Engineering Fluid Mechanics

Ninth Edition

Clayton T. Crowe
WASHINGTON STATE UNIVERSITY, PULLMAN

Donald F. Elger
UNIVERSITY OF IDAHO, MOSCOW

Barbara C. Williams
UNIVERSITY OF IDAHO, MOSCOW

John A. Roberson
WASHINGTON STATE UNIVERSITY, PULLMAN

WILEY
John Wiley & Sons, Inc.

ACQUISITIONS EDITOR   Jennifer Welter
MARKETING MANAGER   Christopher Ruel
PRODUCTION SERVICES MANAGER   Dorothy Sinclair
SENIOR PRODUCTION EDITOR   Sandra Dumas
MEDIA EDITOR   Lauren Sapira
COVER DESIGNER   Jim O'Shea
EDITORIAL ASSISTANT   Mark Owens
MARKETING ASSISTANT   Chelsee Pengal
PRODUCTION MANAGEMENT SERVICES   Publication Services, Inc.
COVER PHOTOGRAPH   ©Bo Tornvig/AgeFotostock America, Inc.

This book was set in Times New Roman by Publication Services, Inc. and printed and bound by
R.R. Donnelley/Jefferson City. The cover was printed by R.R. Donnelley/Jefferson City.

This book is printed on acid-free paper. ∞

Crowe, C. T. (Clayton T.)
    Engineering fluid mechanics/Clayton T. Crowe, Donald F. Elger, Barbara C. Williams, John A. Roberson. —9th ed.

ISBN-13: 978-0470-25977-1

Printed in the United States of America

10 9 8 7 6 5 4 3 2 1

To our spouses, Jeannette and Linda

and in memory of Roy

and to our students past, present, and future

and to those who share our love of learning

Contents

Preface

Audience

This book is written for engineering students of all majors who are taking a first or second course in fluid mechanics. Students should have background knowledge in statics and calculus.

This text is designed to help students develop meaningful and connected knowledge of main concepts and equations as well as develop the skills and approaches that work effectively in professional practice.

Approach

Through innovative ideas and professional skills, engineers can make the world a better place. In particular, fluid mechanics plays a very important role in the design, development, and analysis of systems from microscale applications to giant hydroelectric power generation. For this reason, the study of fluid mechanics is essential to the background of an engineer. The approach in this text is to emphasize both professional practice and technical knowledge.

This text is organized to support the acquisition of deep and connnected knowledge. Each chapter begins by informing students what they should be learning (i.e., learning outcomes) and why this learning is relevant. Topics are linked to previous topics so that students can see how knowledge is building and connecting. Seminal equations, defined as those that are commonly used, are carefully derived and explained. In addition, Table F.2 in the front of the book organizes the main equations.

This text is organized to support the development of skills for problem solving. Example problems are presented with a structured approach that was developed by studying the research literature that describes how experts solve technical problems. This structured approach, labeled as "Engineering Analysis," is presented in Chapter 1. Homework problems are organized by topic, and a variety of types of problems are included.

Organization of Knowledge

Chapters 1 to 11 and 13 are devoted to foundational concepts of fluid mechanics. Relevant content includes fluid properties; forces and pressure variations in static fluids; qualitative descriptions of flow and pressure variations; the Bernoulli equation; the control volume concept; control volume equations for mass, momentum, and energy; dimensional analysis; head loss in conduits; measurements; drag force; and lift force. Nearly all professors cover the material in Chapters 1 to 8 and 10. Chapters 9, 11, and 13 are covered based on instructor preference. Chapters 12, 14, and 15 are devoted to special topics that are optional for a first course in fluid mechanics.

In this 9th edition, there is some reorganization of sections and some additions of new technical material. Chapter 1 provides new material on the nature of fluids, unit practices, and problem solving. Sections in Chapter 4 were reordered to provide a more logical development of the Bernoulli equation. Also, the material on the Eulerian and Lagrangian approaches was moved from Chapter 4 to Chapter 5. Chapter 7 provides new material on energy and power and a new section to describe calculation of power. In Chapter 10, new material on standard pipe sizes was added and new sections were added to describe flow classification and to describe how to solve turbulent flow problems. The open channels flow topics that were in Chapter 10 were moved to Chapter 15. Chapters 11 and 15 were modified by introducing new sections to better organize the material. Also, a list of main equations and a detailed list of unit conversions were added to the front of the book.

Features of this Book*

*Learning Outcomes. Each chapter begins with learning outcomes so students can identify what knowledge they should be gaining by studying the chapter.

*Rationale. Each section describes what content is presented and why this content is relevant so students can connect their learning to what is important to them.

*Visual Approach. The text uses sketches and photographs to help students learn more effectively by connecting images to words and equations.

*Foundational Concepts. This text presents major concepts in a clear and concise format. These concepts form building blocks for higher levels of learning. Concepts are identified by a blue tint.

*Seminal Equations. This text emphasizes technical derivations so that students can learn to do the derivations on their own, increasing their levels of knowledge. Features include

- Derivations of each main equation are presented in a step-by-step fashion.
- The assumptions associated with each main equation are stated during the derivation and after the equation is developed.
- The holistic meaning of main equations is explained using words.
- Main equations are named and listed in Table F.2.

Chapter Summaries. Each chapter concludes with a summary so that students can review the key knowledge in the chapter.

*Engineering Analysis. Example problems and solutions to homework problems are structured with a step-by-step approach. As shown in Fig. 1 (next page), the solution process begins with problem formulation, which involves interpreting the problem before attempting to solve the problem. The plan step involves creating a step-by-step plan prior to jumping into a detailed solution. This structured approach provides students with an approach that can generalize to many types of engineering problems.

*Grid Method. This text presents a systematic process, called the grid method, for carrying and canceling units. Unit practice is emphasized because it helps students spot and fix mistakes and because it helps student put meaning on concepts and equations.

Traditonal and SI Units. Examples and homework problems are presented using both SI and traditional unit systems. This presentation helps students develop unit practices and gain familiarity with units that are used on professional practice.

Example Problems. Each chapter has approximately 10 example problems, each worked out in detail. The purpose of these examples is to show how the knowledge is used in context and to present essential details needed for application.

* Asterisk denotes a new or major modification to this edition.

The first steps model how engineers change a problem statement into a meaningful and clear problem definition.

The next steps model how engineers find a solution path by using main concepts.

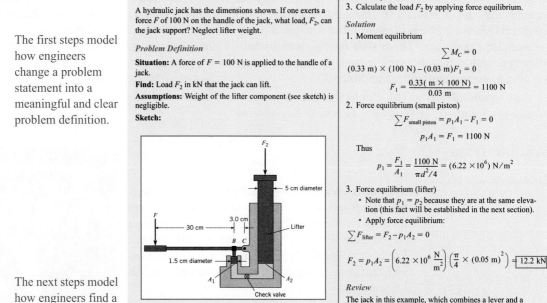

EXAMPLE 3.1 LOAD LIFTED BY A HYDRAULIC JACK

A hydraulic jack has the dimensions shown. If one exerts a force F of 100 N on the handle of the jack, what load, F_2, can the jack support? Neglect lifter weight.

Problem Definition

Situation: A force of $F = 100$ N is applied to the handle of a jack.

Find: Load F_2 in kN that the jack can lift.

Assumptions: Weight of the lifter component (see sketch) is negligible.

Sketch:

Plan

1. Calculate force F_1 acting on the small piston by applying moment equilibrium.

2. Calculate pressure p_1 in the hydraulic fluid by applying force equilibrium.

3. Calculate the load F_2 by applying force equilibrium.

Solution

1. Moment equilibrium

$$\sum M_C = 0$$

$$(0.33 \text{ m}) \times (100 \text{ N}) - (0.03 \text{ m})F_1 = 0$$

$$F_1 = \frac{0.33(\text{ m} \times 100 \text{ N})}{0.03 \text{ m}} = 1100 \text{ N}$$

2. Force equilibrium (small piston)

$$\sum F_{\text{small piston}} = p_1 A_1 - F_1 = 0$$

$$p_1 A_1 = F_1 = 1100 \text{ N}$$

Thus

$$p_1 = \frac{F_1}{A_1} = \frac{1100 \text{ N}}{\pi d^2/4} = (6.22 \times 10^6) \text{ N/m}^2$$

3. Force equilibrium (lifter)
   - Note that $p_1 = p_2$ because they are at the same elevation (this fact will be established in the next section).
   - Apply force equilibrium:

$$\sum F_{\text{lifter}} = F_2 - p_1 A_2 = 0$$

$$F_2 = p_1 A_2 = \left(6.22 \times 10^6 \frac{\text{N}}{\text{m}^2}\right)\left(\frac{\pi}{4} \times (0.05 \text{ m})^2\right) = \boxed{12.2 \text{ kN}}$$

Review

The jack in this example, which combines a lever and a hydraulic machine, provides an output force of 12,200 N from an input force of 100 N. Thus, this jack provides a mechanical advantage of 122 to 1!

Figure 1

Structured problem solving is used throughout the text.

Homework Problems. The text includes several types of end-of-chapter problems, including

- *Preview (or prepare) questions* PQ◄. A preview (or prepare) question is designed to be assigned prior to in-class coverage of the topic. The purpose is so that students come to class with some familiarity with the topics.

- *Analysis problems.* These problems are traditional problems that require a systematic, or step-by-step, approach. They may involve multiple concepts and generally cannot be solved by using a memorization approach. These problems help students learn engineering analysis.

- *Computer problems.* These problems involve use of a computer program. Regarding the choice of software, we have left this open so that instructors may select a program or may allow their students to select a program.

- *Design problems.* These problems have multiple possible solutions and require assumptions and decision making. These problems help students learn to manage the messy, ill-structured problems that typify professional practice.

Solutions Manual. The formatting of the instructor's solutions manual parallels the engineering-analysis approach presented in the text. Each solution includes a description of the situation, statement of the problem goals, statements of main assumptions, summary of the solution approach, a detailed solution, and review comments. In addition, each problem analysis is organized using text labels, such as "momentum equation (*x*-direction)," so that the labels themselves provide a summary of the solution approach.

Image Gallery. The figures from the text are available in PowerPoint format, for easy inclusion in lecture presentations. This resource is available only to instructors. To request access to this password-protected resource, visit the Instructor Companion Site portion of the web site located at www.wiley.com/college/crowe, and register for a password.

Interdisciplinary Approach. Historically, this text was written for the civil engineer. We are retaining this approach while adding material so that the text is also appropriate for other engineering disciplines. For example, the text presents the Bernoulli equation using both head terms (civil engineering approach) and terms with units of pressure (the approach used by chemical and mechanical engineers). We include problems that are relevant to product development as practiced by mechanical and electrical engineers. Some problems feature other disciplines, such as exercise physiology. The reason for this interdisciplinary approach is that the world of today's engineer is becoming more and more interdisciplinary.

Also Available from the Publisher. *Practice Problems with Solutions: A Guide for Learning Engineering Fluid Mechanics*, 9th Edition, by Crow, Elger, and Roberson. ISBN 978 0470 420867. This is a companion manual to the textbook, and presents additional example problems with complete solutions for students seeking additional practice problems and solution guidance.

Visit `www.wiley.com/college/crowe` for ordering information. It is also available in a set with the textbook at a discounted price.

Author Team

Most of the book was originally written by Professor John Roberson, with Clayton Crowe adding the material on compressible flow. Professor Roberson retired from active authorship after the 6th edition, Professor Donald Elger joined on the 7th edition, and Professor Barbara Williams joined on the 9th Edition.

The author team. Left to right: Donald Elger, Barbara Williams, and Clayton Crowe.

Acknowledgments

We wish to thank the following instructors for their help in reviewing and offering feedback on the manuscript for the 9th edition: David Benson, Kettering University; John D. Dietz, University of Central Florida; Howard M. Liljestrand, University of Texas; Nancy Ma, North Carolina State University; Saad Merayyan, California State University, Sacramento; Sergey A. Smirnov, Texas Tech University; and Yaron Sternberg, University of Maryland.

Special recognition is given to our colleagues and mentors. Clayton Crowe acknowledges the many years of professional interaction with his colleagues at Washington State University, and the loving support of his family. Ronald Adams, Engineering Dean at Oregon State University, mentored Donald Elger during his Ph.D. research and introduced him to a new way of thinking about fluid mechanics. Ralph Budwig, a fluid mechanics researcher and educator at The University of Idaho, has provided many hours of delightful discussion. Wilfried Brutsuert at Cornell University and George Bloomsburg at the University of Idaho inspired Barbara Williams's passion for fluid mechanics. Roy Williams, husband and colleague, always pushed for excellence. Charles L. Barker introduced John Roberson to the field of fluid mechanics and motivated him to write a textbook.

Finally, we owe a debt of gratitude to Cody Newbill and John Crepeau, who helped with both the text and the solutions manual. We are also grateful to our families for their encouragement and understanding during the writing and editing of the text.

We welcome feedback and ideas for interesting end-of-chapter problems. Please contact us at the e-mail addresses given below.

Clayton T. Crowe (clayton_crowe@hotmail.com)
Donald F. Elger (delger@uidaho.edu)
Barbara C. Williams (barbwill@uidaho.edu)
John A. Roberson (emeritus)

CHAPTER 1

Introduction

SIGNIFICANT LEARNING OUTCOMES

Conceptual Knowledge
- Describe fluid mechanics.
- Contrast gases and liquids by describing similarities and differences.
- Explain the continuum assumption.

Procedural Knowledge
- Use primary dimensions to check equations for dimensional homogeneity.
- Apply the grid method to carry and cancel units in calculations.
- Explain the steps in the "Structured Approach for Engineering Analysis" (see Table 1.4).

Prior to fluid mechanics, students take courses such as physics, statics, and dynamics, which involve solid mechanics. *Mechanics* is the field of science focused on the motion of material bodies. Mechanics involves force, energy, motion, deformation, and material properties. When mechanics applies to material bodies in the solid phase, the discipline is called *solid mechanics.* When the material body is in the gas or liquid phase, the discipline is called *fluid mechanics.* In contrast to a solid, a *fluid* is a substance whose molecules move freely past each other. More specifically, a fluid is a substance that will continuously deform—that is, flow under the action of a shear stress. Alternatively, a solid will deform under the action of a shear stress but will not flow like a fluid. Both liquids and gases are classified as fluids.

This chapter introduces fluid mechanics by describing gases, liquids, and the continuum assumption. This chapter also presents (a) a description of resources available in the appendices of this text, (b) an approach for using units and primary dimensions in fluid mechanics calculations, and (c) a systematic approach for problem solving.

1.1

Liquids and Gases

This section describes liquids and gases, emphasizing behavior of the molecules. This knowledge is useful for understanding the observable characteristics of fluids.

Liquids and gases differ because of forces between the molecules. As shown in the first row of Table 1.1, a liquid will take the shape of a container whereas a gas will expand to fill a closed container. The behavior of the liquid is produced by strong attractive force between the molecules. This strong attractive force also explains why the density of a liquid is much

Table 1.1   COMPARISON OF SOLIDS, LIQUIDS, AND GASES

Attribute	Solid	Liquid	Gas
Typical Visualization			
Macroscopic Description	Solids hold their shape; no need for a container	Liquids take the shape of the container and will stay in open container	Gases expand to fill a closed container
Mobility of Molecules	Molecules have low mobility because they are bound in a structure by strong intermolecular forces	Liquids typically flow easily even though there are strong intermolecular forces between molecules	Molecules move around freely with little interaction except during collisions; this is why gases expand to fill their container
Typical Density	Often high; e.g., density of steel is 7700 kg/m^3	Medium; e.g., density of water is 1000 kg/m^3	Small; e.g., density of air at sea level is 1.2 kg/m^3
Molecular Spacing	Small—molecules are close together	Small—molecules are held close together by intermolecular forces	Large—on average, molecules are far apart
Effect of Shear Stress	Produces deformation	Produces flow	Produces flow
Effect of Normal Stress	Produces deformation that may associate with volume change; can cause failure	Produces deformation associated with volume change	Produces deformation associated with volume change
Viscosity	NA	High; decreases as temperature increases	Low; increases as temperature increases
Compressibility	Difficult to compress; bulk modulus of steel is 160×10^9 Pa	Difficult to compress; bulk modulus of liquid water is 2.2×10^9 Pa	Easy to compress; bulk modulus of a gas at room conditions is about 1.0×10^5 Pa

higher than the density of gas (see the fourth row). The attributes in Table 1.1 can be generalized by defining a gas and liquid based on the differences in the attractive forces between molecules. A *gas* is a phase of material in which molecules are widely spaced, molecules move about freely, and forces between molecules are minuscule, except during collisions. Alternatively, a *liquid* is a phase of material in which molecules are closely spaced, molecules move about freely, and there are strong attractive forces between molecules.

1.2

The Continuum Assumption

This section describes how fluids are conceptualized as a continuous medium. This topic is important for applying the derivative concept to characterize properties of fluids.

Figure 1.1

When a measuring volume ΔV is large enough for random molecular effects to average out, the continuum assumption is valid

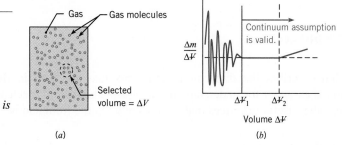

(a)

(b)

While a body of fluid is comprised of molecules, most characteristics of fluids are due to average molecular behavior. That is, a fluid often behaves as if it were comprised of continuous matter that is infinitely divisible into smaller and smaller parts. This idea is called the *continuum assumption*. When the continuum assumption is valid, engineers can apply limit concepts from differential calculus. Recall that a limit concept, for example, involves letting a length, an area, or a volume approach zero. Because of the continuum assumption, fluid parameters such as density and velocity can be considered continuous functions of position with a value at each point in space.

To gain insight into the validity of the continuum assumption, consider a hypothetical experiment to find density. Fig. 1.1a shows a container of gas in which a volume ΔV has been identified. The idea is to find the mass of the molecules ΔM inside the volume and then to calculate density by

$$\rho = \frac{\Delta M}{\Delta V}$$

The calculated density is plotted in Fig. 1.1b. When the measuring volume ΔV is very small (approaching zero), the number of molecules in the volume will vary with time because of the random nature of molecular motion. Thus, the density will vary as shown by the wiggles in the blue line. As volume increases, the variations in calculated density will decrease until the calculated density is independent of the measuring volume. This condition corresponds to the vertical line at ΔV_1. If the volume is too large, as shown by ΔV_2, then the value of density may change due to spatial variations.

In most applications, the continuum assumption is valid. For example, consider the volume needed to contain at least a million (10^6) molecules. Using Avogadro's number of 6×10^{23} molecules/mole, the limiting volume for water is 10^{-13} mm^3, which corresponds to a cube less than 10^{-4} mm on a side. Since this dimension is much smaller than the flow dimensions of a typical problem, the continuum assumption is justified. For an ideal gas (1 atm and 20°C) one mole occupies 24.7 liters. The size of a volume with more than 10^6 molecules would be 10^{-10} mm^3, which corresponds to a cube with sides less than 10^{-3} mm (or one micrometer). Once again this size is much smaller than typical flow dimensions. Thus, the continuum assumption is usually valid in gas flows.

The continuum assumption is invalid for some specific applications. When air is in motion at a very low density, such as when a spacecraft enters the earth's atmosphere, then the spacing between molecules is significant in comparison to the size of the spacecraft. Similarly, when a fluid flows through the tiny passages in nanotechnology devices, then the spacing between molecules is significant compared to the size of these passageways.

Dimensions, Units, and Resources

This section describes the dimensions and units that are used in fluid mechanics. This information is essential for understanding most aspects of fluid mechanics. In addition, this section describes useful resources that are presented in the front and back of this text.

Dimensions

A *dimension* is a category that represents a physical quantity such as mass, length, time, momentum, force, acceleration, and energy. To simplify matters, engineers express dimensions using a limited set that are called *primary dimensions*. Table 1.2 lists one common set of primary dimensions.

Secondary dimensions such as momentum and energy can be related to primary dimensions by using equations. For example, the secondary dimension "force" is expressed in primary dimensions by using Newton's second law of motion, $F = ma$. The primary dimensions of acceleration are L/T^2, so

$$[F] = [ma] = M\frac{L}{T^2} = \frac{ML}{T^2} \tag{1.1}$$

In Eq. (1.1), the square brackets mean "dimensions of." This equation reads "the primary dimensions of force are mass times length divided by time squared." Note that primary dimensions are not enclosed in brackets.

Units

While a dimension expresses a specific type of physical quantity, a unit assigns a number so that the dimension can be measured. For example, measurement of volume (a dimension) can be expressed using units of liters. Similarly, measurement of energy (a dimension) can be expressed using units of joules. Most dimensions have multiple units that are used for measurement. For example, the dimension of "force" can be expressed using units of newtons, pounds-force, or dynes.

Unit Systems

In practice, there are several unit systems in use. The International System of Units (abbreviated SI from the French "Le Système International d'Unités") is based on the meter,

Dimension	Symbol	Unit (SI)
Length	L	meter (m)
Mass	M	kilogram (kg)
Time	T	second (s)
Temperature	θ	kelvin (K)
Electric current	i	ampere (A)
Amount of light	C	candela (cd)
Amount of matter	N	mole (mol)

Table 1.2 PRIMARY DIMENSIONS

kilogram, and second. Although the SI system is intended to serve as an international standard, there are other systems in common use in the United States. The U.S. Customary System (USCS), sometimes called English Engineering System, uses the pound-mass (lbm) for mass, the foot (ft) for length, the pound-force (lbf) for force, and the second (s) for time. The British Gravitational (BG) System is similar to the USCS system that the unit of mass is the slug. To convert between pounds-mass and kg or slugs, the relationships are

$$1.0 \text{ lbm} = \frac{1}{2.2} \text{ kg} = \frac{1}{32.2} \text{ slug}$$

Thus, a gallon of milk, which has mass of approximately 8 lbm, will have a mass of about 0.25 slugs, which is about 3.6 kg.

For simplicity, this text uses two categories for units. The first category is the familiar SI unit system. The second category contains units from both the USCS and the BG systems of units and is called the "Traditional Unit System."

Resources Available in This Text

To support calculations and design tasks, formulas and data are presented in the front and back of this text.

Table F.1 (the notation "F.x" means a table in the front of the text) presents data for converting units. For example, this table presents the factor for converting meters to feet (1 m = 3.281 ft) and the factor for converting horsepower to kilowatts (1 hp = 745.7 W). Notice that a given parameter such as viscosity will have one set of primary dimensions (M/LT) and several possible units, including pascal-second (Pa $\cdot$ s), poise, and lbf $\cdot$ s$/$ft^2. Table F.1 lists *unit conversion formulas*, where each formula is a relationship between units expressed using the equal sign. Examples of unit conversion formulas are 1.0 m = 3.281 ft and 3.281 ft = km$/$1000. Notice that each row of Table F.1 provides multiple conversion formulas. For example, the row for length conversions,

$$1 \text{ m} = 3.281 \text{ ft} = 1.094 \text{ yd} = 39.37 \text{ in} = \frac{\text{km}}{1000} = 10^6 \text{ } \mu\text{m} \tag{1.2}$$

has the usual conversion formulas such as 1 m = 39.37 in, and the less common formulas such as 1.094 yd = 10^6 μm.

Table F.2 presents equations that are commonly used in fluid mechanics. To make them easier to remember, equations are given descriptive names such as the "hydrostatic equation." Also, notice that each equation is given an equation number and page number corresponding to where it is introduced in this text.

Tables F.3, F.4, and F.5 present commonly used constants and fluid properties. Other fluid properties are presented in the appendix. For example, Table A.3 (the notation "A.x" means a table in the appendix) gives properties of air.

Table A.6 lists the variables that are used in this text. Notice that this table gives the symbol, the primary dimensions, and the name of the variable.

1.4

Topics in Dimensional Analysis

This section introduces dimensionless groups, the concept of dimensional homogeneity of an equation, and a process for carrying and canceling units in a calculation. This knowledge is useful in all aspects of engineering, especially for finding and correcting mistakes during calculations and during derivations of equations.

All of topics in this section are part of *dimensional analysis,* which is the process for applying knowledge and rules involving dimensions and units. Other aspects of dimensional analysis are presented in Chapter 8 of this text.

Dimensionless Groups

Engineers often arrange variables so that primary dimensions cancel out. For example, consider a pipe with an inside diameter D and length L. These variables can be grouped to form a new variable L/D, which is an example of a dimensionless group. A *dimensionless group* is any arrangement of variables in which the primary dimensions cancel. Another example of a dimensionless group is the Mach number M, which relates fluid speed V to the speed of sound c:

$$M = \frac{V}{c}$$

Another common dimensionless group is named the Reynolds number and given the symbol Re. The Reynolds number involves density, velocity, length, and viscosity μ :

$$Re = \frac{\rho V L}{\mu} \qquad (1.3)$$

The convention in this text is to use the symbol [-] to indicate that the primary dimensions of a dimensionless group cancel out. For example,

$$[Re] = \left[\frac{\rho V L}{\mu}\right] = [-] \qquad (1.4)$$

Dimensional Homogeneity

When the primary dimensions on each term of an equation are the same, the equation is *dimensionally homogeneous.* For example, consider the equation for vertical position s of an object moving in a gravitational field:

$$s = \frac{gt^2}{2} + v_o t + s_o$$

In the equation, g is the gravitational constant, t is time, v_o is the magnitude of the vertical component of the initial velocity, and s_o is the vertical component of the initial position. This equation is dimensionally homogeneous because the primary dimension of each term is length L. Example 1.1 shows how to find the primary dimension for a group of variables using a step-by-step approach. Example 1.2 shows how to check an equation for dimensional homogeneity by comparing the dimensions on each term.

Since fluid mechanics involves many differential and integral equations, it is useful to know how to find primary dimensions on integral and derivative terms.

To find primary dimensions on a derivative, recall from calculus that a derivative is defined as a ratio:

$$\frac{df}{dy} = \lim_{\Delta y \to 0} \frac{\Delta f}{\Delta y}$$

Thus, the primary dimensions of a derivative can be found by using a ratio:

$$\left[\frac{df}{dy}\right] = \left[\frac{f}{y}\right] = \frac{[f]}{[y]}$$

EXAMPLE 1.1   PRIMARY DIMENSIONS OF THE REYNOLDS NUMBER

Show that the Reynolds number, given in Eq. 1.4, is a dimensionless group.

Problem Definition

Situation: The Reynolds number is given by
$Re = (\rho V L)/\mu$.

Find: Show that Re is a dimensionless group.

Plan

1. Identify the variables by using Table A.6.
2. Find the primary dimensions by using Table A.6.
3. Show that Re is dimensionless by canceling primary dimensions.

Solution

1. Variables
   - mass density, ρ
   - velocity, V
   - Length, L
   - viscosity, μ
2. Primary dimensions

$$[\rho] = M/L^3$$
$$[V] = L/T$$
$$[L] = L$$
$$[\mu] = M/LT$$

3. Cancel primary dimensions:

$$\left[\frac{\rho V L}{\mu}\right] = \left[\frac{M}{L^3}\right]\left[\frac{L}{T}\right][L]\left[\frac{LT}{M}\right] = [\text{-}]$$

Since the primary dimensions cancel, the Reynolds number $(\rho V L)/\mu$ is a dimensionless group.

EXAMPLE 1.2   DIMENSIONAL HOMOGENEITY OF THE IDEAL GAS LAW

Show that the ideal gas law is dimensionally homogeneous.

Problem Definition

Situation: The ideal gas law is given by $p = \rho R T$.

Find: Show that the ideal gas law is dimensionally homogeneous.

Plan

1. Find the primary dimensions of the first term.
2. Find the primary dimensions of the second term.
3. Show dimensional homogeneity by comparing the terms.

Solution

1. Primary dimensions (first term)
   - From Table A.6, the primary dimensions are:

$$[p] = \frac{M}{LT^2}$$

2. Primary dimensions (second term).
   - From Table A.6, the primary dimensions are

$$[\rho] = M/L^3$$
$$[R] = L^2/\theta T^2$$
$$[T] = \theta$$

   - Thus

$$[\rho R T] = \left(\frac{M}{L^3}\right)\left(\frac{L^2}{\theta T^2}\right)(\theta) = \frac{M}{LT^2}$$

3. Conclusion: The ideal gas law is dimensionally homogeneous because the primary dimensions of each term are the same.

The primary dimensions for a higher-order derivative can also be found by using the basic definition of the derivative. The resulting formula for a second-order derivative is

$$\left[\frac{d^2 f}{dy^2}\right] = \lim_{\Delta y \to 0} \frac{\Delta(df/dy)}{\Delta y} = \left[\frac{f}{y^2}\right] = \frac{[f]}{[y^2]} \tag{1.5}$$

Applying Eq. (1.5) to acceleration shows that

$$\left[\frac{d^2 y}{dt^2}\right] = \left[\frac{y}{t^2}\right] = \frac{L}{T^2}$$

To find primary dimensions of an integral, recall from calculus that an integral is defined as a sum:

$$\int f \, dy = \lim_{N \to \infty} \sum_{i=1}^{N} f \, \Delta y_i$$

Thus

$$\left[\int f \, dy\right] = [f][y] \tag{1.6}$$

For example, position is given by the integral of velocity with respect to time. Checking primary dimensions for this integral gives

$$\left[\int V \, dt\right] = [V][t] = \frac{L}{T} \cdot T = L$$

In summary, one can easily find primary dimensions on derivative and integral terms by applying fundamental definitions from calculus. This process is illustrated by Example 1.3

EXAMPLE 1.3 PRIMARY DIMENSIONS OF A DERIVATIVE AND INTEGRAL

Find the primary dimensions of $\mu \dfrac{d^2 u}{dy^2}$, where μ is viscosity, u is fluid velocity, and y is distance. Repeat for $\dfrac{d}{dt}\displaystyle\int_{V} \rho \, dV$ where t is time, V is volume, and ρ is density.

Problem Definition

Situation: A derivative and integral term are specified above.
Find: Primary dimensions on the derivative and the integral.

Plan

1. Find the primary dimensions of the first term by applying Eq. (1.5).
2. Find the primary dimensions of the second term by applying Eqs. (1.5) and (1.6).

Solution

1. Primary dimensions of $\mu \dfrac{d^2 u}{dy^2}$
   - From Table A.6:

$$[\mu] = M/LT$$
$$[u] = L/T$$
$$[x] = L$$

   - Apply Eq. (1.5):

$$\left[\frac{d^2 u}{dy^2}\right] = \left[\frac{u}{y^2}\right] = \frac{L/T}{L^2}$$

   - Combine the previous two steps:

$$\left[\mu \frac{d^2 u}{dy^2}\right] = [\mu]\left[\frac{d^2 u}{dy^2}\right] = \left(\frac{M}{LT}\right)\left(\frac{L/T}{L^2}\right) = \boxed{\frac{M}{L^2 T^2}}$$

2. Primary dimensions of $\dfrac{d}{dt}\displaystyle\int_{\mathcal{V}}\rho\,d\mathcal{V}$

   • Find primary dimensions from Table A.6:

   $$[t] = T$$
   $$[\rho] = M/L^3$$
   $$[\mathcal{V}] = L^3$$

• Apply Eqs. (1.5) and (1.6) together:

$$\left[\frac{d}{dt}\int_{\mathcal{V}}\rho\,d\mathcal{V}\right] = \left[\frac{\rho\mathcal{V}}{t}\right] = \left(\frac{M}{L^3}\right)\left(\frac{L^3}{T}\right) = \boxed{\frac{M}{T}}$$

Sometimes constants have primary dimensions. For example, the hydrostatic equation relates pressure p, density ρ, the gravitational constant g, and elevation z:

$$p + \rho gz = \text{constant} = C$$

For dimensional homogeneity, the constant C needs to have the same primary dimensions as either p or ρgz. Thus the dimensions of C are $[C] = M/LT^2$. Another example involves expressing fluid velocity V as a function of distance y using two constants a and b:

$$V(y) = ay(b - y)$$

For dimensional homogeneity both sides of this equation need to have primary dimensions of $[L/T]$. Thus, $[b] = L$ and $[a] = L^{-1}T^{-1}$.

The Grid Method

Because fluid mechanics involves complex equations and traditional units, this section presents the *grid method*, which is a systematic way to carry and cancel units. For example, Fig. 1.2 shows an estimate of the power P required to ride a bicycle at a speed of $V = 20$ mph. The engineer estimated that the required force to move against wind drag is $F = 4.0$ lbf and applied the equation $P = FV$. As shown, the calculation reveals that the power is 159 watts.

As shown in Fig. 1.2, the grid method involves writing an equation, drawing a grid, and carrying and canceling units. Regarding unit cancellations, the key idea is the use of *unity conversion ratios,* in which unity (1.0) appears on one side of the equation. Examples of unity conversion ratios are

$$1.0 = \frac{1\text{ m/s}}{2.237\text{ mph}} \qquad 1.0 = \frac{1.0\text{ N}}{0.2249\text{ lbf}}$$

Figure 1.2 shows three conversion ratios. Each of these ratios is obtained by using information given in Table F.1. For example, the row in Table F.1 for power shows that $1\text{ W} = (\text{N}\cdot\text{m/s})$. Dividing both sides of this equation by $\text{N}\cdot\text{m/s}$ gives

$$1.0 = \frac{\text{W}\cdot\text{s}}{\text{N}\cdot\text{m}}$$

Table 1.3 shows how to apply the grid method. Notice how the same process steps can apply to different situations.

Figure 1.2

Grid method

$$P = F \times V = \frac{4\text{ lbf}}{}\left|\frac{20\text{ mph}}{}\right|\frac{1.0\text{ m/s}}{2.237\text{ mph}}\left|\frac{1.0\text{ N}}{0.2248\text{ lbf}}\right|\frac{W*s}{N*m}$$

$$P = 159\ W$$

Table 1.3   APPLYING THE GRID METHOD (TWO EXAMPLES)

Process Step	Example 1	Example 2
	Situation: Convert a pressure of 2.00 psi to pascals.	**Situation:** Find the force in newtons that is needed to accelerate a mass of 10 g at a rate of 15 ft/s².
Step 1. **Problem:** Write the term or equation. Include numbers and units.	$p = 2.00$ psi	$F = ma$ $F\,(\mathrm{N}) = (0.01\ \mathrm{kg})(15\ \mathrm{ft/s^2})$
Step 2. **Conversion Ratios:** Look up unit conversion formula(s) in Table F.1 and represent these as unity conversion ratios.	$1.0 = \dfrac{1\ \mathrm{Pa}}{1.45 \times 10^{-4}\ \mathrm{psi}}$	$1.0 = \dfrac{1.0\ \mathrm{m}}{3.281\ \mathrm{ft}}$ \qquad $1.0 = \dfrac{\mathrm{N\cdot s^2}}{\mathrm{kg\cdot m}}$
Step 3. **Algebra:** Multiply variables and cancel units. Fix any errors.	$p = [2.00\ \mathrm{psi}]\left[\dfrac{1\ \mathrm{Pa}}{1.45\times10^{-4}\ \mathrm{psi}}\right]$	$F = [0.01\ \mathrm{kg}]\left[\dfrac{15\ \mathrm{ft}}{\mathrm{s^2}}\right]\left[\dfrac{1.0\ \mathrm{m}}{3.281\ \mathrm{ft}}\right]\left[\dfrac{\mathrm{N\cdot s^2}}{\mathrm{kg\cdot m}}\right]$
Step 4. **Calculations:** Perform the indicated calculations. Round the answer to the correct number of significant figures.	$p = 13.8$ kPa	$F = 0.0457$ N

Since fluid mechanics problems often involve mass units of slugs or pounds-mass (lbm), it is easy to make a mistake when converting units. Thus, it is useful to have a systematic approach. The idea is to relate mass units to force units through $F = ma$. For example, a force of 1.0 N is the magnitude of force that will accelerate a mass of 1.0 kg at a rate of 1.0 m/s². Thus,

$$(1.0\ \mathrm{N}) \equiv (1.0\ \mathrm{kg})(1.0\ \mathrm{m/s^2})$$

Rewriting this expression gives a conversion ratio

$$1.0 = \frac{\mathrm{kg\cdot m}}{\mathrm{N\cdot s^2}} \tag{1.7}$$

When mass is given using slugs, the corresponding conversion ratio is

$$1.0 = \frac{\mathrm{slug\cdot ft}}{\mathrm{lbf\cdot s^2}} \tag{1.8}$$

A force of 1.0 lbf is defined as the magnitude of force that will accelerate a mass of 1.0 lbm at a rate of 32.2 ft/s². Thus,

$$(1.0\ \mathrm{lbf}) \equiv (1.0\ \mathrm{lbm})(32.2\ \mathrm{ft/s^2})$$

Thus, the conversion ratio relating force and mass units becomes

$$1.0 = \frac{32.2\ \mathrm{lbm\cdot ft}}{\mathrm{lbf\cdot s^2}} \tag{1.9}$$

Example 1.4 shows how to apply the grid method. Notice that calculations involving traditional units follow the same process as calculations involving SI units.

EXAMPLE 1.4   GRID METHOD APPLIED TO A ROCKET

A water rocket is fabricated by attaching fins to a 1-liter plastic bottle. The rocket is partially filled with water and the air space above the water is pressurized, causing water to jet out of the rocket and propel the rocket upward. The thrust force T from the water jet is given by $T = \dot{m}V$, where $\dot{m}$ is the rate at which the water flows out of the rocket in units of mass per time and V is the speed of the water jet. (a) Estimate the thrust force in newtons for a jet velocity of $V = 30$ m/s (98.4 ft/s) where the mass flow rate is $\dot{m} = 9$ kg/s (19.8 lbm/s). (b) Estimate the thrust force in units of pounds-force (lbf). Apply the grid method during your calculations.

Problem Definition

Situation:

1. A rocket is propelled by a water jet.
2. The thrust force is given by $T = \dot{m}V$.

Find: Thrust force supplied by the water jet. Present the answer in N and lbf.

Sketch:

- Air
  - Pressure = p
- Water
- Water jet
  - Velocity = V = 30 m/s = 98.4 ft/s
  - Mass flow rate = $\dot{m}$ = 9 kg/s = 19.8 lbm/s

Plan

Find the thrust force by using the process given in Table 1.3. When traditional units are used, apply Eq. (1.9).

Solution

1. Thrust force (SI units)

$$T = \dot{m}V$$

- Insert numbers and units:

$$T\ (\text{N}) = \dot{m}V = (9\ \text{kg/s})(30\ \text{m/s})$$

- Insert conversion ratios and cancel units:

$$T\ (\text{N}) = \left[\frac{9\ \text{kg}}{\text{s}}\right]\left[\frac{30\ \text{m}}{\text{s}}\right]\left[\frac{\text{N} \cdot \text{s}^2}{\text{kg} \cdot \text{m}}\right]$$

$$\boxed{T = 270\ \text{N}}$$

2. Thrust force (traditional units)

$$T = \dot{m}V$$

- Insert numbers and units:

$$T\ (\text{lbf}) = \dot{m}V = (19.8\ \text{lbm/s})(98.4\ \text{ft/s})$$

- Insert conversion ratios and cancel units:

$$T\ (\text{lbf}) = \left[\frac{19.8\ \text{lbm}}{\text{s}}\right]\left[\frac{98.4\ \text{ft}}{\text{s}}\right]\left[\frac{\text{lbf} \cdot \text{s}^2}{32.2\ \text{lbm} \cdot \text{ft}}\right]$$

$$\boxed{T = 60.5\ \text{lbf}}$$

Review

1. Validation. Since 1.0 lbf = 4.45 N, answer (a) is the same as answer (b).
2. Tip! To validate calculations in traditional units, repeat the calculation in SI units.

1.5

Engineering Analysis

In fluid mechanics, many problems are messy and open-ended. Thus, this section presents a structured approach to problem solving.

Engineering analysis is a process for idealizing or representing real-world situations using mathematics and scientific principles and then using calculations to extract useful information. For example, engineering analysis is used to find the power required by a pump, wind force acting on a building, and pipe diameter for a given application. Engineering analysis

Table 1.4   STRUCTURED APPROACH FOR ENGINEERING ANALYSIS

What To Do	Why Do This?	Typical Actions
Problem Definition: This involves figuring out what the problem is, what is involved, and what the end state (i.e., goal state) is. Problem definition is done before trying to solve the problem.	• To visualize the situation (present state). • To visualize the goal (end state).	• Read and interpret the problem statement. • Look up and learn unfamiliar knowledge. • Document your interpretation of the situation. • Interpret and document problem goals. • Make an engineering sketch. • Document main assumptions. • Look up fluid properties; document sources.
Plan: This involves figuring out a solution path or "how to solve the problem." Planning is done prior to jumping into action.	• To find an easy way to solve the problem. • Saves you time.	• Generate multiple ideas for solving the problem • Identify useful equations from Table F.2. • Inventory past solutions. • Analyze equations using a term-by-term approach. • Balancing number of equations with number of unknowns. • Make a step-by-step plan.
Solution: This involves solving the problem by executing the plan.	• To reach the problem goal state.	• Use computer programs. • Perform calculations. • Double-check work. • Carry and cancel units.
Review: This involves validating the solution, exploring implications of the solution, and looking back to learn from the experience you just had.	• Gain confidence that your answer can be trusted. • Increases your understanding. • Gain ideas for applications. • To learn.	• Check the units of answer. • Check that problem goals have been reached. • Validate the answer with a simpler estimate. • Write down knowledge that you want to remember. • List "what worked" and "ideas for improvement."

involves subdividing or organizing a problem into logical parts as described in Table 1.4. Notice that the columns describe what to do, the rationale for this step, and typical actions taken during this step. The approach shown in Table 1.4 is used in example problems throughout this text.

1.6

Applications and Connections

Knowledge in this textbook generalizes to problem solving in many contexts. However, this presents a challenge because it can be hard to understand how this fundamental knowledge relates to everyday problems. Thus, this section describes how knowledge from fluid mechanics connects to other disciplines.

Hydraulics is the study of the flow of water through pipes, rivers, and open-channels. Hydraulics includes pumps and turbines and applications such as hydropower. Hydraulics is important for ecology, policymaking, energy production, recreation, fish and game resources, and water supply.

Hydrology is the study of the movement, distribution, and quality of water throughout the earth. Hydrology involves the hydraulic cycle and water resource issues. Thus, hydrology provides results that are useful for environmental engineering and for policymaking. Hydrology is important nowadays because of global challenges in providing water for human societies.

Aerodynamics is the study of air flow. Topics include lift and drag on objects (e.g., airplanes, automobiles, birds), shock waves associated with flow around a rocket, and the flow through a supersonic or deLaval nozzle. Aerodynamics is important for the design of vehicles, for energy conservation, and for understanding nature.

Bio-fluid mechanics is an emerging field that includes the study of the lungs and circulatory system, blood flow, micro-circulation, and lymph flow. Bio-fluids also includes development of artificial heart valves, stents, vein and dialysis shunts, and artificial organs. Bio-fluid mechanics is important for advancing health care.

Acoustics is the study of sound. Topics include production, control, transmission, reception of sound, and physiological effects of sound. Since sound waves are pressure waves in fluids, acoustics is related to fluid mechanics. In addition, water hammer in a piping system, which involves pressure waves in liquids, involves some of the same knowledge that is used in acoustics.

Microchannel flow is an emerging area that involves the study of flow in tiny passages. The typical size of a microchannel is a diameter in the range of 10 to 200 micrometers. Applications that involve microchannels include microelectronics, fuel cell systems, and advanced heat sink designs.

Computational fluid dynamics (CFD) is the application of numerical methods implemented on computers to model and solve problems that involve fluid flows. Computers perform millions of calculations per second to simulate fluid flow. Examples of flows that are modeled by CFD include water flow in a river, blood flow in the abdominal aorta, and air flow around an automobile.

Petroleum engineering is the application of engineering to the exploration and production of petroleum. Movement of oil in the ground involves flow through a porous medium. Petroleum extraction involves flow of oil through passages in wells. Oil pipelines involve pumps and conduit flow.

Atmospheric science is the study of the atmosphere, its processes, and the interaction of the atmosphere with other systems. Fluid mechanics topics include flow of the atmosphere and applications of CFD to atmospheric modeling. Atmospheric science is important for predicting weather and is relevant to current issues including acid rain, photochemical smog, and global warming.

Electrical engineering problems can involve knowledge from fluid mechanics. For example, fluid mechanics is involved in the flow of solder during a manufacturing process, the cooling of a microprocessor by a fan, sizing of motors to operate pumps, and the production of electrical power by wind turbines.

Environmental engineering involves the application of science to protect or improve the environment (air, water, and/or land resources) or to remediate polluted sites. Environmental engineers design water supply and wastewater treatment systems for communities. Environmental engineers are concerned with local and worldwide environmental issues such as acid rain, ozone depletion, water pollution, and air pollution.

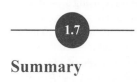

1.7

Summary

Fluid mechanics involves the application of scientific concepts such as position, force, velocity, acceleration, and energy to materials that are in the liquid or gas states. Liquids differ from solids because they can be poured and they flow under the action of shear stress. Similar to liquids, gases also flow when shear stress is nonzero. A significant difference between gases and liquids is that the molecules in liquids experience strong intermolecular forces whereas the molecules in gases move about freely with little or no interactions except during collisions. Thus, gases expand to fill their container while liquids will occupy a fixed volume. In addition, liquids have much larger values of density and liquids exhibit effects such as surface tension.

In fluid mechanics, practices that involve using units and dimensions are known collectively as "dimensional analysis." A dimension is a category associated with a physical quantity such as mass, length, time, or energy. Units are the divisions by which a dimension is measured. For example, the dimension called "mass" may be measured using units of kilograms, slugs, pounds-mass, or ounces.

All dimensions can be expressed using a limited set of primary dimensions. This text mainly uses three primary dimensions: mass (M), length (L), and time (T).

A systematic way to carry and cancel units is called the grid method. The main idea of the grid method is to multiply terms in equations by a ratio (called a unity conversion ratio) that equals 1.0. Examples of unity conversion ratios are $1.0 = (1.0 \text{ kg})/(2.2 \text{ lbm})$ and $1.0 = (1.0 \text{ lbf})/(4.45 \text{ N})$.

Engineering analysis is the process of applying scientific knowledge and mathematical procedures to solve practical problems such as calculating forces on an airplane or estimating the energy requirements of a pump. Engineering analysis involves actions that can be organized into four categories: problem definition, plan, solution, and review.

Problems

*A *Preview Question* ($\mathbb{PQ}$) can be assigned prior to in-class coverage of a topic.

Units, Dimensions, Dimensional Homogeneity

*1.1 $\mathbb{PQ}$◄ For each variable below, list three common units.
a. Volume flow rate (Q), mass flow rate $(\dot{m})$, and pressure (p).
b. Force, energy, power.
c. Viscosity.

*1.2 $\mathbb{PQ}$◄ In Table F.2, find the hydrostatic equation. For each form of the equation that appears, list the name, symbol, and primary dimensions of each variable.

*1.3 $\mathbb{PQ}$◄ For each of the following units, list the primary dimensions: kWh, poise, slug, cfm, cSt.

1.4 The hydrostatic equation is $p/\gamma + z = C$, where p is pressure, γ is specific weight, z is elevation, and C is a constant. Prove that the hydrostatic equation is dimensionally homogeneous.

1.5 Find the primary dimensions of each of the following terms.
a. $(\rho V^2)/2$ (kinetic pressure), where ρ is fluid density and V is velocity.
b. T (torque).
c. P (power).
d. $(\rho V^2 L)/\sigma$ (Weber number), where ρ is fluid density, V is velocity, L is length, and σ is surface tension.

1.6 The power provided by a centrifugal pump is given by $P = \dot{m}gh$, where $\dot{m}$ is mass flow rate, g is the gravitational constant, and h is pump head. Prove that this equation is dimensionally homogeneous.

1.7 Find the primary dimensions of each of the following terms.
a. $\int_A \rho V^2 \, dA$, where ρ is fluid density, V is velocity, and A is area.

b. $\dfrac{d}{dt}\int_{\mathcal{V}} \rho V \, d\mathcal{V}$, where $\dfrac{d}{dt}$ is the derivative with respect to time, ρ is density, and $\mathcal{V}$ is volume.

The Grid Method

*1.8 $\mathbb{PQ}$◄ In your own words, what actions need to be taken in each step of the grid method?

1.9 Apply the grid method to calculate the density of an ideal gas using the formula $\rho = p/RT$. Express your answer in lbm/ft^3. Use the following data: absolute pressure is $p = 35$ psi, the gas constant is $R = 1716$ ft-lbf/slug-°R, and the temperature is $T = 100$ °F.

1.10 The pressure rise Δp associated with wind hitting a window of a building can be estimated using the formula $\Delta p = \rho(V^2/2)$, where ρ is density of air and V is the speed of the wind. Apply the grid method to calculate pressure rise for $\rho = 1.2$ kg/m^3 and $V = 60$ mph.
a. Express your answer in pascals.
b. Express your answer in pounds-force per square inch (psi).
c. Express your answer in inches of water column (in-H$_2$O).

1.11 Apply the grid method to calculate force using $F = ma$.
a. Find force in newtons for $m = 10$ kg and $a = 10$ m/s^2.
b. Find force in pounds-force for $m = 10$ lbm and $a = 10$ ft/s^2.
c. Find force in newtons for $m = 10$ slug and $a = 10$ ft/s^2.

1.12 When a bicycle rider is traveling at a speed of $V = 24$ mph, the power P she needs to supply is given by $P = FV$, where $F = 5$ lbf is the force necessary to overcome aerodynamic drag. Apply the grid method to calculate:
a. power in watts.
b. energy in food calories to ride for 1 hour.

1.13 Apply the grid method to calculate the cost in U.S. dollars to operate a pump for one year. The pump power is 20 hp. The pump operates for 20 hr/day, and electricity costs $0.10 per kWh.

C H A P T E R

Fluid Properties

2

SIGNIFICANT LEARNING OUTCOMES

Conceptual Knowledge
- Define density, specific gravity, viscosity, surface tension, vapor pressure, and bulk modulus of elasticity.
- Describe the differences between absolute viscosity and kinematic viscosity.
- Describe how shear stress, viscosity, and the velocity distribution are related.
- Describe how viscosity, density, and vapor pressure vary with temperature and / or pressure.

Procedural Knowledge
- Look up fluid property values from figures, tables; know when and how to interpolate.
- Calculate gas density using the ideal gas law.

A fluid has certain characteristics by which its physical condition may be described. These characteristics are called *properties* of the fluid. This chapter introduces material properties of fluids and presents key equations, tables, and figures.

2.1

Properties Involving Mass and Weight

Mass and weight properties are needed for most problems in fluid mechanics, including the flow of ground water in aquifers and the pressure acting on a scuba diver or an underwater structure.

Mass Density, ρ

Mass density is defined as the ratio of mass to volume at a point, given by

$$\rho = \lim_{\Delta V \to 0} \frac{\Delta m}{\Delta V} \tag{2.1}$$

Review the continuum assumption developed in Section 1.2 for the meaning of ΔV approaching zero. Mass density has units of kilograms per cubic meter (kg/m^3) or pounds-mass per cubic foot (lbm/ft^3). The mass density of water at 4°C is $1000 \ kg/m^3$ ($62.4 \ lbm/ft^3$), and it decreases slightly with increasing temperature, as shown in Table A.5. The mass density of air at 20°C and standard atmospheric pressure is $1.2 \ kg/m^3$ ($0.075 \ lbm/ft^3$), and it changes significantly with temperature and pressure. Mass density, often simply called density, is represented by the Greek symbol ρ (rho). The densities of common fluids are given in Tables A.2 to A.5.

Specific Weight, γ

The gravitational force per unit volume of fluid, or simply the weight per unit volume, is defined as *specific weight*. It is given the Greek symbol γ (gamma). Water at 20°C has a specific weight of 9790 N/m³ (or 62.4 lbf/ft³ at 50°F). In contrast, the specific weight of air at 20°C and standard atmospheric pressure is 11.8 N/m³. Specific weight and density are related by

$$\gamma = \rho g \tag{2.2}$$

Specific weights of common liquids are given in Table A.4.

Variation in Liquid Density

In practice, engineers need to decide whether or not to model a fluid as constant density or variable density. Usually, a liquid such as water requires a large change in pressure in order to change the density. Thus, for most applications, liquids can be considered incompressible and can be assumed to have constant density. An exception to this occurs when different solutions, such as saline and fresh water, are mixed. A mixture of salt in water changes the density of the water without changing its volume. Therefore in some flows, such as in estuaries, density variations may occur within the flow field even though the fluid is essentially incompressible. A fluid wherein density varies spatially is described as *nonhomogeneous*. This text emphasizes the flow of *homogeneous* fluids, so the term *incompressible,* used throughout the text, implies constant density.

Specific Gravity, S

The ratio of the specific weight of a given fluid to the specific weight of water at the standard reference temperature 4°C is defined as *specific gravity,* S:

$$S = \frac{\gamma_{\text{fluid}}}{\gamma_{\text{water}}} = \frac{\rho_{\text{fluid}}}{\rho_{\text{water}}} \tag{2.3}$$

The specific weight of water at atmospheric pressure is 9790 N/m³. The specific gravity of mercury at 20°C is

$$S_{\text{Hg}} = \frac{133 \ \text{kN/m}^3}{9.79 \ \text{kN/m}^3} = 13.6$$

Because specific gravity is a ratio of specific weights, it is dimensionless and therefore independent of the system of units used.

Ideal Gas Law

The *ideal gas law* relates important thermodynamic properties, and is often used to calculate density.

One form of the law is

$$p V = n R_u T \tag{2.4}$$

where p is the absolute pressure, V is the volume, n is the number of moles, R_u is the universal gas constant (the same for all gases), and T is absolute temperature. Absolute pressure,

---

EXAMPLE 2.1 DENSITY OF AIR

Air at standard sea-level pressure ($p = 101 \text{ kN}/\text{m}^2$) has a temperature of 4°C. What is the density of the air?

Problem Definition

Situation: Air with a known temperature and pressure.

Find: Density (kg/m^3).

Properties: Air, 4°C, p at 101 kN/m²; Table A.2, $R = 287 \text{ J}/\text{kg K}$.

Plan

Apply the ideal gas law, Eq. (2.5), to solve for density, ρ.

Solution

$$\rho = \frac{p}{RT}$$

$$\rho = \frac{101 \times 10^3 \text{ N}/\text{m}^2}{287 \text{ J}/\text{kg K} \times (273 + 4) \text{ K}} = \boxed{1.27 \text{ kg}/\text{m}^3}$$

Review

1. Remember: Use absolute temperatures and pressures with the ideal gas law.
2. Remember: In Eq. (2.5), use R from Table A.2. Do not use R_u.

---

introduced in Chapter 3, is referred to absolute zero. The universal gas constant is 8.314 kJ/kmol-K in the SI system and 1545 ft-lbf/lbmol-°R in the traditional system. Eq. (2.4) can be rewritten as

$$p = \frac{n\mathcal{M}}{\mathcal{V}}\frac{R_u}{\mathcal{M}}T$$

where $\mathcal{M}$ is the molecular weight of the gas. The product of the number of moles and the molecular weight is the mass of the gas. Thus $n\mathcal{M}/\mathcal{V}$ is the mass per unit volume, or density. The quotient $R_u/\mathcal{M}$ is the gas constant, R. Thus a second form of the *ideal gas law* is

$$p = \rho RT \qquad (2.5)$$

Although no gas is ideal, Eq. (2.5) is a valid approximation for most gas flow problems. Values of R for a number of gases are given in Table A.2. To determine the mass density of a gas, solve Eq. (2.5) for ρ:

$$\rho = \frac{p}{RT}$$

2.3

Properties Involving Thermal Energy

Specific Heat, c

The property that describes the capacity of a substance to store thermal energy is called *specific heat*. By definition, it is the amount of thermal energy that must be transferred to a unit mass of substance to raise its temperature by one degree. The specific heat of a gas depends on the process accompanying the change in temperature. If the *specific volume v* of the gas ($v = 1/\rho$) remains constant while the temperature changes, then the specific heat is identified as c_v. However, if the pressure is held constant during the change in state, then the specific heat is identified as c_p. The ratio c_p/c_v is given the symbol k. Values for c_p and k for various gases are given in Table A.2.

Internal Energy

The energy that a substance possesses because of the state of the molecular activity in the substance is termed *internal energy*. Internal energy is usually expressed as a specific quantity—that is, internal energy per unit mass. In the SI system, the *specific internal energy, u,* is given in joules per kilogram; in Traditional Units it is given in Btu / lbm. The internal energy is generally a function of temperature and pressure. However, for an ideal gas, it is a function of temperature alone.

Enthalpy

The combination $u + p/\rho$ is encountered frequently in equations for thermodynamics and compressible flow; it has been given the name *specific enthalpy*. For an ideal gas, u and p/ρ are functions of temperature alone. Consequently their sum, specific enthalpy, is also a function solely of temperature.

2.4

Viscosity

The property of viscosity is important to engineering practice because it leads to significant energy loss when moving fluids contact a solid boundary, or when different zones of fluid are flowing at different velocities.

Viscosity, μ

Viscosity (also called *dynamic viscosity*, or *absolute viscosity*) is a measure of a fluid's resistance to deformation under shear stress. For example, crude oil has a higher resistance to shear than does water. Crude oil will pour *more slowly* than water from an identical beaker held at the same angle. This relative slowness of the oil implies a low "speed" or rate of strain. The symbol used to represent viscosity is μ (mu). To understand the physics of viscosity, it is useful to refer back to solid mechanics and the concepts of shear stress and shear strain. Shear stress, τ, tau, is the ratio of force/area on a surface when the force is aligned parallel to the area. Shear strain is a change in an interior angle of a cubical element, $\Delta\phi$, that was originally a right angle. The shear stress on a material element in solid mechanics is proportional to the strain, and the constant of proportionality is the shear modulus:

$$\{\text{shear stress}\} = \{\text{shear modulus}\} \times \{\text{strain}\}$$

In fluid flow, however, the shear stress on a fluid element is proportional to the rate (speed) of strain, and the constant of proportionality is the viscosity:

$$\{\text{shear stress}\} = \{\text{viscosity}\} \times \{\text{rate of strain}\}$$

Figure 2.1 depicts an initially rectangular element in a parallel flow field. As the element moves downstream, a shear force on the top of the element (and a corresponding shear stress in the opposite direction on the bottom of the element) causes the top surface to move faster (with velocity $V + \Delta V$) than the bottom (at velocity V). The forward and rearward edges become inclined at an angle $\Delta\phi$ with respect to the vertical. The rate at which $\Delta\phi$ changes with time, given by $\dot{\phi}$, is the *rate of strain*, and can be related to the velocity difference between

Figure 2.1

Depiction of strain caused by a shear stress (force per area) in a fluid. The rate of strain is the rate of change of the interior angle of the original rectangle.

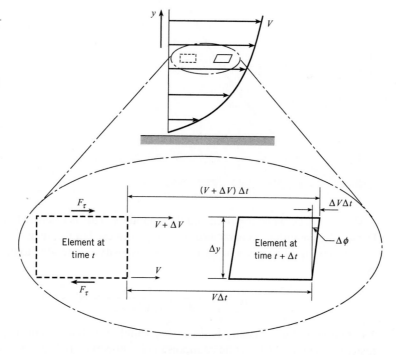

the two surfaces. In time (Δt) the upper surface moves $(V + \Delta V)\Delta t$ while the bottom surface moves $V\Delta t$, so the net difference is $\Delta V\Delta t$. The strain $\Delta\phi$ is

$$\Delta\phi \approx \frac{\Delta V\Delta t}{\Delta y}$$

where Δy is the distance between the two surfaces. The rate of strain is

$$\frac{\Delta\phi}{\Delta t} \approx \frac{\Delta V}{\Delta y}$$

In the limit as $\Delta t \longrightarrow 0$ and $\Delta y \longrightarrow 0$, the rate of strain is related to the velocity gradient by $\dot{\phi} = dV/dy$, so the shear stress (shear force per unit area) is

$$\tau = \mu \frac{dV}{dy} \tag{2.6}$$

For strain in flow near a wall, as shown in Fig. 2.2, the term dV/dy represents the velocity gradient (or change of velocity with distance from the wall), where V is the fluid velocity and y is the distance measured from the wall. The velocity distribution shown is characteristic of flow next to a stationary solid boundary, such as fluid flowing through a pipe. Several observations relating to this figure will help one to appreciate the interaction between viscosity and velocity distribution. First, the velocity gradient at the boundary is finite. The curve of velocity variation cannot be tangent to the boundary because this would imply an infinite velocity gradient and, in turn, an infinite shear stress, which is impossible. Second, a velocity gradient that becomes less steep (dV/dy becomes smaller) with distance from the boundary has a maximum shear stress at the boundary, and the shear stress decreases with distance from the boundary. Also note that the velocity of the fluid is zero at the stationary boundary. That is, at the boundary surface the fluid has the velocity of the boundary—

Figure 2.2

Velocity distribution next to a boundary.

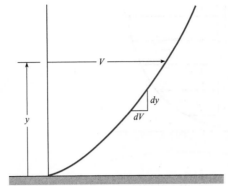

no slip occurs between the fluid and the boundary. This is referred to as the *no-slip condition*. The no-slip condition is characteristic of all flows used in this text.

From Eq. (2.6) it can be seen that the viscosity μ is related to the shear stress and velocity gradient.

$$\mu = \frac{\tau}{dV/dy} \Rightarrow \frac{N/m^2}{(m/s)/m} = N \cdot s/m^2 \tag{2.7}$$

A common unit of viscosity is the *poise,* which is 1 dyne-s/cm^2 or 0.1 N $\cdot$ s/m^2. The viscosity of water at 20°C is one centipoise (10^{-2} poise) or 10^{-3} N $\cdot$ s/m^2. The unit of viscosity in the traditional system is lbf $\cdot$ s/ft^2.

Kinematic Viscosity, ν

Many equations of fluid mechanics include the ratio μ/ρ. Because it occurs so frequently, this ratio has been given the special name *kinematic viscosity.* The symbol used to identify kinematic viscosity is ν (nu). Units of kinematic viscosity ν are m^2/s, as shown.

$$\nu = \frac{\mu}{\rho} \Rightarrow \frac{N \cdot s/m^2}{kg/m^3} = m^2/s \tag{2.8}$$

The units for kinematic viscosity in the traditional system are ft^2/s.

Temperature Dependency

The effect of temperature on viscosity is different for liquids and gases. The viscosity of liquids decreases as the temperature increases, whereas the viscosity of gases increases with increasing temperature; this trend is also true for kinematic viscosity (see Fig. 2.3 and Figs. A.2 and A.3).

To understand the mechanisms responsible for an increase in temperature that causes a decrease in viscosity in a liquid, it is helpful to rely on an approximate theory that has been developed to explained the observed trends (1). The molecules in a liquid form a lattice-like structure with "holes" where there are no molecules, as shown in Fig. 2.4. Even when the liquid is at rest, the molecules are in constant motion, but confined to cells, or "cages." The cage or lattice structure is caused by attractive forces between the molecules. The cages may be thought of as energy barriers. When the liquid is subjected to a rate of strain and thus caused to move, as shown in Fig. 2.4, there is a shear stress, τ, imposed by one layer on another in the fluid. This force/area assists a molecule in overcoming the energy barrier, and it can move into the next hole. The magnitude of these energy barriers is related to viscosity, or resistance

Figure 2.3

Kinematic viscosity for air and crude oil.

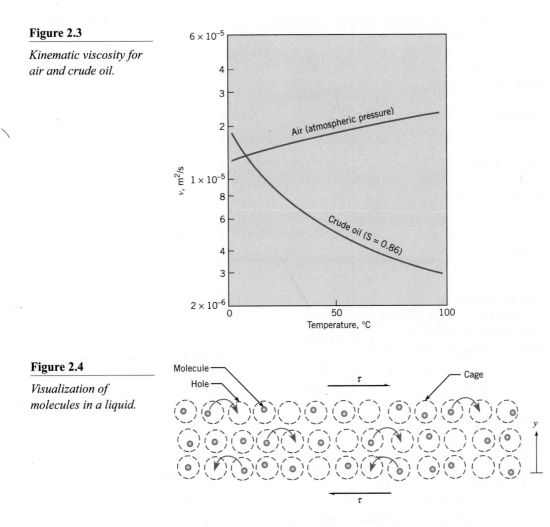

Figure 2.4

Visualization of molecules in a liquid.

to shear deformation. At a higher temperature the size of the energy barrier is smaller, and it is easier for molecules to make the jump, so that the net effect is less resistance to deformation under shear. Thus, an increase in temperature causes a decrease in viscosity for liquids.

An equation for the variation of liquid viscosity with temperature is

$$\mu = C e^{b/T} \tag{2.9}$$

where C and b are empirical constants that require viscosity data at two temperatures for evaluation. This equation should be used primarily for data interpolation. The variation of viscosity (dynamic and kinematic) for other fluids is given in Figs. A.2 and A.3.

As compared to liquids, gases do not have zones or cages to which molecules are confined by intermolecular bonding. Gas molecules are always undergoing random motion. If this random motion of molecules is superimposed upon two layers of gas, where the top layer is moving faster than the bottom layer, periodically a gas molecule will randomly move from one layer to the other. This behavior of a molecule in a low-density gas is analogous to people jumping back and forth between two conveyor belts moving at different speeds as shown in Fig. 2.5. When people jump from the high-speed belt to the low-speed belt, a re-

EXAMPLE 2.2   CALCULATING VISCOSITY OF LIQUID AS A FUNCTION OF TEMPERATURE

The dynamic viscosity of water at 20°C is 1.00×10^{-3} N · s/m², and the viscosity at 40°C is 6.53×10^{-4} N · s/m².
Using Eq. (2.9), estimate the viscosity at 30°C.

Problem Definition

Situation: Viscosity of water is specified at two temperatures.
Find: The viscosity at 30°C by interpolation.
Properties:

a) Water at 20°C, $\mu = 1.00 \times 10^{-3}$ N · s/m².
b) Water at 40°C, $\mu = 6.53 \times 10^{-4}$ N · s/m².

Plan

1. Linearize Eq. (2.9) by taking the logarithm.
2. Interpolate between the two known values of viscosity.
3. Solve for $\ln C$ and b in this linear set of equations.
4. Change back to exponential equation, and solve for μ at 30°C.

Solution

1. Logarithm of Eq. (2.9)
$$\ln \mu = \ln C + b/T$$

2. Interpolation
$$-6.908 = \ln C + 0.00341b$$
$$-7.334 = \ln C + 0.00319b$$

3. Solution for $\ln C$ and b
$$\ln C = -13.51 \qquad b = 1936 \ (K)$$
$$C = e^{-13.51} = 1.357 \times 10^{-6}$$

4. Substitution back in exponential equation
$$\mu = 1.357 \times 10^{-6} e^{1936/T}$$

At 30°C
$$\mu = \boxed{8.08 \times 10^{-4} \ \text{N} \cdot \text{s/m}^2}$$

Review

Note: This value differs by 1% from the reported value in Table A.5, but provides a much better estimate than would be obtained by arithmetically averaging two values on the table.

straining (or braking) force has to be applied to slow the person down (analagous to viscosity). If the people are heavier, or are moving faster, a greater braking force must be applied. This analogy also applies for gas molecules translating between fluid layers where a shear force is needed to maintain the layer speeds. As the gas temperature increases, more of the molecules will be making random jumps. Just as the jumping person causes a braking action on the belt, highly mobile gas molecules have momentum, which must be resisted by the layer to which the molecules jump. Therefore, as the temperature increases, the viscosity, or resistance to shear, also increases.

Figure 2.5

Analogy of people moving between conveyor belts and gas molecules translating between fluid layers.

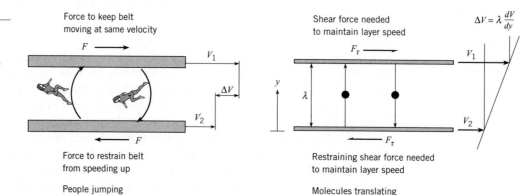

Force to keep belt moving at same velocity

Force to restrain belt from speeding up

People jumping between moving belts

Shear force needed to maintain layer speed

$\Delta V = \lambda \frac{dV}{dy}$

Restraining shear force needed to maintain layer speed

Molecules translating between fluid layers

EXAMPLE 2.3   MODELING A BOARD SLIDING ON A LIQUID LAYER

A board 1 m by 1 m that weighs 25 N slides down an inclined ramp (slope = 20°) with a velocity of 2.0 cm/s. The board is separated from the ramp by a thin film of oil with a viscosity of 0.05 N · s/m². Neglecting edge effects, calculate the space between the board and the ramp.

Problem Definition

Situation: A board is sliding down a ramp, on a thin film of oil.

Find: Space (in m) between the board and the ramp.

Assumptions: A linear velocity distribution in the oil.

Properties: Oil, $\mu = 0.05$ N · s/m².

Sketch:

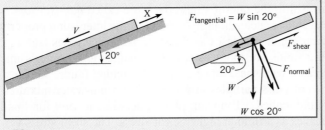

Plan

1. Draw a free body diagram of the board, as shown in "sketch."
   - For a constant sliding velocity, the resisting shear force is equal to the component of weight parallel to the inclined ramp.
   - Relate shear force to viscosity and velocity distribution.

2. With a linear velocity distribution, dV/dy can everywhere be expressed as $\Delta V/\Delta y$, where ΔV is the velocity of the board, and Δy is the space between the board and the ramp.

3. Solve for Δy.

Solution

1. Freebody analysis

$$F_{\text{tangential}} = F_{\text{shear}}$$

$$W \sin 20° = \tau A$$

$$W \sin 20° = \mu \frac{dV}{dy} A$$

2. Substitution of dV/dy as $\Delta V/\Delta y$

$$W \sin 20° = \mu \frac{\Delta V}{\Delta y} A$$

3. Solution for Δy

$$\Delta y = \frac{\mu \Delta V A}{W \sin 20°}$$

$$\Delta y = \frac{0.05 \text{ N} \cdot \text{s/m}^2 \times 0.020 \text{ m/s} \times 1 \text{ m}^2}{25 \text{ N} \times \sin 20°}$$

$$\Delta y = 0.000117 \text{ m}$$

$$\Delta y = \boxed{0.117 \text{ mm}}$$

An estimate for the variation of gas viscosity with temperature is *Sutherland's equation*,

$$\frac{\mu}{\mu_0} = \left(\frac{T}{T_0}\right)^{3/2} \frac{T_0 + S}{T + S} \tag{2.10}$$

where μ_0 is the viscosity at temperature T_0, and S is Sutherland's constant. All temperatures are absolute. Sutherland's constant for air is 111 K; values for other gases are given in Table A.2. Using Sutherland's equation for air yields viscosities with an accuracy of ±2% for temperatures between 170 K and 1900 K. In general, the effect of pressure on the viscosity of common gases is minimal for pressures less than 10 atmospheres.

Newtonian versus Non-Newtonian Fluids

Fluids for which the shear stress is directly proportional to the rate of strain are called *Newtonian fluids*. Because shear stress is directly proportional to the shear strain, dV/dy, a plot relating these variables (see Fig. 2.6) results in a straight line passing through the origin. The slope of this line is the value of the dynamic (absolute) viscosity. For some fluids the shear

Figure 2.6

Shear stress relations for different types of fluids.

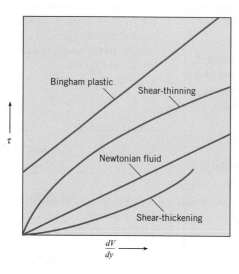

stress may not be directly proportional to the rate of strain; these are called *non-Newtonian fluids*. One class of non-Newtonian fluids, shear-thinning fluids, has the interesting property that the ratio of shear stress to shear strain decreases as the shear strain increases (see Fig. 2.6). Some common shear-thinning fluids are toothpaste, catsup, paints, and printer's ink. Fluids for which the viscosity increases with shear rate are shear-thickening fluids. Some examples of these fluids are mixtures of glass particles in water and gypsum-water mixtures. Another type of non-Newtonian fluid, called a Bingham plastic, acts like a solid for small values of shear stress and then behaves as a fluid at higher shear stress. The shear stress versus shear strain rate for a Bingham plastic is also shown in Fig. 2.6. This book will focus on the theory and applications involving Newtonian fluids. For more information on the theory of flow of non-Newtonian fluids, see references (2) and (3).

2.5

Bulk Modulus of Elasticity

The *bulk modulus of elasticity, E_v*, is a property that relates changes in pressure to changes in volume (e.g., expansion or contraction)

$$E_v = - \frac{dp}{d\Psi / \Psi} = - \frac{\text{change in pressure}}{\text{fractional change in volume}} \qquad (2.11)$$

where dp is the differential pressure change, $d\Psi$ is the differential volume change, and Ψ is the volume of fluid. Because $d\Psi / \Psi$ is negative for a positive dp, a negative sign is used in the definition to yield a positive E_v. The elasticity is often called the compressibility of the fluid.

The fractional change in volume can be related to the change in material density using

$$M = \rho \Psi \qquad (2.12)$$

Since the mass is constant

$$dM = \rho \, d\Psi + \Psi \, d\rho = 0$$

so

$$\mathcal{V}\, d\rho = -\rho\, d\mathcal{V} \qquad or \qquad \frac{d\rho}{\rho} = -\frac{d\mathcal{V}}{\mathcal{V}}$$

and the definition of the bulk modulus of elasticity becomes

$$E_v = \frac{dp}{d\rho/\rho} = \frac{change\ in\ pressure}{fractional\ change\ in\ density} \tag{2.13}$$

The bulk modulus of elasticity of water is approximately 2.2 GN/m^2, which corresponds to a 0.05% change in volume for a change of $1\,MN/m^2$ in pressure. Obviously, the term *incompressible* is justifiably applied to water because it has such a small change in volume for a very large change in pressure.

The elasticity of an ideal gas is proportional to the pressure, according to the ideal gas law. For an isothermal (constant-temperature) process,

$$\frac{dp}{d\rho} = RT$$

so

$$E_v = \rho\,\frac{dp}{d\rho} = \rho RT = p$$

For an adiabatic process, $E_v = kp$, where k is the ratio of specific heats, c_p/c_v.

The elasticity or compressibility of a gas is important in high-speed gas flows where pressure variations can cause significant density changes. As will be shown in Chapter 12, the elasticity of a gas is related to the speed of sound in that gas. The ratio of the flow velocity to the speed of sound is the Mach number, which relates to the importance of elasticity effects.

2.6

Surface Tension

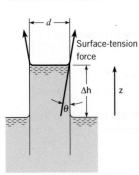

Figure 2.7

Capillary action in a small tube.

Surface tension, σ (sigma), is a material property whereby a liquid at a material interface, usually liquid-gas, exerts a force per unit length along the surface. According to the theory of molecular attraction, molecules of liquid considerably below the surface act on each other by forces that are equal in all directions. However, molecules near the surface have a greater attraction for each other than they do for molecules below the surface because of the presence of a different substance above the surface. This produces a layer of surface molecules on the liquid that acts like a stretched membrane. Because of this membrane effect, each portion of the liquid surface exerts "tension" on adjacent portions of the surface or on objects that are in contact with the liquid surface. This tension acts in the plane of the surface, and is given by:

$$F_\sigma = \sigma L \tag{2.14}$$

where L is the length over which the surface tension acts.

Surface tension for a water–air surface is 0.073 N/m at room temperature. The magnitude of surface tension decreases with increasing temperature; tabulated values for different liquids as a function of temperature are available in the literature and online. The effect of surface tension is illustrated for the case of *capillary action* (rise above a static water level at atmospheric pressure) in a small tube (Fig. 2.7). Here the end of a small-diameter tube is inserted into a reservoir of water, and the characteristic curved water surface profile occurs within the tube. The relatively greater attraction of the water molecules for the glass rather than the air causes the water surface to curve upward in the region of the glass wall. Then the surface tension force acts around the circumference of the tube, in the direction indicated. It may be assumed that the *contact angle* θ (theta) is equal to 0° for water against glass. The surface tension force produces a net upward force on the water that causes the water in the tube to rise above the water surface in the reservoir. A calculation of the surface tension force acting to raise the water in a small-diameter tube is demonstrated in Example 2.4.

Other manifestations of surface tension include the excess pressure (over and above atmospheric pressure) created inside droplets and bubbles because there is necessarily a pressure difference across a curved interface; the breakup of a liquid jet into droplets; the shape and motion of bubbles, the structure of foams, and the binding together of wetted granular material, such as soil.

Surface tension forces for several different cases are shown in Fig. 2.8. Case (*a*) is a spherical droplet of radius *r*. The surface tension force is balanced by the internal pressure.

$$F_\sigma = \sigma L = pA$$

or

$$2\pi r \sigma = p\pi r^2$$

$$p = \frac{2\sigma}{r}$$

Case (*b*) is a bubble of radius *r* that has internal and external surfaces and the surface-tension force acts on both surfaces, so

$$p = \frac{2\sigma}{r}$$

Case (*c*) is a cylinder supported by surface-tension forces. The liquid does not wet the cylinder surface. The maximum weight the surface tension can support is

$$W = 2F_\sigma = 2\sigma L$$

where *L* is the length of the cylinder.

EXAMPLE 2.4 CAPILLARY RISE IN A TUBE

To what height above the reservoir level will water (at 20°C) rise in a glass tube, such as that shown in Fig. 2.7, if the inside diameter of the tube is 1.6 mm?

Problem Definition

Situation: A glass tube of small diameter placed in an open reservoir of water induces capillary rise.

Find: The height the water will rise above the reservoir level.

Sketch: See Figure 2.7.

Properties: Water (20 °C), Table A.5, $\sigma = 0.073$ N/m; $\gamma = 9790$ N/m³.

Plan

1. Perform a force balance on water that has risen in the tube.
2. Solve for Δh.

Solution

1. Force balance: Weight of water (down) is balanced by surface tension force (up).

$$F_{\sigma,z} - W = 0$$

$$\sigma \pi d \cos\theta - \gamma(\Delta h)(\pi d^2/4) = 0$$

Because the contact angle θ for water against glass is so small, it can be assumed to be 0°; therefore $\cos\theta \approx 1$. Therefore:

$$\sigma \pi d - \gamma(\Delta h)\left(\frac{\pi d^2}{4}\right) = 0$$

2. Solve for Δh:

$$\Delta h = \frac{4\sigma}{\gamma d} = \frac{4 \times 0.073 \text{ N/m}}{9790 \text{ N/m}^3 \times 1.6 \times 10^{-3} \text{ m}} = \boxed{18.6 \text{ mm}}$$

Figure 2.8

Surface-tension forces for several different cases.

(a) Spherical droplet

(b) Spherical bubble

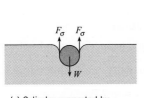
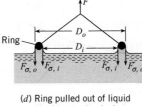

(c) Cylinder supported by surface tension (liquid does not wet cylinder)

(d) Ring pulled out of liquid (liquid wets the ring)

Case (*d*) is a ring being pulled out of a liquid. This is a technique to measure surface tension. The force due to surface tension on the ring is

$$F_\sigma = F_{\sigma,i} + F_{\sigma,o}$$
$$= \pi\sigma(D_i + D_o)$$

2.7

Vapor Pressure

The pressure at which a liquid will vaporize, or boil, at a given temperature, is called its *vapor pressure*. This means that boiling occurs whenever the local pressure equals the vapor pressure. Vapor pressure increases with temperature. Note that there are two ways to boil a liquid. One way is to raise the temperature, assuming that the pressure is fixed. For water at 14.7 psia, this can be accomplished by increasing the temperature of water at sea level to 212°F, thus reaching the temperature where the vapor pressure is equal to the same value. However, boiling can also occur in water at temperatures much below 212°F if the pressure in the water is reduced to the vapor pressure of water corresponding to that lower temperature. For example, the vapor pressure of water at 50°F (10°C) is 0.178 psia (approximately 1% of standard atmospheric pressure). Therefore, if the pressure in water at 50°F is reduced to 0.178 psia, the water boils.*

Such boiling often occurs in localized low-pressure zones of flowing liquids, such as on the suction side of a pump. When localized low-pressure boiling does occur in flowing liquids, vapor bubbles start growing in local regions of very low pressure and then collapse in regions of higher pressure downstream. This phenomenon, which is called *cavitation,* can cause extensive damage to fluids systems, and is discussed in Chapter 5.

Table A.5 gives values of vapor pressure for water.

---

* Actually, boiling can occur at this vapor pressure only if there is a gas–liquid surface present to allow the process to start. Boiling at the bottom of a pot of water is usually initiated in crevices in the material of the pot, in which minute bubbles of air are entrapped even when the pot is filled with water.

Summary

2.8

The commonly used fluid properties are

Mass density (ρ): mass per unit volume.

Specific weight (γ): weight per unit volume.

Specific gravity (S): ratio of specific weight to specific weight of water at reference conditions.

The relationship between pressure, density, and temperature for an ideal gas is

$$p = \rho R T$$

where R is the gas constant, and pressure and temperature must be expressed in absolute values.

In a fluid the shear stress is proportional to the rate of strain, and the constant of proportionality is the viscosity. The shear stress at a wall is given by

$$\tau = \mu \frac{dV}{dy}$$

where dV/dy is the velocity gradient of the fluid evaluated at the wall. In a Newtonian fluid, the viscosity is independent of the rate of strain. A fluid for which the effective viscosity decreases with increasing strain rate is a shear-thinning fluid.

Surface tension is the result of molecular attraction near a free surface, causing the surface to act like a stretched membrane.

The bulk modulus of elasticity relates to the pressure required to change the density of a fluid.

When the local pressure is equal to the vapor pressure at a given temperature, liquid boils.

References

1. Bird, R. B., Stewart, W. E., and Lightfoot, E. N. *Transport Phenomena.* John Wiley & Sons, New York, 1960.

2. Harris, J. *Rheology and Non-Newtonian Flow.* Longman, New York, 1977.

3. Schowalter, W. R. *Mechanics of Non-Newtonian Fluids.* Pergamon Press, New York, 1978.

Problems

*A *Preview Question* ($\mathbb{PQ}$) can be assigned prior to in-class coverage of a topic.

Properties Related to Mass and Weight

*2.1 $\mathbb{PQ}$◀ Describe how density differs from specific weight.

*2.2 $\mathbb{PQ}$◀ For what fluids can we (usually) assume density to be nearly constant? For what fluids must we be careful to calculate density as a function of temperature and pressure?

*2.3 $\mathbb{PQ}$◀ Where in this text can you find density data for such fluids as oil and mercury?

2.4 An engineer living at an elevation of 2500 ft is conducting experiments to verify predictions of glider performance. To process data, density of ambient air is needed. The engineer measures temperature (74.3°F) and atmospheric pressure (27.3 inches of mercury). Calculate density in units of kg/m^3. Compare the calculated value with data from Table A.3 and make a recommendation about the effects of elevation on density; that is, are the effects of elevation significant?

2.5 Calculate the density and specific weight of carbon dioxide at a pressure of 300 kN/m^2 absolute and 60°C.

2.6 Determine the density and specific weight of methane gas at at a pressure of 300 kN/m^2 absolute and 60°C.

2.7 Natural gas is stored in a spherical tank at a temperature of 10°C. At a given initial time, the pressure in the tank is 100 kPa gage, and the atmospheric pressure is 100 kPa absolute. Some time later, after considerably more gas is pumped into the tank, the pressure in the tank is 200 kPa gage, and the temperature is still 10°C. What will be the ratio of the mass of natural gas in the tank when p = 200 kPa gage to that when the pressure was 100 kPa gage?

2.8 At a temperature of 100°C and an absolute pressure of 5 atmospheres, what is the ratio of the density of water to the density of air, ρ_w / ρ_a?

2.9 Find the total weight of a 10 ft^3 tank of oxygen if the oxygen is pressurized to 500 psia, the tank itself weighs 150 lbf, and the temperature is 70°F?

2.10 A 10 m^3 oxygen tank is at 15°C and 800 kPa. The valve is opened, and some oxygen is released until the pressure in the tank drops to 600 kPa. Calculate the mass of oxygen that has been released from the tank if the temperature in the tank does not change during the process.

2.11 What are the specific weight and density of air at an absolute pressure of 600 kPa and a temperature of 50°C?

2.12 Meteorologists often refer to air masses in forecasting the weather. Estimate the mass of 1 mi^3 of air in slugs and kilograms. Make your own reasonable assumptions with respect to the conditions of the atmosphere.

2.13 A bicycle rider has several reasons to be interested in the effects of temperature on air density. The aerodynamic drag force decreases linearly with density. Also, a change in temperature will affect the tire pressure.

a. To visualize the effects of temperature on air density, write a computer program that calculates the air density at atmospheric pressure for temperatures from –10°C to 50°C.

b. Also assume that a bicycle tire was inflated to an absolute pressure of 450 kPa at 20°C. Assume the volume of the tire does not change with temperature. Write a program to show how the tire pressure changes with temperature in the same temperature range, –10°C to 50°C.

Prepare a table or graph of your results for both problems. What engineering insights do you gain from these calculations?

2.14 A design team is developing a prototype CO$_2$ cartridge for a manufacturer of rubber rafts.This cartridge will allow a user to quickly inflate a raft. A typical raft is shown in the sketch. Assume a raft inflation pressure of 3 psi (this means that the absolute pressure is 3 psi greater than local atmospheric pressure). Estimate the volume of the raft and the mass of CO$_2$ in grams in the prototype cartridge.

2.15 A team is designing a helium-filled balloon that will fly to an altitude of 80,000 ft. As the balloon ascends, the upward force (buoyant force) will need to exceed the total weight. Thus, weight is critical. Estimate the weight (in newtons) of the helium inside the balloon. The balloon is inflated at a site where the atmospheric pressure is 0.89 bar and the temperature is 22°C. When inflated prior to launch, the balloon is spherical (radius 1.3 m) and the inflation pressure equals the local atmospheric pressure.

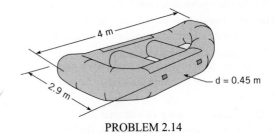

PROBLEM 2.14

2.16 Hydrometers are used in the wine and beer industries to measure the alcohol content of the product. This is accomplished by measuring the specific gravity of the liquid before fermentation, during fermentation, or after fermentation is complete. During fermentation, glucose ($C_6H_{12}O_6$) is converted to ethyl alcohol (CH_3CH_2OH) and carbon dioxide gas, which escapes from the vat.

$$C_6H_{12}O_6 \longrightarrow 2(CH_3CH_2OH) + 2(CO_2)$$

Brewer's yeast tolerates alcohol contents to approximately 5% before fermentation stops, whereas wine yeast tolerates alcohol contents up to 21% depending on the yeast strain. The specific gravity of alcohol is 0.80, and the maximum specific gravity of sugar in solution is 1.59. If a wine has a specific gravity of 1.08 before fermentation, and all the sugar is converted to alcohol, what will be the final specific gravity of the wine and the percent alcohol content by volume? Assume that the initial liquid (the unfermented wine is called must) contains only sugar and water.

Viscosity

*2.17 $\mathbb{PQ}$◀ The following questions relate to viscosity.

a. What are the primary dimensions of viscosity? What are five common units?

b. What is the viscosity of SAE 10W-30 motor oil at 115°F (in traditional units)?

c. How does viscosity of water vary with temperature? Why?

d. How does viscosity of air vary with temperature? Why?

2.18 What is the change in the viscosity and density of water between 10°C and 70°C? What is the change in the viscosity and density of air between 10°C and 70°C? Assume standard atmospheric pressure (p = 101 kN/m^2 absolute).

2.19 Determine the change in the kinematic viscosity of air that is heated from 10°C to 70°C. Assume standard atmospheric pressure.

2.20 Find the dynamic and kinematic viscosities of kerosene, SAE 10W-30 motor oil, and water at a temperature of 38°C (100°F).

2.21 What is the ratio of the dynamic viscosity of air to that of water at standard pressure and a temperature of 20°C? What is the ratio of the kinematic viscosity of air to that of water for the same conditions?

2.22 Using Sutherland's equation and the ideal gas law, develop an expression for the kinematic viscosity ratio v/v_0 in terms of pressures p and p_0 and temperatures T and T_0, where the subscript 0 refers to a reference condition.

2.23 The dynamic viscosity of air at 15°C is 1.78×10^{-5} N $\cdot$ s/m^2. Using Sutherland's equation, find the viscosity at 100°C.

2.24 The kinematic viscosity of methane at 15°C and atmospheric pressure is 1.59×10^{-5} m^2/s. Using Sutherland's equation and the ideal gas law, find the kinematic viscosity at 200°C and 2 atmospheres.

2.25 The dynamic viscosity of nitrogen at 59°F is 3.59×10^{-7} lbf $\cdot$ s/ft^2. Using Sutherland's equation, find the dynamic viscosity at 200°F.

2.26 The kinematic viscosity of helium at 59°F and 1 atmosphere is 1.22×10^{-3} ft^2/s. Using Sutherland's equation and the ideal gas law, find the kinematic viscosity at 30°F and a pressure of 1.5 atmospheres.

2.27 The absolute viscosity of propane at 100°C is 1.00×10^{-5} N $\cdot$ s/m^2 and at 400°C is 1.72×10^{-5} N $\cdot$ s/m^2. Find Sutherland's constant for propane.

2.28 Ammonia is very volatile, so it may be either a gas or a liquid at room temperature. When it is a gas, its absolute viscosity at 68°F is 2.07×10^{-7} lbf $\cdot$ s/ft^2 and at 392°F is 3.46×10^{-7} lbf $\cdot$ s/ft^2. Using these two data points, find Sutherland's constant for ammonia.

2.29 The viscosity of SAE 10W-30 motor oil at 38°C is 0.067 N $\cdot$ s/m^2 and at 99°C is 0.011 N $\cdot$ s/m^2. Using Eq. (2.8) for interpolation, find the viscosity at 60°C. Compare this value with that obtained by linear interpolation.

2.30 The viscosity of grade 100 aviation oil at 100°F is 4.43×10^{-3} lbf $\cdot$ s/ft^2 and at 210°F is 3.9×10^{-4} lbf $\cdot$ s/ft^2. Using Eq. (2.8), find the viscosity at 150°F.

2.31 Two plates are separated by a 1/8-in. space. The lower plate is stationary; the upper plate moves at a velocity of 25 ft/s. Oil (SAE 10W-30, 150°F), which fills the space between the plates, has the same velocity as the plates at the surface of contact. The variation in velocity of the oil is linear. What is the shear stress in the oil?

2.32 Find the kinematic and dynamic viscosities of air and water at a temperature of 40°C (104°F) and an absolute pressure of 170 kPa (25 psia).

2.33 The sliding plate viscometer shown below is used to measure the viscosity of a fluid. The top plate is moving to the right with a constant velocity of 10 m/s in response to a force of 3 N. The bottom plate is stationary. What is the viscosity of the fluid? Assume a linear velocity distribution.

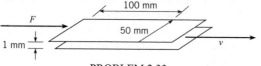

PROBLEM 2.33

2.34 The velocity distribution for water (20°C) near a wall is given by $u = a(y/b)^{1/6}$, where $a = 10$ m/s, $b = 2$ mm, and y is the distance from the wall in mm. Determine the shear stress in the water at $y = 1$ mm.

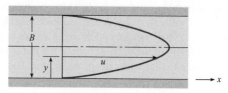

PROBLEMS 2.35, 2.36, 2.37

2.35 The velocity distribution for the flow of crude oil at 100°F ($\mu = 8 \times 10^{-5}$ lbf $\cdot$ s/ft^2) between two walls is shown, and is given by $u = 100y(0.1 - y)$ ft/s, where y is measured in feet and the space between the walls is 0.1 ft. Plot the velocity distribution and determine the shear stress at the walls.

2.36 A liquid flows between parallel boundaries as shown above. The velocity distribution near the lower wall is given in the following table:

y in mm	V in m/s
0.0	0.00
1.0	1.00
2.0	1.99
3.0	2.98

a. If the viscosity of the liquid is 10^{-3} N $\cdot$ s/m^2, what is the maximum shear stress in the liquid?

b. Where will the minimum shear stress occur?

2.37 Suppose that glycerin is flowing $(T = 20°C)$ and that the pressure gradient dp/dx is -1.6 kN/m^3. What are the velocity and shear stress at a distance of 12 mm from the wall if the space B between the walls is 5.0 cm? What are the shear stress and velocity at the wall? The velocity distribution for viscous flow between stationary plates is

$$u = -\frac{1}{2\mu}\frac{dp}{dx}(By - y^2)$$

2.38 A laminar flow occurs between two horizontal parallel plates under a pressure gradient dp/ds (p decreases in the positive s direction). The upper plate moves left (negative) at velocity u_t. The expression for local velocity u is given as

$$u = -\frac{1}{2\mu}\frac{dp}{ds}(Hy - y^2) + u_t\frac{y}{H}$$

a. Is the magnitude of the shear stress greater at the moving plate $(y = H)$ or at the stationary plate $(y = 0)$?

b. Derive an expression for the y position of zero shear stress.

c. Derive an expression for the plate speed u_t required to make the shear stress zero at $y = 0$.

2.39 Consider the ratio μ_{100}/μ_{50}, where μ is the viscosity of oxygen and the subscripts 100 and 50 are the temperatures of the oxygen in degrees Fahrenheit. Does this ratio have a value (a) less than 1, (b) equal to 1, or (c) greater than 1?

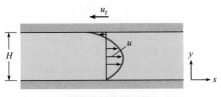

PROBLEM 2.38

2.40 This problem involves a cylinder falling inside a pipe that is filled with oil, as depicted in the figure. The small space between the cylinder and the pipe is lubricated with an oil film that has viscosity μ. Derive a formula for the steady rate of descent of a cylinder with weight W, diameter d, and length ℓ sliding inside a vertical smooth pipe that has inside diameter D. Assume that the cylinder is concentric with the pipe as it falls. Use the general formula to find the rate of descent of a cylinder 100 mm in diameter that slides inside a 100.5 mm pipe. The cylinder is 200 mm long and weighs 15 N. The lubricant is SAE 20W oil at 10°C.

2.41 The device shown consists of a disk that is rotated by a shaft. The disk is positioned very close to a solid boundary. Between the disk and the boundary is viscous oil.

a. If the disk is rotated at a rate of 1 rad/s, what will be the ratio of the shear stress in the oil at $r = 2$ cm to the shear stress at $r = 3$ cm?

b. If the rate of rotation is 2 rad/s, what is the speed of the oil in contact with the disk at $r = 3$ cm?

c. If the oil viscosity is 0.01 N · s/m^2 and the spacing y is 2 mm, what is the shear stress for the conditions noted in part (b)?

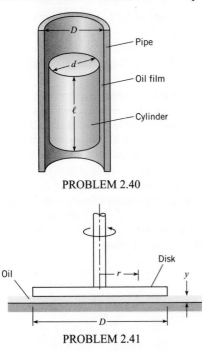

PROBLEM 2.40

PROBLEM 2.41

2.42 Some instruments having angular motion are damped by means of a disk connected to the shaft. The disk, in turn, is im-

mersed in a container of oil, as shown. Derive a formula for the damping torque as a function of the disk diameter D, spacing S, rate of rotation ω, and oil viscosity μ.

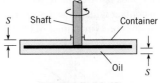

PROBLEM 2.42

2.43 One type of viscometer involves the use of a rotating cylinder inside a fixed cylinder The gap between the cylinders must be very small to achieve a linear velocity distribution in the liquid. (Assume the maximum spacing for proper operation is 0.05 in.). Design a viscometer that will be used to measure the viscosity of motor oil from 50°F to 200°F.

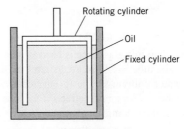

PROBLEM 2.43

Elasticity and Volume Changes

*2.44 $\mathbb{PQ}$◄ The bulk modulus of elasticity of ethyl alcohol is 1.06×10^9 Pa. For water, it is 2.15×10^9 Pa. Which of these liquids is easier to compress? Why?

2.45 A pressure of 2×10^6 N/m^2 is applied to a mass of water that initially filled a 2000 cm^3 volume. Estimate its volume after the pressure is applied.

2.46 Calculate the pressure increase that must be applied to water to reduce its volume by 2%.

2.47 An open vat in a food processing plant contains 400 L of water at 20°C and atmospheric pressure. If the water is heated to 80°C, what will be the percentage change in its volume? If the vat has a diameter of 3 m, how much will the water level rise due to this temperature increase? *Hint:* In this case the volume change is due to change in temperature.

Surface Tension

*2.48 $\mathbb{PQ}$◄ Advanced texts define the surface tension σ as an energy/area. Use primary dimensions to show that energy/area equals force/length.

2.49 Which of the following is the formula for the gage pressure within a very small spherical droplet of water:

(a) $p = \sigma/d$, (b) $p = 4\sigma/d$, or (c) $p = 8\sigma/d$?

2.50 A spherical soap bubble has an inside radius R, a film thickness t, and a surface tension σ. Derive a formula for the pressure within the bubble relative to the outside atmospheric pressure. What is the pressure difference for a bubble with a 4 mm radius? Assume σ is the same as for pure water.

2.51 A water bug is suspended on the surface of a pond by surface tension (water does not wet the legs). The bug has six legs, and each leg is in contact with the water over a length of 5 mm. What is the maximum mass (in grams) of the bug if it is to avoid sinking?

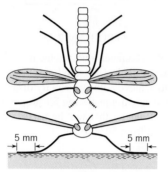

PROBLEM 2.51

2.52 A water column in a glass tube is used to measure the pressure in a pipe. The tube is 1/4 in. (6.35 mm) in diameter. How much of the water column is due to surface-tension effects? What would be the surface-tension effects if the tube were 1/8 in. (3.2 mm) or 1/32 in. (0.8 mm) in diameter?

2.53 Calculate the maximum capillary rise of water between two vertical glass plates spaced 1 mm apart.

PROBLEM 2.52

2.54 What is the pressure within a 1 mm spherical droplet of water relative to the atmospheric pressure outside?

2.55 By measuring the capillary rise in a tube, one can calculate the surface tension. The surface tension of water varies linearly with temperature from 0.0756 N/m at 0°C to 0.0589 N/m at 100°C. Size a tube (specify diameter and length) that uses capillary rise of water to measure temperature in the range from 0°C to 100°C. Is this design for a thermometer a good idea?

2.56 Consider a soap bubble 2 mm in diameter and a droplet of water, also 2 mm in diameter, that are falling in air. If the value

of the surface tension for the film of the soap bubble is assumed to be the same as that for water, which has the greater pressure inside it? (a) the bubble, (b) the droplet, (c) neither—the pressure is the same for both.

2.57 A drop of water at 20°C is forming under a solid surface. The configuration just before separating and falling as a drop is shown in the figure. Assume the forming drop has the volume of a hemisphere. What is the diameter of the hemisphere just before separating?

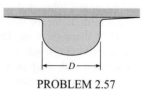

PROBLEM 2.57

2.58 The surface tension of a liquid is being measured with a ring as shown in Fig. 2.6d. The ring has an outside diameter of 10 cm and an inside diameter of 9.5 cm. The mass of the ring is 10 g. The force required to pull the ring from the liquid is the weight corresponding to a mass of 16 g. What is the surface tension of the liquid (in N/m)?

Vapor Pressure

*2.59 PQ◄ If a liquid reaches the vapor pressure, what happens in the liquid?

*2.60 PQ◄ How does vapor pressure change with increasing temperature?

2.61 At a temperature of 60°F, what pressure must be imposed in order for water to boil.

2.62 Water is at 20°C, and the pressure is lowered until bubbles are noticed to be forming. What must the magnitude of the pressure be?

2.63 A student in the laboratory plans to exert a vacuum in the head space above a surface of water in a closed tank. She plans for the the absolute pressure in the tank to be 10,400 Pa. The temperature in the lab is 20°C. Will water bubble into the vapor phase under these circumstances?

2.64 The vapor pressure of water at 100°C is 101 kN/m^2, because water boils under these conditions. The vapor pressure of water decreases approximately linearly with decreasing temperature at a rate of 3.1 $kN/m^2/°C$. Calculate the boiling temperature of water at an altitude of 3000 m, where the atmospheric pressure is 69 kN/m^2 absolute.

Fluid Statics

The first man made structure to exceed the masonry mass of the Great Pyramid of Giza was the Hoover Dam. Design of dams involves calculations of hydrostatic forces.(Photo courtesy of U.S. Bureau of Reclamation, Lower Colorado Region)

SIGNIFICANT LEARNING OUTCOMES

Conceptual Knowledge
- Describe pressure and pressure distribution.
- Describe gage, absolute, and vacuum pressure.
- List the steps used to derive the hydrostatic differential equation.

Procedural Knowledge
- Apply the hydrostatic equation and the manometer equations to predict pressure.
- Apply the panel equations to predict forces and moments.
- Apply the buoyancy equation to predict forces.

Applications (Typical)
- For applications involving the atmosphere, the ocean, manometers, and hydraulic machines, find pressure values and distributions.
- For structures and components subjected to hydrostatic loading, find forces and moments.

Chapters 1 and 2 set the stage by describing fluids and their properties. This chapter begins mechanics of fluids in depth by introducing many concepts related to pressure and by describing how to calculate forces associated with distributions of pressure. This chapter is restricted to fluids that are in hydrostatic equilibrium.

As shown in Fig. 3.1, the hydrostatic condition involves equilibrium of a fluid particle. A *fluid particle,* is defined as a body of fluid having finite mass and internal structure but negligible dimensions. Thus, a fluid particle is very small, but large enough so that the continuum assumption (p. 2) applies. The *hydrostatic condition* means that each fluid particle is in force equilibrium with the net force due to pressure balancing the weight of the fluid particle.

Pressure

This section describes pressure, pressure distribution, and related concepts. This knowledge is foundational for all aspects of fluid mechanics. This section also presents an interesting application, the hydraulic machine.

Figure 3.1

Figure 3.1

The hydrostatic condition.
(a) A fluid particle in a body of fluid.
(b) Forces acting on the fluid paricle.

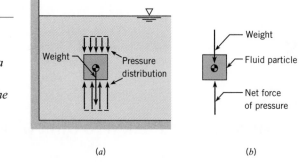

(a)   *(b)*

Definition of Pressure

Pressure is defined as the ratio of normal force to area at a point. For example, Fig. 3.2 shows fluid inside an object such as air inside a soccer ball. The molecules of the fluid interact with the walls to produces a pressure distribution. At each point on the walls, this pressure distribution creates a resultant force ΔF_{normal} that acts on an infinitesimal unit of area ΔA as shown. Pressure is the ratio of normal force magnitude $\left|\Delta \vec{F}_{normal}\right|$ to unit area ΔA at a point:

$$p = \lim_{\Delta A \to 0} \frac{\left|\Delta \vec{F}_{normal}\right|}{\Delta A} = \frac{dF_{normal}}{dA} \tag{3.1}$$

The reason that pressure is defined using a derivative is that pressure often varies from point to point. For example, pressure acting on the windshield of a moving car will vary at different locations on the windshield.

Pressure is a scalar quantity; that is, it has magnitude only. Pressure is not a force; rather it is a scalar that produces resultant force by its action on an area. The resultant force is normal to the area and acts in a direction toward the surface (compressive).

Some units for pressure give a ratio of force to area. Newtons per square meter of area, or pascals (Pa), is the SI unit. The traditional units include psi, which is pounds-force per square inch, and psf, which is pounds-force per square foot. Other units for pressure give the height of a column of liquid. For example, pressure in a balloon will push on a water column upward about 8 inches as shown in Fig. 3.3. Engineers state that the pressure in the balloon is 8 inches of water: $p = 8$ in-H_2O. When pressure is given in units of "height of a fluid column," the pressure value can be directly converted to other units using Table F.1. For example, the pressure in the balloon is

$$p = (8 \text{ in-}H_2O)(248.8 \text{ Pa/in-}H_2O) = 1.99 \text{ kPa}$$

Figure 3.2

Pressure acting on the walls of a sphere.

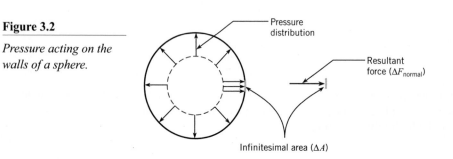

Figure 3.3

Pressure in a balloon causing a column of water to rise 8 inches.

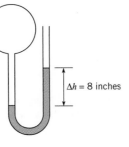

Δh = 8 inches

Standard atmospheric pressure, which is the air pressure at sea level, can be written using multiple units:

$$1.0 \text{ atm} = 101.3 \text{ kPa} = 14.70 \text{ psi} = 33.9 \text{ ft-H}_2\text{O} = 760 \text{ mm-Hg} = 29.92 \text{ in-Hg} = 2116 \text{ psf}$$

Absolute Pressure, Gage Pressure, and Vacuum Pressure

Engineers use several different scales for pressure. Absolute pressure is referenced to regions such as outer space, where the pressure is essentially zero because the region is devoid of gas. The pressure in a perfect vacuum is called absolute zero, and pressure measured relative to this zero pressure is termed *absolute pressure.*

When pressure is measured relative to prevailing local atmospheric pressure, the pressure value is called *gage pressure.* For example, when a tire pressure gage gives a value of 300 kPa (44 psi), this means that the absolute pressure in the tire is 300 kPa greater than local atmospheric pressure. To convert gage pressure to absolute pressure, add the local atmospheric pressure. For example, a gage pressure of 50 kPa recorded in a location where the atmospheric pressure is 100 kPa is expressed as either

$$p = 50 \text{ kPa gage} \quad \text{or} \quad p = 150 \text{ kPa abs} \tag{3.2}$$

Gage and absolute pressures are often identified after the unit as shown in Eq. (3.2). However, engineers sometimes modify the pressure unit. For example, a gage pressure of 10 pounds per square foot is designated as psfg. Other combinations are psfa, psig, psia. The latter two designations are for pounds per square inch gage and pounds per square inch absolute.

When pressure is less than atmospheric, the pressure can be described using vacuum pressure. *Vacuum pressure* is defined as the difference between atmospheric pressure and actual pressure. Vacuum pressure is a positive number and equals the absolute value of gage pressure (which will be negative). For example, if a gage connected to a tank indicates a vacuum pressure of 31.0 kPa, this can also be stated as 70.0 kPa absolute, or –31.0 kPa gage.

Figure 3.4 provides a visual description of the three pressure scales. Notice that $p_B = 7.45$ psia is equivalent to –7.25 psig and +7.25 psi vacuum. Notice that $p_A = $ of 301 kPa abs is equivalent to 200 kPa gage. Gage, absolute, and vacuum pressure can be related using equations labeled as the "pressure equations."

$$p_{\text{abs}} = p_{\text{atm}} + p_{\text{gage}} \tag{3.3a}$$

$$p_{\text{abs}} = p_{\text{atm}} - p_{\text{vacuum}} \tag{3.3b}$$

$$p_{\text{vacuum}} = -p_{\text{gage}} \tag{3.3c}$$

Figure 3.4

Example of pressure relations.

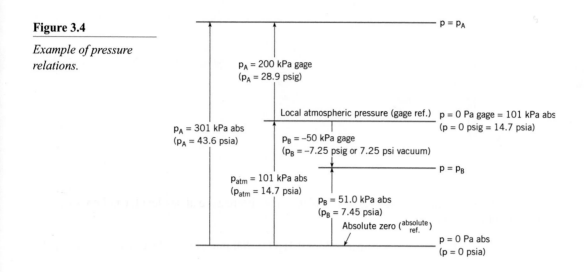

Hydraulic Machines

A *hydraulic machine* uses components such as pistons, pumps, and hoses to transmit forces and energy using fluids. Hydraulic machines are applied, for example, to braking systems, forklift trucks, power steering systems, and airplane control systems (3). Hydraulic machines provide an example of Pascal's law. This law states that pressure applied to an enclosed and continuous body of fluid is transmitted undiminished to every portion of that fluid and to the walls of the containing vessel.

      Hydraulic machines provide mechanical advantage. For example, a person using a hydraulic jack can lift a much larger load, as shown in Example 3.1.

EXAMPLE 3.1   LOAD LIFTED BY A HYDRAULIC JACK

A hydraulic jack has the dimensions shown. If one exerts a force F of 100 N on the handle of the jack, what load, F_2, can the jack support? Neglect lifter weight.

Problem Definition

Situation: A force of $F = 100$ N is applied to the handle of a jack.

Find: Load F_2 in kN that the jack can lift.

Assumptions: Weight of the lifter component (see sketch) is negligible.

Plan

1. Calculate force acting on the small piston by applying moment equilibrium.
2. Calculate pressure p_1 in the hydraulic fluid by applying force equilibrium.
3. Calculate the load F_2 by applying force equilibrium.

Sketch:

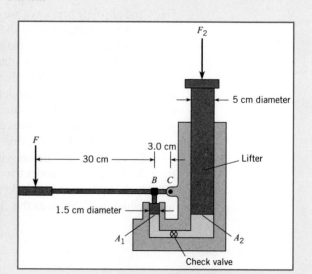

Solution

1. Moment equilibrium

$$\sum M_C = 0$$

$$(0.33 \text{ m}) \times (100 \text{ N}) - (0.03 \text{ m})F_1 = 0$$

$$F_1 = \frac{0.33 \text{ m} \times 100 \text{ N}}{0.03 \text{ m}} = 1100 \text{ N}$$

2. Force equilibrium (small piston)

$$\sum F_{\text{small piston}} = p_1 A_1 - F_1 = 0$$

$$p_1 A_1 = F_1 = 1100 \text{ N}$$

Thus

$$p_1 = \frac{F_1}{A_1} = \frac{1100 \text{ N}}{\pi d^2/4} = 6.22 \times 10^6 \text{ N/m}^2$$

3. Force equilibrium (lifter)
   - Note that $p_1 = p_2$ because they are at the same elevation (this fact will be established in the next section).
   - Apply force equilibrium:

$$\sum F_{\text{lifter}} = F_2 - p_1 A_2 = 0$$

$$F_2 = p_1 A_2 = \left(6.22 \times 10^6 \frac{\text{N}}{\text{m}^2}\right)\left(\frac{\pi}{4} \times (0.05 \text{ m})^2\right) = \boxed{12.2 \text{ kN}}$$

Review

The jack in this example, which combines a lever and a hydraulic machine, provides an output force of 12,200 N from an input force of 100 N. Thus, this jack provides a mechanical advantage of 122 to 1!

3.2

Pressure Variation with Elevation

This section shows how equations for pressure variation are derived and applied. The results are used throughout fluid mechanics.

Hydrostatic Differential Equation

The hydrostatic differential equation is derived by applying force equilibrium to a static body of fluid. To begin the derivation, visualize any region of static fluid (e.g., water behind a dam), isolate a cylindrical body, and then sketch a free-body diagram (FBD) as shown in Fig. 3.5. Notice that the cylindrical body is oriented so that its longitudinal axis is parallel to an arbitrary ℓ direction. The body is $\Delta\ell$ long, ΔA in cross-sectional area, and inclined at an angle α with the horizontal. Apply force equilibrium in the ℓ direction:

$$\sum F_\ell = 0$$

$$F_{\text{Pressure}} - F_{\text{Weight}} = 0$$

$$p\Delta A - (p + \Delta p)\Delta A - \gamma \Delta A \Delta \ell \sin\alpha = 0$$

Figure 3.5

Variation in pressure with elevation.

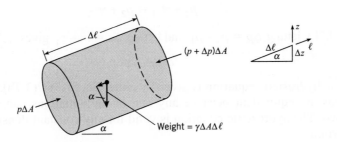

Simplify and divide by the volume of the body $\Delta\ell\Delta A$ to give

$$\frac{\Delta p}{\Delta \ell} = -\gamma \sin \alpha$$

From Fig. 3.5, the sine of the angle is given by

$$\sin \alpha = \frac{\Delta z}{\Delta \ell}$$

Combining the previous two equations and letting Δz approach zero gives

$$\lim_{\Delta z \to 0} \frac{\Delta p}{\Delta z} = -\gamma$$

The final result is

$$\frac{dp}{dz} = -\gamma \qquad \text{(hydrostatic differential equation)} \qquad (3.4)$$

Equation (3.4) is valid in a body of fluid when the force balance shown in Fig. 3.1 is satisfied.

Equation (3.4) means that changes in pressure correspond to changes in elevation. If one travels upward in the fluid (positive z direction), the pressure decreases; if one goes downward (negative z), the pressure increases; if one moves along a horizontal plane, the pressure remains constant. Of course, these pressure variations are exactly what a diver experiences when ascending or descending in a lake or pool.

Hydrostatic Equation

The hydrostatic equation is used to predict pressure variation in a fluid with constant density. This equation is derived by assuming that specific weight γ is constant and then integrating Eq. (3.4) to give

$$p + \gamma z = p_z = \text{constant} \qquad (3.5)$$

where the term z is elevation, which is the height (vertical distance) above a fixed reference point called a datum, and p_z is *piezometric pressure*. Dividing Eq. (3.5) by γ gives

$$\frac{p_z}{\gamma} = \left(\frac{p}{\gamma} + z\right) = h = \text{constant} \qquad (3.6)$$

where h is the *piezometric head*. Since h is constant in Eq. (3.6),

$$\frac{p_1}{\gamma} + z_1 = \frac{p_2}{\gamma} + z_2 \qquad (3.7a)$$

where the subscripts 1 and 2 identify any two points in a static fluid of constant density. Multiplying Eq. (3.7a) by γ gives

$$p_1 + \gamma z_1 = p_2 + \gamma z_2 \qquad (3.7b)$$

In Eq. (3.7b), letting $\Delta p = p_2 - p_1$ and letting $\Delta z = z_2 - z_1$ gives

$$\Delta p = -\gamma \Delta z \qquad (3.7c)$$

The hydrostatic equation is given by either Eq. (3.7a), (3.7b), or (3.7c). These three equations are equivalent because any one of the equations can be used to derive the other two. The hydrostatic equation is valid for any constant density fluid in hydrostatic equilibrium.

Notice that the hydrostatic equation involves

$$\text{piezometric head} = h \equiv \left(\frac{p}{\gamma} + z\right) \tag{3.8}$$

$$\text{piezometric pressure} = p_z \equiv (p + \gamma z) \tag{3.9}$$

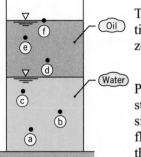

To calculate piezometric head or piezometric pressure, an engineer identifies a specific location in a body of fluid and then uses the value of pressure and elevation at that location. Piezometric pressure and head are related by

$$p_z = h\gamma \tag{3.10}$$

Piezometric head, h, a property that is widely used in fluid mechanics, characterizes hydrostatic equilibrium. When hydrostatic equilibrium prevails in a body of fluid of constant density, then h will be constant at all locations. For example, Fig. 3.6 shows a container with oil floating on water. Since piezometric head is constant in the water, $h_a = h_b = h_c$. Similarly the piezometric head is constant in the oil: $h_d = h_e = h_f$. Notice that piezometric head is not constant when density changes. For example, $h_c \neq h_d$ because points c and d are in different fluids with different values of density. Example 3.2 shows how to find pressure in liquid by applying the idea of constant piezometric head.

Figure 3.6

Oil floating on water.

Example 3.3 shows how to find pressure by applying the idea of "constant piezometric head" to a problem involving several fluids. Notice the continuity of pressure across a planar interface.

EXAMPLE 3.2   WATER PRESSURE IN A TANK

What is the water pressure at a depth of 35 ft in the tank shown?

Problem Definition

Situation: Water is contained in a tank that is 50 ft deep.

Find: Water pressure (psig) at a depth of 35 ft.

Properties: Water (50°F), Table A.5: $\gamma = 62.4$ lbf/ft^3.

Sketch:

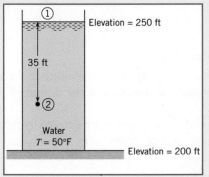

Plan

Use the idea that piezometric head is constant. The steps are

1. Equate piezometric head at elevation 1 with piezometric head at elevation 2 (i.e., apply Eq. 3.7a).

2. Analyze each term in Eq. (3.7a).
3. Solve for the pressure at elevation 2.

Solution

1. Eq. (3.7a):

$$\frac{p_1}{\gamma} + z_1 = \frac{p_2}{\gamma} + z_2$$

2. Term-by-term analysis of Eq. (3.7a) yields:
   - $p_1 = p_{atm} = 0$ psig
   - $z_1 = 250$ ft
   - $z_2 = 215$ ft
3. Combine steps 1 and 2:

$$\frac{p_1}{\gamma} + z_1 = \frac{p_2}{\gamma} + z_2$$

$$0 + 250 \text{ ft} = \frac{p_2}{62.4 \text{ lbf/ft}^3} + 215 \text{ ft}$$

$$p_2 = 2180 \text{ psfg} = \boxed{15.2 \text{ psig}}$$

Review

Remember! Gage pressure at the free surface of a liquid exposed to the atmosphere is zero.

EXAMPLE 3.3 PRESSURE IN TANK WITH TWO FLUIDS

Oil with a specific gravity of 0.80 forms a layer 0.90 m deep in an open tank that is otherwise filled with water. The total depth of water and oil is 3 m. What is the gage pressure at the bottom of the tank?

Problem Definition

Situation: Oil and water are contained in a tank.

Find: Pressure (kPa gage) at the bottom of the tank.

Properties:

1. Oil (10°C), $S = 0.8$.
2. Water (10°C), Table A.5: $\gamma = 9810 \text{ N/m}^3$.

Sketch:

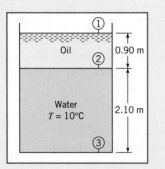

Plan

Use the idea that piezometric head is constant in a body of fluid with constant density. Recognize that pressure across the interface at elevation 2 is constant. The steps are

1. Find p_2 by applying the hydrostatic equation given in Eq. (3.7a).

2. Equate pressures across the oil-water interface.
3. Find p_3 by applying the hydrostatic equation given in Eq. (3.7a).

Solution

1. Hydrostatic equation (oil)

$$\frac{p_1}{\gamma_{\text{oil}}} + z_1 = \frac{p_2}{\gamma_{\text{oil}}} + z_2$$

$$\frac{0 \text{ Pa}}{\gamma_{\text{oil}}} + 3 \text{ m} = \frac{p_2}{0.8 \times 9810 \text{ N/m}^3} + 2.1 \text{ m}$$

$$p_2 = 7.063 \text{ kPa}$$

2. Oil-water interface

$$p_2\big|_{\text{oil}} = p_2\big|_{\text{water}} = 7.063 \text{ kPa}$$

3. Hydrostatic equation (water)

$$\frac{p_2}{\gamma_{\text{water}}} + z_2 = \frac{p_3}{\gamma_{\text{water}}} + z_3$$

$$\frac{7.063 \times 10^3 \text{ Pa}}{9810 \text{ N/m}^3} + 2.1 \text{ m} = \frac{p_3}{9810 \text{ N/m}^3} + 0 \text{ m}$$

$$\boxed{p_3 = 27.7 \text{ kPa gage}}$$

Review

Validation: Since oil is less dense than water, the answer should be slightly smaller than the pressure corresponding to a water column of 3 m. From Table F.1, a water column of 10 m ≈ 1 atm. Thus, a 3 m water column should produce a pressure of about 0.3 atm = 30 kPa. The calculated value appears correct.

Pressure Variation in the Atmosphere

This section describes how to calculate pressure, density and temperature in the atmosphere for applications such as modeling of atmospheric dynamics and the design of gliders, airplanes, balloons, and rockets.

    Equations for pressure variation in the earth's atmosphere are derived by integrating the hydrostatic differential equation (3.4). To begin the derivation, write the ideal gas law (2.5):

$$\rho = \frac{p}{RT} \tag{3.11}$$

Multiply by g:

$$\gamma = \frac{pg}{RT} \tag{3.12}$$

Equation (3.12) requires temperature-versus-elevation data for the atmosphere. It is common practice to use the U.S. Standard Atmosphere (1). The *U.S. Standard Atmosphere* defines values for atmospheric temperature, density, and pressure over a wide range of altitudes. The first model was published in 1958; this was updated in 1962, 1966, and 1976. The U.S. Standard Atmosphere gives average conditions over the United States at 45° N latitude in July.

The U.S. Standard Atmosphere also gives average conditions at sea level. The sea level temperature is 15°C (59°F), the pressure is 101.33 kPa abs (14.696 psia), and the density is 1.225 kg/m³ (0.002377 slugs/ft³).

Temperature data for the U.S. Standard Atmosphere are given in Fig. 3.7 for the lower 30 km of the atmosphere. The atmosphere is about 1000 km thick and is divided into five layers, so Fig. 3.7 only gives data near the earth's surface. In the *troposphere*, defined as the layer between sea level and 13.7 km (45,000 ft), the temperature decreases nearly linearly with increasing elevation at a lapse rate of 5.87 K/km. The *stratosphere* is the layer that begins at the top of the troposphere and extends up to about 50 km. In the lower regions of the stratosphere, the temperature is constant at −57.5°C, to an altitude of 16.8 km (55,000 ft), and then the temperature increases monotonically to −38.5°C at 30.5 km (100,000 ft).

Pressure Variation in the Troposphere

Let the temperature T be given by

$$T = T_0 - \alpha(z - z_0) \tag{3.13}$$

In this equation T_0 is the temperature at a reference level where the pressure is known, and α is the lapse rate. Combine Eq. (3.12) with the hydrostatic differential equation (3.4) to give

$$\frac{dp}{dz} = -\frac{pg}{RT} \tag{3.14}$$

Substituting Eq. (3.13) into Eq. (3.14) gives

$$\frac{dp}{dz} = -\frac{pg}{R[T_0 - \alpha(z - z_0)]}$$

Figure 3.7

Temperature variation with altitude for the U.S. standard atmosphere in July (1).

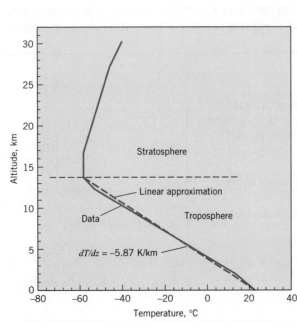

Separate the variables and integrate to obtain

$$\frac{p}{p_0} = \left[\frac{T_0 - \alpha(z - z_0)}{T_0}\right]^{g/\alpha R}$$

Thus, the atmospheric pressure variation in the troposphere is

$$p = p_0\left[\frac{T_0 - \alpha(z - z_0)}{T_0}\right]^{g/\alpha R} \tag{3.15}$$

Example 3.4 shows how to apply Eq. (3.15) to find pressure at a specified elevation in the troposphere.

Pressure Variation in the Lower Stratosphere

In the lower part of the stratosphere (13.7 to 16.8 km above the earth's surface as shown in Fig. 3.7), the temperature is approximately constant. In this region, integration of Eq. (3.14) gives

$$\ln p = -\frac{zg}{RT} + C$$

At $z = z_0, p = p_0$, so the preceding equation reduces to

$$\frac{p}{p_0} = e^{-(z-z_0)g/RT}$$

so the atmospheric pressure variation in the stratosphere takes the form

$$p = p_0 e^{-(z-z_0)g/RT} \tag{3.16}$$

where p_o is pressure at the interface between the troposphere and stratosphere, z_o is the elevation of the interface, and T is the temperature of the stratosphere. Example 3.5 shows how to apply Eq. (3.16) to find pressure at a specified elevation in the troposphere.

EXAMPLE 3.4   PRESSURE IN THE TROPOSPHERE

If at sea level the absolute pressure and temperature are 101.3 kPa and 23°C, what is the pressure at an elevation of 2000 m, assuming that standard atmospheric conditions prevail?

Problem Definition

Situation: Standard atmospheric conditions prevail at an elevation of 2000 m.

Find: Atmospheric pressure (kPa absolute) at an elevation of 2000 m.

Plan

Calculate pressure using Eq. (3.15).

Solution

$$p = p_0\left[\frac{T_0 - \alpha(z - z_0)}{T_0}\right]^{g/\alpha R}$$

where $p_0 = 101{,}300 \text{ N}/\text{m}^2$, $T_0 = 273 + 23 = 296$ K, $\alpha = 5.87 \times 10^{-3}$ K$/$m, $z - z_0 = 2000$ m, and $g/\alpha R = 5.823$. Then

$$p = 101.3\left(\frac{296 - 5.87 \times 10^{-3} \times 2000}{296}\right)^{5.823}$$

$$= \boxed{80.0 \text{ kPa absolute}}$$

EXAMPLE 3.5   PRESSURE IN THE LOWER STRATOSPHERE

If the pressure and temperature are 2.31 psia (p = 15.9 kPa absolute) and –71.5°F (–57.5°C) at an elevation of 45,000 ft (13.72 km), what is the pressure at 55,000 ft (16.77 km), assuming isothermal conditions over this range of elevation?

Problem Definition

Situation: Standard atmospheric conditions prevail at an elevation of 55,000 ft (16.77 km).

Find: Atmospheric pressure (psia and kPa absolute) at 55,000 ft (16.77 km).

Plan

Calculate pressure using Eq. (3.16).

Solution

For isothermal conditions,

$$T = -71.5 + 460 = 388.5°R$$

$$p = p_0 e^{-(z-z_0)g/RT} = 2.31 e^{-(10,000)(32.2)/(1716 \times 388.5)}$$

$$= 2.31 e^{-0.483}$$

Therefore the pressure at 55,000 ft is

$$\boxed{p = 1.43 \text{ psia}}$$

SI units

$$\boxed{p = 9.83 \text{ kPa absolute}}$$

3.3

Pressure Measurements

This section describes five scientific instruments for measuring pressure: the barometer, Bourdon-tube gage, piezometer, manometer, and transducer. This information is used for experimental work, for equipment testing and process monitoring.

Barometer

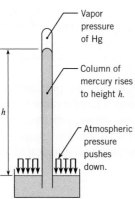

Vapor pressure of Hg

Column of mercury rises to height h.

Atmospheric pressure pushes down.

Figure 3.8

A mercury barometer.

An instrument that is used to measure atmospheric pressure is called a *barometer*. The most common types are the mercury barometer and the aneroid barometer. A mercury barometer is made by inverting a mercury-filled tube in a container of mercury as shown in Fig. 3.8. The pressure at the top of the mercury barometer will be the vapor pressure of mercury, which is very small: $p_v = 2.4 \times 10^{-6}$ atm at 20°C. Thus, atmospheric pressure will push the mercury up the tube to a height h. The mercury barometer is analyzed by applying the hydrostatic equation:

$$p_{atm} = \gamma_{Hg}h + p_v \approx \gamma_{Hg}h \tag{3.17}$$

Thus, by measuring h, local atmospheric pressure can be determined using Eq. (3.17).

An aneroid barometer works mechanically. An aneroid is an elastic bellows that has been tightly sealed after some air was removed. When atmospheric pressure changes, this causes the aneroid to change size, and this mechanical change can be used to deflect a needle to indicate local atmospheric pressure on a scale. An aneroid barometer has some advantages over a mercury barometer because it is smaller and allows data recording over time.

Bourdon-Tube Gage

A *Bourdon-tube* gage, Fig. 3.9, measures pressure by sensing the deflection of a coiled tube. The tube has an elliptical cross section and is bent into a circular arc, as shown in Fig. 3.9b. When atmospheric pressure (zero gage pressure) prevails, the tube is undeflected, and for this

Figure 3.9

Bourdon-tube gage.
(a) View of typical gage.
(b) Internal mechanism
(schematic).

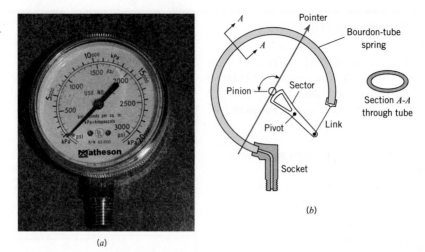

(a)

(b)

condition the gage pointer is calibrated to read zero pressure. When pressure is applied to the gage, the curved tube tends to straighten (much like blowing into a party favor to straighten it out), thereby actuating the pointer to read a positive gage pressure. The Bourdon-tube gage is common because it is low cost, reliable, easy to install, and available in many different pressure ranges. There are disadvantages: dynamic pressures are difficult to read accurately; accuracy of the gage can be lower than other instruments; and the gage can be damaged by excessive pressure pulsations.

Piezometer

A *piezometer* is a vertical tube, usually transparent, in which a liquid rises in response to a positive gage pressure. For example, Fig. 3.10 shows a piezometer attached to a pipe. Pressure in the pipe pushes the water column to a height h, and the gage pressure at the center of the pipe is $p = \gamma h$, which follows directly from the hydrostatic equation (3.7c). The piezometer has several advantages: simplicity, direct measurement (no need for calibration), and accuracy. However, a piezometer cannot easily be used for measuring pressure in a gas, and a piezometer is limited to low pressures because the column height becomes too large at high pressures.

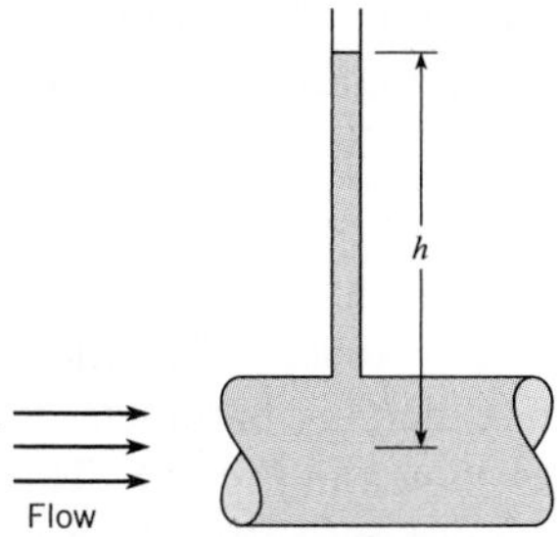

Manometer

A *manometer*, often shaped like the letter "U," is a device for measuring pressure by raising or lowering a column of liquid. For example, Fig. 3.11 shows a U-tube manometer that is being used to measure pressure in a flowing fluid. In the case shown, positive gage pressure in the pipe pushes the manometer liquid up a height Δh. To use a manometer, engineers relate the height of the liquid in the manometer to pressure as illustrated in Example 3.6.

Figure 3.10

Piexometer attached to a pipe.

Figure 3.11

U-tube manometer.

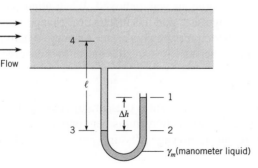

EXAMPLE 3.6   PRESSURE MEASUREMENT (U-TUBE MANOMETER)

Water at 10°C is the fluid in the pipe of Fig. 3.11, and mercury is the manometer fluid. If the deflection Δh is 60 cm and ℓ is 180 cm, what is the gage pressure at the center of the pipe?

Problem Definition

Situation: Pressure in a pipe is being measured using a U-tube manometer.

Find: Gage pressure (kPa) in the center of the pipe.

Properties:

1. Water (10°C), Table A.5, $\gamma = 9810$ N/m^3.
2. Mercury, Table A.4: $\gamma = 133,000$ N/m^3.

Plan

Start at point 1 and work to point 4 using ideas from Eq. (3.7c). When fluid depth increases, add a pressure change. When fluid depth decreases, subtract a pressure change.

Solution

1. Calculate the pressure at point 2 using the hydrostatic equation (3.7c).

$$p_2 = p_1 + \text{ pressure increase between 1 and 2} = 0 + \gamma_m \Delta h_{12}$$

$$= \gamma_m (0.6 \text{ m}) = (133,000 \text{ N/m}^3)(0.6 \text{ m})$$

$$= 79.8 \text{ kPa}$$

2. Find the pressure at point 3.
   - The hydrostatic equation with $z_3 = z_2$ gives

$$p_3\big|_{\text{water}} = p_2\big|_{\text{water}} = 79.8 \text{ kPa}$$

   - When a fluid-fluid interface is flat, pressure is constant across the interface. Thus, at the oil-water interface

$$p_3\big|_{\text{mercury}} = p_3\big|_{\text{water}} = 79.8 \text{ kPa}$$

3. Find the pressure at point 4 using the hydrostatic equation given in Eq. (3.7c).

$$p_4 = p_3 - \text{ pressure decrease between 3 and 4} = p_3 - \gamma_w \ell$$

$$= 79,800 \text{ Pa} - (9810 \text{ N/m}^3)(1.8 \text{ m})$$

$$= 62.1 \text{ kPa gage}$$

Once one is familiar with the basic principle of manometry, it is straightforward to write a single equation rather than separate equations as was done in Example 3.6. The single equation for evaluation of the pressure in the pipe of Fig 3.11 is

$$0 + \gamma_m \Delta h - \gamma \ell = p_4$$

One can read the equation in this way: Zero pressure at the open end, plus the change in pressure from point 1 to 2, minus the change in pressure from point 3 to 4, equals the pressure in the pipe. The main concept to remember is that pressure increases as depth increases and decreases as depth decreases.

The general equation for the pressure difference measured by the manometer is:

$$p_2 = p_1 + \underset{\text{down}}{\sum \gamma_i h_i} - \underset{\text{up}}{\sum \gamma_i h_i} \tag{3.18}$$

where γ_i and h_i are the specific weight and deflection in each leg of the manometer. It does not matter where one starts; that is, where one defines the initial point 1 and final point 2. When liquids and gases are both involved in a manometer problem, it is well within engineering accuracy to neglect the pressure changes due to the columns of gas. This is because $\gamma_{\text{liquid}} \gg \gamma_{\text{gas}}$. Example 3.7 shows how to apply Eq. (3.18) to perform an analysis of a manometer that uses multiple fluids.

EXAMPLE 3.7 MANOMETER ANALYSIS

Sketch: What is the pressure of the air in the tank if $\ell_1 = 40$ cm, $\ell_2 = 100$ cm, and $\ell_3 = 80$ cm?

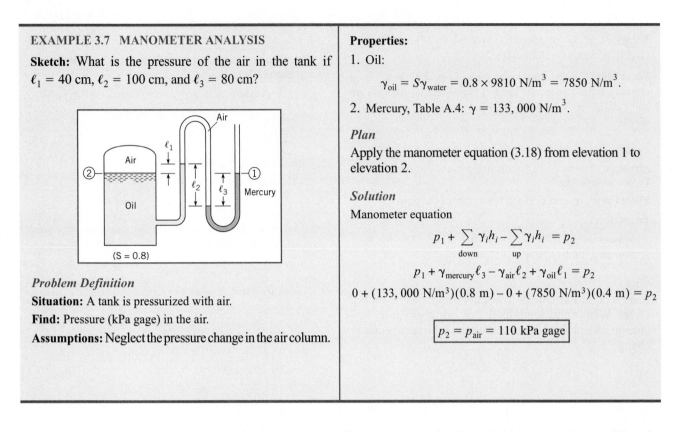

Problem Definition

Situation: A tank is pressurized with air.

Find: Pressure (kPa gage) in the air.

Assumptions: Neglect the pressure change in the air column.

Properties:

1. Oil:
$$\gamma_{oil} = S\gamma_{water} = 0.8 \times 9810 \text{ N/m}^3 = 7850 \text{ N/m}^3.$$

2. Mercury, Table A.4: $\gamma = 133,000 \text{ N/m}^3$.

Plan

Apply the manometer equation (3.18) from elevation 1 to elevation 2.

Solution

Manometer equation

$$p_1 + \underset{\text{down}}{\sum \gamma_i h_i} - \underset{\text{up}}{\sum \gamma_i h_i} = p_2$$

$$p_1 + \gamma_{mercury}\ell_3 - \gamma_{air}\ell_2 + \gamma_{oil}\ell_1 = p_2$$

$$0 + (133,000 \text{ N/m}^3)(0.8 \text{ m}) - 0 + (7850 \text{ N/m}^3)(0.4 \text{ m}) = p_2$$

$$\boxed{p_2 = p_{air} = 110 \text{ kPa gage}}$$

Because the manometer configuration shown in Fig. 3.12 is common, it is useful to derive an equation specific to this application. To begin, apply the manometer equation (3.18) between points 1 and 2:

$$p_1 + \underset{\text{down}}{\sum \gamma_i h_i} - \underset{\text{up}}{\sum \gamma_i h_i} = p_2$$

$$p_1 + \gamma_A(\Delta y + \Delta h) - \gamma_B \Delta h - \gamma_A(\Delta y + z_2 - z_1) = p_2$$

Simplifying gives

$$(p_1 + \gamma_A z_1) - (p_2 + \gamma_A z_2) = \Delta h(\gamma_B - \gamma_A)$$

Dividing through by γ_A gives

$$\left(\frac{p_1}{\gamma_A} + z_1\right) - \left(\frac{p_2}{\gamma_A} + z_2\right) = \Delta h\left(\frac{\gamma_B}{\gamma_A} - 1\right)$$

Recognize that the terms on the left side of the equation are piezometric head and rewrite to give the final result:

$$h_1 - h_2 = \Delta h\left(\frac{\gamma_B}{\gamma_A} - 1\right) \tag{3.19}$$

Equation (3.19) is valid when a manometer is used as shown in Fig. 3.12. Example 3.8 shows how this equation is used.

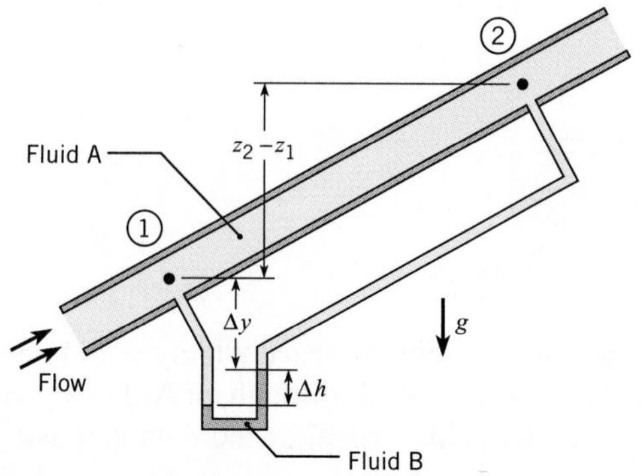

Figure 3.12

Apparatus for determining change in piezometric head corresponding to flow in a pipe.

EXAMPLE 3.8   CHANGE IN PIEZOMETRIC HEAD FOR PIPE FLOW

A differential mercury manometer is connected to two pressure taps in an inclined pipe as shown in Fig. 3.12. Water at 50°F is flowing through the pipe. The deflection of mercury in the manometer is 1 inch. Find the change in piezometric pressure and piezometric head between points 1 and 2.

Problem Definition

Situation: Water is flowing in a pipe.

Find:

1. Change in piezometric head (ft) between points 1 and 2.
2. Change in piezometric pressure (psfg) between 1 and 2.

Properties:

1. Water (50 °F), Table A.5, $\gamma_{water} = 62.4$ lbf/ft^3.
2. Mercury, Table A.4, $\gamma_{Hg} = 847$ lbf/ft^3.

Plan

1. Find difference in the piezometric head using Eq. (3.19).
2. Relate piezometric head to piezometric pressure using Eq. (3.10).

Solution

Difference in piezeometric head

$$h_1 - h_2 = \Delta h \left(\frac{\gamma_{Hg}}{\gamma_{water}} - 1 \right) = \left(\frac{1}{12} \text{ ft} \right) \left(\frac{847 \text{ lbf/ft}^3}{62.4 \text{ lbf/ft}^3} - 1 \right)$$

$$= \boxed{1.05 \text{ ft}}$$

Piezometric pressure

$$p_z = h\gamma_{water}$$

$$= (1.05 \text{ ft})(62.4 \text{ lbf/ft}^3) = \boxed{65.5 \text{ psf}}$$

Pressure Transducers

A *pressure transducer* is a device that converts pressure to an electrical signal. Modern factories and systems that involve flow processes are controlled automatically, and much of their operation involves sensing of pressure at critical points of the system. Therefore, pressure-sensing devices, such as pressure transducers, are designed to produce electronic signals that can be transmitted to oscillographs or digital devices for record-keeping or to control other devices for process operation. Basically, most transducers are tapped into the system with one side of a small diaphragm exposed to the active pressure of the system. When the pressure changes, the diaphragm flexes, and a sensing element connected to the other side of the diaphragm produces a signal that is usually linear with the change in pressure in the system. There are many types of sensing elements; one common type is the resistance-wire strain gage attached to a flexible diaphragm as shown in Fig. 3.13. As the diaphragm flexes, the wires of the strain gage change length, thereby changing the resistance of the wire. This change in resistance is converted into a voltage change that can then be used in various ways.

Another type of pressure transducer used for measuring rapidly changing high pressures, such as the pressure in the cylinder head of an internal combustion engine, is the piezoelectric transducer (2). These transducers operate with a quartz crystal that generates a charge when subjected to a pressure. Sensitive electronic circuitry is required to convert the charge to a measurable voltage signal.

Figure 3.13

Schematic diagram of strain-gage pressure transducer.

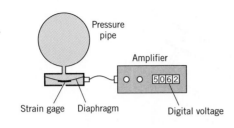

Computer data acquisition systems are used widely with pressure transducers. The analog signal from the transducer is converted (through an A/D converter) to a digital signal that can be processed by a computer. This expedites the data acquisition process and facilitates storing data.

Forces on Plane Surfaces (Panels)

This section explains how to represent hydrostatic pressure distributions on one face of a panel with a resultant force that passes through a point called the center of pressure. This information is relevant to applications such as dams and water towers.

Uniform Pressure Distribution

A plane surface or *panel* is a flat surface of arbitrary shape. A description of the pressure at all points along a surface is called a *pressure distribution*. When pressure is the same at every point, as shown in Fig. 3.14*a*, the pressure distribution is called a uniform pressure distribution. The pressure distribution in Fig. 3.14*a* can be represented by a resultant force as shown in Fig. 3.14*b*. For a uniform pressure distribution, the magnitude of the resultant force is F where

$$F = \int_A p \, dA = \bar{p} A$$

and $\bar{p}$ is the average pressure. The resultant force F passes through a point called the *center of pressure (CP)*. Notice that the CP is represented using a circle with a "plus" inside. For a uniform pressure distribution, the CP is located at the centroid of area of the panel.

Hydrostatic Pressure Distribution

When a pressure distribution is produced by a fluid in hydrostatic equilibrium, as shown in Fig. 3.15*a*, then the pressure distribution is called a *hydrostatic pressure distribution*. Notice

Figure 3.14

(a) Uniform pressure distribution, and (b) equivalent force.

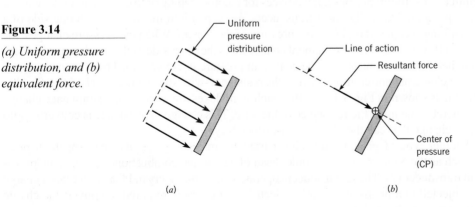

Figure 3.15

(a) Hydrostatic pressure distribution, and (b) resultant force F acting at the center of pressure.

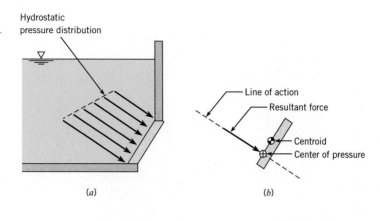

that a hydrostatic pressure distribution is linear and that the arrows representing pressure act normal to the surface. In Fig. 3.15b, the pressure distribution is represented by a resultant force that acts at the CP. Notice that the CP is located below the centroid of area.

Magnitude of Resultant Hydrostatic Force

To derive an equation for the resultant force on a panel under hydrostatic loading, sum forces using an integral. The situation is shown in Fig. 3.16. Line *AB* is the edge view of a panel submerged in a liquid. The plane of this panel intersects the horizontal liquid surface at axis 0-0 with an angle α. The distance from the axis 0-0 to the horizontal axis through the centroid of the area is given by $\overline{y}$. The distance from 0-0 to the differential area dA is y. The pressure on the differential area is

$$p = \gamma y \sin \alpha$$

The differential force is

$$dF = p\, dA = \gamma y \sin \alpha\, dA$$

The total force on the area is

$$F = \int_A p\, dA = \int_A \gamma y \sin \alpha\, dA \tag{3.20}$$

In Eq. (3.20), γ and $\sin \alpha$ are constants. Thus

$$F = \gamma \sin \alpha \int_A y\, dA \tag{3.21}$$

Now the integral in Eq. (3.21) is the first moment of the area. Consequently, this is replaced by its equivalent, $\overline{y}A$. Therefore

$$F = \gamma \overline{y} A \sin \alpha$$

which can be rewritten as

$$F = (\gamma \overline{y} \sin \alpha)A \tag{3.22}$$

The product of the variables within the parentheses of Eq. (3.22) is the pressure at the centroid of the area. Thus

$$F = \overline{p} A \tag{3.23}$$

Figure 3.16

Distribution of hydrostatic pressure on a plane surface.

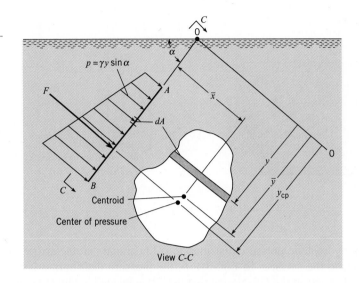

View *C-C*

Equation (3.23), called the panel equation, shows that the hydrostatic force on a panel of arbitrary shape (e.g., rectangular, round, elliptical) is given by the product of panel area and pressure at the centroid of area. Example 3.9 shows how to apply the panel force equation.

Line of Action of the Resultant Force

A general equation for the vertical location of the CP is derived next. The initial situation is shown in Fig. 3.16. The torque due to the resultant force F will balance the torque due to the pressure distribution.

$$y_{cp}F = \int y\, dF$$

The differential force dF is given by $dF = p\, dA$; therefore,

$$y_{cp}F = \int_A yp\, dA$$

EXAMPLE 3.9 HYDROSTATIC FORCE DUE TO CONCRETE

Determine the force acting on one side of a concrete form 2.44 m high and 1.22 m wide (8 ft by 4 ft) that is used for pouring a basement wall. The specific weight of concrete is 23.6 kN/m³ (150 lbf/ft³).

Problem Definition

Situation:

1. Concrete in a liquid state acts on a vertical surface.
2. Vertical wall is 2.44 m high and 1.22 m wide

Find: The resultant force (kN) acting on the wall.

Assumptions: Freshly poured concrete can be represented as a liquid.

Properties: Concrete: $\gamma = 23.6$ kN/m³.

Plan

Apply the panel equation given in Eq. (3.23).

Solution

1. Panel equation
$$F = \bar{p}A$$

2. Term-by-term analysis
   • $\bar{p}$ = pressure at depth of the centroid
   $$\bar{p} = (\gamma_{concrete})(z_{centroid}) = (23.6 \text{ kN/m}^3)(2.44/2 \text{ m})$$
   $$= 28.79 \text{ kPa}$$

   • A = area of panel
   $$A = (2.44 \text{ m})(1.22 \text{ m}) = 2.977 \text{ m}^2$$

3. Resultant force
   $$F = \bar{p}A = (28.79 \text{ kPa})(2.977 \text{ m}^2) = \boxed{85.7 \text{ kN}}$$

Also, $p = \gamma y \sin \alpha$, so

$$y_{cp}F = \int_A \gamma y^2 \sin \alpha \, dA \tag{3.24}$$

Since γ and $\sin \alpha$ are constants,

$$y_{cp}F = \gamma \sin \alpha \int_A y^2 \, dA \tag{3.25}$$

The integral on the right-hand side of Eq. (3.25) is the second moment of the area (often called the area moment of inertia). This shall be identified as I_0. However, for engineering applications it is convenient to express the second moment with respect to the horizontal centroidal axis of the area. Hence by the parallel-axis theorem,

$$I_0 = \bar{I} + \bar{y}^2 A \tag{3.26}$$

Substitute Eq. (3.26) into Eq. (3.25) to give

$$y_{cp}F = \gamma \sin \alpha (\bar{I} + \bar{y}^2 A)$$

However, from Eq. (3.22), $F = \gamma \bar{y} \sin \alpha A$. Therefore,

$$y_{cp}(\gamma \bar{y} \sin \alpha A) = \gamma \sin \alpha (\bar{I} + \bar{y}^2 A)$$

$$y_{cp} = \bar{y} + \frac{\bar{I}}{\bar{y} A} \tag{3.27}$$

$$y_{cp} - \bar{y} = \frac{\bar{I}}{\bar{y} A} \tag{3.28}$$

In Eq. (3.28), the area moment of inertia $\bar{I}$ is taken about a horizontal axis that passes through the centroid of area. Formulas for $\bar{I}$ are presented in Fig. A.1. The slant distance $\bar{y}$ measures the length from the surface of the liquid to the centroid of the panel along an axis that is aligned with the "slant of the panel" as shown in Fig. 3.16.

Equation (3.28) shows that the Center of Pressure (CP) will be situated below the centroid. The distance between the CP and the centroid depends on the depth of submersion, which is characterized by $\bar{y}$, and on the panel geometry, which is characterized by $\bar{I}/A$.

Due to assumptions in the derivations, Eqs. (3.23) and (3.28) have several limitations. First, they only apply to a single fluid of constant density. Second, the pressure at the liquid surface needs to be $p = 0$ gage to correctly locate the CP. Third, Eq. (3.28) gives only the vertical location of the CP, not the lateral location.

Example 3.10 shows how to apply the panel equations. Notice that drawing an FBD makes the analysis easier.

EXAMPLE 3.10   FORCE TO OPEN AN ELLIPTICAL GATE

An elliptical gate covers the end of a pipe 4 m in diameter. If the gate is hinged at the top, what normal force F is required to open the gate when water is 8 m deep above the top of the pipe and the pipe is open to the atmosphere on the other side? Neglect the weight of the gate.

Problem Definition

Situation: Water pressure is acting on an elliptical gate.

Find: Normal force (in newtons) required to open gate.

Sketch:

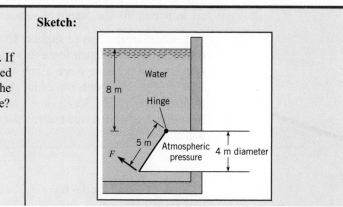

Properties: Water (10°C), Table A.5: $\gamma = 9810$ N/m^3.

Assumptions:

1. Neglect the weight of the gate.
2. Neglect friction between the bottom on the gate and the pipe wall.

Plan

1. Calculate resultant hydrostatic force using $F = \bar{p}A$.
2. Find the location of the center of pressure using Eq. (3.28).
3. Draw an FBD of the gate.
4. Apply moment equilibrium about the hinge.

Solution

1. Hydrostatic (resultant) force
   - $\bar{p}$ = pressure at depth of the centroid

$\bar{p} = (\gamma_{water})(z_{centroid}) = (9810 \text{ N/m}^3)(10 \text{ m}) = 98.1$ kPa

   - A = area of elliptical panel (using Fig. A.1 to find formula)

$A = \pi ab$

$= \pi(2.5 \text{ m})(2 \text{ m}) = 15.71\text{m}^2$

   - Calculate resultant force

$F_p = \bar{p}A = (98.1 \text{ kPa})(15.71 \text{ m}^2) = \boxed{1.54 \text{ MN}}$

2. Center of pressure
   - $\bar{y} = 12.5$ m, where $\bar{y}$ is the slant distance from the water surface to the centroid.
   - Area moment of inertia $\bar{I}$ of an elliptical panel using a formula from Fig. A.1

$\bar{I} = \dfrac{\pi a^3 b}{4} = \dfrac{\pi(2.5 \text{ m})^3(2 \text{ m})}{4} = 24.54$ m^4

   - Finding center of pressure

$y_{cp} - \bar{y} = \dfrac{\bar{I}}{\bar{y}A} = \dfrac{25.54 \text{ m}^4}{(12.5 \text{ m})(15.71\text{m}^2)} = 0.125$ m

3. FBD of the gate:

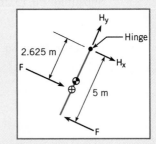

4. Moment equilibrium

$$\sum M_{hinge} = 0$$

$$1.541 \times 10^6 \text{ N} \times 2.625 \text{ m} - F \times 5 \text{ m} = 0$$

$$F = \boxed{809 \text{ kN}}$$

Forces on Curved Surfaces

This section describes how to calculate forces on surfaces that have curvature. These results are important for the design of components such as tanks, pipes, and curved gates.

Consider the curved surface AB in Fig. 3.17a. The goal is to represent the pressure distribution with a resultant force that passes through the center of pressure. One approach is to integrate the pressure force along the curved surface and find the equivalent force. However, it is easier to sum forces for the free body shown in the upper part of Fig. 3.17b. The lower sketch in Fig. 3.17b shows how the force acting on the curved surface relates to the force F acting on the free body. Using the FBD and summing forces in the horizontal direction shows that

$$F_x = F_{AC} \tag{3.29}$$

The line of action for the force F_{AC} is through the center of pressure for side AC, as discussed in the previous section, and designated as y_{cp}.

Figure 3.17

(a) Pressure distribution and equivalent force.
(b) Free-body diagram and action-reaction force pair.

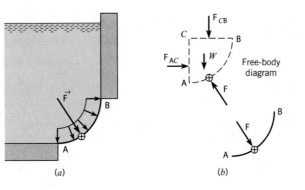

(a)   (b)

The vertical component of the equivalent force is

$$F_y = W + F_{CB} \qquad (3.30)$$

where W is the weight of the fluid in the free body and F_{CB} is the force on the side CB.

The force F_{CB} acts through the centroid of surface CB, and the weight acts through the center of gravity of the free body. The line of action for the vertical force may be found by summing the moments about any convenient axis.

Example 3.11 illustrates how curved surface problems can be solved by applying equilibrium concepts together with the panel force equations.

EXAMPLE 3.11   HYDROSTATIC FORCE ON A CURVED SURFACE

Sketch:

Surface AB is a circular arc with a radius of 2 m and a width of 1 m into the paper. The distance EB is 4 m. The fluid above surface AB is water, and atmospheric pressure prevails on the free surface of the water and on the bottom side of surface AB. Find the magnitude and line of action of the hydrostatic force acting on surface AB.

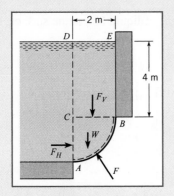

Problem Definition

Situation: A body of water is contained by a curved surface.

Find:

1. Hydrostatic force (in newtons) on the curved surface AB.
2. Line of action of the hydrostatic force.

Properties: Water (10°C), Table A.5: $\gamma = 9810$ N/m^3.

Plan

Apply equilibrium concepts to the body of fluid ABC.

1. Find the horizontal component of F by applying Eq. (3.29).
2. Find the vertical component of F by applying Eq. (3.30).
3. Find the line of action of F by finding the lines of action of components and then using a graphical solution.

Solution

1. Equilibrium in the horizontal direction

$$F_x = F_H = \bar{p}A = (5 \text{ m})(9810 \text{ N/m}^3)(2 \times 1 \text{ m}^2)$$

$$= 98.1 \text{ kN}$$

2. Equilibrium in the horizontal direction
   - Vertical force on side CB

$$F_V = \bar{p}_0 A = 9.81 \text{ kN/m}^3 \times 4 \text{ m} \times 2 \text{ m} \times 1 \text{ m} = 78.5 \text{ kN}$$

   - Weight of the water in volume ABC

$$W = \gamma \Psi_{ABC} = (\gamma)(\tfrac{1}{4}\pi r^2)(w)$$

$$= (9.81 \text{ kN/m}^3) \times (0.25 \times \pi \times 4 \text{ m}^2)(1 \text{ m}) = 30.8 \text{ kN}$$

   - Summing forces

$$F_y = W + F_V = 109.3 \text{ kN}$$

3. Line of action (horizontal force)

$$y_{cp} = \bar{y} + \frac{\bar{I}}{\bar{y}A} = (5 \text{ m}) + \left(\frac{1 \times 2^3/12}{5 \times 2 \times 1} \text{ m}\right)$$

$$y_{cp} = 5.067 \text{ m}$$

4. The line of action (x_{cp}) for the vertical force is found by summing moments about point C:

$$x_{cp}F_y = F_V \times 1\text{ m} + W \times \bar{x}_W$$

The horizontal distance from point C to the centroid of the area ABC is found using Fig. A.1: $\bar{x}_W = 4r/3\pi = 0.849$ m. Thus,

$$x_{cp} = \frac{78.5\text{ kN} \times 1\text{ m} + 30.8\text{ kN} \times 0.849\text{ m}}{109.3\text{ kN}} = 0.957\text{ m}$$

5. The resultant force that acts on the curved surface is shown in the following figure.

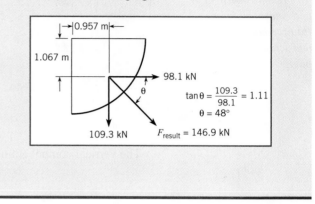

The central idea of this section is that forces on curved surfaces may be found by applying equilibrium concepts to systems comprised of the fluid in contact with the curved surface. Notice how equilibrium concepts are used in each of the following situations.

Consider a sphere holding a gas pressurized to a gage pressure p_i as shown in Fig. 3.18. The indicated forces act on the fluid in volume ABC. Applying equilibrium in the vertical direction gives

$$F = p_i A_{AC} + W$$

Because the specific weight for a gas is quite small, engineers usually neglect the weight of the gas:

$$F = p_i A_{AC} \tag{3.31}$$

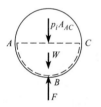

Figure 3.18

Pressurized spherical tank showing forces that act on the fluid inside the marked region.

Another example is finding the force on a curved surface submerged in a reservoir of liquid as shown in Fig. 3.19a. If atmospheric pressure prevails above the free surface and on the outside of surface AB, then force caused by atmospheric pressure cancels out and equilibrium gives

$$F = \gamma V_{ABCD} = W\downarrow \tag{3.32}$$

Hence the force on surface AB equals the weight of liquid above the surface, and the arrow indicates that the force acts downward.

Figure 3.19

Curved surface with (a) liquid above and (b) liquid below. In (a), arrows represent forces acting on the liquid. In (b), arrows represent the pressure distribution on surface ab.

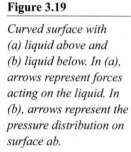
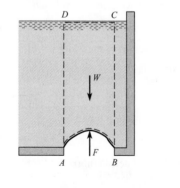
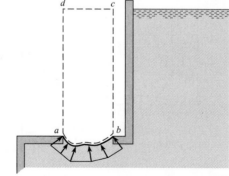

(a)　　　　　　(b)

Now consider the situation where the pressure distribution on a thin curved surface comes from the liquid underneath, as shown in Fig. 3.19b. If the region above the surface, volume $abcd$, were filled with the same liquid, the pressure acting at each point on the upper surface of ab would equal the pressure acting at each point on the lower surface. In other words, there would be no net force on the surface. Thus, the equivalent force on surface ab is given by

$$F = \gamma \Psi_{abcd} = W \uparrow \tag{3.33}$$

where W is the weight of liquid needed to fill a volume that extends from the curved surface to the free surface of the liquid.

Buoyancy

This section describes how to calculate the buoyant force on an object. A *buoyant force* is defined as the upward force that is produced on a body that is totally or partially submerged in a fluid when the fluid is in a gravity field. Buoyant forces are significant for most problems that involve liquids. Examples include surface ships, sediment transport in rivers, and fish migration. Buoyant forces are sometimes significant in problems involving gases, for example, a weather balloon.

The Buoyant Force Equation

The initial situation for the derivation is shown in Fig. 3.20. Consider a body $ABCD$ submerged in a liquid of specific weight γ. The sketch on the left shows the pressure distribution acting on the body. As shown by Eq. (3.33), pressures acting on the lower portion of the body create an upward force equal to the weight of liquid needed to fill the volume above surface ADC. The upward force is

$$F_{up} = \gamma(\Psi_b + \Psi_a)$$

where Ψ_b is the volume of the body (i.e., volume $ABCD$) and Ψ_a is the volume of liquid above the body (i.e., volume $ABCFE$). As shown by Eq. (3.32), pressures acting on the top surface of the body create a downward force equal to the weight of the liquid above the body:

$$F_{down} = \gamma \Psi_a$$

Subtracting the downward force from the upward force gives the net or buoyant force F_B acting on the body:

$$F_B = F_{up} - F_{down} = \gamma \Psi_b \tag{3.34}$$

Hence, the net force or buoyant force (F_B) equals the weight of liquid that would be needed to occupy the volume of the body.

Consider a body that is floating as shown in Fig. 3.21. The marked portion of the object has a volume Ψ_D. Pressure acts on curved surface ADC causing an upward force equal to the weight of liquid that would be needed to fill volume Ψ_D. The buoyant force is given by

$$F_B = F_{up} = \gamma \Psi_D \tag{3.35}$$

Figure 3.20

Two views of a body immersed in a liquid.

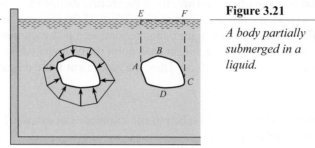

Figure 3.21

A body partially submerged in a liquid.

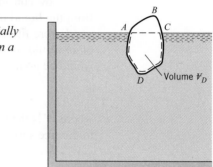

Hence, the buoyant force equals the weight of liquid that would be needed to occupy the volume V_D. This volume is called the displaced volume. Comparison of Eqs. (3.34) and (3.35) shows that one can write a single equation for the buoyant force:

$$F_B = \gamma V_D \qquad (3.36)$$

Although Eq. (3.36) was derived for a liquid, it is equally valid for a gas. If the body is totally submerged, the displaced volume is the volume of the body. If a body is partially submerged, the displaced volume is the portion of the volume that is submerged. For a fluid of uniform density, the line of action of the buoyant force passes through the centroid of the displaced volume.

The general principle of buoyancy embodied in Eq. (3.36) is called *Archimedes' principle:* For an object partially or completely submerged in a fluid, there is a net upward force (buoyant force) equal to the weight of the displaced fluid.

The Hydrometer

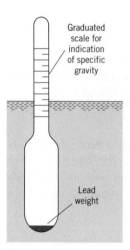

Figure 3.22

Hydrometer

A *hydrometer* (Fig. 3.22) is an instrument for measuring the specific gravity of liquids. It is typically made of a glass bulb that is weighted on one end so the hydrometer floats in an upright position. A stem of constant diameter is marked with a scale, and the specific weight of the liquid is determined by the depth at which the hydrometer floats. The operating principle of the hydrometer is buoyancy. In a heavy liquid (i.e., high γ), the hydrometer will float shallower because a lesser volume of the liquid must be displaced to balance the weight of the hydrometer. In a light liquid, the hydrometer will float deeper.

3.7

Stability of Immersed and Floating Bodies

This section describes how to determine whether an object will tip over or remain in an upright position when placed in a liquid. This topic is important for the design of objects such as ships and buoys.

Immersed Bodies

When a body is completely immersed in a liquid, its stability depends on the relative positions of the center of gravity of the body and the centroid of the displaced volume of fluid,

Figure 3.23

Conditions of stability for immersed bodies. (a) Stable. (b) Neutral. (c) Unstable.

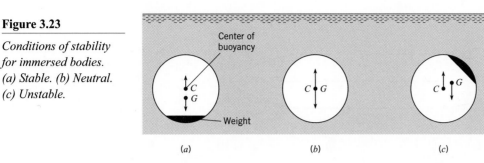

(a)   (b)   (c)

which is called the *center of buoyancy*. If the center of buoyancy is above the center of gravity, such as in Fig. 3.23*a*, any tipping of the body produces a righting couple, and consequently, the body is stable. However, if the center of gravity is above the center of buoyancy, any tipping produces an increasing overturning moment, thus causing the body to turn through 180°. This is the condition shown in Fig. 3.23*c*. Finally, if the center of buoyancy and center of gravity are coincident, the body is neutrally stable—that is, it lacks a tendency for righting or for overturning, as shown in Fig. 3.23*b*.

Floating Bodies

The question of stability is more involved for floating bodies than for immersed bodies because the center of buoyancy may take different positions with respect to the center of gravity, depending on the shape of the body and the position in which it is floating. For example, consider the cross section of a ship shown in Fig. 3.24*a*. Here the center of gravity *G* is above the center of buoyancy *C*. Therefore, at first glance it would appear that the ship is unstable and could flip over. However, notice the position of *C* and *G* after the ship has taken a small angle of heel. As shown in Fig. 3.24*b*, the center of gravity is in the same position, but the center of buoyancy has moved outward of the center of gravity, thus producing a righting moment. A ship having such characteristics is stable.

The reason for the change in the center of buoyancy for the ship is that part of the original buoyant volume, as shown by the wedge shape *AOB*, is transferred to a new buoyant volume *EOD*. Because the buoyant center is at the centroid of the displaced volume, it follows that for this case the buoyant center must move laterally to the right. The point of intersection of the lines of action of the buoyant force before and after heel is called the *metacenter M*, and the distance *GM* is called the *metacentric height*. If *GM* is positive—that is, if *M* is above *G*—the ship is stable; however, if *GM* is negative, the ship is unstable. Quantitative relations involving these basic principles of stability are presented in the next paragraph.

Figure 3.24

Ship stability relations.

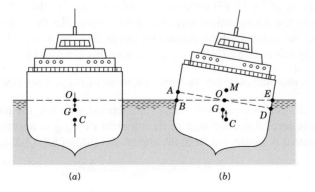

(a)   (b)

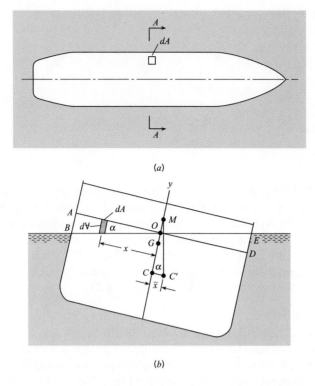

Consider the ship shown in Fig. 3.25, which has taken a small angle of heel α. First evaluate the lateral displacement of the center of buoyancy, CC'; then it will be easy by simple trigonometry to solve for the metacentric height GM or to evaluate the righting moment. Recall that the center of buoyancy is at the centroid of the displaced volume. Therefore, resort to the fundamentals of centroids to evaluate the displacement CC'. From the definition of the centroid of a volume,

$$\bar{x}\,\mathcal{V} = \Sigma x_i \Delta \mathcal{V}_i \tag{3.37}$$

where $\bar{x} = CC'$, which is the distance from the plane about which moments are taken to the centroid of $\mathcal{V}$; $\mathcal{V}$ is the total volume displaced; $\Delta \mathcal{V}_i$ is the volume increment; and x_i is the moment arm of the increment of volume.

Take moments about the plane of symmetry of the ship. Recall from mechanics that volumes to the left produce negative moments and volumes to the right produce positive moments. For the right side of Eq. (3.37) write terms for the moment of the submerged volume about the plane of symmetry. A convenient way to do this is to consider the moment of the volume before heel, subtract the moment of the volume represented by the wedge AOB, and add the moment represented by the wedge EOD. In a general way this is given by the following equation:

$$\bar{x}\,\mathcal{V} = \text{moment of } \mathcal{V} \text{ before heel} - \text{moment of } \mathcal{V}_{AOB} + \text{moment of } \mathcal{V}_{EOD} \tag{3.38}$$

Because the original buoyant volume is symmetrical with y-y, the moment for the first term on the right is zero. Also, the sign of the moment of $\mathcal{V}_{AOB}$ is negative; therefore, when this negative moment is subtracted from the right-hand side of Eq. (3.38), the result is

$$\bar{x}\,\mathcal{V} = \Sigma x_i \Delta \mathcal{V}_{i_{AOB}} + \Sigma x_i \Delta \mathcal{V}_{i_{EOD}} \tag{3.39}$$

EXAMPLE 3.12   BUOYANT FORCE ON A METAL PART

A metal part (object 2) is hanging by a thin cord from a floating wood block (object 1). The wood block has a specific gravity $S_1 = 0.3$ and dimensions of $50 \times 50 \times 10$ mm. The metal part has a volume of 6600 mm³. Find the mass m_2 of the metal part and the tension T in the cord.

Problem Definition

Situation: A metal part is suspended from a floating block of wood.

Find:

1. Mass (in grams) of the metal part.
2. Tension (in newtons) in the cord.

Properties:

1. Water (15°C), Table A.5: $\gamma = 9800$ N/m³.
2. Wood: $S_1 = 0.3$.

Sketch:

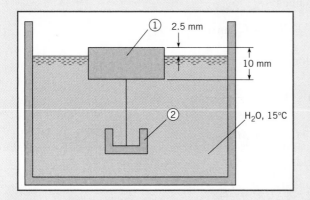

Plan

1. Draw FBDs of the block and the part.
2. Apply equilibrium to the block to find the tension.
3. Apply equilibrium to the part to find the weight of the part.
4. Calculate the mass of the metal part using $W = mg$.

Solution

1. FBDs

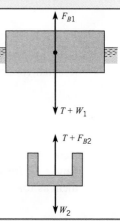

2. Force equilibrium (vertical direction) applied to block

$$T = F_{B1} - W_1$$

- Buoyant force $F_{B1} = \gamma \Psi_{D1}$, where Ψ_{D1} is the submerged volume

$$F_{B1} = \gamma \Psi_{D1}$$
$$= (9800 \text{ N/m}^3)(50 \times 50 \times 7.5 \text{ mm}^3)(10^{-9} \text{ m}^3/\text{mm}^3)$$
$$= 0.184 \text{ N}$$

- Weight of the block

$$W_1 = \gamma S_1 \Psi_1$$
$$= (9800 \text{ N/m}^3)(0.3)(50 \times 50 \times 10 \text{ mm}^3)(10^{-9} \text{ m}^3/\text{mm}^3)$$
$$= 0.0735 \text{ N}$$

- Tension in the cord

$$T = (0.184 - 0.0735) = \boxed{0.110 \text{ N}}$$

3. Force equilibrium (vertical direction) applied to metal part
   - Buoyant force

$$F_{B2} = \gamma \Psi_2 = (9800 \text{ N/m}^3)(6600 \text{ mm}^3)(10^{-9}) = 0.0647 \text{ N}$$

   - Equilibrium equation

$$W_2 = T + F_{B2} = (0.110 \text{ N}) + (0.0647 \text{ N})$$

4. Mass of metal part

$$m_2 = W_2/g = \boxed{17.8 \text{ g}}$$

Review

Notice that tension in the cord (0.11 N) is less than the weight of the metal part (0.18 N). This result is consistent with the common observation that an object will "weigh less in water than in air."

Now, express Eq. (3.39) in integral form:

$$\overline{x}\,\Psi = \int_{AOB} x\,d\Psi + \int_{EOD} x\,d\Psi \tag{3.40}$$

But it may be seen from Fig. 3.25b that $d\Psi$ can be given as the product of the length of the differential volume, $x\tan\alpha$, and the differential area, dA. Consequently, Eq. (3.40) can be written as

$$\overline{x}\,\Psi = \int_{AOB} x^2 \tan\alpha\,dA + \int_{EOD} x^2 \tan\alpha\,dA$$

Here $\tan\alpha$ is a constant with respect to the integration. Also, since the two terms on the right-hand side are identical except for the area over which integration is to be performed, combine them as follows:

$$\overline{x}\,\Psi = \tan\alpha \int_{A_{\text{waterline}}} x^2 dA \tag{3.41}$$

The second moment, or moment of inertia of the area defined by the waterline, is given the symbol I_{00}, and the following is obtained:

$$\overline{x}\,\Psi = I_{00}\tan\alpha$$

Next, replace $\overline{x}$ by CC' and solve for CC':

$$CC' = \frac{I_{00}\tan\alpha}{\Psi}$$

From Fig. 3.25b, $\qquad\qquad CC' = CM\tan\alpha$

Thus eliminating CC' and $\tan\alpha$ yields

$$CM = \frac{I_{00}}{\Psi}$$

However, $\qquad\qquad GM = CM - CG$

Therefore the *metacentric height* is

$$GM = \frac{I_{00}}{\Psi} - CG \tag{3.42}$$

Equation (3.42) is used to determine the stability of floating bodies. As already noted, if GM is positive, the body is stable; if GM is negative, it is unstable.

Note that for small angles of heel α, the righting moment or overturning moment is given as follows:

$$RM = \gamma\,\Psi GM\alpha \tag{3.43}$$

However, for large angles of heel, direct methods of calculation based on these same principles would have to be employed to evaluate the righting or overturning moment.

EXAMPLE 3.13 STABILITY OF A FLOATING BLOCK

A block of wood 30 cm square in cross section and 60 cm long weighs 318 N. Will the block float with sides vertical as shown?

Problem Definition

Situation: A block of wood is floating in water.

Find: Is the block of wood stable?

Sketch:

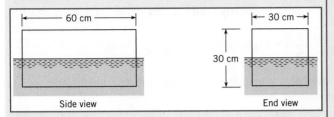

Side view   End view

Plan

1. Apply force equilibrium to find the depth of submergence.
2. Determine if block is stable about the long axis by applying Eq. (3.42).
3. If block is not stable, repeat steps 1 and 2.

Solution

1. Equilibrium (vertical direction)

$$\sum F_y = 0$$

$$-\text{weight} + \text{buoyant force} = 0$$

$$-318 \text{ N} + 9810 \text{ N/m}^3 \times 0.30 \text{ m} \times 0.60 \text{ m} \times d = 0$$

$$d = 0.18 \text{ m} = 18 \text{ cm}$$

2. Stability (longitudinal axis)

$$GM = \frac{I_{00}}{\Psi} - CG = \frac{\frac{1}{12} \times 60 \times 30^3}{18 \times 60 \times 30} - (15 - 9)$$

$$= 4.167 - 6 = -1.833 \text{ cm}$$

Because the metacentric height is negative, the block is not stable about the longitudinal axis. Thus a slight disturbance will make it tip to the orientation shown below.

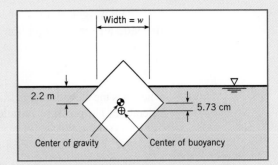

Center of gravity   Center of buoyancy

3. Equilibrium (vertical direction—see above figure)

$$-\text{weight} + \text{buoyant force} = 0$$

$$-(318 \text{ N}) + (9810 \text{ N/m}^3)(\Psi_D) = 0$$

$$\Psi_D = 0.0324 \text{ m}^3$$

4. Find the dimension w.
   (Displaced volume) = (Block volume) – (Volume above the waterline).

$$\Psi_D = 0.0324 \text{ m}^3 = (0.3^2)(0.6) \text{ m}^3 - \frac{w^2}{4}(0.6 \text{ m})$$

$$w = 0.379 \text{ m}$$

5. Moment of inertia at the waterline

$$I_{00} = \frac{bh^3}{12} = \frac{(0.6 \text{ m})(0.379 \text{ m})^3}{12} = 0.00273 \text{ m}^4$$

6. Metacentric height

$$GM = \frac{I_{00}}{\Psi} - CG = \frac{0.00273 \text{ m}^4}{0.0324 \text{ m}^3} - 0.0573 \text{ m} = 0.027 \text{ m}$$

Since the metacentric height is positive, the block will be stable in this position.

Summary

Hydrostatics, or fluid statics, is a situation in which the weight of each fluid particle is balanced by a net force associated with pressure.

Pressure p expresses the magnitude of normal force per unit area at a point. Pressure is a scalar quantity, meaning it has no directional dependence. Absolute pressure p_{abs} is measured relative to absolute zero while gage pressure p_{gage} is measured relative to atmospheric pressure p_{atm}. Gage and absolute pressure are related by

$$p_{abs} = p_{gage} + p_{atm}$$

The weight of a fluid causes pressure to increase with increasing depth, giving the hydrostatic differential equation

$$\frac{dp}{dz} = -\gamma = -\rho g$$

where z is the elevation, γ is fluid weight per volume, ρ is fluid density, and g is gravitational acceleration. The pressure variation for a uniform-density fluid is given by the hydrostatic equation

$$\frac{p}{\gamma} + z = \text{constant}$$

Hence, a fluid with uniform density has a constant value of piezometric head ($p/\gamma + z$) or piezometric pressure ($p + \gamma z$) throughout.

A fluid in contact with a surface produces a pressure distribution on the surface. This pressure distribution can be represented as a statically equivalent force F acting at the center of pressure. For a plane surface, the equivalent force is

$$F = \overline{p} A$$

where $\overline{p}$ is pressure at the centroid of the area A. For a horizontal surface, the center of pressure is at the centroid. Otherwise, the slant distance between the centroid $\overline{y}$ and the center of pressure y_{cp} is given by

$$y_{cp} - \overline{y} = \frac{\overline{I}}{\overline{y} A}$$

where $\overline{I}$ is the moment of area with respect to a horizontal axis through the centroid.

When a surface is curved, one can find the equivalent force by applying force equilibrium to a free body comprised of the fluid in contact with the surface.

When an object is either partially or totally submerged in a fluid, a buoyant force F_B acts. The magnitude is equal to the weight of the displaced volume of fluid:

$$F_B = \gamma \mathcal{V}_D$$

where $\mathcal{V}_D$ is the volume of displaced fluid. For a fluid with a uniform density, the center of buoyancy is the centroid of the displaced volume of fluid.

When an object is floating, it may be unstable or stable. Stable means that if the object is tipped, the buoyant force causes a moment that rotates the object back to its equilibrium position. An object is stable if the metacentric height is positive. In this case tipping the object causes the center of buoyancy to move such that the buoyant force produces a righting moment.

References

1. U.S. Standard Atmosphere, U.S. Government Printing Office, Washington, D.C., 1976.

2. Holman, J. P., and W. J. Gajda, Jr. *Experimental Methods for Engineers.* McGraw-Hill, New York, 1984.

3. Wikipedia contributors, "Hydraulic machinery," Wikipedia, The Free Encyclopedia, http://en.wikipedia.org/w/index.php?title= Hydraulic_machinery&oldid=161288040 (accessed October 4, 2007).

Problems

*A *Preview Question* ($\mathbb{PQ}$) can be assigned prior to in-class coverage of a topic.

Pressure and Related Concepts

*3.1 $\mathbb{PQ}$◀ Apply the grid method (p. 9) to each situation.
a. If pressure is 8 inches of water (vacuum), what is gage pressure in kPa?
b. If the pressure is 120 kPa abs, what is the gage pressure in psi?
c. If gage pressure is 0.5 bar, what is absolute pressure in psi?
d. If a person's blood pressure is 120 mm Hg, what is their blood pressure in kPa abs?

*3.2 $\mathbb{PQ}$◀ A 100 mm diameter sphere contains an ideal gas at 20°C. Apply the grid method (p. 9) to calculate the density in units of kg/m³.
a. Gas is helium. Gage pressure is 20 in H_2O.
b. Gas is methane. Vacuum pressure is 3 psi.

*3.3 $\mathbb{PQ}$◀ Using Section 3.1 and other resources, answer the following questions. Strive for depth, clarity, and accuracy while also combining sketches, words, and equations in ways that enhance the effectiveness of your communication.
a. What are five important facts that engineers need to know about pressure?
b. What are five common instances in which people use gage pressure?
c. What are the most common units for pressure?
d. Why is pressure defined using a derivative?
e. How is pressure similar to shear stress? How does pressure differ from shear stress?

3.4 The Crosby gage tester shown in the figure is used to calibrate or to test pressure gages. When the weights and the piston together weigh 140 N, the gage being tested indicates 200 kPa. If the piston diameter is 30 mm, what percentage of error exists in the gage?

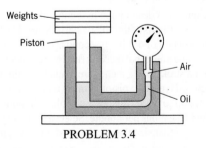

PROBLEM 3.4

3.5 As shown, a mouse can use the mechanical advantage provided by a hydraulic machine to lift up an elephant.
a. Derive an algebraic equation that gives the mechanical advantge of the hydraulic machine shown. Assume the pistons are frictionless and massless.

b. A mouse can have a mass of 25 g and an elephant a mass of 7500 kg. Determine a value of D_1 and D_2 so that the mouse can support the elephant.

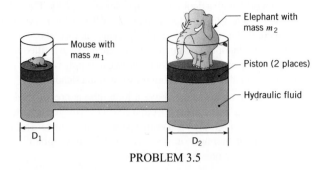

PROBLEM 3.5

3.6 Find a parked automobile for which you have information on tire pressure and weight. Measure the area of tire contact with the pavement. Next, using the weight information and tire pressure, use engineering principles to calculate the contact area. Compare your measurement with your calculation and discuss.

Hydrostatic Equation

*3.7 $\mathbb{PQ}$◀ Apply the grid method (p. 9) with the hydrostatic equation ($\Delta p = \gamma \Delta z$) to each of the following cases.
a. Predict the pressure change Δp in kPa for an elevation change Δz of 10 ft in a fluid with a density of 90 lbm/ft³.
b. Predict the pressure change in psf for a fluid with S = 0.8 and an elevation change of 22 m.
c. Predict pressure change in inches of water for a fluid with a density of 1.2 kg/m³ and an elevation change of 1000 ft.
d. Predict the elevation change in millimeters for a fluid with S = 13 that corresponds to a change in pressure of 1/6 atm.

*3.8 $\mathbb{PQ}$◀ Using Section 3.2 and other resources, answer the following questions. Strive for depth, clarity, and accuracy while also combining sketches, words, and equations in ways that enhance the effectiveness of your communication.
a. What does hydrostatic mean? How do engineers identify whether a fluid is hydrostatic?
b. What are the common forms on the hydrostatic equation? Are the forms equivalent or are they different?
c. What is a datum? How do engineers establish a datum?
d. What are the main ideas of Eq. (3.5)? That is, what is the meaning of this equation?
e. What assumptions need to be satisfied to apply the hydrostatic equation?

3.9 Apply the grid method (p. 11) to each situation.
a. What is the change in air pressure in pascals between the floor and the ceiling of a room with walls that are 8 ft tall.
b. A diver in the ocean (S = 1.03) records a pressure of 2 atm on her depth gage. How deep is she?

c. A hiker starts a hike at an elevation where the air pressure is 940 mbar, and he ascends 1200 ft to a mountain summit. Assuming the density of air is constant, what is the pressure in mbar at the summit?

d. Lake Pend Oreille, in northern Idaho, is one of the deepest lakes in the world, with a depth of 350 m in some locations. This lake is used as a test facility for submarines. What is the maximum pressure that a submarine could experience in this lake?

e. A 60 m tall standpipe (a standpipe is vertical pipe that is filled with water and open to the atmosphere) is used to supply water for fire fighting. What is the maximum pressure in the standpipe?

3.10 As shown, an air space above a long tube is pressurized to 50 kPa vacuum. Water (15°C) from a reservoir fills the tube to a height h. If the pressure in the air space is changed to 25 kPa vacuum, will h increase or descrease and by how much? Assume atmospheric pressure is 100 kPa.

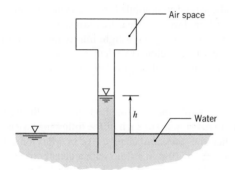

PROBLEM 3.10

3.11 For the closed tank with Bourdon-tube gages tapped into it, what is the specific gravity of the oil and the pressure reading on gage C?

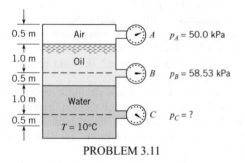

PROBLEM 3.11

3.12 This manometer contains water at room temperature. The glass tube on the left has an inside diameter of 1 mm $(d = 1.0$ mm). The glass tube on the right is three times as large. For these conditions, the water surface level in the left tube will be (a) higher than the water surface level in the right tube, (b) equal to the water surface level in the right tube, or (c) less than the water surface level in the right tube. State your main reason or assumption for making your choice.

3.13 If a 200 N force F_1 is applied to the piston with the 4 cm diameter, what is the magnitude of the force F_2 that can be resisted by the piston with the 10 cm diameter? Neglect the weights of the pistons.

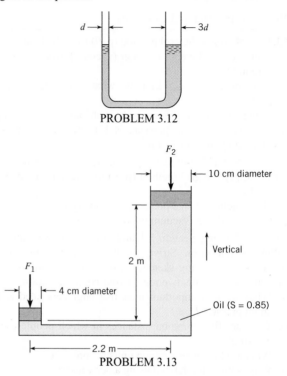

PROBLEM 3.12

PROBLEM 3.13

3.14 Some skin divers go as deep as 50 m. What is the gage pressure at this depth in fresh water, and what is the ratio of the absolute pressure at this depth to normal atmospheric pressure? Assume $T = 20°C$.

3.15 Water occupies the bottom 1.0 m of a cylindrical tank. On top of the water is 0.75 m of kerosene, which is open to the atmosphere. If the temperature is 20°C, what is the gage pressure at the bottom of the tank?

3.16 An engineer is designing a hydraulic lift with a capacity of 10 tons. The moving parts of this lift weigh 1000 lbf. The lift should raise the load to a height of 6 ft in 20 seconds. This will be accomplished with a hydraulic pump that delivers fluid to a cylinder. Hydraulic cylinders with a stroke of 72 inches are available with bore sizes from 2 to 8 inches. Hydraulic piston pumps with an operating pressure range from 200 to 3000 psig are available with pumping capacities of 5, 10, and 15 gallons per minute. Select a hydraulic pump size and a hydraulic cylinder size that can be used for this application.

3.17 A tank with an attached manometer contains water at 20°C. The atmospheric pressure is 100 kPa. There is a stopcock located 1 m from the surface of the water in the manometer. The stopcock is closed, trapping the air in the manometer, and water is added to the tank to the level of the stopcock. Find the increase in elevation of the water in the manometer assuming the air in the manometer is compressed isothermally.

3.18 A tank is fitted with a manometer on the side, as shown. The liquid in the bottom of the tank and in the manometer has a specific gravity (S) of 3.0. The depth of this bottom liquid is 20 cm. A 15 cm layer of water lies on top of the bottom liquid. Find the position of the liquid surface in the manometer.

is connected to a round tube of diameter $D_2 = 5$ mm and oil rises in the tube to height h_2. Find h_2. Assume the oil in the tube is open to atmosphere and neglect the mass of the piston.

3.21 What is the maximum gage pressure in the odd tank shown in the figure? Where will the maximum pressure occur? What is the hydrostatic force acting on the top (CD) of the last chamber on the right-hand side of the tank? Assume $T = 10°C$.

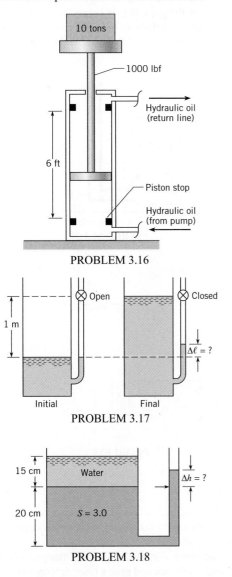

PROBLEM 3.16

PROBLEM 3.17

PROBLEM 3.18

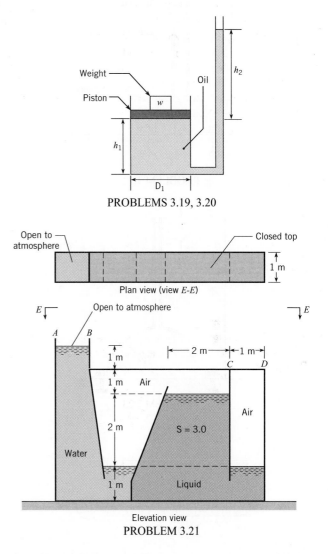

PROBLEMS 3.19, 3.20

Elevation view
PROBLEM 3.21

3.19 As shown, a load acts on a piston of diameter D_1. The piston rides on a reservoir of oil of depth h_1 and specific gravity S. The reservoir is connected to a round tube of diameter D_2 and oil rises in the tube to height h_2. The oil in the tube is open to atmosphere. Derive an equation for the height h_2 in terms of the mass m of the load and other relevant variables. Neglect the mass of the piston.

3.20 As shown, a load of mass 10 kg is situated on a piston of diameter $D_1 = 140$ mm. The piston rides on a reservoir of oil of depth $h_1 = 42$ mm and specific gravity S = 0.8. The reservoir

3.22 The steel pipe and steel chamber shown in the figure together weigh 600 lbf. What force will have to be exerted on the chamber by all the bolts to hold it in place? The dimension ℓ is equal to 2.5 ft. *Note:* There is no bottom on the chamber—only a flange bolted to the floor.

3.23 What force must be exerted through the bolts to hold the dome in place? The metal dome and pipe weigh 6 kN. The dome has no bottom. Here $\ell = 80$ cm.

3.24 Find the vertical component of force in the metal at the base of the spherical dome shown when gage A reads 5 psig. Indicate whether the metal is in compression or tension. The specific gravity of the enclosed fluid is 1.5. The dimension L is 2 ft. Assume the dome weighs 1000 lbf.

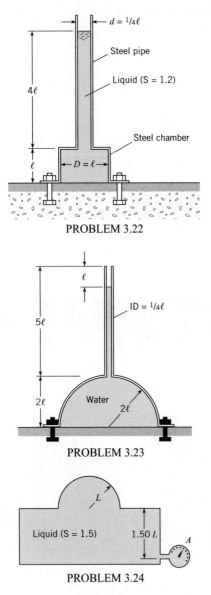

PROBLEM 3.22

PROBLEM 3.23

Liquid (S = 1.5)   1.50 L

PROBLEM 3.24

3.25 The piston shown weighs 10 lbf. In its initial position, the piston is restrained from moving to the bottom of the cylinder by means of the metal stop. Assuming there is neither friction nor leakage between piston and cylinder, what volume of oil (S = 0.85) would have to be added to the 1 in. tube to cause the piston to rise 1 in. from its initial position?

3.26 Consider an air bubble rising from the bottom of a lake. Neglecting surface tension, determine approximately what the

ratio of the density of the air in the bubble will be at a depth of 34 ft to its density at a depth of 8 ft.

3.27 One means of determining the surface level of liquid in a tank is by discharging a small amount of air through a small tube, the end of which is submerged in the tank, and reading the pressure on the gage that is tapped into the tube. Then the level of the liquid surface in the tank can be calculated. If the pressure on the gage is 15 kPa, what is the depth d of liquid in the tank?

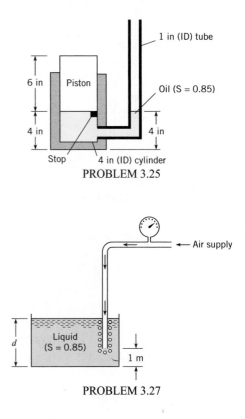

PROBLEM 3.25

PROBLEM 3.27

Manometers

*3.28 $\mathbb{P}\mathbb{Q}\blacktriangleleft$ Using the Internet and other resources, answer the following questions:
a. What are three common types of manometers? For each type, make a sketch and give a brief description.
b. How would you build a manometer from materials that are commonly available? Sketch your design concept.

*3.29 $\mathbb{P}\mathbb{Q}\blacktriangleleft$ As shown, gas at pressure p_g raises a column of liquid to a height h. The gage pressure in the gas is given by $p_g = \gamma_{liquid}h$. Apply the grid method (p. 9) to each situation that follows.
a. The manometer uses a liquid with S = 1.5. Calculate pressure in psia for $h = 1$ ft.
b. The manometer uses mercury. Calculate the column rise in mm for a gage pressure of $1/6$ atm.

c. The liquid has a density of 50 lbm/ft³. Calculate pressure in psfg for $h = 6$ inches.
d. The liquid has a density of 800 kg/m³. Calculate the gage pressure in bar for $h = 3$ m.

3.30 Is the gage pressure at the center of the pipe (a) negative, (b) zero, or (c) positive? Neglect surface tension effects and state your rationale.

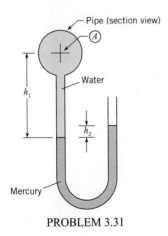

PROBLEM 3.29

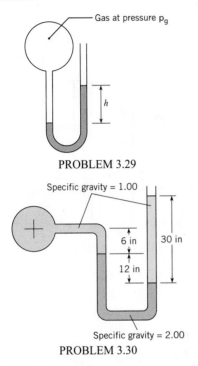

PROBLEM 3.30

3.31 Determine the gage pressure at the center of the pipe (point A) in pounds per square inch when the temperature is 70°F with $h_1 = 16$ in. and $h_2 = 2$ in.

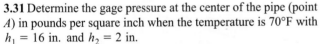

PROBLEM 3.31

3.32 Considering the effects of surface tension, estimate the gage pressure at the center of pipe A for $h = 100$ mm and $T = 20°C$.

3.33 What is the pressure at the center of pipe B?

3.34 The ratio of container diameter to tube diameter is 8. When air in the container is at atmospheric pressure, the free surface in the tube is at position 1. When the container is pressurized, the liquid in the tube moves 40 cm up the tube from position 1 to position 2. What is the container pressure that causes this deflection? The liquid density is 1200 kg/m³.

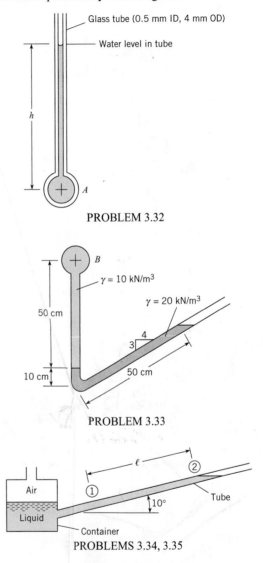

PROBLEM 3.32

PROBLEM 3.33

PROBLEMS 3.34, 3.35

3.35 The ratio of container diameter to tube diameter is 10. When air in the container is at atmospheric pressure, the free surface in the tube is at position 1. When the container is pressurized, the liquid in the tube moves 3 ft up the tube from position 1 to position 2. What is the container pressure that causes this deflection? The specific weight of the liquid is 50 lbf/ft³.

3.36 Determine the gage pressure at the center of pipe A in pounds per square inch and in kilopascals.

3.37 A device for measuring the specific weight of a liquid consists of a U-tube manometer as shown. The manometer tube has an internal diameter of 0.5 cm and originally has water in it. Exactly 2 cm³ of unknown liquid is then poured into one leg of the manometer, and a displacement of 5 cm is measured between the surfaces as shown. What is the specific weight of the unknown liquid?

3.40 Determine (a) the difference in pressure and (b) the difference in piezometric head between points A and B. The elevations z_A and z_B are 10 m and 11 m, respectively, $\ell_1 = 1$ m, and the manometer deflection ℓ_2 is 50 cm.

3.41 The deflection on the manometer is h meters when the pressure in the tank is 150 kPa absolute. If the absolute pressure in the tank is doubled, what will the deflection on the manometer be?

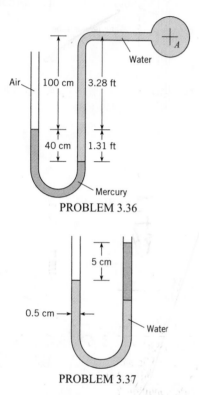

PROBLEM 3.36

PROBLEM 3.37

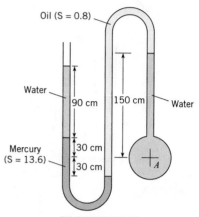

PROBLEM 3.39

3.38 Mercury is poured into the tube in the figure until the mercury occupies 375 mm of the tube's length. An equal volume of water is then poured into the left leg. Locate the water and mercury surfaces. Also determine the maximum pressure in the tube.

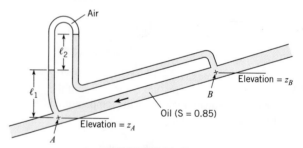

PROBLEM 3.40

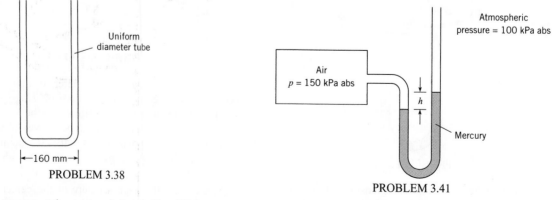

PROBLEM 3.38

PROBLEM 3.41

3.39 Find the pressure at the center of pipe A. $T = 10°C$.

3.42 A vertical conduit is carrying oil (S = 0.95). A differential mercury manometer is tapped into the conduit at points A and B. Determine the difference in pressure between A and B when $h = 3$ in. What is the difference in piezometric head between A and B?

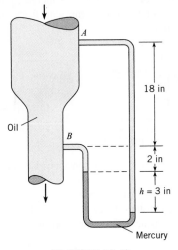

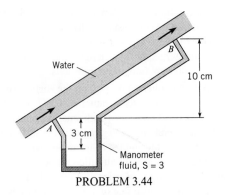

PROBLEM 3.44

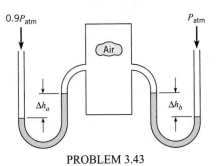

PROBLEM 3.42

3.43 Two water manometers are connected to a tank of air. One leg of the manometer is open to 100 kPa pressure (absolute) while the other leg is subjected to 90 kPa. Find the difference in deflection between both manometers, $\Delta h_a - \Delta h_b$.

Choose the piston size and standpipe diameter. Clearly state the design features you considered. Indicate how you would calibrate the scale on the standpipe. Would the scale be linear?

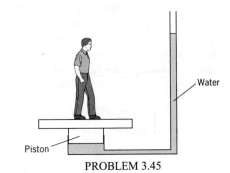

PROBLEM 3.45

Atmospheric Pressure Variation

3.46 The boiling point of water decreases with elevation because of the pressure change. What is the boiling point of water at an elevation of 2000 m and at an elevation of 4000 m for standard atmospheric conditions?

3.47 From a depth of 10 m in a lake to an elevation of 4000 m in the atmosphere, plot the variation of absolute pressure. Assume that the lake water surface elevation is at mean sea level and assume standard atmospheric conditions.

3.48 Assume that a woman must breathe a constant mass rate of air to maintain her metabolic processes. If she inhales and exhales 16 times per minute at sea level, where the temperature is 59°F (15°C) and the pressure is 14.7 psia (101 kPa), what would you expect her rate of breathing at 18,000 ft (5486 m) to be? Use standard atmospheric conditions.

3.49 A pressure gage in an airplane indicates a pressure of 95 kPa at takeoff, where the airport elevation is 1 km and the temperature is 10°C. If the standard lapse rate of 5.87°C/km is assumed, at what elevation is the plane when a pressure of 75 kPa is read? What is the temperature for that condition?

3.50 Denver, Colorado, is called the "mile-high" city. What are the pressure, temperature, and density of the air when standard atmospheric conditions prevail? Give your answer in traditional and SI units.

PROBLEM 3.43

3.44 A manometer is used to measure the pressure difference between points A and B in a pipe as shown. Water flows in the pipe, and the specific gravity of the manometer fluid is 3.0. The distances and manometer deflection are indicated on the figure. Find (a) the pressure differences $p_A - p_B$, and (b) the difference in piezometric pressure, $p_{z,A} - p_{z,B}$. Express both answers in kPa.

3.45 A novelty scale for measuring a person's weight by having the person stand on a piston connected to a water reservoir and stand pipe is shown in the diagram. The level of the water in the stand pipe is to be calibrated to yield the person's weight in pounds force. When the person stands on the scale, the height of the water in the stand pipe should be near eye level so the person can read it. There is a seal around the piston that prevents leaks but does not cause a significant frictional force. The scale should function for people who weigh between 60 and 250 lbf and are between 4 and 6 feet tall.

3.51 An airplane is flying at 10 km altitude in a U.S. standard atmosphere. If the internal pressure of the aircraft interior is 100 kPa, what is the outward force on a window? The window is flat and has an elliptical shape with lengths of 300 mm along the major axis and 200 mm along the minor axis.

3.52 The mean atmospheric pressure on the surface of Mars is 7 mbar, and the mean surface temperature is –63°C. The atmosphere consists primarily of CO_2 (95.3%) with small amounts of nitrogen and argon. The acceleration due to gravity on the surface is 3.72 m/s^2. Data from probes entering the Martian atmosphere show that the temperature variation with altitude can be approximated as constant at –63°C from the Martian surface to 14 km, and then a linear decrease with a lapse rate of 1.5°C/km up to 34 km. Find the pressure at 8 km and 30 km altitude. Assume the atmosphere is pure carbon dioxide. Note that the temperature distribution in the atmosphere of Mars differs from that of Earth because the region of constant temperature is adjacent to the surface and the region of decreasing temperature starts at an altitude of 14 km.

3.53 Design a computer program that calculates the pressure and density for the U.S. standard atmosphere from 0 to 30 km altitude. Assume the temperature profiles are linear and are approximated by the following ranges, where z is the altitude in kilometers:

0–13.72 km	$T = 23.1 - 5.87z$ (°C)
13.7–16.8 km	$T = -57.5°C$
16.8–30 km	$T = -57.5 + 1.387(z - 16.8)°C$

Panel Force Equations

*3.54 PQ◄ Using Section 3.4 and other resources, answer the questions below. Strive for depth, clarity, and accuracy while also combining sketches, words, and equations in ways that enhance the effectiveness of your communication.
a. For hydrostatic conditions, what do typical pressure distributions on a panel look like? Sketch three examples that correspond to different situations.
b. What is a center of pressure? What is a centroid of area?
c. In Eq. (3.23), what does $\bar{p}$ mean? What factors influence the value of $\bar{p}$?
d. What is the relationship between the pressure distribution on a panel and the resultant force?
e. How far is the center of pressure from the centroid of area? What factors influence this distance?

3.55 Consider the two rectangular gates shown in the figure. They are both the same size, but gate A is held in place by a horizontal shaft through its midpoint and gate B is cantilevered to a shaft at its top. Now consider the torque T required to hold the gates in place as H is increased. Choose the valid statement(s): (a) T_A increases with H. (b) T_B increases with H. (c) T_A does not change with H. (d) T_B does not change with H.

3.56 For gate A, choose the statements that are valid: (a) The hydrostatic force acting on the gate increases as H increases. (b)

The distance between the CP on the gate and the centroid of the gate decreases as H increases. (c) The distance between the CP on the gate and the centroid of the gate remains constant as H increases. (d) The torque applied to the shaft to prevent the gate from turning must be increased as H increases. (e) The torque applied to the shaft to prevent the gate from turning remains constant as H increases.

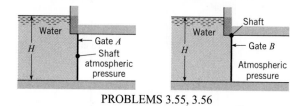

PROBLEMS 3.55, 3.56

3.57 As shown, water (15°C) is in contact with a square panel; $d = 1$ m and $h = 2$ m.
a. Calculate the depth of the centroid
b. Calculate the resultant force on the panel
c. Calculate the distance from the centroid to the CP.

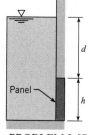

PROBLEM 3.57

3.58 As shown, a round viewing window of diameter $D = 0.8$ m is situated in a large tank of seawater ($S = 1.03$). The top of the window is 1.2 m below the water surface, and the window is angled at 60° with respect to the horizontal. Find the hydrostatic force acting on the window and locate the corresponding CP.

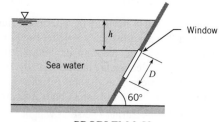

PROBLEM 3.58

3.59 Find the force of the gate on the block. See sketch.

3.60 Assume that wet concrete ($\gamma = 150$ lbf/ft^3) behaves as a liquid. Determine the force per unit foot of length exerted on the forms. If the forms are held in place as shown, with ties between vertical braces spaced every 2 ft, what force is exerted on the bottom tie?

3.61 A rectangular gate is hinged at the water line, as shown. The gate is 4 ft high and 10 ft wide. The specific weight of water is 62.4 lbf/ft³. Find the necessary force (in lbf) applied at the bottom of the gate to keep it closed.

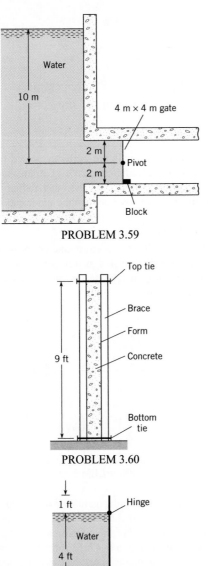

PROBLEM 3.59

PROBLEM 3.60

PROBLEM 3.61

3.62 The gate shown is rectangular and has dimensions 6 m by 4 m. What is the reaction at point A? Neglect the weight of the gate.

3.63 Determine P necessary to just start opening the 2 m–wide gate.

3.64 The square gate shown is eccentrically pivoted so that it automatically opens at a certain value of h. What is that value in terms of ℓ?

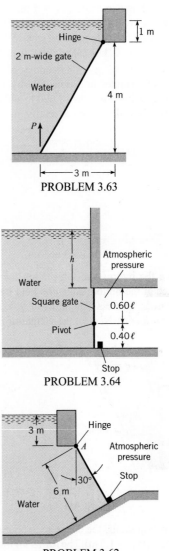

PROBLEM 3.63

PROBLEM 3.64

PROBLEM 3.62

3.65 This 10 ft–diameter butterfly valve is used to control the flow in a 10 ft–diameter outlet pipe in a dam. In the position shown, it is closed. The valve is supported by a horizontal shaft through its center. What torque would have to be applied to the shaft to hold the valve in the position shown?

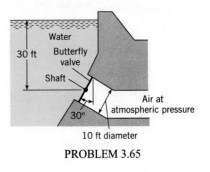

PROBLEM 3.65

3.66 For the gate shown, $\alpha = 45°$, $y_1 = 1$ m, and $y_2 = 4$ m. Will the gate fall or stay in position under the action of the hydrostatic and gravity forces if the gate itself weighs 150 kN and is 1.0 m wide? Assume $T = 10°C$. Use calculations to justify your answer.

3.67 For this gate, $\alpha = 45°$, $y_1 = 3$ ft, and $y_2 = 6$ ft. Will the gate fall or stay in position under the action of the hydrostatic and gravity forces if the gate itself weighs 18,000 lb and is 3 ft wide? Assume $T = 50°F$. Use calculations to justify your answer.

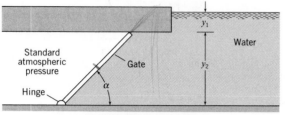

PROBLEMS 3.66, 3.67

3.68 Determine the hydrostatic force F on the triangular gate, which is hinged at the bottom edge and held by the reaction R_T at the upper corner. Express F in terms of γ, h, and W. Also determine the ratio R_T/F. Neglect the weight of the gate.

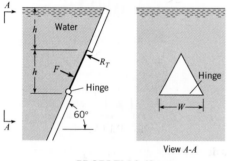

PROBLEM 3.68

3.69 In constructing dams, the concrete is poured in lifts of approximately 1.5 m ($y_1 = 1.5$ m). The forms for the face of the dam are reused from one lift to the next. The figure shows one such form, which is bolted to the already cured concrete. For the new pour, what moment will occur at the base of the form per meter of length (normal to the page)? Assume that concrete acts as a liquid when it is first poured and has a specific weight of 24 kN/m³.

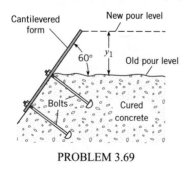

PROBLEM 3.69

3.70 The plane rectangular gate can pivot about the support at B. For the conditions given, is it stable or unstable? Neglect the weight of the gate. Justify your answer with calculations.

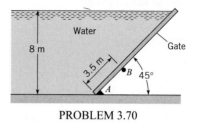

PROBLEM 3.70

Forces on Curved Surfaces

3.71 Two hemispheric shells are perfectly sealed together, and the internal pressure is reduced to 25% of atmospheric pressure. The inner radius is 10.5 cm, and the outer radius is 10.75 cm. The seal is located halfway between the inner and outer radius. If the atmospheric pressure is 101.3 kPa, what force is required to pull the shells apart?

3.72 If exactly 20 bolts of 2.5 cm diameter are needed to hold the air chamber together at A-A as a result of the high pressure within, how many bolts will be needed at B-B? Here $D = 40$ cm and $d = 20$ cm.

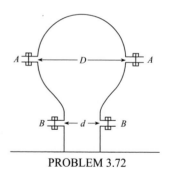

PROBLEM 3.72

3.73 For the plane rectangular gate ($\ell \times w$ in size), figure (*a*), what is the magnitude of the reaction at A in terms of γ_w and the dimensions ℓ and w? For the cylindrical gate, figure (*b*), will the magnitude of the reaction at A be greater than, less than, or the same as that for the plane gate? Neglect the weight of the gates.

3.74 Water is held back by this radial gate. Does the resultant of the pressure forces acting on the gate pass above the pin, through the pin, or below the pin?

3.75 For the curved surface AB:

a. Determine the magnitude, direction, and line of action of the vertical component of hydrostatic force acting on the surface. Here $\ell = 1$ m.

b. Determine the magnitude, direction, and line of action of the horizontal component of hydrostatic force acting on the surface.

c. Determine the resultant hydrostatic force acting on the surface.

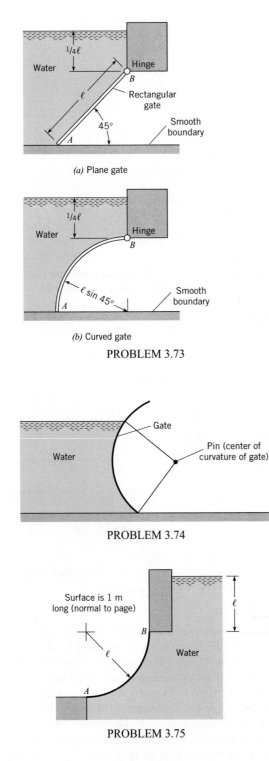

(a) Plane gate

(b) Curved gate

PROBLEM 3.73

PROBLEM 3.74

PROBLEM 3.75

3.76 Determine the hydrostatic force acting on the radial gate if the gate is 40 ft long (normal to the page). Show the line of action of the hydrostatic force acting on the gate.

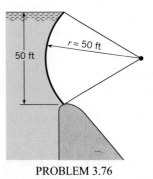

PROBLEM 3.76

3.77 A plug in the shape of a hemisphere is inserted in a hole in the side of a tank as shown in the figure. The plug is sealed by an O-ring with a radius of 0.2 m. The radius of the hemispherical plug is 0.25 m. The depth of the center of the plug is 2 m in fresh water. Find the horizontal and vertical forces on the plug due to hydrostatic pressure.

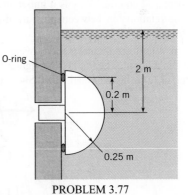

PROBLEM 3.77

3.78 This dome (hemisphere) is located below the water surface as shown. Determine the magnitude and sign of the force components needed to hold the dome in place and the line of action of the horizontal component of force. Here $y_1 = 1$ m and $y_2 = 2$ m. Assume $T = 10°C$.

3.79 Consider the dome of Prob. 3.78. This dome is 10 ft in diameter, but now the dome is not submerged. The water surface is at the level of the center of curvature of the dome. For these conditions, determine the magnitude and direction of the resultant hydrostatic force acting on the dome.

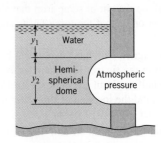

PROBLEMS 3.78, 3.79

Buoyancy

*3.80 $\mathbb{PQ}\blacktriangleleft$ Apply the grid method (p. 11) to each situation below.

a. Determine the buoyant force in newtons on a basketball that is floating in a lake (10°C).

b. Determine the buoyant force in newtons on a 1 mm copper sphere that is immersed in kerosene.

c. Determine the buoyant force in newtons on a 12 inch–diameter balloon. The balloon is filled with helium and situated in ambient air (20°C).

*3.81 $\mathbb{PQ}\blacktriangleleft$ Using Section 3.6 and other resources, answer the following questions. Strive for depth, clarity, and accuracy while also combining sketches, words, and equations in ways that enhance the effectiveness of your communication.

a. Why learn about buoyancy? That is, what are important technical problems that involve buoyant forces?

b. For a buoyant force, where is the CP? Where is the line of action?

c. What is displaced volume? Why is it important?

d. What is the relationship between pressure distribution and buoyant force?

3.82 As shown, a uniform-diameter rod is weighted at one end and is floating in a liquid. The liquid (a) is lighter than water, (b) must be water, or (c) is heavier than water. Show your work.

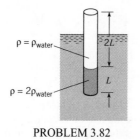

PROBLEM 3.82

3.83 An 800 ft ship has a displacement of 35,000 tons, and the area defined by the waterline is 38,000 ft². Will the ship take more or less draft when steaming from salt water to fresh water? How much will it settle or rise?

3.84 A submerged spherical steel buoy that is 1.2 m in diameter and weighs 1200 N is to be anchored in salt water 20 m below the surface. Find the weight of scrap iron that should be sealed inside the buoy in order that the force on its anchor chain will not exceed 4.5 kN.

3.85 A buoy is designed with a hemispherical bottom and conical top as shown in the figure. The diameter of the hemisphere is 1 m, and the half angle of the cone is 30°. The buoy has a mass of 460 kg. Find the location of the water line on the buoy floating in sea water ($\rho = 1010 \text{ kg/m}^3$).

3.86 A rock weighs 1000 N in air and 609 N in water. Find its volume.

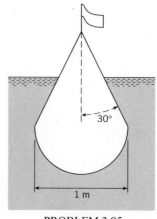

PROBLEM 3.85

3.87 As shown, a cube ($L = 60$ mm) suspended in carbon tetrachloride is exactly balanced by an object of mass $m_1 = 700$ g. Find the mass m_2 of the cube.

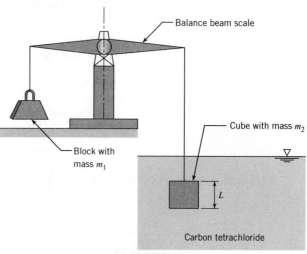

PROBLEM 3.87

3.88 A block of material of unknown volume is submerged in water and found to weigh 300 N (in water). The same block weighs 700 N in air. Determine the specific weight and volume of the material.

3.89 A 1 ft–diameter cylindrical tank is filled with water to a depth of 2 ft. A cylinder of wood 6 in. in diameter and 3 in. long is set afloat on the water. The weight of the wood cylinder is 2 lbf. Determine the change (if any) in the depth of the water in the tank.

3.90 A 90° inverted cone contains water as shown. The volume of the water in the cone is given by $\mathcal{V} = (\pi/3)h^3$. The original depth of the water is 10 cm. A block with a volume of 200 cm³ and a specific gravity of 0.6 is floated in the water. What will be the change (in cm) in water surface height in the cone?

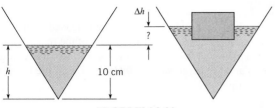

PROBLEM 3.90

3.91 The floating platform shown is supported at each corner by a hollow sealed cylinder 1 m in diameter. The platform itself weighs 30 kN in air, and each cylinder weighs 1.0 kN per meter of length. What total cylinder length L is required for the platform to float 1 m above the water surface? Assume that the specific weight of the water (brackish) is 10,000 N/m^3. The platform is square in plan view.

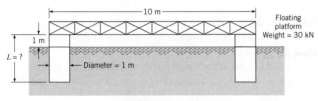

PROBLEM 3.91

3.92 To what depth d will this rectangular block (with density 0.8 times that of water) float in the two-liquid reservoir?

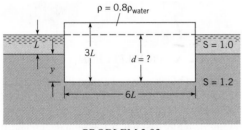

PROBLEM 3.92

3.93 Determine the minimum volume of concrete ($\gamma = 23.6$ kN/m^3) needed to keep the gate (1 m wide) in a closed position, with $\ell = 2$ m. Note the hinge at the bottom of the gate.

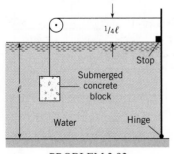

PROBLEM 3.93

3.94 A cylindrical container 4 ft high and 2 ft in diameter holds water to a depth of 2 ft. How much does the level of the water in

the tank change when a 5 lb block of ice is placed in the container? Is there any change in the water level in the tank when the block of ice melts? Does it depend on the specific gravity of the ice? Explain all the processes.

3.95 The partially submerged wood pole is attached to the wall by a hinge as shown. The pole is in equilibrium under the action of the weight and buoyant forces. Determine the density of the wood.

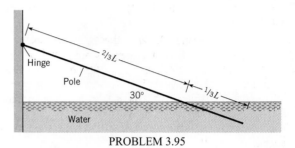

PROBLEM 3.95

3.96 A gate with a circular cross section is held closed by a lever 1 m long attached to a buoyant cylinder. The cylinder is 25 cm in diameter and weighs 200 N. The gate is attached to a horizontal shaft so it can pivot about its center. The liquid is water. The chain and lever attached to the gate have negligible weight. Find the length of the chain such that the gate is just on the verge of opening when the water depth above the gate hinge is 10 m.

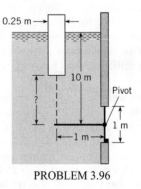

PROBLEM 3.96

3.97 A balloon is to be used to carry meteorological instruments to an elevation of 15,000 ft where the air pressure is 8.3 psia. The balloon is to be filled with helium, and the material from which it is to be fabricated weighs 0.01 lbf/ft^2. If the instruments weigh 10 lbf, what diameter should the spherical balloon have?

3.98 A weather balloon is constructed of a flexible material such that the internal pressure of the balloon is always 10 kPa higher than the local atmospheric pressure. At sea level the diameter of the balloon is 1 m, and it is filled with helium. The balloon material, structure, and instruments have a mass of 100 g. This does not include the mass of the helium. As the balloon rises, it will expand. The temperature of the helium is always equal to the local atmospheric temperature, so it decreases as the balloon gains altitude. Calculate the maximum altitude of the balloon in a standard atmosphere.

Hydrometers

3.99 The hydrometer shown sinks 5.3 cm in water (15°C). The bulb displaces 1.0 cm³, and the stem area is 0.1 cm². Find the weight of the hydrometer.

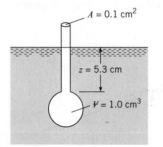

$A = 0.1$ cm²

$z = 5.3$ cm

$\Psi = 1.0$ cm³

PROBLEMS 3.99, 3.100

3.100 The hydrometer of Prob. 3.99 weighs 0.015 N. If the stem sinks 6.3 cm in oil ($z = 6.3$ cm), what is the specific gravity of the oil?

3.101 A common commercial hydrometer for measuring the amount of antifreeze in the coolant system of an automobile engine consists of a chamber with differently colored balls. The system is calibrated to give the range of specific gravity by distinguishing between the balls that sink and those that float. The specific gravity of an ethylene glycol-water mixture varies from 1.012 to 1.065 for 10% to 50% by weight of ethylene glycol. Assume there are six balls, 1 cm in diameter each, in the chamber. What should the weight of each ball be to provide a range of specific gravities between 1.01 and 1.06 with 0.01 intervals?

3.102 A hydrometer with the configuration shown has a bulb diameter of 2 cm, a bulb length of 8 cm, a stem diameter of 1 cm, a length of 8 cm, and a mass of 35 g. What is the range of specific gravities that can be measured with this hydrometer? (*Hint:* Liquid levels range between bottom and top of stem.)

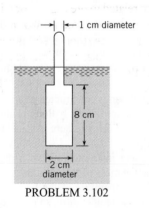

1 cm diameter

8 cm

2 cm diameter

PROBLEM 3.102

Stability

3.103 A barge 20 ft wide and 50 ft long is loaded with rocks as shown. Assume that the center of gravity of the rocks and barge is located along the centerline at the top surface of the barge. If the rocks and the barge weigh 400,000 lbf, will the barge float upright or tip over?

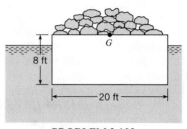

G

8 ft

20 ft

PROBLEM 3.103

3.104 A floating body has a square cross section with side w as shown in the figure. The center of gravity is at the centroid of the cross section. Find the location of the water line, ℓ/w, where the body would be neutrally stable ($GM = 0$). If the body is floating in water, what would be the specific gravity of the body material?

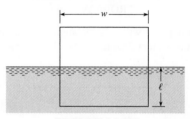

w

ℓ

PROBLEM 3.104

3.105 A cylindrical block of wood 1 m in diameter and 1 m long has a specific weight of 7500 N/m³. Will it float in water with its axis vertical?

3.106 A cylindrical block of wood 1 m in diameter and 1 m long has a specific weight of 5000 N/m³. Will it float in water with the ends horizontal?

3.107 Is the block in this figure stable floating in the position shown? Show your calculations.

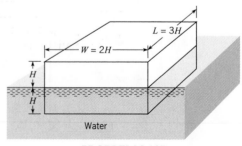

$L = 3H$

$W = 2H$

H

H

Water

PROBLEM 3.107

Flowing Fluids and Pressure Variation

This photograph shows the eye of a hurricane. The motion is the result of pressure variations.

SCIENTIFIC LEARNING OUTCOMES

Conceptual Knowledge

- Distinguish between steady, unsteady, uniform, and nonuniform flows.
- Distinguish between convective and local acceleration.
- Describe the steps to derive the Bernoulli equation from Euler's equation.
- Explain what is meant by rotation and vorticity of a fluid element.
- Describe flow separation.

Procedural Knowledge

- Apply Euler's equation to predict pressure.
- Predict pressure distributions in rotating flows.
- Apply the Bernoulli equation to pressure and velocity variations.
- Evaluate the rotation and vorticity of a fluid element.

Applications (Typical)

- In variable area ducts, relate pressure and velocity distributions.
- Measurement of velocity with stagnation tube or a Pitot-static tube.
- Cyclonic storm pressure distribution.

Many phenomena that affect us in our daily lives are related to pressure in flowing fluids. For example, one indicator of our health, blood pressure, is related to the flow of blood through veins and arteries. The atmospheric pressure readings reported in weather forecasts control atmospheric flow patterns related to local weather conditions. Even the rotary motion generated when we stir a cup of coffee gives rise to pressure variations and flow patterns that enhance mixing.

The relationship between pressure and flow velocity is also important in many engineering applications. In the design of tall structures, the pressure forces from the wind may dictate the design of individual elements, such as windows, as well as the basic structure to withstand wind loads. In aircraft design, the pressure distribution is primarily responsible for lift and contributes to the drag of the aircraft. In the design of flow systems, such as heating and air conditioning, the pressure distribution is responsible for flow in the ducts.

The force balance between pressure and weight in a static fluid was presented in Chapter 3, which lead to an equation for pressure variation with depth. In this chapter the pressure variation in flowing fluids will be addressed. The concepts of pathlines and streamlines help visualize and understand fluid motion. The definition of fluid velocity and acceleration leads to an application of Newton's second law relating forces on a fluid element to the product of mass and acceleration. These relationships lead to the Bernoulli equation, which relates local pressure and elevation to fluid velocity and is fundamental to many fluid mechanic applications. This chapter also introduces the idea of fluid rotation and the concept of irrotationality.

4.1

Descriptions of Fluid Motion

Engineers have developed ways to describe fluid flow patterns and to identify important characteristics of the flow field. This terminology allows engineers to communicate ideas essential to the design of systems such as bridge piers, air-conditioning ducts, airfoils, and structures subjected to wind loads.

Streamlines and Flow Patterns

To visualize the flow field it is desirable to construct lines that show the flow direction. Such a construction is called a flow pattern, and the lines are called streamlines. The *streamline* is defined as a line drawn through the flow field in such a manner that the local velocity vector is tangent to the streamline at every point along the line at that instant. Thus the tangent of the streamline at a given time gives the direction of the velocity vector. A streamline, however, does not indicate the magnitude of the velocity. The flow pattern provided by the streamlines is an instantaneous visualization of the flow field.

An example of streamlines and a flow pattern is shown in Fig. 4.1*a* for water flowing through a slot in the side of a tank. The velocity vectors have been sketched at three different locations: *a*, *b*, and *c*. The streamlines, according to their definition, are tangent to the velocity vectors at these points. Also, the velocities are parallel to the wall in the wall region, so the streamlines adjacent to the wall follow the contour of the wall. The generation of a flow pattern is a very effective way of illustrating the geometric features of the flow field.

Whenever flow occurs around a body, part of it will go to one side and part to the other as shown in Fig. 4.1*b* for flow over an airfoil section. The streamline that follows the flow division (that divides on the upstream side and joins again on the downstream side) is called the *dividing streamline*. At the location where the dividing streamline intersects the body, the velocity will be zero with respect to the body. This is the *stagnation point*.

Figure 4.1

Flow through an opening in a tank and over an airfoil section.

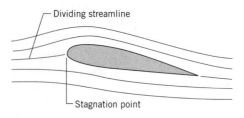

(a)   (b)

Figure 4.2

Predicted streamline pattern over the Volvo ECC prototype. (Courtesy J. Michael Summa, Analytic Methods, Inc.)

Another example of streamlines is shown in Fig. 4.2. These are the streamlines predicted for the flow over an Volvo ECC prototype. Flow patterns of this nature allow the engineer to assess various aerodynamic features of the flow and possibly change the shape to achieve better performance, such as reduced drag.

Having introduced the general concepts of flow patterns, it is convenient to make distinctions between different types of flows. These concepts can be best introduced by expressing the velocity of the fluid in the form

$$V = V(s, t)$$

where s is the distance traveled by a fluid particle along a path, and t is the time, as shown in Fig. 4.3. Flows can be either uniform or nonuniform. In a *uniform flow*, the velocity does not change along a fluid path; that is,

$$\frac{\partial V}{\partial s} = 0$$

It follows that in uniform flow the fluid paths are straight and parallel as shown in Fig. 4.4 for flow in a pipe. In *nonuniform flow*, the velocity changes along a fluid path, so

$$\frac{\partial V}{\partial s} \neq 0$$

For the converging duct in Fig. 4.5a, the magnitude of the velocity increases as the duct converges, so the flow is nonuniform. For the vortex flow shown in Fig. 4.5b, the magnitude of the velocity does not change along the fluid path, but the direction does, so the flow is nonuniform.

Figure 4.3

Fluid particle moving along a pathline.

Figure 4.4

Uniform flow in a pipe.

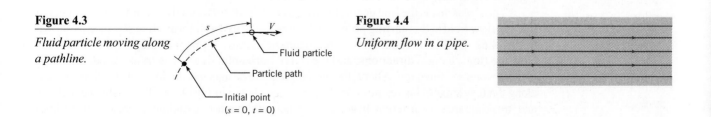

Figure 4.5

*Flow patterns for
nonuniform flow.
(a) Converging flow.
(b) Vortex flow.*

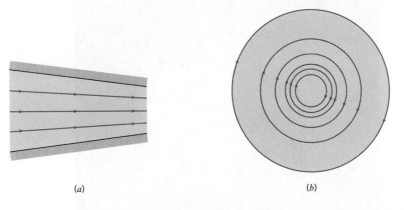

(a)   (b)

Flows can be either steady or unsteady. In a *steady flow* the velocity at a given point on
a fluid path does not change with time:

$$\frac{\partial V}{\partial t} = 0$$

The flow in a pipe, shown in Fig. 4.4, would be an example of steady flow if there
was no change in velocity with time. An *unsteady flow* exists if

$$\frac{\partial V}{\partial t} \neq 0$$

If the flow in the pipe changed with time due to a valve opening or closing, the flow would be
unsteady; that is, the velocity at any point selected on a fluid path would be increasing or decreasing with time. Although unsteady, the flow would still be uniform.

By studying the flow pattern, one can generally decide whether the flow is uniform or
nonuniform. The flow pattern, as represented by streamlines, gives no indication of the steadi-
ness or unsteadiness of the flow because the streamlines are only an instantaneous representation of the flow field.

Pathlines and Streaklines

Besides the streamline described earlier, there are two other approaches commonly used to visualize flow fields; namely, the pathline and streakline.

The *pathline* simply is the path of a fluid particle as it moves through the flow field. In other
words, if a light were attached to a fluid particle, a time exposure photograph taken of the moving
light would be the pathline. For an example of a pathline, consider a two-dimensional flow that
initially has horizontal streamlines as shown in Fig. 4.6. At a given time, t_0, the flow instantly
changes direction, and the flow moves upward to the right at 45° with no further change. The flow
is unsteady because the velocity at a point changes with time. A fluid particle is tracked from the
starting point, and up to time t_0, the pathline is the horizontal line segment shown on Fig. 4.6*a*.
After time t_0, the particle continues to follow the streamline and moves up the right as shown in
Fig. 4.6*b*. Both line segments constitute the pathline.

The *streakline* is the line generated by a tracer fluid, such as a dye, continuously injected
into the flow field at the starting point. Up to time t_0, the dye will form a line segment as shown
in Fig. 4.6*c*. Up to this time, there is no difference between the pathline and the streakline.
Now the flow changes directions, and the initial horizontal dye line is transported, in whole,
in the upward 45° direction. After t_0 the dye continues to be injected and forms a new line segment
along the new streamline, resulting in the streakline shown in Fig. 4.6*d*. Obviously, the pathline
and streakline are very different. In general, neither pathlines nor streaklines represent streamlines

Figure 4.6

Streamlines, pathlines, and streakline for an unsteady flow field.

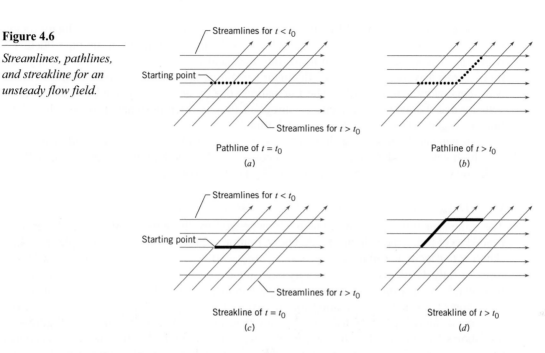

in an unsteady flow. Both the pathline and streakline provide a history of the flow field, and the streamlines indicate the current flow pattern.

In steady flow the pathline, streakline, and streamline are coincident if they pass through the same point.

Laminar and Turbulent Flow

Laminar flow is a well-ordered state of flow in which adjacent fluid layers move smoothly with respect to each other. A typical laminar flow would be the flow of honey or thick syrup from a pitcher. Laminar flow in a pipe has a smooth, parabolic velocity distribution as shown in Fig. 4.7a.

Turbulent flow is an unsteady flow characterized by intense cross-stream mixing. For example, the flow in the wake of a ship is turbulent. The eddies observed in the wake cause intense mixing. The transport of smoke from a smoke stack on a windy day also exemplifies a turbulent flow. The mixing is apparent as the plume widens and disperses.

An instantaneous velocity profile for turbulent flow in a pipe is shown in Fig. 4.7b. A near uniform velocity distribution occurs across the pipe because the high-velocity fluid at

Figure 4.7

*Laminar and turbulent flow in a straight pipe.
(a) Laminar flow.
(b) Turbulent flow.*

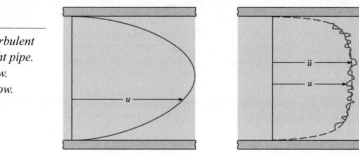

the pipe center is transported by turbulent eddies across the pipe to the low-velocity region near the wall. Because the flow is unsteady, the velocity at any point in the pipe fluctuates with time. The standard approach to treating turbulent flow is to represent the velocity as a time-averaged average value plus a fluctuating quantity, $u = \bar{u} + u'$. The time-averaged value is designated by $\bar{u}$ in Fig. 4.7b. The fluctuation velocity is the difference between the local velocity and the averaged velocity. A turbulent flow is often designated as "steady" if the time-averaged velocity is unchanging with time.

In general, laminar pipe flows are associated with low velocities and turbulent flows with high velocities. Laminar flows can occur in small tubes, highly viscous flows, or flows with low velocities, but turbulent flows are, by far, the most common.

One-Dimensional and Multi-Dimensional Flows

The dimensionality of a flow field is characterized by the number of spatial dimensions needed to describe the velocity field. The definition is best illustrated by example. Fig. 4.8a shows the velocity distribution for an axisymmetric flow in a circular duct. The flow is uniform, or fully developed, so the velocity does not change in the flow direction (z). The velocity depends on only one dimension, namely the radius r, so the flow is one-dimensional. Fig. 4.8b shows the velocity distribution for uniform flow in a square duct. In this case the velocity depends on two dimensions, namely x and y, so the flow is two-dimensional. Figure 4.8c also shows the velocity distribution for the flow in a square duct but the duct cross-sectional area is expanding in the flow direction so the velocity will be dependent on z as well as x and y. This flow is three-dimensional.

Another good example of three-dimensional flow is turbulence, because the velocity components at any one time depend on the three coordinate directions. For example, the velocity component u at a given time depends on x, y, and z; that is, $u(x,y,z)$. Turbulent flow is unsteady, so the velocity components also depend on time.

Another definition frequently used in fluid mechanics is quasi–one–dimensional flow. By this definition it is assumed that there is only one component of velocity in the flow direction and that the velocity profiles are uniformly distributed; that is, constant velocity across the duct cross section.

4.2

Acceleration

Acceleration of a fluid particle as it moves along a pathline, as shown in Fig. 4.9, is the rate of change of the particle's velocity with time. The local velocity of the fluid particle depends on the distance traveled, s, and time, t. The local radius of curvature of the pathline is r. The components of the acceleration vector are shown in Fig. 4.9b. The normal component of acceleration a_n will be present anytime a fluid particle is moving on a curved path (i.e., centripetal acceleration). The tangential component of acceleration a_t will be present if the particle is changing speed.

Using normal and tangential components, the velocity of a fluid particle on a pathline (Fig. 4.9a) may be written as

$$V = V(s, t)\mathbf{e}_t$$

Figure 4.8

*Flow dimensionality,
(a) one-dimensional flow,
(b) two-dimensional flow,
and (c) three-dimensional
flow.*

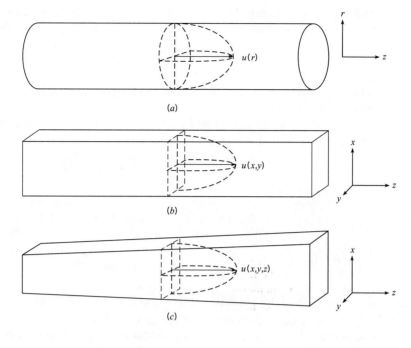

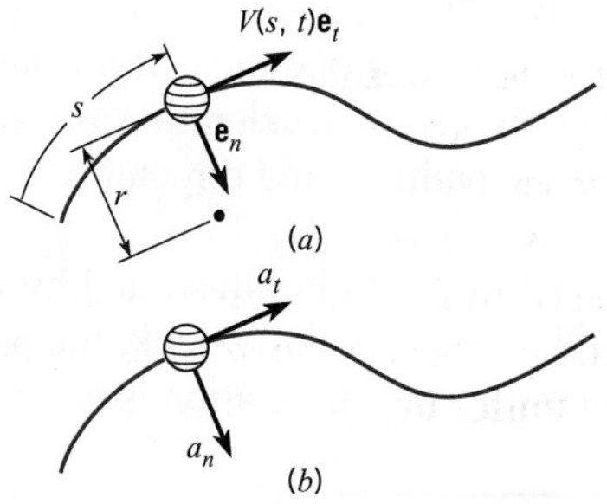

(a)

(b)

(c)

Figure 4.9

*Particle moving on a
pathline. (a) Velocity.
(b) Acceleration.*

where $V(s, t)$ is the speed of the particle, which can vary with distance along the pathline, s, and time, t. The direction of the velocity vector is given by a unit vector $\mathbf{e}_t$.

Using the definition of acceleration,

$$\mathbf{a} = \frac{d\mathbf{V}}{dt} = \left(\frac{dV}{dt}\right)\mathbf{e}_t + V\left(\frac{d\mathbf{e}_t}{dt}\right) \tag{4.1}$$

To evaluate the derivative of speed in Eq. (4.1), the chain rule for a function of two variables can be used.

$$\frac{dV(s, t)}{dt} = \left(\frac{\partial V}{\partial s}\right)\left(\frac{ds}{dt}\right) + \frac{\partial V}{\partial t} \tag{4.2}$$

In a time dt, the fluid particle moves a distance ds, so the derivative ds/dt corresponds to the speed V of the particle, and Eq. (4.2) becomes

$$\frac{dV}{dt} = V\left(\frac{\partial V}{\partial s}\right) + \frac{\partial V}{\partial t} \tag{4.3}$$

In Eq. (4.1), the derivative of the unit vector $d\mathbf{e}_t / dt$ is nonzero because the direction of the unit vector changes with time as the particle moves along the pathline. The derivative is

$$\frac{d\mathbf{e}_t}{dt} = \frac{V}{r}\,\mathbf{e}_n \tag{4.4}$$

where $\mathbf{e}_n$ is the unit vector perpendicular to the pathline and pointing inward toward the center of curvature (1).

Substituting Eqs. (4.3) and (4.4) into Eq. (4.1) gives the acceleration of the fluid particle:

$$\mathbf{a} = \left(V\frac{\partial V}{\partial s} + \frac{\partial V}{\partial t}\right)\mathbf{e}_t + \left(\frac{V^2}{r}\right)\mathbf{e}_n \tag{4.5}$$

The interpretation of this equation is as follows. The acceleration on the left side is the value recorded at a point in the flow field if one were moving with the fluid particle past that point. The terms on the right side represent another way to evaluate the fluid particle acceleration at the same point by measuring the velocity, the velocity gradient, and the velocity change with time at that point and reducing the acceleration according the terms in Eq. (4.5).

Convective, Local, and Centripetal Acceleration

Inspection of Eq. (4.5) reveals that the acceleration component along a pathline depends on two terms. The variation of velocity with time at a point on the pathline, namely $\partial V / \partial t$, is called the *local acceleration*. In steady flow the local acceleration is zero. The other term, $V\partial V / \partial s$, depends on the variation of velocity along the pathline and is called the *convective acceleration*. In a uniform flow, the convective acceleration is zero. The acceleration with magnitude V^2 / r, which is normal to the pathline and directed toward the center of rotation, is the *centripetal acceleration.**

The concept of convective acceleration can be illustrated by use of the cartoon in Fig. 4.10. The carriage represents the fluid particle, and the track, the pathline. It is assumed that the track is stationary. One way to measure the acceleration is to ride on the carriage and read

Figure 4.10

Measuring convective acceleration by two different approaches.

---

* The centripetal acceleration is also a convective acceleration because the acceleration is due to a change in velocity direction as the fluid particle moves along the pathline. Here it will be identified simply as the centripetal acceleration.

the acceleration off an accelerometer. This gives a direct measurement of dV/dt. The other way is to measure the carriage velocity at two locations separated by a distance Δs and calculate the convective acceleration using

$$V\frac{\partial V}{\partial s} \approx V\frac{\Delta V}{\Delta s}$$

Both methods will give the same answer. The centripetal acceleration could also be measured with an accelerometer attached to the carriage or by calculating V^2/r if the local radius of curvature of the track is known.

Example 4.1 illustrates how to find the fluid acceleration by evaluating the local and convective acceleration.

EXAMPLE 4.1 EVALUATING ACCELERATION IN A NOZZLE

A nozzle is designed such that the velocity in the nozzle varies as

$$u(x) = \frac{u_0}{1.0 - 0.5x/L}$$

where the velocity u_0 is the entrance velocity and L is the nozzle length. The entrance velocity is 10 m/s, and the length is 0.5 m. The velocity is uniform across each section. Find the acceleration at the station halfway through the nozzle ($x/L = 0.5$).

Problem Definition

Situation: Given velocity distribution in a nozzle.

Find: Acceleration at nozzle midpoint.

Sketch:

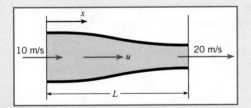

Assumptions: Flow field is quasi–one-dimensional (negligible velocity normal to nozzle centerline).

Plan

1. Select the pathline along the centerline of the nozzle.
2. Evaluate the convective, local, and centripetal accelerations in Eq. (4.5).
3. Calculate the acceleration.

Solution

The distance along the pathline is x, so s in Eq. 4.5 becomes x and V becomes u. The pathline is straight, so $r \longrightarrow \infty$.

1. Evaluation of terms:
   - Convective acceleration

$$\frac{\partial u}{\partial x} = -\frac{u_0}{(1 - 0.5x/L)^2} \times \left(-\frac{0.5}{L}\right)$$

$$= \frac{1}{L}\frac{0.5u_0}{(1 - 0.5x/L)^2}$$

$$u\frac{\partial u}{\partial x} = 0.5\frac{u_0^2}{L}\frac{1}{(1 - 0.5x/L)^3}$$

Evaluation at $x/L = 0.5$:

$$u\frac{\partial u}{\partial x} = 0.5 \times \frac{10^2}{0.5} \times \frac{1}{0.75^3}$$

$$= 237 \text{ m/s}^2$$

   - Local acceleration

$$\frac{\partial u}{\partial t} = 0$$

   - Centripetal acceleration

$$\frac{u^2}{r} = 0$$

2. Acceleration

$$a_x = 237 \text{ m/s}^2 + 0$$

$$= \boxed{237 \text{ m/s}^2}$$

$$a_n \text{ (normal to pathline)} = \boxed{0}$$

Review

Since a_x is positive, the direction of the acceleration is positive; that is, the velocity increases in the x-direction, as expected. Even though the flow is steady, the fluid particles still accelerate.

4.3

Euler's Equation

In Chapter 3 the hydrostatic equations were derived by equating the sum of the forces on a fluid element equal to zero. The same ideas are applied in this section to a moving fluid by equating the sum of the forces acting on a fluid element to the element's acceleration, according to Newton's second law. The resulting equation is Euler's equation, which can be used to predict pressure variation in moving fluids.

Consider the cylindrical element in Fig. 4.11a oriented in an arbitrary direction ℓ with cross-sectional area ΔA in a flowing fluid. The element is oriented at an angle α with respect to the horizontal plane (the x-y plane) as shown in Fig. 4.11b. The element has been isolated from the flow field and can be treated as a "free body" where the presence of the surrounding fluid is replaced by pressure forces acting on the element. Assume that the viscous forces are zero.

Here the element is being accelerated in the ℓ-direction. Note that the coordinate axis z is vertically upward and that the pressure varies along the length of the element. Applying Newton's second law in the ℓ-direction results in

$$\sum F_\ell = ma_\ell$$

$$F_{\text{pressure}} + F_{\text{gravity}} = ma_\ell$$

(4.6)

The mass of the fluid element is

$$m = \rho \Delta A \Delta \ell$$

The net force due to pressure in the ℓ-direction is

$$F_{\text{pressure}} = p\Delta A - (p + \Delta p)\Delta A = -\Delta p \Delta A$$

Any pressure forces acting on the side of the cylindrical element will not contribute to a force in the ℓ-direction. The force due to gravity is the component of weight in the ℓ-direction

$$F_{\text{gravity}} = -\Delta W_\ell = -\Delta W \sin\alpha$$

Figure 4.11

Free-body diagram for fluid element accelerating the ℓ-direction. (a) Fluid element. (b) Orientation of element in coordinate system.

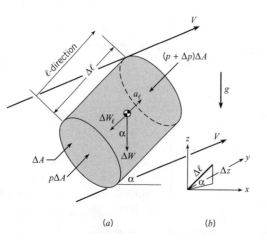

(a)   (b)

where the minus sign occurs because the component of weight is in the negative ℓ-direction. From the diagram in Fig. 4.11b showing the relationship for angle α with respect to $\Delta\ell$, and Δz, one notes that $\sin\alpha = \Delta z/\Delta\ell$, so the force due to gravity can be expressed as

$$F_{\text{gravity}} = -\Delta W\frac{\Delta z}{\Delta\ell}$$

The weight of the element is $\Delta W = \gamma\Delta\ell\Delta A$. Substituting the mass of the element and the forces on the element into Eq. (4.6) yields

$$-\Delta p\Delta A - \gamma\Delta\ell\Delta A\frac{\Delta z}{\Delta\ell} = \rho\Delta\ell\Delta A a_\ell$$

Dividing through by $\Delta A\Delta\ell$ results in

$$-\frac{\Delta p}{\Delta\ell} - \gamma\frac{\Delta z}{\Delta\ell} = \rho a_\ell$$

Taking the limit as $\Delta\ell$ approaches zero (element shrinks to a point) leads to the differential equation for acceleration in the ℓ-direction,

$$-\frac{\partial p}{\partial\ell} - \gamma\frac{\partial z}{\partial\ell} = \rho a_\ell \tag{4.7}$$

For an incompressible flow, γ is constant and Eq. (4.7) reduces to

$$-\frac{\partial}{\partial\ell}(p + \gamma z) = \rho a_\ell \tag{4.8}$$

Equation (4.8) is *Euler's equation* for motion of a fluid. It shows that the acceleration is equal to the change in piezometric pressure with distance, and the minus sign means that the acceleration is in the direction of decreasing piezometric pressure.

In a static body of fluid, Euler's equation reduces to the hydrostatic differential equation, Eq. (3.5). In a static fluid, there are no viscous stresses, which is a condition required in the derivation of Euler's equation. Also there is no motion, so the acceleration is zero in all directions. Thus, Euler's equation reduces to $\partial/\partial\ell(p + \gamma z) = 0$, which yields Eq. (3.4).

Euler's equation can be applied to find the pressure distribution across streamlines in rectilinear flow. Consider the flow with parallel streamlines adjacent a wall shown in Fig. 4.12. In the direction normal to the wall, the n direction, the acceleration is zero. Applying Euler's equation in the n direction gives $\partial/\partial n(p + \gamma z) = 0$, so the piezometric pressure is constant in the normal direction.*

The application of Euler's equation to find the pressure required to accelerate a column of liquid is illustrated in Example 4.2.

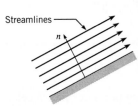

Streamlines

Figure 4.12

Normal direction to parallel stream surfaces.

---

* Euler's equation in this case is applicable in the normal direction because the component of shear stress in this direction is zero for Newtonian fluids.

EXAMPLE 4.2   APPLICATION OF EULER'S EQUATION TO ACCELERATION OF A FLUID

A column water in a vertical tube is being accelerated by a piston in the vertical direction at 100 m/s^2. The depth of the water column is 10 cm. Find the gage pressure on the piston. The water density is 10^3 kg/m^3.

Problem Definition

Situation: A column of water is being accelerated by a piston.

Find: The gage pressure on the piston.

Sketch:

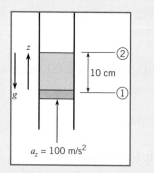

Assumptions:

1. Acceleration is constant.
2. Viscous effects are unimportant.
3. Water is incompressible.

Properties: $\rho = 10^3$ kg/m^3

Plan

1. Apply Euler's equation, Eq. (4.8), in the z-direction.
2. Integrate equation and apply limits at sections 1 and 2.
3. Set pressure equal to zero (gage pressure) at cross-section 2 (atmosphere).
4. Calculate the pressure on piston (cross-section 1).

Solution

1. Because the acceleration is constant there is no dependence on time so the partial derivative in Euler's equation can be replaced by an ordinary derivative. Euler's equation in z-direction:

$$\frac{d}{dz}(p + \gamma z) = -\rho a_z$$

2. Integration between sections 1 and 2:

$$\int_1^2 d(p + \gamma z) = \int_1^2 (-\rho a_z)dz$$

$$(p_2 + \gamma z_2) - (p_1 + \gamma z_1) = -\rho a_z(z_2 - z_1)$$

3. Substitution of limits:

$$p_1 = (\gamma + \rho a_z)\Delta z = \rho(g + a_z)\Delta z$$

4. Evaluation of pressure:

$$p_1 = 10^3 \text{ kg/m}^3 \times (9.81 + 100) \text{ m/s}^2 \times 0.1 \text{ m}$$

$$p_1 = \boxed{10.9 \times 10^3 \text{ Pa} = 10.9 \text{ kPa, gage}}$$

Example 4.3 shows how to apply Euler's equation for predicting pressures in a decelerating tank of liquid.

EXAMPLE 4.3   PRESSURE IN A DECELERATING TANK OF LIQUID

The tank on a trailer truck is filled completely with gasoline, which has a specific weight of 42 lbf/ft^3 (6.60 kN/m^3). The truck is decelerating at a rate of 10 ft/s^2 (3.05 m/s^2).

a. If the tank on the trailer is 20 ft (6.1 m) long and if the pressure at the top rear end of the tank is atmospheric, what is the pressure at the top front?

b. If the tank is 6 ft (1.83 m) high, what is the maximum pressure in the tank?

Sketch:

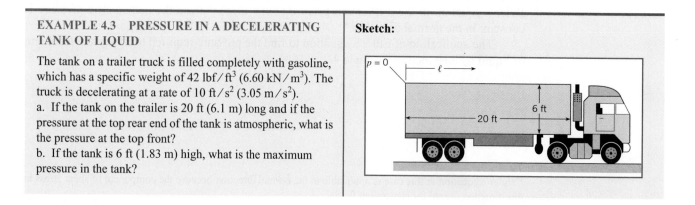

Problem Definition

Situation: Decelerating tank of gasoline with pressure equal to zero gage at top rear end.

Find:

1. Pressure (psfg and kPa, gage) at top front of tank.
2. Maximum pressure (psfg and kPa, gage) in tank.

Assumptions:

1. Deceleration is constant.
2. Gasoline is incompressible.

Properties: $\gamma = 42 \; \text{lbf}/\text{ft}^3 \; (6.60 \; \text{kN}/\text{m}^3)$

Plan

1. Apply Euler's equation, Eq. (4.8), along top of tank. Elevation, z, is constant.
2. Evaluate pressure at top front.
3. Maximum pressure will be at front bottom. Apply Euler's equation from top to bottom at front of tank.
4. Using result from step 2, evaluate pressure at front bottom.

Solution

1. Euler's equation along the top of the tank

$$\frac{dp}{d\ell} = -\rho a_\ell$$

Integration from back (1) to front (2)

$$p_2 - p_1 = -\rho a_\ell \Delta\ell = -\frac{\gamma}{g} a_\ell \Delta\ell$$

2. Evaluation of p_2 with $p_1 = 0$

$$p_2 = -\left(\frac{42 \; \text{lbf/ft}^3}{32.2 \; \text{ft/s}^2}\right) \times (-10 \; \text{ft/s}^2) \times 20 \; \text{ft}$$

$$= \boxed{261 \; \text{psfg}}$$

In SI units

$$p_2 = -\left(\frac{6.60 \; \text{kN/m}^3}{9.81 \; \text{m/s}^2}\right) \times (-3.05 \; \text{m/s}^2) \times 6.1 \; \text{m}$$

$$= \boxed{12.5 \; \text{kPa, gage}}$$

3. Euler's equation in vertical direction

$$\frac{d}{dz}(p + \gamma z) = -\rho a_z$$

4. For vertical direction, $a_z = 0$. Integration from top of tank (2) to bottom (3):

$$p_2 + \gamma z_2 = p_3 + \gamma z_3$$

$$p_3 = p_2 + \gamma(z_2 - z_3)$$

$$p_3 = 261 \; \text{lbf/ft}^2 + 42 \; \text{lbf/ft}^3 \times 6 \; \text{ft} = \boxed{513 \; \text{psfg}}$$

In SI units

$$p_3 = 12.5 \; \text{kN/m}^2 + 6.6 \; \text{kN/m}^3 \times 1.83 \; \text{m}$$

$$p_3 = \boxed{24.6 \; \text{kPa, gage}}$$

4.4

Pressure Distribution in Rotating Flows

Situations in which a fluid rotates as a solid body are found in many engineering applications. One common application is the centrifugal separator. The centripetal accelerations resulting from rotating a fluid separate the heavier elements from the lighter elements as the heavier elements move toward the outside and the lighter elements are displaced toward the center. A milk separator operates in this fashion, as does a cyclone separator for removing particulates from an air stream.

To learn how pressure varies in a rotating, incompressible flow, apply Euler's equation in the direction normal to the streamlines and outward from the center of rotation. In this case the fluid elements rotate as the spokes of a wheel, so the direction ℓ in Euler's equation, Eq. (4.8), is replaced by r giving

$$-\frac{d}{dr}(p + \gamma z) = \rho a_r \tag{4.9}$$

where the partial derivative has been replaced by an ordinary derivative since the flow is steady and a function only of the radius r. From Eq. (4.5), the acceleration in the radial direction (away from the center of curvature) is

$$a_r = -\frac{V^2}{r}$$

and Euler's equation becomes

$$-\frac{d}{dr}(p + \gamma z) = -\rho\frac{V^2}{r} \tag{4.10}$$

For a liquid rotating as a rigid body,

$$V = \omega r$$

Substituting this velocity distribution into Euler's equation results in,

$$\frac{d}{dr}(p + \gamma z) = \rho r \omega^2 \tag{4.11}$$

Integrating Eq. (4.11) with respect to r gives

$$p + \gamma z = \frac{\rho r^2 \omega^2}{2} + \text{const} \tag{4.12}$$

or

$$\frac{p}{\gamma} + z - \frac{\omega^2 r^2}{2g} = C \tag{4.13a}$$

This equation can also be written as

$$p + \gamma z - \rho\frac{\omega^2 r^2}{2} = C \tag{4.13b}$$

These equivalent equations describe the *pressure variation in rotating flow*.

The equation for pressure variation in a rotating flow is used in Example 4.4 to predict the surface profile of a liquid in a rotating tank.

EXAMPLE 4.4 SURFACE PROFILE OF ROTATING LIQUID

A cylindrical tank of liquid shown in the figure is rotating as a solid body at a rate of 4 rad/s. The tank diameter is 0.5 m. The line AA depicts the liquid surface before rotation, and the line $A'A'$ shows the surface profile after rotation has been established. Find the elevation difference between the liquid at the center and the wall during rotation.

Sketch:

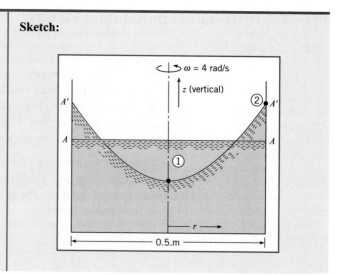

Problem Definition

Situation: Liquid rotating in a cylindrical tank.

Find: Elevation difference (in meters) between liquid at center and at the wall.

Assumptions: Fluid is incompressible.

Plan

Pressure at liquid surface is constant (atmospheric).

1. Apply equation for pressure variation in rotating flow, Eq. (4.13a), between points 1 and 2.
2. Evaluate elevation difference.

Solution

1. Equation (4.13a) applied between points 1 and 2.

$$\frac{p_1}{\gamma} + z_1 - \frac{\omega^2 r_1^2}{2g} = \frac{p_2}{\gamma} + z_2 - \frac{\omega^2 r_2^2}{2g}$$

The pressure at both points is atmospheric, so $p_1 = p_2$ and the pressure terms cancel out. At point 1, $r_1 = 0$, and at point 2, $r = r_2$. The equation reduces to

$$z_2 - \frac{\omega^2 r_2^2}{2g} = z_1$$

$$z_2 - z_1 = \frac{\omega^2 r_2^2}{2g}$$

2. Evaluation of elevation difference:

$$z_2 - z_1 = \frac{(4 \text{ rad/s})^2 \times (0.25 \text{ m})^2}{2 \times 9.81 \text{ m/s}^2}$$

$$= \boxed{0.051 \text{ m or } 5.1 \text{ cm}}$$

Review

Notice that the surface profile is parabolic.

Example 4.5 illustrates the application of the equation for pressure variation in rotating flows to a rotating manometer.

EXAMPLE 4.5   ROTATING MANOMETER TUBE

When the U-tube is not rotated, the water stands in the tube as shown. If the tube is rotated about the eccentric axis at a rate of 8 rad/s, what are the new levels of water in the tube?

Problem Definition

Situation: Manometer tube is rotated around an eccentric axis.

Find: Levels of water in each leg.

Assumptions: Liquid is incompressible.

Sketch:

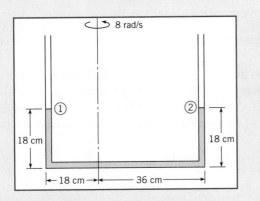

Plan

The total length of the liquid in the manometer must be the same before and after rotation, namely 90 cm. Assume, to start with, that liquid remains in the bottom leg. The pressure at the top of the liquid in each leg is atmospheric.

1. Apply the equation for pressure variation in rotating flows, Eq. (4.13a), to evaluate difference in elevation in each leg.
2. Using constraint of total liquid length, find the level in each leg.

Solution

1. Application of Eq. (4.13a) between top of leg on left (1) and on right (2):

$$z_1 - \frac{r_1^2 \omega^2}{2g} = z_2 - \frac{r_2^2 \omega^2}{2g}$$

$$z_2 - z_1 = \frac{\omega^2}{2g}(r_2^2 - r_1^2)$$

$$= \frac{(8 \text{ rad/s})^2}{2 \times 9.81 \text{ m/s}^2}(0.36^2 \text{ m}^2 - 0.18^2 \text{ m}^2) = 0.317 \text{ m}$$

2. The sum of the heights in each leg is 36 cm.

$$z_2 + z_1 = 0.36 \text{ m}$$

Solution for the leg heights:

$$z_2 = 0.338 \text{ m}$$

$$z_1 = 0.022 \text{ m}$$

Review

If the result was a negative height in one leg, it would mean that one end of the liquid column would be in the horizontal leg, and the problem would have to be reworked to reflect this configuration.

4.5

The Bernoulli Equation Along a Streamline

Derivation

From the dynamics of particles in solid-body mechanics, one knows that integrating Newton's second law for particle motion along a path provides a relationship between the change in kinetic energy and the work done on the particle. Integrating Euler's equation along a pathline in the steady flow of an incompressible fluid yields an equivalent relationship called the Bernoulli equation.

The Bernoulli equation is developed by applying Euler's equation along a pathline with the direction ℓ replaced by s, the distance along the pathline, and the acceleration a_ℓ replaced by a_t, the direction tangent to the pathline. Euler's equation becomes

$$-\frac{\partial}{\partial s}(p + \gamma z) = \rho a_t \tag{4.14}$$

The tangential component of acceleration is given by Eq. (4.15), namely

$$a_t = V\frac{\partial V}{\partial s} + \frac{\partial V}{\partial t} \tag{4.15}$$

For a steady flow, the local acceleration is zero and the pathline becomes a streamline. Also, the properties along a streamline depend only on the distance s, so the partial derivatives become ordinary derivatives. Euler's equation now becomes

$$-\frac{d}{ds}(p + \gamma z) = \rho V\frac{dV}{ds} = \rho \frac{d}{ds}\left(\frac{V^2}{2}\right) \tag{4.16}$$

Moving all the terms to one side yields

$$\frac{d}{ds}\left(p + \gamma z + \rho \frac{V^2}{2}\right) = 0 \tag{4.17}$$

or

$$p + \gamma z + \rho \frac{V^2}{2} = C \tag{4.18a}$$

where C is a constant. This is known as the *Bernoulli equation,* which states that the sum of the piezometric pressure $(p + \gamma z)$ and kinetic pressure $(\rho V^2/2)$ * is constant along a streamline for the *steady* flow of an *incompressible, inviscid* fluid. Dividing Eq. (4.18a) by the specific weight yields the equivalent form of the Bernoulli equation along a streamline, namely

$$\frac{p}{\gamma} + z + \frac{V^2}{2g} = h + \frac{V^2}{2g} = C \tag{4.18b}$$

---

* Another term, dynamic pressure, which is closely associated with kinetic pressure, is equal to the difference between the total pressure (pressure at a point of stagnation) and the static pressure. Under conditions where the Bernoulli equation applies, kinetic and dynamic pressure are equal. However, in high-speed gas flow, where compressibility effects are important, the two may have significantly different values.

Figure 4.13

Piezometric and velocity head variation for flow through a venturi section.

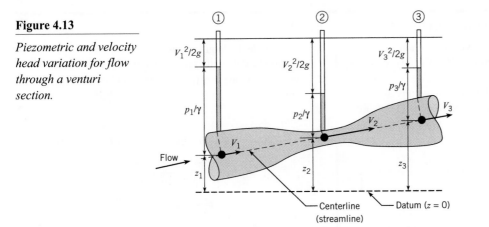

where h is the piezometric head and $(V^2/2g)$ is the velocity head. In words,

$$\left(\begin{array}{c}\text{Pressure}\\\text{head}\end{array}\right) + \left(\begin{array}{c}\text{Elevation}\\\text{head}\end{array}\right) + \left(\begin{array}{c}\text{Velocity}\\\text{head}\end{array}\right) = \left(\begin{array}{c}\text{Constant along}\\\text{streamline}\end{array}\right)$$

The concept underlying the Bernoulli equation can be illustrated by considering the flow through the inclined venturi (contraction-expansion) section as shown in Fig. 4.13. This configuration is often used as a flow metering device. The reduced area of the throat section leads to an increased velocity and attendant pressure change. The streamline is the centerline of the venturi. Piezometers are tapped into the wall at three locations, and the height of the liquid in the tube above the centerline is p/γ. The elevation of the centerline (streamline) above a datum is z. The location of the datum line is arbitrary. The constant in the Bernoulli equation is the same at all three locations, so

$$\frac{p_1}{\gamma} + z_1 + \frac{V_1^2}{2g} = \frac{p_2}{\gamma} + z_2 + \frac{V_2^2}{2g} = \frac{p_3}{\gamma} + z_3 + \frac{V_3^2}{2g}$$

Even though the elevation, pressure head, and velocity head vary through the venturi section, the sum of the three heads is the same. The higher velocity at the throat leads to a higher velocity head at this location and a corresponding decrease in pressure head. The elevations of the liquid in the piezometers above the datum are the piezometric heads. So as the velocity increases, the piezometric head decreases as the velocity head increases, maintaining a constant sum through the venturi.

The fact that the Bernoulli equation has been derived for an inviscid fluid does not limit its application here. Even though the real fluid is viscous, the effects of viscosity are small for short distances. Also, the effects of viscosity on pressure change are negligible compared to the pressure change due to velocity variation.

Application of the Bernoulli Equation

The Bernoulli equation is often used to calculate the velocity in venturi configurations given the pressure difference between the upstream section and the throat section, as shown in Example 4.6.

Example 4.7 shows the application of the Bernoulli equation to the efflux of a liquid from a tank. This analysis is important to the engineer in calculating the draining time.

EXAMPLE 4.6   VELOCITY IN A VENTURI SECTION

Piezometric tubes are tapped into a venturi section as shown in the figure. The liquid is incompressible. The upstream piezometric head is 1 m, and the piezometric head at the throat is 0.5 m. The velocity in the throat section is twice large as in the approach section. Find the velocity in the throat section.

Sketch:

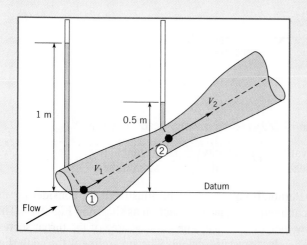

Problem Definition

Situation: Incompressible flow in venturi section. Piezometric heads and velocity ratio given.

Find: Velocity (in m/s) in venturi section.

Assumptions: Viscosity effects are negligible, and the Bernoulli equation is applicable.

Plan

1. Write out the Bernoulli equation, Eq. (4.18b), incorporating velocity ratio and solve for throat velocity.
2. Substitute in piezometric heads to calculate throat velocity.

Solution

1. The Bernoulli equation with $V_2 = 2V_1$ gives

$$h_1 - h_2 = \frac{V_2^2 - V_1^2}{2g} = \frac{3V_1^2}{2g}$$

$$V_1^2 = \frac{2g}{3}(h_1 - h_2)$$

$$V_2 = 2\sqrt{\frac{2g}{3}(h_1 - h_2)}$$

2. Substitution of values and velocity calculation:

$$V_2 = 2\sqrt{\frac{2 \times 9.81 \text{ m/s}^2(1 - 0.5) \text{ m}}{3}}$$

$$= \boxed{3.62 \text{ m/s}}$$

Review

A piezometric tube could not be used to measure the piezometric head if the pressure anywhere in the line were subatmospheric. In this case, pressure gages or manometers would have to be used.

EXAMPLE 4.7   OUTLET VELOCITY FROM DRAINING TANK

A open tank filled with water and drains through a port at the bottom of the tank. The elevation of the water in the tank is 10 m above the drain. The drain port is at atmospheric pressure. Find the velocity of the liquid in the drain port.

Problem Definition

Situation: Tank draining through port at bottom.

Find: Velocity (in m/s) in drain port.

Sketch:

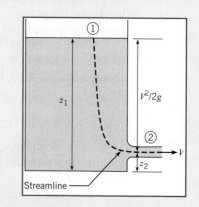

Assumptions:

1. Flow is steady.
2. Viscous effects are unimportant.
3. Velocity at liquid surface is much less than velocity in drain port.

Plan

Since the flow is steady and viscous effects are unimportant, the Bernoulli equation is applicable along a streamline. The streamline chosen is shown in the sketch with point 1 at the liquid surface and point 2 at the drain port.

1. Apply the Bernoulli equation, Eq. (4.18b), between points 1 and 2.
2. Reduce the equation to yield velocity in drain port.
3. Calculate velocity.

Solution

1. The Bernoulli equation between points 1 and 2 on streamline:

$$\frac{p_1}{\gamma} + z_1 + \frac{V_1^2}{2g} = \frac{p_2}{\gamma} + z_2 + \frac{V_2^2}{2g}$$

2. The pressure at the outlet and the tank surface are the same (atmospheric), so $p_1 = p_2$. The velocity at the tank surface is much less than in the drain port so $V_2^2 \gg V_1^2$. Solution for V_2:

$$z_1 - z_2 = \frac{V_2^2}{2g}$$

$$V_2 = \sqrt{2g(z_1 - z_2)}$$

3. Velocity calculation:

$$V_2 = \sqrt{2 \times 9.81 \text{ m/s}^2 \times 10 \text{ m}}$$

$$= \boxed{14 \text{ m/s}}$$

Review

1. Notice that the answer is independent of liquid properties. This would be true for all liquids so long as viscous effects are unimportant. Also note that the same velocity would be calculated for an object dropped from the same elevation as the liquid in the tank.
2. Selection of point 1 is not critical; it can be taken at any point on liquid surface.
3. The assumption of the small velocity at the liquid surface is generally valid. It will be shown in Chapter 5 that the ratio of the velocity at the liquid surface to the drain port velocity is

$$\frac{V_1}{V_2} = \frac{A_2}{A_1}$$

where A_2 is the cross-sectional area of the drain port and A_1 is the cross-sectional area of the tank. For example with $A_2/A_1 = 0.1$, $V_1^2 = 0.01 V_2^2$.

Application of the Bernoulli Equation to Velocity Measurement Devices

The Bernoulli equation can be used to reduce data for flow velocity measurements from a stagnation tube and a Pitot-static tube.

Stagnation Tube

A *stagnation tube* (sometimes call a total head tube) is an open-ended tube directed upstream in a flow and connected to a pressure sensor. Because the velocity is zero at the tube opening, the pressure measured corresponds to stagnation conditions.

Consider the stagnation tube shown in Fig. 4.14. In this case the pressure sensor is a piezometer. The rise of the liquid in the vertical leg is a measure of the pressure. When the Bernoulli equation is written between points 0 and 1 on the streamline, one notes that $z_0 \cong z_1$. Therefore, the Bernoulli equation reduces to

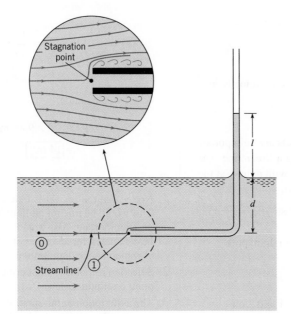

$$p_1 + \frac{\rho V_1^2}{2} = p_0 + \frac{\rho V_0^2}{2} \tag{4.19}$$

The velocity at point 1 is zero (stagnation point). Hence, Eq. (4.19) simplifies to

$$V_0^2 = \frac{2}{\rho}(p_1 - p_0) \tag{4.20}$$

By the equations of hydrostatics (there is no acceleration normal to the streamlines where the streamlines are straight and parallel), $p_0 = \gamma d$ and $p_1 = \gamma(l + d)$. Therefore, Eq. (4.20) can be written as

$$V_0^2 = \frac{2}{\rho}(\gamma(l + d) - \gamma d)$$

which reduces to

$$V_0 = \sqrt{2gl} \tag{4.21}$$

This equation will be referred to as the *stagnation tube equation*. Thus it is seen that a very simple device such as this curved tube can be used to measure the velocity of flow.

Pitot-Static Tube

The *Pitot-static tube*, named after the eighteenth-century French hydraulic engineer who invented it, is based on the same principle as the stagnation tube, but it is much more versatile than the stagnation tube. The Pitot-static tube, shown in Fig. 4.15, has a pressure tap at the upstream end of the tube for sensing the stagnation pressure. There are also ports located several tube diameters downstream of the front end of the tube for sensing the static

Figure 4.15

Pitot-static tube.

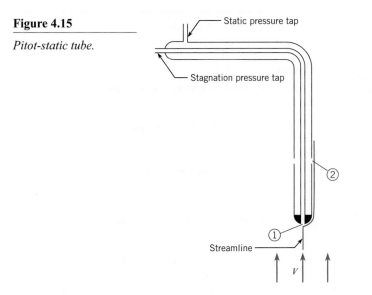

pressure in the fluid where the velocity is essentially the same as the approach velocity. When the Bernoulli equation, Eq. (4.18a), is applied between points 1 and 2 along the streamline shown in Fig. 4.15, the result is

$$p_1 + \gamma z_1 + \frac{\rho V_1^2}{2} = p_2 + \gamma z_2 + \frac{\rho V_2^2}{2}$$

But $V_1 = 0$, so solving that equation for V_2 gives the *Pitot-static tube equation*

$$V_2 = \left[\frac{2}{\rho} (p_{z,1} - p_{z,2}) \right]^{1/2} \qquad (4.22)$$

Here $V_2 = V$, where V is the velocity of the stream and $p_{z,1}$ and $p_{z,2}$ are the piezometric pressures at points 1 and 2, respectively.

By connecting a pressure gage or manometer between the pressure taps shown in Fig. 4.15 that lead to points 1 and 2, one can easily measure the flow velocity with the Pitot-static tube. A major advantage of the Pitot-static tube is that it can be used to measure velocity in a pressurized pipe; a simple stagnation tube is not convenient to use in such a situation.

If a differential pressure gage is connected across the taps, the gage measures the difference in piezometric pressure directly. Therefore Eq. (4.22) simplifies to

$$V = \sqrt{2 \Delta p / \rho}$$

where Δp is the pressure difference measured by the gage.

More information on Pitot-static tubes and flow measurement is available in the *Flow Measurement Engineering Handbook* (2).

Example 4.8 illustrates the application of the Pitot-static tube to measuring liquid velocity in a pipe using a manometer.

Example 4.9 shows the Pitot-static tube application with a pressure gage.

EXAMPLE 4.8 APPLICATION OF PITOT EQUATION WITH MANOMETER

A mercury manometer is connected to the Pitot-static tube in a pipe transporting kerosene as shown. If the deflection on the manometer is 7 in., what is the kerosene velocity in the pipe? Assume that the specific gravity of the kerosene is 0.81.

Sketch:

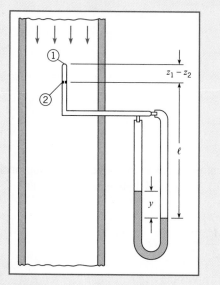

Problem Definition

Situation: Pitot-static tube is mounted in a pipe and connected to a manometer.

Find: Flow velocity (in m/s).

Assumptions: Pitot-static tube equation is applicable.

Properties: $S_{kero} = 0.81$, from Table A.4, $S_{Hg} = 13.55$.

Plan

1. Find difference in piezometric pressure using the manometer equation, Eq. (3.18).

2. Substitute in Pitot-static tube equation.
3. Evaluate velocity.

Solution

1. Manometer equation between points 1 and 2 on Pitot-static tube:

$$p_1 + (z_1 - z_2)\gamma_{kero} + \ell\gamma_{kero} - y\gamma_{Hg} - (\ell - y)\gamma_{kero} = p_2$$

or

$$p_1 + \gamma_{kero}z_1 - (p_2 + \gamma_{kero}z_2) = y(\gamma_{Hg} - \gamma_{kero})$$

$$p_{z,1} - p_{z,2} = y(\gamma_{Hg} - \gamma_{kero})$$

2. Substitution into the Pitot-static tube equation:

$$V = \left[\frac{2}{\rho_{kero}} y(\gamma_{Hg} - \gamma_{kero}) \right]^{1/2}$$

$$= \left[2gy\left(\frac{\gamma_{Hg}}{\gamma_{kero}} - 1 \right) \right]^{1/2}$$

3. Velocity evaluation:

$$V = \left[2 \times 32.2 \ \text{ft/s}^2 \times \frac{7}{12} \ \text{ft} \left(\frac{13.55}{0.81} - 1 \right) \right]^{1/2}$$

$$= \left[2 \times 32.2 \times \frac{7}{12}(16.7 - 1)\text{ft}^2/\text{s}^2 \right]^{1/2}$$

$$= \boxed{24.3 \ \text{ft/s}}$$

Review

The -1 in the quantity $(16.7 - 1)$ reflects the effect of the column of kerosene in the right leg of the manometer, which tends to counterbalance the mercury in the left leg. Thus with a gas–liquid manometer, the counterbalancing effect is negligible.

EXAMPLE 4.9 PITOT TUBE APPLICATION WITH PRESSURE GAGE

A differential pressure gage is connected across the taps of a Pitot-static tube. When this Pitot-static tube is used in a wind tunnel test, the gage indicates a Δp of 730 Pa. What is the air velocity in the tunnel? The pressure and temperature in the tunnel are 98 kPa absolute and 20°C, respectively.

Problem Definition

Situation: Differential pressure gage attached to Pitot-static tube for velocity measurement in wind tunnel.

Find: Air velocity (in m/s).

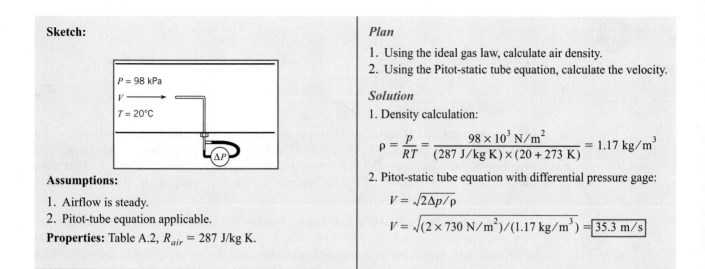

Sketch:

$P = 98$ kPa
V —→
$T = 20°C$
ΔP

Assumptions:

1. Airflow is steady.
2. Pitot-tube equation applicable.

Properties: Table A.2, $R_{air} = 287$ J/kg K.

Plan

1. Using the ideal gas law, calculate air density.
2. Using the Pitot-static tube equation, calculate the velocity.

Solution

1. Density calculation:

$$\rho = \frac{p}{RT} = \frac{98 \times 10^3 \text{ N/m}^2}{(287 \text{ J/kg K}) \times (20 + 273 \text{ K})} = 1.17 \text{ kg/m}^3$$

2. Pitot-static tube equation with differential pressure gage:

$$V = \sqrt{2\Delta p/\rho}$$

$$V = \sqrt{(2 \times 730 \text{ N/m}^2)/(1.17 \text{ kg/m}^3)} = \boxed{35.3 \text{ m/s}}$$

Application of the Bernoulli Equation to Flow of Gases

In the flow of gases, the contribution of pressure change due to elevation difference is generally very small compared with the change in kinetic pressure. Thus it is reasonable when applying the Bernoulli equation to gas flow (such as air) to use the simpler formulation

$$p + \frac{1}{2}\rho V^2 = C \tag{4.23}$$

This is the form used by aerodynamicists in studying the flow over airfoils and aircraft components.

Applicability of the Bernoulli Equation to Rotating Flows

The Bernoulli equation (4.18a) relates pressure, elevation, and kinetic pressure along streamlines in steady, incompressible flows where viscous effects are negligible. The question arises as to whether it can be used across streamlines; that is, could it be applied between two points on adjacent streamlines? The answer is provided by the form of the equation for pressure variation in a rotating flow, Eq. (4.13b), which can be written as

$$p + \gamma z - \frac{1}{2}\rho V^2 = C$$

where ωr has been replaced by the velocity, V. Obviously the sign on the kinetic pressure term is different than the Bernoulli equation, so the Bernoulli equation does not apply across streamlines in a rotating flow.

In the next section the concept of flow rotation is introduced. There is a situation in which flows have concentric, circular streamlines and yet the fluid elements do not rotate. In this "irrotational" flow, the Bernoulli equation is applicable across streamlines as well a along streamlines.

Rotation and Vorticity

Concept of Rotation

In many applications of fluid mechanics it is important to establish the rotationality of a flow field. There are many closed-form solutions for fluid flow fields based on irrotational flow. To make use of these solutions, the engineer must be able to determine the degree of flow rotation in his or her application. The purpose of this section is to introduce the concept of rotation.

The idea of fluid rotation is clear when a fluid rotates as a solid body. However, in other flow configurations it may not be so obvious. Consider fluid flow between two horizontal plates, Fig. 4.16, where the bottom plate is stationary and the top is moving to the right with a velocity V. The velocity distribution is linear; therefore, an element of fluid will deform as shown. Here it is seen that the element face that was initially vertical rotates clockwise, whereas the horizontal face does not. It is not clear whether this is a case of rotational motion or not.

Rotation is defined as the average rotation of two initially mutually perpendicular faces of a fluid element. The test is to look at the rotation of the line that bisects both faces (*a-a* and *b-b* in Fig. 4.16). The angle between this line and the horizontal axis is the rotation, θ.

The general relationship between θ and the angles defining the sides is shown in Fig. 4.17, where θ_A is the angle of one side with the *x*-axis and the angle θ_B is the angle of the

Figure 4.16

Rotation of a fluid element in flow between a moving and stationary parallel plate.

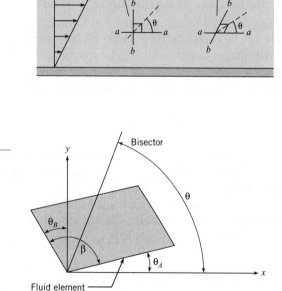

Figure 4.17

Orientation of rotated fluid element.

other side with the y-axis. The angle between the sides is $\beta = \frac{\pi}{2} + \theta_B - \theta_A$, so the orientation of the element with respect to the x-axis is

$$\theta = \frac{1}{2}\beta + \theta_A = \frac{\pi}{4} + \frac{1}{2}(\theta_A + \theta_B)$$

The rotational rate of the element is

$$\dot{\theta} = \frac{1}{2}(\dot{\theta}_A + \dot{\theta}_B) \tag{4.24}$$

If $\dot{\theta} = 0$, the flow is *irrotational*.

An expression will now be derived that will give the rate of rotation of the bisector in terms of the velocity gradients in the flow. Consider the element shown in Fig. 4.18. The sides of the element are initially perpendicular with lengths Δx and Δy. Then the element moves with time and deforms as shown with point 0 going to 0', point 1 to 1', and point 2 to 2'. The lengths of the sides are unchanged. After time Δt the horizontal side has rotated counter-clockwise by $\Delta\theta_A$ and the vertical side clockwise (negative direction) by $-\Delta\theta_B$.

The y velocity component of point 1 is $v + (\partial v/\partial x)\Delta x$, and the x component of point 2 is $u + (\partial u/\partial y)\Delta y$. The net displacements of points 1 and 2 are*

$$\Delta y_1 \sim \left[\left(v + \frac{\partial v}{\partial x}\Delta x\right)\Delta t - v\Delta t\right] = \frac{\partial v}{\partial x}\Delta x \Delta t$$

$$\Delta x_2 \sim \left[\left(u + \frac{\partial u}{\partial y}\Delta y\right)\Delta t - u\Delta t\right] = \frac{\partial u}{\partial y}\Delta y \Delta t \tag{4.25}$$

Referring to Fig. 4.18, the angles $\Delta\theta_A$ and $\Delta\theta_B$ are given by

$$\Delta\theta_A = \operatorname{asin}\left(\frac{\Delta y_1}{\Delta x}\right) \sim \frac{\Delta y_1}{\Delta x} \sim \frac{\partial v}{\partial x}\Delta t$$

$$-\Delta\theta_B = \operatorname{asin}\left(\frac{\Delta x_2}{\Delta y}\right) \sim \frac{\Delta x_2}{\Delta y} \sim \frac{\partial u}{\partial y}\Delta t \tag{4.26}$$

Figure 4.18

Translation and deformation of a fluid element.

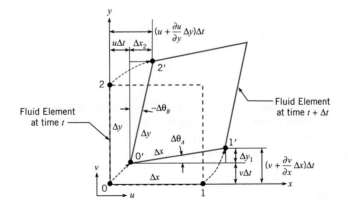

* The symbol ~ means that the quantities are approximately equal but become exactly equal as the quantities approach zero.

Dividing the angles by Δt and taking the limit as $\Delta t \longrightarrow 0$,

$$
\begin{aligned}
\dot{\theta}_A &= \lim_{\Delta t \to 0} \frac{\Delta \theta_A}{\Delta t} = \frac{\partial v}{\partial x} \\[2mm]
\dot{\theta}_B &= \lim_{\Delta t \to 0} \frac{\Delta \theta_B}{\Delta t} = -\frac{\partial u}{\partial y}
\end{aligned}
\tag{4.27}
$$

Substituting these results into Eq. (4.24) gives the rotational rate of the element about the z-axis (normal to the page),

$$
\dot{\theta} = \frac{1}{2}\left(\frac{\partial v}{\partial x} - \frac{\partial u}{\partial y}\right)
$$

This component of rotational velocity is defined as Ω_z, so

$$
\Omega_z = \frac{1}{2}\left(\frac{\partial v}{\partial x} - \frac{\partial u}{\partial y}\right)
\tag{4.28a}
$$

Likewise, the rotation rates about the other axes are

$$
\Omega_x = \frac{1}{2}\left(\frac{\partial w}{\partial y} - \frac{\partial v}{\partial z}\right)
\tag{4.28b}
$$

$$
\Omega_y = \frac{1}{2}\left(\frac{\partial u}{\partial z} - \frac{\partial w}{\partial x}\right)
\tag{4.28c}
$$

The rate-of-rotation vector is

$$
\Omega = \Omega_x \mathbf{i} + \Omega_y \mathbf{j} + \Omega_z \mathbf{k}
\tag{4.29}
$$

An irrotational flow $(\Omega = 0)$ requires that

$$
\frac{\partial v}{\partial x} = \frac{\partial u}{\partial y}
\tag{4.30a}
$$

$$
\frac{\partial w}{\partial y} = \frac{\partial v}{\partial z}
\tag{4.30b}
$$

$$
\frac{\partial u}{\partial z} = \frac{\partial w}{\partial x}
\tag{4.30c}
$$

The most extensive application of these equations is in ideal flow theory. An ideal flow is the flow of an irrotational, incompressible fluid. Flow fields in which viscous effects are small can often be regarded as irrotational. In fact, if a flow of an incompressible, inviscid fluid is initially irrotational, it will remain irrotational.

Vorticity

Another property used frequently in fluid mechanics is *vorticity,* which is a vector equal to twice the rate-of-rotation vector. The magnitude of the vorticity indicates the rotationality of a flow and is very important in flows where viscous effects dominate, such as boundary layer, separated, and wake flows. The vorticity equation is

$$\omega = 2\Omega$$

$$= \left(\frac{\partial w}{\partial y} - \frac{\partial v}{\partial z}\right)\mathbf{i} + \left(\frac{\partial u}{\partial z} - \frac{\partial w}{\partial x}\right)\mathbf{j} + \left(\frac{\partial v}{\partial x} - \frac{\partial u}{\partial y}\right)\mathbf{k} \qquad (4.31)$$

$$= \nabla \times V$$

where $\nabla \times \mathbf{V}$, from vector calculus means the curl of the vector V.
An irrotational flow signifies that the vorticity vector is everywhere zero.
Example 4.10 illustrates how to evaluate the rotationality of a flowfield.

EXAMPLE 4.10   EVALUATION OF ROTATION OF VELOCITY FIELD

The vector $V = 10x\mathbf{i} - 10y\mathbf{j}$ represents a two-dimensional velocity field. Is the flow irrotational?

Problem Definition

Situation: Velocity field given.

Find: If flow is irrotational.

Plan

Flow is two-dimensional, so $w = 0$ and $\frac{\partial}{\partial z} = 0$. Use Eq. (4.30a) to evaluate rotationality.

Solution

Velocity components and derivatives

$$u = 10x \qquad \frac{\partial u}{\partial y} = 0$$

$$v = -10y \qquad \frac{\partial v}{\partial x} = 0$$

Thus flow is irrotational.

The calculation to determine the amount of rotation of a fluid element in a given time is shown in Example 4.11.

EXAMPLE 4.11   ROTATION OF A FLUID ELEMENT

A fluid exists between stationary and moving parallel flat plates, and the velocity is linear as shown. The distance between the plates is 1 cm, and the upper plate moves at 2 cm/s. Find the amount of rotation that the fluid element located at 0.5 cm will undergo after it has traveled a distance of 1 cm.

Sketch:

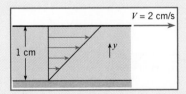

Problem Definition

Situation: Flow between moving, parallel, flat plates.

Find: Rotation of fluid element at midpoint after traveling 1 cm.

Assumptions: Planar flow ($w = 0$ and $\frac{\partial}{\partial z} = 0$).

Plan

1. Use Eq. (4.28a) to evaluate rotational rate with $v = 0$.

2. Find time for element to travel 1 cm.

3. Calculate amount of rotation.

Solution

1. Velocity distribution

$$u = 0.02 \text{ m/s} \times \frac{y}{0.01 \text{ m}} = 2y \text{ (1/s)}$$

Rotational rate

$$\Omega_z = \frac{1}{2}\left(\frac{\partial v}{\partial x} - \frac{\partial u}{\partial y}\right) = -1 \text{ rad/s}$$

2. Time to travel 1 cm:

$$u = 2 \text{ (1/s)} \times 0.005 \text{ m} = 0.01 \text{ m/s}$$

$$\Delta t = \frac{\Delta x}{u} = \frac{0.01 \text{ m}}{0.01 \text{ m/s}} = 1 s$$

3. Amount of rotation

$$\Delta\theta = \Omega_z \times \Delta t = -1 \times 1 = -1 \text{ rad}$$

Review

Note that the rotation is negative (in clockwise direction).

Rotation in Flows with Concentric Streamlines

It is interesting to realize that a flow field rotating with circular streamlines can be irrotational; that is, the fluid elements do not rotate. Consider the two-dimensional flow field shown in Fig. 4.19. The circumferential velocity on the circular streamline is V, and its radius is r. The z-axis is perpendicular to the page. As before, the rotation of the element is quantified by the rotation of the bisector, Eq. (4.24), which is

$$\dot{\theta} = \frac{1}{2}(\dot{\theta}_A + \dot{\theta}_B)$$

From geometry, the angle $\Delta\theta_B$ is equal to the angle $\Delta\phi$. The rotational rate of angle ϕ is V/r, so

$$\dot{\theta}_B = \frac{V}{r}$$

Using the same analysis as for $\dot{\theta}_A$, Eq. (4.27), with x replaced by r yields

$$\dot{\theta}_A = \frac{\partial V}{\partial r}$$

Since V is a function of r only, the partial derivative can be replaced by the ordinary derivative. Therefore, the rotational rate about the z-axis is

$$\Omega_z = \frac{1}{2}\left(\frac{dV}{dr} + \frac{V}{r}\right) \qquad (4.32)$$

As a check on this equation, apply it to a flow rotating as a solid body. The velocity distribution is $V = \omega r$, so the rate of rotation is

$$\Omega_z = \frac{1}{2}\left[\frac{d}{dr}(\omega r) + \omega\right]$$

$$= \omega$$

as expected. This type of circular motion is called a "forced" vortex.

If the flow is irrotational, then

$$\frac{dV}{dr} = -\frac{V}{r} \qquad (4.33)$$

Figure 4.19

Deformation of element in flow with concentric, circular streamlines.

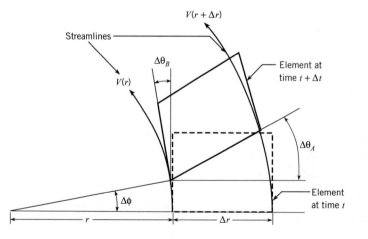

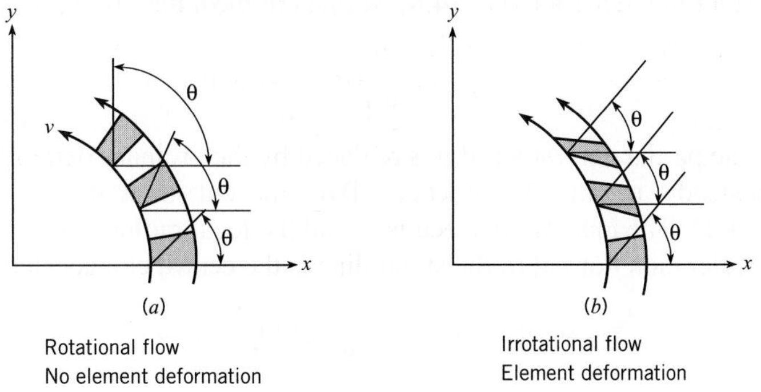

Figure 4.20

Element deformation and rotation in flow with circular streamlines. (a) Rotational flow. (b) Irrotational flow.

Rotational flow
No element deformation

Irrotational flow
Element deformation

or

$$\frac{dV}{V} = -\frac{dr}{r}$$

Integrating this equation leads to

$$V = \frac{C}{r} \qquad (4.34)$$

where C is a constant. In this case, the circumferential velocity varies inversely with r, so the velocity decreases with increasing radius. This flow field is known as a "free" vortex. The fluid elements go around in circles, but do not rotate.

Although the condition for irrotationality given by Eq. (4.33) was derived using concentric circular streamlines, it is valid for any point along any streamline in an irrotational flow where r is the local radius of curvature of the streamline (or stream surface generated by the streamlines).

The difference between element rotation and deformation for flows with circular streamlines is shown in Fig. 4.20. For the rotational flow shown in Fig. 4.20a the fluid elements rotate but they do not deform. In the irrotational flow shown in Fig. 4.20b, the elements continuously deform but do not rotate. In other words, the elements deform to maintain a constant orientation (no rotation).

In a general flow there is both deformation and rotation. An ideal fluid is one that has no viscosity and is incompressible. If the flow of an ideal fluid is initially irrotational. it will remain irrotational. This is the foundation for many classical studies of flow fields in fluid mechanics.

4.7

The Bernoulli Equation in Irrotational Flow

In Section 4.5 the Bernoulli equation was developed for pressure variation between any two points along a streamline in steady flow with no viscous effects. In an irrotational flow, the Bernoulli equation is not limited to flow along streamlines but can be applied between any two points in the flow field. This feature of the Bernoulli equation is used extensively in classical hydrodynamics, the aerodynamics of lifting surfaces (wings), and atmospheric winds.

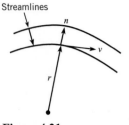

Streamlines

Figure 4.21

*Two adjacent stream-
lines showing direction* n
between lines.

The Euler equation, Eq. (4.8), applied in the n direction (normal to the streamline) is

$$-\frac{d}{dn}(p + \gamma z) = \rho a_n \qquad (4.35)$$

where the partial derivative of n is replaced by the ordinary derivative because the flow is assumed steady (no time dependence). Two adjacent streamlines and the direction n is shown in Fig. 4.21. The local fluid speed is V, and the local radius of curvature of the streamline is r. The acceleration normal to the streamline is the centripetal acceleration, so

$$a_n = -\frac{V^2}{r} \qquad (4.36)$$

where the negative sign occurs because the direction n is outward from the center of curvature and the centripetal acceleration is toward the center of curvature. Using the irrotationality condition, the acceleration can be written as

$$a_n = -\frac{V^2}{r} = -V\left(\frac{V}{r}\right) = V\frac{dV}{dr} = \frac{d}{dr}\left(\frac{V^2}{2}\right) \qquad (4.37)$$

Also the derivative with respect to r can be expressed as a derivative with respect to n by

$$\frac{d}{dr}\left(\frac{V^2}{2}\right) = \frac{d}{dn}\left(\frac{V^2}{2}\right)\frac{dn}{dr} = \frac{d}{dn}\left(\frac{V^2}{2}\right)$$

because the direction of n is the same as r so $dn/dr = 1$. Equation (4.37) can be rewritten as

$$a_n = \frac{d}{dn}\left(\frac{V^2}{2}\right) \qquad (4.38)$$

Substituting the expression for acceleration into Euler's equation, Eq. (4.35), and assuming constant density results in

$$\frac{d}{dn}\left(p + \gamma z + \rho\frac{V^2}{2}\right) = 0 \qquad (4.39)$$

or

$$p + \gamma z + \rho\frac{V^2}{2} = C \qquad (4.40)$$

which is the Bernoulli equation, and C is constant in the n direction (across streamlines). Thus for an irrotational flow, the constant C in the Bernoulli equation is the same across streamlines as well as along streamlines, so it is the same everywhere in the flow field. Equivalently, the sum of the piezometric head and velocity head

$$\frac{p}{\gamma} + z + \frac{V^2}{2g} = C \qquad (4.41)$$

is constant everywhere in the flow field if the flow is steady, incompressible, inviscid, and irrotational. Thus for any two points 1 and 2 in the flow field,

$$\frac{p_1}{\gamma} + z_1 + \frac{V_1^2}{2g} = \frac{p_2}{\gamma} + z_2 + \frac{V_2^2}{2g} \qquad (4.42)$$

EXAMPLE 4.12   VELOCITY AND PRESSURE DISTRIBUTION IN A FREE VORTEX

A free vortex in air rotates in a horizontal plane and has a velocity of 40 m/s at a radius of 4 km from the vortex center. Find the velocity at 10 km from the center and the pressure difference between the two locations. The air density is $1.2\,kg/m^3$.

Problem Definition

Situation: Free vortex in horizontal plane.

Find:

1. Velocity (in m/s) 10 km from center.
2. Pressure difference (Pa) between two radii.

Assumptions: Flow is incompressible and steady.

Properties: $\rho = 1.2\ kg/m^3$.

Plan

1. Apply Eq. (4.34) to calculate velocity.
2. Since flow is irrotational, use the Bernoulli equation, Eq. (4.40), for pressure difference.

Solution

1. Velocity distribution

$$V = \frac{C}{r}$$

$$\frac{V_{10km}}{V_{4km}} = \frac{r_{4km}}{r_{10km}} = 0.4$$

$$V_{10km} = 0.4 \times 40$$

$$= \boxed{16\ m/s}$$

2. The Bernoulli equation for a horizontal plane

$$p_{4km} + \rho\frac{V_{4km}^2}{2} = p_{10km} + \rho\frac{V_{10km}^2}{2}$$

$$p_{10km} - p_{4km} = \frac{\rho}{2}(V_{4km}^2 - V_{10km}^2)$$

$$= \frac{1.2\ kg/m^3}{2}(40^2 - 16^2)\ (m/s)^2$$

$$= \boxed{806\ Pa}$$

Note that the V in the Bernoulli equation is the speed of the fluid and not a velocity component.

The calculation of pressure and velocity in a free vortex is illustrated in Example 4.12.

Pressure Variation in a Cyclonic Storm

A cyclonic storm is characterized by rotating winds with a low-pressure region in the center. Tornadoes and hurricanes are examples of cyclonic storms. A simple model for the flow field in a cyclonic storm is a forced vortex at the center surrounded by a free vortex, as shown in Fig. 4.22. This model is used in several applications of vortex flows. In practice, however, there will be no discontinuity in the slope of the velocity distribution as shown in Fig. 4.22, but rather a smooth transition between the inner forced vortex and the outer free vortex. Still, the model can be used to make reasonable predictions of the pressure field.

The model for the cyclonic storm is an illustration of where the Bernoulli equation can and cannot be used across streamlines. The Bernoulli equation cannot be used across streamlines in the vortex at the center because the flow is rotational. The pressure distribution in the forced vortex is given by Eq. (4.13b). The Bernoulli equation can be used across streamlines in the free vortex since the flow is irrotational.

Take point 1 as the center of the forced vortex and point 2 at the junction of the forced and free vortices, where the velocity is maximum. Let point 3 be at the extremity of the free vortex, where the velocity is essentially zero and the pressure is atmospheric pressure p_0. Applying the Bernoulli equation, Eq. (4.42), between any arbitrary point in the free vortex and point 3, one can write

$$p + \gamma z + \rho\frac{V^2}{2} = p_3 + \gamma z_3 + \rho\frac{V_3^2}{2} \tag{4.43}$$

Figure 4.22

Combination of forced and free vortex to model a cyclonic storm.

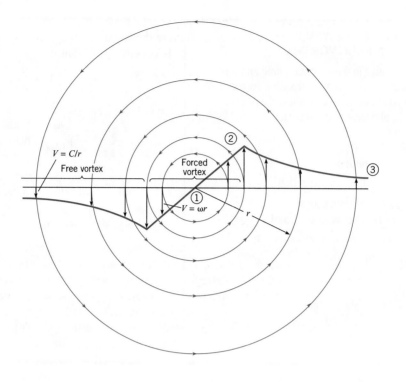

Figure 4.23

The "secondary" flow produced by the pressure difference in a tornado.

Neglecting any elevation change, setting $p_0 = p_3$, and taking V_3 as zero gives

$$p - p_0 = -\rho \frac{V^2}{2} \qquad (4.44)$$

which shows that the pressure decreases toward the center of the vortex. This decreasing pressure provides the centripetal force to keep the flow moving along circular streamlines. The pressure at point 2 is

$$p_2 - p_0 = -\rho \frac{V_{max}^2}{2} \qquad (4.45)$$

Applying the equation for pressure variation in rotating flow, Eq. (4.13b), across the center, forced vortex region yields

$$p + \gamma z - \rho \frac{\omega^2 r^2}{2} = p_2 + \gamma z_2 - \rho \frac{\omega^2 r_2^2}{2} \qquad (4.46)$$

Once again there is no elevation change, so $z = z_2$. At point 2, ωr_2 is the maximum speed V_{max}, and ωr is the speed of the fluid in the forced vortex. Solving for the pressure, one finds

$$p = p_2 - \rho\frac{V_{max}^2}{2} + \rho\frac{V^2}{2} \tag{4.47}$$

Substituting in the expression for p_2 from Eq. (4.45) gives

$$p = p_0 - \rho V_{max}^2 + \rho\frac{V^2}{2} \tag{4.48}$$

The pressure difference between the center of the cyclonic storm where the speed is zero and the outer edge of the storm is

$$p_1 - p_0 = -\rho V_{max}^2 \tag{4.49}$$

The minimum pressure at the vortex center can give rise to a "secondary" flow as shown in Fig. 4.23. In this case the secondary flow is produced by the pressure gradient in the primary (vortex) flow. In the region near the ground, the wind velocity is decreased due to the friction provided by the ground. However, the pressure difference in the radial direction causes a radially inward flow adjacent to the ground and an upward flow at the vortex center.

More information on cyclonic storms is available in Moran and Morgan. (3).

The pressure change across a tornado is shown in Example (4.13).

The Pressure Coefficient

Describing the pressure distribution is important because pressure gradients influence flow patterns and pressure distributions acting on bodies create resultant forces. A common dimensionless group for describing the pressure distribution is called the *pressure coefficient*:

$$C_p = \frac{p_z - p_{zo}}{\rho V_o^2/2} = \frac{h - h_o}{V_o^2/(2g)} \tag{4.50}$$

EXAMPLE 4.13   PRESSURE DIFFERENCE IN TORNADO

Assume that a tornado is modeled as the combination of a forced and a free vortex. The maximum wind speed in the tornado is 150 mph. What is the pressure difference, in inches of mercury, between the center and the outer edge of the tornado? The density of the air is 0.075 lbm/ft^3.

Problem Definition

Situation: Tornado with 150 mph winds.

Find: Pressure difference (inches Hg) between center and edge.

Assumptions: Tornado modeled as forced and free vortex.

Properties: $\rho = 0.075$ lbm/ft^3.

Plan

1. Use Eq. (4.49) to calculate pressure difference.
2. Convert result to inches Hg.

Solution

1. Convert velocity to ft/s:

$$150\frac{\text{mi}}{\text{hr}} \times \frac{5280\text{ ft}}{\text{mi}} \times \frac{\text{hr}}{3600\text{ s}} = 220\frac{\text{ft}}{\text{s}}$$

Convert density to slug/ft^3:

$$0.075\frac{\text{lbm}}{\text{ft}^3} \times \frac{\text{slug}}{32.2\text{ lbm}} = 0.00233\frac{\text{slug}}{\text{ft}^3}$$

Pressure difference

$$p_1 - p_0 = -\rho(V_{max})^2$$

$$p_1 - p_0 = -0.00233\frac{\text{slugs}}{\text{ft}^3} \times 220^2\ \frac{\text{ft}^2}{\text{s}^2} = -112.8\text{ psf}$$

2. Convert to inches Hg:

$$p_1 - p_0 = -112.8\text{ psf} \times \frac{29.92\text{ in Hg}}{2116\text{ psf}} = -1.59\text{ in Hg}$$

In the next section, the pressure coefficient is used to describe the pressure distribution around a circular cylinder.

Pressure Distribution Around a Circular Cylinder—Ideal Fluid

If a fluid is nonviscous and incompressible (an *ideal fluid*) and if the flow is initially irrotational, then the flow will be irrotational throughout the entire flow field.* Then, if the flow is also steady, the Bernoulli equation will apply everywhere because all the restrictions for the Bernoulli equation will have been satisfied. The flow pattern about a circular cylinder with such restrictions is shown in Fig. 4.24a.

Because the flow pattern is symmetrical with either the vertical or the horizontal axis through the center of the cylinder, the pressure distribution on the surface of the cylinder, obtained by application of the Bernoulli equation, is also symmetrical as shown in Fig. 4.24b. The pressure coefficient reduces to

$$C_p = \frac{p - p_0}{\frac{1}{2}\rho V^2_0} \tag{4.51}$$

where p is the local pressure and p_0 and V_0 are the free-stream pressure and velocity. The pressure coefficient is plotted outward (negative) or inward (positive) from the surface of the cylinder, depending on the sign of the relative pressure and on a line normal to the surface of the cylinder. The points at the front and rear of the cylinder, points B and D, are points of stagnation ($C_p = +1.0$), and the minimum pressure ($C_p = -3.0$) occurs at the midsection, point C, where the velocity is highest.

A useful concept in understanding the flow around a cylinder is the definition of the favorable and adverse pressure gradient. From Euler's equation for pressure gradient and acceleration along a pathline, Eq. (4.14), neglecting gravitational effects, one has

$$\rho a_t = -\frac{\partial p}{\partial s}$$

Figure 4.24

Irrotational flow past a cylinder.
(a) Streamline pattern.
(b) Pressure distribution.

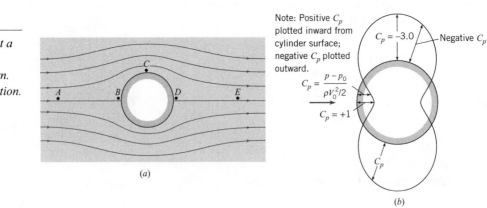

Note: Positive C_p plotted inward from cylinder surface; negative C_p plotted outward.

$$C_p = \frac{p - p_0}{\rho V_0^2/2}$$

(a)

(b)

* This can be seen in a qualitative sense if one visualizes a small spherical element of fluid within a nonviscous flow field. Since pressure forces act normal to the surface of the spherical element, the element deforms if the pressure is not of equal intensity over the entire surface. However, the element cannot rotate (irrotational situation) because there is no shear stress (viscosity is zero) on the surface of the sphere to possibly cause rotation.

One notes that $a_t > 0$ if $\partial p/\partial s < 0$; that is, the fluid particle accelerates if the pressure decreases with distance along a pathline. This is a *favorable pressure gradient*. On the other hand, $a_t < 0$ if $\partial p/\partial p > 0$, so the fluid particle decelerates if the pressure increases along a pathline. This is an *adverse pressure gradient*. The definitions of pressure gradient are summarized in the table.

Favorable pressure gradient	$\partial p/\partial s < 0$	$a_t > 0$ (acceleration)
Adverse pressure gradient	$\partial p/\partial s > 0$	$a_t < 0$ (deceleration)

Visualize the motion of a fluid particle in Fig. 4.24a as it travels around the cylinder from A to B to C to D and finally to E. Notice that it first decelerates from the free-stream velocity to zero velocity at the forward stagnation point as it travels in an adverse pressure gradient. Then as it passes from B to C, it is in a favorable pressure gradient, and it accelerated to its highest speed. From C to D the pressure increases again toward the rearward stagnation point, and the particle decelerates but has enough momentum to reach D. Finally, the pressure decreases from D to E, and this favorable pressure gradient accelerates the particle back to the free-stream velocity. Understanding this qualitative description of how the fluid particle travels from one point to another will be helpful when the phenomenon of separation is explained in the next section.

Separation

Flow separation occurs when the fluid pathlines adjacent to body deviate from the contour of the body and produce a wake. This flow condition is very common. It tends to increase drag, reduce lift, and produce unsteady forces that can lead to structural failure.

Consider the flow of a real (viscous) fluid past a cylinder as shown in Fig. 4.25. The flow pattern upstream of the midsection is very similar to the pattern for an ideal fluid. However, in a viscous fluid the velocity at the surface is zero (no-slip condition), whereas

Figure 4.25

Flow of a real fluid past a circular cylinder.
(a) Flow pattern.
(b) Pressure distribution

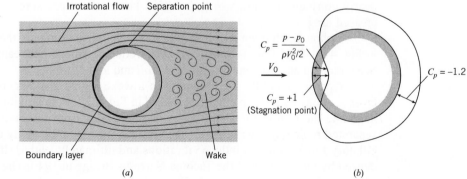

$$C_p = \frac{p - p_0}{\rho V_0^2/2}$$

V_0

$C_p = +1$
(Stagnation point)

$C_p = -1.2$

Irrotational flow   Separation point

Boundary layer   Wake

(a)   (b)

Figure 4.26

Smoke traces showing separation on an airfoil section at a large angle of attack. (Courtesy of Education Development Center, Inc. Newton, MA)

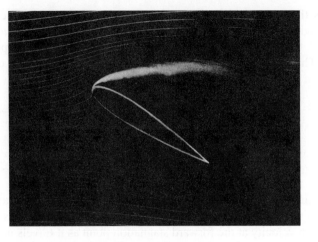

Figure 4.27

Flow pattern past a square rod illustrating separation at the edges.

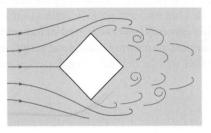

with the flow of an inviscid fluid the surface velocity need not be zero. Because of viscous effects, a thin layer, called a boundary layer, forms next to the surface. The velocity changes from zero at the surface to the free-stream velocity across the boundary layer. Over the forward section of the cylinder, where the pressure gradient is favorable, the boundary layer is quite thin.

Downstream of the midsection, the pressure gradient is adverse and the fluid particles in the boundary layer, slowed by viscous effects, can only go so far and then are forced to detour away from the surface. This is called the *separation point*. A recirculatory flow called a *wake* develops behind the cylinder. The flow in the wake region is called *separated flow*. The pressure distribution on the cylinder surface in the wake region is nearly constant, as shown in Fig. 4.25b. The reduced pressure in the wake leads to increased drag.

A photograph of an airfoil section with flow separation near the leading edge is shown in Fig. 4.26. This flow is visualized by introducing smoke upstream of the airfoil section. Separation on an airfoil surface leads to *stall* and loss of lift.

Separation and the development of a wake region also occurs on blunt objects and cross sections with sharp edges, as shown in Fig. 4.27. In these situations, the flow cannot negotiate the turn at the sharp edges and separates from the body, generating eddies, a separated region, and wake flow. The vortices shed from the body can produce lateral oscillatory forces than can induce vibrations and ultimately lead to structural failure, as evidenced by the collapse of the Tacoma Narrows Bridge in 1940. The prediction and control of separation is a continuing challenge for engineers involved with the design of fluid systems.

Summary

The streamline is a curve everywhere tangent to the local velocity vector. The configuration of streamlines in a flow field is called the flow pattern. The pathline is the line traced out by a particle. A streakline is the line produced by a dye introduced at a point in the field. Pathlines, streaklines, and streamlines are coincident in steady flow if they share a common point but differ in unsteady flows.

In a uniform flow, the velocity does not change along a streamline. In a steady flow, the velocity does not change with time at any location.

The tangential acceleration of a fluid element along a pathline is

$$a_t = \frac{\partial V}{\partial t} + V \frac{\partial V}{\partial s}$$

where the first term is the local acceleration and the second term is the convective acceleration. The acceleration normal to the pathline and toward the center of rotation is

$$a_n = \frac{V^2}{r}$$

where r is the local radius of curvature of the pathline.

Applying Newton's second law to a fluid element in the flow of an incompressible, inviscid fluid results in *Euler's equation*,

$$-\frac{\partial}{\partial \ell}(p + \gamma z) = \rho a_\ell$$

where ℓ is an arbitrary direction. Integrating Euler's equation in the radial direction for a rotating flow results in

$$p + \gamma z - \rho \frac{\omega^2 r^2}{2} = C$$

Integrating Euler's equation along a streamline in steady flow results in the *Bernoulli equation*,

$$p + \gamma z + \rho \frac{V^2}{2} = C$$

in terms of pressure or

$$\frac{p}{\gamma} + z + \frac{V^2}{2g} = C$$

in terms of head where V is the speed of the fluid and C is a constant along a streamline. The value of C may vary from streamline to streamline.

The rotation of a fluid element is defined as the average rotation of two initially perpendicular lines defining the sides of the element. If every fluid element in a flow does not rotate, the flow is irrotational and the value for C in the Bernoulli equation is the same for every streamline.

The *vorticity vector* is defined as

$$\omega = \left(\frac{\partial w}{\partial y} - \frac{\partial v}{\partial z}\right)\mathbf{i} + \left(\frac{\partial u}{\partial z} - \frac{\partial w}{\partial x}\right)\mathbf{j} + \left(\frac{\partial v}{\partial x} - \frac{\partial u}{\partial y}\right)\mathbf{k}$$

and is equal to twice the fluid rotation vector. In an irrotational flow, the vorticity is zero.

Separation occurs when streamlines move away from the surface of the body and create a local recirculation zone or wake. Typically the pressure in the recirculation zone assumes the value at the point of separation.

References

1. Hibbeler, R.C. *Dynamics*. Prentice Hall, Englewood Cliffs, New Jersey, 1995

2. Miller, R.W. (ed) *Flow Measurement Engineering Handbook*, McGraw-Hill, New York, 1996.

3. Moran, J.M., and M.D. Morgan. *Meteorology: The Atmosphere and Science of Weather*. Prentice Hall, Englewood Cliff, New Jersey, 1997.

Problems

Flow Descriptions

4.1 $\mathbb{PQ}\blacktriangleleft$ Identify five examples of an unsteady flow and explain what features classify them as an unsteady flow?

4.2 $\mathbb{PQ}\blacktriangleleft$ You are pouring a heavy syrup on your pancakes. As the syrup spreads over the pancake, would the thin film of syrup be a laminar or turbulent flow? Explain.

4.3 $\mathbb{PQ}\blacktriangleleft$ Breathe in and out of your mouth. Try to sense the air flow patterns near your face while doing this. Discuss the type of flow associated with these flow processes. If you were to blow out a candle, you would do it while exhaling (at least most people do). Why is it easier to do this by exhaling than by inhaling?

4.4 In the system in the figure, the valve at C is gradually opened in such a way that a constant rate of increase in discharge is produced. How would you classify the flow at B while the valve is being opened? How would you classify the flow at A?

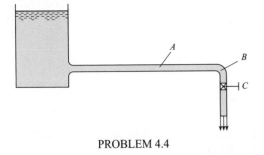

PROBLEM 4.4

4.5 Water flows in the passage shown. If the flow rate is decreasing with time, the flow is classified as (a) steady, (b) unsteady, (c) uniform, or (d) nonuniform.

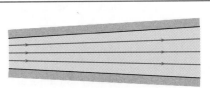

PROBLEM 4.5

4.6 If a flow pattern has converging streamlines, how would you classify the flow?

4.7 Consider flow in a straight conduit. The conduit is circular in cross section. Part of the conduit has a constant diameter, and part has a diameter that changes with distance. Then, relative to flow in that conduit, correctly match the items in column A with those in column B.

A	B
Steady flow	$\partial V_s/\partial s = 0$
Unsteady flow	$\partial V_s/\partial s \neq 0$
Uniform flow	$\partial V_s/\partial t = 0$
Nonuniform flow	$\partial V_s/\partial t \neq 0$

4.8 Classify each of the following as a one-dimensional, two-dimensional, or three-dimensional flow.
a. Water flow over the crest of a long spillway of a dam.
b. Flow in a straight horizontal pipe.
c. Flow in a constant-diameter pipeline that follows the contour of the ground in hilly country.
d. Airflow from a slit in a plate at the end of a large rectangular duct.
e. Airflow past an automobile.
f. Air flow past a house.
g. Water flow past a pipe that is laid normal to the flow across the bottom of a wide rectangular channel.

Pathlines, Streamlines and Streaklines

4.9 $\mathbb{PQ}\blacktriangleleft$ If somehow you could attach a light to a fluid particle and take a time exposure, would the image you photographed be a pathline or streakline? Explain from definition of each.

4.10 $\mathbb{PQ}\blacktriangleleft$ The pattern produced by smoke rising from a chimney on a windy day is analogous to a pathline or streakline? Explain from the definition of each.

4.11 At time $t = 0$, dye was injected at point A in a flow field of a liquid. When the dye had been injected for 4 s, a pathline for a particle of dye that was emitted at the 4 s instant was started. The streakline at the end of 10 s is shown below. Assume that the speed (but not the velocity) of flow is the same throughout the 10 s period. Draw the pathline of the particle that was emitted at $t = 4$ s. Make your own assumptions for any missing information.

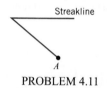

Streakline

A

PROBLEM 4.11

4.12 For a given hypothetical flow, the velocity from time $t = 0$ to $t = 5$ s was $u = 2$ m/s, $v = 0$. Then, from time $t = 5$ s to $t = 10$ s, the velocity was $u = +3$ m/s, $v = -4$ m/s. A dye streak was started at a point in the flow field at time $t = 0$, and the path of a particle in the fluid was also traced from that same point starting at the same time. Draw to scale the streakline, pathline of the particle, and streamlines at time $t = 10$ s.

4.13 At time $t = 0$, a dye streak was started at point A in a flow field of liquid. The speed of the flow is constant over a 10 s period, but the flow direction is not necessarily constant. At any particular instant the velocity in the entire field of flow is the same. The streakline produced by the dye is shown above. Draw (and label) a streamline for the flow field at $t = 8$ s.

Draw (and label) a pathline that one would see at $t = 10$ s for a particle of dye that was emitted from point A at $t = 2$ s.

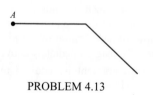

A

PROBLEM 4.13

Acceleration

4.14 $\mathbb{PQ}\blacktriangleleft$ Acceleration is the rate of change of velocity with time. Is the acceleration vector always aligned with the velocity vector? Explain.

4.15 $\mathbb{PQ}\blacktriangleleft$ For a rotating body, is the acceleration toward the center of rotation a centripetal or centrifugal acceleration? Look up word meanings and word roots.

4.16 Figure 4.24 on p. 110 shows the flow pattern for flow past a circular cylinder. Assume that the approach velocity at A is constant (does not vary with time).

a. Is the flow past the cylinder steady or unsteady?

b. Is this a case of one-dimensional, two-dimensional, or three-dimensional flow?

c. Are there any regions of the flow where local acceleration is present? If so, show where they are and show vectors representing the local acceleration in the regions where it occurs.

d. Are there any regions of flow where convective acceleration is present? If so, show vectors representing the convective acceleration in the regions where it occurs.

4.17 The velocity along a pathline is given by V (m/s) $= s^2 t^{1/2}$ where s is in meters and t is in seconds. The radius of curvature is 0.5 m. Evaluate the acceleration along and normal to the path at $s = 2$ m and $t = 0.5$ seconds.

4.18 Tests on a sphere are conducted in a wind tunnel at an air speed of U_0. The velocity of flow toward the sphere along the longitudinal axis is found to be $u = -U_0 (1 - r_0^3/x^3)$, where r_0 is the radius of the sphere and x the distance from its center. Determine the acceleration of an air particle on the x-axis upstream of the sphere in terms of x, r_0, and U_0.

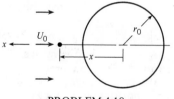

x ← U_0

r_0

x

PROBLEM 4.18

4.19 In this flow passage the velocity is varying with time. The velocity varies with time at section A-A as

$$V = 5 \text{ m/s} - 2.55\frac{t}{t_0} \text{ m/s}$$

At time $t = 0.50$ s, it is known that at section A-A the velocity gradient in the s direction is $+2$ m/s per meter. Given that t_0 is 0.5s and assuming quasi–one-dimensional flow, answer the following questions for time $t = 0.5$ s.

a. What is the local acceleration at A-A?

b. What is the convective acceleration at A-A?

4.20 The nozzle in the figure is shaped such that the velocity of flow varies linearly from the base of the nozzle to its tip. Assuming quasi–one-dimensional flow, what is the convective acceleration midway between the base and the tip if the velocity is 1 ft/s at the base and 4 ft/s at the tip? Nozzle length is 18 inches.

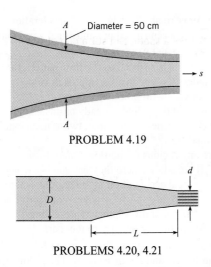

PROBLEM 4.19

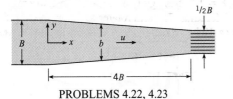

PROBLEMS 4.20, 4.21

4.21 In Prob. 4.20 the velocity varies linearly with time throughout the nozzle. The velocity at the base is $1t$ (ft/s) and at the tip is $4t$ (ft/s). What is the local acceleration midway along the nozzle when $t = 2$ s?

4.22 Liquid flows through this two-dimensional slot with a velocity of $V = 2(q_0/b)(t/t_0)$, where q_0 and t_0 are reference values. What will be the local acceleration at $x = 2B$ and $y = 0$ in terms of B, t, t_0, and q_0?

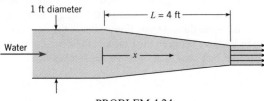

PROBLEMS 4.22, 4.23

4.23 What will be the convective acceleration for the conditions of Prob. 4.22?

4.24 The velocity of water flow in the nozzle shown is given by the following expression:

$$V = 2t/(1 - 0.5x/L)^2,$$

where V = velocity in feet per second, t = time in seconds, x = distance along the nozzle, and L = length of nozzle = 4 ft. When $x = 0.5L$ and $t = 3$ s, what is the local acceleration along the centerline? What is the convective acceleration? Assume quasi–one-dimensional flow prevails.

PROBLEM 4.24

Euler's Equation

4.25 $\mathbb{P}\mathbb{Q}\blacktriangleleft$ State Newton's second law of motion as used in dynamics. Are there any limitations on the use of Newton's second law? Explain.

4.26 $\mathbb{P}\mathbb{Q}\blacktriangleleft$ What is the difference between a force due to weight and a force due to pressure? Explain.

4.27 A pipe slopes upward in the direction of liquid flow at an angle of 30° with the horizontal. What is the pressure gradient in the flow direction along the pipe in terms of the specific weight of the liquid if the liquid is decelerating (accelerating opposite to flow direction) at a rate of $0.3g$?

4.28 What pressure gradient is required to accelerate kerosene ($S = 0.81$) vertically upward in a vertical pipe at a rate of $0.3\,g$?

4.29 The hypothetical liquid in the tube shown in the figure has zero viscosity and a specific weight of 10 kN/m³. If $p_B - p_A$ is equal to 12 kPa, one can conclude that the liquid in the tube is being accelerated (a) upward, (b) downward, or (c) neither: acceleration = 0.

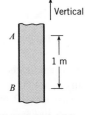

PROBLEM 4.29

4.30 If the piston and water ($\rho = 62.4$ lbm/ft³) are accelerated upward at a rate of $0.5g$, what will be the pressure at a depth of 2 ft in the water column?

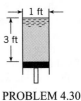

PROBLEM 4.30

4.31 Water ($\rho = 62.4$ lbm/ft³) stands at a depth of 10 ft in a vertical pipe that is open at the top and closed at the bottom by a piston. What upward acceleration of the piston is necessary to create a pressure of 8 psig immediately above the piston?

4.32 What pressure gradient is required to accelerate water ($\rho = 1000$ kg/m³) in a horizontal pipe at a rate of 6 m/s²?

4.33 Water ($\rho = 1000$ kg/m³) is accelerated from rest in a horizontal pipe that is 100 m long and 30 cm in diameter. If the acceleration rate (toward the downstream end) is 5 m/s², what is the pressure at the upstream end if the pressure at the downstream end is 90 kPa gage?

4.34 Water ($\rho = 62.4$ lbm/ft³) stands at a depth of 10 ft in a vertical pipe that is closed at the bottom by a piston. Assuming that the vapor pressure is zero (abs), determine the maximum downward acceleration that can be given to the piston without causing the water immediately above it to vaporize.

4.35 A liquid with a specific weight of 100 lbf/ft³ is in the conduit. This is a special kind of liquid that has zero viscosity. The pressures at points A and B are 170 psf and 100 psf, respectively. Which one (or more) of the following conclusions can one draw with certainty? (a) The velocity is in the positive ℓ direction. (b) The velocity is in the negative ℓ direction. (c) The acceleration is in the positive ℓ direction. (d) The acceleration is in the negative ℓ direction.

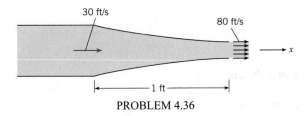

PROBLEM 4.35

4.36 If the velocity varies linearly with distance through this water nozzle, what is the pressure gradient, dp/dx, halfway through the nozzle? ($\rho = 62.4$ lbm/ft³).

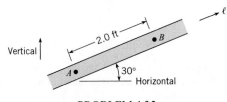

PROBLEM 4.36

4.37 The closed tank shown, which is full of liquid, is accelerated downward at $1.5g$ and to the right at $0.9g$. Here $L = 3$ ft, $H = 4$ ft, and the specific gravity of the liquid is 1.2. Determine $p_C - p_A$ and $p_B - p_A$.

4.38 The closed tank shown, which is full of liquid, is accelerated downward at $\frac{2}{3}g$ and to the right at $1g$. Here $L = 2.5$ m, $H = 3$ m, and the liquid has a specific gravity of 1.3. Determine $p_C - p_A$ and $p_B - p_A$.

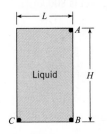

PROBLEMS 4.37, 4.38

Pressure Distribution in Rotating Flows

4.39 PQ◄ Take a spoon and stir a cup of liquid. Report on the contour of the surface. Provide an explanation for the observed shape.

4.40 PQ◄ A cyclonic separator is a device for separating solid particles from a gas stream by inducing a spin in the gas stream. Explain the mechanism by which the particles are separated from the gas.

4.41 This closed tank, which is 4 ft in diameter, is filled with water ($\rho = 62.4$ lbm/ft³) and is spun around its vertical centroidal axis at a rate of 10 rad/s. An open piezometer is connected to the tank as shown so that it is also rotating with the tank. For these conditions, what is the pressure at the center of the bottom of the tank?

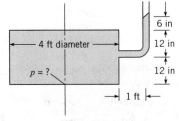

PROBLEM 4.41

4.42 A tank of liquid (S = 0.80) that is 1 ft in diameter and 1.0 ft high ($h = 1.0$ ft) is rigidly fixed (as shown) to a rotating arm having a 2 ft radius. The arm rotates such that the speed at point A is 20 ft/s. If the pressure at A is 25 psf, what is the pressure at B?

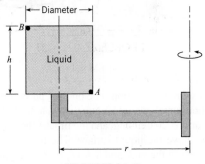

PROBLEM 4.42

4.43 Separators are used to separate liquids of different densities, such as cream from skim milk, by rotating the mixture at high speeds. In a cream separator the skim milk goes to the outside while the cream migrates toward the middle. A factor of merit for the centrifuge is the centrifugal acceleration force (RCF), which is the radial acceleration divided by the acceleration due to gravity. A cream separator can operate at 9000 rpm (rev/min). If the bowl of the separator is 20 cm in diameter, what is the centripetal acceleration if the liquid rotates as a solid body and what is the RCF?

4.44 A closed tank of liquid ($S = 1.2$) is rotated about a vertical axis (see the figure), and at the same time the entire tank is accelerated upward at 4 m/s². If the rate of rotation is 10 rad/s, what is the difference in pressure between points A and B ($p_B - p_A$)? Point B is at the bottom of the tank at a radius of 0.5 m from the axis of rotation, and point A is at the top on the axis of rotation.

PROBLEM 4.44

4.45 A U-tube is rotated about one leg, as shown. Before being rotated the liquid in the tube fills 0.25 m of each leg. The length of the base of the U-tube is 0.5 m, and each leg is 0.5 m long. What would be the maximum rotation rate (in rad/s) to ensure that no liquid is expelled from the outer leg?

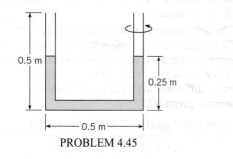

PROBLEM 4.45

4.46 An arm with a stagnation tube on the end is rotated at 100 rad/s in a horizontal plane 10 cm below a liquid surface as shown. The arm is 20 cm long, and the tube at the center of rotation extends above the liquid surface. The liquid in the tube is the same as that in the tank and has a specific weight of 10,000 N/m³. Find the location of the liquid surface in the central tube.

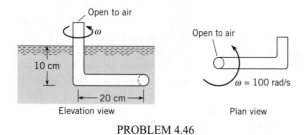

PROBLEM 4.46

4.47 A U-tube is rotated at 50 rev/min about one leg. The fluid at the bottom of the U-tube has a specific gravity of 3.0. The distance between the two legs of the U-tube is 1 ft. A 6 in. height of another fluid is in the outer leg of the U-tube. Both legs are open to the atmosphere. Calculate the specific gravity of the other fluid.

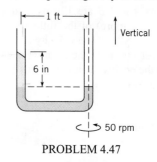

PROBLEM 4.47

4.48 A manometer is rotated around one leg, as shown. The difference in elevation between the liquid surfaces in the legs is 20 cm. The radius of the rotating arm is 10 cm. The liquid in the manometer is oil with a specific gravity of 0.8. Find the number of g's of acceleration in the leg with greatest amount of oil.

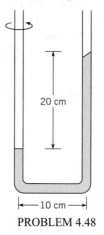

PROBLEM 4.48

4.49 A fuel tank for a rocket in space under a zero-g environment is rotated to keep the fuel in one end of the tank. The system is rotated at 3 rev/min. The end of the tank (point A) is 1.5 m from the axis of rotation, and the fuel level is 1 m from the rotation axis. The pressure in the nonliquid end of the tank is 0.1 kPa, and the density of the fuel is 800 kg/m³. What is the pressure at the exit (point A)?

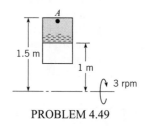

PROBLEM 4.49

4.50 Water stands in these tubes as shown when no rotation occurs. Derive a formula for the angular speed at which water will just begin to spill out of the small tube when the entire system is rotated about axis *A-A*.

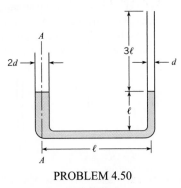

PROBLEM 4.50

4.51 Water ($\rho = 1000$ kg/m^3) fills a slender tube 1 cm in diameter, 40 cm long, and closed at one end. When the tube is rotated in the horizontal plane about its open end at a constant speed of 50 rad/s, what force is exerted on the closed end?

4.52 Water ($\rho = 1000$ kg/m^3) stands in the closed-end U-tube as shown when there is no rotation. If $\ell = 10$ cm and if the entire system is rotated about axis *A-A*, at what angular speed will water just begin to spill out of the open tube? Assume that the temperature for the system is the same before and after rotation and that the pressure in the closed end is initially atmospheric.

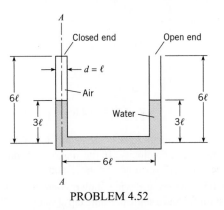

PROBLEM 4.52

4.53 A simple centrifugal pump consists of a 10 cm disk with radial ports as shown. Water is pumped from a reservoir through a central tube on the axis. The wheel spins at 3000 rev/min, and the liquid discharges to atmospheric pressure. To establish the maximum height for operation of the pump, assume that the flow rate is zero and the pressure at the pump intake is atmospheric pressure. Calculate the maximum operational height z for the pump.

4.54 A closed cylindrical tank of water ($\rho = 1000$ kg/m^3) is rotated about its horizontal axis as shown. The water inside the tank rotates with the tank ($V = r\omega$). Derive an equation for

dp/dz along a vertical-radial line through the center of rotation. What is dp/dz along this line for $z = -1$ m, $z = 0$, and $z = +1$ m when $\omega = 5$ rad/s? Here $z = 0$ at the axis.

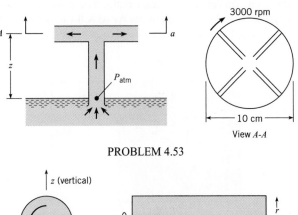

PROBLEM 4.53

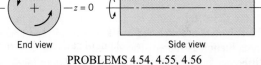

PROBLEMS 4.54, 4.55, 4.56

4.55 For the conditions of Prob. 4.54, derive an equation for the maximum pressure difference in the tank as a function of the significant variables.

4.56 The tank shown is 4 ft in diameter and 12 ft long and is closed and filled with water ($\rho = 62.4$ lbm/ft^3). It is rotated about its horizontal-centroidal axis, and the water in the tank rotates with the tank ($V = r\omega$). The maximum velocity is 25 ft/s. What is the maximum difference in pressure in the tank? Where is the point of minimum pressure?

The Bernoulli Equation Along a Streamline

4.57 $\mathbb{PQ}$◀ Often, in high winds, a roof will be lifted from a house. Applying the Bernoulli equation, explain how this might happen.

4.58 $\mathbb{PQ}$◀ Describe in your own words how an aspirator works.

4.59 A water jet issues vertically from a nozzle, as shown. The water velocity as it exits the nozzle is 20 ft/s. Calculate how high h the jet will rise. (*Hint:* Apply the Bernoulli equation along the centerline.)

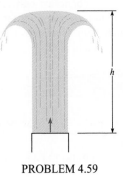

PROBLEM 4.59

4.60 A pressure of 10 kPa, gage, is applied to the surface of water in an enclosed tank. The distance from the water surface to the outlet is 0.5 m. The temperature of the water is 20°C. Find the velocity (m/s) of water at the outlet. The speed of the water surface is much less than the water speed at the outlet.

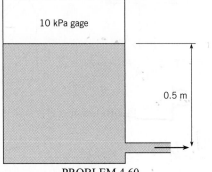

10 kPa gage

0.5 m

PROBLEM 4.60

4.61 Water flows through a vertical contraction (venturi) section. Piezometers are attached to the upstream pipe and minimum area section as shown. The velocity in the pipe is 10 ft/s. The difference in elevation between the two water levels in the piezometers is 6 inches. The water temperature is 68°F. What is the velocity (ft/s) at the minimum area?

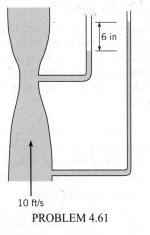

6 in

10 ft/s

PROBLEM 4.61

4.62 Kerosene at 20°C flows through a contraction section as shown. A pressure gage connected between the upstream pipe and throat section shows a pressure difference of 20 kPa. The gasoline velocity in the throat section is 10 m/s. What is the velocity (m/s) in the upstream pipe?

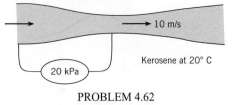

10 m/s

Kerosene at 20° C

20 kPa

PROBLEM 4.62

4.63 A Pitot-static tube is mounted on an airplane to measure airspeed. At an altitude of 10,000 ft, where the temperature is 23°F and the pressure is 10 psia, a pressure difference corresponding to 10 in of water is measured. What is the airspeed?

4.64 A glass tube is inserted into a flowing stream of water with one opening directed upstream and the other end vertical. If the water velocity is 4 m/s, how high will the water rise in the vertical leg relative to the level of the water surface of the stream?

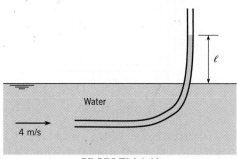

ℓ

Water

4 m/s

PROBLEM 4.64

4.65 A Bourdon-tube gage is tapped into the center of a disk as shown. Then for a disk that is about 1 ft in diameter and for an approach velocity of air (V_0) of 40 ft/s, the gage would read a pressure intensity that is (a) less than $\rho V_0^2/2$, (b) equal to $\rho V_0^2/2$, or (c) greater than $\rho V_0^2/2$.

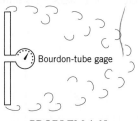

Bourdon-tube gage

PROBLEM 4.65

4.66 An air-water manometer is connected to a Pitot-static tube used to measure air velocity. If the manometer deflects 2 in., what is the velocity? Assume $T = 60°F$ and $p = 15$ psia.

4.67 The flow-metering device shown consists of a stagnation probe at station 2 and a static pressure tap at station 1. The velocity at station 2 is twice that at station 1. Air with a density of 1.2 kg/m^3 flows through the duct. A water manometer is connected between the stagnation probe and the pressure tap, and a deflection of 10 cm is measured. What is the velocity at station 2?

4.68 The "spherical" Pitot probe shown is used to measure the flow velocity in water ($\rho = 1000$ kg/m^3). Pressure taps are located at the forward stagnation point and at 90° from the forward stagnation point. The speed of fluid next to the surface of the sphere varies as $1.5V_0 \sin\theta$, where V_0 is the free-stream velocity and θ is measured from the forward stagnation point. The pressure taps are at the same level; that is, they are in the same horizontal plane. The piezometric pressure difference between the two taps is 2 kPa. What is the free-stream velocity V_0?

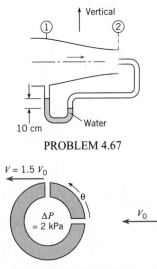

PROBLEM 4.67

PROBLEM 4.68

4.69 A device used to measure the velocity of fluid in a pipe consists of a cylinder, with a diameter much smaller than the pipe diameter, mounted in the pipe with pressure taps at the forward stagnation point and at the rearward side of the cylinder. Data show that the pressure coefficient at the rearward pressure tap is -0.3. Water with a density of 1000 kg/m³ flows in the pipe. A pressure gage connected by lines to the pressure taps shows a pressure difference of 500 Pa. What is the velocity in the pipe?

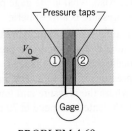

PROBLEM 4.69

4.70 Explain how you might design a spherical Pitot-static probe to provide the direction and velocity of a flowing stream. The Pitot-static probe will be mounted on a sting that can be oriented in any direction.

4.71 Two Pitot-static tubes are shown. The one on the top is used to measure the velocity of air, and it is connected to an air-water manometer as shown. The one on the bottom is used to measure the velocity of water, and it too is connected to an air-water manometer as shown. If the deflection h is the same for both manometers, then one can conclude that (a) $V_A = V_w$, (b) $V_A > V_w$, or (c) $V_A < V_w$.

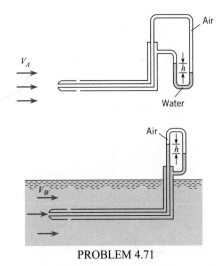

PROBLEM 4.71

4.72 A Pitot-static tube is used to measure the velocity at the center of a 12 in. pipe. If kerosene at 68°F is flowing and the deflection on a mercury-kerosene manometer connected to the Pitot tube is 4 in., what is the velocity?

4.73 A Pitot-static tube used to measure air velocity is connected to a differential pressure gage. If the air temperature is 20°C at standard atmospheric pressure at sea level, and if the differential gage reads a pressure difference of 3 kPa, what is the air velocity?

4.74 A Pitot-static tube used to measure air velocity is connected to a differential pressure gage. If the air temperature is 60°F at standard atmospheric pressure at sea level, and if the differential gage reads a pressure difference of 11 psf, what is the air velocity?

4.75 A Pitot-static tube is used to measure the gas velocity in a duct. A pressure transducer connected to the Pitot tube registers a pressure difference of 1.0 psi. The density of the gas in the duct is 0.12 lbm/ft³. What is the gas velocity in the duct?

4.76 A sphere moves horizontally through still water at a speed of 11 ft/s. A short distance directly ahead of the sphere (call it point A), the velocity, with respect to the earth, induced by the sphere is 1 ft/s in the same direction as the motion of the sphere. If p_0 is the pressure in the undisturbed water at the same depth as the center of the sphere, then the value of the ratio p_A/p_0 will be (a) less than unity, (b) equal to unity, or (c) greater than unity.

4.77 Body A travels through water at a constant speed of 13 m/s as shown. Velocities at points B and C are induced by the moving body and are observed to have magnitudes of 5 m/s and 3 m/s, respectively. What is $p_B - p_C$?

4.78 Water in a flume is shown for two conditions. If the depth d is the same for each case, will gage A read greater or less than gage B? Explain.

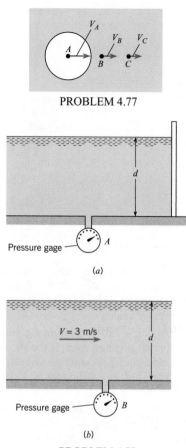

PROBLEM 4.77

(a)

(b)

PROBLEM 4.78

4.79 The apparatus shown in the figure is used to measure the velocity of air at the center of a duct having a 10 cm diameter. A tube mounted at the center of the duct has a 2 mm diameter and is attached to one leg of a slant-tube manometer. A pressure tap in the wall of the duct is connected to the other end of the slant-tube manometer. The well of the slant-tube manometer is sufficiently large that the elevation of the fluid in it does not change significantly when fluid moves up the leg of the manometer. The air in the duct is at a temperature of 20°C, and the pressure is 150 kPa. The manometer liquid has a specific gravity of 0.7, and the slope of the leg is 30°. When there is no flow in the duct, the liquid surface in the manometer lies at 2.3 cm on the slanted scale. When there is flow in the duct, the liquid moves up to 6.7 cm on the slanted scale. Find the velocity of the air in the duct. Assuming a uniform velocity profile in the duct, calculate the rate of flow of the air.

4.80 A rugged instrument used frequently for monitoring gas velocity in smoke stacks consists of two open tubes oriented to the flow direction as shown and connected to a manometer. The pressure coefficient is 1.0 at A and -0.3 at B. Assume that water, at 20°C, is used in the manometer and that a 5 mm deflection is noted. The pressure and temperature of the stack gases are

101 kPa and 250°C. The gas constant of the stack gases is 200 J/kg K. Determine the velocity of the stack gases.

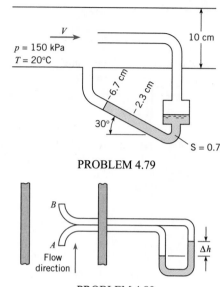

PROBLEM 4.79

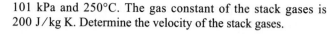

PROBLEM 4.80

4.81 The pressure in the wake of a bluff body is approximately equal to the pressure at the point of separation. The velocity distribution for flow over a sphere is $V = 1.5\ V_0 \sin \theta$, where V_0 is the free-stream velocity and θ is the angle measured from the forward stagnation point. The flow separates at $\theta = 120°$. If the free-stream velocity is 100 m/s and the fluid is air ($\rho = 1.2$ kg/m^3), find the pressure coefficient in the separated region next to the sphere. Also, what is the gage pressure in this region if the free-stream pressure is atmospheric?

4.82 A Pitot-static tube is used to measure the airspeed of an airplane. The Pitot tube is connected to a pressure-sensing device calibrated to indicate the correct airspeed when the temperature is 17°C and the pressure is 101 kPa. The airplane flies at an altitude of 3000 m, where the pressure and temperature are 70 kPa and -6.3°C. The indicated airspeed is 70 m/s. What is the true airspeed?

4.83 An aircraft flying at 10,000 feet uses a Pitot-static tube to measure speed. The instrumentation on the aircraft provides the differential pressure as well as the local static pressure and the local temperature. The local static pressure is 9.8 psig, and the air temperature is 25°F. The differential pressure is 0.5 psid. Find the speed of the aircraft in mph.

4.84 You need to measure air flow velocity. You order a commercially available Pitot-static tube, and the accompanying instructions state that the airflow velocity is given by

$$V \text{ (ft/min)} = 1096.7 \sqrt{\frac{h_v}{d}}$$

where h_v is the "velocity pressure" in inches of water and d is the density in pounds per cubic foot. The velocity pressure is the deflection measured on a water manometer attached to the static and total pressure ports. The instructions also state the density d can be calculated using

$$d \ (\text{lbm}/\text{ft}^3) = 1.325 \frac{P_a}{T}$$

where P_a is the barometric pressure in inches of mercury and T is the absolute temperature in degrees Rankine. Before you use the Pitot tube you want to confirm that the equations are correct. Determine if they are correct.

4.85 Consider the flow of water over the surfaces shown. For each case the depth of water at section D-D is the same (1 ft), and the mean velocity is the same and equal to 10 ft/s. Which of the following statements are valid?

a. $p_C > p_B > p_A$

b. $p_B > p_C > p_A$

c. $p_A = p_B = p_C$

d. $p_B < p_C < p_A$

e. $p_A < p_B < p_C$

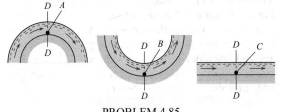

PROBLEM 4.85

Rotation of Fluid Elements

4.86 $\mathbb{PQ}◀$ What is meant by rotation of a fluid element? Use a sketch to explain.

4.87 $\mathbb{PQ}◀$ Consider a spherical fluid element in an inviscid fluid (no shear stresses). If pressure and gravitational forces are the only forces acting on the element, can they cause the element to rotate? Explain.

4.88 The vector $\mathbf{V} = 10x\mathbf{i} - 10y\mathbf{j}$ represents a two-dimensional velocity field. Is the flow irrotational?

4.89 The u and v velocity components of a flow field are given by $u = -\omega y$ and $v = \omega x$. Determine the vorticity and the rate of rotation of flow field.

4.90 The velocity components for a two-dimensional flow are

$$u = \frac{Cx}{(y^2 + x^2)} \qquad v = \frac{Cy}{(x^2 + y^2)}$$

where C is a constant. Is the flow irrotational?

4.91 A two-dimensional flow field is defined by $u = x^2 - y^2$ and $v = -2xy$. Is the flow rotational or irrotational?

4.92 Fluid flows between two parallel stationary plates. The distance between the plates is 1 cm. The velocity profile between the two plates is a parabola with a maximum velocity at the centerline of 2 cm/s. The velocity is given by

$$u = 2(1 - 4y^2)$$

where y is measured from the centerline. The cross-flow component of velocity, v, is zero. There is a reference line located 1 cm downstream. Find an expression, as a function of y, for the amount of rotation (in radian) a fluid element will undergo when it travels a distance of 1 cm downstream.

4.93 A combination of a forced and a free vortex is represented by the velocity distribution

$$v_\theta = \frac{1}{r}[1 - \exp(-r^2)]$$

For $r \longrightarrow 0$ the velocity approaches a rigid body rotation, and as r becomes large, a free-vortex velocity distribution is approached. Find the amount of rotation (in radians) that a fluid element will experience in completing one circuit around the center as a function of r. *Hint:* The rotation rate in a flow with concentric streamlines is given by

$$2\dot{\theta} = \frac{dv_\theta}{dr} + \frac{v_\theta}{r} = \frac{1}{r}\frac{d}{dr}(v_\theta r)$$

Evaluate the rotation for $r = 0.5, 1.0,$ and 1.5.

The Bernoulli Equation in Irrotational Flow

4.94 Liquid flows with a free surface around a bend. The liquid is inviscid and incompressible, and the flow is steady and irrotational. The velocity varies with the radius across the flow as $V = 1/r$ m/s, where r is in meters. Find the difference in depth of the liquid from the inside to the outside radius. The inside radius of the bend is 1 m and the outside radius is 3 m.

4.95 The velocity in the outlet pipe from this reservoir is 16 ft/s and $h = 15$ ft. Because of the rounded entrance to the pipe, the flow is assumed to be irrotational. Under these conditions, what is the pressure at A?

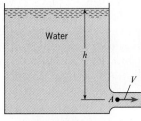

PROBLEMS 4.95. 4.96

4.96 The velocity in the outlet pipe from this reservoir is 6 m/s and $h = 15$ m. Because of the rounded entrance to the pipe, the flow is assumed to be irrotational. Under these conditions, what is the pressure at A?

4.97 The maximum velocity of the flow past a circular cylinder, as shown, is twice the approach velocity. What is Δp between the point of highest pressure and the point of lowest pressure in a 40 m/s wind? Assume irrotational flow and standard atmospheric conditions.

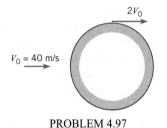

PROBLEM 4.97

4.98 The velocity and pressure are given at two points in the flow field. Assume that the two points lie in a horizontal plane and that the fluid density is uniform in the flow field and is equal to 1000 kg/m³. Assume steady flow. Then, given these data, determine which of the following statements is true. (a) The flow in the contraction is nonuniform and irrotational. (b) The flow in the contraction is uniform and irrotational. (c) The flow in the contraction is nonuniform and rotational. (d) The flow in the contraction is uniform and rotational.

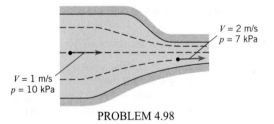

PROBLEM 4.98

4.99 Water ($\rho = 62.4$ lbm/ft³) flows from the large orifice at the bottom of the tank as shown. Assume that the flow is irrotational. Point B is at zero elevation, and point A is at 1 ft elevation. If $V_A = 8$ ft/s at an angle of 45° with the horizontal and if $V_B = 20$ ft/s vertically downward, what is the value of $p_A - p_B$?

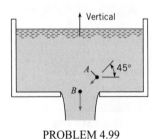

PROBLEM 4.99

4.100 Ideal flow theory will yield a flow pattern past an airfoil similar to that shown. If the approach air velocity V_0 is 80 m/s, what is the pressure difference between the bottom and the top

of this airfoil at points where the velocities are $V_1 = 85$ m/s and $V_2 = 75$ m/s? Assume ρ_{air} is uniform at 1.2 kg/m³.

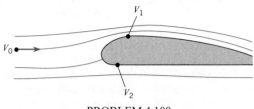

PROBLEM 4.100

4.101 Consider the flow of water between two parallel plates in which one plate is fixed as shown. The distance between the plates is h, and the speed of the moving plate is V. A person wishes to calculate the pressure difference between the plates and applies the Bernoulli equation between points 1 and 2,

$$z_1 + \frac{p_1}{\gamma} + \frac{V_1^2}{2g} = z_2 + \frac{p_2}{\gamma} + \frac{V_2^2}{2g}$$

and concludes that

$$p_1 - p_2 = \gamma(z_2 - z_1) + \rho\frac{V_2^2}{2}$$

$$= \gamma h + \rho\frac{V^2}{2}$$

Is this correct? Provide the reason for your answer.

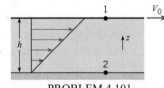

PROBLEM 4.101

Cyclonic Storms

4.102 During the fall of 2005, Hurricane Wilma passed through the Gulf of Mexico. When the storm was classified as a category 5 hurricane, the pressure at the center of the hurricane was measured at 902 mbars. The highest wind velocity was 175 mph. Assuming the pressure far from the center of the hurricane was 1 bar and the air density was 1.2 kg/m³, estimate the pressure at the center of the hurricane. Comment on the difference in your prediction and the measured value and provide some rationale for the discrepancy.

4.103 On June 24, 2003, a violent tornado occurred near Manchester, ND. A pressure drop of 100 mbar was recorded in the tornado. Assume an air density of 1.2 kg/m³ and estimate the maximum velocity.

4.104 A whirlpool is modeled as the combination of a free and a forced vortex. The maximum velocity in the whirlpool is 10 m/s, and the radius at the juncture of the free and the forced vortex is 10 m. The pressure is atmospheric (gage pressure = 0) at the free surface. Plot the shape of the water surface from the center to a radius of 50 m. The elevation is zero for the vortex center, where the velocity is zero.

4.105 The intensity of a tornado is measured by the Fujita Tornado Intensity Scale (F-scale). An intense tornado with an F-scale reading of 4 has a maximum wind velocity of 350 km/hr. The tornado is modeled as a combination of a free and a forced vortex, and the radius of the forced vortex is 50 m. The atmospheric pressure is 100 kPa. Plot the variation of pressure with radius from the center. There is no elevation change.

4.106 A weather balloon is caught in a tornado modeled as a combination free-forced vortex. Will it move toward the center or away from the center? Think carefully about pressure gradients and buoyancy. Provide a rationale for your answer.

4.107 The pressure distribution in a tornado is predicted using the Bernoulli equation, which is based on a constant density. However, the density will decrease as the pressure decreases in the tornado. Does the Bernoulli equation overpredict or underpredict the pressure drop in the tornado? Explain.

Separation

4.108 The velocity distribution over the surface of a sphere upstream of the separation point is $u_\theta = 1.5U\sin\theta$, where U is the free stream velocity and θ is the angle measured from the forward stagnation point. A pressure of -2.5 in H_2O gage is measured at the point of separation on a sphere in a 100 ft/s airflow with a density of 0.07 lbm/ft^3. The pressure far upstream of the sphere in atmospheric. Estimate the location of the stagnation point (θ). Separation occurs on the windward side of the sphere.

4.109 Knowing the speed at point 1 of a fluid upstream of a sphere and the average speed at point 2 in the wake of in the sphere, can one use the Bernoulli equation to find the pressure difference between the two points? Provide the rationale for your decision.

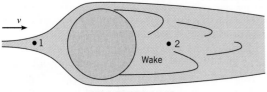

PROBLEM 4.109

General

4.110 Euler's equations for a planar (two-dimensional) flow in the xy-plane are

$$u\frac{\partial u}{\partial x} + v\frac{\partial u}{\partial y} = -g\frac{\partial h}{\partial x} \qquad x = \text{direction}$$

$$u\frac{\partial v}{\partial x} + v\frac{\partial v}{\partial y} = -g\frac{\partial h}{\partial y} \qquad y = \text{direction}$$

a. The slope of a streamline is given by

$$\frac{dy}{dx} = \frac{v}{u}$$

Using this relation in Euler's equation, show that

$$d\left(\frac{u^2 + v^2}{2g} + h\right) = 0$$

or

$$d\left(\frac{V^2}{2g} + h\right) = 0$$

which means that $V^2/2g + h$ is constant along a streamline.

b. For an irrotational flow,

$$\frac{\partial u}{\partial y} = \frac{\partial v}{\partial x}$$

Substituting this equation into Euler's equation, show that

$$\frac{\partial}{\partial x}\left(\frac{V^2}{2g} + h\right) = 0$$

$$\frac{\partial}{\partial y}\left(\frac{V^2}{2g} + h\right) = 0$$

which means that $V^2/2g + h$ is constant in all directions.

Control Volume Approach and Continuity Equation

The engineer can find flow properties (pressure and velocity) in a flow field in one of two ways. One approach is to generate a series of pathlines or streamlines through the field and determine flow properties at any point along the lines by applying Euler's equation or the Bernoulli equation developed in Chapter 4. This is called the Lagrangian approach. The other way is to solve a set of equations for flow properties at any point in the flow field. This is called the Eulerian approach. Both approaches are discussed in this chapter. The foundational concepts for the Eulerian approach, or control volume approach, are developed and applied to the conservation of mass. This leads to the continuity equation, a fundamental and widely used equation in fluid mechanics.

Lagrangian and Eulerian Approach

Engineers are often concerned about evaluating the pressure and velocities at arbitrary locations in a flow field. They may need to find the local velocity around a bridge pier to assess the possibility of erosion. Or they may need to evaluate the lowest pressure points in a flow

Figure 5.1

The Lagrangian and Eulerian approaches for quantifying the flow field in sudden contraction.

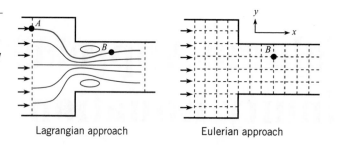

Lagrangian approach   Eulerian approach

field to determine whether local boiling (cavitation) may occur. In the design of lifting surfaces, such as airfoils or hydrofoils, they may be interested in the pressure gradient to predict the onset of separation. It is essential to have some scheme to find flow properties at a point in a flow field. The Lagrangian and Eulerian descriptions of the flow field represent two different ways to obtain this information.

The flow into a sudden contraction, shown in Fig. 5.1, can be used to illustrate the difference between the Lagrangian and Eulerian approach to quantify flow properties in a flow field. It is desired to evaluate the pressure at point B. The pressure and velocity are known at the inlet. One approach is to locate the pathline that starts at the inlet, point A, and passes through point B as shown in Fig. 5.1. As the fluid particle moves along this pathline, the pressure changes with velocity according to Euler's equation.* Integrating Euler's equation from A to B would yield the pressure at point B. If the flow is steady, the Bernoulli equation could be used between the two points. This is called the *Lagrangian approach*. Obviously, it is an enormous task to keep track of all the pathlines required to evaluate the flow properties at a given point in the flow field. The problem is further compounded for unsteady flows where different pathlines will pass through the same point at different times.

The other way to describe the flow is to develop a solution to the flow field that provides the flow properties at any point. Thus if the pressure were available as a function of location, $p(x,y)$, then the pressure at point B in Fig. 5.1 would simply be obtained by substituting in the values of the coordinates at that point. Solving the fluid flow equations to yield the flow properties at any point in the field is known as the *Eulerian approach.*

In order to use the Eulerian approach, the basic equations must be recast in Eulerian form. In solid body mechanics, the fundamental equations are developed using the "free body" concept in which an element in the field is isolated and the effect of the surroundings is replaced by forces acting on the surface of element. An analogous approach is used in fluid mechanics as shown in Fig. 5.2. The volume enclosing a point is identified as a "control volume." The effects of the surroundings are replaced by forces due to pressure and shear stress acting on the surface of the control volume. In addition to the forces like those applied to the "free body," there is a flow through the control volume that has to be taken into account. Because of the use of a control volume, the Eulerian approach is often called the *control volume approach*.

The basic equations, such as Newton's second law and the energy equation, are applicable for a body of given mass, or a fluid particle, moving in the field; in other words, they are in Lagrangian form. This chapter introduces a procedure for converting the Lagrangian form to the Eulerian form so they can be applied to a control volume. In the limit as the control volume approaches zero volume, the Eulerian forms of the differential equations are derived. The solution of these equations provides the Eulerian description of the flow field.

* If viscous effects were important, Euler's equation would have to be extended to include such effects.

Figure 5.2

A control volume in a flow field.

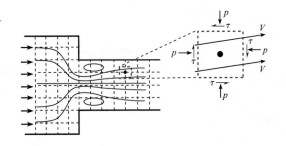

The Eulerian form of the equations also provides the basic equations for numerical solutions of the flow field, commonly known as CFD (computational fluid dynamics). The equations are written as a set of algebraic equations that are solved by various numerical schemes and provide flow properties at discrete points in the field. CFD is used extensively for industrial design.

5.1

Rate of Flow

In order to develop the Eulerian, or control volume approach, there is a need to be able to calculate the flow rates through a control volume. Also, the capability to calculate flow rates is important in analyzing water supply systems, natural gas distribution networks, and river flows. The equations for calculating flow rates are developed in this section.

Discharge

The *discharge*, Q, often called the *volume flow rate*, is the volume of fluid that passes through an area per unit time. For example, when filling the gas tank of an automobile, the discharge or volume flow rate would be the gallons per minute flowing through the nozzle. Typical units for discharge are ft^3/s (cfs), ft^3/min (cfm), gpm, m^3/s, and L/s.

The discharge or volume flow rate in a pipe is related to the flow velocity and cross-sectional area. Consider the idealized flow of fluid in a pipe as shown in Fig. 5.3 in which the velocity is constant across the pipe section. Suppose a marker is injected over the cross section at section *A-A* for a period of time Δt. The fluid that passes *A-A* in time Δt is represented by the marked volume. The length of the marked volume is $V\Delta t$ so the volume is $\Delta \mathcal{V} = AV\Delta t$, where A is the cross-sectional area of the pipe. The volume flow per unit time past *A-A* is

$$\frac{\Delta \mathcal{V}}{\Delta t} = VA$$

Taking the limit as $\Delta t \longrightarrow 0$ gives

$$Q = \lim_{\Delta t \to 0} \frac{\Delta \mathcal{V}}{\Delta t} = VA \tag{5.1}$$

which will be referred to as the *discharge* or *volume flow rate equation*. It is important to realize that discharge refers to a volume flow rate.

The discharge given by Eq. (5.1) is based on a constant flow velocity over the cross-sectional area. In general, the velocity varies across the section such as shown in Fig. 5.4.

Figure 5.3

Volume of fluid in flow with uniform velocity distribution that passes section A-A in time Δt.

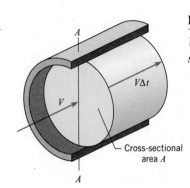

Cross-sectional area A

Figure 5.4

Volume of fluid that passes section A-A in time Δt.

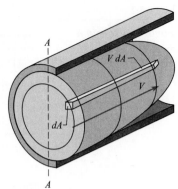

The volume flow rate through a differential area of the section is $V\,dA$, and the total volume flow rate is obtained by integration over the entire cross-section:

$$Q = \int_A V\,dA \tag{5.2}$$

In many problems—for example, those involving flow in pipes—one may know the discharge and need to find the mean (average) velocity without knowing the actual velocity distribution across the pipe section. By definition, the *mean velocity* is the discharge divided by the cross-sectional area,

$$\overline{V} = \frac{Q}{A} \tag{5.3}$$

For laminar flows in circular pipes, the velocity profile is parabolic like the case shown in Fig. 5.4. In this case, the mean velocity is half the centerline velocity. However, for turbulent pipe flow as shown in Fig. 4.7b, the time-averaged velocity profile is nearly uniformly distributed across the pipe, so the mean velocity is fairly close to the velocity at the pipe center. It is customary to leave the bar off the velocity symbol and simply indicate the mean velocity with V.

The volume flow rate equation can be generalized by using the concept of the dot product. In Fig. 5.5 the flow velocity vector is not normal to the surface but is oriented at an angle θ with respect to the direction that is normal to the surface. The only component of velocity that contributes to the flow through the differential area dA is the component normal to the area, V_n. The differential discharge through area dA is

$$dQ = V_n dA$$

Figure 5.5

Velocity vector oriented at angle θ with respect to normal.

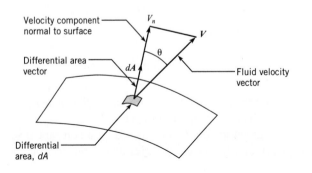

If the vector, $d\mathbf{A}$, is defined with magnitude equal to the differential area, dA, and direction normal to the surface, then $V_n dA = |\mathbf{V}| \cos \theta \, dA = \mathbf{V} \cdot d\mathbf{A}$ where $\mathbf{V} \cdot d\mathbf{A}$ is the dot product of the two vectors. Thus a more general equation for the discharge or volume flow rate through a surface A is

$$Q = \int_A \mathbf{V} \cdot d\mathbf{A} \qquad (5.4)$$

If the velocity is constant over the area and the area is a planar surface, then the discharge is given as

$$Q = \mathbf{V} \cdot \mathbf{A}$$

If, in addition, the velocity and area vectors are aligned, then

$$Q = VA$$

which reverts to the original equation developed for discharge, Eq. (5.1).

Mass Flow Rate

The *mass flow rate*, $\dot{m}$, is the mass of fluid passing through a cross-sectional area per unit time. The common units for mass flow rate are kg/s, lbm/s, and slugs/s. Using the same approach as for volume flow rate, the mass of the fluid in the marked volume in Fig. 5.3 is $\Delta m = \rho \Delta V\!\!\!-$, where ρ is the average density. The *mass flow rate equation* is

$$\dot{m} = \lim_{\Delta t \to 0} \frac{\Delta m}{\Delta t} = \rho \lim_{\Delta t \to 0} \frac{\Delta V\!\!\!-}{\Delta t} = \rho Q$$
$$= \rho A V \qquad (5.5)$$

The generalized form of the mass flow equation corresponding to Eq. (5.4) is

$$\dot{m} = \int_A \rho \mathbf{V} \cdot d\mathbf{A} \qquad (5.6)$$

where both the velocity and fluid density can vary over the cross-sectional area. If the density is constant, then Eq. (5.5) is recovered. Also if the velocity vector is aligned with the area vector, such as integrating over the cross-sectional area of a pipe, the mass flow equation reduces to

$$\dot{m} = \int_A \rho V dA \qquad (5.7)$$

In summary, Eqs. (5.1) to (5.7) can be combined to create several useful formulas for volume flow rate (discharge):

$$Q = \bar{V}A = \frac{\dot{m}}{\rho} = \int_A V \, dA = \int_A \mathbf{V} \, d\mathbf{A} \qquad (5.8)$$

Useful formulas for mass flow rate are:

$$\dot{m} = \rho A \bar{V} = \rho Q = \int_A \rho V dA = \int_A \rho \mathbf{V} \, d\mathbf{A} \qquad (5.9)$$

The equations for discharge and mass flow rate are summarized in Table F.2.

Example 5.1 shows how to calculate the discharge and mean velocity using the mass flow rate, fluid density, and pipe diameter.

EXAMPLE 5.1 VOLUME FLOW RATE AND MEAN VELOCITY

Air that has a mass density of $1.24 \text{ kg}/\text{m}^3$ ($0.00241 \text{ slugs}/\text{ft}^3$) flows in a pipe with a diameter of 30 cm (0.984 ft) at a mass rate of flow of $3 \text{ kg}/\text{s}$ ($0.206 \text{ slugs}/\text{s}$). What are the mean velocity and discharge in this pipe for both systems of units?

Problem Definition

Situation: Airflow in pipe with 30 cm diameter at 3 kg/s.

Find:

1. Discharge (m^3/s and ft^3/s).
2. Mean velocity (m/s and ft/s).

Assumptions: Properties are uniformly distributed across section.

Properties: $\rho = 1.24 \text{ kg}/\text{m}^3$ ($0.00241 \text{ slugs}/\text{ft}^3$).

Plan

1. Find the volume flow rate using the volume flow rate equation, Eq. (5.5).

2. Calculate the mean velocity using Eq. (5.3).

Solution

1. Discharge:

$$Q = \frac{\dot{m}}{\rho} = \frac{3 \text{ kg/s}}{1.24 \text{ kg/m}^3} = \boxed{2.42 \text{ m}^3/\text{s}}$$

$$Q = 2.42 \text{ m}^3/\text{s} \times \left(\frac{35.31 \text{ ft}^3}{1 \text{ m}^3} \right) = \boxed{85.5 \text{ cfs}}$$

2. Mean velocity

$$V = \frac{Q}{A} = \frac{2.42 \text{ m}^3/\text{s}}{(\frac{1}{4}\pi) \times (0.30 \text{ m})^2} = \boxed{34.2 \text{ m}/\text{s}}$$

$$V = 34.2 \text{ m/s} \times \left(\frac{1 \text{ ft}}{0.3048 \text{ m}} \right) = \boxed{112 \text{ ft/s}}$$

Example 5.2 shows how to evaluate the discharge when the velocity vector is not normal to the cross section area by using the dot product.

Example 5.3 illustrates how to evaluate the volume flow rate for a nonuniform velocity distribution by integration.

EXAMPLE 5.2 FLOW IN SLOPING CHANNEL

Water flows in a channel that has a slope of $30°$. If the velocity is assumed to be constant, 12 m/s, and if a depth of 60 cm is measured along a vertical line, what is the discharge per meter of width of the channel?

Sketch:

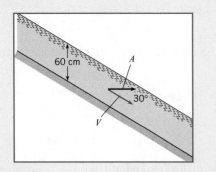

Problem Definition

Situation: Channel slope of $30°$. Velocity is 12 m/s and vertical depth is 60 cm.

Find: Discharge per meter width (m^2/s).

Assumptions: Velocity is uniformly distributed across channel.

Plan

Use Eq. (5.7) with area based on 1 meter width.

Solution

$$Q = \mathbf{V} \cdot \mathbf{A} = V\cos 30 \times A$$
$$= 12 \text{ m/s} \times \cos 30° \times 0.6 \text{ m}$$
$$= \boxed{6.24 \text{ m}^3/\text{s per meter}}$$

Review

The discharge per unit width is usually designated as q.

EXAMPLE 5.3   DISCHARGE IN CHANNEL WITH NON-UNIFORM VELOCITY DISTRIBUTION

The water velocity in the channel shown in the accompanying figure has a distribution across the vertical section equal to $u/u_{max} = (y/d)^{1/2}$. What is the discharge in the channel if the water is 2 m deep ($d = 2$ m), the channel is 5 m wide, and the maximum velocity is 3 m/s?

Problem Definition

Situation: Water flows in a 2 m by 5 m channel with a given velocity distribution.

Find: Discharge (in m^3/s).

Sketch:

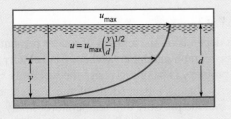

Plan

Find the discharge by using Eq. (5.2).

Solution

Discharge equation

$$Q = \int_0^d u \, dA$$

Channel is 5 m wide, so differential area is $dA = 5dy$. Using given velocity distribution,

$$Q = \int_0^2 u_{max}(y/d)^{1/2} 5 \, dy$$

$$= \frac{5u_{max}}{d^{1/2}} \int_0^2 y^{1/2} \, dy$$

$$= \frac{5u_{max}}{d^{1/2}} \frac{2}{3} y^{3/2} \Big|_0^2$$

$$= \frac{5 \times 3}{2^{1/2}} \times \frac{2}{3} \times 2^{3/2} = \boxed{20 \text{ m}^3/\text{s}}$$

5.2

Control Volume Approach

The control volume (or Eulerian) approach is the method whereby a volume in the flow field is identified and the governing equations are solved for the flow properties associated with this volume. A scheme is needed that allows one to rewrite the equations for a moving fluid particle in terms of flow through a control volume. Such a scheme is the Reynolds transport theorem introduced in this section. This is a very important theorem because it is used to derive many of the basic equations used in fluid mechanics.

System and Control Volume

A *system* is a continuous mass of fluid that always contains the same matter. A system moving through a flow field is shown in Fig. 5.6. The shape of the system may change with time, but the mass is constant since it always consists of the same matter. The fundamental equations, such as Newton's second law and the first law of thermodynamics, apply to a system.

A *control volume* is volume located in space and through which matter can pass, as shown in Fig. 5.6. The system can pass through the control volume. The selection of the control volume position and shape is problem-dependent. The control volume is enclosed by the control surface as shown in Fig. 5.6. Fluid mass enters and leaves the control volume through the control surface. The control volume can deform with time as well as move and rotate in space and the mass in the control volume can change with time.

Figure 5.6

System, control surface, and control volume in a flow field.

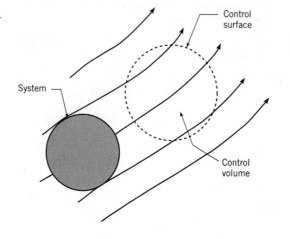

Intensive and Extensive Properties

An *extensive property* is any property that depends on the amount of matter present. The extensive properties of a system include mass, m, momentum, $m\mathbf{v}$ (where $\mathbf{v}$ is velocity), and energy, E. Another example of an extensive property is weight because the weight is mg.

An *intensive property* is any property that is independent of the amount of matter present. Examples of intensive properties include pressure and temperature. Many intensive properties are obtained by dividing the extensive property by the mass present. The intensive property for momentum is velocity $\mathbf{v}$, and for energy is e, the energy per unit mass. The intensive property for weight is g.

In this section an equation for a general extensive property, B, will be developed. The corresponding intensive property will be b. The amount of extensive property B contained in a control volume at a given instant is

$$B_{cv} = \int_{cv} b\,dm = \int_{cv} b\rho\,d\mathcal{V} \qquad (5.10)$$

where dm and $d\mathcal{V}$ are the differential mass and differential volume, respectively, and the integral is carried out over the control volume.

Property Transport Across the Control Surface

When fluid flows across a control surface, properties such as mass, momentum, and energy are transported with the fluid either into or out of the control volume. Consider the flow through the control volume in the duct in Fig. 5.7. If the velocity is uniformly distributed across the control surface, the mass flow rate through each cross section is given by

$$\dot{m}_1 = \rho_1 A_1 V_1 \qquad \dot{m}_2 = \rho_2 A_2 V_2$$

The net mass flow rate out* of the control volume, that is, the outflow rate minus the inflow rate, is

$$\text{net mass outflow rate} = \dot{m}_2 - \dot{m}_1 = \rho_2 A_2 V_2 - \rho_1 A_1 V_1$$

The same control volume is shown in Fig. 5.8 with each control surface area represented by a vector, $\mathbf{A}$, oriented outward from the control volume and with magnitude equal to the

---

* Another term used often for net flow rate out is net efflux.

Figure 5.7

Flow through control volume in a duct.

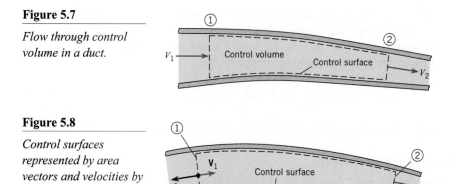

Figure 5.8

Control surfaces represented by area vectors and velocities by velocity vectors.

cross-sectional area. The velocity is represented by a vector, **V**. Taking the dot product of the velocity and area vectors at both stations gives

$$V_1 \cdot A_1 = -V_1 A_1 \qquad V_2 \cdot A_2 = V_2 A_2$$

because at station 1 the velocity and area have the opposite directions while at station 2 the velocity and area vectors are in the same direction. Now the net mass outflow rate can be written as

$$\text{net mass outflow rate} = \rho_2 V_2 A_2 - \rho_1 V_1 A_1$$

$$= \rho_2 \mathbf{V}_2 \cdot \mathbf{A}_2 + \rho_1 \mathbf{V}_1 \cdot \mathbf{A}_1$$

$$= \sum_{cs} \rho \mathbf{V} \cdot \mathbf{A} \tag{5.11}$$

Equation (5.11) states that if the dot product $\rho V \cdot A$ is summed for all flows into and out of the control volume, the result is the net mass flow rate out of the control volume, or the net mass efflux. If the summation is positive, the net mass flow rate is out of the control volume. If it is negative, the net mass flow rate is into the control volume. If the inflow and outflow rates are equal, then

$$\sum_{cs} \rho V \cdot A = 0$$

In a similar manner, to obtain the net rate of flow of an extensive property B out of the control volume, the mass flow rate is multiplied by the intensive property b:

$$\dot{B}_{\text{net}} = \sum_{cs} b \overbrace{\rho V \cdot A}^{\dot{m}} \tag{5.12}$$

To reinforce the validity of Eq. (5.12) one may consider the dimensions involved. Equation (5.12) states that the flow rate of B is given by

$$\overbrace{\left(\frac{B}{\text{mass}}\right)}^{b}\overbrace{\left(\frac{\text{mass}}{\text{time}}\right)}^{\dot{m}} = \left(\frac{B}{\text{time}}\right) = \dot{B}$$

Equation (5.12) is applicable for all flows where the properties are uniformly distributed across the area. If the properties vary across a flow section, then it becomes necessary to integrate across the section to obtain the rate of flow. A more general expression for the net rate of flow of the extensive property from the control volume is thus

$$\dot{B}_{net} = \int_{cs} b\rho V \cdot dA \tag{5.13}$$

Equation (5.13) will be used in the most general form of the Reynolds transport theorem.

Reynolds Transport Theorem

The Reynolds transport theorem, fundamental to the control volume approach, is developed in this section. It relates the Eulerian and Lagrangian approaches. The Reynolds transport theorem is derived by considering the rate of change of an extensive property of a system as it passes through a control volume.

A control volume with a system moving through it is shown in Fig. 5.9. The control volume is enclosed by the control surface identified by the dashed line. The system is identified by the darker shaded region. At time t the system consists of the material inside the control volume and the material going in, so the property B of the system at this time is

$$B_{sys}(t) = B_{cv}(t) + \Delta B_{in} \tag{5.14}$$

At time $t + \Delta t$ the system has moved and now consists of the material in the control volume and the material passing out, so B of the system is

$$B_{sys}(t + \Delta t) = B_{cv}(t + \Delta t) + \Delta B_{out} \tag{5.15}$$

The rate of change of the property B is

$$\frac{dB_{sys}}{dt} = \lim_{\Delta t \to 0}\left[\frac{B_{sys}(t + \Delta t) - B_{sys}(t)}{\Delta t}\right] \tag{5.16}$$

Substituting in Eqs. (5.14) and (5.15) results in

$$\frac{dB_{sys}}{dt} = \lim_{\Delta t \to 0}\left[\frac{B_{cv}(t + \Delta t) - B_{cv}(t) + \Delta B_{out} - \Delta B_{in}}{\Delta t}\right] \tag{5.17}$$

Rearranging terms yields

$$\frac{dB_{sys}}{dt} = \lim_{\Delta t \to 0}\left[\frac{B_{cv}(t + \Delta t) - B_{cv}(t)}{\Delta t}\right] + \lim_{\Delta t \to 0}\frac{\Delta B_{out}}{\Delta t} - \lim_{\Delta t \to 0}\frac{\Delta B_{in}}{\Delta t} \tag{5.18}$$

Figure 5.9

Progression of a system through a control volume.

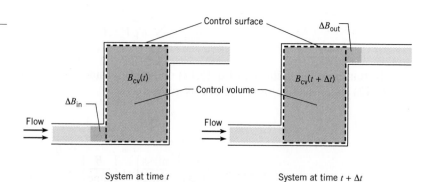

System at time t System at time $t + \Delta t$

The first term on the right side of Eq. (5.18) is the rate of change of the property B inside the control volume, or

$$\lim_{\Delta t \to 0} \left[\frac{B_{cv}(t + \Delta t) - B_{cv}(t)}{\Delta t} \right] = \frac{dB_{cv}}{dt} \qquad (5.19)$$

The remaining terms are

$$\lim_{\Delta t \to 0} \frac{\Delta B_{out}}{\Delta t} = \dot{B}_{out} \qquad \text{and} \qquad \lim_{\Delta t \to 0} \frac{\Delta B_{in}}{\Delta t} = \dot{B}_{in}$$

These two terms can be combined to give

$$\dot{B}_{net} = \dot{B}_{out} - \dot{B}_{in} \qquad (5.20)$$

or the net efflux, or net outflow rate, of the property B through the control surface. Equation (5.18) can now be written as

$$\frac{dB_{sys}}{dt} = \frac{d}{dt} B_{cv} + \dot{B}_{net}$$

Substituting in Eq. (5.13) for $\dot{B}_{net}$ and Eq. (5.10) for B_{cv} results in the most general form of the *Reynolds transport theorem*:

$$\underbrace{\frac{dB_{sys}}{dt}}_{\text{Lagrangian}} = \underbrace{\frac{d}{dt} \int_{cv} b\rho \, d\mathcal{V} + \int_{cs} b\rho V \cdot dA}_{\text{Eulerian}} \qquad (5.21)$$

This equation may be expressed in words as

$$\left\{ \begin{array}{c} \text{Rate of change} \\ \text{of property } B \\ \text{of system} \end{array} \right\} = \left\{ \begin{array}{c} \text{Rate of change} \\ \text{of property } B \\ \text{in control volume} \end{array} \right\} + \left\{ \begin{array}{c} \text{Net outflow} \\ \text{of property } B \\ \text{through control surface} \end{array} \right\}$$

The left side of the equation is the Lagrangian form; that is, the rate of change of property B evaluated moving with the system. The right side is the Eulerian form; that is, the change of property B evaluated in the control volume and the flux measured at the control surface. This equation applies at the instant the system occupies the control volume and provides the connection between the Lagrangian and Eulerian descriptions of fluid flow. The application of this equation is called the *control volume approach*. The velocity V is always measured with respect to the control surface because it relates to the mass flux across the surface.

A simplified form of the Reynolds transport theorem can be written if the mass crossing the control surface occurs through a number of inlet and outlet ports, and the velocity, density and intensive property b are uniformly distributed (constant) across each port. Then

$$\frac{dB_{sys}}{dt} = \int_{cv} b\rho \, d\mathcal{V} + \sum_{cs} \rho b V \cdot A \qquad (5.22)$$

where the summation is carried out for each port crossing the control surface.

An alternative form can be written in terms of the mass flow rates:

$$\frac{dB_{sys}}{dt} = \int_{cv} \rho b \, d\Psi + \sum_{cs} \dot{m}_o b_o - \sum_{cs} \dot{m}_i b_i \qquad (5.23)$$

where the subscripts i and o refer to the inlet and outlet ports, respectively, located on the control surface. This form of the equation does not require that the velocity and density be uniformly distributed across each inlet and outlet port, but the property b must be.

Continuity Equation

The continuity equation derives from the conservation of mass, which, in Lagrangian form, simply states that the mass of the system is constant.

$$m_{sys} = \text{constant}$$

The Eulerian form is derived by applying the Reynolds transport theorem. In this case the extensive property of the system is its mass, $B_{cv} = m_{sys}$. The corresponding value for b is the mass per unit mass, or simply, unity.

$$b = \frac{m_{sys}}{m_{sys}} = 1$$

General Form of the Continuity Equation

The general form of the continuity equation is obtained by substituting the properties for mass into the Reynolds transport theorem, Eq. (5.21), resulting in

$$\frac{dm_{sys}}{dt} = \frac{d}{dt}\int_{cv} \rho \, d\Psi + \int_{cs} \rho V \cdot dA$$

However, $dm_{sys}/dt = 0$, so the general, or integral, form of the *continuity equation* is

$$\frac{d}{dt}\int_{cv} \rho \, d\Psi + \int_{cs} \rho V \cdot dA = 0 \qquad (5.24)$$

This equation can be expressed in words as

$$\left\{ \begin{array}{c} \text{The accumulation rate} \\ \text{of mass in the} \\ \text{control volume} \end{array} \right\} + \left\{ \begin{array}{c} \text{The net outflow rate} \\ \text{of mass through} \\ \text{the control surface} \end{array} \right\} = 0$$

If the mass crosses the control surface through a number of inlet and exit ports, the continuity equation simplifies to

$$\frac{d}{dt}m_{cv} + \sum_{cs} \dot{m}_o - \sum_{cs} \dot{m}_i = 0 \qquad (5.25)$$

where m_{cv} is the mass of fluid in the control volume. Note that the two summation terms represent the net mass outflow through the control surface.

Example 5.4 shows an application of the continuity equation to calculating the mass accumulation rate in a tank.

EXAMPLE 5.4 MASS ACCUMULATION IN A TANK

A jet of water discharges into an open tank, and water leaves the tank through an orifice in the bottom at a rate of 0.003 m^3/s. If the cross-sectional area of the jet is 0.0025 m^2 where the velocity of water is 7 m/s, at what rate is water accumulating in (or evacuating from) the tank?

Problem Definition

Situation: Jet of water (7 m/s at 0.0025 m^2) entering tank and water leaving at 0.003 m^3/s through orifice.

Find: Rate of accumulation (or evacuation) in tank (kg/s).

Sketch:

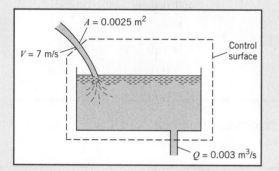

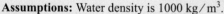

Assumptions: Water density is 1000 kg/m^3.

Plan

A control volume is drawn around the tank as shown. There is one inlet and one outlet.

1. Develop equation for accumulation rate by applying the continuity equation, Eq. (5.25).

2. Analyze equation term by term.
3. Calculate the accumulation rate.

Solution

1. Continuity equation

$$\frac{d}{dt}m_{cv} + \sum_{cs} \dot{m}_o - \sum_{cs} \dot{m}_i = 0$$

Because there is only one inlet and outlet, the equation reduces to

$$\frac{d}{dt}m_{cv} = \dot{m}_i - \dot{m}_o$$

2. Term-by-term analysis
   - The inlet mass flow rate is calculated using Eq. (5.5)
   $$\dot{m}_i = \rho VA$$
   $$= 1000 \text{ kg/m}^3 \times 7 \text{ m/s} \times 0.0025 \text{ m}^2$$
   $$= 17.5 \text{ kg/s}$$
   - Outlet flow rate is
   $$\dot{m}_o = \rho Q = 1000 \text{ kg/m}^3 \times 0.003 \text{ m}^3/s = 3 \text{ kg/s}$$

3. Accumulation rate:
   $$\frac{dm_{cv}}{dt} = 17.5 \text{ kg/s} - 3 \text{ kg/s}$$
   $$= \boxed{14.5 \text{ kg/s}}$$

Review

Note that the result is positive so water is accumulating in the tank.

The rate of water level rise in a reservoir is an often-used application of the continuity equation. Example 5.5 illustrates this application.

EXAMPLE 5.5 RATE OF WATER RISE IN RESERVOIR

A river discharges into a reservoir at a rate of 400,000 ft^3/s (cfs), and the outflow rate from the reservoir through the flow passages in the dam is 250,000 cfs. If the reservoir surface area is 40 mi^2, what is the rate of rise of water in the reservoir?

Problem Definition

Situation: Reservoir with 400,000 cfs entering and 250,000 cfs leaving. Area is 40 mi^2.

Find: Rate of water rise (ft/hr) in reservoir.

Sketch:

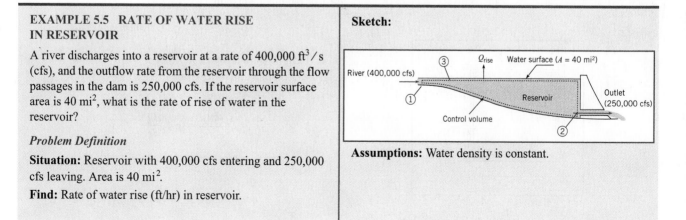

Assumptions: Water density is constant.

Plan

The control volume selected is shown in the sketch. There is an inlet from the river at location 1 and an outlet at location 2. The control surface 3 is just below the water surface and is stationary. Mass passes through control surface 3 as the water level in the reservoir rises (or falls). The mass in the control volume is constant.

1. Apply the continuity equation, Eq. (5.25).
2. Analyze term by term.
3. Evaluate rise rate using Eq. (5.1).

Solution

Continuity equation:

$$\frac{d}{dt}m_{cv} + \sum_{cs} \dot{m}_o - \sum_{cs} \dot{m}_i = 0$$

2. Term-by-term analysis:
   • Mass in the control volume is constant, so $dm_{cv}/dt = 0$.

• Inlet port 1 is river flow rate, $\sum_{cs} \dot{m}_i = \rho Q_1$.

• Outlets are reservoir surface and dam outlet,

$$\sum_{cs} \dot{m}_o = \rho Q_2 + \rho Q_{rise}.$$

Substitution of terms back into continuity equation:

$$\rho Q_2 + \rho Q_{rise} = \rho Q_1$$
$$Q_{rise} = Q_1 - Q_2$$

3. Rise rate calculation using Eq. (5.1):

$$V_{rise} = \frac{Q_{rise}}{A_3} = \frac{Q_1 - Q_2}{A_3}$$

$$V_{rise} = \frac{400,000 \text{ cfs} - 250,000 \text{ cfs}}{40 \text{ mi}^2 \times (5280 \text{ ft/mi})^2}$$

$$= 1.34 \times 10^{-4} \text{ ft/s} = \boxed{0.482 \text{ ft/hr}}$$

Example 5.6 illustrates how to predict the emptying rate of a tank. In this example, a control volume of varying size is chosen.

EXAMPLE 5.6 WATER LEVEL DROP RATE IN DRAINING TANK

A 10 cm jet of water issues from a 1 m diameter tank. Assume that the velocity in the jet is $\sqrt{2gh}$ m/s where h is the elevation of the water surface above the outlet jet. How long will it take for the water surface in the tank to drop from $h_0 = 2$ m to $h_f = 0.50$ m?

Problem Definition

Situation: Water draining by a 10 cm jet from 1 m diameter tank.
Find: Time (in seconds) to drain from depth of 2 m to 0.5 m.
Sketch:

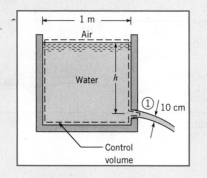

Plan

The control selected is shown in the sketch. The control surface is located at and moves with the water surface. Water crosses control surface at location 1.

1. Apply the continuity equation, Eq. (5.25).
2. Analyze term by term.
3. Solve the equation for elapsed time.
4. Calculate time to change levels.

Solution

1. Continuity equation

$$\frac{d}{dt}m_{cv} + \sum_{cs} \dot{m}_o - \sum_{cs} \dot{m}_i = 0$$

2. Term-by-term analysis
   • Accumulation rate term

$$dm_{cv} = \rho A_T dh$$

$$\frac{dm_{cv}}{dt} = \rho A_T \frac{dh}{dt}$$

where A_T is cross-sectional area of tank.

- Inlet mass flow rate with no inflow is

$$\sum_{cs} \dot{m}_i = 0$$

- Outlet mass flow rate

$$\sum_{cs} \dot{m}_o = \rho A_1 V_1 = \rho \sqrt{2gh} A_1$$

Substitution of terms in continuity equation:

$$-\rho V_1 A_1 = \frac{d(\rho A_T h)}{dt}$$

$$-\sqrt{2gh} A_1 = A_T \frac{dh}{dt}$$

3. Equation for elapsed time:
- Separating variables

$$dt = \frac{-A_T}{\sqrt{2g} A_1} \frac{dh}{\sqrt{h}} \quad \text{or} \quad dt = \frac{-A_T}{\sqrt{2g} A_1} h^{-1/2} dh$$

- Integrating

$$t = \frac{-2A_T}{\sqrt{2g} A_1} h^{1/2} + C$$

- Substituting in initial condition, $h(0) = h_0$, and final condition, $h(t) = h_f$, and solving for time

$$t = \frac{2A_T}{\sqrt{2g} A_1} (h_0^{1/2} - h_f^{1/2})$$

4. Time calculation:
- Evaluating tank and outlet areas

$$A_1 = \frac{\pi}{4}(0.10 \text{ m})^2 = 0.01\left(\frac{\pi}{4}\right) \text{m}^2$$

$$A_T = \frac{\pi}{4} D^2 = \frac{\pi}{4} \times (1 \text{ m})^2 = \frac{\pi}{4} \text{ m}^2$$

- Elapsed time

$$t = \frac{2(\pi/4) \text{ m}^2}{\sqrt{2 \times 9.81 \text{ m/s}^2}(\pi/4 \times 0.01 \text{ m}^2)}(\sqrt{2} \text{ m} - \sqrt{0.5} \text{ m})$$

$$= \boxed{31.9 \text{ s}}$$

Example 5.7 shows the application of the continuity equation to predict the time for depressurization of a tank where gas leaks out through a small hole. In this case a control volume with constant volume is selected, but the mass in the control volume changes with time as the density changes.

EXAMPLE 5.7   DEPRESSURIZATION OF GAS IN TANK

Methane escapes through a small (10^{-7} m^2) hole in a 10 m^3 tank. The methane escapes so slowly that the temperature in the tank remains constant at 23°C. The mass flow rate of methane through the hole is given by $\dot{m} = 0.66 \, pA/\sqrt{RT}$, where p is the pressure in the tank, A is the area of the hole, R is the gas constant, and T is the temperature in the tank. Calculate the time required for the absolute pressure in the tank to decrease from 500 to 400 kPa.

Problem Definition

Situation: Methane leaks through a 10^{-7} m^2 hole in 10 m^3 tank.

Find: Time (in seconds) for pressure to decrease from 500 kPa to 400 kPa.

Assumptions:

1. Gas temperatures constant at 23°C during leakage.
2. Ideal gas law is applicable.

Properties: Table A.2, $R = 518$ J/kgK.

Sketch:

Plan

Control volume selected encloses whole tank, the tank shell, and the methane in the tank. There are no inlets and only one outlet.

1. Apply continuity equation, Eq. (5.25).
2. Analyze term by term.
3. Solve equation for elapsed time.
4. Calculate time.

Solution

1. Continuity equation

$$\frac{d}{dt}m_{cv} + \sum_{cs}\dot{m}_o - \sum_{cs}\dot{m}_i = 0$$

2. Term-by-term analysis.
   - Rate of accumulation term. The mass in the control volume is the sum of the mass of the tank shell, M_{shell}, and the mass of methane in the tank,

$$m_{cv} = m_{shell} + \rho V$$

where V is the internal volume of the tank which is constant. The mass of the tank shell is constant, so

$$\frac{dm_{cv}}{dt} = V\frac{d\rho}{dt}$$

   - There is no mass inflow:

$$\sum_{cs}\dot{m}_i = 0$$

   - Mass out flow rate is

$$\sum_{cs}\dot{m}_o = 0.66\frac{pA}{\sqrt{RT}}$$

Substituting terms into continuity equation

$$V\frac{d\rho}{dt} = -0.66\frac{pA}{\sqrt{RT}}$$

3. Equation for elapsed time:
   - Use ideal gas law (Eq. 2.5) for ρ,

$$V\frac{d}{dt}\left(\frac{p}{RT}\right) = -0.66\frac{pA}{\sqrt{RT}}$$

   - Because R, T, A, and V are constant,

$$\frac{dp}{dt} = -0.66\frac{pA\sqrt{RT}}{V}$$

   or

$$\frac{dp}{p} = -0.66\frac{A\sqrt{RT}\,dt}{V}$$

   - Integrating equation and substituting limits for initial and final pressure

$$t = \frac{1.52V}{A\sqrt{RT}}\ln\frac{p_0}{p_f}$$

4. Elapsed time

$$t = \frac{1.52(10\text{ m}^3)}{(10^{-7}\text{ m}^2)\left(518\dfrac{\text{J}}{\text{kg}\cdot\text{K}}\times 300\text{ K}\right)^{1/2}}\ln\frac{500}{400} = \boxed{8.6\times 10^4\text{ s}}$$

Review

1. The time corresponds to approximately one day.
2. Since the ideal gas law is used, the pressure and temperature have to be in absolute values.

Continuity Equation for Flow in a Pipe

Several simplified forms of the continuity equation are used by engineers for flow in a pipe. The equation is developed by positioning a control volume inside a pipe, as shown in Fig. 5.10. Mass enters through station 1 and exits through station 2. The control volume is fixed to the pipe walls, and its volume is constant. If the flow is steady, then m_{cv} is constant so the mass flow formulation of the continuity equation reduces to

$$\dot{m}_2 = \dot{m}_1$$

For flow with a uniform velocity and density distribution, the continuity equation for steady flow in a pipe is

$$\rho_2 A_2 V_2 = \rho_1 A_1 V_1 \tag{5.26}$$

If the flow is incompressible, then

$$A_2 V_2 = A_1 V_1 \tag{5.27}$$

Figure 5.10

*Flow through a pipe
section.*

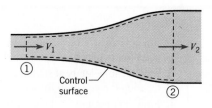

or, equivalently,

$$Q_2 = Q_1$$

This equation is valid for both steady and unsteady incompressible flow.

Equations (5.26) and (5.27) are very common forms of the continuity equation and are used in numerous applications. If the flow is not uniformly distributed, the mass flow must be calculated using Eq. (5.5).

If there are more than two ports, then the general form of the continuity equation for steady flow is

$$\sum_{cs} \dot{m}_i = \sum_{cs} \dot{m}_o \tag{5.28}$$

If the flow is incompressible, Eq. (5.28) can be written in terms of discharge:

$$\sum_{cs} Q_i = \sum_{cs} Q_o \tag{5.29}$$

Example 5.8 illustrates the application of the continuity equation to calculate velocity in a variable-area pipe.

The venturimeter is a device commonly used in engineering applications. Example 5.9 illustrates the application of both the Bernoulli equation and the continuity equation to the venturimeter, which is used routinely in industry to measure flow rates.

EXAMPLE 5.8 VELOCITY IN A VARIABLE-AREA PIPE

A 120 cm pipe is in series with a 60 cm pipe. The speed of the water in the 120 cm pipe is 2 m/s. What is the water speed in the 60 cm pipe?

Problem Definition

Situation: Two pipes connected in series.
Find: Velocity in 60 cm pipe.
Sketch:

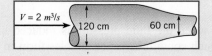

Plan

Flow rate is the same for each section, $Q_{120} = Q_{60}$. Use Eq. (5.27) to calculate velocity in the 60 cm pipe.

Solution

Equation (5.27) for V_{60}

$$V_{60} = V_{120}\frac{A_{120}}{A_{60}}$$

Calculation for V_{60}:

$$V_{60} = 2\,\text{m/s} \times \frac{(120\ \text{cm})^2}{(60\ \text{cm})^2} = \boxed{8\ \text{m/s}}$$

EXAMPLE 5.9 WATER FLOW THROUGH A VENTURIMETER

Water with a density of 1000 kg/m^3 flows through a vertical venturimeter as shown. A pressure gage is connected across two taps in the pipe (station 1) and the throat (station 2). The area ratio $A_{\text{throat}}/A_{\text{pipe}}$ is 0.5. The velocity in the pipe is 10 m/s. Find the pressure difference recorded by the pressure gage. Assume the flow has a uniform velocity distribution and that viscous effects are not important.

Problem Definition

Situation: Water flow in venturimeter with gage connected between upstream and throat. Area ratio is 0.5 and pipe velocity is 10 m/s.

Find: Pressure difference measured by gage.

Assumptions:

1. Velocity distribution is uniform.
2. Viscous effects are unimportant.

Properties: $\rho = 1000$ kg/m^3.

Sketch:

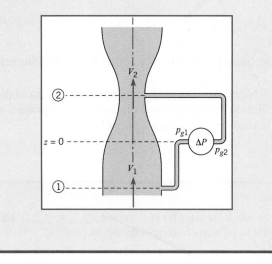

Plan

1. Since viscous effects unimportant, apply the Bernoulli equation between stations 1 and 2.
2. Find mean velocity at station 2 by applying Eq. (5.27), and develop the equation for piezometric pressure.
3. Find the pressure on the gage by applying the hydrostatic equation, Eq. (3.7a).

Solution

1. The Bernoulli equation

$$p_1 + \gamma z_1 + \rho \frac{V_1^2}{2} = p_2 + \gamma z_2 + \rho \frac{V_2^2}{2}$$

Rewrite the equation in terms of piezometric pressure.

$$p_{z_1} - p_{z_2} = \frac{\rho}{2}(V_2^2 - V_1^2)$$

$$= \frac{\rho V_1^2}{2}\left(\frac{V_2^2}{V_1^2} - 1\right)$$

2. Continuity equation $V_2/V_1 = A_1/A_2$

$$p_{z_1} - p_{z_2} = \frac{\rho V_1^2}{2}\left(\frac{A_1^2}{A_2^2} - 1\right)$$

$$= \frac{1000 \text{ kg/m}^3}{2} \times (10 \text{ m/s})^2 \times (2^2 - 1)$$

$$= 150 \text{ kPa}$$

3. Gage is located at zero elevation. Apply hydrostatic equation through static fluid in gage line between gage attachment point where the pressure is p_{g_1} and station 1 where the gage line is tapped into the pipe,

$$p_{z_1} = p_{g_1}$$

Also $p_{z_2} = p_{g_2}$ so

$$\Delta p_{\text{gage}} = p_{g_1} - p_{g_2} = p_{z_1} - p_{z_2} = \boxed{150 \text{ kPa}}$$

5.4

Cavitation

Cavitation is the phenomenon that occurs when the fluid pressure is reduced to the local vapor pressure and boiling occurs. Under such conditions vapor bubbles form in the liquid, grow, and then collapse, producing shock waves, noise, and dynamic effects that lead to decreased equipment performance and, frequently, equipment failure. Engineers are often

Figure 5.11

Flow through pipe restriction: variation of pressure for three different flow rates.

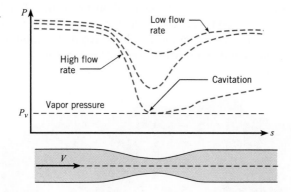

concerned about the possibility of cavitation, and they must design flow systems to avoid potential problems.

Besides its deleterious effects on machinery, cavitation can also be beneficial. Cavitation is responsible for the effectiveness of ultrasonic cleaning. Supercavitating torpedoes have been developed in which a large bubble envelops the torpedo, significantly reducing the contact area with the water and leading to significantly faster speeds. Cavitation plays a medical role in shock wave lithotripsy for the destruction of kidney stones.

Cavitation typically occurs at locations where the velocity is high. Consider the water flow through the pipe restriction shown in Fig. 5.11. The pipe area decreases, so the velocity increases according to the continuity equation and, in turn, the pressure decreases as dictated by the Bernoulli equation. For low flow rates, there is a relatively small drop in pressure at the restriction, so the water remains well above the vapor pressure and boiling does not occur. However, as the flow rate increases, the pressure at the restriction becomes progressively lower until a flow rate is reached where the pressure is equal to the vapor pressure as shown in Fig. 5.11. At this point, the liquid boils to form bubbles and cavitation ensues. The onset of cavitation can also be affected by the presence of contaminant gases, turbulence and viscosity.

The formation of vapor bubbles at the restriction is shown in Fig. 5.12a. The vapor bubbles form and then collapse as they move into a region of higher pressure and are swept downstream with the flow. When the flow velocity is increased further, the minimum pressure is still the local vapor pressure, but the zone of bubble formation is extended as shown in Fig. 5.12b. In this case, the entire vapor pocket may intermittently grow and collapse, producing serious

Figure 5.12

Formation of vapor bubbles in the process of cavitation.
(a) Cavitation.
(b) Cavitation—higher flow rate.

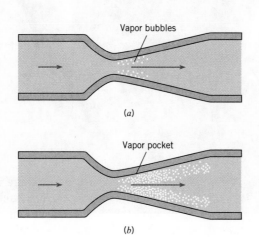

vibration problems. Severe damage that occurred on a centrifugal pump impeller is shown in Fig. 5.13, and serious erosion produced by cavitation in a spillway tunnel of Hoover Dam is shown in Fig. 5.14. Obviously, cavitation should be avoided or minimized by proper design of equipment and structures and by proper operational procedures.

Figure 5.13

Cavitation damage to impeller of a centrifugal pump.

Figure 5.14

Cavitation damage to a hydroelectric power dam spillway tunnel.

Experimental studies reveal that very high intermittent pressure, as high as 800 MPa (115,000 psi), develops in the vicinity of the bubbles when they collapse (1). Therefore, if bubbles collapse close to boundaries such as pipe walls, pump impellers, valve casings, and dam slipway floors, they can cause considerable damage. Usually this damage occurs in the form of fatigue failure brought about by the action of millions of bubbles impacting (in effect, imploding) against the material surface over a long period of time, thus producing a material pitting in the zone of cavitation.

The world's largest and most technically advanced water tunnel for studying cavitation is located in Memphis, Tennessee—the William P. Morgan Large Cavitation Tunnel. This facility is used to test large-scale models of submarine systems and full-scale torpedoes as well as applications in the maritime shipping industry.

More detailed discussions of cavitation can be found in Brennen (2) and Young (3).

Differential Form of the Continuity Equation

In the analysis of fluid flows and the development of numerical models, one of the fundamental independent equations needed is the differential form of the continuity equation. This equation is derived in this section. The derivation is accomplished by applying the integral form of the continuity equation to a small control volume and taking the limit as the volume approaches zero.

A small control volume defined by the x, y, z coordinate system is shown in Fig. 5.15. The integral form of the continuity equation, Eq. (5.24), is

$$\frac{d}{dt}\int_{\text{cv}} \rho \, d\mathcal{V} + \int_{\text{cs}} \rho V \cdot dA = 0$$

where V is the velocity measured with respect to the local control surface. Applying the Leibnetz theorem for differentiation of an integral allows the unsteady term to be expressed as

Figure 5.15

Elemental control volume.

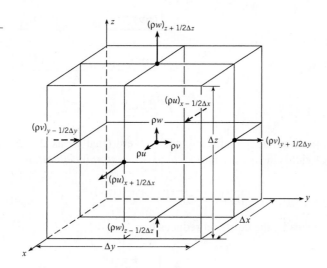

$$\frac{d}{dt}\int_{cv}\rho\, d\mathcal{V} = \int_{cv}\frac{\partial\rho}{\partial t}d\mathcal{V} + \int_{cs}\rho V_s \cdot d\mathbf{A}$$

where V_s is the local velocity of the control surface with respect to the reference frame. For a control volume with stationary sides, as shown in Fig. 5.15, $V_s = 0$, so the continuity equation for the control volume can be written as

$$\int_{cs}\rho V \cdot d\mathbf{A} + \int_{cv}\frac{\partial\rho}{\partial t}d\mathcal{V} = 0$$

Because the volume is very small (infinitesimal), one can assume that the velocity and densities are uniformly distributed across each face (control surface), and the mass flux term becomes

$$\int_{cs}\rho V \cdot d\mathbf{A} = \sum_{cs}\rho V \cdot A$$

and the continuity equation assumes the form

$$\sum_{cs}\rho V \cdot A + \int_{cv}\frac{\partial\rho}{\partial t}d\mathcal{V} = 0$$

Considering the flow rates through the six faces of the cubical element and applying those to the foregoing form of the continuity equation, results in

$$\begin{aligned}
&[(\rho u)_{x+(1/2)\Delta x}]\Delta y\Delta z - [(\rho u)_{x-(1/2)\Delta x}]\Delta y\Delta z \\
&+ [(\rho v)_{y+(1/2)\Delta y}]\Delta z\Delta x - [(\rho v)_{y-(1/2)\Delta y}]\Delta z\Delta x \\
&+ [(\rho w)_{z+(1/2)\Delta z}]\Delta x\Delta y - [(\rho w)_{z-(1/2)\Delta z}]\Delta x\Delta y \\
&+ \frac{\partial\rho}{\partial t}\Delta x\Delta y\Delta z = 0
\end{aligned} \tag{5.30}$$

Dividing Eq. (5.30) by the volume of the element ($\Delta x\Delta y\Delta z$) yields

$$\frac{[(\rho u)_{x+(1/2)\Delta x}] - [(\rho u)_{x-(1/2)\Delta x}]}{\Delta x}$$

$$+ \frac{[(\rho v)_{y+(1/2)\Delta y}] - [(\rho v)_{y-(1/2)\Delta y}]}{\Delta y}$$

$$+ \frac{[(\rho w)_{z+(1/2)\Delta z}] - [(\rho w)_{z-(1/2)\Delta z}]}{\Delta z}$$

$$+ \frac{\partial\rho}{\partial t} = 0$$

Taking the limit as the volume approaches zero (that is, as Δx, Δy, and Δz uniformly approach zero) yields the *differential form of the continuity equation*

$$\frac{\partial}{\partial x}(\rho u) + \frac{\partial}{\partial y}(\rho v) + \frac{\partial}{\partial z}(\rho w) = -\frac{\partial\rho}{\partial t} \tag{5.31}$$

If the flow is steady, the equation reduces to

$$\frac{\partial}{\partial x}(\rho u) + \frac{\partial}{\partial y}(\rho v) + \frac{\partial}{\partial z}(\rho w) = 0 \tag{5.32}$$

EXAMPLE 5.10   APPLICATION OF DIFFERENTIAL FORM OF CONTINUITY EQUATION

The expression $V = 10x\mathbf{i} - 10y\mathbf{j}$ is said to represent the velocity for a two-dimensional (planar) incompressible flow. Check to see if the continuity equation is satisfied.

Problem Definition

Situation: Velocity field is given.

Find: Determine if continuity equation is satisfied.

Plan

Reduce Eq. (5.33) to two-dimensional flow ($w = 0$ and substitute velocity components into equation).

Solution

Continuity equation for two-dimensional flow

$$\frac{\partial u}{\partial x} + \frac{\partial v}{\partial y} = 0$$

$$u = 10x; \qquad \frac{\partial u}{\partial x} = 10$$

$$v = -10y; \qquad \frac{\partial v}{\partial y} = -10$$

$$\frac{\partial u}{\partial x} + \frac{\partial v}{\partial y} = 10 - 10 = 0$$

Continuity is satisfied.

And if the fluid is incompressible, the equation further simplifies to

$$\frac{\partial u}{\partial x} + \frac{\partial v}{\partial y} + \frac{\partial w}{\partial z} = 0 \qquad (5.33)$$

for both steady and unsteady flow.

In vector notation, Eq. (5.33) is given as

$$\nabla \cdot V = 0 \qquad (5.34)$$

where ∇ is the del operator, defined as

$$\nabla = \mathbf{i}\frac{\partial}{\partial x} + \mathbf{j}\frac{\partial}{\partial y} + \mathbf{k}\frac{\partial}{\partial z}$$

Summary

There are two ways to describe a flow field, the Lagrangian and Eulerian approaches. The Eulerian, in which flow properties are available at points in the field, is the most appropriate for fluid mechanics.

Flow rate refers to either the volume per unit time or the mass per unit time passing through a surface. The volume flow rate, or discharge, is given by

$$Q = \int_A V \cdot dA$$

where dA is the vector normal to the surface with magnitude equal the differential surface area and V is the velocity vector. If the area vector and velocity vector are aligned, then

$$Q = \int_A V\, dA = \overline{V}A$$

where $\overline{V}$ is the average velocity. The corresponding mass flow rate is

$$\dot{m} = \int_A \rho V\, dA$$

If the density is uniformly distributed across the area,

$$\dot{m} = \rho Q$$

A fluid system is a given quantity of matter consisting always of the same matter. A control volume (cv) is a geometric volume defined in space and enclosed by a control surface (cs). Mass can cross the control surface.

The Reynolds transport theorem relates the time rate of change of an extensive property of a system to the rate of change of the property in the control volume plus the net outflow of the property across the control surface. It provides a link between the Lagrangian and Eulerian forms for the rate of property changes in a fluid.

The continuity equation derives from the application of the Reynolds transport theorem to the conservation of mass principle and is expressed as

$$\frac{d}{dt}\bigg|_{cv} \rho\, d\forall + \int_{cs} \rho V \cdot dA = 0$$

where V is the velocity with respect to the control surface and dA is the differential area directed outward from the control volume. An alternative form of the continuity equation is

$$\frac{d}{dt}M_{cv} = \sum \dot{m}_i - \sum \dot{m}_o$$

where M_{cv} is the mass in the control volume and $\dot{m}_i$ and $\dot{m}_o$ are the mass flow rates of flows entering and leaving the control volume, respectively.

For steady, one-dimensional flow in a pipe, the continuity equation reduces to

$$\rho_1 A_1 V_1 = \rho_2 A_2 V_2$$

where the subscripts 1 and 2 refer to the inlet and outlet of the pipe. If, in addition, the flow is incompressible, then

$$A_1 V_1 = A_2 V_2$$

The differential form of the continuity equation for incompressible flow is $\nabla \cdot V = 0$

Cavitation occurs when the pressure drops to the local vapor pressure of the liquid and bubbles appear due to liquid boiling. The presence of cavitation can cause serious equipment failures.

References

1. Knapp, R.T., J. W. Daily, and F. G. Hammitt. *Cavitation.* McGraw-Hill, New York, 1970.

2. Brennen, C. E. *Cavitation and Bubble Dynamics.* Oxford University Press, New York, 1995.

3. Young, F. R. *Cavitation.* McGraw-Hill, New York, 1989.

Problems

Flow Rate (Discharge)

5.1 PQ◀ Consider filling the gasoline tank of an automobile at a gas station. (a) Estimate the discharge in gpm. (b) Using the same nozzle, estimate the time to put 50 gallons in the tank. (c) Estimate the cross-sectional area of the nozzle and calculate the velocity at the nozzle exit.

5.2 PQ◀ The average flow rate (release) through Grand Coulee Dam is 110,000 ft³/s. The width of the river downstream of the dam is 100 yards. Making a reasonable estimate of the river velocity, estimate the river depth.

5.3 PQ◀ Taking a jar of known volume, fill it with water from your household tap and measure the time to fill. Calculate the discharge from the tap. Estimate the cross-sectional area of the faucet outlet, and calculate the water velocity issuing from the tap.

5.4 PQ◀ A liquid flows through a pipe with a constant velocity. If a pipe twice the size is used with the same velocity, will the flow rate be (a) halved, (b) doubled, (c) quadrupled. Explain.

5.5 The discharge of water in a 25 cm diameter pipe is 0.05 m^3/s. What is the mean velocity?

5.6 A pipe with a 16 in. diameter carries water having a velocity of 3 ft/s. What is the discharge in cubic feet per second and in gallons per minute (1 cfs is equivalent to 449 gpm)?

5.7 A pipe with a 2 m diameter carries water having a velocity of 4 m/s. What is the discharge in cubic meters per second and in cubic feet per second?

5.8 A pipe whose diameter is 8 cm transports air with a temperature of 20°C and pressure of 200 kPa absolute at 20 m/s. Determine the mass flow rate.

5.9 Natural gas (methane) flows at 20 m/s through a pipe with a 1 m diameter. The temperature of the methane is 15°C, and the pressure is 150 kPa gage. Determine the mass flow rate.

5.10 An aircraft engine test pipe is capable of providing a flow rate of 200 kg/s at altitude conditions corresponding to an absolute pressure of 50 kPa and a temperature of –18°C. The velocity of air through the duct attached to the engine is 240 m/s. Calculate the diameter of the duct.

5.11 A heating and air-conditioning engineer is designing a system to move 1000 m^3 of air per hour at 100 kPa abs, and 30°C. The duct is rectangular with cross-sectional dimensions of 1 m by 20 cm. What will be the air velocity in the duct?

5.12 The hypothetical velocity distribution in a circular duct is

$$\frac{v}{V_0} = 1 - \frac{r}{R}$$

where r is the radial location in the duct, R is the duct radius, and V_0 is the velocity on the axis. Find the ratio of the mean velocity to the velocity on the axis.

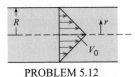

PROBLEM 5.12

5.13 Water flows in a two-dimensional channel of width W and depth D as shown in the diagram. The hypothetical velocity profile for the water is

$$V(x, y) = V_s\left(1 - \frac{4x^2}{W^2}\right)\left(1 - \frac{y^2}{D^2}\right)$$

where V_s is the velocity at the water surface midway between the channel walls. The coordinate system is as shown; x is measured from the center plane of the channel and y downward from the water surface. Find the discharge in the channel in terms of V_s, D, and W.

5.14 Water flows in a pipe that has a 4 ft diameter and the following hypothetical velocity distribution: The velocity is maximum at the centerline and decreases linearly with r to a minimum at the pipe wall. If $V_{max} = 15$ ft/s and $V_{min} = 12$ ft/s, what is the discharge in cubic feet per second and in gallons per minute?

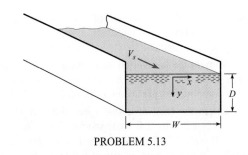

PROBLEM 5.13

5.15 In Prob. 5.14, if $V_{max} = 8$ m/s, $V_{min} = 6$ m/s, and $D = 2$ m, what is the discharge in cubic meters per second and the mean velocity?

5.16 Air enters this square duct at section 1 with the velocity distribution as shown. Note that the velocity varies in the y direction only (for a given value of y, the velocity is the same for all values of z).
a. What is the volume rate of flow?
b. What is the mean velocity in the duct?
c. What is the mass rate of flow if the mass density of the air is 1.2 kg/m^3?

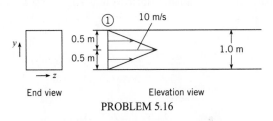

End view   Elevation view

PROBLEM 5.16

5.17 The velocity at section A-A is 18 ft/s, and the vertical depth y at the same section is 4 ft. If the width of the channel is 30 ft, what is the discharge in cubic feet per second?

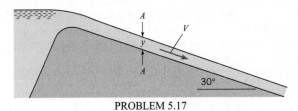

PROBLEM 5.17

5.18 The rectangular channel shown is 1.5 m wide. What is the discharge in the channel?

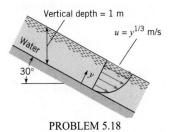

PROBLEM 5.18

5.19 If the velocity in the channel of Prob. 5.18 is given as $u = 10[\exp(y) - 1]$ m/s and the channel width is 2 m, what is the discharge in the channel and what is the mean velocity?

5.20 Water from a pipe is diverted into a weigh tank for exactly 20 min. The increased weight in the tank is 20 kN. What is the discharge in cubic meters per second? Assume $T = 20°C$.

5.21 Water enters the lock of a ship canal through 180 ports, each port having a 2 ft by 2 ft cross section. The lock is 900 ft long and 105 ft wide. The lock is designed so that the water surface in it will rise at a maximum rate of 6 ft/min. For this condition, what will be the mean velocity in each port?

5.22 An empirical equation for the velocity distribution in a horizontal, rectangular, open channel is given by $u = u_{max}(y/d)^n$, where u is the velocity at a distance y feet above the floor of the channel. If the depth d of flow is 1.2 m, $u_{max} = 3$ m/s, and $n = 1/6$, what is the discharge in cubic meters per second per meter of width of channel? What is the mean velocity?

5.23 The hypothetical water velocity in a V-shaped channel (see the accompanying figure) varies linearly with depth from zero at the bottom to maximum at the water surface. Determine the discharge if the maximum velocity is 6 ft/s.

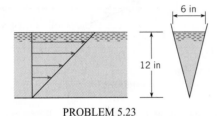

PROBLEM 5.23

5.24 The velocity of flow in a circular pipe varies according to the equation $V/V_c = (1 - r^2/r_0^2)^n$, where V_c is the centerline velocity, r_0 is the pipe radius, and r is the radial distance from the centerline. The exponent n is general and is chosen to fit a given profile ($n = 1$ for laminar flow). Determine the mean velocity as a function of V_c and n.

5.25 Plot the velocity distribution across the pipe, and determine the discharge of a fluid flowing through a pipe 1 m in diameter that has a velocity distribution given by $V = 12(1 - r^2/r_0^2)$ m/s. Here r_0 is the radius of the pipe, and r is the radial distance from the centerline. What is the mean velocity?

5.26 Water flows through a 2.0 in.–diameter pipeline at 80 lb/min. Calculate the mean velocity. Assume $T = 60°F$.

5.27 Water flows through a 20 cm pipeline at 1000 kg/min. Calculate the mean velocity in meters per second if $T = 20°C$.

5.28 Water from a pipeline is diverted into a weigh tank for exactly 15 min. The increased weight in the tank is 4765 lbf. What is the average flow rate in gallons per minute and in cubic feet per second? Assume $T = 60°F$.

5.29 A shell and tube heat exchanger consists of a one pipe inside another pipe as shown. The liquid flows in opposite directions in each pipe. If the speed of the liquid is the same in each pipe, what is the ratio of the outer pipe diameter to the inner pipe diameter if the discharge in each pipe is the same?

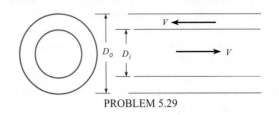

PROBLEM 5.29

5.30 The cross section of a heat exchanger consists of three circular pipes inside a larger pipe. The internal diameter of the three smaller pipes is 2.5 cm, and the pipe wall thickness is 3 mm. The inside diameter of the larger pipe is 8 cm. If the velocity of the fluid in region between the smaller pipes and larger pipe is 10 m/s, what is the discharge in m^3/s?

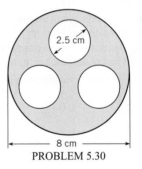

PROBLEM 5.30

5.31 The mean velocity of water in a 4 in. pipe is 10 ft/s. Determine the flow in slugs per second, gallons per minute, and cubic feet per second if $T = 60°F$.

Control Volume Approach

5.32 $\mathbb{PQ}$◄ What is a control surface and a control volume? Can mass pass through a control surface?

5.33 $\mathbb{PQ}$◄ What is the difference between an intensive and extensive property? Give an example of each.

5.34 $\mathbb{PQ}$◄ Explain the differences between the Eulerian and Lagrangian descriptions of a flow field.

5.35 $\mathbb{PQ}$◄ What are the shortcomings of describing a flow field using the Lagrangian description?

5.36 $\mathbb{PQ}$◄ What is the purpose of the Reynolds transport theorem?

5.37 Gas flows into and out of the chamber as shown. For the conditions shown, which of the following statement(s) are true of the application of the control volume equation to the continuity principle?

a. $B_{sys} = 0$

b. $dB_{\text{sys}}/dt = 0$

c. $\sum_{\text{cs}} b\rho \mathbf{V} \cdot \mathbf{A} = 0$

d. $\dfrac{d}{dt}\displaystyle\int_{\text{cv}} \rho \, d\Psi = 0$

e. $b = 0$

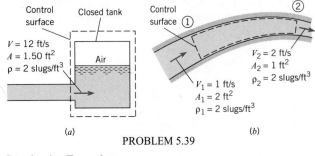

(a)   (b)

PROBLEM 5.39

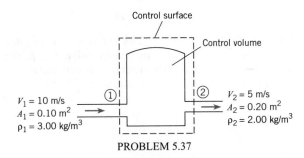

PROBLEM 5.37

5.38 The piston in the cylinder is moving up. Assume that the control volume is the volume inside the cylinder above the piston (the control volume changes in size as the piston moves). A gaseous mixture exists in the control volume. For the given conditions, indicate which of the following statements are true.

a. $\sum_{\text{cs}} \rho \mathbf{V} \cdot \mathbf{A}$ is equal to zero.

b. $\dfrac{d}{dt}\displaystyle\int_{\text{cv}} \rho \, d\Psi$ is equal to zero.

c. The mass density of the gas in the control volume is increasing with time.

d. The temperature of the gas in the control volume is increasing with time.

e. The flow inside the control volume is unsteady.

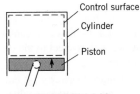

PROBLEM 5.38

5.39 For cases *a* and *b* shown in the figure, respond to the following questions and statements concerning the application of the Reynolds transport theorem to the continuity equation.

a. What is the value of b?

b. Determine the value of dB_{sys}/dt.

c. Determine the value of $\sum_{\text{cs}} b\rho \mathbf{V} \cdot \mathbf{A}$.

d. Determine the value of $d/dt \displaystyle\int_{\text{cv}} b\rho \, d\Psi$.

Continuity Equation

5.40 PQ◄ A pipe flows full with water. Is it possible for the volume flow rate into the pipe to be different than the flow rate out of the pipe? Explain.

5.41 PQ◄ Air is pumped into one end of a tube at a certain mass flow rate. Is it necessary that the same mass flow rate of air comes out the other end of the tube? Explain.

5.42 PQ◄ If an automobile tire develops a leak, how does the mass of air and density change inside the tire with time? Assuming the temperature remains constant, how is the change in density related to the tire pressure?

5.43 PQ◄ Two pipes are connected together in series. The diameter of one pipe is twice the diameter of the second pipe. With liquid flowing in the pipes, the velocity in the large pipe is 5 m/s. What is the velocity in the smaller pipe?

5.44 Both pistons are moving to the left, but piston *A* has a speed twice as great as that of piston *B*. Then the water level in the tank is (a) rising, (b) not moving up or down, or (c) falling?

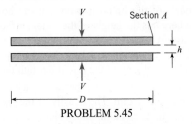

PROBLEM 5.44

5.45 Two parallel disks of diameter *D* are brought together, each with a normal speed of *V*. When their spacing is *h*, what is the radial component of convective acceleration at the section just inside the edge of the disk (section *A*) in terms of *V*, *h*, and *D*? Assume uniform velocity distribution across the section.

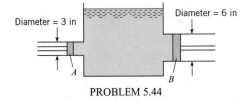

PROBLEM 5.45

5.46 Two streams discharge into a pipe as shown. The flows are incompressible. The volume flow rate of stream *A* into the pipe is given by $Q_A = 0.02t \text{ m}^3/\text{s}$ and that of stream *B* by $Q_B = 0.008t^2 \text{ m}^3/\text{s}$, where t is in seconds. The exit area of the pipe is 0.01 m². Find the velocity and acceleration of the flow at the exit at $t = 1$ s.

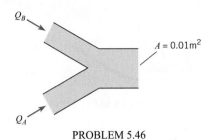

PROBLEM 5.46

5.47 Air discharges downward in the pipe and then outward between the parallel disks. Assuming negligible density change in the air, derive a formula for the acceleration of air at point A, which is a distance r from the center of the disks. Express the acceleration in terms of the constant air discharge Q, the radial distance r, and the disk spacing h. If $D = 10$ cm, $h = 0.6$ cm, and $Q = 0.380$ m³/s, what are the velocity in the pipe and the acceleration at point A where $r = 20$ cm?

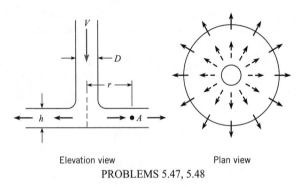

Elevation view   Plan view
PROBLEMS 5.47, 5.48

5.48 All the conditions of Prob. 5.47 are the same except that $h = 1$ cm and the discharge is given as $Q = Q_0(t/t_0)$, where $Q_0 = 0.1$ m³/s and $t_0 = 1$ s. For the additional conditions, what will be the acceleration at point A when $t = 2$ s and $t = 3$ s?

5.49 A tank has a hole in the bottom with a cross-sectional area of 0.0025 m² and an inlet line on the side with a cross-sectional area of 0.0025 m², as shown. The cross-sectional area of the tank is 0.1 m². The velocity of the liquid flowing out the bottom hole is $V = \sqrt{2gh}$, where h is the height of the water surface in the tank above the outlet. At a certain time the surface level in the tank is 1 m and rising at the rate of 0.1 cm/s. The liquid is incompressible. Find the velocity of the liquid through the inlet.

5.50 A mechanical pump is used to pressurize a bicycle tire. The inflow to the pump is 1 cfm. The density of the air entering the pump is 0.075 lbm/ft³. The inflated volume of a bicycle tire is 0.045 ft³. The density of air in the inflated tire is 0.4 lbm/ft³. How many seconds does it take to pressurize the tire if there initially was no air in the tire?

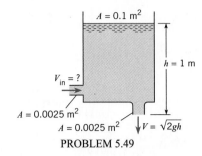

PROBLEM 5.49

5.51 A 6 in.–diameter cylinder falls at a rate of 4 ft/s in an 8 in.–diameter tube containing an incompressible liquid. What is the mean velocity of the liquid (with respect to the tube) in the space between the cylinder and the tube wall?

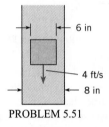

PROBLEM 5.51

5.52 This circular tank of water is being filled from a pipe as shown. The velocity of flow of water from the pipe is 10 ft/s. What will be the rate of rise of the water surface in the tank?

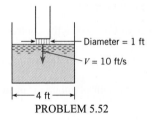

PROBLEM 5.52

5.53 A sphere 8 inches in diameter falls at 4 ft/s downward axially through water in a 1 ft–diameter container. Find the upward speed of the water with respect to the container wall at the midsection of the sphere.

5.54 A rectangular air duct 20 cm by 60 cm carries a flow of 1.44 m³/s. Determine the velocity in the duct. If the duct tapers to 10 cm by 40 cm, what is the velocity in the latter section? Assume constant air density.

5.55 A 30 cm pipe divides into a 20 cm branch and a 15 cm branch. If the total discharge is 0.30 m³/s and if the same mean velocity occurs in each branch, what is the discharge in each branch?

5.56 The conditions are the same as in Prob. 5.55 except that the discharge in the 20 cm branch is twice that in the 15 cm branch. What is the mean velocity in each branch?

5.57 Water flows in a 10 in. pipe that is connected in series with a 6 in. pipe. If the rate of flow is 898 gpm (gallons per minute), what is the mean velocity in each pipe?

5.58 What is the velocity of the flow of water in leg B of the tee shown in the figure?

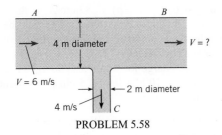

PROBLEM 5.58

5.59 For a steady flow of gas in the conduit shown, what is the mean velocity at section 2?

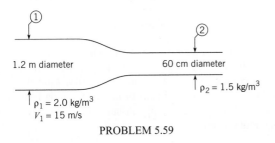

PROBLEM 5.59

5.60 Two pipes, A and B, are connected to an open water tank. The water is entering the bottom of the tank from pipe A at 10 cfm. The water level in the tank is rising at 1.0 in./min, and the surface area of the tank is 80 ft². Calculate the discharge in a second pipe, pipe B, that is also connected to the bottom of the tank. Is the flow entering or leaving the tank from pipe B?

5.61 Is the tank in the figure filling or emptying? At what rate is the water level rising or falling in the tank?

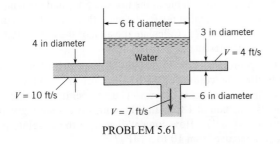

PROBLEM 5.61

5.62 Given: Flow velocities as shown in the figure and water surface elevation (as shown) at $t = 0$ s. At the end of 22 s, will the water surface in the tank be rising or falling, and at what speed?

5.63 A lake with no outlet is fed by a river with a constant flow of 1200 ft³/s. Water evaporates from the surface at a constant rate of 13 ft³/s per square mile surface area. The area varies with depth h (feet) as A (square miles) $= 4.5 + 5.5h$. What is the equilibrium depth of the lake? Below what river discharge will the lake dry up?

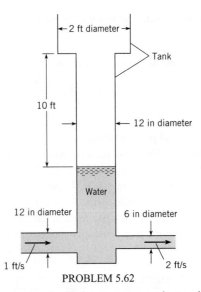

PROBLEM 5.62

5.64 A stationary nozzle discharges water against a plate moving toward the nozzle at half the jet velocity. When the discharge from the nozzle is 5 cfs, at what rate will the plate deflect water?

5.65 An open tank has a constant inflow of 20 ft³/s. A 1.0 ft–diameter drain provides a variable outflow velocity V_{out} equal to $\sqrt{(2gh)}$ ft/s. What is the equilibrium height h_{eq} of the liquid in the tank?

5.66 Assuming that complete mixing occurs between the two inflows before the mixture discharges from the pipe at C, find the mass rate of flow, the velocity, and the specific gravity of the mixture in the pipe at C.

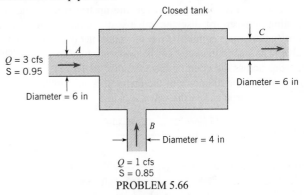

PROBLEM 5.66

5.67 Oxygen and methane are mixed at 250 kPa absolute pressure and 100°C. The velocity of the gases into the mixer is 5 m/s. The density of the gas leaving the mixer is 2.2 kg/m³. Determine the exit velocity of the gas mixture.

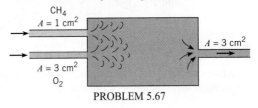

PROBLEM 5.67

5.68 A pipe with a series of holes as shown in the figure is used in many engineering systems to distribute gas into a system. The volume flow rate through each hole depends on the pressure difference across the hole and is given by

$$Q = 0.67 A_o \left(\frac{2\Delta p}{\rho}\right)^{1/2}$$

where A_o is the area of the hole, Δp is the pressure difference across the hole and ρ is the density of the gas in the pipe. If the pipe is sufficiently large, the pressure will be uniform along the pipe. A distribution pipe for air at 20° C is 0.5 meters in diameter and 10 m long. The gage pressure in the pipe is 100 Pa. The pressure outside the pipe is atmospheric at 1 bar. The hole diameter is 2.5 cm and there are 50 holes per meter length of pipe. The pressure is constant in the pipe. Find the velocity of the air entering the pipe.

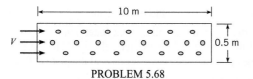

PROBLEM 5.68

5.69 The globe valve shown in the figure is a very common device to control flow rate. The flow comes through the pipe at the left and then passes through a minimum area formed by the disc and valve seat. As the valve is closed, the area for flow between the disc and valve is reduced. The flow area can be approximated by the annular region between the disc and the seat. The pressure drop across the valve can be estimated by application of the Bernoulli equation between the upstream pipe and the opening between the disc and valve seat. Assume there is a 10 gpm (gallons per minute) flow of water at 60°F through the valve. The inside diameter of the upstream pipe is 1 inch. The distance across the opening from the disc to the seat is 1/8th of an inch, and the diameter of the opening is 1/2 inch. What is the pressure drop across the valve in psid?

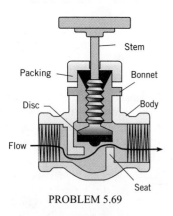

PROBLEM 5.69

5.70 In the flow through an orifice shown in the diagram the flow goes through a minimum area downstream of the orifice.

This is called the "vena contracta." The ratio of the flow area at the vena contracta to the area of the orifice is 0.64.

a. Derive an equation for the discharge through the orifice in the form $Q = CA_o(2\Delta p/\rho)^{1/2}$, where A_o is the area of the orifice, Δp is the pressure difference between the upstream flow and the vena contracta, and ρ is the fluid density. C is a dimensionless coefficient.

b. Evaluate the discharge for water at 1000 kg/m³ and a pressure difference of 10 kPa for a 1.5 cm orifice centered in a 2.5 cm diameter pipe.

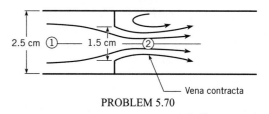

PROBLEM 5.70

5.71 A compressor supplies gas to a 10 m³ tank. The inlet mass flow rate is given by $\dot{m} = 0.5\rho_0/\rho$ (kg/s), where ρ is the density in the tank and ρ_0 is the initial density. Find the time it would take to increase the density in the tank by a factor of 2 if the initial density is 2 kg/m³. Assume the density is uniform throughout the tank.

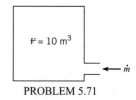

PROBLEM 5.71

5.72 A slow leak develops in a tire (assume constant volume), in which it takes 3 hr for the pressure to decrease from 30 psig to 25 psig. The air volume in the tire is 0.5 ft³, and the temperature remains constant at 60°F. The mass flow rate of air is given by $\dot{m} = 0.68\,pA/\sqrt{RT}$. Calculate the area of the hole in the tire. Atmospheric pressure is 14 psia.

5.73 Oxygen leaks slowly through a small orifice in an oxygen bottle. The volume of the bottle is 0.1 m³, and the diameter of the orifice is 0.12 mm. The temperature in the tank remains constant at 18°C, and the mass-flow rate is given by $\dot{m} = 0.68pA/\sqrt{RT}$. How long will it take the absolute pressure to decrease from 10 to 5 MPa?

5.74 How long will it take the water surface in the tank shown to drop from $h = 3$ m to $h = 50$ cm?

5.75 A cylindrical drum of water, lying on its side, is being emptied through a 2 in.–diameter short pipe at the bottom of the drum. The velocity of the water out of the pipe is $V = \sqrt{2gh}$, where g is the acceleration due to gravity and h is the height of the water surface above the outlet of the tank. The tank is 4 ft long and 2 ft in diameter. Initially the tank is half full. Find the time for the tank to empty.

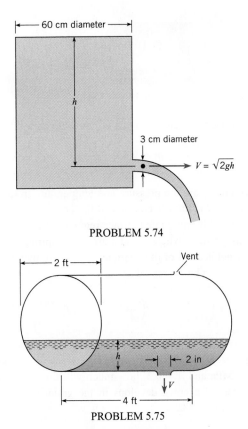

PROBLEM 5.74

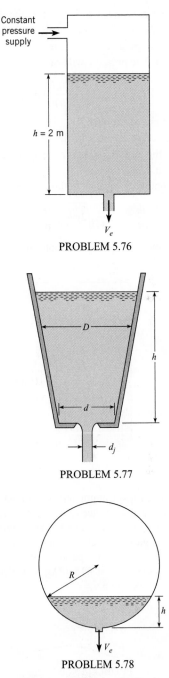

PROBLEM 5.76

PROBLEM 5.75

5.76 Water is draining from a pressurized tank as shown in the figure. The exit velocity is given by

$$V_e = \sqrt{\frac{2p}{\rho} + 2gh}$$

where p is the pressure in the tank, ρ is the water density, and h is the elevation of the water surface above the outlet. The depth of the water in the tank is 2 m. The tank has a cross-sectional area of 1 m^2, and the exit area of the pipe is 10 cm^2. The pressure in the tank is maintained at 10 kPa. Find the time required to empty the tank. Compare this value with the time required if the tank is not pressurized.

PROBLEM 5.77

5.77 For the type of tank shown, the tank diameter is given as $D = d + C_1 h$, where d is the bottom diameter and C_1 is a constant. Derive a formula for the time of fall of liquid surface from $h = h_0$ to $h = h$ in terms of d_j, d, h_0, h, and C_1. Solve for t if $h_0 = 1$ m, $h = 20$ cm, $d = 20$ cm, $C_1 = 0.3$, and $d_j = 5$ cm. The velocity of water in the liquid jet exiting the tank is $V_e = \sqrt{2gh}$.

5.78 A spherical tank with a diameter of 1 m is half filled with water. A port at the bottom of the tank is opened to drain the tank. The hole diameter is 1 cm, and the velocity of the water draining from the hole is $V_e = \sqrt{2gh}$, where h is the elevation of the water surface above the hole. Find the time required for the tank to empty.

PROBLEM 5.78

5.79 A tank containing oil is to be pressurized to decrease the draining time. The tank, shown in the figure, is 2 m in diameter and 6 m high. The oil is originally at a level of 5 m. The oil has a density of 880 kg/m^3. The outlet port has a diameter of 2 cm, and the velocity at the outlet is given by

$$V_e = \sqrt{2gh + \frac{2p}{\rho}}$$

where p is the gage pressure in the tank, ρ is the density of the oil, and h is the elevation of the surface above the hole. Assume during the emptying operation that the temperature of the air in the tank is constant. The pressure will vary as

$$p = (p_0 + p_{atm})\frac{(L - h_0)}{(L - h)} - p_{atm}$$

where L is the height of the tank, p_{atm} is the atmospheric pressure, and the subscript 0 refers to the initial conditions. The initial pressure in the tank is 300 kPa gage, and the atmospheric pressure is 100 kPa.

Applying the continuity equation to this problem, one finds

$$\frac{dh}{dt} = -\frac{A_e}{A_T}\sqrt{2gh + \frac{2p}{\rho}}$$

Integrate this equation to predict the depth of the oil with time for a period of one hour.

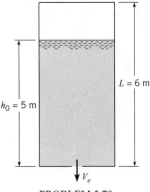

$L = 6$ m

$h_0 = 5$ m

V_e

PROBLEM 5.79

5.80 An end-burning rocket motor has a chamber diameter of 10 cm and a nozzle exit diameter of 8 cm. The density of the propellant is 1750 kg/m³, and the surface regresses at the rate of 1 cm/s. The gases crossing the nozzle exit plane have a pressure of 10 kPa abs and a temperature of 2200°C. The gas constant of the exhaust gases is 415 J/kg K. Calculate the gas velocity at the nozzle exit plane.

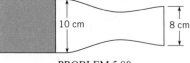

10 cm   8 cm

PROBLEM 5.80

5.81 A cylindrical-port rocket motor has a grain design consisting of a cylindrical shape as shown. The curved internal surface and both ends burn. The propellant surface regresses uniformly at 1.2 cm/s. The propellant density is 2200 kg/m³. The inside diameter of the motor is 20 cm. The propellant grain is 40 cm long and has an inside diameter of 12 cm. The diameter of the nozzle exit plane is 20 cm. The gas velocity at the exit plane is 2000 m/s. Determine the gas density at the exit plane.

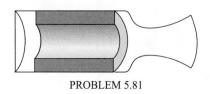

PROBLEM 5.81

5.82 The mass flow rate through a nozzle is given by

$$\dot{m} = 0.65\frac{p_c A_t}{\sqrt{RT_c}},$$

where p_c and T_c are the pressure and temperature in the rocket chamber and R is the gas constant of the gases in the chamber. The propellant burning rate (surface regression rate) can be expressed as $\dot{r} = ap_c^n$, where a and n are two empirical constants. Show, by application of the continuity equation, that the chamber pressure can be expressed as

$$p_c = \left(\frac{a\rho_p}{0.65}\right)^{1/(1-n)}\left(\frac{A_g}{A_t}\right)^{1/(1-n)}(RT_c)^{1/[2(1-n)]}$$

where ρ_p is the propellant density and A_g is the grain surface burning area. If the operating chamber pressure of a rocket motor is 3.5 MPa and $n = 0.3$, how much will the chamber pressure increase if a crack develops in the grain, increasing the burning area by 20%?

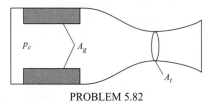

p_c A_g

A_t

PROBLEM 5.82

5.83 The piston shown is moving up during the exhaust stroke of a four-cycle engine. Mass escapes through the exhaust port at a rate given by

$$\dot{m} = 0.65\frac{p_c A_v}{\sqrt{RT_c}}$$

where p_c and T_c are the cylinder pressure and temperature, A_v is the valve opening area, and R is the gas constant of the exhaust gases. The bore of the cylinder is 10 cm, and the piston is moving upward at 30 m/s. The distance between the piston and the head is 10 cm. The valve opening area is 1 cm², the chamber pressure is 300 kPa abs, the chamber temperature is 600°C, and the gas constant is 350 J/kg K. Applying the continuity equation, determine the rate at which the gas density is changing in the cylinder. Assume the density and pressure are uniform in the cylinder and the gas is ideal.

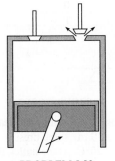

PROBLEM 5.83

5.84 The flow pattern through the pipe contraction is as indicated, and the discharge of water is 70 cfs. For $d = 2$ ft and $D = 6$ ft, what will be the pressure at point B if the pressure at point A is 3500 psf?

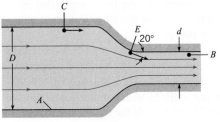

PROBLEM 5.84

5.85 Water flows through a rigid contraction section of circular pipe in which the outlet diameter is one-half the inlet diameter. The velocity of the water at the inlet varies with time as $V_{in} = (10 \text{ m/s})[1 - \exp(-t/10)]$. How will the velocity vary with time at the outlet?

5.86 The annular venturimeter is useful for metering flows in pipe systems for which upstream calming distances are limited. The annular venturimeter consists of a cylindrical section mounted inside a pipe as shown in the figure. The pressure difference is measured between the upstream pipe and at the region adjacent to the cylindrical section. Air at standard conditions flows in the system. The pipe diameter is 4 inches. The ratio of the cylindrical section diameter to the inside pipe diameter is 0.8. A pressure difference of 3 inches of water is measured. Find the volume flow rate in the system. Assume the flow is incompressible, inviscid, and steady and that the velocity is uniformly distributed across the pipe sections.

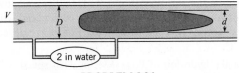

PROBLEM 5.86

5.87 Venturi-type applicators are frequently used to spray liquid fertilizers. Water flowing through the venturi creates a subatmospheric pressure at the throat, which in turn causes the liquid

fertilizer to flow up the feed tube and mix with the water in the throat region. The venturi applicator shown in the figure uses water at 20°C to spray a liquid fertilizer with the same density. The venturi exhausts to the atmosphere, and the exit diameter is 1 cm. The ratio of exit area to throat area (A_2/A_1) is 2. The flow rate of water through the venturi is 10 lpm (liters per minute). The bottom of the feed tube in the reservoir is 5 cm below the liquid fertilizer surface and 10 cm below the centerline of the venturi. The pressure at the liquid fertilizer surface is atmospheric. The flow rate through the feed tube between the reservoir and venturi throat is

$$Q_l \text{ (lpm)} = 0.5\sqrt{\Delta h}$$

where Δh is the drop in piezometric head (in meters) between the feed tube entrance and the venturi centerline. Find the flow rate of liquid fertilizer in the feed tube, Q_l. Also find the concentration of liquid fertilizer in the mixture, $[Q_l/(Q_l + Q_w)]$, at the end of the sprayer.

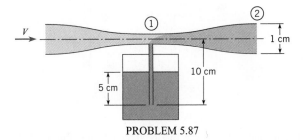

PROBLEM 5.87

5.88 Air with a density of 0.0644 lbm/ft³ is flowing upward in the vertical duct, as shown. The velocity at the inlet (station 1) is 80 ft/s, and the area ratio between stations 1 and 2 is 0.5 $(A_2/A_1 = 0.5)$. Two pressure taps, 10 ft apart, are connected to a manometer, as shown. The specific weight of the manometer liquid is 120 lbf/ft³. Find the deflection, Δh, of the manometer.

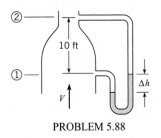

PROBLEM 5.88

5.89 An atomizer utilizes a constriction in an air duct as shown. Design an operable atomizer making your own assumptions regarding the air source.

5.90 A suction device is being designed based on the venturi principle to lift objects submerged in water. The operating water temperature is 15°C. The suction cup is located 1 m below the water surface, and the venturi throat is located 1 m above the water. The atmospheric pressure is 100 kPa. The ratio of the throat area to the

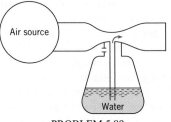

PROBLEM 5.89

exit area is $\frac{1}{4}$, and the exit area is 0.001 m^2. The area of the suction cup is 0.1 m^2.

a. Find the velocity of the water at the exit for maximum lift condition.
b. Find the discharge through the system for maximum lift condition.
c. Find the maximum load the suction cup can support.

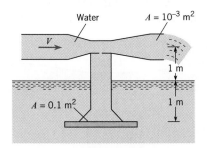

PROBLEM 5.90

5.91 A design for a hovercraft is shown in the figure. A fan brings air at 60°F into a chamber, and the air is exhausted between the skirts and the ground. The pressure inside the chamber is responsible for the lift. The hovercraft is 15 ft long and 7 ft wide. The weight of the craft including crew, fuel, and load is 2000 lbf. Assume that the pressure in the chamber is the stagnation pressure (zero velocity) and the pressure where the air exits around the skirt is atmospheric. Assume the air is incompressible, the flow is steady, and viscous effects are negligible. Find the airflow rate necessary to maintain the skirts at a height of 3 inches above the ground.

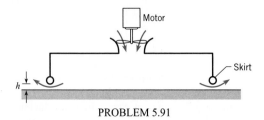

PROBLEM 5.91

5.92 Water is forced out of this cylinder by the piston. If the piston is driven at a speed of 6 ft/s, what will be the speed of efflux of the water from the nozzle if $d = 2$ in. and $D = 4$ in.? Neglecting friction and assuming irrotational flow, determine the force F that will be required to drive the piston. The exit pressure is atmospheric pressure.

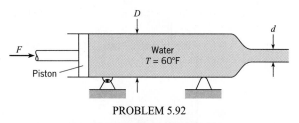

PROBLEM 5.92

5.93 Air flows through a constant-area heated pipe. At the entrance, the velocity is 10 m/s, the pressure is 100 kPa absolute and the temperature is 20°C. At the outlet, the pressure is 80 kPa absolute, and the temperature is 50°C. What is the velocity at the outlet? Can the Bernoulli equation be used to relate the pressure and velocity changes? Explain.

Cavitation

5.94 PQ◄ Sometimes driving your car on a hot day, you may encounter a problem with the fuel pump called pump cavitation. What is happening to the gasoline? How does this affect the operation of the pump?

5.95 PQ◄ What is cavitation? Why does the tendency for cavitation in a liquid increase with increased temperatures?

5.96 When gage A indicates a pressure of 120 kPa gage, then cavitation just starts to occur in the venturi meter. If $D = 40$ cm and $d = 10$ cm, what is the water discharge in the system for this condition of incipient cavitation? The atmospheric pressure is 100 kPa gage, and the water temperature is 10°C. Neglect gravitational effects.

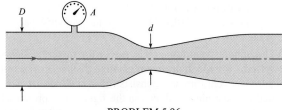

PROBLEM 5.96

5.97 A sphere 1 ft in diameter is moving horizontally at a depth of 12 ft below a water surface where the water temperature is 50°F. $V_{max} = 1.5V_o$, where V_o is the free stream velocity and occurs at the maximum sphere width. At what speed in still water will cavitation first occur?

5.98 When the hydrofoil shown was tested, the minimum pressure on the surface of the foil was found to be 70 kPa absolute when the foil was submerged 1.80 m and towed at a speed of 8 m/s. At the same depth, at what speed will cavitation first occur? Assume irrotational flow for both cases and $T = 10°C$.

5.99 For the hydrofoil of Prob. 5.98, at what speed will cavitation begin if the depth is increased to 3 m?

5.100 When the hydrofoil shown was tested, the minimum pressure on the surface of the foil was found to be 2.5 psi vacuum when the foil was submerged 4 ft and towed at a speed of 25

ft/s. At the same depth, at what speed will cavitation first occur? Assume irrotational flow for both cases and $T = 50°F$.

5.101 For the conditions of Prob. 5.100, at what speed will cavitation begin if the depth is increased to 10 ft?

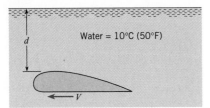

PROBLEMS 5.98, 5.99, 5.100, 5.101

5.102 A sphere is moving in water at a depth where the absolute
pressure is 18 psia. The maximum velocity on a sphere occurs
90° from the forward stagnation point and is 1.5 times the free-
stream velocity. The density of water is 62.4 lbm/ft³. Calculate
the speed of the sphere at which cavitation will occur. $T = 50°F$.

5.103 The minimum pressure on a cylinder moving horizontally
in water ($T = 10°C$) at 5 m/s at a depth of 1 m is 80 kPa absolute. At what velocity will cavitation begin? Atmospheric pressure is 100 kPa absolute.

Differential Form of the Continuity Equation

5.104 It is predicted that a flow field will have the following velocity components:

$$u = V(x^3 + xy^2) \qquad v = V(y^3 + yx^2) \qquad w = 0$$

V is a constant. Is such a flow field possible? (Does it satisfy
continuity?)

5.105 The velocity components of a flow field are given by

$$u = \frac{y}{(x^2 + y^2)^{3/2}} \qquad v = \frac{-x}{(x^2 + y^2)^{3/2}}$$

Is continuity satisfied? Is the flow irrotational?

5.106 A u component of velocity is given by $u = Axy$, where A
is a constant. What is a possible v component? What must the v
component be if the flow is irrotational?

5.107 The continuity equation can be expressed in vector notation as

$$\frac{\partial \rho}{\partial t} + \nabla(\rho \cdot \mathbf{V}) = 0.$$

Show that this equation can also be expressed as

$$\frac{D\rho}{Dt} + \rho(\nabla \cdot V) = 0$$

where the operator D/Dt is defined as

$$\frac{D}{Dt} = \frac{\partial}{\partial t} + V \cdot \nabla.$$

CHAPTER 6

Momentum Equation

SIGNIFICANT LEARNING OUTCOMES

Conceptual Knowledge
- Explain the steps in deriving the momentum equation.
- Define an inertial reference frame.
- Identify the accumulation and momentum flux terms in the momentum equation.
- Explain the steps in deriving the moment-of-momentum equation.
- Explain the water-hammer effect.

Procedural Knowledge
- Apply the component form of the momentum equation to stationary and moving control volumes.
- Apply the vector form of the momentum equation.
- Apply the moment-of-momentum equation to stationary and rotating control volumes.

Typical Applications
- For jets, vanes, nozzles and pipe sections, calculate forces and moments.
- For water hammer effects, calculate pressure rise.
- For radial turbines, calculate power output.

The analysis of forces on vanes and pipe bends, the thrust produced by a rocket or turbojet, and torque produced by a hydraulic turbine are all examples of the application of the momentum equation. In Chapter 5, the Reynolds transport theorem was introduced, which enables one to take fundamental equations for a system (given mass) and write the equivalent equations in Eulerian form suitable for the control volume approach. In this chapter the Reynolds transport theorem is applied to Newton's second law of motion, $\boldsymbol{F} = m\boldsymbol{a}$, to develop the Eulerian form of the momentum equation. Application of this equation allows the engineer to analyze forces and moments produced by flowing fluids.

6.1

Momentum Equation: Derivation

Newton's second law of motion is a familiar concept in mechanics. In this section the principle that the rate of change of momentum of a body is equal to the force acting on the body is rewritten in Eulerian form using the Reynolds transport theorem. The result is momentum equation in the form suitable for the control volume approach.

When forces act on a particle, the particle accelerates according to Newton's second law of motion:

$$\sum \mathbf{F} = m\mathbf{a} \tag{6.1}$$

By definition, the mass (m) is constant, so the equation may be written using momentum:

$$\sum \mathbf{F} = \frac{d(m\boldsymbol{v})}{dt} \tag{6.2}$$

Although Eqs. (6.1) and (6.2) apply to a single particle, the law can also be formulated for a system composed of a group of particles, for example, a fluid system. In this case, the law may be written as

$$\sum \mathbf{F} = \frac{d(\mathbf{Mom}_{\text{sys}})}{dt} \tag{6.3}$$

The term $\mathbf{Mom}_{\text{sys}}$ denotes the total momentum of all mass comprising the system.

Equation (6.3) is a Lagrangian equation. To derive an Eulerian equation, the Reynolds transport theorem, Eq. (5.21),

$$\frac{dB_{\text{sys}}}{dt} = \frac{d}{dt}\int_{\text{cv}} b\rho \, d\mathcal{V} + \int_{\text{cs}} b\rho \mathbf{V} \cdot d\mathbf{A}$$

is applied where V is fluid velocity relative to the control surface at the location where the flow crosses the surface. The extensive property B_{sys} becomes the momentum of the system: $B = \mathbf{Mom}_{\text{sys}}$. The corresponding intensive property b becomes the momentum per unit mass within the system. The momentum of any fluid particle of mass m in the system is $m\boldsymbol{v}$, and so $b = (m\boldsymbol{v})/m = \boldsymbol{v}$. The velocity $\boldsymbol{v}$ must be relative to an inertial reference frame, that is, a frame that does not rotate and can either be stationary or moving at a constant velocity. Substituting for B_{sys} and b into Eq. (5.21) gives

$$\frac{d(\mathbf{Mom}_{\text{sys}})}{dt} = \frac{d}{dt}\int_{\text{cv}} \boldsymbol{v}\rho \, d\mathcal{V} + \int_{\text{cs}} \boldsymbol{v}\rho \mathbf{V} \cdot d\mathbf{A} \tag{6.4}$$

Combining Eqs. (6.3) and (6.4) gives the *integral form of the momentum equation*:

$$\sum \mathbf{F} = \frac{d}{dt}\int_{\text{cv}} \boldsymbol{v}\rho \, d\mathcal{V} + \int_{\text{cs}} \boldsymbol{v}\rho \mathbf{V} \cdot d\mathbf{A} \tag{6.5}$$

This equation can be expressed in words as

$$\begin{bmatrix} \text{sum of forces} \\ \text{acting on the matter} \\ \text{in control volume} \end{bmatrix} = \begin{bmatrix} \text{time rate of} \\ \text{change of momentum} \\ \text{in control volume} \end{bmatrix} + \begin{bmatrix} \text{net outflow rate} \\ \text{of momentum} \\ \text{through control surface} \end{bmatrix}$$

It is important to remember that the momentum equation is a vector equation; that is, there is a direction associated with the each term in the equation.

If the flow crossing the control surface occurs through a series of inlet and outlet ports and if the velocity $\boldsymbol{v}$ is uniformly distributed across each port, then a simplified form of the Reynolds transport theorem, Eq. (5.23), can be used, and the momentum equation becomes

$$\sum \mathbf{F} = \frac{d}{dt}\int_{\text{cv}} \rho\boldsymbol{v} \, d\mathcal{V} + \sum_{\text{cs}} \dot{m}_o \boldsymbol{v}_o - \sum_{\text{cs}} \dot{m}_i \boldsymbol{v}_i \tag{6.6}$$

where the subscripts o and i refer to the outlet and inlet ports, respectively. This form of the momentum equation will be identified as the *vector form*. Notice that the product of $\dot{m}\boldsymbol{v}$ corresponds to the mass per unit time times velocity, or momentum per unit time, which has the same units as force.

As long as v is uniformly distributed across control surface, Eq. (6.6) applies to any control volume, including one that is moving, deforming, or both. In all cases, $\dot{m}$ is the rate at which mass is passing across the control surface, and v is velocity evaluated at the control surface with respect to the inertial reference frame that is selected.

The three components of Eq. (6.6) for the Cartesian coordinate (x,y,z) system are

$$x\text{-direction: } \sum F_x = \frac{d}{dt}\int_{\text{cv}} v_x\,\rho\,d\Psi + \sum_{\text{cs}} \dot{m}_o v_{ox} - \sum_{\text{cs}} \dot{m}_i v_{ix} \qquad (6.7a)$$

$$y\text{-direction: } \sum F_y = \frac{d}{dt}\int_{\text{cv}} v_y\,\rho\,d\Psi + \sum_{\text{cs}} \dot{m}_o v_{oy} - \sum_{\text{cs}} \dot{m}_i v_{iy} \qquad (6.7b)$$

$$z\text{-direction: } \sum F_z = \frac{d}{dt}\int_{\text{cv}} v_z\,\rho\,d\Psi + \sum_{\text{cs}} \dot{m}_o v_{oz} - \sum_{\text{cs}} \dot{m}_i v_{iz} \qquad (6.7c)$$

where the subscripts x, y, and z refer to the force and velocity components in the coordinate directions. These equations will be identified as the *component form of the momentum equation*.

When velocity v varies across the control surface, the general form of the momentum equation, Eq. (6.5), must always be used.

Momentum Equation: Interpretation

Application of the momentum equation to fluid flow problems is analogous to the use of the free-body approach in solid mechanics. In solid mechanics, a system of interest is isolated from its surroundings, thereby creating a free body, and forces are applied to replace the influence of the surroundings. These forces are then summed and equated with the product of mass and acceleration. In fluid mechanics, the system of interest is the material contained within the control volume, and forces are applied to this system to represent the effect of the surroundings. Forces are then summed and equated to momentum changes of the flow.

Figure 6.1

Forces associated with flow in a pipe: (a) pipe schematic, (b) control volume situated inside the pipe, and (c) control volume surrounding the pipe.

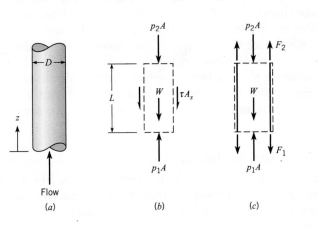

Force Terms

In Eq. (6.5), the force term is the sum of all the forces acting on the matter in the control volume. For example, consider flow inside a vertical pipe, as shown in Fig. 6.1a. One possible control volume is a cylinder with diameter D and length L located just inside the pipe wall. As shown in Fig. 6.1b, the fluid within the control volume has been isolated from its surroundings, and the effect of the surroundings are shown as forces. The effect of the wall is replaced by a force equal to the shear stress (τ) times the pipe surface area ($A_s = \pi DL$). The force due to pressure is given by pressure (p) times the section area ($A = \pi D^2/4$) and always acts toward the control surface (a compressive force). The weight of the fluid is given by $W = \gamma(\pi D^2/4)L$. Thus, the net force acting in the z-direction is given by

$$\sum F_z = (p_1 - p_2)\frac{\pi}{4}D^2 - \tau\pi DL - \gamma(\pi D^2/4)L \tag{6.8}$$

Another possible control volume has a length L and a diameter that is larger than the pipe's outside diameter. As shown in Fig. 6.1c, this control volume cuts through the pipe wall. Comparing Figs. 6.1b and 6.1c shows that the pressure forces are the same. However, in Fig. 6.1c, there is no force associated with shear stress, but there are two new forces, F_1 and F_2, which represent the forces due to the pipe wall. Also, the weight of matter within the control volume now includes the weight of the fluid and the pipe wall (W_p). The net z-direction force is

$$\sum F_z = (p_1 - p_2)\frac{\pi}{4}D^2 - F_1 + F_2 - (W_p + \gamma(\pi D^2/4)L) \tag{6.9}$$

The choice of control volume depends on what information being sought. To relate the pressure change between sections 1 and 2 to wall shear stress, Eq. (6.8) would be best. To find the tensile force carried by the pipe wall, Eq. (6.9) would be used.

The sketches shown in Figs. 6.1b and 6.1c are identified as *force diagrams* (FD). A force diagram shows the forces acting on the matter contained within a control volume. A force diagram is equivalent to a free-body diagram at the instant in time when the momentum equation is applied.

In Fig. 6.1b, the force of gravity (weight) acts on each mass element in the control volume (with the resultant force acting at the mass center). A force that acts on mass elements within the body is defined as a *body force*. A body force can act at a distance without any physical contact. Examples of body forces include gravitational, electrostatic, and magnetic forces.

Except for the body force (weight), all forces shown in Figs. 6.1b and 6.1c are surface forces. A *surface force* is defined as a force that requires physical contact, meaning that surface forces act at the control surface. For example, $p_1 A_1$ acts at the control surface and requires contact between the fluid outside the control volume and the fluid inside the control volume. With respect to pressure, the net force is obtained by integrating the pressure over the area of the surface. For example, if the pressure varies hydrostatically, the magnitude of the force and the line of action would be determined using the methods presented in Chapter 3. When evaluating pressure forces, engineers commonly use gage pressure. In this case, the force associated with atmospheric pressure acting over a surface is zero. In addition to pressure, surface forces can be caused by shear stress, for example the force τA_s shown in Fig. 6.1b.

Momentum Accumulation

The first term on the right side of Eq. (6.5) represents the rate at which the momentum of the material inside the control volume is changing with time. In particular, the mass of a volume element in the control volume is $\rho\, d\mathcal{V}$, so the product $v\,\rho\,\mathcal{V}$ is the momentum of a volume

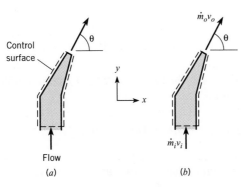

element. Integrating over the control volume gives total momentum of the material in the control volume. Taking the time derivative gives the rate at which the momentum is changing. This term may be described as the net rate of momentum accumulation, and it will be referred to it as the *momentum accumulation* term. The units are momentum per unit time, which are equivalent to the units for force.

In many problems, the momentum accumulation is zero. For example, consider steady flow through the control volume surrounding the nozzle shown in Fig. 6.2a. The fluid inside the control volume has momentum because it is moving. However, the velocity and density at each point do not change with time, so the total momentum in the control volume is constant, and the momentum accumulation term is zero. The evaluation of the momentum accumulation term is completed by considering the structural elements (i.e., the nozzle walls). Since the structural elements are stationary, there is no momentum change, so the momentum accumulation rate is zero.

In summary, the momentum of the material inside a control volume is evaluated by integrating the momentum of each volume element over the control volume. If the momentum in each differential volume is constant with time (e.g., steady flow, a stationary structural part), the momentum accumulation rate is zero.

Momentum Diagram

The momentum terms on the right side of Eq. (6.5) may be visualized with a *momentum diagram* (MD). The momentum diagram is created by sketching a control volume and then drawing a vector to represent the momentum accumulation term and a vector to represent momentum flow at each section where mass crosses the control surface.

Although the momentum diagram applies to the integral form of the momentum principle, Eq. (6.5), the diagram takes on a simple form when the velocity **v** is uniformly distributed across each inlet and outlet port, and Eq. (6.6) applies. For example, consider steady flow through the nozzle shown in Fig. 6.2. For the control volume indicated, the momentum accumulation term is zero, and this vector is omitted from the diagram. If the velocity is assumed to be uniform across the inlet and exit sections, the outlet momentum flow is given by $\dot{m}_o v_o$, and the inlet momentum flow is given by $\dot{m}_i v_i$, as shown in Fig. 6.2b. To evaluate the momentum flow, one can use the diagram to see that

$$\sum_{cs} \dot{m}_o \mathbf{v}_o = [\dot{m}_o v_o \cos\theta]\mathbf{i} + [\dot{m}_o v_o \sin\theta]\mathbf{j}$$

and

$$\sum_{cs} \dot{m}_i \mathbf{v}_i = [\dot{m}_i v_i]\mathbf{j}$$

Recognizing that $\dot{m}_i = \dot{m}_o = \dot{m}$, the above equations can be combined to show that the net outward flow of momentum is

$$\sum_{cs} \dot{m}_o \mathbf{v}_o - \sum_{cs} \dot{m}_i \mathbf{v}_i = [\dot{m} v_o \cos\theta]\mathbf{i} + [\dot{m} v_o \sin\theta - \dot{m} v_i]\mathbf{j}$$

Using the momentum diagram is a straightforward way to evaluate the momentum terms.

Systematic Approach

A systematic approach is recommended for using the momentum equation. One such approach is summarized here.

Problem Setup

• Select an appropriate control volume. Sketch the control volume and coordinate axes. Select an inertial reference frame.
• Identify governing equations. This will include either the vector or component form of the momentum equation. Other equations, such as the Bernoulli equation and/or the continuity equation, may be needed.

Force Analysis and Diagram

• Sketch body force(s) (usually only gravitational force) on the force diagram.
• Sketch surface forces on the force diagram; these are forces caused by pressure distribution, shear stress distribution, and supports and structures.

Momentum Analysis and Diagram

• Evaluate the momentum accumulation term. If the flow is steady and other materials in the control volume are stationary, the momentum accumulation is zero. Otherwise, the momentum accumulation term is evaluated by integration, and an appropriate vector is added to the momentum diagram.
• Sketch momentum flow vectors on the momentum diagram. For uniform velocity, each vector is $\dot{m}\mathbf{v}$.

6.3

Common Applications

This section discusses four common applications of the momentum equation: fluid jets, nozzles, vanes, and pipe bends.

Fluid Jets

A fluid jet is created by a high-speed stream of fluid leaving a nozzle as shown in Fig. 6.3. Examples would include a firehose or an ink jet in a printer. Provided that the jet is "free" (not confined by walls), the pressure is constant across any cross section of the jet, such as

Figure 6.3

A fluid jet exiting a nozzle. Pressure of the ambient fluid is p_s. Letters indicate cross sections.

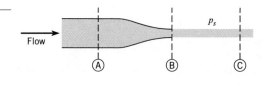

sections B and C, and equal to the pressure of the surrounding fluid p_s.* Thus $p_B = p_C = p_s$. It is convenient to use gage pressures when calculating pressure forces so the exit pressure is zero and so there is no surface force at the exit plane. Finally, it is typically assumed that fluid velocity is uniform across the cross section of a jet.

Example 6.1 shows how to calculate the thrust on a rocket by applying the momentum equation.

EXAMPLE 6.1   THRUST OF ROCKET

The sketch below shows a 40 g rocket, of the type used for model rocketry, being fired on a test stand in order to evaluate thrust. The exhaust jet from the rocket motor has a diameter of $d = 1$ cm, a speed of $v = 450$ m/s, and a density of $\rho = 0.5$ kg/m^3. Assume the pressure in the exhaust jet equals ambient pressure, and neglect any momentum changes inside the rocket motor. Find the force F_b acting on the beam that supports the rocket.

Sketch:

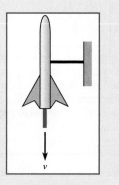

Problem Definition

Situation: Model rocket supported by a beam. Exit conditions of model rocket provided.

Find: Force (in newtons) on beam.

Assumptions: Pressure is atmospheric at the nozzle exit plane.

Properties: $\rho = 0.5$ kg/m^3.

Plan

1. Choose a control volume such that the force acting on the control surface is the force on the beam and where information is available for the momentum flux crossing the control surface.
2. Sketch the force diagram.
3. Sketch the momentum diagram.
4. Because this problem involves only one direction, the component form of the momentum equation in the z-direction, Eq. (6.7c), will be used.
5. Evaluate the sum of the forces from the force diagram.
6. Evaluate momentum terms.
7. Calculate the force.

Solution

1. The control volume chosen is shown below. The control volume is stationary.

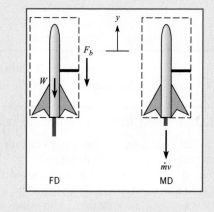

---

* This assumption is valid if a jet is subsonic, meaning the speed of the jet is less than the local speed of sound in the fluid. Otherwise, the exit pressure can be higher than atmospheric pressure. Supersonic jets are discussed in Chapter 12.

2. From the force diagram, the force on the control surface exerted by the beam is chosen as downward (negative z-direction) with magnitude F_b. (The corresponding force exerted by the rocket on the beam is upward.) The weight also acts downward. Also there is no pressure force at the nozzle exit plane because exit pressure is atmospheric.

3. The momentum diagram shows only one momentum outflow and no inflow.

4. Momentum equation in z-direction:

$$\sum F_z = \frac{d}{dt}\int_{cv} v_z \rho \, dV + \sum_{cs} \dot{m}_o v_{oz} - \sum_{cs} \dot{m}_i v_{iz}$$

5. Sum of forces.

$$\sum F_z = (-F_b - W)$$
$$= -F_b - mg$$

6. Evaluation of momentum terms.
   - Accumulation term: No changes in control volume,
   $$\frac{d}{dt}\int_{cv} v_z \rho \, dV = 0.$$

   - Momentum inflow: No inflow, $\sum_{cs} \dot{m}_i v_{iz} = 0$.

   - Momentum outflow: $\sum_{cs} \dot{m}_o v_{oz} = \dot{m}(-v) = -\rho A v^2$

7. Force on beam:

$$-F_b - mg = -\rho A v^2$$
$$F_b = \rho A v^2 - mg$$
$$= (0.5 \text{ kg/m}^3)(\pi \times 0.01^2 \text{ m}^2 / 4)(450^2 \text{ m}^2/\text{s}^2)$$
$$- (0.04 \text{ kg})(9.81 \text{ m/s}^2)$$
$$F_b = \boxed{7.56 \text{ N}}$$

The direction of F_b (on the beam) is upward.

Review

1. The thrust force of the rocket motor is $\dot{m}v = 7.95$ N (1.79 lbf); this value is typical of a small motor used for model rocketry.

2. The force F_b acts downward at the control surface, and an equal and opposite force acts upward on the support beam, as shown in the sketch below. This is an example of an action and reaction force, as described by Newton's third law of motion.

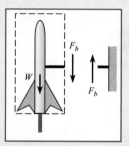

3. For solving this problem, two separate diagrams were used: the force and the momentum diagrams. The rationale for this approach is to regard forces and momentum flows as separate phenomena. This facilitates writing the equations and provides a systematic approach to more complex problems.

Example 6.2 shows how to calculate the forces due to a jet flowing into a cart. The component form of the momentum equation is utilized.

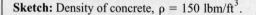

EXAMPLE 6.2 CONCRETE FLOWING INTO CART

As shown in the sketch, concrete flows into a cart sitting on a scale. The stream of concrete has a density of $\rho = 150$ lbm/ft^3, an area of $A = 1$ ft^2, and a speed of $v = 10$ ft/s. At the instant shown, the weight of the cart plus the concrete is 800 lbf. Determine the tension in the cable and the weight recorded by the scale. Assume steady flow.

Sketch: Density of concrete, $\rho = 150$ lbm/ft^3.

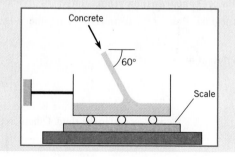

Problem Definition

Situation: Concrete flowing into cart held by cable and mounted on a scale.

Find:

1. Force (in lbf) on cable.
2. Weight (in lbf) recorded on scale.

Assumptions: The velocity of the concrete in the cart is zero.

Plan

1. Select a control volume that provides force on cable and weight on scale.
2. Sketch the force diagram.
3. Sketch the momentum diagram.
4. Since this problem involves two directions, the component form of the momentum equations in the x- and z-directions, Eqs. (6.7a) and (6.7c), will be used.
5. Evaluate forces from force diagram.
6. Evaluate momentum terms.
7. Calculate tension in cable and weight on scale.

Solution

1. Control volume selected is shown on diagram. Control volume is stationary.

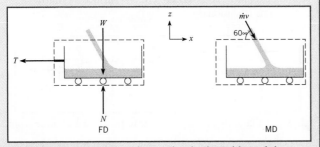

2. Force diagram shows the tension in the cable and the weight on the scale.
3. Momentum diagram shows only an inflow of momentum. Velocity of the concrete in the tank is neglected.
4. Component momentum equations
   - Momentum equation in x-direction

$$\sum F_x = \frac{d}{dt}\int_{cv} v_x \rho \, d\Psi + \sum_{cs} \dot{m}_o v_{ox} - \sum_{cs} \dot{m}_i v_{ix} \qquad \text{(a)}$$

   - Momentum equation in z-direction

$$\sum F_z = \frac{d}{dt}\int_{cv} v_z \rho \, d\Psi + \sum_{cs} \dot{m}_o v_{oz} - \sum_{cs} \dot{m}_i v_{iz} \qquad \text{(b)}$$

5. Forces from the force diagram

$$\sum F_x = -T$$

$$\sum F_z = N - W$$

6. Evaluation of momentum terms
   - Momentum accumulation: $v_x = 0$, $v_z = 0$, so

$$\frac{d}{dt}\int_{cv} v_x \rho \, d\Psi = 0, \quad \frac{d}{dt}\int_{cv} v_z \rho \, d\Psi = 0.$$

   - Momentum inflow

$$\sum_{cs} \dot{m}_i v_{ix} = \dot{m}v\cos 60° = \rho A v^2 \cos 60°$$

$$\sum_{cs} \dot{m}_i v_{iz} = \dot{m}(-v\sin 60°) = -\rho A v^2 \sin 60°$$

   - Momentum outflow: No outflow, so,

$$\sum_{cs} \dot{m}_o v_{ox} = 0, \text{ and } \sum_{cs} \dot{m}_o v_{oz} = 0.$$

7. Evaluate tension in cable using (a).

$$-T = -\rho A v^2 \cos 60°$$

$$T = (150 \text{ lbm/ft}^3)\left(\frac{\text{slugs}}{32.2 \text{ lbm}}\right)(1 \text{ ft}^2)(10 \text{ ft/s})^2 \cos 60°$$

$$= \boxed{233 \text{ lbf}}$$

Evaluate force on scale using (b).

$$N - W = -(-\rho A v^2 \sin 60°)$$

$$N = W + \rho A v^2 \sin 60°$$

$$= 800 \text{ lbf} + 403 \text{ lbf} = \boxed{1200 \text{ lbf}}$$

Review

1. The weight recorded by the scale is larger than the weight of the cart because of the momentum carried by the fluid jet.
2. Notice that unit conversions are usually needed when using English units.
3. There will be some velocity in the cart due to mixing, but the momentum associated with those velocities would be insignificant, so the momentum accumulation term can be neglected.
4. Answers are expressed with three significant figures.

Nozzles

Nozzles are flow devices used to accelerate a fluid stream by reducing the cross-sectional area of the flow. When a fluid flows through a nozzle as shown in Fig. 6.3, it is reasonable to assume the velocity is uniform across sections A and B. Hence, the momentum flows will have magnitude $\dot{m}v$. If the nozzle exhausts into the atmosphere, the pressure at section B is atmospheric. Applying the Bernoulli equation between sections A and B will provide an equation for the pressure at section A. This pressure will exert a force of magnitude pA, where p is the pressure at the centroid of section A.

In many applications involving finding the force on a nozzle, the Bernoulli equation is used along with the momentum equation. Example 6.3 illustrates one such application.

EXAMPLE 6.3   FORCE ON A NOZZLE

The sketch shows air flowing through a nozzle. The inlet pressure is $p_1 = 105$ kPa abs, and the air exhausts into the atmosphere, where the pressure is 101.3 kPa abs. The nozzle has an inlet diameter of 60 mm and an exit diameter of 10 mm, and the nozzle is connected to the supply pipe by flanges. Find the air speed at the exit of the nozzle and the force required to hold the nozzle stationary. Assume the air has a constant density of 1.22 kg/m^3. Neglect the weight of the nozzle.

Sketch:

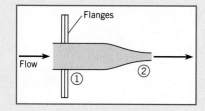

Problem Definition

Situation: Air flows through nozzle with contraction ratio of 6:1. Nozzle attached to pipe by flange.

Find:

1. Exit velocity (in m/s) of nozzle.
2. Force on flange (in newtons) required to hold nozzle.

Assumptions:

1. Density is constant.
2. Viscous effects are negligible.
3. Flow is steady.

Properties: $\rho = 1.22$ kg/m^3.

Plan

1. Select a control surface that includes the force on the flange.

2. Sketch the force diagram.
3. Sketch the momentum diagram.
4. Apply the Bernoulli equation and the continuity equation to find exit (and inlet) velocity.
5. Apply the component form of the momentum equation in the x-direction, Eq. (6.7a).
6. Evaluate forces (use gage pressures for pressure force).
7. Evaluate momentum terms.
8. Calculate force on flange.

Solution

1. Select control volume (and control surface). Control volume is stationary.

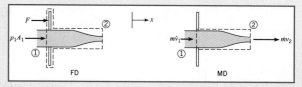

2. Force diagram shows force due to pressure and force from flange.
3. Momentum diagram shows a momentum inflow and outflow.
4. Application of the Bernoulli equation between sections 1 and 2

$$p_1 + \gamma z_1 + \frac{1}{2}\rho v_1^2 = p_2 + \gamma z_2 + \frac{1}{2}\rho v_2^2$$

• Set $z_1 = z_2$.

• Set $p_2 = 0$ kPa gage, and now

$$p_1 = 105 \text{ kPa} - 101.3 \text{ kPa} = 3.7 \text{ kPa gage.}$$

The Bernoulli equation simplifies to

$$p_1 + \rho v_1^2/2 = \rho v_2^2/2$$

From the continuity equation,

$$v_1 A_1 = v_2 A_2$$
$$v_1 d_1^2 = v_2 d_2^2$$

Substitute into the Bernoulli equation and solve for v_2:

$$v_2 = \sqrt{\frac{2p_1}{\rho(1-(d_2/d_1)^4)}}$$

Evaluate exit velocity:

$$v_2 = \sqrt{\frac{2\times 3.7\times 1000\ \text{Pa}}{(1.22\ \text{kg/m}^3)(1-(10/60)^4)}} = \boxed{77.9\ \text{m/s}}$$

Inlet velocity is

$$v_1 = v_2\left(\frac{d_2}{d_1}\right)^2$$
$$= 77.9\ \text{m/s}\times\left(\frac{1}{6}\right)^2 = 2.16\ \text{m/s}$$

5. Momentum equation

$$\sum F_x = \frac{d}{dt}\int_{cv} v_x \rho\, d\Psi + \sum_{cs} \dot{m}_o v_{ox} - \sum_{cs} \dot{m}_i v_{ix}$$

6. Sum of forces in x-direction

$$\sum F_x = F + p_1 A_1$$

7. Term-by-term evaluation of momentum terms
   • Accumulation term: Flow is steady,

   $$\frac{d}{dt}\int_{cv} v_x \rho\, d\Psi = 0.$$

• Momentum outflux with one outflow at section 2,

$$\sum_{cs} \dot{m}_o v_{ox} = \dot{m} v_2.$$

• Momentum influx with one inlet at section 1,

$$\sum_{cs} \dot{m}_i v_{ix} = \dot{m} v_1.$$

8. Force on flange

$$F + p_1 A_1 = \dot{m}(v_2 - v_1)$$
$$F = \rho A_1 v_1 (v_2 - v_1) - p_1 A_1$$
$$= (1.22\ \text{kg/m}^3)\left(\frac{\pi}{4}\right)(0.06\ \text{m})^2(2.16\ \text{m/s})$$
$$\times (77.9 - 2.16)(\text{m/s})$$
$$-3.7\times 1000\ \text{N/m}^2\times\left(\frac{\pi}{4}\right)(0.06\ \text{m})^2$$
$$= 0.564\ \text{N} - 10.46\ \text{N} = \boxed{-9.90\ \text{N}}$$

Because F is negative, the direction is opposite to the direction assumed on the force diagram. Hence, the force on the control surface acts in the negative x-direction, but the force on the flange will be in the positive direction.

$$\text{Force on flange} = \boxed{9.90\ \text{N}}$$

The tension in the bolts holding the flange will be increased.

Review

1. The direction initially assumed for the force on the flange was arbitrary. If the answer for the force is negative, then the force is opposite the chosen direction.

2. By choosing to use gage pressure there are no pressure forces on any surfaces exposed to the atmosphere. Also there is no force due to pressure across the nozzle exit plane.

Vanes

A *vane* is a structural component, typically thin, that is used to turn a fluid jet or is turned by a fluid jet. Examples include a blade in a turbine, a sail on a ship, and a thrust reverser on an aircraft engine. Figure 6.4 shows a flat vane impacted by a jet of fluid. A typical control volume is also shown. In analyzing flow over a vane, it is common to neglect the pressure change due to elevation difference. Since the pressure is constant (atmospheric pressure or surrounding pressure), the Bernoulli equation shows the speed is constant $v_1 = v_2 = v_3$. Another common assumption is that viscous forces are negligible compared to pressure forces. Thus when a vane is flat, as in Fig. 6.4, the force needed to hold the vane stationary is normal to the vane.

Figure 6.4

Fluid jet striking a flat vane.

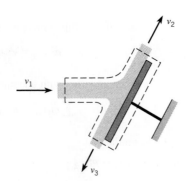

Example 6.4 illustrates how to calculate the force on a vane used to deflect a stream of water. This example utilizes the vector form of the momentum equation.

EXAMPLE 6.4 WATER DEFLECTED BY A VANE

A water jet is deflected 60° by a stationary vane as shown in the figure. The incoming jet has a speed of 100 ft/s and a diameter of 1 in. Find the force exerted by the jet on the vane. Neglect the influence of gravity.

Sketch:

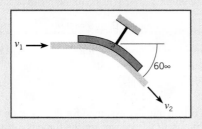

Problem Definition

Situation: Water deflected by a vane.

Find: Force (lbf) on vane due to jet.

Assumptions:

1. Viscous effects are negligible.
2. Neglect gravitational effects.

Properties: $\rho = 62.4 \ \text{lbm/ft}^3 = 1.94 \ \text{slug/ft}^3$.

Plan

From the Bernoulli equation, since the pressure is constant, the inlet and outlet speeds are the same. Also, from continuity, $\dot{m}_1 = \dot{m}_2 = \dot{m}$.

1. Select a control volume such that the control surface includes the force on the vane and flux of momentum.
2. Sketch the force diagram.
3. Sketch the momentum diagram.

4. Use the vector form of momentum equation, Eq. (6.6).
5. Evaluate force terms.
6. Evaluate momentum terms
7. Evaluate mass flow rate.
8. Calculate force.

Solution

1. The control volume selected is shown in the sketch. The control volume is stationary.

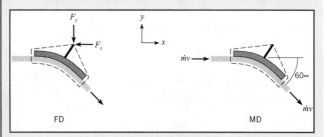

2. The force diagram shows only the reaction force.
3. The momentum diagram shows an inflow and outflow.
4. Vector form of momentum equation.

$$\sum \boldsymbol{F} = \frac{d}{dt}\bigg|_{\text{cv}} \int \rho \boldsymbol{v} \, d\mathcal{V} + \sum_{\text{cs}} \dot{m}_o \boldsymbol{v}_o - \sum_{\text{cs}} \dot{m}_i \boldsymbol{v}_i$$

5. Force vector is

$$\sum \boldsymbol{F} = - F_x \mathbf{i} - F_y \mathbf{j}$$

6. Evaluation of momentum terms

   - Control volume is stationary, $\dfrac{d}{dt}\displaystyle\int_{cv} \rho\mathbf{v}\,d\mathcal{V} = 0$

   - Momentum outflow vector,

     $$\sum_{cs} \dot{m}_o\mathbf{v}_o = [(\dot{m}v\cos 60°)\mathbf{i} - (\dot{m}v\sin 60°)\mathbf{j}].$$

   - Momentum inflow vector, $\displaystyle\sum_{cs} \dot{m}_i\mathbf{v}_i = \dot{m}v\mathbf{i}.$

7. Mass flow rate

   $$\dot{m} = \rho A v$$

   $$= (1.94\ \text{slug/ft}^3)(\pi \times 0.0417^2\ \text{ft}^2)(100\ \text{ft/s})$$

   $$= 1.06\ \text{slug/s}$$

8. Force

   $$-F_x\mathbf{i} - F_y\mathbf{j} = (\dot{m}v\cos 60° - \dot{m}v)\mathbf{i} - (\dot{m}v\sin 60°)\mathbf{j}$$

   For each component,

   $$-F_x = \dot{m}v\cos 60° - \dot{m}v$$

   $$-F_y = -\dot{m}v\sin 60°$$

   Force in x-direction

   $$F_x = \dot{m}v(1 - \cos 60°)$$

   $$= (1.06\ \text{slug/s})(100\ \text{ft/s})(1 - \cos 60°)$$

   $$F_x = \boxed{53.0\ \text{lbf}}$$

   Force in y-direction

   $$F_y = \dot{m}v\sin 60°$$

   $$= (1.06\ \text{slug/s})(100\ \text{ft/s})\sin 60°$$

   $$F_y = \boxed{91.8\ \text{lbf}}$$

   The force of the jet on the vane ($\mathbf{F}_{\text{jet}}$) is opposite in direction to the force required to hold the vane stationary ($\mathbf{F}$). Therefore,

   $$\mathbf{F}_{\text{jet}} = (53.0\ \text{lbf})\mathbf{i} + (91.8\ \text{lbf})\mathbf{j}$$

Example 6.5 shows how to calculate the force on an axisymmetric vane, which redirects the flow in the radial direction.

EXAMPLE 6.5   FORCE ON AN AXISYMMETRIC VANE

As shown in the figure, an incident jet of fluid with density ρ, speed v, and area A is deflected through an angle β by a stationary, axisymmetric vane. Find the force required to hold the vane stationary. Express the answer using ρ, v, A, and β. Neglect the influence of gravity.

Sketch: Gravitational effects are negligible.

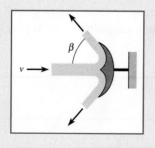

Problem Definition

Situation: Fluid deflected by axisymmetric vane.

Find: Force required to hold vane stationary.

Assumptions:

1. Flow is steady.
2. Fluid is incompressible.
3. Viscous effects are negligible.

Plan

Because the pressure is constant, the Bernoulli equation shows the inlet and outlet speeds are the same. Application of the continuity equation shows the inlet and outlet mass flows are also the same.

1. Select a control volume with the constraining force on control surface.
2. Sketch the force diagram.
3. Sketch the momentum diagram.
4. Apply the component form of the momentum equation in x-direction, Eq. (6.7a).
5. Evaluate force terms.
6. Evaluate momentum terms.
7. Calculate force.

Solution

1. Control volume selected is shown. Control volume is stationary.

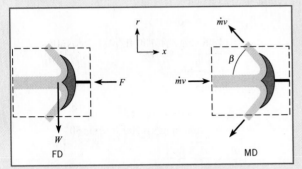

FD

MD

2. The force diagram shows only one force.

3. The momentum diagram shows one momentum flux in and one axisymmetric flux out. The net radial flux of momentum is zero, so only the component in the axial direction contributes to the momentum flux.

4. Momentum equation in x-direction.

$$\sum F_x = \frac{d}{dt}\int_{cv} v_x \rho \, d\Psi + \sum_{cs} \dot{m}_o v_{ox} - \sum_{cs} \dot{m}_i v_{ix}$$

5. Sum of forces

$$\sum F_x = -F$$

6. Evaluation of momentum terms
   • Accumulation term for stationary control volume is

   $$\frac{d}{dt}\int_{cv} v_x \rho \, d\Psi = 0.$$

   • Momentum outflow is $\sum_{cs} \dot{m}_o v_{ox} = -\dot{m}v\cos\beta$.

   • Momentum inflow is $\sum_{cs} \dot{m}_i v_{ix} = \dot{m}v$.

7. Force on vane

   $$-F = -\dot{m}v(1 + \cos\beta)$$
   $$F = \dot{m}v(1 + \cos\beta)$$

   Apply mass flow rate equation, $\dot{m} = \rho A v$,

   $$\boxed{F = \rho A v^2 (1 + \cos\beta)}$$

   and the direction of this force is to the left, as shown in the force diagram.

Review

This type of reverse flow vane is used to reverse thrust on aircraft engines.

Pipe Bends

Calculating the force on pipe bends is important in engineering applications using large pipes to design the support system. Because flow in a pipe is usually turbulent, it is common practice to assume that velocity is nearly constant across each cross section of the pipe. Also, the force acting on a pipe cross section is given by pA, where p is the pressure at the centroid of area and A is area.

Example 6.6 illustrates how to calculate a restraining force on a pipe bend using the vector form of the momentum equation. In contrast to vanes, the pressure forces play a key role in the solution.

EXAMPLE 6.6   FORCES ACTING ON A PIPE BEND

A 1 m–diameter pipe bend shown in the diagram is carrying crude oil $(S = 0.94)$ with a steady flow rate of 2 m^3/s. The bend has an angle of 30° and lies in a horizontal plane. The volume of oil in the bend is 1.2 m^3, and the empty weight of the bend is 4 kN. Assume the pressure along the centerline of the bend is constant with a value of 75 kPa gage. Find the net force required to hold the bend in place.

Sketch:

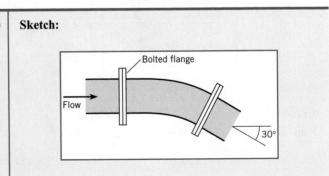

Problem Definition

Situation: Crude oil flows through a 30° pipe bend.

Find: Force (in kN) required to hold bend in place.

Assumptions: Pressure is constant through bend.

Properties: $S_{\text{oil}} = 0.94$.

Plan

From the continuity equation, the inlet and outlet mass flows are the same.

1. Select a control volume that accommodates the pressure forces and reaction forces at the flanges.
2. Sketch the force diagram.
3. Sketch the momentum diagram.
4. Use the vector form of the momentum equation, Eq. (6.6).
5. Evaluate the sum of the forces.
6. Evaluate the momentum terms.
7. Calculate the reaction force.

Solution

1. The control volume selected is shown. The control volume is stationary. The z-direction is outward from the page.

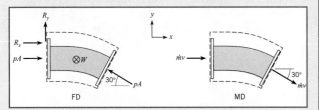

FD   MD

2. The force diagram shows pressure forces and the component reaction forces.
3. Vector form of momentum equation

$$\sum \boldsymbol{F} = \frac{d}{dt}\bigg|_{\text{cv}} \int \rho \boldsymbol{v}\, d\mathcal{V} + \sum_{\text{cs}} \dot{m}_o \boldsymbol{v}_o - \sum_{\text{cs}} \dot{m}_i \boldsymbol{v}_i$$

4. Sum of the forces: The weight of the pipe and fluid therein is W and acts in the negative z-direction.

$$\sum \boldsymbol{F} = (R_x + pA - pA\cos 30°)\mathbf{i} + (R_y + pA\sin 30°)\mathbf{j}$$
$$+ (R_z - W)\mathbf{k}$$

5. Momentum terms
   - Accumulation term for stationary control volume is
   $$\frac{d}{dt}\bigg|_{\text{cv}} \int \rho \boldsymbol{v}\, d\mathcal{V} = 0.$$
   - Momentum outflow is
   $$\sum_{\text{cs}} \dot{m}_o \boldsymbol{v}_o = (\dot{m}v\cos 30°)\mathbf{i} - (\dot{m}v\sin 30°)\mathbf{j}.$$
   - Momentum inflow is $\sum_{\text{cs}} \dot{m}_i \boldsymbol{v}_i = (\dot{m}v)\mathbf{i}$.

6. Reaction force
$$(R_x + pA - pA\cos 30°)\mathbf{i} + (R_y + pA\sin 30°)\mathbf{j} + (R_z - W)\mathbf{k}$$
$$= [\dot{m}v(\cos 30 - 1)]\mathbf{i} - (\dot{m}v\sin 30°)\mathbf{j}$$

   - Equating components
   $$R_x + pA - pA\cos 30° = \dot{m}v\cos 30° - \dot{m}v$$
   $$R_y + pA\sin 30° = -\dot{m}v\sin 30°$$
   $$R_z - W = 0$$

   - Pressure force
   $$pA = (75\ \text{kN/m}^2)(\pi \times 0.5^2\ \text{m}^2) = 58.9\ \text{kN}$$

   - Fluid speed
   $$v = Q/A = \frac{(2\ \text{m}^3/\text{s})}{(\pi \times 0.5^2\ \text{m}^2)} = 2.55\ \text{m/s}$$

   - Momentum flux
   $$\dot{m}v = \rho Q v = (0.94 \times 1000\ \text{kg/m}^3)(2\ \text{m}^3/\text{s})(2.55\ \text{m/s})$$
   $$= 4.79\ \text{kN}$$

Reaction force in x-direction
$$R_x = -(pA + \dot{m}v)(1 - \cos 30°)$$
$$= -(58.9 + 4.79)(\text{kN})(1 - \cos 30°) = \boxed{-8.53\ \text{kN}}$$

Reaction force in y-direction
$$R_y = -(pA + \dot{m}v)\sin 30°$$
$$= -(58.9 + 4.79)(\text{kN})(\sin 30°) = \boxed{-31.8\ \text{kN}}$$

Reaction force in z-direction. (The bend weight includes the oil plus the empty pipe).

$$W = \gamma \mathcal{V} + 4\ \text{kN}$$
$$= (0.94 \times 9.81\ \text{kN/m}^3)(1.2\ \text{m}^3) + 4\ \text{kN} = \boxed{15.1\ \text{kN}}$$

Reaction force vector
$$\mathbf{R} = (-8.53\ \text{kN})\mathbf{i} + (-31.8\ \text{kN})\mathbf{j} + (15.1\ \text{kN})\mathbf{k}$$

Example 6.7 illustrates how to calculate the force required to restrain a reducing bend, a bend in which the fluid speed and direction both change, by application of the component form of the momentum equation.

EXAMPLE 6.7   WATER FLOW THROUGH REDUCING BEND

Water flows through a 180° reducing bend, as shown. The discharge is 0.25 m³/s, and the pressure at the center of the inlet section is 150 kPa gage. If the bend volume is 0.10 m³, and it is assumed that the Bernoulli equation is valid, what force is required to hold the bend in place? The metal in the bend weighs 500 N. The water density is 1000 kg/m³. The bend is in the vertical plane.

Sketch:

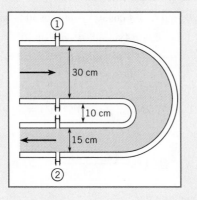

Problem Definition

Situation: Water flow through reducing bend.

Find: Force (in newtons) required to hold bend in place.

Assumptions:

1. The Bernoulli equation is valid.
2. Neglect pipe wall thickness.

Properties: $\rho = 1000$ kg/m³.

Plan

The flow is steady, so $Q_1 = Q_2 = Q$.

1. Select control volume that encloses bend and the reaction force acts on the control surface.
2. Sketch the force diagram.
3. Sketch the momentum diagram.
4. Apply the component form of the momentum equation in the x- and z-directions, Eqs. (6.7a) and (6.7c).
5. Evaluate the force terms.
6. Evaluate the momentum terms.
7. Solve momentum equations for reaction forces.
8. Calculate the inlet and outlet speed.

9. Apply the Bernoulli equation to find the outlet pressure.
10. Calculate the reaction force.

Solution

1. The control volume selected is shown. The control volume is stationary.

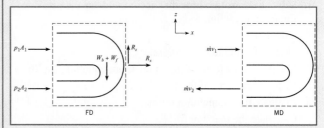

2. There are two forces due to pressure and a reaction force component in the x-direction, and there are weight and reaction forces component in the z-direction.

3. There is inlet and outlet momentum flux in x-direction

4. Momentum equations in x- and z-directions

$$\sum F_x = \frac{d}{dt}\int_{cv} v_x \rho \, d\Psi + \sum_{cs} \dot{m}_o v_{ox} - \sum_{cs} \dot{m}_i v_{ix}$$

$$\sum F_z = \frac{d}{dt}\int_{cv} v_z \rho \, d\Psi + \sum_{cs} \dot{m}_o v_{oz} - \sum_{cs} \dot{m}_i v_{iz}$$

5. Summation of forces in x- and z-directions

$$\sum F_x = p_1 A_1 + p_2 A_2 + R_x$$

$$\sum F_z = R_z - W_b - W_f$$

6. Evaluation of momentum terms
   - Accumulation terms, steady flow

   $$\frac{d}{dt}\int_{cv} v_x \rho \, d\Psi = 0, \text{ and } \frac{d}{dt}\int_{cv} v_z \rho \, d\Psi = 0.$$

   - Momentum outflow

   $$\sum_{cs} \dot{m}_o v_{ox} = \dot{m}(-v_2) = -\rho Q v_2, \text{ and } \sum_{cs} \dot{m}_o v_{oz} = 0.$$

   - Momentum inflow

   $$\sum_{cs} \dot{m}_i v_{ix} = \dot{m} v_1 = \rho Q v_1, \text{ and } \sum_{cs} \dot{m}_i v_{iz} = 0.$$

7. Solution for reaction forces
   - x-direction

   $$p_1 A_1 + p_2 A_2 + R_x = -\rho Q(v_2 + v_1)$$
   $$R_x = -(p_1 A_1 + p_2 A_2) - \rho Q(v_2 + v_1)$$

   - z-direction

   $$R_z = W_b + W_f$$

8. Inlet and outlet speeds

   $$v_1 = \frac{Q}{A_1} = \frac{0.25 \text{ m}^3/\text{s}}{\pi/4 \times 0.3^2 \text{ m}^2} = 3.54 \text{ m/s}$$

   $$v_2 = \frac{Q}{A_2} = \frac{0.25 \text{ m}^3/\text{s}}{\pi/4 \times 0.15^2 \text{ m}^2} = 14.15 \text{ m/s}$$

9. Outlet pressure (the Bernoulli equation between sections 1 and 2)

   $$p_1 + \frac{\rho v_1^2}{2} + \gamma z_1 = p_2 + \frac{\rho v_2^2}{2} + \gamma z_2$$

   From diagram, neglecting pipe wall thickness, $z_1 - z_2 = 0.325$ m.

   $$p_2 = p_1 + \frac{\rho(v_1^2 - v_2^2)}{2} + \gamma(z_1 - z_2)$$

   $$= 150 \text{ kPa} + \frac{(1000)(3.54^2 - 14.15^2)\text{Pa}}{2}$$

   $$+ (9810)(0.325)\text{Pa}$$

   $$= 59.3 \text{ kPa}$$

10. Reaction force
    - Pressure forces

    $$p_1 A_1 + p_2 A_2 = (150 \text{ kPa})(\pi \times 0.3^2/4 \text{ m}^2)$$
    $$+ (59.3 \text{ kPa})(\pi \times 0.15^2/4 \text{ m}^2)$$
    $$= 11.6 \text{ kN}$$

    - Momentum flux

    $$\rho Q(v_2 + v_1) = (1000 \text{ kg/m}^3)(0.25 \text{ m}^3)$$
    $$\times (14.15 + 3.54)(\text{m/s})$$
    $$= 4420 \text{ N}$$

    - Reaction force components

    $$R_x = -(11.6 \text{ kN}) - (4.42 \text{ kN})$$
    $$= \boxed{-16.0 \text{ kN}}$$
    $$R_z = W_b + W_f$$
    $$= 500 \text{ N} + (9810 \text{ N/m}^3)(0.1 \text{ m}^3)$$
    $$= \boxed{1.48 \text{ kN}}$$

Additional Applications

Evaluation of Drag Force on Wind Tunnel Model

The previous examples in this chapter were cases in which the velocity across each flow section was constant. However in many applications the velocity is not uniformly distributed across the control surface, and the momentum flux must be evaluated by integration.

Example 6.8 involves calculating the force on a model in a wind tunnel by application on the integral form of the momentum equation. The downstream velocity profile is not uniform and requires integration to establish the momentum outflow.

EXAMPLE 6.8   DRAG FORCE ON WIND-TUNNEL MODEL

The drag force of a bullet-shaped device may be measured using a wind tunnel. The tunnel is round with a diameter of 1 m, the pressure at section 1 is 1.5 kPa gage, the pressure at section 2 is 1.0 kPa gage, and air density is 1.0 kg/m³. At the inlet, the velocity is uniform with a magnitude of 30 m/s. At the exit, the velocity varies linearly as shown in the sketch. Determine the drag force on the device and support vanes. Neglect viscous resistance at the wall, and assume pressure is uniform across sections 1 and 2.

Sketch:

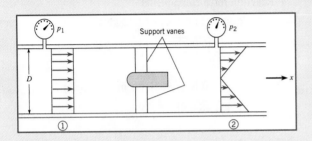

Problem Definition

Situation: Wind tunnel test on suspended model with upstream and downstream velocity distributions and pressures measured.

Find: Drag force (in newtons) on model.

Assumptions: Steady flow.

Properties: $\rho = 1.0 \text{ kg/m}^3$.

Plan

1. Select a control volume that encloses the model.
2. Sketch the force diagram.
3. Sketch the momentum diagram.
4. The downstream velocity profile is not uniformly distributed. Apply the integral form of the momentum equation, Eq. (6.5).
5. Evaluate the sum of forces.
6. Determine velocity profile at section 2 by application of continuity equation.
7. Evaluate the momentum terms.
8. Calculate drag force on model.

Solution

1. The control volume selected is shown. The control volume is stationary.

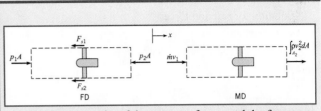

2. The forces consist of the pressure forces and the force on the model support struts cut by the control surface. The drag force on the model is equal and opposite to the force on the support struts: $F_D = F_{s1} + F_{s2}$.

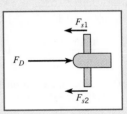

3. There is inlet and outlet momentum flux.
4. Integral form of momentum equation in x-direction

$$\sum F_x = \frac{d}{dt}\int_{cv} \rho v_x \, d\!V + \int_{cs} \rho v_x (V \cdot dA)$$

On cross-section 1, $V \cdot dA = -v_x dA$, and on cross-section 2, $V \cdot dA = v_x dA$, so

$$\sum F_x = \frac{d}{dt}\int_{cv} \rho v_x \, d\!V - \int_1 \rho v_x^2 \, dA + \int_2 \rho v_x^2 \, dA$$

5. Evaluation of force terms.

$$\sum F_x = p_1 A - p_2 A - (F_{s1} + F_{s2})$$

$$= p_1 A - p_2 A - F_D$$

6. Velocity profile at section 2.
   Velocity is linear in radius, so choose
   $v_2 = v_1 K(r/r_o)$, where r_o is the tunnel radius and K is a proportionality factor to be determined.

$$Q_1 = Q_2$$

$$A_1 v_1 = \int_{A_2} v_2(r) \, dA = \int_0^{r_o} v_1 K(r/r_o) 2\pi r \, dr$$

$$\pi r_o^2 v_1 = 2\pi v_1 K \frac{1}{3} r_o^2$$

$$K = \frac{3}{2}$$

7. Evaluation of momentum terms

   • Accumulation term for steady flow is $\dfrac{d}{dt}\displaystyle\int_{cv} \rho v_x \, d\!V = 0$

- Momentum at cross-section 1 with $v_x = v_1$ is

$$\int_1 \rho v_x^2 \, dA = \rho v_1^2 A = \dot{m} v_1$$

- Momentum at cross-section 2 is

$$\int_2 \rho v_x^2 \, dA = \int_0^{r_o} \rho \left[\frac{3}{2} v_1 \left(\frac{r}{r_o} \right) \right]^2 2 \pi r dr = \frac{9}{8} \dot{m} v_1$$

8. Drag force

$$p_1 A - p_2 A - F_D = \dot{m} v_1 \left(\frac{9}{8} - 1 \right)$$

$$F_D = (p_1 - p_2)A - \frac{1}{8} \rho A v_1^2$$

$$= (\pi \times 0.5^2 \text{ m}^2)(1.5 - 1.0)(10^3) \text{ N/m}^2$$

$$- \frac{1}{8}(1 \text{ kg/m}^3)(\pi \times 0.5^2 \text{ m}^2)(30 \text{ m/s})^2$$

$$F_D = \boxed{304 \text{ N}}$$

Force on a Sluice Gate

Sluice gates are often used in hydroelectric facilities to control the flow rate of water from the dam onto a spillway. Example 6.9 provides another example on the use of the momentum equation to predict the force on a sluice gate by accounting for the momentum flux.

EXAMPLE 6.9   FORCE ON A SLUICE GATE

A sluice gate is used to control the water flow rate over a dam. The gate is 20 ft wide, and the depth of the water above the bottom of the sluice gate is 16 ft. The depth of the water upstream of the gate is 20 ft, and the depth downstream is 3 ft. Estimate the flow rate under the gate and the force on the gate. The water density is 62.4 lbm/ft³.

Sketch:

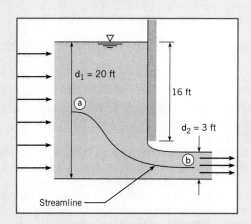

$d_1 = 20$ ft

a

16 ft

$d_2 = 3$ ft

b

Streamline

Problem Definition

Situation: Sluice gate to control water flow over dam.

Find:

1. Flow rate (cfs) under the gate.
2. Force (tons) on gate.

Assumptions:

1. Velocity profiles are uniformly distributed.
2. Streamlines are straight at stations 1 and 2.
3. Viscous effects are negligible.

Properties: $\rho = 62.4$ lbm/ft³ $= 1.94$ slugs/ft³, and $\gamma = 62.4$ lbf/ft³.

Plan

1. Select a control volume such that the control surface includes the force on the gate and passes through the outlet flow where the velocity profile is uniformly distributed.
2. Sketch the force diagram.
3. Sketch the momentum diagram.
4. Use the component form of the momentum equation in x-direction, Eq. (6.7a).
5. Use the Bernoulli and continuity equations to find the velocities at the inlet and outlet.
6. Evaluate the sum of the forces.
7. Evaluate the momentum terms.
8. Calculate the force on the gate.

Solution

1. The control volume selected is shown. The control volume is stationary.

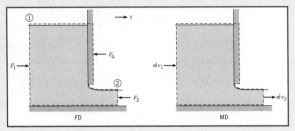

2. The force diagram shows forces due to pressure and the force on the gate.
3. The momentum diagram shows an influx and outflux of momentum.
4. Component form of momentum equation in x-direction.

$$\sum F_x = \frac{d}{dt}\int_{cv} v_x \rho\, d\Psi + \sum_{cs} \dot{m}_o v_{ox} - \sum_{cs} \dot{m}_i v_{ix}$$

5. The Bernoulli equation between points a and b along the streamline.

$$\frac{p_a}{\gamma} + z_a + \frac{v_1^2}{2g} = \frac{p_b}{\gamma} + z_b + \frac{v_2^2}{2g}$$

The piezometric pressure is constant across sections 1 and 2, so $\frac{p_a}{\gamma} + z_a = d_1$ and $\frac{p_b}{\gamma} + z_b = d_2$. From continuity equation, Eq. (5.27), $v_1 d_1 w = v_2 d_2 w$ where w is the flow width. Combine the Bernoulli and continuity equations.

$$2g(d_1 - d_2) = v_2^2 - v_1^2 = v_2^2\left(1 - \frac{d_2^2}{d_1^2}\right)$$

$$v_2 = \frac{1}{\sqrt{1 - \frac{d_2^2}{d_1^2}}}\sqrt{2g(d_1 - d_2)}$$

Velocities and discharge

$$v_2 = \frac{1}{\sqrt{1 - \left(\frac{3}{20}\right)^2}}\sqrt{2 \times 32.2 \text{ ft/s}^2 \times (20 - 3) \text{ ft}} = 33.5 \text{ ft/s}$$

$$v_1 = \frac{d_2}{d_1} v_2 = 33.5 \text{ ft/s} \times \frac{3 \text{ ft}}{20 \text{ ft}} = 5.02 \text{ ft/s}$$

$$Q = v_2 d_2 w = 33.5 \text{ ft/s} \times 3 \text{ ft} \times 20 \text{ ft} = \boxed{2010 \text{ ft}^3/\text{s}}$$

6. Sum of the forces from force diagram

$$\sum F_x = F_1 - F_2 - F_G$$

From equation for force on planar surface, Eq. (3.23), $F = \bar{p}A$,

$$F_1 = \frac{\gamma d_1}{2} d_1 w$$

$$F_2 = \frac{\gamma d_2}{2} d_2 w$$

7. Evaluation of momentum terms

- Accumulation term for steady flow is $\dfrac{d}{dt}\displaystyle\int_{cv} v_x \rho\, d\Psi = 0.$

- Momentum inflow with one inlet is $\displaystyle\sum_{cs} \dot{m}_i v_{ix} = \dot{m}v_1.$

- Momentum outflow with one outlet is $\displaystyle\sum_{cs} \dot{m}_o v_{ox} = \dot{m}v_2.$

8. Force on sluice gate

$$\frac{\gamma}{2}d_1^2 w - \frac{\gamma}{2}d_2^2 w - F_G = \dot{m}(v_2 - v_1)$$

$$F_G = \frac{\gamma}{2}w(d_1^2 - d_2^2) + \rho Q(v_1 - v_2)$$

$$F_G = \frac{62.4 \text{ lbf/ft}^3}{2} \times 20 \text{ ft} \times (20^2 - 3^2)(\text{ft})^2$$

$$+ 1.94 \text{ slug/ft}^3 \times 2010 \text{ ft}^3/\text{s}$$

$$\times (5.02 - 33.5)(\text{ft/s})$$

$$= 1.33 \times 10^5 \text{ lbf} \times \frac{1 \text{ ton}}{2000 \text{ lbf}} = \boxed{66.5 \text{ tons}}$$

Review

1. The sluice gate is often used in hydroelectric facilities to control the flow rate of water from the dam onto a spillway.
2. Note that if a hydrostatic pressure distribution were assumed over the sluice gate, the force would be 79.9 tons, which is larger than predicted using the momentum equation. This is because the gage pressure at the bottom of the sluice gate will be zero and not the pressure predicted using the hydrostatic equation.

Moving Control Volumes

All the applications of the momentum equation up to this point have involved a stationary control volume. However, in some problems it may be more useful to attach the control volume to a moving body. The purpose of this section is to illustrate how to apply the momentum equation to nonstationary control volumes.

As discussed in Section 6.1, the velocity $\mathbf{v}$ in the momentum equation must be relative to an inertial reference frame. When applying Eq. (6.6) or Eqs. (6.7a–6.7c) each mass flow rate is calculated using the velocity with respect to the control surface, but the velocity $\mathbf{v}$ must be evaluated with respect to an inertial reference frame.

Example 6.10 illustrates the use of a moving control volume to find the force on moving block due to an impinging water jet. The example emphasizes the selection of the reference frame. It shows that two different inertial reference frames can be used to address the same problem.

EXAMPLE 6.10   JET IMPINGING ON MOVING BLOCK

A stationary nozzle produces a water jet with a speed of 50 m/s and a cross-sectional area of 5 cm². The jet strikes a moving block and is deflected 90° relative to the block. The block is sliding with a constant speed of 25 m/s on a surface with friction. The density of the water is 1000 kg/m³. Find the frictional force F acting on the block.

Solve the problem using two different inertial reference frames: (a) the moving block and (b) the stationary nozzle.

Sketch:

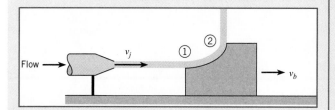

Problem Definition

Situation: Jet impinges on block moving at constant velocity.

Find: The force (in newtons) on the block using

(a) the block as the inertial reference frame.
(b) the nozzle as the inertial reference frame.

Plan

Two different inertial reference frames will be used. Case (a) will use the moving cart, which is a valid inertial frame because it moves at a constant velocity. Case (b) will use the stationary nozzle location.

1. Select a control volume that moves with the block.
2. Sketch the force diagram.

3. Sketch the momentum diagram.
4. Apply the component form of the momentum equation in the x-direction, Eq. (6.7a).
5. Evaluate the sum of the forces.
6. Evaluate the momentum terms using (a) the moving block and (b) the stationary nozzle as inertial reference frames.
7. Evaluate mass flow rate.
8. Calculate force on cart.

Solution

1. The control volume selected is shown in the sketch. The control volume is not stationary.

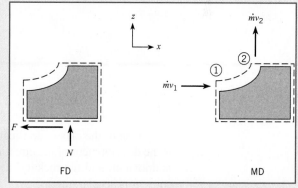

2. The force diagram shows one force in the horizontal direction.
3. The momentum diagram shows an influx and outflux of momentum.

4. Momentum equation

$$\sum F_x = \frac{d}{dt} \int_{cv} v_x \rho \, d\Psi + \sum_{cs} \dot{m}_o v_{ox} - \sum_{cs} \dot{m}_i v_{ix}$$

5. The sum of the forces

$$\sum F_x = -F$$

6. Evaluation of terms in momentum equation
   (a) Inertial reference frame on cart

- Accumulation term with $v_x = 0$ is $\dfrac{d}{dt} \int_{cv} v_x \rho \, d\Psi = 0$.

- Momentum inflow for x-component of velocity at station 1, $v_{ix} = v_j - v_b$, is

$$\sum_{cs} \dot{m}_i v_{ix} = \dot{m}(v_j - v_b)$$

- Momentum outflow for x-component of velocity at station 2, $v_{ox} = 0$, is

$$\sum_{cs} \dot{m}_o v_{ox} = 0$$

(b) Inertial reference frame at nozzle

- Accumulation term with $v_x = v_b = $ constant, is

$$\frac{d}{dt} \int_{cv} v_x \rho \, d\Psi = 0.$$

- Momentum inflow for x-component of velocity at station 1, $v_{xi} = v_j$, is

$$\sum_{cs} \dot{m}_i v_{ix} = \dot{m} v_j$$

- Momentum outflow for x-component of velocity at station 2, $v_{xo} = v_b$, is

$$\sum_{cs} \dot{m}_o v_{ox} = \dot{m} v_b$$

7. Mass flow rate. Since flow is steady with respect to the block, $\dot{m}_i = \dot{m}_o = \dot{m}$.

$$\dot{m} = \rho A (v_j - v_b)$$

8. Evaluate force.
   (a) Moving block as inertial reference frame

$$-F = -\rho A (v_j - v_b)^2$$

$$F = \rho A (v_j - v_b)^2$$

(b) Stationary nozzle as inertial reference frame

$$-F = \dot{m} v_b - \dot{m} v_j = -\dot{m}(v_j - v_b)$$

$$F = \rho A (v_j - v_b)^2$$

Force on cart

$$F_x = (1000 \text{ kg/m}^3)(5 \times 10^{-4} \text{ m}^2)(50 - 25)^2 (\text{m/s})^2$$

$$F_x = \boxed{312 \text{ N}}$$

Review

Note that the same answer for force is obtained independent of the inertial reference frame chosen.

One of the classic applications of the momentum equation to a moving control volume is the development of the equation of motion for a rocket. In this situation the control volume is drawn around the rocket, but the rocket cannot be used as an inertial reference frame because it is accelerating.

Up to this point, the accumulation term in the momentum equation has been zero. In accelerating control volumes, the accumulation term is not zero and plays a very important role in the analysis.

A rocket moving vertically upward with a speed v_r measured with respect to the ground is shown in Fig. 6.5. Exhaust gases leave the engine nozzle (area A_e) at a speed V_e relative to the rocket nozzle with a gage pressure of p_e. The goal is to obtain the equation of motion of the rocket which can be used to predict performance.

The control volume is drawn around and accelerates with the rocket. The force and momentum diagrams are shown in Fig. 6.6. There is a drag force of D and a weight of W acting

Figure 6.5

Vertical launch of rocket.

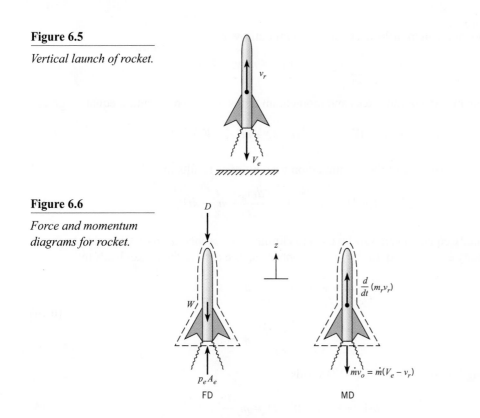

Figure 6.6

Force and momentum diagrams for rocket.

downward. There is a pressure force of $p_e A_e$ on the nozzle exit plane acting upward. The summation of the forces in the z-direction is

$$\sum F_z = p_e A_e - W - D \tag{6.10}$$

There is only one momentum flux out of the rocket nozzle, $\dot{m}v_o$. The speed v_o must be referenced to an inertial reference frame, which in this case is chosen as the ground. The speed of the exit gases with respect to the ground is

$$v_o = (v_r - V_e) \tag{6.11}$$

since the rocket is moving upward with speed v_r with respect to the ground, and the exit gases are moving downward at speed V_e with respect to the rocket. There is also a momentum accumulation rate in the rocket of $d/(m_r v_r)dt$.

The component momentum equation, Eq. (6.7c), in the z-direction is

$$\sum F_z = \frac{d}{dt}\int_{cv} v_z \rho \, d\Psi + \sum_{cs} \dot{m}_o v_{oz} - \sum_{cs} \dot{m}_i v_{iz}$$

The velocity inside the control volume is the speed of the rocket, v_r, so the accumulation term becomes

$$\frac{d}{dt}\left(\int_{cv} v_z \rho \, d\Psi\right) = \frac{d}{dt}\left[v_r\int_{cv} \rho \, d\Psi\right] = \frac{d}{dt}(m_r v_r)$$

There is no momentum inflow. The momentum outflow is

$$\sum_{cs} \dot{m}_o v_{oz} = \dot{m} v_0 = \dot{m}(v_r - V_e)$$

Substituting the sum of the forces and momentum terms into the momentum equation gives

$$p_e A_e - W - D = \frac{d}{dt}(m_r v_r) + \dot{m}(v_r - V_e) \tag{6.12}$$

By taking the derivative of the accumulation term by parts results in

$$p_e A_e - W - D = m_r \frac{dv_r}{dt} + v_r\left(\frac{dm_r}{dt} + \dot{m}\right) - \dot{m}V_e \tag{6.13}$$

The continuity equation can now be used to eliminate the second term on the right. Applying the continuity equation, Eq. (5.25), to the control surface around the rocket leads to

$$\frac{d}{dt}\int_{cv} \rho d\Psi + \sum \dot{m}_o - \sum \dot{m}_i = 0$$

$$\frac{dm_r}{dt} + \dot{m} = 0 \tag{6.14}$$

Substituting Eq. (6.14) into Eq. (6.13) yields

$$\dot{m}V_e + p_e A_e - W - D = m_r \frac{dv_r}{dt} \tag{6.15}$$

The sum of the momentum outflow and the pressure force at the nozzle exit is identified as the thrust of the rocket

$$T = \dot{m}V_e + p_e A_e = \rho_e A_e V_e^2 + p_e A_e$$

so Eq. (6.15) simplifies to

$$m_r \frac{dv_r}{dt} = T - D - W \tag{6.16}$$

which is the equation used to predict and analyze rocket performance.

Integration of Eq. (6.16) leads to one of the fundamental equations for rocketry: the burnout velocity or the velocity achieved when all the fuel is burned. Neglecting the drag and weight, the equation of motion reduces to

$$T = m_r \frac{dv_r}{dt} \tag{6.17}$$

The instantaneous mass of the rocket is given by $m_r = m_i - \dot{m}t$, where m_i is the initial rocket mass and t is the time from ignition. Substituting the expression for mass into Eq. (6.17) and integrating with the initial condition $v_r(0) = 0$ results in

$$v_{bo} = \frac{T}{\dot{m}} \ln \frac{m_i}{m_f} \tag{6.18}$$

where v_{bo} is the burnout velocity and m_f is the final (or payload) mass. The ratio $T/\dot{m}$ is known as the specific impulse, I_{sp}, and has units of velocity.

EXAMPLE 6.11   PROPELLANT MASS RATIO FOR ACHIEVING ORBITAL VELOCITY

A single-stage rocket utilizing a liquid oxygen/kerosene propellant has a specific impulse of 3200 m/s. The orbital velocity for an earth satellite is 7600 m/s. What would be the ratio of propellant mass to total initial mass to achieve orbital velocity?

Problem Definition

Situation: Rocket launch to achieve orbital velocity.

Find: Ratio of propellent mass to initial mass.

Plan

1. Use Eq. (6.18) to calculate initial/final mass ratio.
2. Calculate the propellant/initial mass ratio using $m_p = m_i - m_f$.

Solution

1. From Eq. (6.18)

$$v_{bo} = I_{sp} \ln \frac{m_i}{m_f}$$

$$\frac{m_i}{m_f} = \exp\left(\frac{v_{bo}}{I_{sp}}\right) = \exp\left(\frac{7600}{3200}\right) = 10.7$$

2. Solve for propellant/initial mass ratio:

$$\frac{m_p}{m_i} = 1 - \frac{m_f}{m_i}$$

$$= 1 - \frac{1}{10.7} = \boxed{0.906}$$

Review

For single-stage rockets, a very large fraction of the initial mass must be propellant to achieve orbital speeds. For this reason, multi-stage rockets are used in space applications.

Example 6.11 illustrates the use of the equation for rocket burnout velocity for conditions necessary to achieve orbital velocity for a earth satellite.

Water Hammer: Physical Description

Whenever a valve is closed in a pipe, a positive pressure wave is created upstream of the valve and travels up the pipe at the speed of sound. In this context a positive pressure wave is defined as one for which the pressure is greater than the existing steady-state pressure. This pressure wave may be great enough to cause pipe failure. Therefore, a basic understanding of this process, which is called *water hammer,* is necessary for the proper design and operation of such systems. The simplest case of water hammer will be considered here. For a more comprehensive treatment of the subject, the reader is referred to Chaudhry (1) and Streeter and Wylie (2).

Consider flow in the pipe shown in Fig. 6.7. Initially the valve at the end of the pipe is only partially open (Fig. 6.7a); consequently, an initial velocity V and initial pressure p_0 exist in the pipe. At time $t = 0$ it is assumed that the valve is instantaneously closed, thus creating a pressure increase behind the valve and a pressure wave that travels from the valve toward the reservoir at the speed of sound, c. All the water between the pressure wave and the upper end of the pipe will have the initial velocity V, but all the water on the other side of the pressure wave (between the wave and the valve) will be at rest. This condition is shown in Fig. 6.7b. Once the pressure wave reaches the upper end of the pipe (after time $t = L/c$), it can be visualized that all of the water in the pipe will be under a pressure $p_0 + \Delta p$; however, the pressure in the reservoir at the end of the pipe is only p_0. This imbalance of pressure at the reservoir end causes the water to flow from the pipe back into the reservoir with a velocity V. Thus a new pressure wave is formed that travels toward the valve end of the pipe (Fig. 6.7c), and the pressure on the reservoir side of the wave is reduced to p_0. When this wave finally reaches the valve, all the water in the pipe is flowing toward the reservoir with a velocity V. This condition is only momentary, however, because the closed valve prevents any sustained flow.

Figure 6.7

Water hammer process.
(a) Initial condition.
(b) Condition during
time $0 < t < L/c$.
(c) Condition during time
$L/c < t < 2L/c$.
(d) Condition during time
$2L/c < t < 3L/c$.
(e) Condition during time
$3L/c < t < 4L/c$.
(f) Condition at time
$t = 4L/c$.

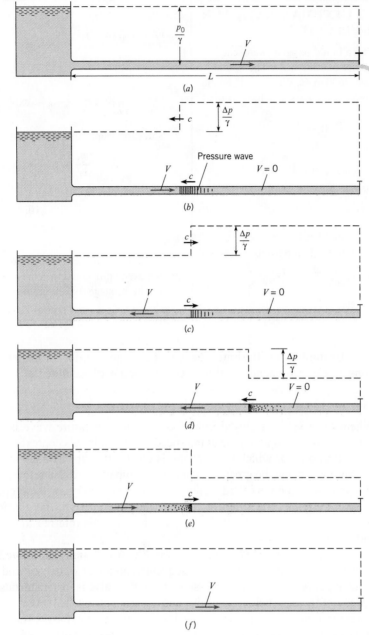

Next, during time $2L/c < t < 3L/c$, a rarefied wave of pressure $(p < p_0)$ travels up to the reservoir, as shown in Fig. 6.7d. When the wave reaches the reservoir, all the water in the pipe has a pressure less than that in the reservoir. This imbalance of pressure causes flow to be established again in the entire pipe, as shown in Fig. 6.7f, and the condition is exactly the same as in the initial condition (Fig. 6.7a). Hence the process will repeat itself in a periodic manner.

Figure 6.8

Variation of water hammer pressure with time at two points in a pipe. (a) Location: adjacent to valve. (b) Location: at midpoint of pipe. (c) Actual variation of pressure near valve.

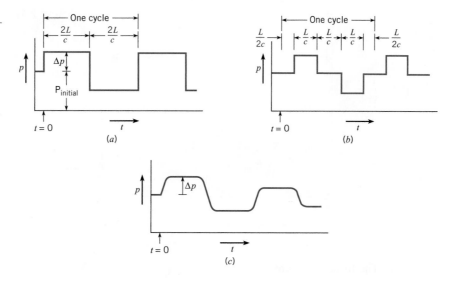

From this description, it may be seen that the pressure in the pipe immediately upstream of the valve will be alternately high and low, as shown in Fig. 6.7a. A similar observation for the pressure at the midpoint of the pipe reveals a more complex variation of pressure with time, as shown in Fig. 6.8b. Obviously, a valve cannot be closed instantaneously, and viscous effects, which were neglected here, will have a damping effect on the process. Therefore, a more realistic pressure–time trace for the point just upstream of the valve is given in Fig. 6.8c. The finite time of closure erases the sharp discontinuities in the pressure trace that were present in Fig. 6.8a. However, it should be noted that the maximum pressure developed at the valve will be virtually the same as for instantaneous closure if the time of closure is less than $2L/c$. That is, the change in pressure will be the same for a given change in velocity unless the negative wave from the reservoir mitigates the positive pressure, and it takes a time $2L/c$ before this negative wave can reach the valve. The value $2L/c$ is called the *critical time of closure* and is given the symbol t_c.

Magnitude of Water Hammer Pressure and Speed of Pressure Wave

The quantitative relations for water hammer can be analyzed with the momentum equation by letting the control volume either move with the pressure wave, thus creating steady motion, or be fixed, thus retaining the inherently unsteady character of the process. To illustrate the use of the momentum equation with unsteady motion, the latter approach will be taken. Consider a pressure wave in a rigid pipe, as shown in Fig. 6.9. The density, pressure, and velocity of the fluid on the reservoir side of the pressure wave are ρ, p, and V, respectively, and the similar quantities on the valve side of the wave are $\rho + \Delta\rho$, $p + \Delta p$, and 0. Because the wave in this case is traveling from the valve to the reservoir, its distance from the valve at any time t is given as ct. The momentum equation can now be applied to the flow in the control volume. Let the x-direction be along the pipe. The equation for x-momentum, Eq. 6.7a, simplifies to

$$\sum F_x = \frac{d}{dt}\int_{cv} v_x \rho \, d\mathcal{V} - \dot{m}v_i$$

Figure 6.9

Pressure wave in a pipe.

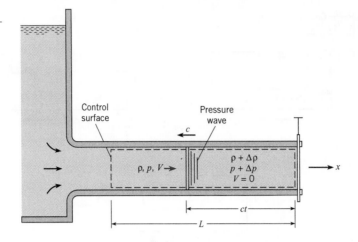

The force terms are given by

$$\sum F_x = pA - (p + \Delta p)A$$

The inlet momentum flow is given by $\dot{m}v_i = \rho A V^2$. The momentum within the control volume decreases with time because fluid that is in motion stops as the pressure wave passes by. Evaluation of the momentum accumulation term gives

$$\frac{d}{dt}\int_{cv} v_x \rho \, d\Psi = \frac{d}{dt}[V\rho(L - ct)A]$$

$$= -V\rho cA$$

When force and momentum terms are substituted into the momentum equation, one obtains

$$pA - (p + \Delta p)A = -\rho V^2 A - \rho V cA$$

This reduces to

$$\Delta p = \rho V^2 + \rho V c$$

In this equation the first term on the right-hand side is usually negligible with respect to the second term on the right, because for liquids c is much greater than V. Consequently, the equation simplifies to

$$\Delta p = \rho V c \tag{6.19}$$

The speed of the pressure wave can be obtained by applying the continuity equation to the control volume in Fig. 6.9. The continuity equation is

$$0 = \sum \dot{m}_o - \sum \dot{m}_i + \frac{d}{dt}\int_{cv} \rho \, d\Psi$$

and when applied to Fig. 6.9 results in

$$0 = \dot{m}_i + \frac{d}{dt}[\rho(L - ct)A + (\rho + \Delta \rho)ctA]$$

because there is no mass flow out of the control volume. The mass flow rate is given by $\dot{m}_i = \rho V A$, so the continuity equation reduces to

$$\frac{\Delta \rho}{\rho} = \frac{V}{c}$$

or
$$c = \frac{V}{\Delta \rho / \rho} \tag{6.20}$$

However, by definition $E_v = \Delta p / (\Delta \rho / \rho)$. Therefore,

$$\frac{\Delta \rho}{\rho} = \frac{\Delta p}{E_v} \tag{6.21}$$

Now when $\Delta \rho / \rho$ is eliminated between Eqs. (6.20) and (6.21), the result is

$$c = \frac{V E_v}{\Delta p} \tag{6.22}$$

From Eq. (6.19), $\Delta p = \rho V c$. Therefore, Eq. (6.22) becomes

$$c = \sqrt{\frac{E_v}{\rho}} \tag{6.23}$$

Thus, by application of the momentum and continuity equations, expressions for both Δp and c have been derived.

Example 6.12 illustrates how to calculate the pressure rise due to the water hammer effect.

EXAMPLE 6.12   PRESSURE RISE DUE TO WATER HAMMER EFFECT

A rigid pipe leading from a reservoir is 3000 ft long, and water is flowing through it with a velocity of 4 ft/s. If the initial pressure at the downstream end is 40 psig, what maximum pressure will develop at the downstream end when a rapid-acting valve at that end is closed in 1 s?

Problem Definition

Situation: Water flowing in pipe and valve closed quickly.

Find: Maximum pressure (psig) at downstream end.

Assumptions: Water temperature is 60°F.

Properties: From Table A.5, $E_v = 3.2 \times 10^5$ lbf/in^2, and $\rho = 1.94$ slugs/ft^3.

Plan

1. Calculate the speed of sound in the water from Eq. (6.23).
2. Calculate the critical closure time, t_c.
3. Check to ensure that valve closure time is less than t_c.
4. Calculate pressure rise using Eq. (6.19) and add initial pipe pressure.

Solution

1. Calculation for sound speed:

$$c = \sqrt{\frac{E_v}{\rho}} = \sqrt{\frac{320{,}000 \text{ lbf/in}^2 \times 144 \text{ in}^2/\text{ft}^2}{1.94 \text{ slugs/ft}^3}} = 4874 \text{ ft/s}$$

2. Calculation for critical closure time:

$$t_c = 2L/c$$
$$= 2(3000 \text{ ft}/4874 \text{ ft/s}) = 1.23 \text{ s}$$

3. Closure time of 1 s is less than 1.23 s.

4. Pressure rise calculation:

$$\Delta p = \rho V c$$
$$= 1.94 \text{ slugs/ft}^3 \times 4 \text{ ft/s} \times 4874 \text{ ft/s}$$
$$= 37{,}820 \text{ lbf/ft}^2 \times \frac{1 \text{ ft}^2}{144 \text{ in}^2} = 263 \text{ psi}$$

Maximum pressure is

$$p_{max} = 40 + 263 = \boxed{303 \text{ psig}}$$

As indicated by Example 6.12, water hammer pressures can be quite large. Therefore, engineers must design piping systems to keep the pressure within acceptable limits. This is done by installing an accumulator near the valve and/or operating the valve in such a way that rapid closure is prevented. Accumulators may be in the form of air chambers for relatively small systems, or surge tanks (a surge tank is a large open tank connected by a branch pipe to the main pipe) for large systems. Another way to eliminate excessive water-hammer pressures is to install pressure-relief valves at critical points in the pipe system. These valves are pressure-activated so that water is automatically diverted out of the system when the water-hammer pressure reaches excessive levels.

6.5

Moment-of-Momentum Equation

The moment-of-momentum equation is very useful for situations that involve torques. Examples include analyses of rotating machinery such as pumps, turbines, fans, and blowers.

Torques acting on a control volume are related to changes in angular momentum through the moment-of-momentum equation. Development of this equation parallels the development of the momentum equation as presented in Section 6.1. When forces act on a system of particles, used to represent a fluid system, Newton's second law of motion can be used to derive an equation for rotational motion:

$$\sum M = \frac{d(H_{sys})}{dt} \tag{6.24}$$

where M is a moment and H_{sys} is the total angular momentum of all mass forming the system.

Equation (6.24) is a Lagrangian equation, which can be converted to an Eulerian form using the Reynolds transport theorem from Eq. (5.21). The extensive property B_{sys} becomes the angular momentum of the system: $B_{sys} = H_{sys}$. The intensive property b becomes the angular momentum per unit mass. The angular momentum of an element is $r \times mv$, and so $b = r \times v$. Substituting for B_{sys} and b into Eq. (5.21) gives

$$\frac{d(H_{sys})}{dt} = \frac{d}{dt} \int_{cv} (r \times v)\rho \, d\mathcal{V} + \int_{cs} (r \times v)\rho V \cdot dA \tag{6.25}$$

Combining Eqs. (6.24) and (6.25) gives the integral form of the *moment-of-momentum equation*:

$$\sum M = \frac{d}{dt} \int_{cv} (r \times v)\rho \, d\mathcal{V} + \int_{cs} (r \times v)\rho V \cdot dA \tag{6.26}$$

where r is a position vector that extends from the moment center, V is flow velocity relative to the control surface, and v is flow velocity relative to the inertial reference frame selected.

The moment-of-momentum equation has the following physical interpretation: The sum of moments acting on the material within the control volume equals the rate of change of angular momentum within the control volume plus the net rate at which angular momentum flows out of the control volume.

If the mass crosses the control surface through a series of inlet and outlet ports with uniformly distributed properties across each port, the moment-of-momentum equation becomes

$$\sum M = \frac{d}{dt}\int_{cv}(r \times v)\rho \, d\mathcal{V} + \sum_{cs} r_o \times (\dot{m}_o v_o) - \sum_{cs} r_i \times (\dot{m}_i v_i) \qquad (6.27)$$

The methods used for applying the moment-of-momentum equation parallel the methods described in Section 6.2. The origin for evaluating moments may be selected at any convenient location.

Example 6.13 shows how to apply the moment-of-momentum equation to find the resisting moment required to support the flow through a 180° reducing bend.

EXAMPLE 6.13   RESISTING MOMENT ON REDUCING BEND

The reducing bend shown in the figure is supported on a horizontal axis through point A. Water flows through the bend at 0.25 m³/s. The inlet pressure at cross-section 1 is 150 kPa gage, and the outlet pressure at section 2 is 59.3 kPa gage. A weight of 1420 N acts 20 cm to the right of point A. Find the moment the support system must resist. The diameters of the inlet and outlet pipes are 30 cm and 10 cm, respectively.

Sketch:

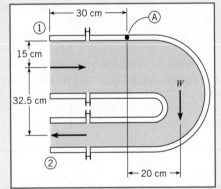

Situation: Reverse bend supported on a horizontal axis.

Find: Resisting moment (in N · m) by support system.

Assumptions:

1. Water flow is steady.
2. Water temperature is 20°C.

Properties: From Table A.5, $\rho = 998$ kg/m³.

Plan

The reducing bend is stationary, so it serves as an inertial reference frame. Flow is steady, so $\dot{m}_1 = \dot{m}_2 = \dot{m}$.

1. Select a control volume surrounding the reducing bend.
2. Sketch the force (and moment) diagram.
3. Sketch the momentum diagram.
4. Apply the moment-of-momentum equation, Eq. (6.27). The bend lies in the x-z plane so the y-direction, represented by unit vector **j**, goes into the page. A positive moment in the x-z plane is in the clockwise direction.
5. Evaluate the sum of moments.
6. Evaluate the moments of momentum.
7. Calculate resisting moment.

Solution

1. The control volume selected is shown in diagram.

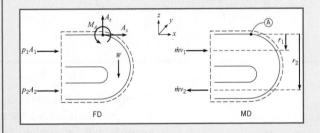

2. The force diagram shows two pressure forces contributing to the moment and the resisting moment.
3. The momentum diagram shows an influx and outflux of momentum.
4. Apply the moment-of-momentum equation.

$$\sum M_A = \frac{d}{dt}\int_{cv}(r \times v)\rho \, d\mathcal{V} + \sum_{cs} r_o \times (\dot{m}_o v_o) - \sum_{cs} r_i \times (\dot{m}_i v_i)$$

5. Sum of moments about axis A.

$$\sum M_A = -(M_A + 0.15p_1A_1 + 0.475p_2A_2 - 0.2W)\mathbf{j}$$

6. Evaluate the moment-of-momentum terms.
   • Accumulation term, steady flow, is

   $$\frac{d}{dt}\int_{cv}(r \times v)\rho \, d\rlap{\diagup}V = 0 \,.$$

   • Inflow term is $\sum_{cs} r_i \times (\dot{m}_i v_i) = r_1 \times (\dot{m}v_1) = -r_1\dot{m}v_1\mathbf{j}$.

   • Outflow term is

   $$\sum_{cs} r_o \times (\dot{m}_o v_o) = r_2 \times (\dot{m}v_2) = r_2\dot{m}v_2\mathbf{j}.$$

7. Resisting moment at A

   $$-(M_A + 0.15p_1A_1 + 0.475p_2A_2 - 0.2W)\mathbf{j} = \dot{m}(r_2v_2 + r_1v_1)\mathbf{j}$$

   $$M_A = -0.15p_1A_1 - 0.475p_2A_2 + 0.2W - \dot{m}(r_2v_2 + r_1v_1)$$

   • Torque due to pressure

   $$0.15p_1A_1 = (0.15 \text{ m})(150 \times 1000 \text{ N/m}^2)(\pi \times 0.3^2/4 \text{ m}^2)$$

   $$= 1590 \text{ N} \cdot \text{m}$$

   $$0.475p_2A_2 = (0.475 \text{ m})(59.3 \times 1000 \text{ N/m}^2)(\pi \times 0.15^2/4 \text{ m}^2)$$

   $$= 498 \text{ N} \cdot \text{m}$$

• Net moment-of-momentum flow

$$\dot{m} = \rho Q = (998 \text{ kg/m}^3)(0.25 \text{ m}^3/\text{s})$$

$$= 250 \text{ kg/s}$$

$$v_1 = \frac{Q}{A_1} = \frac{0.25 \text{ m}^3/\text{s}}{\pi \times 0.15^2 \text{m}^2} = 3.54 \text{ m/s}$$

$$v_2 = \frac{Q}{A_2} = \frac{0.25 \text{ m}^3/\text{s}}{\pi \times 0.075^2 \text{m}^2} = 14.15 \text{ m/s}$$

$$\dot{m}(r_2v_2 + r_1v_1) = (250 \text{ kg/s})$$

$$\times (0.475 \times 14.15 + 0.15 \times 3.54)(\text{m}^2/\text{s})$$

$$= 1813 \text{ N} \cdot \text{m}$$

• Moment exerted by support

$$M_A = -0.15p_1A_1 - 0.475p_2A_2 + 0.2W - \dot{m}(r_2v_2 + r_1v_1)$$

$$= -(1590 \text{ N} \cdot \text{m}) - (498 \text{ N} \cdot \text{m})$$

$$+ (0.2 \text{ m} \times 1420 \text{ N}) - (1813 \text{ N} \cdot \text{m})$$

$$M_A = \boxed{-3.62 \text{ kN} \cdot \text{m}}$$

Thus, a moment of 3.62 kN · m acting in the **j**, or clockwise, direction is needed to hold the bend stationary. Stated differently, the support system must be designed to withstand a counterclockwise moment of 3.62 kN · m.

Review

1. Care must be taken in addressing moment-of-momentum problems that the "right-hand-rule" to find the direction of the moment be applied correctly.
2. This type of problem may be encountered when the flanges are flexible and provide no force on the system.

Example 6.14 illustrates how the moment-of-momentum equation can be applied to predict the power delivered by turbomachinery. The momentum enters and exits radially. This analysis can be applied to both power-producing machines (turbines) and power-absorbing machines (pumps and compressors), which are addressed in Chapter 14.

EXAMPLE 6.14   POWER DELIVERED BY A FRANCIS TURBINE

A Francis turbine is shown in the diagram. Water is directed by guide vanes into the rotating wheel (runner) of the turbine. The guide vanes have a 70° angle from the radial direction. The water exits with only a radial component of velocity with respect to the environment. The outer diameter of the wheel is 1 m, and the inner diameter is 0.5 m. The distance across the runner is 4 cm. The discharge is 0.5 m³/s, and the rotational rate of the wheel is 1200 rpm. The water density is 1000 kg/m³. Find the power (kW) produced by the turbine.

Sketch:

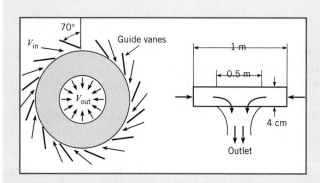

Problem Definition

Situation: Water flows through a Francis turbine and exits with no circumferential velocity.

Find: The power (kW) produced by the turbine.

Properties: $\rho = 1000 \text{ kg/m}^3$.

Plan

1. Select a control volume enclosing the turbine wheel.
2. Sketch the force diagram (in this case, moment diagram).
3. Sketch the momentum diagram.
4. Apply the moment-of-momentum equation, Eq. (6.27).
5. Evaluate the sum of moments. The counterclockwise direction is the positive direction.
6. Evaluate the moment-of-momentum terms.
7. Calculate the torque and power.

Solution

1. The control volume selected is shown in the sketch. The control volume is stationary.

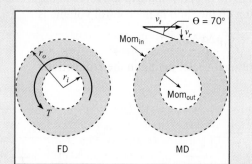

2. The force diagram shows positive torque acting on control surface.
3. The moment diagram shows an inflow and outflow of momentum.

4. Moment of momentum equation

$$\sum M_z = \frac{d}{dt} \int_{cv} (\mathbf{r} \times \mathbf{v})_z \rho \, d\mathcal{V} + \sum_{cs} [\mathbf{r}_o \times (\dot{m} \mathbf{v}_o)]_z$$

$$- \sum_{cs} [\mathbf{r}_i \times (\dot{m}_i \mathbf{v}_i)]_z$$

5. Sum of moments

$$\sum M_z = T$$

6. Moment-of-momentum terms
   - Accumulation term for steady flow is

   $$\frac{d}{dt} \int_{cv} (\mathbf{r} \times \mathbf{v})_z \rho \, d\mathcal{V} = 0.$$

   - Inlet momentum flux is $\sum_{cs} [\mathbf{r}_i \times (\dot{m}_i \mathbf{v}_i)]_z = -\dot{m} r_o v_t$.

   - Outlet momentum flux with $v_t = 0$ at outlet, is

   $$\sum_{cs} [\mathbf{r}_o \times (\dot{m} \mathbf{v}_o)]_z = 0.$$

7. Torque and power

$$T = \dot{m} r_o v_t = \rho Q r_o v_t$$

$$P = T\omega$$

   - Radial velocity component, v_r

   $$Q = \pi D w v_r$$

   $$v_r = Q/(\pi D w) = (0.5 \text{ m}^3/\text{s})/(\pi \times 1 \text{ m} \times 0.04 \text{ m})$$

   $$= 3.98 \text{ m/s}$$

   - Tangential velocity, v_t, from velocity triangle,

   $$\frac{v_t}{v_r} = \tan 70°$$

   $$v_t = 3.98 \text{ m/s} \times \tan 70° = 10.9 \text{ m/s}$$

   - Torque

   $$T = 1000 \text{ kg/m}^3 \times 0.5 \text{m}^3/\text{s} \times 0.5 \text{ m} \times 10.9 \text{ m/s}$$

   $$= 2725 \text{ N} \cdot \text{m}$$

   - Power

   $$\omega = \frac{1200 \text{ rev}}{\text{min}} \times \frac{1 \text{ min}}{60 \text{ s}} \times \frac{2\pi \text{ rad}}{1 \text{ rev}} = 126 \text{ rad/s}$$

   $$P = T\omega = 2725 \text{ N} \cdot \text{m} \times 126 \ (1/\text{s}) = 343000 \text{ N} \cdot \text{m/s}$$

   $$\boxed{= 343 \text{ kW}}$$

Navier-Stokes Equation

In Chapter 5, the continuity equation at a point in the flow is derived using a control volume of infinitesimal size. The resulting differential equation is an independent equation in the analysis of fluid flow. The same approach can be applied to the momentum equation, yielding the differential equation for momentum at a point in the flow. For simplicity, the derivation will be restricted to a two-dimensional planar flow, and the extension to three dimensions will be outlined.

Consider the infinitesimal control volume shown in Fig. 6.10a. The dimensions of the control volume are Δx and Δy, and the dimension in the third direction (normal to page) is taken as unity. Assume that the center of the control volume is fixed with respect to the coordinate system and that the coordinate system is an inertial reference frame. Also assume that the control surfaces are fixed with respect to the coordinate system. The x-direction momentum equation is Eq. (6.7a), namely

$$\sum F_x = \frac{d}{dt}\int_{cv} \rho v_x \, d\Psi + \sum_{cs} \dot{m}_o v_{o_x} - \sum_{cs} \dot{m}_i v_{i_x}$$

where v_x is the x-component of velocity of the fluid with respect to an inertial reference frame. In this derivation, the component velocities in the x- and y-directions are u and v, respectively. These velocities are referenced to the coordinate system, which is an inertial reference frame. The velocities at the control surface are also u and v since the control surfaces are fixed with respect to the coordinate system.

The forces acting on the fluid are due to pressure, stress, and body force, as shown in Fig. 6.10b. The pressure on the east face (the face to the right of the center of the element) is $p|_{x+\Delta x/2}$ and on the west face is $p|_{x-\Delta x/2}$. The net force acting in the x-direction due to pressure is

$$F_{x,p} = (p|_{x-\Delta x/2} - p|_{x+\Delta x/2})\Delta y \qquad (6.28)$$

The pressures on the north and south faces do not contribute to the force in the x-direction.

The body force due to gravity acting on the fluid in the control volume is

$$F_{x,g} = g_x \rho \Delta x \Delta y \qquad (6.29)$$

where g_x is the component of the gravitational vector acting in the x-direction.

Figure 6.10

Infinitesimal control volume.

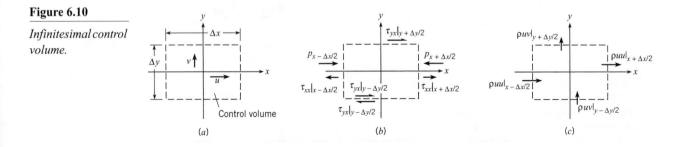

The evaluation of the net shear stress forces is somewhat more complicated. The shear stress acting in the x-direction on the north face is $\tau_{yx}|_{y+\Delta y/2}$. The subscripts on τ refer to the face on which the force acts and to the direction of the force. Thus τ_{yx} is the shear stress that acts on the y face in the x-direction. The face of a control surface is defined in the same way as the area vector in Section 5.2; that is, the y face corresponds to the face with the area vector in the y-direction. The shear stress on the south face inside the control volume is $\tau_{yx}|_{y-\Delta y/2}$. However, the stress that acts on the south face outside the control volume is equal and opposite to the stress on the inside face. Thus, the stress on the fluid on the south face of the control volume is $-\tau_{yx}|_{y-\Delta y/2}$.

There is also a normal stress (other than pressure) that acts on the east and west faces. This stress is proportional to strain rate of the fluid in the control volume in the x-direction. The stress on the east face is $\tau_{xx}|_{x+\Delta x/2}$. The stress acting on the outside west face of the control volume is $-\tau_{xx}|_{x-\Delta x/2}$ using the same argument as used for the shear stress on the south face. The net force in the x-direction due to shear and normal stresses is

$$F_{\tau,x} = (\tau_{yx}|_{y+\Delta y/2} - \tau_{yx}|_{y-\Delta y/2})\Delta x + (\tau_{xx}|_{x+\Delta x/2} - \tau_{xx}|_{x-\Delta x/2})\Delta y \tag{6.30}$$

Applying Liebnitz theorem for the differentiation of an integral in the same fashion as done for the continuity equation, the rate of change of momentum in the x-direction of the fluid in the control volume can be expressed as

$$\frac{d}{dt}\int_{cv}\rho u\,d\forall = \int_{cv}\frac{\partial}{\partial t}(\rho u)\,d\forall + \sum_{cs}\rho u(\mathbf{V}_c \cdot \mathbf{A}) \tag{6.31}$$

where $\mathbf{V}_c$ is the velocity of the control surface with respect to the coordinate system. For the control volume used in this derivation, $\mathbf{V}_c = 0$. Thus the rate of change of momentum of the fluid in the control volume becomes*

$$\frac{d}{dt}\int_{cv}\rho u\,d\forall = \int_{cv}\frac{\partial}{\partial t}(\rho u)\,dx\,dy \cong \frac{\partial}{\partial t}(\rho u)\Delta x\Delta y \tag{6.32}$$

The net efflux of momentum is obtained by summing the momentum flow from all four faces as shown in Fig. 6.10c. The flux of momentum outward from the east face is $\dot{m}_{x+\Delta x/2}u_{x+\Delta x/2}$ or $(\rho u_{x+\Delta x/2}\Delta y)u_{x+\Delta x/2}$. The momentum flux outward from the north face is $\dot{m}_{y+\Delta y/2}u_{x+\Delta y/2}$ or $(\rho v_{y+\Delta y/2}\Delta x)u_{y+\Delta y/2}$. The moment flux inward from the west and south faces is calculated in the same fashion. Finally, the net efflux of momentum is

$$\sum_{cs}\dot{m}_o v_{o_x} - \sum_{cs}\dot{m}_i v_{i_x} = (\rho uu|_{x+\Delta x/2} - \rho uu|_{x-\Delta x/2})\Delta y$$

$$+ (\rho vu|_{y+\Delta y/2} - \rho vu|_{y-\Delta y/2})\Delta x \tag{6.33}$$

Collecting all the terms that Eq. (6.7a) comprises and dividing through by the product $\Delta x \Delta y$ results in

---

* In the limit, as Δx and Δy approach zero, $\displaystyle\lim_{\Delta x,\,\Delta y \to 0}\frac{1}{\Delta x\Delta y}\int_{cv}\frac{\partial}{\partial t}(\rho u)\,dx\,dy = \frac{\partial}{\partial t}(\rho u)$

$$\frac{1}{\Delta x \Delta y}\int_{cv}\frac{\partial}{\partial t}(\rho u)\,dx\,dy + \frac{\rho uu|_{x+\Delta x/2} - \rho uu|_{x-\Delta x/2}}{\Delta x}$$

$$+ \frac{\rho vu|_{y+\Delta y/2} - \rho vu|_{y-\Delta y/2}}{\Delta y} = \frac{p|_{x+\Delta x/2} - p|_{x+\Delta x/2}}{\Delta x}$$

$$+ \frac{\tau_{yx}|_{y+\Delta y/2} - \tau_{yx}|_{y-\Delta y/2}}{\Delta y} + \frac{\tau_{xx}|_{x+\Delta x/2} - \tau_{xx}|_{x-\Delta x/2}}{\Delta x} + \rho g_x \qquad (6.34)$$

Taking the limit as Δx and Δy approach zero yields the differential form of the momentum equation in the x-direction:

$$\frac{\partial}{\partial t}(\rho u) + \frac{\partial}{\partial x}(\rho uu) + \frac{\partial}{\partial y}(\rho uv) = -\frac{\partial p}{\partial x} + \frac{\partial \tau_{xx}}{\partial x} + \frac{\partial \tau_{yx}}{\partial y} + \rho g_x \qquad (6.35)$$

One further step is to use the differential form of the continuity equation for two-dimensional flow [Eq. (5.31) with $\partial(\rho w)/\partial z = 0$] to convert the equation to a different form. Through differentiation by parts, the left side of Eq. (6.35) can be written as

$$\frac{\partial}{\partial t}(\rho u) + \frac{\partial}{\partial x}(\rho uu) + \frac{\partial}{\partial y}(\rho uv) = \rho\frac{\partial u}{\partial t} + \rho u\frac{\partial u}{\partial x} + \rho v\frac{\partial u}{\partial y}$$

$$+ u\left[\frac{\partial \rho}{\partial t} + \frac{\partial}{\partial x}(\rho u) + \frac{\partial}{\partial y}(\rho v)\right] \qquad (6.36)$$

Note that the last term is zero because of the continuity equation (Eq. 5.23), so the momentum equation in the x-direction at a point becomes

$$\rho\frac{\partial u}{\partial t} + \rho u\frac{\partial u}{\partial x} + \rho v\frac{\partial u}{\partial y} = -\frac{\partial p}{\partial x} + \frac{\partial \tau_{xx}}{\partial x} + \frac{\partial \tau_{yx}}{\partial y} + \rho g_x \qquad (6.37)$$

The left side of this equation is the product of the density and the acceleration of a fluid element in the x-direction and is usually written in more compact form as $\rho Du/Dt$, so the equation can be expressed as

$$\rho\frac{Du}{Dt} = -\frac{\partial p}{\partial x} + \frac{\partial \tau_{xx}}{\partial x} + \frac{\partial \tau_{yx}}{\partial y} + \rho g_x \qquad (6.38)$$

By applying Eq. (6.7b) to the same control volume, the momentum equation in the y-direction at a point can be derived. The result is

$$\rho\frac{\partial v}{\partial t} + \rho u\frac{\partial v}{\partial x} + \rho v\frac{\partial v}{\partial y} = \rho\frac{Dv}{Dt} = -\frac{\partial p}{\partial y} + \frac{\partial \tau_{yy}}{\partial y} + \frac{\partial \tau_{xy}}{\partial x} + \rho g_y \qquad (6.39)$$

The same approach can be used to derive the momentum equation in three dimensions. In this case the two-dimensional infinitesimal volume is extended to a three-dimensional volume with depth Δz and with a velocity component w in the z-direction.

In order to complete the development of these equations, a relationship is needed between the shear and normal stresses and the rate of strain of the fluid elements. These relationships are called the "constitutive equations." For a Newtonian fluid, by definition, the stress is proportional to the *rate* of strain. The rate of shear strain of a fluid element can be related to the gradients in velocity in the same way as the rate of rotation in Section 4.6.

With reference to Fig. 4.18, the shear strain of an element is $\Delta\theta_A - \Delta\theta_B$, so the rate of shear strain is

$$\lim_{\Delta t \to 0} \left(\frac{\Delta\theta_A}{\Delta t} - \frac{\Delta\theta_B}{\Delta t} \right) = \frac{1}{2}\left(\frac{\partial u}{\partial y} + \frac{\partial v}{\partial x} \right) \tag{6.40}$$

The constant of proportionality between the shear stress and the rate of shear strain is 2μ, where μ is the coefficient of dynamic viscosity, so the shear stress τ_{yx} is related to the velocity gradients by

$$\tau_{yx} = \mu\left(\frac{\partial u}{\partial y} + \frac{\partial v}{\partial x} \right) \tag{6.41}$$

The normal stress in the x-direction for an incompressible fluid* is given by

$$\tau_{xx} = 2\mu \frac{\partial u}{\partial x} \tag{6.42}$$

Substituting the foregoing constitutive relations into Eq. (6.39) yields

$$\rho\frac{Du}{Dt} = -\frac{\partial p}{\partial x} + \frac{\partial}{\partial x}\left(2\mu \frac{\partial u}{\partial x} \right) + \frac{\partial}{\partial y}\left[\mu\left(\frac{\partial u}{\partial y} + \frac{\partial v}{\partial x} \right) \right] + \rho g_x \tag{6.43}$$

If the dynamic viscosity is constant, the normal and shear stress terms can be written as

$$\frac{\partial}{\partial x}\left(2\mu \frac{\partial u}{\partial x} \right) + \frac{\partial}{\partial y}\left[\mu\left(\frac{\partial u}{\partial y} + \frac{\partial v}{\partial x} \right) \right] = \mu\left[\frac{\partial^2 u}{\partial x^2} + \frac{\partial^2 u}{\partial y^2} \right] + \mu\frac{\partial}{\partial x}\left(\frac{\partial u}{\partial x} + \frac{\partial v}{\partial y} \right) \tag{6.44}$$

From the continuity equation for the planar flow of an incompressible fluid, the last term is zero, so Eq. (6.43) reduces to

$$\rho\frac{Du}{Dt} = -\frac{\partial p}{\partial x} + \mu\left[\frac{\partial^2 u}{\partial x^2} + \frac{\partial^2 u}{\partial y^2} \right] + \rho g_x \tag{6.45}$$

The corresponding equation in the y-direction is

$$\rho\frac{Dv}{Dt} = -\frac{\partial p}{\partial y} + \mu\left[\frac{\partial^2 v}{\partial x^2} + \frac{\partial^2 v}{\partial y^2} \right] + \rho g_y \tag{6.46}$$

The three-dimensional form for these equations can be found in Schlichting (3). These are called the Navier-Stokes equations after L. M. Navier (1785–1836) and G. G. Stokes (1819–1903), who are credited with their development.

Example 6.15 shows how the Navier-Stokes equations can be used to solve for the vertical flow of fluid between two plates. The equations are reduced to the only terms that apply, then integrated.

---

* An additional term is needed for the normal stress equation for a compressible fluid, which is discussed in Schlichting (3).

EXAMPLE 6.15 APPLICATION OF THE NAVIER-STOKES EQUATIONS

A two-dimensional (planar) flow is generated between a fixed and moving vertical plate as shown in the figure. The left plate is fixed, and the right plate moves with a velocity U. The distance between the plates is δ. The flow is steady and fully developed; that is, there is no change in the x-direction. The fluid is incompressible, and the pressure is constant. Use the Navier-Stokes equations to find the velocity distribution between the plates.

Sketch:

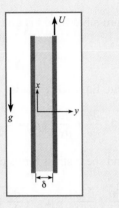

Problem Definition

Situation: Flow between a fixed and moving vertical plate.

Find: The velocity distribution between the plates, using the Navier-Stokes equations.

Plan

The following conditions are given:

- Pressure is constant.
- Flow is steady, $\partial u / \partial t = \partial v / \partial t = 0$.
- Flow is fully developed in x-direction, $\dfrac{\partial u}{\partial x} = 0$ and

$$\frac{\partial}{\partial x}\left(\frac{\partial u}{\partial x}\right) = 0.$$

- Boundary conditions are

$$u(0) = 0$$
$$u(\delta) = U$$
$$v(0) = v(\delta) = 0$$

- Gravitational acceleration is $g_x = -g$.

1. Use continuity equation, Eq. (5.31), to evaluate velocity component v.
2. Write out the x-component of Navier-Stokes equation, Eq. (6.45), and eliminate the terms that are zero.
3. Integrate equation and substitute in boundary conditions for the velocity u.

Solution

1. Continuity equation

$$\frac{\partial u}{\partial x} + \frac{\partial v}{\partial y} = 0$$

But $\partial u / \partial x = 0$, so $\partial v / \partial y = 0$, which means that v does not change with y. Thus $v = v(0) = 0$ everywhere.

2. The Navier-Stokes equation in x-direction

$$\rho \frac{\partial u}{\partial t} + \rho u \frac{\partial u}{\partial x} + \rho v \frac{\partial u}{\partial y} = -\frac{\partial p}{\partial x} + \mu \left[\frac{\partial^2 u}{\partial x^2} + \frac{\partial^2 u}{\partial y^2}\right] - \rho g$$

Eliminate terms that are zero:

$$\mu \frac{\partial^2 u}{\partial y^2} = \rho g$$

Since u is only a function of y, the partial derivative can be replaced by total derivatives.

$$\frac{d^2 u}{dy^2} = \frac{\rho g}{\mu}$$

3. Integrate and solve for u.

$$\frac{du}{dy} = \frac{\rho g}{\mu} y + C_1$$

$$u = \frac{\rho g y^2}{2\mu} + C_1 y + C_2$$

Substitute in boundary conditions.

$$\boxed{\frac{u}{U} = \frac{y}{\delta} - \frac{g\delta^2}{2\nu U}\left(1 - \frac{y^2}{\delta^2}\right)}$$

Review

Note that, due to gravity, the velocity profile lies below the linear profile $u/U = (y/\delta)$. The linear profile is approached for closer plate spacing, higher kinematic viscosity, and larger velocity.

6.7

Summary

The momentum principle is used to analyze problems involving forces and flow. It is expressed as

$$\sum_{cs} F = \frac{d}{dt}\bigg|_{cv} v\rho \, d\Psi + \int_{cs} v\rho V \cdot dA$$

where V is flow velocity relative to the control surface, and v is flow velocity relative to an inertial (nonaccelerating) reference frame.

The physical interpretation of the momentum principle is that the sum of forces equals the rate of momentum change inside the control volume plus the net rate at which momentum flows out of the control volume. To apply the momentum equation, one selects a control volume and then evaluates the forces, momentum accumulation, and momentum flow terms. These terms may be represented visually by using a force diagram and a momentum diagram.

The force term represents all external forces that act on the material inside the control volume. These forces can be either body forces or surface forces. For most problems, the only body force is weight. There are three common types of surface forces: those caused by structural elements, those caused by pressure, and those caused by shear stress distributions.

The momentum accumulation term $(d/dt|_{cv} v\rho \, d\Psi)$ gives the rate at which the momentum inside the control volume is changing with time. If flow is steady, and other mass in the control volume is stationary, the momentum accumulation is zero. Otherwise, the momentum accumulation is evaluated by integration.

The momentum flow term $(\int_{cs} v\rho V \cdot dA)$ gives the net rate at which momentum is flowing outward across the control surface. If velocity varies across the control surface, integration is needed to find the momentum flow.

If mass enters and leaves the control volume through a number of ports, and if the velocity v is uniformly distributed across each port, the momentum equation simplifies to

$$\sum_{cs} F = \frac{d}{dt}\bigg|_{cv} v\rho \, d\Psi + \sum_{cs} \dot{m}_o v_o - \sum_{cs} \dot{m}_i v_i$$

where subscripts o and i denote out and in, respectively. In this equation, $\dot{m}$ is the rate at which mass is crossing the control surface, and v is flow velocity evaluated at the control surface with respect to the inertial reference frame selected.

Water hammer is due to a pressure wave in a duct that travels at the speed of sound. The pressure rise across the wave is

$$\Delta p = \rho V c$$

where V is the duct flow velocity and c is the speed of sound, given by

$$c = \sqrt{\frac{E_v}{\rho}}$$

Problems involving moments may be analyzed by applying the moment-of-momentum principle. If mass crosses the control surface through a number of inlet and outlet ports, and if the properties are uniformly distributed across each port, the moment-of-momentum principle is expressed as

$$\sum_{cs} \boldsymbol{M} = \frac{d}{dt} \int_{cv} \rho(\boldsymbol{r} \times \boldsymbol{v}) \, d\Psi + \sum_{cs} \boldsymbol{r}_o \times (\dot{m}_o \boldsymbol{v}_o) - \sum_{cs} \boldsymbol{r}_i \times (\dot{m}_i \boldsymbol{v}_i)$$

where $\boldsymbol{r}$ is a position vector from the moment center. The physical interpretation is that the sum of moments equals the rate of angular-momentum change inside the control volume plus the net rate at which angular momentum flows out of the control volume. Application of the moment-of-momentum equation parallels the approaches used for the momentum equation.

The Navier-Stokes equation is a differential form of Newton's second law that applies to motion of a fluid element. The Navier-Stokes equation is a nonlinear, partial-differential equation that is widely used in advanced studies of fluid mechanics phenomena.

References

1. Chaudhry, M. H. *Applied Hydraulic Transients,* 2nd ed. Van Nostrand Reinhold, New York, 1987.

2. Streeter, V. L., and E. B. Wylie. *Fluid Transients.* FEB Press, Ann Arbor, Michigan, 1983.

3. Schlichting, H. *Boundary Layer Theory.* McGraw-Hill, New York, 1979.

Problems

Momentum Equation

6.1 $\mathbb{PQ}\blacktriangleleft$ Using the Internet or some other source as reference, define in your own words the meaning of "inertial reference frame."

6.2 $\mathbb{PQ}\blacktriangleleft$ The surface of the earth is not a true inertial reference frame because there is a centripetal acceleration due to the earth's rotation. The earth rotates once every 24 hours and has a diameter of 8000 miles. What is the centripetal acceleration on the surface of the earth, and how does it compare to the gravitational acceleration?

6.3 $\mathbb{PQ}\blacktriangleleft$ Newton's second law can be stated that the force is equal to the rate of change of momentum, $F = d(mv)/dt$. Taking the derivative by parts yields $F = m(dv) / (dt) + v(dm) / (dt)$. This does not correspond to $F = ma$. What is the source of the discrepancy?

Jets

6.4 $\mathbb{PQ}\blacktriangleleft$ Give five examples of jets and how they are used in practice.

6.5 A "balloon rocket" is a balloon suspended from a taut wire by a hollow tube (drinking straw) and string. The nozzle is formed of a 1 cm–diameter tube, and an air jet exits the nozzle with a speed of 40 m/s and a density of 1.2 kg/m³. Find the force F needed to hold the balloon stationary. Neglect friction.

6.6 The balloon rocket is held in place by a force F. The pressure inside the balloon is 8 in-H₂O, the nozzle diameter is 1.0 and the air density is 1.2 kg/m³. Find the exit velocity v

and the force F. Neglect friction and assume the air flow is inviscid and irrotational.

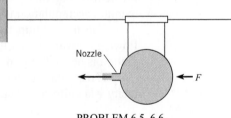

PROBLEM 6.5, 6.6

6.7 A water jet of diameter 30 mm and speed $v = 20$ m/s is filling a tank. The tank has a mass of 20 kg and contains 20 liters of water at the instant shown. The water temperature is 15°C. Find the force acting on the bottom of the tank and the force acting on the stop block. Neglect friction.

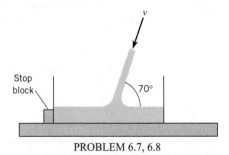

PROBLEM 6.7, 6.8

6.8 A water jet of diameter 2 inches and speed $v = 50$ ft/s is filling a tank. The tank has a mass of 25 lbm and contains 5 gallons of water at the instant shown. The water temperature is 70°F. Find the minimum coefficient of friction such that the force acting on the stop block is zero.

6.9 A design contest features a submarine that will travel at a steady speed of $V_{sub} = 1$ m/s in 15°C water. The sub is powered by a water jet. This jet is created by drawing water from an inlet of diameter 25 mm, passing this water through a pump and then accelerating the water through a nozzle of diameter 5 mm to a speed of V_{jet}. The hydrodynamic drag force (F_D) can be calculated using

$$F_D = C_D \left(\frac{\rho V_{sub}^2}{2} \right) A_p$$

where the coefficient of drag is $C_D = 0.3$ and the projected area is $A_p = 0.28$ m^2. Specify an acceptable value of V_{jet}.

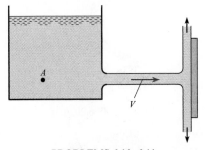

PROBLEM 6.9

6.10 A horizontal water jet at 70°F impinges on a vertical-perpendicular plate. The discharge is 2 cfs. If the external force required to hold the plate in place is 200 lbf, what is the velocity of the water?

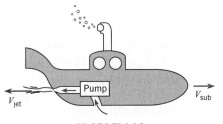

PROBLEMS 6.10, 6.11

6.11 A horizontal water jet at 70°F issues from a circular orifice in a large tank. The jet strikes a vertical plate that is normal to the axis of the jet. A force of 600 lbf is needed to hold the plate in place against the action of the jet. If the pressure in the tank is 25 psig at point A, what is the diameter of the jet just downstream of the orifice?

6.12 An engineer, who is designing a water toy, is making preliminary calculations. A user of the product will apply a force F_1 that moves a piston ($D = 80$ mm) at a speed of $V_{piston} = 300$ mm/s. Water at 20°C jets out of a converging nozzle of diameter $d = 15$ mm. To hold the toy stationary, the user applies a force F_2 to the handle. Which force (F_1 versus F_2) is larger? Explain your answer using concepts of the momentum principle. Then calculate F_1 and F_2. Neglect friction between the piston and the walls.

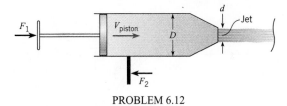

PROBLEM 6.12

6.13 A firehose on a boat is producing a 3 in.–diameter water jet with a speed of $V = 70$ mph. The boat is held stationary by a cable attached to a pier, and the water temperature is 50°F. Calculate the tension in the cable.

6.14 A boat is held stationary by a cable attached to a pier. A firehose directs a spray of 5°C water at a speed of $V = 50$ m/s. If the allowable load on the cable is 5 kN, calculate the mass flow rate of the water jet. What is the corresponding diameter of the water jet?

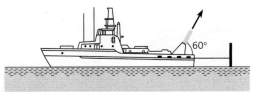

PROBLEMS 6.13, 6.14

6.15 A tank of water (15°C) with a total weight of 200 N (water plus the container) is suspended by a vertical cable. Pressurized air drives a water jet ($d = 12$ mm) out the bottom of the tank such that the tension in the vertical cable is 10 N. If $H = 425$ mm, find the required air pressure in units of atmospheres (gage). Assume the flow of water is irrotational.

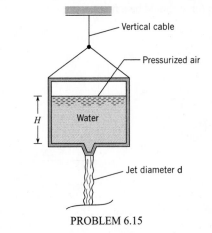

PROBLEM 6.15

6.16 A jet of water (60°F) is discharging at a constant rate of 2.0 cfs from the upper tank. If the jet diameter at section 1 is 4 in., what forces will be measured by scales A and B? Assume the empty tank weighs 300 lbf, the cross-sectional area of the tank is 4 ft², $h = 1$ ft, and $H = 9$ ft.

PROBLEM 6.16

6.17 A conveyor belt discharges gravel into a barge as shown at a rate of 50 yd³/min. If the gravel weighs 120 lbf/ft³, what is the tension in the hawser that secures the barge to the dock?

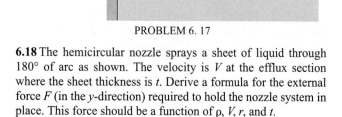

PROBLEM 6. 17

6.18 The hemicircular nozzle sprays a sheet of liquid through 180° of arc as shown. The velocity is V at the efflux section where the sheet thickness is t. Derive a formula for the external force F (in the y-direction) required to hold the nozzle system in place. This force should be a function of ρ, V, r, and t.

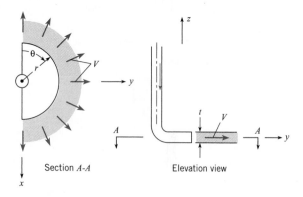

PROBLEM 6.18

6.19 The expansion section of a rocket nozzle is often conical in shape; and because the flow diverges, the thrust derived from

the nozzle is less than it would be if the exit velocity were everywhere parallel to the nozzle axis. By considering the flow through the spherical section suspended by the cone and assuming that the exit pressure is equal to the atmospheric pressure, show that the thrust is given by

$$T = \dot{m} V_e \frac{(1 + \cos\alpha)}{2}$$

where $\dot{m}$ is the mass flow through the nozzle, V_e is the exit velocity, and α is the nozzle half-angle.

PROBLEM 6.19

Vanes

6.20 Determine the external reactions in the x- and y-directions needed to hold this fixed vane, which turns the oil jet in a horizontal plane. Here V_1 is 18 m/s, $V_2 = 17$ m/s, and $Q = 0.15$ m³/s.

6.21 Solve Prob. 6.20 for $V_1 = 90$ ft/s, $V_2 = 85$ ft/s, and $Q = 2$ cfs.

6.22 This planar water jet (60°F) is deflected by a fixed vane. What are the x- and y-components of force per unit width needed to hold the vane stationary? Neglect gravity.

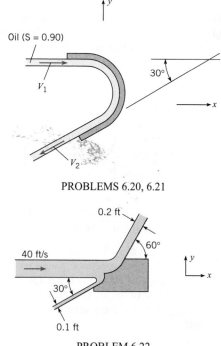

PROBLEMS 6.20, 6.21

PROBLEM 6.22

6.23 A water jet with a speed of 20 ft/s and a mass flow rate of 25 lbm/s is turned 30° by a fixed vane. Find the force of the water jet on the vane. Neglect gravity.

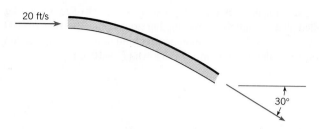

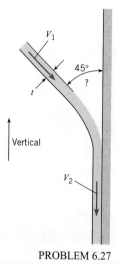

you think it will take, and explain your reasons for drawing it that way.

PROBLEM 6.23

6.24 Water ($\rho = 1000$ kg/m^3) strikes a block as shown and is deflected 30°. The flow rate of the water is 1.5 kg/s, and the inlet velocity is $V = 10$ m/s. The mass of the block is 1 kg. The coefficient of static friction between the block and the surface is 0.1 (friction force/normal force). If the force parallel to the surface exceeds the frictional force, the block will move. Determine the force on the block and whether the block will move. Neglect the weight of the water.

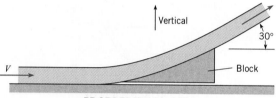

PROBLEMS 6.24, 6.25

6.25 For the situation described in Prob. 6.24, find the maximum inlet velocity (V) such that the block will not slip.

6.26 Plate A is 50 cm in diameter and has a sharp-edged orifice at its center. A water jet (at 10°C) strikes the plate concentrically with a speed of 30 m/s. What external force is needed to hold the plate in place if the jet issuing from the orifice also has a speed of 30 m/s? The diameters of the jets are $D = 5$ cm and $d = 2$ cm.

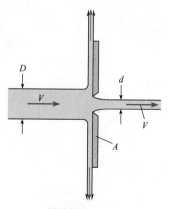

PROBLEM 6.26

6.27 A two-dimensional liquid jet impinges on a vertical wall. Assuming that the incoming jet speed is the same as the exiting jet speed ($V_1 = V_2$), derive an expression for the force per unit width of jet exerted on the wall. What form do you think the upper liquid surface will take next to the wall? Sketch the shape

PROBLEM 6.27

6.28 A cone that is held stable by a wire is free to move in the vertical direction and has a jet of water (at 10°C) striking it from below. The cone weighs 30 N. The initial speed of the jet as it comes from the orifice is 15 m/s, and the initial jet diameter is 2 cm. Find the height to which the cone will rise and remain stationary. *Note:* The wire is only for stability and should not enter into your calculations.

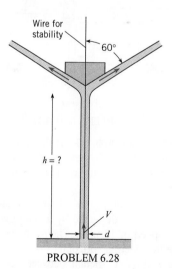

PROBLEM 6.28

6.29 A horizontal jet of water (at 10°C) that is 6 cm in diameter and has a velocity of 20 m/s is deflected by the vane as shown. If the vane is moving at a rate of 7 m/s in the x-direction, what components of force are exerted on the vane by the water in the x- and y-directions? Assume negligible friction between the water and the vane.

6.30 A vane on this moving cart deflects a 10 cm water ($\rho = 1000$ kg/m^3) jet as shown. The initial speed of the water in the jet is 20 m/s, and the cart moves at a speed of 3 m/s. If the vane splits the jet so that half goes one way and half the other, what force is exerted on the vane by the jet?

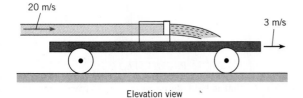

PROBLEM 6.29

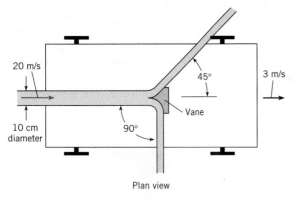

Elevation view

Plan view

PROBLEMS 6.30, 6.31

6.31 Refer to the cart of Prob. 6.30. If the cart speed is constant at 5 ft/s, and if the initial jet speed is 60 ft/s, and jet diameter = 0.15 ft, what is the rolling resistance of the cart? (ρ = 62.4 lbm/ft^3)

6.32 The water (ρ = 1000 kg/m^3) in this jet has a speed of 25 m/s to the right and is deflected by a cone that is moving to the left with a speed of 13 m/s. The diameter of the jet is 10 cm. Determine the external horizontal force needed to move the cone. Assume negligible friction between the water and the vane.

6.33 This two-dimensional water (at 50°F) jet is deflected by the two-dimensional vane, which is moving to the right with a speed of 60 ft/s. The initial jet is 0.30 ft thick (vertical dimension), and its speed is 100 ft/s. What power per foot of the jet (normal to the page) is transmitted to the vane?

6.34 Assume that the scoop shown, which is 20 cm wide, is used as a braking device for studying deceleration effects, such as

those on space vehicles. If the scoop is attached to a 1000 kg sled that is initially traveling horizontally at the rate of 100 m/s, what will be the initial deceleration of the sled? The scoop dips into the water 8 cm (d = 8 cm). (T = 10°C)

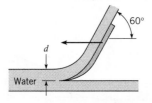

PROBLEMS 6.32, 6.33

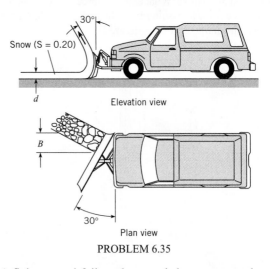

PROBLEM 6.34

6.35 This snowplow "cleans" a swath of snow that is 4 in. deep (d = 4 in.) and 2 ft wide (B = 2 ft). The snow leaves the blade in the direction indicated in the sketches. Neglecting friction between the snow and the blade, estimate the power required for just the snow removal if the speed of the snowplow is 40 ft/s.

Elevation view

Plan view

PROBLEM 6.35

6.36 A finite span airfoil can be regarded as a vane as shown in the figure. The cross section of air affected is equal to the circle with the diameter of the wing span, b. The wing deflects the air by an angle α and produces a force normal to the free-stream velocity, the lift L, and in the free-stream direction, the drag D.

The airspeed is unchanged. Calculate the lift and drag for a 30 ft wing span in a 300 ft/s airstream at 14.7 psia and 60°F for flow deflection of 2°.

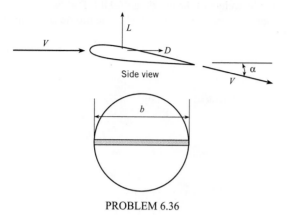

PROBLEM 6.36

6.37 The "clam shell" thrust reverser sketched in the figure is often used to decelerate aircraft on landing. The sketch shows normal operation (a) and when deployed (b). The vanes are oriented 20° with respect to the vertical. The mass flow through the engine is 150 lbm/s, the inlet velocity is 300 ft/s and the exit velocity is 1400 ft/s. Assume that when the thrust reverser is deployed, the exit velocity of the exhaust is unchanged. Assume the engine is stationary. Calculate the thrust under normal operation (lbf) and when the thrust reverser is deployed.

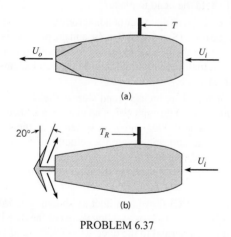

PROBLEM 6.37

Pipes

6.38 A hot gas stream enters a uniform-diameter return bend as shown. The entrance velocity is 100 ft/s, the gas density is 0.02 lbm/ft³, and the mass flow rate is 1 lbm/s. Water is sprayed into the duct to cool the gas down. The gas exits with a density of 0.06 lbm/ft³. The mass flow of water into the gas is negligible. The pressures at the entrance and exit are the same and equal to the atmospheric pressure. Find the force required to hold the bend.

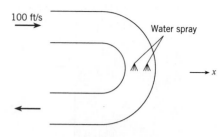

PROBLEM 6.38

6.39 Assume that the gage pressure p is the same at sections 1 and 2 in the horizontal bend shown in the figure. The fluid flowing in the bend has density ρ, discharge Q, and velocity V. The cross-sectional area of the pipe is A. Then the magnitude of the force (neglecting gravity) required at the flanges to hold the bend in place will be (a) pA, (b) $pA + \rho QV$, (c) $2pA + \rho QV$, or (d) $2pA + 2\rho QV$.

6.40 The pipe shown has a 180° vertical bend in it. The diameter D is 1 ft, and the pressure at the center of the upper pipe is 15 psig. If the flow in the bend is 20 cfs, what external force will be required to hold the bend in place against the action of the water? The bend weighs 200 lbf, and the volume of the bend is 3 ft³. Assume the Bernoulli equation applies. ($\rho = 62.4$ lbm/ft³)

6.41 The pipe shown has a 180° horizontal bend in it as shown, and D is 20 cm. The discharge of water ($\rho = 1000$ kg/m³) in the pipe and bend is 0.30 m³/s, and the pressure in the pipe and bend is 100 kPa gage. If the bend volume is 0.10 m³ and the bend itself weighs 500 N, what force must be applied at the flanges to hold the bend in place?

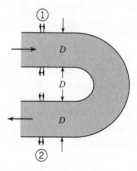

PROBLEMS 6.39, 6.40, 6.41

6.42 Water (at 50°F) flows in the horizontal bend at a rate of 12 cfs and discharges into the atmosphere past the downstream flange. The pipe diameter is 1 ft. What force must be applied at the upstream flange to hold the bend in place? Assume that the volume of water downstream of the upstream flange is 4 ft³ and that the bend and pipe weigh 100 lbf. Assume the pressure at the inlet section is 4 psig.

6.43 The gage pressure throughout the horizontal 90° pipe bend is 300 kPa. If the pipe diameter is 1 m and the water (at 10°C) flow rate is 10 m³/s, what x-component of force must be applied to the bend to hold it in place against the water action?

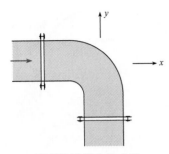

PROBLEMS 6.42, 6.43

6.44 This 30° vertical bend in a pipe with a 2 ft diameter carries water ($\rho = 62.4$ lbm/ft³) at a rate of 31.4 cfs. If the pressure p_1 is 10 psi at the lower end of the bend, where the elevation is 100 ft, and p_2 is 8.5 psi at the upper end, where the elevation is 103 ft, what will be the vertical component of force that must be exerted by the "anchor" on the bend to hold it in position? The bend itself weighs 300 lb, and the length L is 4 ft.

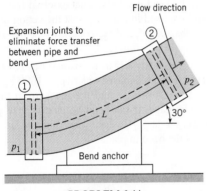

PROBLEM 6.44

6.45 This bend discharges water ($\rho = 1000$ kg/m³) into the atmosphere. Determine the force components at the flange required to hold the bend in place. The bend lies in a horizontal plane. Assume viscous forces are negligible. The interior volume of the bend is 0.25 m³, $D_1 = 60$ cm, $D_2 = 30$ cm, and $V_2 = 10$ m/s. The mass of the bend material is 250 kg.

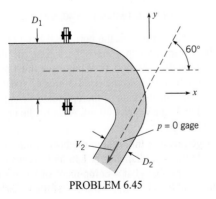

PROBLEM 6.45

6.46 This nozzle bends the flow from vertically upward to 30° with the horizontal and discharges water ($\gamma = 62.4$ lbf/ft³) at a speed of $V = 130$ ft/s. The volume within the nozzle itself is 1.8 ft³, and the weight of the nozzle is 100 lbf. For these conditions, what *vertical* force must be applied to the nozzle at the flange to hold it in place?

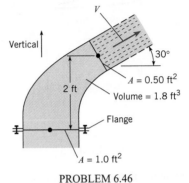

PROBLEM 6.46

6.47 A pipe 1 ft in diameter bends through an angle of 135°. The velocity of flow of gasoline ($S = 0.8$) is 20 ft/s, and the pressure is 10 psig in the bend. What external force is required to hold the bend against the action of the gasoline? Neglect the gravitational force.

6.48 A 6 in. horizontal pipe has a 180° bend in it. If the rate of flow of water (60°F) in the bend is 6 cfs and the pressure therein is 20 psig, what external force in the original direction of flow is required to hold the bend in place?

6.49 A pipe 15 cm in diameter bends through 135°. The velocity of flow of gasoline ($S = 0.8$) is 8 m/s, and the pressure is 100 kPa gage throughout the bend. Neglecting gravitational force, determine the external force required to hold the bend against the action of the gasoline.

6.50 A horizontal reducing bend turns the flow of water ($\rho = 1000$ kg/m³) through 60°. The inlet area is 0.001 m², and the outlet area is 0.0001 m². The water from the outlet discharges into the atmosphere with a velocity of 50 m/s. What horizontal force (parallel to the initial flow direction) acting through the metal of the bend at the inlet is required to hold the bend in place?

6.51 Water (at 10°C) flows in a duct as shown. The inlet water velocity is 10 m/s. The cross-sectional area of the duct is 0.1 m². Water is injected normal to the duct wall at the rate of 500 kg/s midway between stations 1 and 2. Neglect frictional forces on the duct wall. Calculate the pressure difference $(p_1 - p_2)$ between stations 1 and 2.

6.52 For this wye fitting, which lies in a horizontal plane, the cross-sectional areas at sections 1, 2, and 3 are 1 ft², 1 ft², and 0.25 ft², respectively. At these same respective sections the pressures are 1000 psfg, 900 psfg, and 0 psfg, and the water discharges are 20 cfs to the right, 12 cfs to the right, and exits to atmosphere at 8 cfs. What x-component of force would have to be applied to the wye to hold it in place?

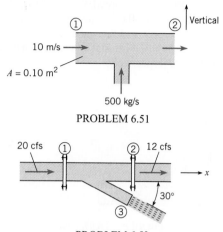

PROBLEM 6.51

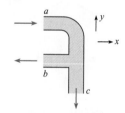

PROBLEM 6.52

6.53 Water ($\rho = 62.4$ lbm/ft^3) flows through a horizontal bend and T section as shown. The mass flow rate entering at section a is 12 lbm/s, and those exiting at sections b and c are 6 lbm/s each. The pressure at section a is 5 psig. The pressure at the two outlets is atmospheric. The cross-sectional areas of the pipes are the same: 5 in^2. Find the x-component of force necessary to restrain the section.

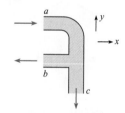

PROBLEMS 6.53, 6.54

6.54 Water ($\rho = 1000$ kg/m^3) flows through a horizontal bend and T section as shown. At section a the flow enters with a velocity of 6 m/s, and the pressure is 4.8 kPa. At both sections b and c the flow exits the device with a velocity of 3 m/s, and the pressure at these sections is atmospheric ($p = 0$). The cross-sectional areas at a, b, and c are all the same: 0.20 m^2. Find the x- and y-components of force necessary to restrain the section.

6.55 For this horizontal T through which water ($\rho = 1000$ kg/m^3) is flowing, the following data are given: $Q_1 = 0.25$ m^3/s, $Q_2 = 0.10$ m^3/s, $p_1 = 100$ kPa, $p_2 = 70$ kPa, $p_3 = 80$ kPa, $D_1 = 15$ cm, $D_2 = 7$ cm, and $D_3 = 15$ cm. For these conditions, what external force in the x-y plane (through the bolts or other supporting devices) is needed to hold the T in place?

Nozzles

6.56 PQ◄ Firehoses are fitted with special nozzles. Use the Internet or contact your local fire department to find information on operational conditions and typical hose and nozzle sizes used.

6.57 High-speed water jets are used for speciality cutting applications. The pressure in the chamber is approximately 60,000

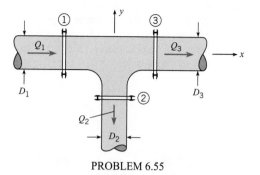

PROBLEM 6.55

psig. Using the Bernoulli equation, estimate the water speed exiting the nozzle exhausting to atmospheric pressure. Neglect compressibility effects and assume a water temperature of 60°F.

6.58 Water at 60°F flows through a nozzle that contracts from a diameter of 3 in. to 1 in. The pressure at section 1 is 2500 psfg, and atmospheric pressure prevails at the exit of the jet. Calculate the speed of the flow at the nozzle exit and the force required to hold the nozzle stationary. Neglect weight.

6.59 Water at 15°C flows through a nozzle that contracts from a diameter of 10 cm to 2 cm. The exit speed is $v_2 = 25$ m/s, and atmospheric pressure prevails at the exit of the jet. Calculate the pressure at section 1 and the force required to hold the nozzle stationary. Neglect weight.

6.60 Write a computer program that models the nozzle described in Prob. 6.59. Apply your computer program to solve Probs. 6.59 and 6.15 and compare answers.

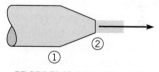

PROBLEMS 6.58, 6.59, 6.60

6.61 Water (at 50°F) flows through this nozzle at a rate of 15 cfs and discharges into the atmosphere. $D_1 = 12$ in., and $D_2 = 9$ in. Determine the force required at the flange to hold the nozzle in place. Assume irrotational flow. Neglect gravitational forces.

6.62 Solve Prob. 6.61 using the following values: $Q = 0.30$ m^3/s, $D_1 = 30$ cm, and $D_2 = 10$ cm. ($\rho = 1000$ kg/m^3)

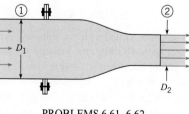

PROBLEMS 6.61, 6.62

6.63 This "double" nozzle discharges water ($\rho = 62.4$ lbm/ft^3) into the atmosphere at a rate of 16 cfs. If the nozzle is lying in a horizontal plane, what x-component of force acting through the flange bolts is required to hold the nozzle in place? *Note:* Assume irrotational flow, and assume the water speed in each jet to be the same. Jet A is 4 in. in diameter, jet B is 4.5 in. in diameter, and the pipe is 1 ft in diameter.

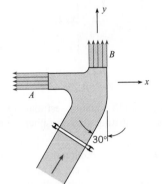

PROBLEMS 6.63, 6.64

6.64 This "double" nozzle discharges water (at 10°C) into the atmosphere at a rate of 0.50 m³/s. If the nozzle is lying in a horizontal plane, what x-component of force acting through the flange bolts is required to hold the nozzle in place? *Note:* Assume irrotational flow, and assume the water speed in each jet to be the same. Jet A is 10 cm in diameter, jet B is 12 cm in diameter, and the pipe is 30 cm in diameter.

6.65 A rocket-nozzle designer is concerned about the force required to hold the nozzle section on the body of a rocket. The nozzle section is shaped as shown in the figure. The pressure and velocity at the entrance to the nozzle are 1.5 MPa and 100 m/s. The exit pressure and velocity are 80 kPa and 2000 m/s. The mass flow through the nozzle is 220 kg/s. The atmospheric pressure is 100 kPa. The rocket is not accelerating. Calculate the force on the nozzle-chamber connection. *Note:* The given pressures are absolute.

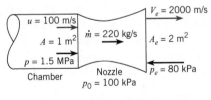

PROBLEM 6.65

6.66 A 15 cm nozzle is bolted with six bolts to the flange of a 30 cm pipe. If water ($\rho = 1000$ kg/m³) discharges from the nozzle into the atmosphere, calculate the tension load in each bolt when the pressure in the pipe is 200 kPa. Assume irrotational flow.

6.67 Water ($\rho = 62.4$ lbm/ft³) is discharged from the two-dimensional slot shown at the rate of 8 cfs per foot of slot. Determine the pressure p at the gage and the water force per foot on the vertical end plates A and C. The slot and jet dimensions B and b are 8 in. and 4 in., respectively.

6.68 Water (at 10°C) is discharged from the two-dimensional slot shown at the rate of 0.40 m³/s per meter of slot. Determine the pressure p at the gage and the water force per meter on the vertical end plates A and C. The slot and jet dimensions B and b are 20 cm and 7 cm, respectively.

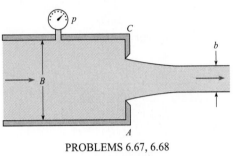

PROBLEMS 6.67, 6.68

6.69 This spray head discharges water ($\rho = 62.4$ lbm/ft³) at a rate of 4 ft³/s. Assuming irrotational flow and an efflux speed of 65 ft/s in the free jet, determine what force acting through the bolts of the flange is needed to keep the spray head on the 6 in. pipe. Neglect gravitational forces.

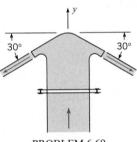

PROBLEM 6.69

6.70 Two circular water ($\rho = 62.4$ lbm/ft³) jets of 1 in. diameter ($d = 1$ in.) issue from this unusual nozzle. If the efflux speed is 80.2 ft/s, what force is required at the flange to hold the nozzle in place? The pressure in the 4 in. pipe ($D = 4$ in.) is 43 psig.

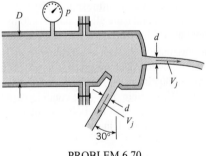

PROBLEM 6.70

6.71 Liquid (S = 1.2) enters the "black sphere" through a 2 in. pipe with velocity of 50 ft/s and a pressure of 60 psig. It leaves the sphere through two jets as shown. The velocity in the vertical jet is 100 ft/s, and its diameter is 1 in. The other jet's diameter is also 1 in. What force through the 2 in. pipe wall is required in the x- and y-directions to hold the sphere in place? Assume the sphere plus the liquid inside it weighs 200 lbf.

6.72 Liquid ($S = 1.5$) enters the "black sphere" through a 5 cm pipe with a velocity of 10 m/s and a pressure of 400 kPa. It leaves the sphere through two jets as shown. The velocity in the vertical jet is 30 m/s, and its diameter is 25 mm. The other jet's diameter is also 25 mm. What force through the 5 cm pipe wall is required in the x- and y-directions to hold the sphere in place? Assume the sphere plus the liquid inside it weighs 600 N.

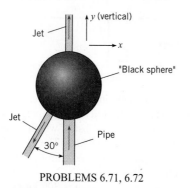

PROBLEMS 6.71, 6.72

Applications (Stationary)

6.73 Assume that you have access to a laboratory that includes a sluice gate as shown in Example (6.9). Design an experiment utilizing the sluice gate to verify the analytic approach. In your design you may attach instruments or drill holes in the sluice gate, etc., to facilitate the taking of data. In any case describe what is to be done to achieve a workable experiment.

6.74 Neglecting viscous resistance, determine the force of the water per unit of width acting on a sluice gate for which the upstream depth is 5 ft and the downstream depth is 0.6 ft.

6.75 For laminar flow in a pipe, wall shear stress (τ_0) causes the velocity distribution to change from uniform to parabolic as shown. At the fully developed section (section 2), the velocity is distributed as follows: $u = u_{max}[1 - (r/r_0)^2]$. Derive a formula for the force on the wall due to shear stress, F_τ, between 1 and 2 as a function of U (the mean velocity in the pipe), ρ, p_1, p_2, and D (the pipe diameter).

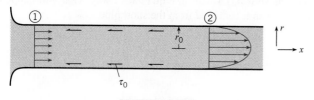

PROBLEM 6.75

6.76 The propeller on a swamp boat produces a slipstream 3 ft in diameter with a velocity relative to the boat of 100 ft/s. If the air temperature is 80°F, what is the propulsive force when the boat is not moving and also when its forward speed is 30 ft/s? *Hint:* Assume that the pressure, except in the immediate vicinity of the propeller, is atmospheric.

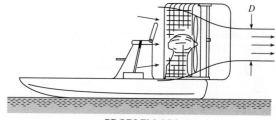

PROBLEM 6.76

6.77 A windmill is operating in a 10 m/s wind that has a density of 1.2 kg/m³. The diameter of the windmill is 4 m. The constant-pressure (atmospheric) streamline has a diameter of 3 m upstream of the windmill and 4.5 m downstream. Assume that the velocity distributions are uniform and the air is incompressible. Determine the thrust on the windmill.

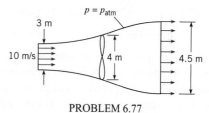

PROBLEM 6.77

6.78 The figure illustrates the principle of the jet pump. Derive a formula for $p_2 - p_1$ as a function of D_j, V_j, D_0, V_0, and ρ. Assume that the fluid from the jet and the fluid initially flowing in the pipe are the same, and assume that they are completely mixed at section 2, so that the velocity is uniform across that section. Also assume that the pressures are uniform across both sections 1 and 2. What is $p_2 - p_1$ if the fluid is water, $A_j/A_0 = 1/3$, $V_j = 15$ m/s, and $V_0 = 2$ m/s? Neglect shear stress.

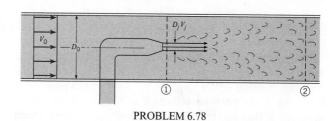

PROBLEM 6.78

6.79 Jet-type pumps are sometimes used for special purposes, such as to circulate the flow in basins in which fish are being reared. The use of a jet-type pump eliminates the need for mechanical machinery that might be injurious to the fish. The accompanying figure shows the basic concept for this type of application. For this type of basin the jets would have to increase the water surface elevation by an amount equal to $6V^2/2g$, where V is the average velocity in the basin (1 ft/s as shown in this example). Propose a basic design for a jet system that would make such a recirculating system work for a channel 8 ft wide and 4 ft deep. That is, determine the speed, size, and number of jets.

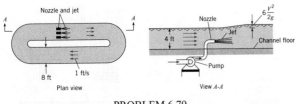

PROBLEM 6.79

6.80 An engineer is measuring the lift and drag on an airfoil section mounted in a two-dimensional wind tunnel. The wind tunnel is 0.5 m high and 0.5 m deep (into the paper). The upstream wind velocity is uniform at 10 m/s, and the downstream velocity is 12 m/s and 8 m/s as shown. The vertical component of velocity is zero at both stations. The test section is 1 m long. The engineer measures the pressure distribution in the tunnel along the upper and lower walls and finds

$$p_u = 100 - 10x - 20x(1-x)\,(\text{Pa gage})$$
$$p_l = 100 - 10x + 20x(1-x)\,(\text{Pa gage})$$

where x is the distance in meters measured from the beginning of the test section. The gas density is homogeneous throughout and equal to 1.2 kg/m³. The lift and drag are the vectors indicated on the figure. The forces acting on the fluid are in the opposite direction to these vectors. Find the lift and drag forces acting on the airfoil section.

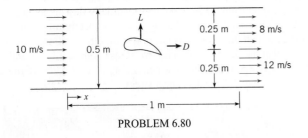

PROBLEM 6.80

6.81 A torpedo-like device is tested in a wind tunnel with an air density of 0.0026 slugs/ft³. The tunnel is 3 ft in diameter, the upstream pressure is 0.24 psig, and the downstream pressure is 0.10 psig. If the mean air velocity V is 120 ft/s, what are the mass rate of flow and the maximum velocity at the downstream section at C? If the pressure is assumed to be uniform across the sections at A and C, what is the drag of the device and support vanes? Assume viscous resistance at the walls is negligible.

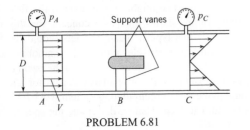

PROBLEM 6.81

6.82 A ramjet operates by taking in air at the inlet, providing fuel for combustion, and exhausting the hot air through the exit. The mass flow at the inlet and outlet of the ramjet is 60 kg/s (the mass flow rate of fuel is negligible). The inlet velocity is 225 m/s. The density of the gases at the exit is 0.25 kg/m³, and the exit area is 0.5 m². Calculate the thrust delivered by the ramjet. The ramjet is not accelerating, and the flow within the ramjet is steady.

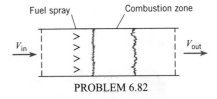

PROBLEM 6.82

6.83 A modern turbofan engine in a commercial jet takes in air, part of which passes through the compressors, combustion chambers, and turbine, and the rest of which bypasses the compressor and is accelerated by the fans. The mass flow rate of bypass air to the mass flow rate through the compressor-combustor-turbine path is called the "bypass ratio." The total flow rate of air entering a turbofan is 300 kg/s with a velocity of 300 m/s. The engine has a bypass ratio of 2.5. The bypass air exits at 600 m/s, whereas the air through the compressor-combustor-turbine path exits at 1000 m/s. What is the thrust of the turbofan engine? Clearly show your control volume and application of momentum equation.

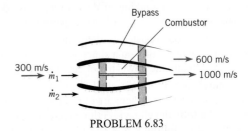

PROBLEM 6.83

Applications (Nonstationary)

6.84 PQ◄ A large tank of liquid is resting on a frictionless plane as shown. Explain in a qualitative way what will happen after the cap is removed from the short pipe.

PROBLEM 6.84

6.85 Consider a tank of water ($\rho = 1000$ kg/m³) in a container that rests on a sled. A high pressure is maintained by a compressor so that a jet of water leaving the tank horizontally from an orifice does so at a constant speed of 25 m/s relative to the tank. If there is 0.10 m³ of water in the tank at time t and the diameter of the jet is 15 mm, what will be the acceleration of the sled at time t if the empty tank and compressor have a weight of 350 N and the coefficient of friction between the sled and the ice is 0.05?

6.86 A cart is moving along a track at a constant velocity of 5 m/s as shown. Water ($\rho = 1000$ kg/m^3) issues from a nozzle at 10 m/s and is deflected through 180° by a vane on the cart. The cross-sectional area of the nozzle is 0.0012 m^2. Calculate the resistive force on the cart.

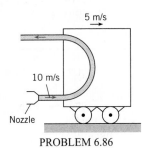

PROBLEM 6.86

6.87 A water jet is used to accelerate a cart as shown. The discharge (Q) from the jet is 0.1 m^3/s, and the velocity of the jet (V_j) is 10 m/s. When the water hits the cart, it is deflected normally as shown. The mass of the cart (M) is 10 kg. The density of water (ρ) is 1000 kg/m^3. There is no resistance on the cart, and the initial velocity of the cart is zero. The mass of the water in the jet is much less than the mass of the cart. Derive an equation for the acceleration of the cart as a function of Q, ρ, V_c, M, and V_j. Evaluate the acceleration of the cart when the velocity is 5 m/s.

6.88 A water jet strikes a cart as shown. After striking the cart, the water is deflected vertically with respect to the cart. The cart is initially at rest and is accelerated by the water jet. The mass in the water jet is much less than that of the cart. There is no resistance on the cart. The mass flow rate from the jet is 10 kg/s. The mass of the cart is 100 kg. Find the time required for the cart to achieve a speed one-half of the jet speed.

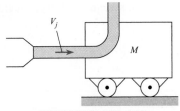

PROBLEMS 6.87, 6.88

6.89 It is common practice in rocket trajectory analyses to neglect the body-force term and drag, so the velocity at burnout is given by

$$v_{bo} = \frac{T}{\lambda} \ln \frac{M_0}{M_f}$$

Assuming a thrust-to-mass-flow ratio of 3000 N · s/kg and a final mass of 50 kg, calculate the initial mass needed to establish the rocket in an earth orbit at a velocity of 7200 m/s.

6.90 A very popular toy on the market several years ago was the water rocket. Water (at 10°C) was loaded into a plastic rocket and pressurized with a hand pump. The rocket was released and would travel a considerable distance in the air. Assume that a water rocket has a mass of 50 g and is charged with 100 g of water. The pressure inside the rocket is 100 kPa gage. The exit area is one-tenth of the chamber cross-sectional area. The inside diameter of the rocket is 5 cm. Assume that Bernoulli's equation is valid for the water flow inside the rocket. Neglecting air friction, calculate the maximum velocity it will attain.

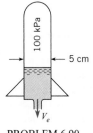

PROBLEM 6.90

Water Hammer

6.91 A valve at the end of a gasoline pipeline is rapidly closed (assume it is closed instantaneously). If the gasoline velocity was initially 12 m/s, what will be the water-hammer pressure rise? The bulk modulus of elasticity of the gasoline is 715 MPa, and the density of gasoline is 680 kg/m^3.

6.92 Estimate the maximum water-hammer pressure that is generated in a rigid pipe if the initial water ($\rho = 1000$ kg/m^3) velocity is 4 m/s and the pipe is 10 km long with a valve at the downstream end that is closed in 10 s.

6.93 The length of a 20 cm rigid pipe carrying 0.15 m^3/s of water (at 10°C) is estimated by instantaneously closing a valve at the downstream end and noting the time required for the pressure fluctuation to complete a cycle. If the time interval is 3 s, what is the pipe length?

6.94 Estimate the maximum water-hammer pressure that is generated in a rigid pipe if the initial water ($\rho = 62.4$ lbm/ft^3) velocity is 8 ft/s and the pipe is 5 mi long with a valve at the downstream end that is closed in 10 s.

6.95 A rigid pipe 4 km long and 12 cm in diameter discharges water ($\rho = 1000$ kg/m^3) at the rate of 0.03 m^3/s. If a valve at the end of the pipe is closed in 3 s, what is the maximum force that will be exerted on the valve as a result of the pressure rise? Assume that the water temperature is 10°C.

6.96 By letting the control volume move with the water-hammer wave, steady-flow conditions are established. Using the momentum and continuity equations and the steady-flow approach, derive Eq. (6.19).

6.97 The 60-cm pipe carries water (at 10°C) with an initial velocity, V_0, of 0.10 m/s. If the valve at C is instantaneously closed at time $t = 0$, what will the pressure-versus-time trace look like at point B for the next 5 s? Graph your results and indicate significant quantitative relations or values from $t = 0$ to $t = 5$ s. What does the pressure versus the position along the pipe look like at $t = 1.5$ s? Plot your results and indicate the velocity or velocities in the pipe.

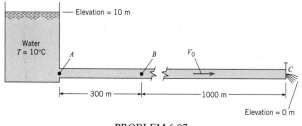

PROBLEM 6.97

6.98 Steady flow initially occurs in this 1 m steel pipe. There is a rapid-acting valve at the end of the pipe at point B, and there are pressure transducers at both points A and B. If the valve is closed at B and the p-versus-t traces are made as shown, estimate the initial discharge and the length L from A to B.

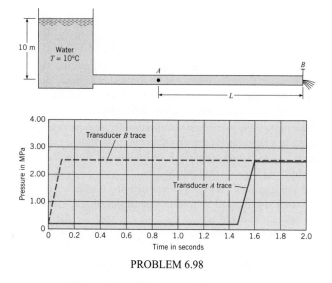

PROBLEM 6.98

Moment-of-Momentum

6.99 Water ($\rho = 1000$ kg/m^3) is discharged from the slot in the pipe as shown. If the resulting two-dimensional jet is 100 cm long and 15 mm thick, and if the pressure at section A-A is 30 kPa, what is the reaction at section A-A? In this calculation, do not consider the weight of the pipe.

6.100 Two small liquid-propellant rocket motors are mounted at the tips of a helicopter rotor to augment power under emergency conditions. The diameter of the helicopter rotor is 7 m, and it rotates at 1 rev/s. The air enters at the tip speed of the rotor, and exhaust gases exit at 500 m/s with respect to the rocket motor. The intake area of each motor is 20 cm^2, and the air density is 1.2 kg/m^3. Calculate the power provided by the rocket motors. Neglect the mass rate of flow of fuel in this calculation.

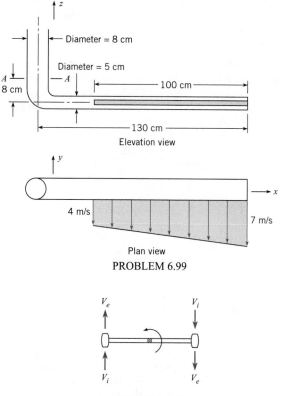

Elevation view

Plan view

PROBLEM 6.99

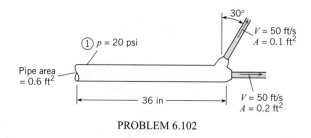

PROBLEM 6.100

6.101 Design a rotating lawn sprinkler to deliver 0.25 in. of water per hour over a circle of 50 ft radius. Make the simplifying assumptions that the pressure to the sprinkler is 50 psig and that frictional effects involving the flow of water through the sprinkler flow passages are negligible (the Bernoulli equation is applicable). However, do not neglect the friction between the rotating element and the fixed base of the sprinkler.

6.102 What is the force and moment reaction at section 1? Water (at 50°F) is flowing in the system. Neglect gravitational forces.

PROBLEM 6.102

6.103 What is the reaction at section 1? Water ($\rho = 1000$ kg/m^3) is flowing, and the axes of the two jets lie in a vertical plane. The pipe and nozzle system weighs 90 N.

6.104 A reducing pipe bend is held in place by a pedestal as shown. There are expansion joints at sections 1 and 2, so no force is transmitted through the pipe past these sections. The

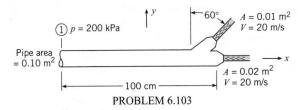

PROBLEM 6.103

pressure at section 1 is 20 psig, and the rate of flow of water ($\rho = 62.4$ lbm/ft^3) is 2 cfs. Find the force and moment that must be applied at section 3 to hold the bend stationary. Assume the flow is irrotational and neglect the influence of gravity.

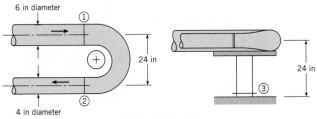

PROBLEM 6.104

6.105 A centrifugal fan is used to pump air. The fan rotor is 1 ft in diameter, and the blade spacing is 2 in. The air enters with no angular momentum and exits radially with respect to the fan rotor. The discharge is 1500 cfm. The rotor spins at 3600 rev/min. The air is at atmospheric pressure and a temperature of 60°F. Neglect the compressibility of the air. Calculate the power (hp) required to operate the fan.

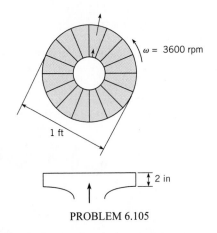

PROBLEM 6.105

Navier-Stokes Equation

6.106 Show how the momentum equation can be applied to derive Euler's equation for the flow of inviscid fluids. *Hint:* Select an arbitrary control volume of length Δs enclosed by a stream tube in an unsteady, nonuniform flow as shown. The volume of the control volume is $[A + (\partial A/\partial s)(\Delta s/2)]\Delta s$. First derive the continuity equation by applying the continuity principle to the flow through the control volume. Then apply the momentum

equation along the stream-tube direction, and use the continuity equation to reduce it to Euler's equation.

6.107 Using a three-dimensional, infinitesimal parallelepiped with dimensions Δx, Δy, and Δz and velocity components u, v, and w, show that, as the volume approaches zero, (a) the rate of momentum change in the x-direction per unit volume in the control volume plus the next efflux of momentum from the six surfaces per unit volume is

$$\frac{\partial}{\partial t}(\rho u) + \frac{\partial}{\partial x}(\rho uu) + \frac{\partial}{\partial y}(\rho uv) + \frac{\partial}{\partial z}(\rho uw)$$

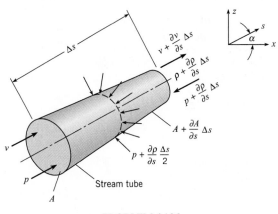

PROBLEM 6.106

(b) the forces due to pressure, normal, and shear stresses per unit volume are

$$\frac{\partial p}{\partial x} + \frac{\partial \tau_{xx}}{\partial x} + \frac{\partial \tau_{yx}}{\partial y} + \frac{\partial \tau_{zx}}{\partial z}$$

and (c) the body force per unit volume is ρg_x. Assemble these three components to obtain the momentum equation in the x-direction at a point and use the continuity equation to arrive at the form

$$\rho\frac{\partial u}{\partial t} + u\frac{\partial u}{\partial x} + v\frac{\partial u}{\partial y} + w\frac{\partial u}{\partial z}$$

$$= -\frac{\partial p}{\partial x} + \frac{\partial \tau_{xx}}{\partial x} + \frac{\partial \tau_{yx}}{\partial y} + \frac{\partial \tau_{zx}}{\partial z} + \rho g_x$$

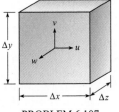

PROBLEM 6.107

6.108 Using the constitutive relations (stress proportional to rate of strain) for an incompressible liquid,

$$\tau_{xx} = 2\mu\frac{\partial u}{\partial x};$$

$$\tau_{yx} = \mu\left(\frac{\partial u}{\partial y} + \frac{\partial v}{\partial x}\right)$$

$$\tau_{zx} = \mu\left(\frac{\partial u}{\partial z} + \frac{\partial w}{\partial x}\right)$$

show for a liquid with constant dynamic viscosity that

$$\frac{\partial \tau_{xx}}{\partial x} + \frac{\partial \tau_{yx}}{\partial y} + \frac{\partial \tau_{zx}}{\partial z} = \mu\left(\frac{\partial^2 u}{\partial x^2} + \frac{\partial^2 u}{\partial y^2} + \frac{\partial^2 u}{\partial z^2}\right)$$

6.109 A flow with rectilinear streamlines adjacent to a wall is shown. The y-direction is normal to the wall. Apply the Navier-Stokes equation in the y-direction, Eq. (6.46), to show that the piezometric pressure is constant in the y-direction. The relationship between Δy and Δz is shown on the figure.

PROBLEM 6.109

CHAPTER 7

The Energy Equation

SIGNIFICANT LEARNING OUTCOMES

Conceptual Knowledge

- Explain the meaning of energy, work, and power.
- Describe various type of head terms (pressure head, pump head, velocity head, turbine head, etc.).
- Explain the meaning of pump, turbine, efficiency, and head loss.
- List the steps used to derive the energy equation.

Procedural Knowledge

- Apply the energy equation to predict variables such as pressure drop and head loss.
- Apply the power equation to find the power required for a pump or power supplied by a turbine.
- Sketch an Energy Grade Line (EGL) or a Hydraulic Grade Line (HGL) and explain the trends.

Applications (Typical)

- For a centrifugal pump or an axial fan, determine the requirements (e.g., energy, head, flow rate).
- For a turbine, establish how much power can be produced.
- For a piping system, identify locations of cavitation by sketching an HGL.

Chapters 4 to 6 have introduced important (i.e., commonly applied) equations such as the flow rate equations, the continuity equation, the momentum equation, and the Bernoulli equation. To complete this list, this chapter introduces the energy equation, which is used in many applications. This chapter also introduces equations for determining power and efficiency.

7.1

Energy, Work, and Power

The energy equation involves energy, work, and power as well as machines that interact with flowing fluids. These topics are introduced in this section.

When matter has *energy*, the matter can be used to do work. A fluid can have several forms of energy. For example a fluid jet has kinetic energy, water behind a dam has gravitational potential energy, and hot steam has thermal energy. *Work* is force acting through a distance when the force is parallel to the direction of motion. For example, for the spray bottle shown in Fig. 7.1, work is done when a finger exerts a force that acts through a distance as the trigger is depressed. Similarly, work is done when the piston exerts a pressure force that acts on the liquid over a distance. Another example of work involves wind passing over the blades of a wind turbine as shown in Figure 7.2. The wind exerts a force on the blades; this force produces a torque and work is given by

$$\text{work} = \text{force} \times \text{distance} = \text{torque} \times \text{angular displacement}$$

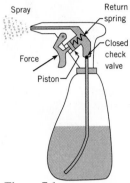

Figure 7.1

In a spray bottle, a piston pump does work on the fluid thereby increasing the energy in the liquid.

A machine is any device that transmits or modifies energy, typically to perform or assist in a human task. In fluid mechanics, a *turbine* is a machine that is used to extract energy from a flowing fluid.* Examples of turbines include the horizontal-axis wind turbine shown in Fig. 7.2, the gas turbine, the Kaplan turbine, the Francis turbine, and the Pelton wheel. Similarly, a *pump* is a machine that is used to provide energy to a flowing fluid. Examples of pumps include the piston pump shown in Fig. 7.1, the centrifugal pump, the diaphragm pump, and the gear pump.

Work and energy both have the same primary dimensions, and the same units, and both characterize an amount or quantity. For example, 1 calorie is the amount of thermal energy needed to raise the temperature of 1 gram of water by 1°C. Other common units include the joule (J), newton-meter, kilowatt-hour (kWh), foot-pound-force (ft-lbf), calorie (cal), and the British thermal unit (Btu).

Power, which expresses a rate of work or energy, is defined by

$$P \equiv \frac{\text{quantity of work (or energy)}}{\text{interval of time}} = \lim_{\Delta t \to 0} \frac{\Delta W}{\Delta t} = \dot{W} \tag{7.1}$$

Equation (7.1) uses a derivative because power can vary with each instant in time. To derive an equation for power, let the amount of work be given by the product of force and displacement $\Delta W = F\Delta x$:

$$P = \lim_{\Delta t \to 0} \frac{F\Delta x}{\Delta t} = FV \tag{7.2a}$$

where V is velocity of a moving body. When a shaft is rotating (e.g., Fig. 7.2), the amount of work is given by the product of torque and angular displacement $\Delta W = T\Delta\theta$. In this case, the power equation is

$$P = \lim_{\Delta t \to 0} \frac{T\Delta\theta}{\Delta t} = T\omega \tag{7.2b}$$

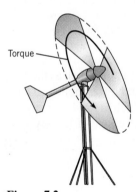

Figure 7.2

When air passes across the rotor of a wind turbine, the air exerts forces that result in a net torque. This torque does work on the blades.

where ω is the angular speed. Common units are radians per second (s^{-1}), revolutions per minute (rpm), and revolutions per second (rps).

Equations (7.2a) and (7.2b) may be combined to show the relationships between power associated with linear motion and power associated with rotational motion:

$$P = FV = T\omega \tag{7.3}$$

Common units for power are the watt (W), horsepower (hp), and the ft-lbf/s. Other units are given in Table F.1. Watts and horsepower are related by 1 kW = 1.34 hp. Similarly, 1 hp = 550 ft-lbf/s. It is useful to know some typical values of power. A 60-watt light bulb uses 60 J/s of electrical energy. A well-conditioned athlete can sustain a power output of about 300 W = 0.4 hp for one hour. A 1970 Volkswagen bug has an engine that is rated at about 50 hp. The Bonneville Dam on the Columbia River 40 miles east of Portland, Oregon, has a rated power of about 1080 MW.

---

* The engine on a jet, which is called a gas turbine, is a notable exception. The jet engine adds energy to a flowing fluid, thereby increasing the momentum of a fluid jet and producing thrust.

Energy Equation: General Form

This section derives the energy equation for a control volume by applying the Reynolds transport theorem to the system equation. The energy equation for a system is (1, 2):

$$\dot{Q} - \dot{W} = \frac{dE}{dt} \tag{7.4}$$

Equation (7.4), also called the first law of thermodynamics, can be stated in words:

$$\left\{ \begin{array}{c} \text{net rate of} \\ \text{thermal energy} \\ \text{entering system} \end{array} \right\} - \left\{ \begin{array}{c} \text{net rate at which} \\ \text{system does work} \\ \text{on environment} \end{array} \right\} = \left\{ \begin{array}{c} \text{rate of change of} \\ \text{energy of the matter} \\ \text{within the system} \end{array} \right\}$$

Recall that a *system* is a body of matter that is under consideration. By definition, a system always contains the same matter. An imaginary boundary separates the system from all other matter, which is called the *environment* (or surroundings).

Equation (7.4) involves sign conventions. Thermal energy is positive when there is an addition of thermal energy to the system and negative when there is a removal. Work is positive when the system is doing work on the environment and negative when work is done on the system.

To extend Eq. (7.4) to a control volume, apply the Reynolds transport theorem Eq. (5.21). Let the extensive property be energy ($B_{\text{sys}} = E$) and let $b = e$ to obtain

$$\dot{Q} - \dot{W} = \frac{d}{dt} \int_{\text{cv}} e\rho \, d\Psi + \int_{\text{cs}} e\rho \mathbf{V} \cdot d\mathbf{A} \tag{7.5}$$

where e is energy per mass in the fluid. Let $e = e_k + e_p + u$ where e_k is the kinetic energy per unit mass, e_p is the gravitational potential energy per unit mass, and u is the thermal energy (or internal energy) per unit mass.

$$\dot{Q} - \dot{W} = \frac{d}{dt} \int_{\text{cv}} (e_k + e_p + u)\rho \, d\Psi + \int_{\text{cs}} (e_k + e_p + u)\rho \mathbf{V} \cdot d\mathbf{A} \tag{7.6}$$

To simplify Eq. (7.6), let*

$$e_k = \frac{\text{kinetic energy of a fluid particle}}{\text{mass of this fluid particle}} = \frac{mV^2/2}{m} = \frac{V^2}{2} \tag{7.7}$$

Similarly, let

$$e_p = \frac{\text{gravitational potential energy of a fluid particle}}{\text{mass of this fluid particle}} = \frac{mgz}{m} = gz \tag{7.8}$$

---

* It is assumed that the control surface is not accelerating, so V, which is referenced to the control surface, is also referenced to an inertial reference frame.

where z is the elevation measured relative to a datum. When Eqs. (7.7) and (7.8) are substituted into Eq. (7.6), the result is

$$\dot{Q} - \dot{W} = \frac{d}{dt} \int_{cv} \left(\frac{V^2}{2} + gz + u\right) \rho \, d\Psi + \int_{cs} \left(\frac{V^2}{2} + gz + u\right) \rho \mathbf{V} \cdot d\mathbf{A} \qquad (7.9)$$

Shaft and Flow Work

Work is classified into two categories:

$$(\text{work}) = (\text{flow work}) + (\text{shaft work})$$

Each work term involves force acting over a distance. When this force is associated with a pressure distribution, then the work is called *flow work*. Alternatively, *shaft work* is any work that is not associated with a pressure force. Shaft work is usually done through a shaft (from which the term originates) and is commonly associated with a pump or turbine. According to the sign convention for work (see p. 219), pump work is negative. Similarly, turbine work is positive. Thus,

$$\dot{W}_{\text{shaft}} = \dot{W}_{\text{turbines}} - \dot{W}_{\text{pumps}} = \dot{W}_t - \dot{W}_p \qquad (7.10)$$

To derive an equation for flow work, use the idea that work equals force times distance. For example, Fig. 7.3 defines a control volume that is situated inside a converging pipe. At section 2, the fluid that is inside the control volume will push on the fluid that is outside of the control volume. The magnitude of the pushing force is $p_2 A_2$. During a time interval Δt, the displacement of the fluid at section 2 is $\Delta x_2 = V_2 \Delta t$. Thus, the amount of work is

$$\Delta W_2 = (F_2)(\Delta x_2) = (p_2 A_2)(V_2 \Delta t) \qquad (7.11)$$

Convert the amount of work given by Eq. (7.11) into a rate of work:

$$\dot{W}_2 = \lim_{\Delta t \to 0} \frac{\Delta W_2}{\Delta t} = p_2 A_2 V_2 = \left(\frac{p_2}{\rho}\right)(\rho A_2 V_2) = \dot{m}\left(\frac{p_2}{\rho}\right) \qquad (7.12)$$

This work is positive because the fluid inside the control volume is doing work on the environment. In a similar manner, the flow work at section 1 is negative and is given by

$$\dot{W}_1 = -\dot{m}\left(\frac{p_1}{\rho}\right)$$

Figure 7.3

Sketch for deriving flow work.

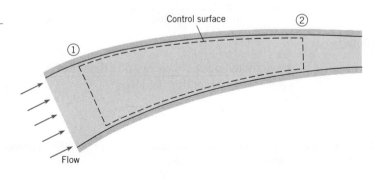

Control surface

① ②

Flow

The net flow work for the situation pictured in Fig. 7.3 is

$$\dot{W}_{\text{flow}} = \dot{W}_2 + \dot{W}_1 = \dot{m}\left(\frac{p_2}{\rho}\right) - \dot{m}\left(\frac{p_1}{\rho}\right) \tag{7.13}$$

Equation (7.13) can be generalized to a situation involving multiple streams of fluid passing across a control surface:

$$\dot{W}_{\text{flow}} = \sum_{\text{outlets}} \dot{m}_{\text{out}}\left(\frac{p_{\text{out}}}{\rho}\right) - \sum_{\text{inlets}} \dot{m}_{\text{in}}\left(\frac{p_{\text{in}}}{\rho}\right) \tag{7.14}$$

To develop a general form of flow work, use integrals to account for velocity and pressure variation across the control surface. Also, use the dot product to account for flow direction. The general equation for flow work is

$$\dot{W}_{\text{flow}} = \int_{\text{cs}} \left(\frac{p}{\rho}\right) \rho \mathbf{V} \cdot d\mathbf{A} \tag{7.15}$$

In summary, the work term is the sum of flow work [Eq. (7.15)] and shaft work [Eq. (7.10)]:

$$\dot{W} = \dot{W}_{\text{flow}} + \dot{W}_{\text{shaft}} = \left(\int_{\text{cs}} \left(\frac{p}{\rho}\right) \rho \mathbf{V} \cdot d\mathbf{A}\right) + \dot{W}_{\text{shaft}} \tag{7.16}$$

Final Steps in the Derivation of the Energy Equation

Introduce the work term from Eq. (7.16) into Eq. (7.9) and let $\dot{W}_{\text{shaft}} = \dot{W}_s$:

$$\begin{aligned}
\dot{Q} - \dot{W}_s - \int_{\text{cs}} \frac{p}{\rho} \rho \mathbf{V} \cdot d\mathbf{A} \\
= \frac{d}{dt}\int_{\text{cv}} \left(\frac{V^2}{2} + gz + u\right) \rho \, d\Psi + \int_{\text{cs}} \left(\frac{V^2}{2} + gz + u\right) \rho \mathbf{V} \cdot d\mathbf{A}
\end{aligned} \tag{7.17}$$

In Eq. (7.17), combine the last term on the left side with the last term on the right side:

$$\dot{Q} - \dot{W}_s = \frac{d}{dt}\int_{\text{cv}} \left(\frac{V^2}{2} + gz + u\right) \rho \, d\Psi + \int_{\text{cs}} \left(\frac{V^2}{2} + gz + u + \frac{p}{\rho}\right) \rho \mathbf{V} \cdot d\mathbf{A} \tag{7.18}$$

Replace $p/\rho + u$ by the specific enthalpy, h. The integral form of the energy principle is:

$$\dot{Q} - \dot{W}_s = \frac{d}{dt}\int_{\text{cv}} \left(\frac{V^2}{2} + gz + u\right) \rho \, d\Psi + \int_{\text{cs}} \left(\frac{V^2}{2} + gz + h\right) \rho \mathbf{V} \cdot d\mathbf{A} \tag{7.19}$$

If the flow crossing the control surface occurs through a series of inlet and outlet ports and if the velocity $\mathbf{V}$ is uniformly distributed across each port, then a simplified form of the Reynolds transport theorem, Eq. (5.21), can be used to derive the following form of the energy equation:

$$\dot{Q} - \dot{W}_s = \frac{d}{dt}\bigg]_c \int \left(\frac{V^2}{2} + gz + u\right)\rho\, d\forall$$

$$+ \sum_{cs} \dot{m}_o\left(\frac{V_o^2}{2} + gz_o + h_o\right) - \sum_{cs} \dot{m}_i\left(\frac{V_i^2}{2} + gz_i + h_i\right)$$

(7.20)

where the subscripts o and i refer to the outlet and inlet ports, respectively.

Energy Equation: Pipe Flow

Section 7.2 showed how to derive the general form of the energy equation for a control volume. Now, this section will simplify this general form to give an equation that is much easier to use yet still applicable to most situations. The first step is to develop a way to account for the kinetic energy distribution in the flowing fluid.

Kinetic Energy Correction Factor

Figure 7.4 shows fluid that is pumped through a pipe. At sections 1 and 2, kinetic energy is transported across the control surface by the flowing fluid. To derive an equation for this kinetic energy, start with the mass flow rate equation.

$$\dot{m} = \rho A \bar{V} = \int_A \rho V\, dA$$

This integral can be conceptualized as adding up the mass of each fluid particle that is crossing the section area and then dividing by the time interval associated with this crossing. To convert this integral to kinetic energy (KE), multiply the mass of each fluid particle by $(V^2/2)$.

Figure 7.4

Flow carries kinetic energy into and out of a control surface.

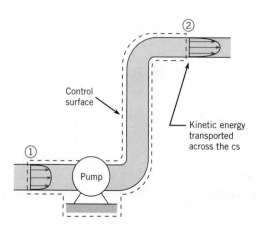

$$\left\{\begin{array}{c} \text{Rate of KE} \\ \text{transported} \\ \text{across a section} \end{array}\right\} = \int_A \rho V \left(\frac{V^2}{2}\right) dA = \int_A \frac{\rho V^3 \, dA}{2}$$

The *kinetic energy correction factor* is defined as

$$\alpha = \frac{\text{actual KE/time that crosses a section}}{\text{KE/time by assuming a uniform velocity distribution}} = \frac{\displaystyle\int_A \frac{\rho V^3 dA}{2}}{\displaystyle\frac{\rho \overline{V}^3 A}{2}}$$

For a constant density fluid, this equation simplifies to

$$\alpha = \frac{1}{A}\int_A \left(\frac{V}{\overline{V}}\right)^3 dA \tag{7.21}$$

In most cases, α takes on a value of 1 or 2. When the velocity profile in a pipe is uniformly distributed, then $\alpha = 1$. When flow is laminar, the velocity distribution is parabolic and $\alpha = 2$. When flow is turbulent, the velocity profile is plug-like and $\alpha \approx 1.05$. For turbulent flow it is common practice to let $\alpha = 1$.

To establish a value for α, integrate the velocity profile using Eq. (7.21). This approach is illustrated in Example 7.1.

Derivation of a Simplified Form of the Energy Equation

Now that the KE correction factor is available for representing the distribution of kinetic energy, the derivation may by completed. Begin by applying Eq. (7.18) to the control volume shown in Fig. 7.4. Assuming steady flow, Eq. (7.18). simplifies to:

$$\dot{Q} - \dot{W}_s + \int_{A_1}\left(\frac{p_1}{\rho} + gz_1 + u_1\right)\rho V_1 \, dA_1 + \int_{A_1}\frac{\rho V_1^3}{2} \, dA_1$$

$$= \int_{A_2}\left(\frac{p_2}{\rho} + gz_2 + u_2\right)\rho V_2 \, dA_2 + \int_{A_2}\frac{\rho V_2^3}{2} \, dA_2 \tag{7.22}$$

As explained in Chapter 4, (see p. 87), piezometric head $p/\rho + gz$ is constant across sections 1 and 2 because the streamlines are straight and parallel. If temperature is also assumed constant across each section, then $p/\rho + gz + u$ can be taken outside the integral to yield

$$\dot{Q} - \dot{W}_s + \left(\frac{p_1}{\rho} + gz_1 + u_1\right)\int_{A_1}\rho V_1 \, dA_1 + \int_{A_1}\rho\frac{V_1^3}{2} \, dA_1$$

$$= \left(\frac{p_2}{\rho} + gz_2 + u_2\right)\int_{A_2}\rho V_2 \, dA_2 + \int_{A_2}\rho\frac{V_2^3}{2} \, dA_2 \tag{7.23}$$

EXAMPLE 7.1   KINETIC ENERGY CORRECTION FACTOR FOR LAMINAR FLOW

The velocity distribution for laminar flow in a pipe is given by the equation

$$V = V_{max}\left[1 - \left(\frac{r}{r_0}\right)^2\right]$$

Here r_0 is the radius of the pipe and r is the radial distance from the center. Find the kinetic-energy correction factor α.

Problem Definition

Situation: Laminar flow in a round pipe. Velocity profile is given.

Find: Kinetic-energy correction factor α (no units).

Sketch:

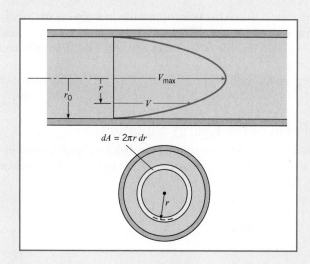

Plan

1. Find dA using the above sketch.
2. Find the mean velocity $\overline{V}$ using the flow rate equation (5.8).
3. Find α using Eq. (7.21).

Solution

1. Differential area
   - From the sketch, dA can be visualized as a rectangular strip of length $2\pi r$ and width dr. Thus, $dA = 2\pi r\, dr$.

2. Mean velocity

$$\overline{V} = \frac{1}{A}\left[\int_A V\, dA\right] = \frac{1}{\pi r_0^2}\left[\int_0^{r_0} V_{max}\left(1 - \frac{r^2}{r_0^2}\right)2\pi r\, dr\right]$$

$$= \frac{2V_{max}}{r_0^2}\left[\int_0^{r_0}\left(1 - \frac{r^2}{r_0^2}\right)r\, dr\right] = \frac{2V_{max}}{r_0^2}\left[\int_0^{r_0}\left(r - \frac{r^3}{r_0^2}\right)dr\right]$$

$$= \frac{2V_{max}}{r_0^2}\left[\left(\frac{r^2}{2} - \frac{r^4}{4r_0^2}\right)\Bigg|_0^{r_0}\right] = \frac{2V_{max}}{r_0^2}\left[\frac{r_0^2}{2} - \frac{r_0^2}{4}\right] = V_{max}/2$$

Interpretation: For laminar flow in a round pipe, the mean velocity is one-half of the maximum (centerline) velocity.

3. Kinetic-energy correction factor (α).

$$\alpha = \frac{1}{A}\left[\int_A\left(\frac{V}{\overline{V}}\right)^3 dA\right] = \frac{1}{\pi r_0^2 \overline{V}^3}\left[\int_0^{r_0} V^3 2\pi r\, dr\right]$$

$$= \frac{1}{\pi r_0^2(V_{max}/2)^3}\left[\int_0^{r_0}\left[V_{max}\left(1 - \frac{r^2}{r_0^2}\right)\right]^3 2\pi r\, dr\right]$$

$$= \frac{16}{r_0^2}\left[\int_0^{r_0}\left(1 - \frac{r^2}{r_0^2}\right)^3 r\, dr\right]$$

To evaluate the integral, make a change of variable by letting $u = (1 - r^2/r_0^2)$. The integral becomes

$$\alpha = \left(\frac{16}{r_0^2}\right)\left(-\frac{r_0^2}{2}\right)\left(\int_1^0 u^3\, du\right) = 8\left(\int_0^1 u^3\, du\right)$$

$$= 8\left(\frac{u^4}{4}\Bigg|_0^1\right) = 8\left(\frac{1}{4}\right)$$

$$\boxed{\alpha = 2}$$

Next, factor out $\int \rho V \, dA = \rho \overline{V} A = \dot{m}$ from each term in Eq. (7.23). Since $\dot{m}$ does not appear as a factor of $\int (\rho V^3/2) \, dA$, express $\int (\rho V^3/2) \, dA$ as $\alpha(\rho \overline{V}^3/2)A$, where α is the kinetic energy correction factor:

$$\dot{Q} - \dot{W}_s + \left(\frac{p_1}{\rho} + gz_1 + u_1 + \alpha_1 \frac{\overline{V}_1^2}{2} \right) \dot{m} = \left(\frac{p_2}{\rho} + gz_2 + u_2 + \alpha_2 \frac{\overline{V}_2^2}{2} \right) \dot{m} \qquad (7.24)$$

Divide through by $\dot{m}$:

$$\frac{1}{\dot{m}} (\dot{Q} - \dot{W}_s) + \frac{p_1}{\rho} + gz_1 + u_1 + \alpha_1 \frac{\overline{V}_1^2}{2} = \frac{p_2}{\rho} + gz_2 + u_2 + \alpha_2 \frac{\overline{V}_2^2}{2} \qquad (7.25)$$

Introduce Eq. (7.10) into Eq. (7.25):

$$\frac{\dot{W}_p}{\dot{m}g} + \frac{p_1}{\gamma} + z_1 + \alpha_1 \frac{\overline{V}_1^2}{2g} = \frac{\dot{W}_t}{\dot{m}g} + \frac{p_2}{\gamma} + z_2 + \alpha_2 \frac{\overline{V}_2^2}{2g} + \frac{u_2 - u_1}{g} - \frac{\dot{Q}}{\dot{m}g} \qquad (7.26)$$

Introduce *pump head* and *turbine head*:

$$\textit{Pump head} = h_p = \frac{\dot{W}_p}{\dot{m}g} = \frac{\text{work/time done by pump on flow}}{\text{weight/time of flowing fluid}}$$

$$\textit{Turbine head} = h_t = \frac{\dot{W}_t}{\dot{m}g} = \frac{\text{work/time done by flow on turbine}}{\text{weight/time of flowing fluid}} \qquad (7.27)$$

Equation (7.26) becomes

$$\frac{p_1}{\gamma} + \alpha_1 \frac{\overline{V}_1^2}{2g} + z_1 + h_p = \frac{p_2}{\gamma} + \alpha_2 \frac{\overline{V}_2^2}{2g} + z_2 + h_t + \left[\frac{1}{g}(u_2 - u_1) - \frac{\dot{Q}}{\dot{m}g} \right] \qquad (7.28)$$

Equation (7.28) is separated into terms that represent mechanical energy (nonbracketed terms) and terms that represent thermal energy (the bracketed term). This bracketed term is always positive because of the second law of thermodynamics. This term is called head loss and is represented by h_L. *Head loss* is the conversion of useful mechanical energy to waste thermal energy through viscous action between fluid particles. Head loss is analogous to thermal energy (heat) that is produced by Coulomb friction. When the bracketed term is replace by head loss h_L, Eq. (7.28) becomes the *energy equation*.

$$\left(\frac{p_1}{\gamma} + \alpha_1 \frac{\overline{V}_1^2}{2g} + z_1 \right) + h_p = \left(\frac{p_2}{\gamma} + \alpha_2 \frac{\overline{V}_2^2}{2g} + z_2 \right) + h_t + h_L \qquad (7.29)$$

Equation (7.29) is based on three main assumptions: (a) the flow is steady; (b) the control volume has one inlet port and one exit port; and (c) the density of the flow is constant.

In words, Eq. (7.29) can be stated as

$$\left(\begin{array}{c} \text{head} \\ \text{carried by} \\ \text{flow into the cv} \end{array} \right) + \left(\begin{array}{c} \text{head} \\ \text{added by} \\ \text{pumps} \end{array} \right) = \left(\begin{array}{c} \text{head} \\ \text{carried by} \\ \text{flow out of the cv} \end{array} \right) + \left(\begin{array}{c} \text{head} \\ \text{extracted by} \\ \text{turbines} \end{array} \right) + \left(\begin{array}{c} \text{head} \\ \text{loss due to} \\ \text{viscous effects} \end{array} \right)$$

Terms in Eq. (7.29) use the concept of head. A head term has a primary dimension of length, and it represents an energy or work concept. *Head* is related to energy or work:

$$\text{head} \approx \frac{\text{energy/time or work/time}}{\text{weight/time of flowing fluid}}$$

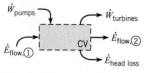

Figure 7.5

Equation (7.29) involves a balance of work and energy terms.

Thus, another way to interpret Eq. (7.29) is that work and energy flows are balanced as shown in Fig. 7.5. Energy can enter the control volume in two ways: energy can be transported across the control surface by the flowing fluid* or a pump can do work on the fluid and thereby add energy to the fluid. Energy can leave the cv in three ways. Energy within the flow can be used to do work on a turbine, energy can be transported across the control surface by the flowing fluid, or mechanical energy can be converted to waste heat via head loss.

    The recommended way to apply the energy equation is to start with Eq. (7.29) and then analyze each term in a stepwise fashion as shown in Example 7.2.

EXAMPLE 7.2 PRESSURE IN A PIPE

A horizontal pipe carries cooling water at 10°C for a thermal power plant from a reservoir as shown. The head loss in the pipe is

$$h_L = \frac{0.02(L/D)V^2}{2g}$$

where L is the length of the pipe from the reservoir to the point in question, V is the mean velocity in the pipe, and D is the diameter of the pipe. If the pipe diameter is 20 cm and the rate of flow is 0.06 m³/s, what is the pressure in the pipe at $L = 2000$ m. Assume $\alpha_2 = 1$.

Problem Definition

Situation: Cooling water for a power plant is flowing in a horizontal pipe.

Find: Pressure (kPa) in the pipe at section 2.

Sketch:

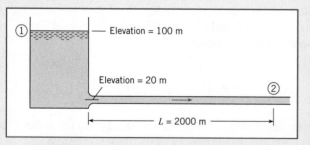

Assumptions: $\alpha_2 = 1.0$.

Properties: Water (10°C), Table A.5: $\gamma = 9810$ N/m³.

Plan

1. Write the energy equation from Eq. (7.29) between section 1 and section 2.
2. Analyze each term in the energy equation.
3. Solve for p_2.

Solution

1. Energy equation (general form)

$$\frac{p_1}{\gamma} + \alpha_1 \frac{\overline{V}_1^2}{2g} + z_1 + h_p = \frac{p_2}{\gamma} + \alpha_2 \frac{\overline{V}_2^2}{2g} + z_2 + h_t + h_L$$

2. Term-by-term analysis

- $p_1 = 0$ because the pressure at top of a reservoir is $p_{atm} = 0$ gage.
- $V_1 \approx 0$ because the level of the reservoir is constant or changing very slowly.
- $z_1 = 100$ m; $z_2 = 20$ m.
- $h_p = h_t = 0$ because there are no pumps or turbines in the system.
- Find V_2 using the flow rate equation (5.3).

$$V_2 = \frac{Q}{A} = \frac{0.06 \text{ m}^3/\text{s}}{(\pi/4)(0.2 \text{ m})^2} = 1.910 \text{ m/s}$$

- Head loss is

$$h_L = \frac{0.02(L/D)V^2}{2g} = \frac{0.02(2000 \text{ m}/0.2 \text{ m})(1.910 \text{ m/s})^2}{2(9.81 \text{ m/s}^2)}$$

$$= 37.2 \text{ m}$$

3. Combine steps 1 and 2.

$$(z_1 - z_2) = \frac{p_2}{\gamma} + \alpha_2 \frac{\overline{V}_2^2}{2g} + h_L$$

$$80 \text{ m} = \frac{p_2}{\gamma} + 1.0 \frac{(1.910 \text{ m/s})^2}{2(9.81 \text{ m/s}^2)} + 37.2 \text{ m}$$

$$80 \text{ m} = \frac{p_2}{\gamma} + (0.186 \text{ m}) + (37.2 \text{ m})$$

$$p_2 = \gamma(42.6 \text{ m}) = (9810 \text{ N/m}^3)(42.6 \text{ m}) = \boxed{418 \text{ kPa}}$$

---

* The term "$\dot{E}_{\text{flow}}$" includes a work term, namely flow work.

7.4

Power Equation

The last section showed how to find the head of a pump or turbine. This section shows how to relate head to power and efficiency. These parameters are used for applications such as selecting a motor for operating a centrifugal pump, calculating the amount of power that can supplied by a proposed hydroelectric plant, and estimating the pump size for a piping system.

An equation for pump power follows from the definition of pump head given in Eq. (7.27):

$$\dot{W}_p = \gamma Q h_p = \dot{m} g h_p \tag{7.30a}$$

Similarly, the power delivered from a flow to a turbine is

$$\dot{W}_t = \gamma Q h_t = \dot{m} g h_t \tag{7.30b}$$

Equations (7.30a) and (7.30b) can be generalized to give an equation called the power equation:

$$P = \dot{m} g h = \gamma Q h \tag{7.31}$$

Both pumps and turbines lose energy due to factors such as mechanical friction, viscous dissipation, and leakage. These losses are accounted for by the *efficiency*, which is defined as the ratio of power output to power input:

$$\eta \equiv \frac{\text{power output from a machine or system}}{\text{power input to a machine or system}} = \frac{P_{\text{output}}}{P_{\text{input}}} \tag{7.32}$$

If the mechanical efficiency of the pump is η_p, the power output $\dot{W}_p$ delivered by the pump to the flow is

$$\dot{W}_p = \eta_p \dot{W}_s \tag{7.33a}$$

where $\dot{W}_s$ is the power supplied to the pump, usually by a rotating shaft that is connected to a motor. For a turbine, the output power W_{shaft} is usually delivered by a rotating shaft to a generator. If the mechanical efficiency of the turbine is η_t, the output power supplied by the turbine is

$$\dot{W}_s = \eta_t \dot{W}_t \tag{7.33b}$$

where $\dot{W}_t$ is the power input to the turbine from the flow. Example 7.3 shows how to apply the power equation.

EXAMPLE 7.3   POWER NEEDED BY A PUMP

A pipe 50 cm in diameter carries water (10°C) at a rate of 0.5 m³/s. A pump in the pipe is used to move the water from an elevation of 30 m to 40 m. The pressure at section 1 is 70 kPa gage and the pressure at section 2 is 350 kPa gage. What power in kilowatts and in horsepower must be supplied to the flow by the pump? Assume $h_L = 3$ m of water and $\alpha_1 = \alpha_2 = 1$.

Problem Definition

Situation: A pump is used to increase the elevation and pressure of water.

Find: Power (kW and hp) that is supplied to the water.

Properties: Water (10°C), Table A.5: $\gamma = 9810$ N/m³.

Sketch:

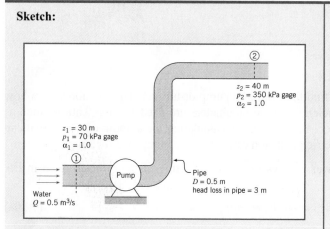

$z_2 = 40$ m
$p_2 = 350$ kPa gage
$\alpha_2 = 1.0$

$z_1 = 30$ m
$p_1 = 70$ kPa gage
$\alpha_1 = 1.0$

Pipe
$D = 0.5$ m
head loss in pipe = 3 m

Water
$Q = 0.5$ m³/s

Pump

Plan

1. Write the energy equation between section 1 and section 2.
2. Analyze each term in the energy equation.
3. Calculate the head of the pump h_p.
4. Find the power by applying the power equation (7.30a).

Solution

1. Energy equation (general form)

$$\frac{p_1}{\gamma} + \alpha_1 \frac{\overline{V}_1^2}{2g} + z_1 + h_p = \frac{p_2}{\gamma} + \alpha_2 \frac{\overline{V}_2^2}{2g} + z_2 + h_t + h_L$$

2. Term-by-term analysis
   - Velocity head cancels because $V_1 = V_2$.
   - $h_t = 0$ because there are no turbines in the system.
   - All other head terms are given.
   - Inserting terms into the general equation gives

$$\frac{p_1}{\gamma} + z_1 + h_p = \frac{p_2}{\gamma} + z_2 + h_L$$

3. Pump head (from step 2)

$$h_p = \left(\frac{p_2 - p_1}{\gamma} \right) + (z_2 - z_1) + h_L$$

$$= \left(\frac{(350{,}000 - 70{,}000) \text{ N/m}^2}{9810 \text{ N/m}^3} \right) + (10 \text{ m}) + (3 \text{ m})$$

$$= (28.5 \text{ m}) + (10 \text{ m}) + (3 \text{ m}) = 41.5 \text{ m}$$

Interpretation: The head provided by the pump (41.5 m) is balanced by the increase in pressure head (28.5 m) plus the increase in elevation head (10 m) plus the head loss (3 m).

4. Power equation

$$P = \gamma Q h_p$$

$$= (9810 \text{ N/m}^3)(0.5 \text{ m}^3/\text{s})(41.5 \text{ m})$$

$$= \boxed{204 \text{ kW}} = (204 \text{ kW})\left(\frac{1.0 \text{ hp}}{0.746 \text{ kW}} \right) = \boxed{273 \text{ hp}}$$

Example 7.4 shows how efficiency enters into a calculation of power.

EXAMPLE 7.4 POWER PRODUCED BY A TURBINE

At the maximum rate of power generation, a small hydroelectric power plant takes a discharge of 14.1 m³/s through an elevation drop of 61 m. The head loss through the intakes, penstock, and outlet works is 1.5 m. The combined efficiency of the turbine and electrical generator is 87%. What is the rate of power generation?

Problem Definition

Situation: A small hydroelectric plant is producing electrical power.

Find: Electrical power generation (in kW).

Properties: Water (10°C), Table A.5: $\gamma = 9810$ N/m³.

Sketch:

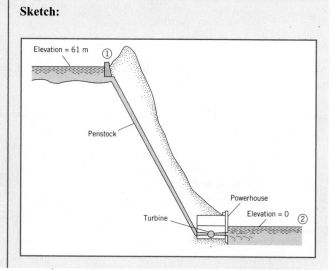

Elevation = 61 m ①

Penstock

Powerhouse

Elevation = 0 ②

Turbine

Plan

1. Write the energy equation (7.29) between section 1 and section 2.
2. Analyze each term in the energy equation.
3. Solve for the head of the turbine h_t.
4. Find the input power to the turbine using the power equation (7.30a).
5. Find the output power from generator by using the efficiency equation (7.32).

Solution

1. Energy equation (general form)

$$\frac{p_1}{\gamma} + \alpha_1 \frac{\overline{V}_1^2}{2g} + z_1 + h_p = \frac{p_2}{\gamma} + \alpha_2 \frac{\overline{V}_2^2}{2g} + z_2 + h_t + h_L$$

2. Term-by-term analysis
   - Velocity heads are negligible because $V_1 \approx 0$ and $V_2 \approx 0$.
   - Pressure heads are zero because $p_1 = p_2 = 0$ gage.

- $h_p = 0$ because there is no pump in the system.
- Elevation head terms are given.

3. Combine steps 1 and 2:

$$h_t = (z_1 - z_2) - h_L$$

$$= (61 \text{ m}) - (1.5 \text{ m}) = 59.5 \text{ m}$$

Interpretation: Head supplied to the turbine (59.5 m) is equal to the net elevation change of the dam (61 m) minus the head loss (1.5 m).

4. Power equation

$$P_{\text{input to turbine}} = \gamma Q h_t = (9810 \text{ N/m}^3)(14.1 \text{ m}^3/\text{s})(59.5 \text{ m})$$

$$= 8.23 \text{ MW}$$

5. Efficiency equation

$$P_{\text{output from generator}} = \eta P_{\text{input to turbine}} = 0.87(8.23 \text{ MW})$$

$$= \boxed{7.16 \text{ MW}}$$

7.5

Contrasting the Bernoulli Equation and the Energy Equation

While the Bernoulli equation given in Eq. (4.18b) and the energy equation from Eq. (7.29) have a similar form and several terms in common, they are not the same equation. This section explains the differences between these two equations. This information is important for conceptual understanding of these two very important equations.

The Bernoulli equation and the energy equation are derived in different ways. The Bernoulli equation was derived by applying Newton's second law to a particle and then integrating the resulting equation along a streamline. The energy equation was derived by starting with the first law of thermodynamics and then using the Reynolds transport theorem. Consequently, the Bernoulli equation involves only mechanical energy, whereas the energy equation includes both mechanical and thermal energy.

The two equations have different methods of application. The Bernoulli equation is applied by selecting two points on a streamline and then equating terms at these points:

$$\frac{V_1^2}{2} + \frac{p_1}{\gamma} + z_1 = \frac{V_2^2}{2} + \frac{p_2}{\gamma} + z_2$$

In addition, these two points can be anywhere in the flow field for the special case of irrotational flow. The energy equation is applied by selecting an inlet section and an outlet section in a pipe and then equating terms as they apply to the pipe:

$$\alpha_1 \frac{V_1^2}{2} + \frac{p_1}{\gamma} + z_1 + h_p = \alpha_2 \frac{V_2^2}{2} + \frac{p_2}{\gamma} + z_2 + h_t + h_L$$

The two equations have different assumptions. The Bernoulli equation applies to steady, incompressible, and inviscid flow. The energy equation applies to steady, viscous, incompressible flow in a pipe with additional energy being added through a pump or extracted through a turbine.

Under special circumstances the energy equation can be reduced to the Bernoulli equation. If the flow is inviscid, there is no head loss; that is, $h_L = 0$. If the "pipe" is regarded as a small stream tube enclosing a streamline, then $\alpha = 1$. There is no pump or turbine along a streamline, so $h_p = h_t = 0$. In this case the energy equation is identical to the Bernoulli equation. Note that the energy equation cannot be developed starting with the Bernoulli equation.

In summary, the energy equation is *not* the Bernoulli equation.

7.6

Transitions

The purpose of this section is to illustrate how the energy, momentum, and continuity equations can be used together to analyze (a) head loss for an abrupt expansion and (b) forces on transitions. These results are useful for designing systems, especially those with large pipes such as the penstock in a dam.

Abrupt Expansion

An *abrupt or sudden expansion* in a pipe or duct is a change from a smaller section area to a larger section area as shown in Fig. 7.6. Notice that a confined jet of fluid from the smaller pipe discharges into the larger pipe and creates a zone of separated flow. Because the streamlines in the jet are initially straight and parallel, the piezometric pressure distribution across the jet at section 1 will be uniform. This same uniform pressure distribution will also occur in the zone of separated flow. Apply the energy equation between sections 1 and 2:

$$\frac{p_1}{\gamma} + \alpha_1 \frac{V_1^2}{2g} + z_1 = \frac{p_2}{\gamma} + \alpha_2 \frac{V_2^2}{2g} + z_2 + h_L \tag{7.34}$$

Figure 7.6

Flow through an abrupt expansion.

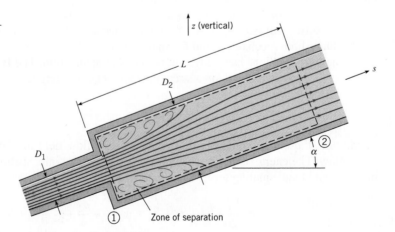

Assume turbulent flow conditions so $\alpha_1 = \alpha_2 \approx 1$. The momentum equation for the fluid in the large pipe between section 1 and section 2, written for the s direction, is

$$\sum F_s = \dot{m}V_2 - \dot{m}V_1$$

Neglect the force due to shear stress to give

$$p_1A_2 - p_2A_2 - \gamma A_2 L \sin\alpha = \rho V_2^2 A_2 - \rho V_1^2 A_1$$

or
$$\frac{p_1}{\gamma} - \frac{p_2}{\gamma} - (z_2 - z_1) = \frac{V_2^2}{g} - \frac{V_1^2}{g}\frac{A_1}{A_2} \qquad (7.35)$$

The continuity equation simplifies to

$$V_1 A_1 = V_2 A_2 \qquad (7.36)$$

Combining Eqs. (7.34) to (7.36) gives an equation for the head loss h_L caused by a sudden expansion:

$$h_L = \frac{(V_1 - V_2)^2}{2g} \qquad (7.37)$$

If a pipe discharges fluid into a reservoir, then $V_2 = 0$ and the sudden-expansion head loss simplifies to

$$h_L = \frac{V^2}{2g}$$

which is the velocity head of the liquid in the pipe. This energy is dissipated by the viscous action of the fluid in the reservoir.

Forces on Transitions

To find forces on transitions in pipes, apply the momentum equation in combination with the energy equation, the flow rate equations, and head loss equations. This approach is illustrated by Example 7.5.

EXAMPLE 7.5   FORCE ON A CONTRACTION IN A PIPE

A pipe 30 cm in diameter carries water (10°C, 250 kPa) at a rate of 0.707 m³/s. The pipe contracts to a diameter of 20 cm. The head loss through the contraction is given by

$$h_L = 0.1\frac{V_2^2}{2g}$$

where V_2 is the velocity in the 20 cm pipe. What horizontal force is required to hold the transition in place? Assume $\alpha_1 = \alpha_2 = 1$.

Problem Definition

Situation: Water flows through a contraction in a round pipe.
Find: The horizontal force F_x (newtons) required to hold the contraction stationary.
Sketch:

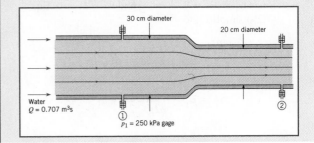

Assumptions: $\alpha_1 = \alpha_2 = 1$.

Properties: Water (10°C), Table A.5: $\gamma = 9810$ N/m^3.

Plan

1. Develop an equation for F_x by applying the momentum equation (6.6).
2. Develop an equation for p_2 by applying the energy equation (7.29).
3. Calculate p_2.
4. Calculate F_x.

Solution

1. Momentum equation (horizontal direction)
   - General equation

$$\sum \boldsymbol{F} = \frac{d}{dt}\bigg|_{cv} \int \rho \boldsymbol{v}\, d\Psi + \sum_{cs} \dot{m}_o \boldsymbol{v}_o - \sum_{cs} \dot{m}_i \boldsymbol{v}_i$$

   - Force terms (see force diagram)

$$\sum F_x = p_1 A_1 + F_x - p_2 A_2.$$

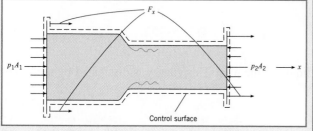

   - Momentum accumulation is zero because flow is steady.
   - Momentum efflux is

$$\sum_{cs} \dot{m}_o \boldsymbol{v}_o - \sum_{cs} \dot{m}_i \boldsymbol{v}_i = (\dot{m} V_2 - \dot{m} V_1)\hat{\boldsymbol{i}}$$

   - Substitute force and momentum terms into the momentum equation.

$$p_1 A_1 - p_2 A_2 + F_x = \dot{m} V_2 - \dot{m} V_1$$

$$F_x = \rho Q(V_2 - V_1) + p_2 A_2 - p_1 A_1$$

2. Energy equation (from section 1 to section 2)
   - Since $\alpha_1 = \alpha_2 = 1$, $z_1 = z_2$, and $h_p = h_t = 0$, Eq. (7.29) simplifies to

$$\frac{p_1}{\gamma} + \frac{V_1^2}{2g} = \frac{p_2}{\gamma} + \frac{V_2^2}{2g} + h_L$$

$$p_2 = p_1 - \gamma\left(\frac{V_2^2}{2g} - \frac{V_1^2}{2g} + h_L\right)$$

3. Pressure at section 2
   - Find velocities using the flow rate equation.

$$V_1 = \frac{Q}{A_1} = \frac{0.707 \text{ m}^3/\text{s}}{(\pi/4) \times (0.3 \text{ m})^2} = 10 \text{ m/s}$$

$$V_2 = \frac{Q}{A_2} = \frac{0.707 \text{ m}^3/\text{s}}{(\pi/4) \times (0.2 \text{ m})^2} = 22.5 \text{ m/s}$$

   - Calculate head loss.

$$h_L = \frac{0.1 V_2^2}{2g} = \frac{0.1 \times (22.5 \text{ m/s})^2}{2 \times (9.81 \text{ m/s}^2)} = 2.58 \text{ m}$$

   - Calculate pressure.

$$p_2 = p_1 - \gamma\left(\frac{V_2^2}{2g}\right) - \frac{V_1^2}{2g} + h_L$$

$$= 250 \text{ kPa} - 9.81 \text{ kN/m}^3$$

$$\times \left(\frac{(22.5 \text{ m/s})^2}{2(9.81 \text{ m/s}^2)} - \frac{(10 \text{ m/s})^2}{2(9.81 \text{ m/s}^2)} + 2.58 \text{ m}\right)$$

$$= 21.6 \text{ kPa}$$

4. Calculate F_x.

$$F_x = \rho Q(V_2 - V_1) + p_2 A_2 - p_1 A_1$$

$$= (1000 \text{ kg/m}^3)(0.707 \text{ m}^3/\text{s})(22.5 - 10)(\text{m/s})$$

$$+ (21,600 \text{ Pa})\left(\frac{\pi(0.2 \text{ m})^2}{4}\right) - (250,000 \text{ Pa})$$

$$\times \left(\frac{\pi(0.3 \text{ m})^2}{4}\right)$$

$$= (8837 + 677 - 17,670)\text{N} = -8.16 \text{ kN}$$

$\boxed{F_x = 8,16 \text{ kN applied in the negative } x \text{ direction}}$

Hydraulic and Energy Grade Lines

This section introduces the hydraulic grade line (HGL) and the energy grade line (EGL), which are graphical representations that show head in a system. This visual approach provides insights and helps one locate and correct trouble spots in the system (usually points of low pressure).

The *EGL*, shown in Fig. 7.7, is a line that indicates the total head at each location in a system. The *EGL* is related to terms in the energy equation by

$$\text{EGL} = \left(\begin{array}{c}\text{velocity}\\\text{head}\end{array}\right) + \left(\begin{array}{c}\text{pressure}\\\text{head}\end{array}\right) + \left(\begin{array}{c}\text{elevation}\\\text{head}\end{array}\right) = \alpha\frac{V^2}{2g} + \frac{p}{\gamma} + z = \left(\begin{array}{c}\text{total}\\\text{head}\end{array}\right) \qquad (7.38)$$

Notice that *total head*, which characterizes the energy that is carried by a flowing fluid, is the sum of velocity head, the pressure head, and the elevation head.

The *HGL,* shown in Fig. 7.7, is a line that indicates the piezometric head at each location in a system:

$$\text{HGL} = \left(\begin{array}{c}\text{pressure}\\\text{head}\end{array}\right) + \left(\begin{array}{c}\text{elevation}\\\text{head}\end{array}\right) = \frac{p}{\gamma} + z = \left(\begin{array}{c}\text{piezometric}\\\text{head}\end{array}\right) \qquad (7.39)$$

Since the HGL gives piezometric head, the HGL will be coincident with the liquid surface in a piezometer as shown in Fig. 7.7. Similarly, the EGL will be coincident with the liquid surface in a stagnation tube.

Tips for Drawing HGLs and EGLs

1. In a lake or reservoir, the HGL and EGL will coincide with the liquid surface. Also, both the HGL and EGL will indicate piezometric head. For example, see Fig. 7.7.
2. A pump causes an abrupt rise in the EGL and HGL by adding energy to the flow. For example, see Fig. 7.8.
3. For steady flow in a pipe of constant diameter and wall roughness, the slope ($\Delta h_L / \Delta L$) of the EGL and the HGL will be constant. For example, see Fig. 7.7.

Figure 7.7

EGL and HGL in a straight pipe.

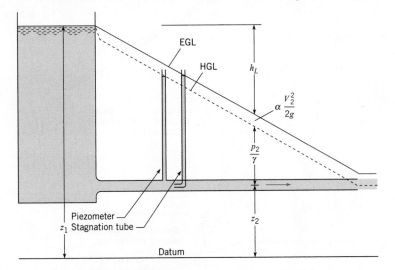

4. Locate the HGL below the EGL by a distance of the velocity head ($\alpha V^2/2g$).

5. Height of the EGL decreases in the flow direction unless a pump is present.

6. A turbine causes an abrupt drop in the EGL and HGL by removing energy from the flow. For example, see Fig. 7.9.

7. Power generated by a turbine can be increased by using a gradual expansion at the turbine outlet. As shown in Fig. 7.9, the expansion converts kinetic energy to pressure. If the outlet to a reservoir is an abrupt expansion, as in Fig. 7.11, this kinetic energy is lost.

8. When a pipe discharges into the atmosphere the HGL is coincident with the system because $p/\gamma = 0$ at these points. For example, in Figures 7.10 and 7.12, the HGL in the liquid jet is drawn through the jet itself.

9. When a flow passage changes diameter, the distance between the EGL and the HGL will change (see Fig. 7.10 and Fig. 7.11) because velocity changes. In addition, the slope on the EGL will change because the head loss per length will be larger in the conduit with the larger velocity (see Fig. 7.11).

10. If the HGL falls below the pipe, then p/γ is negative, indicating subatmospheric pressure (see Fig. 7.12) and a potential location of cavitation.

Figure 7.8

Rise in EGL and HGL due to pump.

Figure 7.9

Drop in EGL and HGL due to turbine.

Figure 7.10

Change in HGL and EGL due to flow through a nozzle.

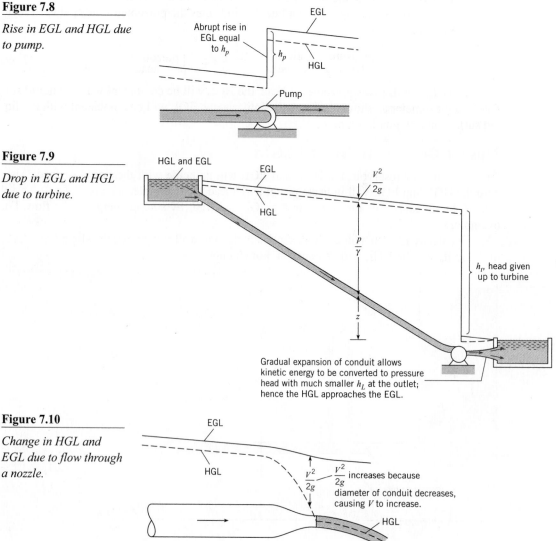

Figure 7.11

Change in EGL and HGL due to change in diameter of pipe.

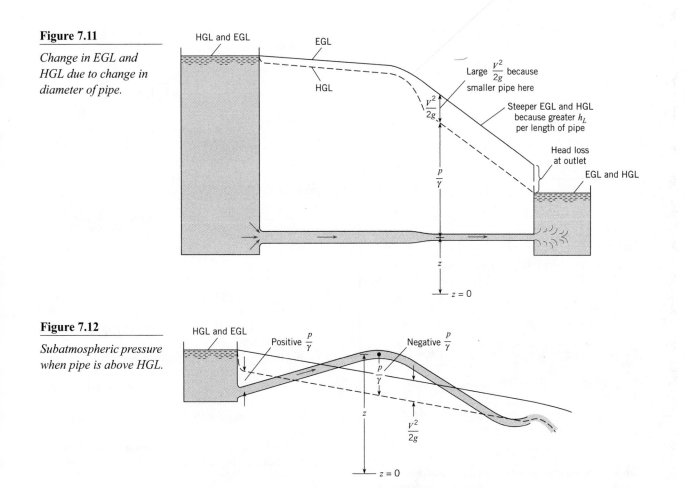

The recommended procedure for drawing an EGL and HGL is shown in Example 7.6. Notice how the tips from pp. 233–234 are applied.

Figure 7.12

Subatmospheric pressure when pipe is above HGL.

EXAMPLE 7.6   EGL AND HGL FOR A SYSTEM

A pump draws water (50°F) from a reservoir, where the water-surface elevation is 520 ft, and forces the water through a pipe 5000 ft long and 1 ft in diameter. This pipe then discharges the water into a reservoir with water-surface elevation of 620 ft. The flow rate is 7.85 cfs, and the head loss in the pipe is given by

$$h_L = 0.01 \left(\frac{L}{D} \right) \left(\frac{V^2}{2g} \right)$$

Determine the head supplied by the pump, h_p, and the power supplied to the flow, and draw the HGL and EGL for the system. Assume that the pipe is horizontal and is 510 ft in elevation.

Problem Definition

Situation: Water is pumped from a lower reservoir to a higher reservoir.

Find:

1. Pump head (in ft).
2. Power (in hp) supplied to the flow.
3. Draw HGL. Draw EGL.

Properties: Water (50°F), Table A.5: $\gamma = 62.4$ lbf/ft^3.

Sketch:

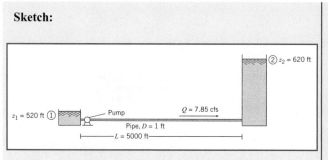

Plan

1. Apply the energy equation (7.29) between sections 1 and section 2.
2. Calculate terms in the energy equation.
3. Find the power by applying the power equation (7.30a).
4. Draw the HGL and EGL by using the tips given on p. 270.

Solution

1. Energy equation (general form)

$$\frac{p_1}{\gamma} + \alpha_1\frac{\overline{V}_1^2}{2g} + z_1 + h_p = \frac{p_2}{\gamma} + \alpha_2\frac{\overline{V}_2^2}{2g} + z_2 + h_t + h_L$$

- Velocity heads are negligible because $V_1 \approx 0$ and $V_2 \approx 0$.
- Pressure heads are zero because $p_1 = p_2 = 0$ gage.
- $h_t = 0$ because there are no turbines in the system.

$$h_p = (z_2 - z_1) + h_L$$

Interpretation: Head supplied by the pump provides the energy to lift the fluid to a higher elevation plus the energy to overcome head loss.

2. Calculations of terms in the energy equation
- Calculate V using the flow rate equation.

$$V = \frac{Q}{A} = \frac{7.85\,\text{ft}^3/\text{s}}{(\pi/4)(1\,\text{ft})^2} = 10\,\text{ft/s}$$

- Calculate head loss.

$$h_L = 0.01\left(\frac{L}{D}\right)\left(\frac{V^2}{2g}\right) = 0.01\left(\frac{5000\,\text{ft}}{1.0\,\text{ft}}\right)\left(\frac{(10\,\text{ft/s})^2}{2\times(32.2\,\text{ft/s}^2)}\right)$$

$$= 77.6\,\text{ft}$$

- Calculate h_p.

$$h_p = (z_2 - z_1) + h_L = (620\,\text{ft} - 520\,\text{ft}) + 77.6\,\text{ft} = \boxed{178\,\text{ft}}$$

3. Power

$$\dot{W}_p = \gamma Q h_p = \left(\frac{62.4\,\text{lbf}}{\text{ft}^3}\right)\left(\frac{7.85\,\text{ft}^3}{\text{s}}\right)(178\,\text{ft})\left(\frac{\text{hp}\cdot\text{s}}{550\,\text{ft}\cdot\text{lbf}}\right)$$

$$= \boxed{159\,\text{hp}}$$

4. HGL and EGL
- From Tip 1 on p. 233, locate the HGL and EGL along the reservoir surfaces.
- From Tip 2, sketch in a head rise of 178 ft corresponding to the pump.
- From Tip 3, sketch the EGL from the pump outlet to the reservoir surface. Use the fact that the head loss is 77.6 ft. Also, sketch EGL from the reservoir on the left to the pump inlet. Show a small head loss.
- From Tip 4, sketch the HGL below the EGL by a distance of $V^2/2g \approx 1.6$ ft.
- From Tip 5, check the sketches to ensure that EGL and HGL are decreasing in the direction of flow (except at the pump).

Sketch: HGL (dashed black line) and EGL (solid blue line)

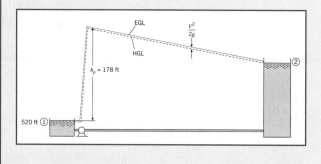

7.8

Summary

The energy equation relates the rate of change of energy of a system to the rate of heat transfer to the system and the rate at which the system does work on the surroundings.

Applying the energy equation to a control volume with steady and uniform flow at the control surfaces results in

$$\dot{Q} - \dot{W}_s = \sum_{cs} \dot{m}_o \left(\frac{V_o^2}{2} + h_o + gz_o \right) - \sum_{cs} \dot{m}_i \left(\frac{V_i^2}{2} + h_i + gz_i \right)$$

where $\dot{Q}$ is the rate of heat transfer to the control volume, $\dot{W}_s$ is the rate at which shaft work is done on the surroundings, and h is the enthalpy of the fluid.

Further simplification of the energy equation for the flow of an incompressible fluid in a pipe yields

$$\frac{p_1}{\gamma} + \alpha_1 \frac{V_1^2}{2g} + z_1 + h_p = \frac{p_2}{\gamma} + \alpha_2 \frac{V_2^2}{2g} + z_2 + h_t + h_L$$

where α is the kinetic-energy correction factor, h_p is the head provided by a pump, h_t is the head removed by a turbine, and h_L is the head loss. Station 1 is always upstream, and station 2 is downstream. For laminar flow, $\alpha = 2$ and for turbulent flow, $\alpha \approx 1$. The increase in head across a pump is related to the pump power by

$$\dot{W}_p = \gamma Q h_p$$

and the power delivered by a turbine is given by

$$\dot{W}_t = \gamma Q h_t$$

The actual power required by the pump is $\dot{W}_s = \dot{W}_p / \eta_p$, and the actual power delivered by a turbine is $\dot{W}_s = \eta_t \dot{W}_t$, where η_p and η_t are the pump and turbine efficiencies.

The head loss is always positive and represents the irreversible conversion of mechanical energy to thermal energy through the viscous action of the fluid. The head loss due to a sudden expansion is

$$h_L = \frac{(V_1 - V_2)^2}{2g}$$

where V_1 and V_2 are the upstream and downstream velocities.

The hydraulic grade line (HGL) is the profile of the piezometric head, $p/\gamma + z$, along a pipe. The energy grade line (EGL) is a plot of the total head, $V^2/2g + p/\gamma + z$, along a pipe. If the hydraulic grade line falls below the elevation of a pipe, subatmospheric pressure exists in the pipe at that location, giving rise to the possibility of cavitation or leakage into the pipe.

References

1. Cengel, Y. A., and M. A. Boles. *Thermodynamics: An Engineering Approach*. McGraw-Hill, New York, 1998.

2. Moran, M. J., and H. N. Shapiro. *Fundamentals of Engineering Thermodynamics*. John Wiley, New York, 1992.

Problems

Energy and Power

7.1 $\mathbb{P}\mathbb{Q}\blacktriangleleft$ From the list below, select one topic that is interesting to you. Then, use references such as the Internet to research your topic and prepare one page of documentation that you could use to present your topic to your peers.

a. Explain how hydroelectric power is produced.
b. Explain how a Kaplan turbine works, how a Francis turbine works, and the differences between these two types of turbines.
c. Explain how a horizontal-axis wind turbine is used to produce electrical power.
d. Explain how a steam turbine is used to produce electrical power.

7.2 PQ◀ Skim this chapter and identify three real-world applications that are motivating to you. For each application, write a paragraph that describes what you already know about the application and why this application has appeal to you.

7.3 PQ◀ Using Section 7.1 and other resources, answer the following questions. Strive for depth, clarity, and accuracy. Also, strive for effective use of sketches, words, and equations.

a. What are the common forms of energy? Which of these forms are relevant to fluid mechanics?

b. What is work? Describe three example of work that are relevant to fluid mechanics.

c. What are the most common units of power?

d. List three significant differences between power and energy.

7.4 PQ◀ Apply the grid method to each situation.

a. Calculate the energy in joules used by a 1 hp pump that is operating for 6 hours. Also, calculate the cost of electricity for this time period. Assume that electricity costs $0.15 per kW-hr.

b. A motor is being to used to turn the shaft of a centrifugal pump. Apply Eq. (7.2b) to calculate the power in watts corresponding to a torque of 100 lbf-in and a rotation speed of 850 rpm.

c. A turbine produces a power of 7500 ft-lbf/s. Calculate the power in hp and in watts.

7.5 PQ◀ Estimate the power required to spray water out of the spray bottle that is pictured in Fig. 7.1. *Hint:* Make appropriate assumptions about the number of sprays per unit time and the force exerted by the finger.

7.6 The sketch shows a common consumer product called the Water Pik. This device uses a motor to drive a piston pump that produces a jet of water ($d = 3$ mm, $T = 10°C$) with a speed of 25 m/s. Estimate the minimum electrical power in watts that is required by the device. *Hints:* (a) Assume that the power is used only to produce the kinetic energy of the water in the jet; and (b) in a time interval Δt, the amount of mass that flows out the nozzle is Δm, and the corresponding amount of kinetic energy is $(\Delta m V^2/2)$.

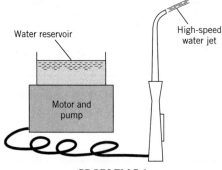

PROBLEM 7.6

7.7 An engineer is considering the development of a small wind turbine ($D = 1.25$ m) for home applications. The design wind speed is 15 mph at $T = 10°C$ and $p = 0.9$ bar. The efficiency of the turbine is $\eta = 20\%$, meaning that 20% of the kinetic energy in the wind can be extracted. Estimate the power in watts that

can be produced by the turbine. *Hint:* In a time interval Δt, the amount of mass that flows through the rotor is $\Delta m = \dot{m}\Delta t$ and the corresponding amount of kinetic energy in this flow is $(\Delta m V^2/2)$.

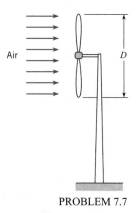

PROBLEM 7.7

Kinetic Energy Correction Factor (α)

7.8 PQ◀ Using Section 7.3 and other resources, answer the questions below. Strive for depth, clarity, and accuracy while also combining sketches, words, and equations in ways that enhance the effectiveness of your communication.

a. What is the kinetic-energy correction factor? Why do engineers use this term?

b. What is the meaning of each variable ($\alpha, A, V, \bar{V}$) that appears in Eq. 7.21?

c. What values of α are commonly used?

7.9 For this hypothetical velocity distribution in a wide rectangular channel, evaluate the kinetic-energy correction factor α.

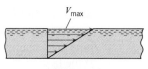

PROBLEM 7.9

7.10 For these velocity distributions in a round pipe, indicate whether the kinetic-energy correction factor α is greater than, equal to, or less than unity.

7.11 Calculate α for case (c) in Prob. 7.10.

7.12 Calculate α for case (d) in Prob. 7.10.

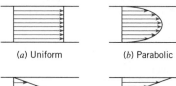

(a) Uniform   (b) Parabolic

(c) Linear   (d) Linear

PROBLEMS 7,10, 7.11, 7.12

7.13 An approximate equation for the velocity distribution in a pipe with turbulent flow is

$$\frac{V}{V_{max}} = \left(\frac{y}{r_0}\right)^n$$

where V_{max} is the centerline velocity, y is the distance from the wall of the pipe, r_0 is the radius of the pipe, and n is an exponent that depends on the Reynolds number and varies between $1/6$ and $1/8$ for most applications. Derive a formula for α as a function of n. What is α if $n = 1/7$?

7.14 An approximate equation for the velocity distribution in a rectangular channel with turbulent flow is

$$\frac{u}{u_{max}} = \left(\frac{y}{d}\right)^n$$

where u_{max} is the velocity at the surface, y is the distance from the floor of the channel, d is the depth of flow, and n is an exponent that varies from about $1/6$ to $1/8$ depending on the Reynolds number. Derive a formula for α as a function of n. What is the value of α for $n = 1/7$?

7.15 The following data were taken for turbulent flow in a circular pipe with a radius of 3.5 cm. Evaluate the kinetic energy correction factor. The velocity at the pipe wall is zero.

r (cm)	V (m/s)	r (cm)	V (m/s)
0.0	32.5	2.8	22.03
0.5	32.44	2.9	21.24
1.0	32.27	3.0	20.49
1.5	31.22	3.1	19.6
2.0	28.21	3.2	18.69
2.25	26.51	3.25	18.16
2.5	24.38	3.3	17.54
2.6	23.7	3.35	17.02
2.7	22.88	3.4	16.14

Energy Equation

7.16 $\mathbb{PQ}$◀ Using Section 7.3 and other resources, answer the questions below. Strive for depth, clarity, and accuracy. Also, strive for effective use of sketches, words, and equations.
a. What is conceptual meaning of the first law of thermodynamics for a system?
b. What is flow work? How is the equation for flow work (Eq. 7.15) derived?
c. What is shaft work? How is shaft work different than flow work?

7.17 $\mathbb{PQ}$◀ Using Section 7.3 and other resources, answer the questions below. Strive for depth, clarity, and accuracy. Also, strive for effective use of sketches, words, and equations.
a. What is head? How is head related to energy? To power?

b. What is head of a turbine?
c. How is head of a pump related to power? To energy?
d. What is head loss?

7.18 $\mathbb{PQ}$◀ Using Sections 7.3 and 7.6 and using other resources, answer the following questions. Strive for depth, clarity, and accuracy. Also, strive for effective use of sketches, words and equations.
a. What are the five main terms in the energy equation (7.29)? What does each term mean?
b. How are terms in the energy equation related to energy? To power?
c. What assumptions are required for using the energy equation (7.29)?
d. How is the energy equation (7.29) similar to the Bernoulli equation? How is it different? Give three important similarities and three important differences.

7.19 Using the energy equation (7.29), prove that fluid in a pipe will flow from a location with high piezometric head to a location with low piezometric head. Assume there are no pumps or turbines and that the pipe has a constant diameter.

7.20 Water flows at a steady rate in this vertical pipe. The pressure at A is 10 kPa, and at B it is 98.1 kPa. Then the flow in the pipe is (a) upward, (b) downward, or (c) no flow. (*Hint:* see problem 7.19).

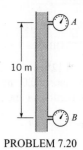

PROBLEM 7.20

7.21 Determine the discharge in the pipe and the pressure at point B. Neglect head losses. Assume $\alpha = 1.0$ at all locations.

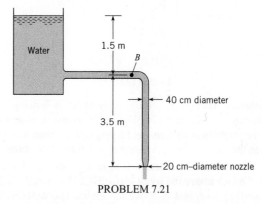

PROBLEM 7.21

7.22 A pipe drains a tank as shown. If $x = 10$ ft, $y = 4$ ft, and head losses are neglected, what is the pressure at point A and what is the velocity at the exit? Assume $\alpha = 1.0$ at all locations.

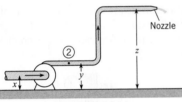

PROBLEMS 7.22, 7.23

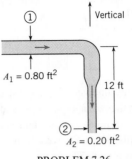

PROBLEM 7.26

7.23 A pipe drains a tank as shown. If $x = 10$ m, $y = 1.5$ m, and head losses are neglected, what is the pressure at point A and what is the velocity at the exit? Assume $\alpha = 1.0$ at all locations.

7.24 For this system, the discharge of water is 0.1 m³/s, $x = 1.0$ m, $y = 2.0$ m, $z = 7.0$ m, and the pipe diameter is 30 cm. Neglecting head losses, what is the pressure head at point 2 if the jet from the nozzle is 10 cm in diameter? Assume $\alpha = 1.0$ at all locations.

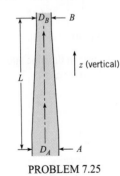

PROBLEM 7.24

7.25 If $D_A = 20$ cm, $D_B = 12$ cm, and $L = 1$ m, and if crude oil ($S = 0.90$) is flowing at a rate of 0.06 m³/s, determine the difference in pressure between sections A and B. Neglect head losses.

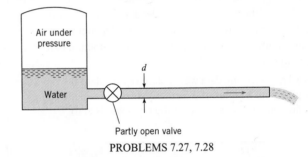

PROBLEM 7.25

7.26 Gasoline having a specific gravity of 0.8 is flowing in the pipe shown at a rate of 5 cfs. What is the pressure at section 2 when the pressure at section 1 is 18 psig and the head loss is 6 ft between the two sections? Assume $\alpha = 1.0$ at all locations.

7.27 Water flows from a pressurized tank as shown. The pressure in the tank above the water surface is 100 kPa gage, and the water surface level is 8 m above the outlet. The water exit velocity is 10 m/s. The head loss in the system varies as $h_L = K_L V^2/2g$, where K_L is the minor-loss coefficient. Find the value for K_L. Assume $\alpha = 1.0$ at all locations.

7.28 A reservoir with water is pressurized as shown. The pipe diameter is 1 in. The head loss in the system is given by $h_L = 5V^2/2g$. The height between the water surface and the pipe outlet is 10 ft. A discharge of 0.10 ft³/s is needed. What must the pressure in the tank be to achieve such a flow rate? Assume $\alpha = 1.0$ at all locations.

PROBLEMS 7.27, 7.28

7.29 In the figure for Probs. 7.27 and 7.28, suppose that the reservoir is open to the atmosphere at the top. The valve is used to control the flow rate from the reservoir. The head loss across the valve is given as $h_L = 5V^2/2g$, where V is the velocity in the pipe. The cross-sectional area of the pipe is 9 cm². The head loss due to friction in the pipe is negligible. The elevation of the water level in the reservoir above the pipe outlet is 11 m. Find the discharge in the pipe. Assume $\alpha = 1.0$ at all locations.

7.30 As shown a microchannel is being designed to transfer fluid in a MEMS (microelectricalmechanical system) application. The channel is 200 micrometers in diameter and is 5 cm long. Ethyl alcohol is driven through the system at the rate of 0.1 microliters/s (μL/s) with a syringe pump, which is essentially a moving piston. The pressure at the exit of the channel is atmospheric. The flow is laminar, so $\alpha = 2$. The head loss in the channel is given by

$$h_L = \frac{32 \mu L V}{\gamma D^2}$$

where L is the channel length, D the diameter, V the mean velocity, μ the viscosity of the fluid, and γ the specific weight of the fluid. Find the pressure in the syringe pump. The velocity head associated with the motion of the piston in the syringe pump is negligible.

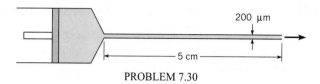

PROBLEM 7.30

7.31 Fire-fighting equipment requires that the exit velocity of the firehose be 40 m/s at an elevation of 50 m above the hydrant. The nozzle at the end of the hose has a contraction ratio of 4:1 $(A_e/A_{hose} = 1/4)$. The head loss in the hose is $10V^2/2g$, where V is the velocity in the hose. What must the pressure be at the hydrant to meet this requirement? The pipe supplying the hydrant is much larger than the firehose.

7.32 The discharge in the siphon is 2.80 cfs, $D = 8$ in., $L_1 = 3$ ft, and $L_2 = 3$ ft. Determine the head loss between the reservoir surface and point C. Determine the pressure at point B if three-quarters of the head loss (as found above) occurs between the reservoir surface and point B. Assume $\alpha = 1.0$ at all locations.

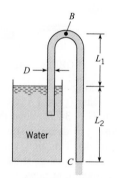

PROBLEM 7.32

7.33 For this siphon the elevations at A, B, C, and D are 30 m, 32 m, 27 m, and 26 m, respectively. The head loss between the inlet and point B is three-quarters of the velocity head, and the head loss in the pipe itself between point B and the end of the pipe is one-quarter of the velocity head. For these conditions, what is the discharge and what is the pressure at point B? The pipe diameter = 25 cm. Assume $\alpha = 1.0$ at all locations.

7.34 For this system, point B is 10 m above the bottom of the upper reservoir. The head loss from A to B is $1.8V^2/2g$, and the pipe area is 10^{-4} m^2. Assume a constant discharge of 8×10^{-4} m^3/s. For these conditions, what will be the depth of water in the upper reservoir for which cavitation will begin at point B? Vapor pressure = 1.23 kPa and atmospheric pressure = 100 kPa. Assume $\alpha = 1.0$ at all locations.

7.35 In this system, $d = 6$ in., $D = 12$ in., $\Delta z_1 = 6$ ft, and $\Delta z_2 = 12$ ft. The discharge of water in the system is 10 cfs. Is the machine a pump or a turbine? What are the pressures at points A and B? Neglect head losses. Assume $\alpha = 1.0$ at all locations.

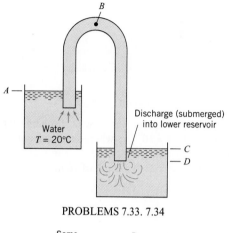

PROBLEMS 7.33. 7.34

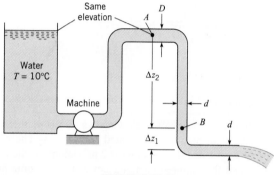

PROBLEM 7.35

7.36 The pipe diameter D is 30 cm, d is 15 cm, and the atmospheric pressure is 100 kPa. What is the maximum allowable discharge before cavitation occurs at the throat of the venturi meter if $H = 5$ m? Assume $\alpha = 1.0$ at all locations.

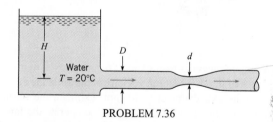

PROBLEM 7.36

7.37 In this system $d = 25$ cm, $D = 40$ cm, and the head loss from the venturi meter to the end of the pipe is given by $h_L = 0.9V^2/2g$, where V is the velocity in the pipe. Neglecting all other head losses, determine what head H will first initiate cavitation if the atmospheric pressure is 100 kPa absolute. What will be the discharge at incipient cavitation? Assume $\alpha = 1.0$ at all locations.

7.38 A pump is used to fill a tank 5 m in diameter from a river as shown. The water surface in the river is 2 m below the bottom of the tank. The pipe diameter is 5 cm, and the head loss in the pipe is given by $h_L = 10V^2/2g$, where V is the mean velocity in the pipe. The flow in the pipe is turbulent, so $\alpha = 1$. The

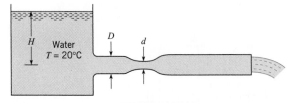

PROBLEM 7.37

head provided by the pump varies with discharge through the pump as $h_p = 20 - 4 \times 10^4 Q^2$, where the discharge is given in cubic meters per second (m³/s) and h_p is in meters. How long will it take to fill the tank to a depth of 10 m?

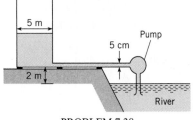

PROBLEM 7.38

Power Equation

7.39 As shown, water at 15°C is flowing in a 15 cm–diameter by 60 m–long run of pipe that is situated horizontally. The mean velocity is 2 m/s, and the head loss is 2 m. Determine the pressure drop and the required pumping power to overcome head loss in the pipe.

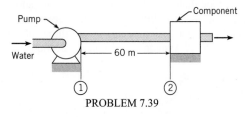

PROBLEM 7.39

7.40 A fan produces a pressure rise of 6 mm of water to move air through a hair dryer. The mean velocity of the air at the exit is 10 m/s, and the exit diameter is 44 mm. Estimate the electrical power in watts that needs to be supplied to operate the fan. Assume that the fan/motor combination has an efficiency of 60%.

PROBLEM 7.40

7.41 As shown in the figure, the pump supplies energy to the flow such that the upstream pressure (12 in. pipe) is 5 psi and the downstream pressure (6 in. pipe) is 90 psi when the flow of water is 5.0 cfs. What horsepower is delivered by the pump to the flow? Assume $\alpha = 1.0$ at all locations.

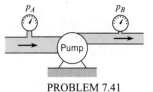

PROBLEM 7.41

7.42 A water discharge of 8 m³/s is to flow through this horizontal pipe, which is 1 m in diameter. If the head loss is given as $7V^2/2g$ (V is velocity in the pipe), how much power will have to be supplied to the flow by the pump to produce this discharge? Assume $\alpha = 1.0$ at all locations.

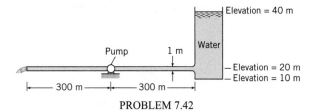

PROBLEM 7.42

7.43 An engineer is making an estimate for a home owner. This owner has a small stream ($Q = 1.4$ cfs, $T = 40°F$) that is located at an elevation $H = 34$ ft above the owner's residence. The owner is proposing to dam the stream, diverting the flow through a pipe (penstock). This flow will spin a hydraulic turbine, which in turn will drive a generator to produce electrical power. Estimate the maximum power in kilowatts that can be generated if there is no head loss and both the turbine and generator are 100% efficient. Also, estimate the power if the head loss is 5.5 ft, the turbine is 70% efficient, and the generator is 90% efficient.

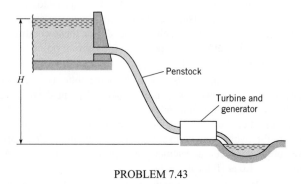

PROBLEM 7.43

7.44 A pump draws water through an 8 in. suction pipe and discharges it through a 4 in. pipe in which the velocity is 12 ft/s. The 4 in. pipe discharges horizontally into air at C. To what height h above the water surface at A can the water be raised if 25 hp is delivered to the pump? Assume that the pump operates at 60% efficiency and that the head loss in the pipe between A and C is equal to $2V_C^2/2g$. Assume $\alpha = 1.0$ at all locations.

7.45 A pump draws water (20°C) through a 20 cm suction pipe and discharges it through a 10 cm pipe in which the velocity is 3 m/s. The 10 cm pipe discharges horizontally into air at point C. To what height h above the water surface at A can the water be raised if 35 kW is delivered to the pump? Assume that the pump operates at 60% efficiency and that the head loss in the pipe between A and C is equal to $2V_C^2/2g$. Assume $\alpha = 1.0$ at all locations.

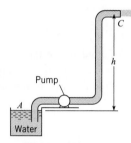

PROBLEMS 7.44, 7.45

7.46 An engineer is designing a subsonic wind tunnel. The test section is to have a cross-sectional area of 4 m² and an airspeed of 60 m/s. The air density is 1.2 kg/m³. The area of the tunnel exit is 10 m². The head loss through the tunnel is given by $h_L = (0.025)(V_T^2/2g)$, where V_T is the airspeed in the test section. Calculate the power needed to operate the wind tunnel. *Hint:* Assume negligible energy loss for the flow approaching the tunnel in region A, and assume atmospheric pressure at the outlet section of the tunnel. Assume $\alpha = 1.0$ at all locations.

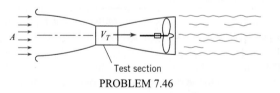

Test section

PROBLEM 7.46

7.47 Neglecting head losses, determine what horsepower the pump must deliver to produce the flow as shown. Here the elevations at points A, B, C, and D are 110 ft, 200 ft, 110 ft, and 90 ft, respectively. The nozzle area is 0.10 ft².

7.48 Neglecting head losses, determine what power the pump must deliver to produce the flow as shown. Here the elevations at points A, B, C, and D are 40 m, 65 m, 35 m, and 30 m, respectively. The nozzle area is 25 cm².

7.49 A pumping system is to be designed to pump crude oil a distance of 1 mile in a 1 foot-diameter pipe at a rate of 3500 gpm. The pressures at the entrance and exit of the pipe are atmospheric, and the exit of the pipe is 200 feet higher than the entrance. The pressure loss in the system due to pipe friction is 60 psi. The specific weight of the oil is 53 lbf/ft³. Find the power, in horsepower, required for the pump.

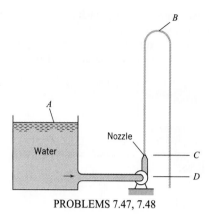

PROBLEMS 7.47, 7.48

7.50 Water (10°C) is flowing at a rate of 0.35 m³/s, and it is assumed that $h_L = 2V^2/2g$ from the reservoir to the gage, where V is the velocity in the 30-cm pipe. What power must the pump supply? Assume $\alpha = 1.0$ at all locations.

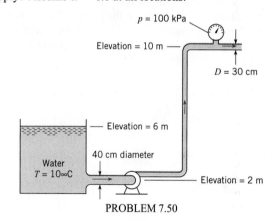

PROBLEM 7.50

7.51 In the pump test shown, the rate of flow is 6 cfs of oil (S = 0.88). Calculate the horsepower that the pump supplies to the oil if there is a differential reading of 46 in. of mercury in the U-tube manometer. Assume $\alpha = 1.0$ at all locations.

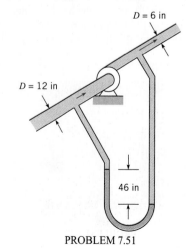

PROBLEM 7.51

7.52 If the discharge is 500 cfs, what power output may be expected from the turbine? Assume that the turbine efficiency is 90% and that the overall head loss is $1.5V^2/2g$, where V is the velocity in the 7 ft penstock Assume $\alpha = 1.0$ at all locations.

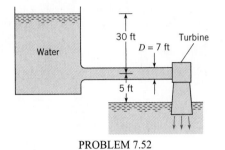

PROBLEM 7.52

7.53 A small-scale hydraulic power system is shown. The elevation difference between the reservoir water surface and the pond water surface downstream of the reservoir, H, is 15 m. The velocity of the water exhausting into the pond is 5 m/s, and the discharge through the system is 1 m³/s. The head loss due to friction in the penstock is negligible. Find the power produced by the turbine in kilowatts.

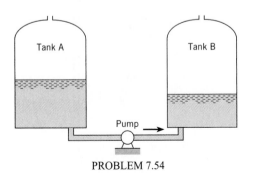

PROBLEM 7.53

7.54 A pump is used to transfer SAE-30 oil from tank A to tank B as shown. The tanks have a diameter of 12 m. The initial depth of the oil in tank A is 20 m, and in tank B the depth is 1 m. The pump delivers a constant head of 60 m. The connecting pipe has a diameter of 20 cm, and the head loss due to friction in the pipe is $20V^2/2g$. Find the time required to transfer the oil from tank A to B; that is, the time required to fill tank B to 20 m depth.

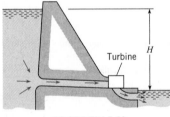

PROBLEM 7.54

7.55 A pump is used to pressurize a tank to 300 kPa abs. The tank has a diameter of 2 m and a height of 4 m. The initial level of water in the tank is 1 m, and the pressure at the water surface is 0 kPa gage. The atmospheric pressure is 100 kPa. The pump operates with a constant head of 50 m. The water is drawn from a source that is 4 m below the tank bottom. The pipe connecting the source and the tank is 4 cm in diameter and the head loss, including the expansion loss at the tank, is $10V^2/2g$. The flow is turbulent.

Assume the compression of the air in the tank takes place isothermally, so the tank pressure is given by

$$p_T = \frac{3}{4 - z_t}p_0$$

where z_t is the depth of fluid in the tank in meters. Write a computer program that will show how the pressure varies in the tank with time, and find the time to pressurize the tank to 300 kPa abs.

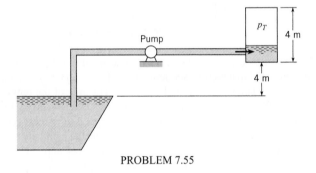

PROBLEM 7.55

Sudden Expansions and Other Components

7.56 What is the head loss at the outlet of the pipe that discharges water into the reservoir at a rate of 10 cfs if the diameter of the pipe is 12 in.?

7.57 What is the head loss at the outlet of the pipe that discharges water into the reservoir at a rate of 0.5 m³/s if the diameter of the pipe is 50 cm?

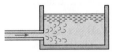

PROBLEMS 7.56, 7.57

7.58 An 8 cm pipe carries water with a mean velocity of 2 m/s. If this pipe abruptly expands to a 15 cm pipe, what will be the head loss due to the abrupt expansion?

7.59 A 6 in. pipe abruptly expands to a 12 in. size. If the discharge of water in the pipes is 5 cfs, what is the head loss due to abrupt expansion?

7.60 Water is draining from tank A to tank B. The elevation difference between the two tanks is 10 m. The pipe connecting the two tanks has a sudden-expansion section as shown. The cross-sectional area of the pipe from A is 8 cm^2, and the area of the pipe into B is 25 cm^2. Assume the head loss in the system consists only of that due to the sudden-expansion section and the loss due to flow into tank B. Find the discharge between the two tanks.

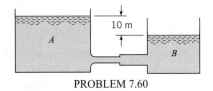

PROBLEM 7.60

7.61 A 40 cm pipe abruptly expands to a 60 cm size. These pipes are horizontal, and the discharge of water from the smaller size to the larger is 1.0 m^3/s. What horizontal force is required to hold the transition in place if the pressure in the 40 cm pipe is 70 kPa gage? Also, what is the head loss? Assume $\alpha = 1.0$ at all locations.

7.62 Water ($\gamma = 62.4$ lbf/ft^3) flows through a horizontal constant diameter pipe with a cross-sectional area of 9 in^2. The velocity in the pipe is 15 ft/s, and the water discharges to the atmosphere. The head loss between the pipe joint and the end of the pipe is 3 ft. Find the force on the joint to hold the pipe. The pipe is mounted on frictionless rollers. Assume $\alpha = 1.0$ at all locations.

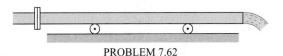

PROBLEM 7.62

7.63 This abrupt expansion is to be used to dissipate the high-energy flow of water in the 5 ft–diameter penstock. Assume $\alpha = 1.0$ at all locations.
a. What power (in horsepower) is lost through the expansion?
b. If the pressure at section 1 is 5 psig, what is the pressure at section 2?
c. What force is needed to hold the expansion in place?

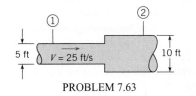

PROBLEM 7.63

7.64 This rough aluminum pipe is 6 in. in diameter. It weighs 1.5 lb per foot of length, and the length L is 50 ft. If the discharge of water is 6 cfs and the head loss due to friction from section 1 to the end of the pipe is 10 ft, what is the longitudinal force transmitted across section 1 through the pipe wall?

7.65 Water flows in this bend at a rate of 5 m^3/s, and the pressure at the inlet is 650 kPa. If the head loss in the bend is 10 m, what will the pressure be at the outlet of the bend?   Also estimate the

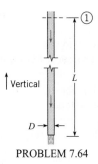

PROBLEM 7.64

force of the anchor block on the bend in the x direction required to hold the bend in place. Assume $\alpha = 1.0$ at all locations.

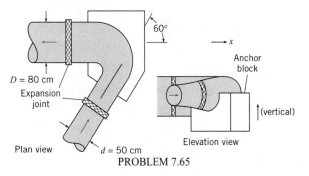

PROBLEM 7.65

7.66 Fluid flowing along a pipe of diameter D accelerates around a disk of diameter d as shown in the figure. The velocity far upstream of the disk is U, and the fluid density is ρ. Assuming incompressible flow and that the pressure downstream of the disk is the same as that at the plane of separation, develop an expression for the force required to hold the disk in place in terms of U, D, d, and ρ. Using the expression you developed, determine the force when $U = 10$ m/s, $D = 5$ cm, $d = 4$ cm, and $\rho = 1.2$ kg/m^3. Assume $\alpha = 1.0$ at all locations.

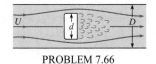

PROBLEM 7.66

EGL and HGL

7.67 $\mathbb{PQ}$◀ Using resources such as the Internet and Section 7.7, identify two real-world applications of HGLs and EGLs that are interesting to you. For each application, write a paragraph in which you describe how and why HGLs and EGLs are used.

7.68 $\mathbb{PQ}$◀ Using Section 7.7 and other resources, answer the following questions. Strive for depth, clarity, and accuracy while also combining sketches, words, and equations in ways that enhance the effectiveness of your communication.
a. What are three important reasons that engineers use the HGL and the EGL?
b. What factors influence the magnitude of the HGL? What factors influence the magnitude of the EGL?

c. How are the EGL and HGL related to the piezometer? To the stagnation tube?

d. How is the EGL related to the energy equation?

e. How can you use an HGL or an EGL to determine the direction of flow?

7.69 The energy grade line for steady flow in a uniform-diameter pipe is shown. Which of the following could be in the "black box"? (a) a pump, (b) a partially closed valve, (c) an abrupt expansion, or (d) a turbine. Choose all valid answer(s) and state your rationale.

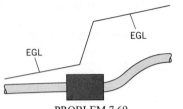

PROBLEM 7.69

7.70 If the pipe shown has constant diameter, is this type of HGL possible? If so, under what additional conditions? If not, why not?

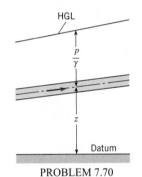

PROBLEM 7.70

7.71 For the system shown,

a. What is the flow direction?

b. What kind of machine is at A?

c. Do you think both pipes, AB and CA, are the same diameter?

d. Sketch in the EGL for the system.

e. Is there a vacuum at any point or region of the pipes? If so, identify the location.

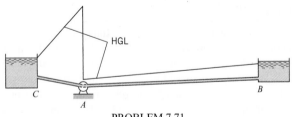

PROBLEM 7.71

7.72 The HGL and the EGL are as shown for a certain flow system.

a. Is flow from A to E or from E to A?

b. Does it appear that a reservoir exists in the system?

c. Does the pipe at E have a uniform or a variable diameter?

d. Is there a pump in the system?

e. Sketch the physical setup that could yield the conditions shown between C and D.

f. Is anything else revealed by the sketch?

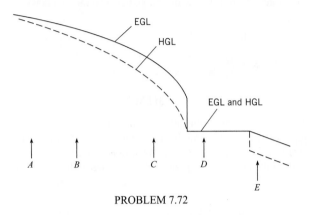

PROBLEM 7.72

7.73 Sketch the HGL and the EGL for this conduit, which tapers uniformly from the left end to the right end.

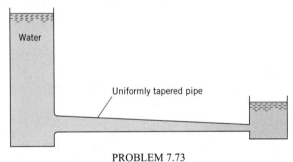

PROBLEM 7.73

7.74 The HGL and the EGL for a pipeline are shown in the figure.

a. Indicate which is the HGL and which is the EGL.

b. Are all pipes the same size? If not, which is the smallest?

c. Is there any region in the pipes where the pressure is below atmospheric pressure? If so, where?

d. Where is the point of maximum pressure in the system?

e. Where is the point of minimum pressure in the system?

f. What do you think is located at the end of the pipe at point E?

g. Is the pressure in the air in the tank above or below atmospheric pressure?

h. What do you think is located at point B?

7.75 Assume that the head loss in the pipe is given by $h_L = 0.014(L/D)(V^2/2g)$, where L is the length of pipe and D is the pipe diameter. Assume $\alpha = 1.0$ at all locations.

a. Determine the discharge of water through this system.

b. Draw the HGL and the EGL for the system.

c. Locate the point of maximum pressure.

d. Locate the point of minimum pressure.

e. Calculate the maximum and minimum pressures in the system.

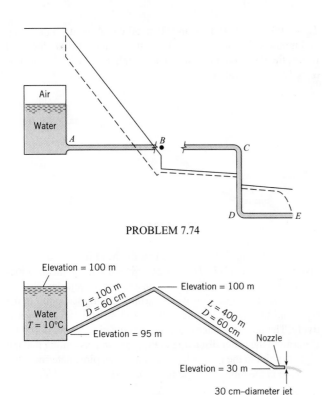

PROBLEM 7.74

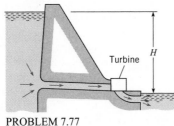

PROBLEM 7.75

7.76 Sketch the HGL and the EGL for the reservoir and pipe of Example 7.2.

7.77 The discharge of water through this turbine is 1000 cfs. What power is generated if the turbine efficiency is 85% and the total head loss is 4 ft? $H = 100$ ft. Also, carefully sketch the EGL and the HGL.

PROBLEM 7.77

7.78 Water flows from the reservoir through a pipe and then discharges from a nozzle as shown. The head loss in the pipe itself is given as $h_L = 0.025(L/D)(V^2/2g)$, where L and D are the length and diameter of the pipe and V is the velocity in the pipe. What is the discharge of water? Also draw the HGL and EGL for the system. Assume $\alpha = 1.0$ at all locations.

7.79 Refer to Fig. 7.11. Assume that the head loss in the pipes is given by $h_L = 0.02(L/D)(V^2/2g)$, where V is the mean velocity in the pipe, D is the pipe diameter, and L is the pipe length.

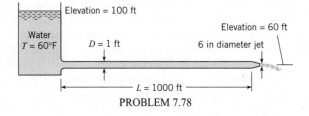

PROBLEM 7.78

The water surface elevations of the upper and lower reservoirs are 100 m and 70 m, respectively. The respective dimensions for upstream and downstream pipes are $D_u = 30$ cm, and $L_u = 200$ m, and $D_d = 15$ cm, and $L_d = 100$ m. Determine the discharge of water in the system.

7.80 What horsepower must be supplied to the water to pump 3.0 cfs at 68°F from the lower to the upper reservoir? Assume that the head loss in the pipes is given by $h_L = 0.018 (L/D)(V^2/2g)$, where L is the length of the pipe in feet and D is the pipe diameter in feet. Sketch the HGL and the EGL.

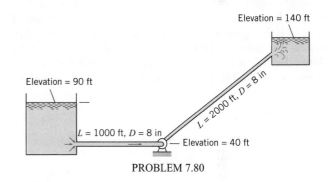

PROBLEM 7.80

7.81 Water flows from reservoir A to reservoir B. The water temperature in the system is 10°C, the pipe diameter D is 1 m, and the pipe length L is 300 m. If $H = 16$ m, $h = 2$ m, and the pipe head loss is given by $h_L = 0.01(L/D)(V^2/2g)$, where V is the velocity in the pipe, what will be the discharge in the pipe? In your solution, include the head loss at the pipe outlet, and sketch the HGL and the EGL. What will be the pressure at point P halfway between the two reservoirs? Assume $\alpha = 1.0$ at all locations.

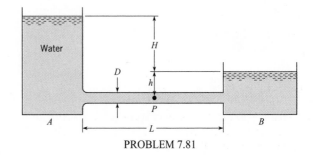

PROBLEM 7.81

7.82 Water flows from the reservoir on the left to the reservoir on the right at a rate of 16 cfs. The formula for the head losses in the pipes is $h_L = 0.02(L/D)(V^2/2g)$. What elevation in the left reservoir is required to produce this flow? Also carefully sketch the HGL and the EGL for the system. *Note:* Assume the head-loss formula can be used for the smaller pipe as well as for the larger pipe. Assume $\alpha = 1.0$ at all locations.

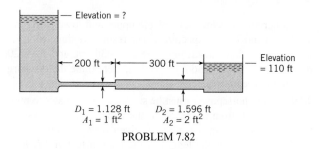

PROBLEM 7.82

7.83 What power is required to pump water at a rate of 3 m^3/s from the lower to the upper reservoir? Assume the pipe head loss is given by $h_L = 0.018(L/D)(V^2/2g)$, where L is the length of pipe, D is the pipe diameter, and V is the velocity in the pipe. The water temperature is 10°C, the water surface elevation in the lower reservoir is 150 m, and the surface elevation in the upper reservoir is 250 m. The pump elevation is 100 m, $L_1 = 100$ m,

$L_2 = 1000$ m, $D_1 = 1$ m, and $D_2 = 50$ cm. Assume the pump and motor efficiency is 74%. In your solution, include the head loss at the pipe outlet and sketch the HGL and the EGL. Assume $\alpha = 1.0$ at all locations.

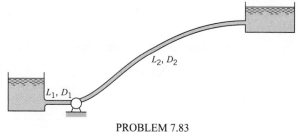

PROBLEM 7.83

7.84 Refer to Fig. 7.12. Assume that the head loss in the pipe is given by $h_L = 0.02(L/D)(V^2/2g)$, where V is the mean velocity in the pipe, D is the pipe diameter, and L is the pipe length. The elevations of the reservoir water surface, the highest point in the pipe, and the pipe outlet are 200 m, 200 m, and 185 m, respectively. The pipe diameter is 30 cm, and the pipe length is 200 m. Determine the water discharge in the pipe, and, assuming that the highest point in the pipe is halfway along the pipe, determine the pressure in the pipe at that point. Assume $\alpha = 1.0$ at all locations.

Dimensional Analysis and Similitude

SIGNIFICANT LEARNING OUTCOMES

Conceptual Knowledge
- State the Buckingham Π theorem.
- Identify and explain the significance of the common π-groups.
- Distinguish between model and prototype.
- Explain the concepts of dynamic and geometric similitude.

Procedural Knowledge
- Apply the Buckingham Π theorem to determine number of dimensionless variables.
- Apply the step-by-step procedure to determine the dimensionless π-groups.
- Apply the exponent method to determine the dimensionless π-groups.
- Distinguish the significant π-groups for a given a flow problem.

Typical Applications
- Drag force on a blimp from model testing.
- Ship model tests to evaluate wave and friction drag.
- Pressure drop in a prototype nozzle from model measurements.

Because of the complexity of fluid mechanics, the design of fluid systems relies heavily on experimental results. Tests are typically carried out on a subscale model, and the results are extrapolated to the full-scale system (prototype). The principles underlying the correspondence between the model and the prototype are addressed in this chapter.

The ideas of primary dimensions, dimensionless groups, and dimensional homogeneity were introduced in Chapter 1. In this chapter, the procedures for obtaining the significant dimensionless groups are outlined, and the requirements for model testing to have correspondence between the model and prototype are presented.

8.1

Need for Dimensional Analysis

Fluid mechanics is more heavily involved with experimental testing than other disciplines because the analytical tools currently available to solve the momentum and energy equations are not capable of providing accurate results. This is particularly evident in turbulent, separating flows. The solutions obtained by utilizing techniques from computational fluid dynamics with the largest computers available yield only fair approximations for turbulent flow problems—hence the need for experimental evaluation and verification.

For analyzing model studies and for correlating the results of experimental research, it is essential that researchers employ dimensionless groups. To appreciate the advantages of using dimensionless groups, consider the flow of water through the unusual orifice illustrated in Fig. 8.1. Actually, this is much like a nozzle used for flow metering except that the flow is in the opposite direction. An orifice operating in this flow condition will have a much different performance than one operating in the normal mode. However, it is not unlikely that a firm or city water department might have such a situation where the flow may occur the "right way" most of the time and the "wrong way" part of the time—hence the need for such knowledge.

Because of size and expense it is not always feasible to carry out tests on a full-scale prototype. Thus engineers will test a subscale model and measure the pressure drop across the model. The test procedure may involve testing several orifices, each with a different throat diameter d_0. For purposes of discussion, assume that three nozzles are to be tested. The Bernoulli equation, introduced in Chapter 4, suggests that the pressure drop will depend on flow velocity and fluid density. It may also depend on the fluid viscosity.

The test program may be carried out with a range of velocities and possibly with fluids of different density (and viscosity). The pressure drop, $p_1 - p_2$, is a function of the velocity V_1, density ρ, and diameter d_0. By carrying out numerous measurements at different values of V_1 and ρ for the three different nozzles, the data could be plotted as shown in Fig. 8.2a for tests using water. In addition, further tests could be planned with different fluids at considerably more expense.

The material introduced in this chapter leads to a much better approach. Through dimensional analysis it can be shown that the pressure drop can be expressed as

$$\frac{p_1 - p_2}{\left(\rho V^2\right)/2} = f\left(\frac{d_0}{d_1}, \frac{\rho V_1 d_0}{\mu}\right) \tag{8.1}$$

which means that dimensionless group for pressure, $(p_1 - p_2)/(\rho V^2/2)$, is a function of the dimensionless throat/pipe diameter ratio d_0/d_1 and the dimensionless group, $(\rho V_1 d_0)/\mu$, which will be identified later as the Reynolds number. The purpose of the experimental program is to establish the functional relationship. As will be shown later, if the Reynolds number is sufficiently large, the results are independent of Reynolds number. Then

$$\frac{p_1 - p_2}{(\rho V^2)/2} = f\left(\frac{d_0}{d_1}\right) \tag{8.2}$$

Thus for any specific orifice design (same d_0/d_1) the pressure drop, $p_1 - p_2$, divided by $\rho V_1^2/2$ for the model is same for the prototype. Therefore the data collected from the model

Figure 8.1

Flow through inverted flow nozzle.

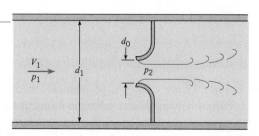

Figure 8.2

Relations for pressure, velocity, and diameter. (a) Using dimensional variables. (b) Using dimensionless groups.

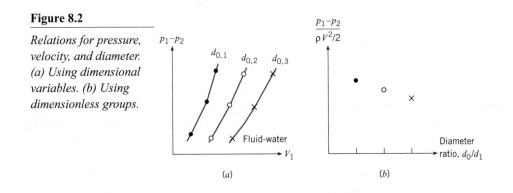

(a)   (b)

tests can be applied directly to the prototype. Only one test is needed for each orifice design. Consequently only three tests are needed, as shown in Fig. 8.2b. The fewer tests result in considerable savings in effort and expense.

The identification of dimensionless groups that provide correspondence between model and prototype data is carried out through *dimensional analysis.*

Buckingham Π Theorem

In 1915 Buckingham (1) showed that the number of independent dimensionless groups of variables (dimensionless parameters) needed to correlate the variables in a given process is equal to $n - m$, where n is the number of variables involved and m is the number of basic dimensions included in the variables.

Buckingham referred to the dimensionless groups as Π, which is the reason the theorem is called the Π theorem. Henceforth dimensionless groups will be referred to as π-*groups*. If the equation describing a physical system has n dimensional variables and is expressed as

$$y_1 = f(y_2, y_3, \ldots y_n)$$

then it can be rearranged and expressed in terms of $(n - m)$ π-groups as

$$\pi_1 = \varphi(\pi_2, \pi_3, \ldots \pi_{n-m})$$

Thus if the drag force F of a fluid flowing past a sphere is known to be a function of the velocity V, mass density ρ, viscosity μ, and diameter D, then five variables (F, V, ρ, μ, and D) and three basic dimensions (L, M, and T) are involved.* By the Buckingham Π theorem there will be $5 - 3 = 2$ π-groups that can be used to correlate experimental results in the form

$$\pi_1 = \varphi(\pi_2)$$

* Note that only three basic dimensions will be considered here. Temperature will not be included.

8.3

Dimensional Analysis

Dimensional analysis is the process used to obtain the π-groups for experimental design. There are two methods: the step-by-step method and the exponent method. Both are addressed in this section.

The Step-by-Step Method

Several methods may be used to carry out the process of finding the π-groups, but the step-by-step approach, very clearly presented by Ipsen (2), is one of the easiest and reveals much about the process. The procedure for the step-by-step method follows in Table 8.1.

The final result can be expressed as a functional relationship of the form

$$\pi_1 = f(\pi_2, \pi_2, ... \pi_n) \tag{8.3}$$

The selection of the dependent and independent π-groups depends on the application. Also the selection of variables used to eliminate dimensions is arbitrary.

	Table 8.1 THE STEP-BY-STEP APPROACH
Step	**Action Taken During This Step**
1	Identify the significant dimensional variables and write out the primary dimensions of each.
2	Apply the Buckingham Π theorem to find the number of π-groups.*
3	Set up table with the number of rows equal to the number of dimensional variables and the number of columns equal to the number of basic dimensions plus one ($m + 1$).
4	List all the dimensional variables in the first column with primary dimensions.
5	Select a dimension to be eliminated, choose a variable with that dimension in the first column, and combine with remaining variables to eliminate the dimension. List combined variables in the second column with remaining primary dimensions.
6	Select another dimension to be eliminated, choose from variables in the second column that has that dimension, and combine with the remaining variables. List the new combinations with remaining primary dimensions in the third column
7	Repeat Step 6 until all dimensions are eliminated. The remaining dimensionless groups are the π-groups. List the π-groups in the last column

* Note that, in rare instances, the number of π-groups may be one more than predicted by the Buckingham Π theorem. This anomaly can occur because it is possible that two-dimensional categories can be eliminated when dividing (or multiplying) by a given variable. See Ipsen (2) for an example of this.

Example 8.1 shows how to use the step-by-step method to find the π-groups for a body falling in a vacuum.

EXAMPLE 8.1 Π-GROUPS FOR BODY FALLING IN A VACUUM

There are three significant dimensional variables for a body falling in a vacuum (no viscous effects): the velocity V; the acceleration due to gravity, g; and the distance through which the body falls, h. Find the π-groups using the step-by-step method.

Problem Definition

Situation: Body falling in vacuum, $V = f(g, h)$.

Find: π-groups.

Plan

Follow procedure for step-by-step method in Table 8.1.

Solution

1. Significant variables and dimensions

$$[V] = L/T$$

$$[g] = L/T^2$$

$$[h] = L$$

There are only two dimensions, L and T.

2. From the Buckingham Π theorem, there is only one (three variables–two dimensions) π-group.
3. Set up table with three rows (number of variables) and three (dimensions + 1) columns.
4. List variables and primary dimensions in first column.

Variable	[ ]	Variable	[ ]	Variable	[ ]
V	$\dfrac{L}{T}$	$\dfrac{V}{h}$	$\dfrac{1}{T}$	$\dfrac{V}{\sqrt{gh}}$	0
g	$\dfrac{L}{T^2}$	$\dfrac{g}{h}$	$\dfrac{1}{T^2}$		
h	L				

5. Select h to eliminate L. Divide g by h, enter in second column with dimension $1/T^2$. Divide V by h, enter in second column with dimension $1/T$.
6. Select g/h to eliminate T. Divide V/h by $\sqrt{g/h}$ and enter in third column.
   As expected, there is only one π-group,

$$\pi = \frac{V}{\sqrt{gh}}$$

The final functional form of equation of the equation is

$$\frac{V}{\sqrt{gh}} = C$$

Review

1. From basic physics $C = \sqrt{2}$.
2. The proper relationship between V, h, and g is found with dimensionless analysis. If the value of C were not known from basic physics, it could be determined from experiment.

Example 8.2 illustrates the application of the step-by-step method for finding π-groups for a problem with five variables and three primary dimensions.

EXAMPLE 8.2 π-GROUPS FOR DRAG ON A SPHERE USING STEP-BY-STEP METHOD

The drag F_D of a sphere in a fluid flowing past the sphere is a function of the viscosity μ, the mass density ρ, the velocity of flow V, and the diameter of the sphere D. Use the step-by-step method to find the π-groups.

Problem Definition

Situation: Given $F_D = f(V, \rho, \mu, D)$.

Find: The π-groups using the step-by-step method.

Plan

Use the step-by-step procedure from Table 8.1.

Solution

1. Dimensions of significant variables

$$F = \frac{ML}{T^2}, \; V = \frac{L}{T}, \; \rho = \frac{M}{L^3}, \; \mu = \frac{M}{LT}, \; D = L$$

2. Number of π-groups, $5 - 3 = 2$.
3. Set up table with five rows and four columns.
4. Write variables and dimensions in first column.

Variable	[ ]	Variable	[ ]	Variable	[ ]	Variable	[ ]
F_D	$\dfrac{ML}{T^2}$	$\dfrac{F_D}{D}$	$\dfrac{M}{T^2}$	$\dfrac{F_D}{\rho D^4}$	$\dfrac{1}{T^2}$	$\dfrac{F_D}{\rho V^2 D^2}$	0
V	$\dfrac{L}{T}$	$\dfrac{V}{D}$	$\dfrac{1}{T}$	$\dfrac{V}{D}$	$\dfrac{1}{T}$		
ρ	$\dfrac{M}{L^3}$	ρD^3	M				
μ	$\dfrac{M}{LT}$	μD	$\dfrac{M}{T}$	$\dfrac{\mu}{\rho D^2}$	$\dfrac{1}{T}$	$\dfrac{\mu}{\rho V D}$	0
D	L						

5. Eliminate L using D and write new variable combinations with corresponding dimensions in the second column.
6. Eliminate M using ρD^3 and write new variable combinations with dimensions in the third column.
7. Eliminate T using V/D and write new combinations in the fourth column.

The final two π-groups are

$$\pi_1 = \frac{F_D}{\rho V^2 D^2} \quad \text{and} \quad \pi_2 = \frac{\mu}{\rho V D}$$

The functional equation can be written as

$$\frac{F_D}{\rho V^2 D^2} = f\left(\frac{\mu}{\rho V D}\right)$$

Review

The functional relationship between the π-groups is obtained from experiment.

The form of the π-groups obtained will depend on the variables selected to eliminate dimensions. For example, if in Example 8.2, $\mu/\rho D^2$ had been used to eliminate the time dimension, the two π-groups would have been

$$\pi_1 = \frac{\rho F_D}{\mu^2} \quad \text{and} \quad \pi_2 = \frac{\mu}{\rho V D}$$

The result is still valid but may not be convenient to use. The form of any π-group can be altered by multiplying or dividing by another π-group. Multiplying the π_1 by the square of π_2 yields the original π_1 in Example 8.2.

$$\frac{\rho F_D}{\mu^2} \times \left(\frac{\mu}{\rho V D}\right)^2 = \frac{F_D}{\rho V^2 D^2}$$

By so doing the two π-groups would be the same as in Example 8.2.

The Exponent Method

An alternative method for finding the π-groups is the exponent method. This method involves solving a set of algebraic equations to satisfy dimensional homogeneity. The procedural steps for the exponent method follow.

Table 8.2 THE EXPONENT METHOD	
Step	**Action Taken During This Step**
1	Identify the significant dimensional variables, y_i, and write out the primary dimensions of each, $[y_i]$.
2	Apply the Buckingham Π theorem to find the number of π-groups.
3	Write out the product of the primary dimensions in the form $$[y_1] = [y_2]^a \times [y_3]^b \times \ldots \times [y_n]^k$$ where n is the number of dimensional variables and a, b, etc. are exponents.
4	Find the algebraic equations for the exponents that satisfy dimensional homogeneity (same power for dimensions on each side of equation).
5	Solve the equations for the exponents.
6	Express the dimensional equation in the form $y_1 = y_2^a y_3^b \ldots y_n^k$ and identify the π-groups.

Example 8.3 illustrates how to apply the exponent method to find the π-groups of the same problem addressed in Example 8.2.

EXAMPLE 8.3 π-GROUPS FOR DRAG ON A SPHERE USING EXPONENT METHOD

The drag of a sphere, F_D, in a flowing fluid is a function of the velocity V, the fluid density ρ, the fluid viscosity μ, and the sphere diameter D. Find the π-groups using the exponent method.

Problem Definition

Situation: Given $F_D = f(V, \rho, \mu, D)$.

Find: The π-groups using exponent method.

Plan

Follow the procedure for the exponent method in Table 8.2.

Solution

1. Dimensions of significant variables are

$$[F] = \frac{ML}{T^2}, \ [V] = \frac{L}{T}, \ [\rho] = \frac{M}{L^3}, \ [\mu] = \frac{M}{LT}, \ [D] = L$$

2. Number of π-groups is $5 - 3 = 2$.
3. Form product with dimensions.

$$\frac{ML}{T^2} = \left[\frac{L}{T}\right]^a \times \left[\frac{M}{L^3}\right]^b \times \left[\frac{M}{LT}\right]^c \times [L]^d$$

$$= \frac{L^{a-3b-c+d}M^{b+c}}{T^{a+c}}$$

4. Dimensional homogeneity. Equate powers of dimensions on each side.

$$L: \quad a - 3b - c + d = 1$$
$$M: \quad b + c = 1$$
$$T: \quad a + c = 2$$

5. Solve for exponents a, b, and c in terms of d.

$$\begin{pmatrix} 1 & -3 & -1 \\ 0 & 1 & 1 \\ 1 & 0 & 1 \end{pmatrix} \begin{pmatrix} a \\ b \\ c \end{pmatrix} = \begin{pmatrix} 1-d \\ 1 \\ 2 \end{pmatrix}$$

The value of the determinant is −1 so a unique solution is achievable. Solution is $a = d$, $b = d - 1$, $c = 2 - d$

6. Write dimensional equation with exponents.

$$F = V^d \rho^{d-1} \mu^{2-d} D^d$$

$$F = \frac{\mu^2}{\rho} \left(\frac{\rho VD}{\mu}\right)^d$$

$$\frac{F\rho}{\mu^2} = \left(\frac{\rho VD}{\mu}\right)^d$$

There are two π-groups:

$$\pi_1 = \frac{F\rho}{\mu^2} \text{ and } \pi_2 = \frac{\rho VD}{\mu}$$

By dividing π_1 by the square of π_2, the π_1 group can be written as $F_D/(\rho V^2 D^2)$, so the functional form of the equation can be written as

$$\frac{F}{\rho V^2 D^2} = f\left(\frac{\rho VD}{\mu}\right)$$

Review

1. The group of variables raised to the power forms a π-group.
2. The functional relationship between the two π-groups is obtained from experiment.

Selection of Significant Variables

All the foregoing procedures deal with straightforward situations. However, some problems do occur. In order to apply dimensional analysis one must first decide which variables are significant. If the problem is not sufficiently well understood to make a good choice of the significant variables, dimensional analysis seldom provides clarification.

A serious shortcoming might be the omission of a significant variable. If this is done, one of the significant π-groups will likewise be missing. In this regard, it is often best to identify a list of variables that one regards as significant to a problem and to determine if only one dimensional category (such as M or L or T) occurs. When this happens, it is likely that there is an error in choice of significant variables because it is not possible to combine two variables to eliminate the lone dimension. Either the variable with the lone dimension should not have been included in the first place (it is not significant), or another variable should have been included.

How does one know if a variable is significant for a given problem? Probably the truest answer is by experience. After working in the field of fluid mechanics for several years, one develops a feel for the significance of variables to certain kinds of applications. However, even the inexperienced engineer will appreciate the fact that free-surface effects have no significance in closed-conduit flow; consequently, surface tension, σ, would not be included as a variable. In closed-conduit flow, if the velocity is less than approximately one-third the speed of sound, compressibility effects are usually negligible. Such guidelines, which have been observed by previous experimenters, help the novice engineer develop confidence in her or his application of dimensional analysis and similitude.

8.4

Common π-Groups

The most common π-groups can be found by applying dimensional analysis to all the variables that might be significant in a general flow situation, The purpose of this section is to develop these common π-groups and discuss their significance.

Variables that have significance in a general flow field are the velocity V, the density ρ, the viscosity μ, and the acceleration due to gravity g. In addition, if fluid compressibility were likely, then the bulk modulus of elasticity, E_v, should be included. If there is a liquid-gas interface, the surface tension effects may also be significant. Finally the flow field will be affected by a general length, L, such as the width of a building or the diameter of a pipe. These variables will be regarded as the independent variables. The primary dimensions of the significant independent variables are

$$[V] = L/T \qquad [\rho] = M/L^3 \qquad [\mu] = M/LT$$

$$[g] = L/T^2 \qquad [E_v] = M/LT^2 \qquad [\sigma] = M/T^2 \qquad [L] = L$$

There are several other independent variables that could be identified for thermal effects, such as temperature, specific heat, and thermal conductivity. Inclusion of these variables is beyond the scope of this text.

Products that result from a flowing fluid are pressure distributions (p), shear stress distributions (τ), and forces on surfaces and objects (F) in the flow field. These will be identified as the dependent variables. The primary dimensions of the dependent variables are

$$[p] = M/LT^2 \qquad [\tau] = [\Delta p] = M/LT^2 \qquad [F] = (ML)/T^2$$

There are other dependent variables not included here, but they will be encountered and introduced for specific applications.

Altogether there are 10 significant variables, which, by application of the Buckingham Π theorem, means there are seven π-groups. Utilizing either the step-by-step method or the exponent method yields

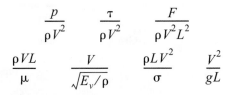

$$\frac{p}{\rho V^2} \qquad \frac{\tau}{\rho V^2} \qquad \frac{F}{\rho V^2 L^2}$$

$$\frac{\rho V L}{\mu} \qquad \frac{V}{\sqrt{E_v/\rho}} \qquad \frac{\rho L V^2}{\sigma} \qquad \frac{V^2}{gL}$$

The first three groups, the dependent π-groups, are identified by specific names. For these groups it is common practice to use the kinetic pressure, $\rho V^2/2$, instead of ρV^2. In most applications one is concerned with a pressure difference, so the pressure π-group is expressed as

$$C_p = \frac{p - p_0}{\frac{1}{2}\rho V^2}$$

where C_p is called the pressure coefficient and p_0 is a reference pressure. The pressure coefficient was introduced earlier in Chapter 4 and Section 8.1. The π-group associated with shear stress is called the shear-stress coefficient and defined as

$$c_f = \frac{\tau}{\frac{1}{2}\rho V^2}$$

where the subscript f denotes "friction." The π-group associated with force is referred to, here, as a force coefficient and defined as

$$C_F = \frac{F}{\frac{1}{2}\rho V^2 L^2}$$

This coefficient will be used extensively in Chapter 11 for lift and drag forces on airfoils and hydrofoils.

The independent π-groups are named after earlier contributors to fluid mechanics. The π-group $VL\rho/\mu$ is called the Reynolds number, after Osborne Reynolds, and designated by Re. The group $V/(\sqrt{E_v/\rho})$ is rewritten as (V/c), since $\sqrt{E_v/\rho}$ is the speed of sound, c. This π-group is called the Mach number and designated by M. The π-group $\rho L V^2/\sigma$ is called the Weber number and designated by We. The remaining π-group is usually expressed as $V/\sqrt{gL}$ and identified as the Froude (rhymes with "food") number* and written as Fr.

The general functional form for all the π-groups is

$$C_p, c_f, C_F = f(\text{Re}, \text{M}, \text{We}, \text{Fr}) \qquad (8.4)$$

which means that either of the three dependent π-groups are functions of the four independent π-groups; that is, the pressure coefficient, the shear-stress coefficient, or the force coefficient are functions of the Reynolds number, Mach number, Weber number, and Froude number.

The π-groups, their symbols, and their names are summarized in Table 8.3. Each independent π-group has an important interpretation as indicated by the ratio column. The Reynolds number can be viewed as the ratio of kinetic to viscous forces. The kinetic forces are the forces associated with fluid motion. The Bernoulli equation indicates that the pressure

---

* Sometimes the Froude number is written as $V/\sqrt{(\Delta \gamma gL)/\gamma}$ and called the densimetric Froude number. It has application in studying the motion of fluids in which there is density stratification, such as between salt water and fresh water in an estuary or heated-water effluents associated with thermal power plants.

π-Group	Symbol	Name	Ratio
$\dfrac{p - p_0}{(\rho V^2)/2}$	C_p	Pressure coefficient	$\dfrac{\text{Pressure difference}}{\text{Kinetic pressure}}$
$\dfrac{\tau}{(\rho V^2)/2}$	c_f	Shear-stress coefficient	$\dfrac{\text{Shear stress}}{\text{Kinetic pressure}}$
$\dfrac{F}{(\rho V^2 L^2)/2}$	C_F	Force coefficient	$\dfrac{\text{Force}}{\text{Kinetic force}}$
$\dfrac{\rho L V}{\mu}$	Re	Reynolds number	$\dfrac{\text{Kinetic force}}{\text{Viscous force}}$
$\dfrac{V}{c}$	M	Mach number	$\dfrac{\text{Kinetic force}}{\text{Compressive force}}$
$\dfrac{\rho L V^2}{\sigma}$	We	Weber number	$\dfrac{\text{Kinetic force}}{\text{Surface-tension force}}$
$\dfrac{V}{\sqrt{gL}}$	Fr	Froude number	$\dfrac{\text{Kinetic force}}{\text{Gravitational force}}$

Table 8.3 COMMON Π-GROUPS

difference required to bring a moving fluid to rest is the kinetic pressure, $\rho V^2/2$, so the kinetic forces,* F_k, should be proportional to

$$F_k \propto \rho V^2 L^2$$

The shear force due to viscous effects, F_v, is proportional to the shear stress and area

$$F_v \propto \tau A \propto \tau L^2$$

and the shear stress is proportional to

$$\tau \propto \mu \frac{dV}{dy} \propto \frac{\mu V}{L}$$

so $F_v \propto \mu V L$. Taking the ratio of the kinetic to the viscous forces

$$\frac{F_k}{F_v} \propto \frac{\rho V L}{\mu} = \text{Re}$$

yields the Reynolds number. The magnitude of the Reynolds number provides important information about the flow. A low Reynolds number implies viscous effects are important; a high Reynolds number implies kinetic forces predominate. The Reynolds number is one of the most widely used π-groups in fluid mechanics. It is also often written using kinematic viscosity, $\text{Re} = \rho V L/\mu = V L/\nu$.

The ratios of the other independent π-groups have similar significance. The Mach number is an indicator of how important compressibility effects are in a fluid flow. If the Mach number is small, then the kinetic force associated with the fluid motion does not cause

* Traditionally the kinetic force has been identified as the "inertial" force.

a significant density change, and the flow can be treated as incompressible (constant density). On the other hand, if the Mach number is large, there are often appreciable density changes that must be considered in model studies.

The Weber number is an important parameter in liquid atomization. The surface tension of the liquid at the surface of a droplet is responsible for maintaining the droplet's shape. If a droplet is subjected to an air jet and there is a relative velocity between the droplet and the gas, kinetic forces due to this relative velocity cause the droplet to deform. If the Weber number is too large, the kinetic force overcomes the surface-tension force to the point that the droplet shatters into even smaller droplets. Thus a Weber-number criterion can be useful in predicting the droplet size to be expected in liquid atomization. The size of the droplets resulting from liquid atomization is a very significant parameter in gas-turbine and rocket combustion.

The Froude number is unimportant when gravity causes only a hydrostatic pressure distribution, such as in a closed conduit. However, if the gravitational force influences the pattern of flow, such as in flow over a spillway or in the formation of waves created by a ship as it cruises over the sea, the Froude number is a most significant parameter.

8.5

Similitude

Scope of Similitude

Similitude is the theory and art of predicting prototype performance from model observations. Whenever it is necessary to perform tests on a model to obtain information that cannot be obtained by analytical means alone, the rules of similitude must be applied. The theory of similitude involves the application of π-groups, such as the Reynolds number or the Froude number, to predict prototype performance from model tests. The art of similitude enters the problem when the engineer must make decisions about model design, model construction, performance of tests, or analysis of results that are not included in the basic theory.

Present engineering practice makes use of model tests more frequently than most people realize. For example, whenever a new airplane is being designed, tests are made not only on the general scale model of the prototype airplane but also on various components of the plane. Numerous tests are made on individual wing sections as well as on the engine pods and tail sections.

Models of automobiles and high-speed trains are also tested in wind tunnels to predict the drag and flow patterns for the prototype. Information derived from these model studies often indicates potential problems that can be corrected before the prototype is built, thereby saving considerable time and expense in development of the prototype.

In civil engineering, model tests are always used to predict flow conditions for the spillways of large dams. In addition, river models assist the engineer in the design of flood-control structures as well as in the analysis of sediment movement in the river. Marine engineers make extensive tests on model ship hulls to predict the drag of the ships. Much of this type of testing is done at the David Taylor Model Basin, Naval Surface Warfare Center, Carderock Division, near Washington, D.C. (see Fig. 8.3). Tests are also regularly performed on models of tall buildings to help predict the wind loads on the buildings, the stability characteristics of the buildings, and the airflow patterns in their vicinity. The latter information is used by the architects to design walkways and passageways that are safer and more comfortable for pedestrians to use.

Figure 8.3

Ship-model test at the David Taylor Model Basin, Naval Surface Warfare Center, Carderock Division.

Geometric Similitude

Geometric similitude means that the model is an exact geometric replica of the prototype.* Consequently, if a 1:10 scale model is specified, all linear dimensions of the model must be 1/10 of those of the prototype. In Fig. 8.4 if the model and prototype are geometrically similar, the following equalities hold:

$$\frac{\ell_m}{\ell_p} = \frac{w_m}{w_p} = \frac{c_m}{c_p} = L_r \tag{8.5}$$

Here ℓ, w, and c are specific linear dimensions associated with the model and prototype, and L_r is the scale ratio between model and prototype. It follows that the ratio of corresponding

Figure 8.4

(a) Prototype. (b) Model.

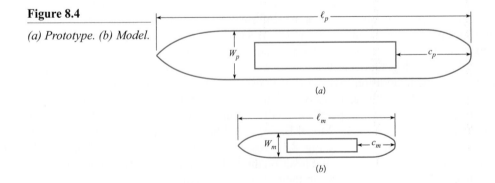

* For most model studies this is a basic requirement. However, for certain types of problems, such as river models, distortion of the vertical scale is often necessary to obtain meaningful results.

areas between model and prototype will be the square of the length ratio: $A_r = L_r^2$. The ratio of corresponding volumes will be given by $V_m/V_p = L_r^3$.

Dynamic Similitude

Dynamic similitude means that the forces that act on corresponding masses in the model and prototype are in the same ratio $(F_m/F_p = \text{constant})$ throughout the entire flow field. For example, the ratio of the kinetic to viscous forces must be the same for the model and the prototype. Since the forces acting on the fluid elements control the motion of those elements, it follows that dynamic similarity will yield similarity of flow patterns. Consequently, the flow patterns for the model and the prototype will be the same if geometric similitude is satisfied and if the relative forces acting on the fluid are the same in the model as in the prototype. This latter condition requires that the appropriate π-groups introduced in Section 8.4 be the same for the model and prototype, because these π-groups are indicators of relative forces within the fluid.

A more physical interpretation of the force ratios can be illustrated by considering the flow over the spillway shown in Fig. 8.5a. Here corresponding masses of fluid in the model and prototype are acted on by corresponding forces. These forces are the force of gravity F_g, the pressure force F_p, and the viscous resistance force F_v. These forces add vectorially as shown in Fig. 8.5 to yield a resultant force F_R, which will in turn produce an acceleration of the volume of fluid in accordance with Newton's second law of motion. Hence, because the force polygons in the prototype and model are similar, the magnitudes of the forces in the prototype and model will be in the same ratio as the magnitude of the vectors representing mass times acceleration:

$$\frac{m_m a_m}{m_p a_p} = \frac{F_{gm}}{F_{gp}}$$

or

$$\frac{\rho_m L_m^3 (V_m/t_m)}{\rho_p L_p^3 (V_p/t_p)} = \frac{\gamma_m L_m^3}{\gamma_p L_p^3}$$

which reduces to

$$\frac{V_m}{g_m t_m} = \frac{V_p}{g_p t_p}$$

But

$$\frac{t_m}{t_p} = \frac{L_m/V_m}{L_p/V_p}$$

so

$$\frac{V_m^2}{g_m L_m} = \frac{V_p^2}{g_p L_p} \tag{8.6}$$

Taking the square root of each side of Eq. (8.6) gives

$$\frac{V_m}{\sqrt{g_m L_m}} = \frac{V_p}{\sqrt{g_p L_p}} \quad \text{or} \quad \text{Fr}_m = \text{Fr}_p \tag{8.7}$$

Thus the Froude number for the model must be equal to the Froude number for the prototype to have the same ratio of forces on the model and the prototype.

Figure 8.5

Model-prototype relations: prototype view (a) and model view (b).

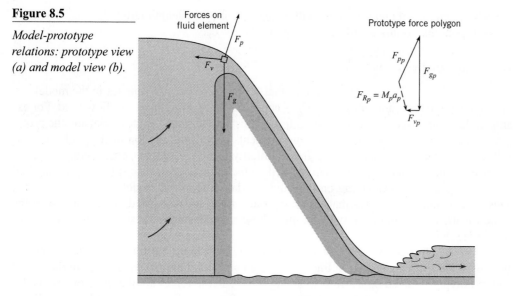

(a)

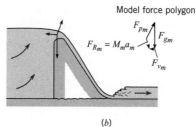

(b)

Equating the ratio of the forces producing acceleration to the ratio of viscous forces,

$$\frac{m_m a_m}{m_p a_p} = \frac{F_{vm}}{F_{vp}} \tag{8.8}$$

where $F_v \propto \mu V L$ leads to

$$\text{Re}_m = \text{Re}_p$$

The same analysis can be carried out for the Mach number and the Weber number. To summarize, if the independent π-groups for the model and prototype are equal, then the condition for dynamic similitude is satisfied.

Referring back to Eq. (8.4) for the general functional relationship,

$$C_p, c_f, C_F = f(\text{Re}, \text{M}, \text{We}, \text{Fr})$$

if the independent π-groups are the same for the model and the prototype, then dependent π-groups must also be equal so

$$C_{p,m} = C_{p,p} \qquad c_{f,m} = c_{f,p} \qquad C_{F,m} = C_{F,p} \tag{8.9}$$

To have complete similitude between the model and the prototype, it is necessary to have both geometric and dynamic similitude.

In many situations it may not be possible nor necessary to have all the independent π-groups the same for the model and the prototype to carry out useful model studies. For the flow of a liquid in a horizontal pipe, for example, in which the fluid completely fills the pipe (no free surface), there would be no surface tension effects, so the Weber number would be inappropriate. Compressibility effects would not be important, so the Mach number would not be needed. In addition, gravity would not be responsible for the flow, so the Froude number would not have to be considered. The only significant π-group would be the Reynolds number; thus dynamic similitude would be achieved by matching the Reynolds number between the model and the prototype.

On the other hand if a model test were to be done for the flow over a spillway, the Froude number would be a significant π-group because gravity is responsible for the motion of the fluid. Also, the action of viscous stresses due to the spillway surface could possibly affect the flow pattern, so the Reynolds number may be a significant π-group. In this situation, dynamic similitude may require that both the Froude number and the Reynolds number be the same for the model and prototype.

The choice of significant π-groups for dynamic similitude and their actual use in predicting prototype performance are considered in the next two sections.

8.6

Model Studies for Flows Without Free-Surface Effects

Free-surface effects are absent in the flow of liquids or gases in closed conduits, including control devices such as valves, or in the flow about bodies (e.g., aircraft) that travel through air or are deeply submerged in a liquid such as water (submarines). Free-surface effects are also absent where a structure such as a building is stationary and wind flows past it. In all these cases, given relatively low Mach numbers, the Reynolds-number criterion is the most significant for dynamic similarity. That is, the Reynolds number for the model must equal the Reynolds number for the prototype.

Example 8.4 illustrates the application of Reynolds-number similitude for the flow over a blimp.

EXAMPLE 8.4 REYNOLDS-NUMBER SIMILITUDE

The drag characteristics of a blimp 5 m in diameter and 60 m long are to be studied in a wind tunnel. If the speed of the blimp through still air is 10 m/s, and if a 1/10 scale model is to be tested, what airspeed in the wind tunnel is needed for dynamically similar conditions? Assume the same air pressure and temperature for both model and prototype.

Problem Definition

Situation: Wind tunnel test of a 1/10 scale model blimp. Prototype speed is 10 m/s.

Find: Speed (in m/s) in wind tunnel for dynamic similitude.

Assumptions: Same air pressure and temperature for model and prototype, therefore $\nu_m = \nu_p$.

Plan

The only π-group that is appropriate is the Reynolds number (there are no compressibility effects, free-surface effects, or gravitation effects). Thus equating the model and prototype Reynolds number satisfies dynamic similitude.

1. Equate the Reynolds number of the model and the prototype.
2. Calculate model speed.

Solution

1. Reynolds-number similitude

$$\mathrm{Re}_m = \mathrm{Re}_p$$

$$\frac{V_m L_m}{\nu_m} = \frac{V_p L_p}{\nu_p}$$

2. Model velocity

$$V_m = V_p \frac{L_p}{L_m} \frac{\nu_m}{\nu_p} = 10 \text{ m/s} \times 10 \times 1 = \boxed{100 \text{ m/s}}$$

Example 8.4 shows that the airspeed in the wind tunnel must be 100 m/s for true Reynolds-number similitude. This speed is quite large, and in fact Mach-number effects may start to become important at such a speed. However, it will be shown in Section 8.9 that it is not always necessary to operate models at true Reynolds-number similitude to obtain useful results.

If the engineer feels that it is essential to maintain Reynolds-number similitude, then only a few alternatives are available. One way to produce high Reynolds numbers at nominal airspeeds is to increase the density of the air. A NASA wind tunnel at the Ames Research Center at Moffett Field in California is one such facility. It has a 12 ft–diameter test section; it can be pressurized up to 90 psia (620 kPa); it can be operated to yield a Reynolds number per foot up to 1.2×10^7, and the maximum Mach number at which a model can be tested in this wind tunnel is 0.6. The airflow in this wind tunnel is produced by a single-stage, 20-blade axial-flow fan, which is powered by a 15,000-horsepower, variable-speed, synchronous electric motor (3). There are several problems that are peculiar to a pressurized tunnel. First, a shell (essentially a pressurized bottle) must surround the entire tunnel and its components, adding to the cost of the tunnel. Second, it takes a long time to pressurize the tunnel in preparation for operation, increasing the time from the start to the finish of runs. In this regard it should be noted that the original pressurized wind tunnel at the Ames Research Center was built in 1946; however, because of extensive use, the tunnel's pressure shell began to deteriorate, so a new facility (the one previously described) was built and put in operation in 1995. Improvements over the old facility include a better data collection system, very low turbulence, and capability of depressurizing only the test section instead of the entire 620,000 ft^3 wind tunnel circuit when installing and removing models. The original pressurized wind tunnel was used to test most models of U.S. commercial aircraft over the past half-century, including the Boeing 737, 757, and 767; Lockheed L-1011; and McDonnell Douglas DC-9 and DC-10.

The Boeing 777 was tested in the low-speed, pressurized 5 m–by–5 m tunnel in Farnborough, England. This tunnel, operated by the Defence Evaluation and Research Agency (DERA) of Great Britain, can operate at three atmospheres with Mach numbers up to 0.2. Approximately 15,000 hours of total testing time was required for the Boeing 777 (4).

Another method of obtaining high Reynolds numbers is to build a tunnel in which the test medium (gas) is at a very low temperature, thus producing a relatively high-density–low-viscosity fluid. NASA has built such a tunnel and operates it at the Langley Research Center. This tunnel, called the National Transonic Facility, can be pressurized up to 9 atmospheres. The test medium is nitrogen, which is cooled by injecting liquid nitrogen into the system. In this wind tunnel it is possible to reach Reynolds numbers of 10^8 based on a model size of 0.25 m (5). Because of its sophisticated design, its initial cost of approximately $100,000,000 (6), and its operating expenses are high.

Another modern approach in wind-tunnel technology is the development of magnetic or electrostatic suspension of models. The use of the magnetic suspension with model airplanes has been studied (6), and the electrostatic suspension for the study of single-particle aerodynamics has been reported (7).

The use of wind tunnels for aircraft design has grown significantly as the size and sophistication of aircraft have increased. For example, in the 1930s the DC-3 and B-17 each had about 100 hours of wind-tunnel tests at a rate of $100 per hour of run time. By contrast the F-15 fighter required about 20,000 hours of tests at a cost of $20,000 per hour (6). The latter test time is even more staggering when one realizes that a much greater volume of data per hour at higher accuracy is obtained from the modern wind tunnels because of the high-speed data acquisition made possible by computers.

Example 8.5 illustrates the use of Reynolds-number similitude to design a test for a valve.

EXAMPLE 8.5   REYNOLDS-NUMBER SIMILITUDE OF A VALVE

The valve shown is the type used in the control of water in large conduits. Model tests are to be done, using water as the fluid, to determine how the valve will operate under wide-open conditions. The prototype size is 6 ft in diameter at the inlet. What flow rate is required for the model if the prototype flow is 700 cfs? Assume that the temperature for model and prototype is 60°F and that the model inlet diameter is 1 ft.

Sketch:

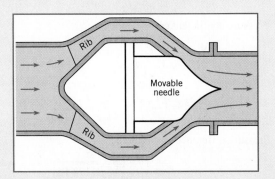

Situation: A 1/6 scale model of valve tested in water tunnel. Prototype flow rate is 700 cfs.

Find: Flow rate through model.

Assumptions:

1. No compressibility, free surface or gravitational effects.
2. Temperature of water in model and prototype is the same. Therefore kinematic viscosity of model and prototype are equal.

Plan

Dynamic similitude is obtained by equating the model and prototype Reynolds number. The model/prototype area ratio is the square of the scale ratio.

1. Equate Reynolds number of model and prototype.
2. Calculate the velocity ratio.
3. Calculate the discharge ratio using model/prototype area ratio.

Solution

1. Reynolds-number similitude

$$\mathrm{Re}_m = \mathrm{Re}_p$$

$$\frac{V_m L_m}{\nu_m} = \frac{V_p L_p}{\nu_p}$$

2. Velocity ratio

$$\frac{V_m}{V_p} = \frac{L_p}{L_m}\frac{\nu_m}{\nu_p}$$

Since $\nu_p = \nu_m$,

$$\frac{V_m}{V_p} = \frac{L_p}{L_m}$$

3. Discharge

$$\frac{Q_m}{Q_p} = \frac{V_m A_m}{V_p A_p} = \frac{L_p}{L_m}\left(\frac{L_m}{L_p}\right)^2 = \frac{L_m}{L_p}$$

$$Q_m = 700 \text{ cfs} \times \frac{1}{6} = \boxed{117 \text{ cfs}}$$

Review

This discharge is very large and serves to emphasize that very few model studies are made that completely satisfy the Reynolds-number criterion. This subject will be discussed further in the next sections.

8.7

Model-Prototype Performance

Geometric (scale model) and dynamic (same π-groups) similitude mean that the dependent π-groups are the same for both the model and the prototype. For this reason, measurements made with the model can be applied directly to the prototype. Such correspondence is illustrated in this section.

Example 8.6 shows how the pressure difference measured in a model test can be used to find the pressure difference between the corresponding two points on the prototype.

EXAMPLE 8.6 APPLICATION OF PRESSURE COEFFICIENT

A 1/10 scale model of a blimp is tested in a wind tunnel under dynamically similar conditions. The speed of the blimp through still air is 10 m/s. A 17.8 kPa pressure difference is measured between two points on the model. What will be the pressure difference between the two corresponding points on the prototype? The temperature and pressure in the wind tunnel is the same as the prototype.

Problem Definition

Situation: A 1/10 scale of a blimp is tested in a wind tunnel under dynamically similar conditions. A pressure difference of 17.8 kPa is measured on model.

Find: Corresponding pressure difference (Pa) on prototype.

Properties: Pressure and temperature are the same for wind tunnel test and prototype, so $v_m = v_p$.

Plan

Reynolds number is the only significant π-group. Thus Eq. (8.4) reduces to

$$C_p = f(\mathrm{Re})$$

For dynamic similitude, $\mathrm{Re}_m = \mathrm{Re}_p$. Then with geometric similitude, $C_{pm} = C_{pp}$.

1. Calculate the model/prototype velocity ratio.
2. Calculate pressure difference on prototype.

Solution

1. Reynolds-number similitude

$$\mathrm{Re}_m = \mathrm{Re}_p$$

$$\frac{V_m L_m}{v_m} = \frac{V_p L_p}{v_p}$$

$$\frac{V_p}{V_m} = \frac{L_m}{L_p} = \frac{1}{10}$$

2. Pressure coefficient correspondence

$$\frac{\Delta p_m}{\frac{1}{2}\rho_m V_m^2} = \frac{\Delta p_p}{\frac{1}{2}\rho_p V_p^2}$$

$$\frac{\Delta p_p}{\Delta p_m} = \left(\frac{V_p}{V_m}\right)^2 = \left(\frac{L_m}{L_p}\right)^2 = \frac{1}{100}$$

Pressure difference on prototype

$$\Delta p_p = \frac{\Delta p_m}{100} = \frac{17.8\ \text{kPa}}{100} = \boxed{178\ \text{Pa}}$$

Example 8.7 illustrates calculating the fluid dynamic force on a prototype blimp from wind tunnel data using similitude.

EXAMPLE 8.7 DRAG FORCE FROM WIND TUNNEL TESTING

A 1/10 scale of a blimp is tested in a wind tunnel under dynamically similar conditions. If the drag force on the model blimp is measured to be 1530 N, what corresponding force could be expected on the prototype? The air pressure and temperature are the same for both model and prototype.

Problem Definition

Situation: A 1/10 scale model of blimp is tested in a wind tunnel, and a drag force of 1530 N is measured.

Find: The drag force (in newtons) on the prototype.

Properties: Pressure and temperature are the same, $v_m = v_p$.

Plan

Reynolds number is the only significant π-group, so Eq. (8.4) reduces to $C_F = f(\mathrm{Re})$. For dynamic similitude, $\mathrm{Re}_m = \mathrm{Re}_p$. Thus with geometric similitude $C_{F,m} = C_{F,p}$.

1. Find velocity ratio by equating Reynolds numbers.
2. Find the force ratio from force coefficient.

Solution

1. Reynolds-number similitude

$$\mathrm{Re}_m = \mathrm{Re}_p$$

$$\frac{V_m L_m}{v_m} = \frac{V_p L_p}{v_p}$$

$$\frac{V_p}{V_m} = \frac{L_m}{L_p} = \frac{1}{10}$$

2. Force coefficient correspondence

$$\frac{F_p}{\frac{1}{2}\rho_p V_p^2 L_p^2} = \frac{F_m}{\frac{1}{2}\rho_m V_m^2 L_m^2}$$

$$\frac{F_p}{F_m} = \frac{V_p^2 L_p^2}{V_m^2 L_m^2} = \frac{L_m^2 L_p^2}{L_p^2 L_m^2} = 1$$

Therefore

$$F_p = 1530\text{N}$$

Review

The result that the model force is the same as the prototype force is interesting. When Reynolds-number similitude is used, and the fluid properties are the same fluid for model and prototype, the forces on the model will always be the same as the forces on the prototype.

8.8

Approximate Similitude at High Reynolds Numbers

The primary justification for model tests is that it is more economical to get answers needed for engineering design by such tests than by any other means. However, as revealed by Examples 8.3, 8.4, and 8.6, Reynolds-number similitude requires expensive model tests (high-pressure facilities, large test sections, or using different fluids). This section shows that approximate similitude is achievable even though high Reynolds numbers cannot be reached in model tests.

Consider the size and power required for wind-tunnel tests of the blimp in Example 8.3. The wind tunnel would probably require a section at least 2 m by 2 m to accommodate the model blimp. With a 100 m/s airspeed in the tunnel, the power required for producing continuously a stream of air of this size and velocity is in the order of 4 MW. Such a test is not prohibitive, but it is very expensive. It is also conceivable that the 100 m/s airspeed would introduce Mach-number effects not encountered with the prototype, thus generating concern over the validity of the model data. Furthermore, a force of 1530 N is generally larger than that usually associated with model tests. Therefore, especially in the study of problems involving non-free-surface flows, it is desirable to perform model tests in such a way that large magnitudes of forces or pressures are not encountered.

For many cases, it is possible to obtain all the needed information from abbreviated tests. Often the Reynolds-number effect (relative viscous effect) either becomes insignificant at high Reynolds numbers or becomes independent of the Reynolds number. The point where testing can be stopped often can be detected by inspection of a graph of the pressure coefficient C_p versus the Reynolds number Re. Such a graph for a venturi meter in a pipe is shown in Fig. 8.6. In this meter, Δp is the pressure difference between the points shown, and V is the velocity in the restricted section of the venturi meter. Here it is seen that viscous forces affect the value of C_p below a Reynolds number of approximately 50,000. However, for higher Reynolds numbers, C_p is virtually constant. Physically this means that at low Reynolds numbers (relatively high viscous forces), a significant part of the change in pressure comes from viscous resistance, and the remainder comes from the acceleration (change in kinetic energy) of the fluid as it passes through the venturi meter. However, with high Reynolds numbers (resulting from either small viscosity or a large product of V, D, and ρ), the viscous resistance is negligible compared with the force required to accelerate the fluid. Since the ratio of Δp to the kinetic pressure does not change (constant C_p) for high Reynolds numbers, there is no need to carry out tests at higher Reynolds numbers. This is true in general, so long as the flow pattern does not change with the Reynolds number.

Figure 8.6

C_p for a venturi meter as a function of the Reynolds number.

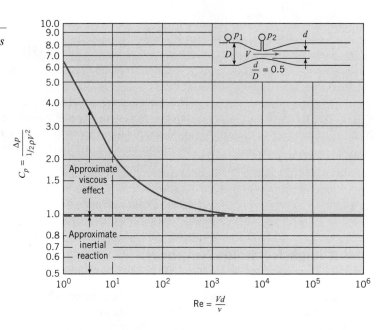

In a practical sense, whoever is in charge of the model test will try to predict from previous works approximately what maximum Reynolds number will be needed to reach the point of insignificant Reynolds-number effect and then will design the model accordingly. After a series of tests has been made on the model, C_p versus Re will be plotted to see whether the range of constant C_p has indeed been reached. If so, then no more data are needed to predict the prototype performance. However, if C_p has not reached a constant value, the test program has to be expanded or results extrapolated. Thus the results of some model tests can be used to predict prototype performance even though the Reynolds numbers are not the same for the model and the prototype. This is especially valid for angular-shaped bodies, such as model buildings, tested in wind tunnels.

In addition, the results of model testing can be combined with analytic results. Computational fluid dynamics (CFD) may predict the change in performance with Reynolds number but may not be reliable to predict the performance level. In this case, the model testing would be used to establish the level and of performance, and the trends predicted by CFD would be used to extrapolate the results to other conditions.

Example 8.8 is an illustration on the approximate similitude at high Reynolds number for flow through a constriction.

EXAMPLE 8.8 MEASURING HEAD LOSS IN NOZZLE IN REVERSE FLOW

Tests are to be performed to determine the head loss in a nozzle under a reverse-flow situation. The prototype operates with water at 50°F and with a nominal reverse-flow velocity of 5 ft/s. The diameter of the prototype is 3 ft. The tests are done in a 1/12 scale model facility with water at 60°F. A head loss (pressure drop) of 1 psid is measured with a velocity of 20 ft/s. What will be the head loss in the actual nozzle?

Sketch:

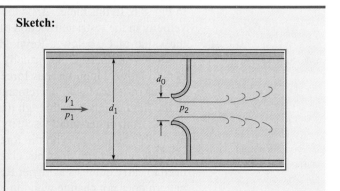

Problem Definition

Situation: A 1/12 scale model tests for head loss in a reverse-flow nozzle. A pressure difference of 1 psid is measured with model at 20 ft/s.

Find: Pressure drop (psid) in actual nozzle.

Properties: Table F.5.: Water at 50°F, $\rho = 1.94$ slugs/ft^3, $\nu = 1.41 \times 10^{-5}$ ft^2/s; water at 60°F, $\rho = 1.94$ slugs/ft^3, $\nu = 1.22 \times 10^{-5}$ ft^2/s.

Plan

The only significant π-group is the Reynolds number, so Eq. (8.4) reduces to $C_p = f(\text{Re})$. Dynamic similitude achieved if $\text{Re}_m = \text{Re}_p$, then $C_{p,m} = C_{p,p}$. From Fig. 8.6, if Re_m, $\text{Re}_p > 10^3$, then $C_{p,m} = C_{p,p}$.

1. Calculate Reynolds number for model and prototype.
2. Check if both exceed 10^3. If not, model tests need to be reevaluated.
3. Calculate pressure coefficient.
4. Evaluate pressure drop in prototype.

Solution

1. Reynolds numbers

$$\text{Re}_m = \frac{VD}{\nu} = \frac{20 \text{ ft/s} \times (3/12 \text{ ft})}{1.22 \times 10^{-5} \text{ ft}^2/\text{s}} = 4.10 \times 10^5$$

$$\text{Re}_p = \frac{5 \text{ ft/s} \times 3 \text{ ft}}{1.41 \times 10^{-5} \text{ ft}^2/\text{s}} = 1.06 \times 10^6$$

2. Both Reynolds numbers exceed 10^3. Therefore $C_{p,m} = C_{p,p}$. The test is valid.

3. Pressure coefficient from model tests

$$C_{p,m} = \frac{\Delta p}{\frac{1}{2}\rho V^2} = \frac{1 \text{ lbf/in}^2 \times 144 \text{ in}^2/\text{ft}^2}{\frac{1}{2} \times 1.94 \text{ slug/ft}^3 \times (20 \text{ ft/s})^2} = 0.371$$

4. Pressure drop in prototype

$$\Delta p_p = 0.371 \times \frac{1}{2}\rho V^2 = 0.371 \times 0.5 \times 1.94 \text{ slug/ft}^3 \times (5 \text{ ft/s})^2$$

$$= 9.0 \text{ lbf/ft}^2 = \boxed{0.0625 \text{ psid}}$$

Review

1. Because the Reynolds numbers are so much greater than 10^3, the equation for pressure drop is valid over a wide range of velocities.

2. This example justifies the independence of Reynolds number referred to in Section 8.1.

In some situations viscous and compressibility effects may both be important, but it is not possible to have dynamic similitude with both π-groups. Which π-group is chosen for similitude depends a great deal on what information the engineer is seeking. If the engineer is interested in the viscous motion of fluid near a wall in shock-free supersonic flow, then the Reynolds number should be selected as the significant π-group. However, if the shock wave pattern over a body is of interest, then the Mach number should be selected for similitude. A useful rule of thumb is that compressibility effects are unimportant for M < 0.3.

Example 8.9 shows the difficulty in having Reynolds-number similitude and avoiding Mach-number effects in wind tunnel tests of an automobile.

EXAMPLE 8.9   MODEL TESTS FOR DRAG FORCE ON AN AUTOMOBILE

A 1/10 scale of an automobile is tested in a wind tunnel with air at atmospheric pressure and 20°C. The automobile is 4 m long and travels at a velocity of 100 km/hr in air at the same conditions. What should the wind-tunnel speed be such that the measured drag can be related to the drag of the prototype? Experience shows that the dependent π-groups are independent of Reynolds numbers for values exceeding 10^5. The speed of sound is 1235 km/hr.

Problem Definition

Situation: A 1/10 scale model of a 4 m–long automobile moving at 100 km/hr is tested in wind tunnel.

Find: The wind tunnel speed to achieve similitude.

Properties: Air (20°C), Table A.3, $\rho = 1.2$ kg/m^3, $\nu = 1.51 \times 10^{-5}$ N $\cdot$ s/m^2.

Plan

Mach number of the prototype is about 0.08 (100/1235), so Mach-number effects are unimportant. Dynamic similitude is achieved with Reynolds numbers, $\text{Re}_m = \text{Re}_p$. With dynamic similitude, $C_{F,m} = C_{F,p}$, and model measurements can be applied to prototype.

1. Determine the model speed for dynamic similitude.
2. Evaluate the model speed. If it is not feasible, continue to next step.
3. Calculate the prototype Reynolds number. If $\text{Re}_p > 10^5$, then $\text{Re}_m \geq 10^5$ for $C_{F,m} = C_{F,p}$.
4. Find the speed for which $\text{Re}_m \geq 10^5$.

Solution

1. Velocity from Reynolds-number similitude

$$\left(\frac{VL}{\nu}\right)_m = \left(\frac{VL}{\nu}\right)_p$$

$$\frac{V_m}{V_p} = \frac{L_p}{L_m} = 10$$

$$V_m = 10 \times 100 \text{ km/hr} = 1000 \text{ km/hr}$$

2. With this velocity, $M = 1000/1235 = 0.81$. This is too high for model tests because it would introduce unwanted compressibility effects.

3. Reynolds number of prototype

$$\text{Re}_p = \frac{VL\rho}{\mu} = \frac{100 \text{ km/hr} \times 0.278 \text{ (m/s)/(km/hr)} \times 4 \text{ m}}{1.51 \times 10^{-5} \text{ m}^2/\text{s}}$$

$$= 7.4 \times 10^6$$

Therefore $C_{F,m} = C_{F,p}$ if $\text{Re}_m \geq 10^5$.

4. Wind tunnel speed

$$V_m \geq \text{Re}_m \frac{\nu_m}{L_m} = 10^5 \times \frac{1.51 \times 10^{-5} \text{ m}^2/\text{s}}{0.4 \text{ m}}$$

$$\geq \boxed{3.8 \text{ m/s}}$$

Review

The wind-tunnel speed must exceed 3.8 m/s. From a practical point of view, the speed will be chosen to provide sufficiently large forces for reliable and accurate measurements.

8.9

Free-Surface Model Studies

Spillway Models

The flow over a spillway is a classic case of a free-surface flow. The major influence, besides the spillway geometry itself, on the flow of water over a spillway is the action of gravity. Hence the Froude-number similarity criterion is used for such model studies. It can be appreciated for large spillways with depths of water on the order of 3 m or 4 m and velocities on the order of 10 m/s or more, that the Reynolds number is very large. At high values of the Reynolds number, the relative viscous forces are often independent of the Reynolds number, as noted in the foregoing section (Sec. 8.8). However, if the reduced-scale model is made too small, the viscous forces as well as the surface-tension forces would have a larger relative effect on the flow in the model than in the prototype. Therefore, in practice, spillway models are made large enough so that the viscous effects have about the same relative effect in the model as in the prototype (i.e., the viscous effects are nearly independent of the Reynolds number). Then the Froude number is the significant π-group. Most model spillways are made at least 1 m high, and for precise studies, such as calibration of individual spillway bays, it is not uncommon to design and construct model spillway sections that are 2 m or 3 m high. Figures 8.7 and 8.8 show a comprehensive model and spillway model for Hell's Canyon Dam in Idaho.

Figure 8.7

Comprehensive model for Hell's Canyon Dam. Tests were made at the Albrook Hydraulic Laboratory, Washington State University.

Figure 8.8

Spillway model for Hell's Canyon Dam. Tests were made at the Albrook Hydraulic Laboratory, Washington State University.

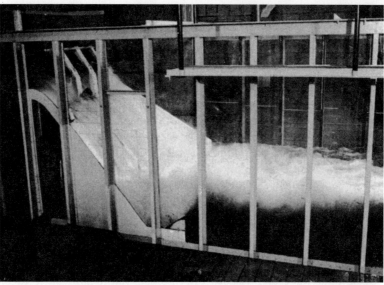

Example 8.10 is an application of Froude-number similitude in modeling discharge over a spillway.

EXAMPLE 8.10 MODELING FLOOD DISCHARGE OVER A SPILLWAY

A 1/49 scale model of a proposed dam is used to predict prototype flow conditions. If the design flood discharge over the spillway is 15,000 m³/s, what water flow rate should be established in the model to simulate this flow? If a velocity of 1.2 m/s is measured at a point in the model, what is the velocity at a corresponding point in the prototype?

Problem Definition

Situation: A 1/49 scale model of spillway with discharge of 15,000 m³/s.

Find:

1. Flow rate over model.
2. Velocity on prototype at point where velocity is 1.2 m/s on model.

Plan

Gravity is responsible for the flow, so significant π-group is the Froude number. For dynamic similitude, $\text{Fr}_m = \text{Fr}_p$.

1. Calculate velocity ratio from Froude-number similitude.
2. Calculate discharge ratio using scale ratio and calculate model discharge.
3. Use velocity ratio from step 1 to find velocity at point on prototype.

Solution

1. Froude-number similitude

$$\text{Fr}_m = \text{Fr}_p$$

$$\frac{V_m}{\sqrt{g_m L_m}} = \frac{V_p}{\sqrt{g_p L_p}}$$

The acceleration due to gravity is the same, so

$$\frac{V_m}{V_p} = \sqrt{\frac{L_m}{L_p}}$$

2. Discharge ratio

$$\frac{Q_m}{Q_p} = \frac{A_m V_m}{A_p V_p} = \frac{L_m^2}{L_p^2}\sqrt{\frac{L_m}{L_p}} = \left(\frac{L_m}{L_p}\right)^{5/2}$$

Discharge for model

$$Q_m = Q_p\left(\frac{1}{49}\right)^{5/2} = 15{,}000\,\frac{\text{m}^3}{\text{s}} \times \frac{1}{16{,}800} = \boxed{0.89\ \text{m}^3/\text{s}}$$

3. Velocity on prototype

$$\frac{V_p}{V_m} = \sqrt{\frac{L_p}{L_m}}$$

$$V_p = \sqrt{49} \times 1.2\ \text{m/s} = \boxed{8.4\ \text{m/s}}$$

Ship Model Tests

The largest facility for ship testing in the United States is the David Taylor Model Basin, Naval Surface Warfare Center, Carderock Division, near Washington, D.C. Two of the core facilities are the towing basins and the rotating arm facility. In the rotating arm facility, models are suspended from the end of a rotating arm in a larger circular basin. Forces and moments can be measured on ship models up to 9 m in length at steady-state speeds as high as 15.4 m/s (30 knots). In the high-speed towing basin, models 1.2 m to 6.1 m can be towed at speeds up to 16.5 m/s (32 knots).

The aim of the ship model testing is to determine the resistance that the propulsion system of the ship must overcome. This resistance is the sum of the wave resistance and the surface resistance of the hull. The wave resistance is a free-surface, or Froude-number, phenomenon, and the hull resistance is a viscous, or Reynolds-number, phenomenon. Because both wave and viscous effects contribute significantly to the overall resistance, it would appear that both the Froude and Reynolds criteria should be used. However, it is impossible to satisfy both if the model liquid is water (the only practical test liquid), because the Reynolds-number similitude dictates a higher velocity for the model than for the prototype [equal to $V_p(L_p/L_m)$], whereas the Froude-number similitude dictates a lower velocity for the model [equal to $V_p(\sqrt{L_m}/\sqrt{L_p})$]. To circumvent such a dilemma, the procedure is to model for the phenomenon that is the most difficult to predict analytically and to account for the other resistance by analytical means. Since the wave resistance is the most difficult problem, the model is operated according to the Froude-number similitude, and the hull resistance is accounted for analytically.

To illustrate how the test results and the analytical solutions for surface resistance are merged to yield design data, the following necessary sequential steps are indicated:

1. Make model tests according to Froude-number similitude, and the total model resistance is measured. This total model resistance will be equal to the wave resistance plus the surface resistance of the hull of the model.

2. Estimate the surface resistance of the model by analytical calculations.

3. Subtract the surface resistance calculated in step 2 from the total model resistance of step 1 to yield the wave resistance of the model.

4. Using the Froude-number similitude, scale the wave resistance of the model up to yield the wave resistance of the prototype.

5. Estimate the surface resistance of the hull of the prototype by analytical means.

6. The sum of the wave resistance of the prototype from step 4 and the surface resistance of the prototype from step 5 yields the total prototype resistance, or drag.

8.10

Summary

Dimensional analysis involves combining dimensional variables to form dimensionless groups. These groups, called π-groups, can be regarded as the scaling parameters for fluid flow. The Buckingham Π theorem states that the number of independent π-groups is $n - m$, where n is the number of dimensional variables and m is the number of basic dimensions included in the variables. In fluid mechanics the three basic dimensions are mass (M), length (L), and time (T).

The π-groups can be found by either the step-by-step method or the exponent method. In the step-by-step method each dimension is removed by successively using a dimensional variable until the π-groups are obtained. In the exponent method, each variable is raised to a power, they are multiplied together, and three simultaneous algebraic equations formulated for dimensional homogeneity are solved to yield the π-groups.

Four common independent π-groups are

$$\text{Reynolds number Re} = \frac{\rho VL}{\mu} \qquad \text{Mach number M} = \frac{V}{c}$$

$$\text{Weber number We} = \frac{\rho V^2 L}{\sigma} \qquad \text{Froude number Fr} = \frac{V}{\sqrt{gL}}$$

Three common dependent π-groups are

$$\text{Pressure coefficient, } C_p = \frac{\Delta p}{(\rho V^2)/2}$$

$$\text{Shear stress coefficient, } c_f = \frac{\tau}{(\rho V^2)/2}$$

$$\text{Force coefficient, } C_F = \frac{F}{(\rho V^2 L^2)/2}$$

The general functional form of the common π-groups is

$$C_F, c_f, C_p = f(\text{Re, M, We, Fr})$$

Experimental testing is often performed with a small-scale replica (model) of the full-scale structure (prototype). Similitude is the art and theory of predicting prototype performance from model observations. To achieve exact similitude, the model must be a scale model of the prototype (geometric similitude), and the values of the π-groups must be the same for the model and the prototype (dynamic similitude). In practice, it is not always possible to have complete dynamic similitude, so only the most important π-groups are matched.

References

1. Buckingham, E. "Model Experiments and the Forms of Empirical Equations." *Trans. ASME*, 37 (1915), 263.

2. Ipsen, D. C. *Units, Dimensions and Dimensionless Numbers.* McGraw-Hill, New York, 1960.

3. NASA publication available from the U.S. Government Printing Office: No. 1995-685-893.

4. Personal communication. Mark Goldhammer, Manager, Aerodynamic Design of the 777.

5. Kilgore, R. A., and D. A. Dress. "The Application of Cryogenics to High Reynolds-Number Testing in Wind Tunnels, Part 2. Development and Application of the Cryogenic Wind Tunnel Concept." *Cryogenics,* Vol. 24, no. 9, September 1984.

6. Baals, D. D., and W. R. Corliss. *Wind Tunnels of NASA.* U.S. Govt. Printing Office, Washington, D.C., 1981.

7. Kale, S., et al. "An Experimental Study of Single-Particle Aerodynamics." *Proc. of First Nat. Congress on Fluid Dynamics,* Cincinnati, Ohio, July 1988.

Problems

Dimensional Analysis

8.1 $\mathbb{PQ}$◀ Find the primary dimensions of density ρ, viscosity μ, and pressure p.

8.2 $\mathbb{PQ}$◀ According to the Buckingham Π theorem, if there are six dimensional variables and three primary dimensions, how many dimensionless variables will there be?

8.3 $\mathbb{PQ}$◀ Explain what is meant by dimensional homogeneity.

8.4 $\mathbb{PQ}$◀ Determine which of the following equations are dimensionally homogeneous:

a. $Q = \frac{2}{3}CL\sqrt{2g}H^{3/2}$

where Q is discharge, C is a pure number, L is length, g is acceleration due to gravity, and H is head.

b. $V = \frac{1.49}{n}R^{2/3}S^{1/2}$

where V is velocity, n is length to the one-sixth power, R is length, and S is slope.

c. $h_f = f\,\frac{L}{D}\frac{V^2}{2g}$

where h_f is head loss, f is a dimensionless resistance coefficient, L is length, D is diameter, V is velocity, and g is acceleration due to gravity.

d. $D = \frac{0.074}{\mathrm{Re}^{0.2}}\frac{Bx\rho V^2}{2}$

where D is drag force, Re is Vx/v, B is width, x is length, ρ is mass density, v is the kinematic viscosity, and V is velocity.

8.5 $\mathbb{PQ}$◀ Determine the dimensions of the following variables and combinations of variables in terms of primary dimensions.

a. T (torque)
b. $\rho V^2/2$, where V is velocity and ρ is mass density
c. $\sqrt{\tau/\rho}$, where τ is shear stress
d. Q/ND^3, where Q is discharge, D is diameter, and N is angular speed of a pump

8.6 It takes a certain length of time for the liquid level in a tank of diameter D to drop from position h_1 to position h_2 as the tank is being drained through an orifice of diameter d at the bottom. Determine the π-groups that apply to this problem. Assume that the liquid is nonviscous. Express your answer in the functional form.

$$\frac{\Delta h}{d} = f(\pi_1, \pi_2, \pi_3)$$

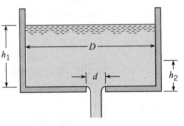

PROBLEM 8.6

8.7 The maximum rise of a liquid in a small capillary tube is a function of the diameter of the tube, the surface tension, and the specific weight of the liquid. What are the significant π-groups for the problem?

8.8 For very low velocities it is known that the drag force F_D of a small sphere is a function solely of the velocity V of flow past the sphere, the diameter d of the sphere, and the viscosity μ of the fluid. Determine the π-groups involving these variables.

8.9 Observations show that the side thrust F, for a rough spinning ball in a fluid is a function of the ball diameter D, the free-stream velocity V_0, the density ρ, the viscosity μ, the roughness height k_s, and the angular velocity of spin ω. Determine the dimensionless parameter(s) that would be used to correlate the experimental results of a study involving the variables noted above. Express your answer in the functional form

$$\frac{F}{V_0^2 D^2} = f(\pi_1, \pi_2, \pi_3)$$

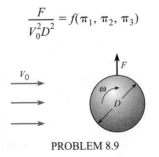

PROBLEM 8.9

8.10 Consider steady viscous flow through a small horizontal tube. For this type of flow, the pressure gradient along the tube, $\Delta p / \Delta \ell$ should be a function of the viscosity μ, the mean velocity V, and the diameter D. By dimensional analysis, derive a functional relationship relating these variables.

8.11 A flow-metering device, called a vortex meter, consists of a square element mounted inside a pipe. Vortices are generated by the element, which gives rise to an oscillatory pressure measured on the leeward side of the element. The fluctuation frequency is related to the flow velocity. The discharge in the pipe is a function of the frequency of the oscillating pressure ω, the pipe diameter D, the size of the element l, the density ρ, and the viscosity μ. Thus

$$Q = f(\omega, D, l, \rho, \mu)$$

Find the π-groups in the form

$$\frac{Q}{\omega D^3} = f(\pi_1, \pi_2)$$

8.12 It is known that the pressure developed by a centrifugal pump, Δp, is a function of the diameter D of the impeller, the speed of rotation n, the discharge Q, and the fluid density ρ. By dimensional analysis, determine the π-groups relating these variables.

8.13 The force on a satellite in the earth's upper atmosphere depends on the mean path of the molecules λ (a length), the density ρ, the diameter of the body D, and the molecular speed c: $F = f(\lambda, \rho, D, c)$. Find the nondimensional form of this equation.

8.14 A general study is to be made of the height of rise of liquid in a capillary tube as a function of time after the start of a test. Other significant variables include surface tension, mass density, specific weight, viscosity, and diameter of the tube. Determine the dimensionless parameters that apply to the problem. Express your answer in the functional form

$$\frac{h}{d} = f(\pi_1, \pi_2, \pi_3)$$

8.15 An engineer is using an experiment to characterize the power P consumed by a fan (see photo) to be used in an electronics cooling application. Power depends on four variables: $P = f(\rho, D, Q, n)$, where ρ is the density of air, D is the diameter of the fan impeller, Q is the flow rate produced by the fan, and n is the rotation rate of the fan. Find the relevant π-groups and suggest a way to plot the data.

8.16 By dimensional analysis, determine the π-groups for the change in pressure that occurs when water or oil flows through a horizontal pipe with an abrupt contraction as shown. Express your answer in the functional form

$$\frac{\Delta p d^4}{\rho Q^2} = f(\pi_1, \pi_2)$$

PROBLEM 8.15

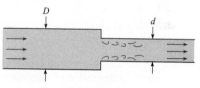

PROBLEM 8.16

8.17 A solid particle falls through a viscous fluid. The falling velocity V, is believed to be a function of the fluid density ρ_f, the particle density ρ_p, the fluid viscosity μ, the particle diameter D, and the acceleration due to gravity g:

$$V = f(\rho_f, \rho_p, \mu, D, g)$$

By dimensional analysis, develop the π-groups for this problem. Express your answer in the form

$$\frac{V}{\sqrt{gD}} = f(\pi_1, \pi_2)$$

8.18 An experimental test program is being set up to calibrate a new flow meter. The flow meter is to measure the mass flow rate of liquid flowing through a pipe. It is assumed that the mass flow rate is a function of the following variables:

$$\dot{m} = f(\Delta p, D, \mu, \rho)$$

where Δp is the pressure difference across the meter, D is the pipe diameter, μ is the liquid viscosity, and ρ is the liquid density. Using dimensional analysis, find the π-groups. Express your answer in the form

$$\frac{\dot{m}}{\sqrt{\rho \Delta p D^4}} = f(\pi)$$

8.19 A torpedo-like device is being designed to travel just below the water surface. Which dimensionless numbers in Section 8.4 would be significant in this problem? Give a rationale for your answer.

8.20 Experiments are to be done on the drag forces on an oscillating fin in a water tunnel. It is assumed that the drag force F_D, is a function of the liquid density ρ, the fluid velocity V, the surface area of the fin S, and the frequency of oscillation ω:

$$F_D = f(\rho, V, S, \omega)$$

By dimensional analysis, find the dimensionless parameters for this problem. Express your answer in the form *

$$\frac{F_D}{\rho V^2 S} = f(\pi)$$

8.21 Flow situations in biofluid mechanics involve the flow through tubes that change in size with time (such as blood vessels) or are supplied by an oscillatory source. The volume flow rate Q in the tube will be a function of the frequency ω, the tube diameter D, the fluid density ρ, viscosity μ, and the pressure gradient $(\Delta p)/(\Delta l)$. Find the π-groups for this situation in the form

$$\frac{Q}{\omega D^3} = f(\pi_1, \pi_2)$$

8.22 The rise velocity V_b of a bubble with diameter D in a liquid of density ρ_l and viscosity μ depends on the acceleration due to gravity, g, and the density difference between the bubble and the fluid, $\rho_l - \rho_b$. Find the π-groups in the form

$$\frac{V_b}{\sqrt{gD}} = f(\pi_1, \pi_2)$$

8.23 The discharge of a centrifugal pump is a function of the rotational speed of the pump N, the diameter of the impeller D, the head across the pump h_p, the viscosity of the fluid μ, the density of the fluid ρ, and the acceleration due to gravity g. The functional relationship is

$$Q = f(N, D, h_p, \mu, \rho, g)$$

By dimensional analysis, find the dimensionless parameters. Express your answer in the form

$$\frac{Q}{ND^3} = f(\pi_1, \pi_2, \pi_3)$$

8.24 Drag tests show that the drag of a square plate placed normal to the free-stream velocity is a function of the velocity V, the density ρ, the plate dimensions B, the viscosity μ, the free-stream turbulence root mean square velocity u_{rms}, and the turbulence length scale L_x. Here u_{rms} and L_x are in ft/s and ft, respectively. By dimensional analysis, develop the π-groups that could be used to correlate the experimental results. Express your answer in the functional form

$$\frac{F_D}{\rho V^2 B^2} = f(\pi_1, \pi_2, \pi_3)$$

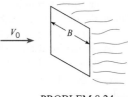

PROBLEM 8.24

Similitude

8.25 PQ◀ What is meant by geometric similitude?

8.26 PQ◀ Many automobile companies advertise products with low drag for improved performance. Gather all the information you can find on wind-tunnel testing of automobiles and summarize your findings in a concise, informative manner on two pages or less.

8.27 PQ◀ One of the shortcomings of mounting a model of an automobile in a wind tunnel and measuring drag is that the effect of the road is not included. Give some thought as to your impressions of what the effect of the road may be on automobile drag and your reasoning. Also list some variables that may influence the effect of the ground on automobile drag.

8.28 PQ◀ One of the largest wind tunnels in the United States is the NASA facility in Moffat Field, California. Look up information on this facility (size, test section velocity, etc.) and summarize your findings in a concise, informative manner.

8.29 PQ◀ The hydrodynamic drag on a sailboat is very important to the performance of the craft, especially in competitive races such as the America's Cup. Investigate on the Internet or other sources the extent and types of simulations that have been carried out on high performance sailboats.

8.30 The drag on a submarine moving below the free surface is to be determined by a test on a 1/15 scale model in a water tunnel. The velocity of the prototype in sea water ($\rho = 1015$ kg/m^3, $v = 1.4 \times 10^{-6}$ m^2/s) is 2 m/s. The test is done in pure water at 20°C. Determine the speed of the water in the water tunnel for dynamic similitude and the ratio of the drag force on the model to the drag force on the prototype.

8.31 Water with a kinematic viscosity of 10^{-6} m^2/s flows through a 4 cm pipe. What would the velocity of water have to be for the water flow to be dynamically similar to oil ($v = 10^{-5}$ m^2/s) flowing through the same pipe at a velocity of 0.5 m/s?

8.32 Oil with a kinematic viscosity of 4×10^{-6} m^2/s flows through a smooth pipe 15 cm in diameter at 2 m/s. What velocity should water have at 20°C in a smooth pipe 5 cm in diameter to be dynamically similar?

8.33 A large venturi meter is calibrated by means of a 1/10 scale model using the prototype liquid. What is the discharge ratio Q_m/Q_p for dynamic similarity? If a pressure difference of 400 kPa is measured across ports in the model for a given discharge, what pressure difference will occur between similar ports in the prototype for dynamically similar conditions?

8.34 A 1/5 scale model of an experimental bathosphere that will operate at great depths is to be tested to determine its drag characteristic by towing it behind a submarine. For true similitude, what should be the towing speed relative to the speed of the prototype?

8.35 A spherical balloon that is to be used in air at 60°F and atmospheric presssure is tested by towing a 1/4 scale model in a lake. The model is 1 ft in diameter, and a drag of 15 lbf is measured when the model is being towed in deep water at 5 ft/s. What drag (in pounds force and newtons) can be expected for the prototype in air under dynamically similar conditions? Assume that the water temperature is 60°F.

8.36 An engineer needs a value of lift force for an airplane that has a coefficient of lift (C_L) of 0.4. The π-group is defined as

$$C_L = 2\frac{F_L}{\rho V^2 S}$$

where F_L is the lift force, ρ is the density of ambient air, V is the speed of the air relative to the airplane, and S is the area of the wings from a top view. Estimate the lift force in newtons for a speed of 80 m/s, an air density of 1.1 kg/m³, and a wing area (planform area) of 15 m².

PROBLEM 8.36

8.37 An airplane travels in air ($p = 100$ kPa, $T = 10°C$) at 150 m/s. If a 1/5 scale model of the plane is tested in a wind tunnel at 25°C, what must the density of the air in the tunnel be so that both the Reynolds-number and the Mach-number criteria are satisfied? The speed of sound varies with the square root of the absolute temperature. (*Note:* The dynamic viscosity is independent of pressure.)

8.38 The new Airbus A380-300 has a wing span of 79.8 m. The cruise altitude is 10,000 m in a standard atmosphere. Assume you are designing a wind tunnel to operate with air at 20°C. The span of the scale model A380 in the wind tunnel is 1 m. Assume Mach number correspondence between model and prototype. Both the speed of sound and the dynamic viscosity vary linearly with the square root of the absolute temperature. What would the pressure of the air in the wind tunnel have to be to have Reynolds-number similitude? Use the properties for a standard atmosphere in Chapter 3 to find properties at 10,000 m altitude.

8.39 The new Boeing 787-3 Dreamliner has a wing span of 52 m. It flies at a cruise Mach number of 0.85, which corresponds to a velocity of 945 km/hr at an altitude of 10,000 m. You are going to estimate the drag on the prototype by measuring the drag on a 1 m wing span scale model in a wind tunnel with air where the speed of sound is 340 m/s and the density is 0.98 kg/m³. What is the ratio of the force on the prototype to the force on the model. Only Mach-number similitude is considered. Use the properties of the standard atmosphere in Chapter 3 to evaluate the density of air for the prototype.

8.40 Flow in a given pipe is to be tested with air and then with water. Assume that the velocities (V_A and V_W) are such that the flow with air is dynamically similar to the flow with water. Then for this condition, the magnitude of the ratio of the velocities, V_A/V_W, will be (a) less than unity, (b) equal to unity, or (c) greater than unity.

8.41 A smooth pipe designed to carry crude oil (diameter = 48 in., $\rho = 1.75$ slugs/ft³, and $\mu = 4 \times 10^{-4}$ lbf-s/ft²) is to be modeled with a smooth pipe 4 in. in diameter carrying water ($T = 60°F$). If the mean velocity in the prototype is 4 ft/s, what should be the mean velocity of water in the model to ensure dynamically similar conditions?

8.42 A student is competing in a contest to design a radio-controlled blimp. The drag force acting on the blimp depends on the Reynolds number, $Re = (\rho V D)/\mu$, where V is the speed of the blimp, D is the maximum diameter, ρ is the density of air, and μ is the viscosity of air. This blimp has a coefficient of drag (C_D) of 0.3. This π-group is defined as

$$C_D = 2\frac{F_D}{\rho V^2 A_p}$$

where F_D is the drag force ρ is the density of ambient air, V is the speed of the air relative to the blimp, and $A_p = \pi D^2/4$ is the maximum section area of the blimp from a front view. Calculate the Reynolds number, the drag force in newtons, and the power in watts required to move the blimp through the air. Blimp speed is 800 mm/s, and the maximum diameter is 475 mm. Assume that ambient air is at 20°C.

PROBLEM 8.42

8.43 Colonization of the moon will require an improved understanding of fluid flow under reduced gravitational forces. The gravitational force on the moon is 1/5 that on the surface of the earth. An engineer is designing a model experiment for flow in a conduit on the moon. The important scaling parameters are the Froude number and the Reynolds number. The model will be full-scale. The kinematic viscosity of the fluid to be used on the moon is 0.5×10^{-5} m^2/s. What should be the kinematic viscosity of the fluid to be used for the model on earth?

8.44 A drying tower at an industrial site is 10 m in diameter. The air inside the tower has a kinematic viscosity of 4×10^{-5} m^2/s and enters at 12 m/s. A 1/15 scale model of this tower is fabricated to operate with water that has a kinematic viscosity of 10^{-6} m^2/s. What should the entry velocity of the water be to achieve Reynolds-number scaling?

8.45 A discharge meter to be used in a 40 cm pipeline carrying oil ($v = 10^{-5}$ m^2/s, $\rho = 860$ kg/m^3) is to be calibrated by means of a model (1/4 scale) carrying water ($T = 20°C$ and standard atmospheric pressure). If the model is operated with a velocity of 1 m/s, find the velocity for the prototype based on Reynolds-number scaling. For the given conditions, if the pressure difference in the model was measured as 3.0 kPa, what pressure difference would you expect for the discharge meter in the oil pipeline?

8.46 Water at 10°C flowing through a rough pipe 10 cm in diameter is to be simulated by air (20°C) flowing through the same pipe. If the velocity of the water is 1.5 m/s, what will the air velocity have to be to achieve dynamic similarity? Assume the absolute air pressure in the pipe to be 150 kPa. If the pressure difference between two sections of the pipe during air flow was measured as 780 Pa, what pressure difference occurs between these two sections when water is flowing under dynamically similar conditions?

8.47 The "noisemaker" B is towed behind the mine-sweeper A to set off enemy acoustic mines such as that shown at C. The drag force of the "noisemaker" is to be studied in a water tunnel at a 1/5 scale (the model is 1/5 the size of the full scale). If the full-scale towing speed is 5 m/s, what should be the water velocity in the water tunnel for the two tests to be exactly similar? What will be the prototype drag force if the model drag force is found to be 2400 N? Assume that sea water at the same temperature is used in both the full-scale and the model tests.

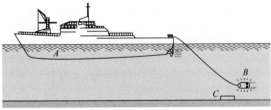

PROBLEM 8.47

8.48 An experiment is being designed to measure aerodynamic forces on a building. The model is a 1/100 scale replica of the prototype. The wind velocity on the prototype is 30 ft/s, and the density is 0.0024 slugs/ft^3. The maximum velocity in the wind tunnel is 300 ft/s. The viscosity of the air flowing for the model and the prototype is the same. Find the density needed in the wind tunnel for dynamic similarity. A force of 50 lbf is measured on the model. What will the force be on the prototype?

8.49 A 60 cm valve is designed for control of flow in a petroleum pipeline. A 1/3 scale model of the full-size valve is to be tested with water in the laboratory. If the prototype flow rate is to be 0.5 m^3/s, what flow rate should be established in the laboratory test for dynamic similitude to be established? Also, if the pressure coefficient C_p in the model is found to be 1.07, what will be the corresponding C_p in the full-scale valve? The relevant fluid properties for the petroleum are S $= 0.82$ and $\mu = 3 \times 10^{-3}$ N $\cdot$ s/m^2. The viscosity of water is 10^{-3} N $\cdot$ s/m^2.

8.50 The moment acting on a submarine rudder is studied by a 1/50 scale model. If the test is made in a water tunnel and if the moment measured on the model is 2 m $\cdot$ N when the fresh-water speed in the tunnel is 10 m/s, what are the corresponding moment and speed for the prototype? Assume the prototype operates in sea water. Assume $T = 10°C$ for both the fresh water and the sea water.

8.51 A model hydrofoil is tested in a water tunnel. For a given angle of attack, the lift of the hydrofoil is measured to be 25 kN when the water velocity is 15 m/s in the tunnel. If the prototype hydrofoil is to be twice the size of the model, what lift force would be expected for the prototype for dynamically similar conditions? Assume a water temperature of 20°C for both model and prototype.

8.52 A 1/8 scale model of an automobile is tested in a pressurized wind tunnel. The test is to simulate the automobile traveling at 80 km/h in air at atmospheric pressure and 25°C. The wind tunnel operates with air at 25°C. At what pressure in the test section must the tunnel operate to have the same Mach and Reynolds numbers? The speed of sound in air at 25°C is 345 m/s.

8.53 If the tunnel in Prob. 8.52 were to operate at atmospheric pressure and 25°C, what speed would be needed to achieve the same Reynolds number for the prototype? At this speed, would you conclude that Mach-number effects were important?

8.54 Experimental studies have shown that the condition for breakup of a droplet in a gas stream is

$$We/Re^{1/2} = 0.5$$

where Re is the Reynolds number and We is the Weber number based on the droplet diameter. What diameter water droplet would break up in a 25 m/s air stream at 20°C and standard atmospheric pressure? The surface tension of water is 7.3×10^{-2} N/m.

8.55 Water is sprayed from a nozzle at 30 m/s into air at atmospheric pressure and 20°C. Estimate the size of the droplets produced if the Weber number for breakup is 6.0 based on the droplet diameter.

8.56 Determine the relationship between the kinematic viscosity ratio v_m/v_P and the scale ratio if both the Reynolds-number and the Froude-number criteria are to be satisfied in a given model test.

8.57 A hydraulic model, 1/20 scale, is built to simulate the flow conditions of a spillway of a dam. For a particular run, the waves downstream were observed to be 8 cm high. How high would be similar waves on the full-scale dam operating under the same conditions? If the wave period in the model is 2 s, what would the wave period in the prototype be?

8.58 The scale ratio between a model dam and its prototype is 1/25. In the model test, the velocity of flow near the crest of the spillway was measured to be 2.5 m/s. What is the corresponding prototype velocity? If the model discharge is 0.10 m^3/s, what is the prototype discharge?

8.59 A seaplane model is built at a 1/10 scale. To simulate take-off conditions at 125 km/h, what should be the corresponding model speed to achieve Froude-number scaling?

8.60 If the scale ratio between a model spillway and its prototype is 1/36, what velocity and discharge ratio will prevail between model and prototype? If the prototype discharge is 3000 m^3/s, what is the model discharge?

8.61 The depth and velocity at a point in a river are measured to be 20 ft and 15 ft/s, respectively. If a 1/64 scale model of this river is constructed and the model is operated under dynamically similar conditions to simulate the free-surface conditions, then what velocity and depth can be expected in the model at the corresponding point?

8.62 A 1/25 scale model of a spillway is tested in a laboratory. If the model velocity and discharge are 7.87 ft/s and 3.53 cfs, respectively, what are the corresponding values for the prototype?

8.63 Flow around a bridge pier is studied using a model at 1/12 scale. When the velocity in the model is 0.9 m/s, the standing wave at the pier nose is observed to be 2.5 cm in height. What are the corresponding values of velocity and wave height in the prototype?

8.64 A 1/25 scale model of a spillway is tested. The discharge in the model is 0.1 m^3/s. To what prototype discharge does this correspond? If it takes 1 min for a particle to float from one point to another in the model, how long would it take a similar particle to traverse the corresponding path in the prototype?

8.65 A tidal estuary is to be modeled at 1/250 scale. In the actual estuary, the maximum water velocity is expected to be 4 m/s, and the tidal period is approximately 12.5 h. What corresponding velocity and period would be observed in the model?

8.66 The maximum wave force on a 1/36 model sea wall was found to be 80 N. For a corresponding wave in the full-scale wall, what full-scale force would you expect? Assume fresh water is used in the model study. Assume $T = 10°C$ for both model and prototype water.

8.67 A model of a spillway is to be built at 1/25 scale. If the prototype has a discharge of 200 m^3/s, what must be the water discharge in the model to ensure dynamic similarity? The total force on part of the model is found to be 22 N. To what prototype force does this correspond?

8.68 A newly designed dam is to be modeled in the laboratory. The prime objective of the general model study is to determine the adequacy of the spillway design and to observe the water velocities, elevations, and pressures at critical points of the structure. The reach of the river to be modeled is 1200 m long, the width of the dam (also the maximum width of the reservoir upstream) is to be 300 m, and the maximum flood discharge to be modeled is 5000 m^3/s. The maximum laboratory discharge is limited to 0.90 m^3/s, and the floor space available for the model construction is 50 m long and 20 m wide. Determine the largest feasible scale ratio (model/prototype) for such a study.

8.69 A ship model 5 ft long is tested in a towing tank at a speed that will produce waves that are dynamically similar to those observed around the prototype. The test speed is 5 ft/s. What should the prototype speed be, given that the prototype length is 150 ft? Assume both the model and the prototype are to operate in fresh water.

8.70 The wave resistance of a model of a ship at 1/25 scale is 2 lbf at a model speed of 5 ft/s. What are the corresponding velocity and wave resistance of the prototype?

8.71 A 1/20 scale model building that is rectangular in plan view and is three times as high as it is wide is tested in a wind tunnel. If the drag of the model in the wind tunnel is measured to be 200 N for a wind speed of 20 m/s, then the prototype building in a 40 m/s wind (same temperature) should have a drag of about (a) 40 kN, (b) 80 kN, (c) 230 kN, or (d) 320 kN.

8.72 A model of a high-rise office building at 1/250 scale is tested in a wind tunnel to estimate the pressures and forces on the full-scale structure. The wind-tunnel air speed is 20 m/s at 20°C and atmospheric pressure, and the full-scale structure is expected to withstand winds of 150 km/h (10°C). If the extreme values of the pressure coefficient are found to be 1.0, –2.7, and –0.8 on the windward wall, side wall, and leeward wall of the model, respectively, what corresponding pressures could be expected to act on the prototype? If the lateral wind force (wind force on building normal to wind direction) was measured as 20 N in the model, what lateral force might be expected in the prototype in the 150 km/h wind?

8.73 Experiments were carried out in a water tunnel and a wind tunnel to measure the drag force on an object. The water tunnel was operated with fresh water at 20°C, and the wind tunnel was operated at 20°C and atmospheric pressure. Three models were used with dimensions of 5 cm, 8 cm, and 15 cm. The drag force on each model was measured at different velocities. The following data were obtained.

Data for the water tunnel

Model Size, cm	Velocity, m/s	Force, N
5	1.0	0.064
5	4.0	0.69
5	8.0	2.20
8	1.0	0.135
8	4.0	1.52
8	8.0	4.52

Data for the wind tunnel

Model Size, cm	Velocity, m/s	Force, N
8	10	0.025
8	40	0.21
8	80	0.64
15	10	0.06
15	40	0.59
15	80	1.82

The drag force is a function of the density, viscosity, velocity, and model size,

$$F_D = f(\rho, \mu, V, D)$$

Using dimensional analysis, express this equation using π-groups and then write a computer program or use a spread sheet to reduce the data. Plot the data using the dimensionless parameters.

8.74 Experiments are performed to measure the pressure drop in a pipe with water at 20°C and crude oil at the same temperature. Data are gathered with pipes of two diameters, 5 cm and 10 cm. The following data were obtained for pressure drop per unit length.

For water

Pipe Diameter, cm	Velocity, m/s	Pressure Drop, N/m^3
5	1	210
5	2	730
5	5	3750
10	1	86
10	2	320
10	5	1650

For crude oil

Pipe Diameter, cm	Velocity, m/s	Pressure Drop, N/m^3
5	1	310
5	2	1040
5	5	5300
10	1	130
10	2	450
10	5	2210

The pressure drop per unit length is assumed to be a function of the pipe diameter, liquid density and viscosity, and the velocity,

$$\frac{\Delta p}{L} = f(\rho, \mu, V, D)$$

Perform a dimensional analysis to obtain the π-groups and then write a computer program or use a spreadsheet to reduce the data. Plot the results using the dimensionless parameters.

9

Surface Resistance

SIGNIFICANT LEARNING OUTCOMES

Conceptual Knowledge

- Identify Couette and Hele-Shaw flows.
- Distinguish between the laminar and turbulent boundary layer.
- Sketch the development of a boundary layer on a flat plate showing main features.
- Explain the meaning of the boundary-layer thickness.
- Distinguish between the local shear stress and average shear stress coefficients.
- Explain the process of boundary-layer separation.

Procedural Knowledge

- Calculate shear stress in Couette flow.
- Calculate flow rate or pressure gradient in Hele-Shaw flow.
- Calculate the boundary-layer thickness.
- Calculate local shear stress and overall resistance for laminar and turbulent boundary layers.

Typical Applications

- For Couette flow, calculate surface resistance.
- For Hele-Shaw flow, determine the flow rate.
- For flow over a flat plate, find the shear stress as a function of Reynolds number.

Viscous stresses create resistance to motion, or drag, as a body travels through a fluid. Aeronautical engineers and naval architects are vitally interested in the drag on an airplane or the surface resistance of a ship because the success or failure of the craft is directly related to its resistive force. If the force is too large, the craft may be an economic failure because of the excessive costs (initial and operational) of the propulsion system.

The phenomena responsible for shear stress at a surface and the prediction of shear force, or surface resistance, on a flat plate are addressed in this chapter. The discussion will build on the ideas of shear stress, viscosity, and velocity gradients presented in Chapter 2. In addition the concepts of the boundary layer and separation, introduced in Chapter 4, will be further expanded.

9.1

Surface Resistance with Uniform Laminar Flow

In this section two cases of one-dimensional laminar flow with parallel streamlines are introduced, flow between two parallel plates with one plate stationary and the other moving and flow between two stationary parallel plates. The flows are uniform and steady. These flows have important practical applications and illustrate the connections between velocity gradient and shear stress.

Differential Equation for Uniform Laminar Flow

Consider the control volume shown in Fig. 9.1, which is aligned with the flow direction s. The streamlines are inclined at an angle θ with respect to the horizontal plane. The control volume has dimensions $\Delta s \times \Delta y \times$ unity; that is, the control volume has a unit length into the page. By application of the momentum equation, the sum of the forces acting in the s-direction is equal to the net outflow of momentum from the control volume. The flow is uniform, so the outflow of momentum is equal to the inflow and the momentum equation reduces to

$$\sum F_s = 0 \tag{9.1}$$

There are three forces acting on the matter in the control volume: the forces due to pressure, shear stress, and gravity. The net pressure force is

$$p\Delta y - \left(p + \frac{dp}{ds}\Delta s\right)\Delta y = -\frac{dp}{ds}\Delta s\Delta y$$

The net force due to shear stress is

$$\left(\tau + \frac{d\tau}{dy}\Delta y\right)\Delta s - \tau\Delta s = \frac{d\tau}{dy}\Delta y\Delta s$$

The component of gravitational force is $\rho g\Delta s\Delta y\sin\theta$. However, $\sin\theta$ can be related to the rate at which the elevation, z, decreases with increasing s and is given by $-dz/ds$. Thus the gravitational force becomes

$$\rho g\Delta s\Delta y\,\sin\theta = -\gamma\Delta y\Delta s\frac{dz}{ds}$$

Summing all the forces to zero and dividing through by $\Delta s\Delta y$ results in

$$\frac{d\tau}{dy} = \frac{d}{ds}(p + \gamma z) \tag{9.2}$$

Figure 9.1

Control volume for analysis of uniform flow with parallel streamlines.

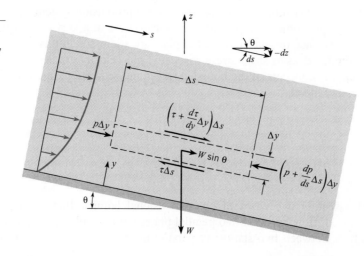

where it is noted that the gradient of the shear stress is equal to the gradient in piezometric pressure in the flow direction. The shear stress is equal to $\mu\, du/dy$, so the basic equation becomes

$$\frac{d^2u}{dy^2} = \frac{1}{\mu}\frac{d}{ds}(p + \gamma z) \tag{9.3}$$

where μ is constant. This equation is now applied to the two flow configurations.

Flow Produced by a Moving Plate (Couette Flow)

Consider the flow between the two plates shown in Fig. 9.2. The lower plate is fixed, and the upper plate is moving with a speed U. The plates are separated by a distance L. In this problem there is no pressure gradient in the flow direction $(dp/ds = 0)$, and the streamlines are in the horizontal direction $(dz/ds = 0)$, so Eq. (9.3) reduces to

$$\frac{d^2u}{dy^2} = 0$$

The two boundary conditions are

$$u = 0 \quad \text{at} \quad y = 0$$
$$u = U \quad \text{at} \quad y = L$$

Integrating this equation twice gives

$$u = C_1 y + C_2$$

Applying the boundary conditions results in

$$u = \frac{y}{L}U \tag{9.4}$$

which shows that the velocity profile is linear between the two plates. The shear stress is constant and equal to

$$\tau = \mu \frac{du}{dy} = \mu \frac{U}{L} \tag{9.5}$$

This flow is known as a *Couette flow* after a French scientist, M. Couette, who did pioneering work on the flow between parallel plates and rotating cylinders. It has application in the design of lubrication systems.

Example 9.1 illustrates the application of Couette flow in calculating surface resistance.

Figure 9.2

Flow generated by a moving plate (Couette flow).

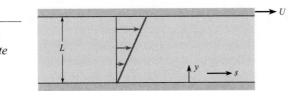

EXAMPLE 9.1   SHEAR STRESS IN COUETTE FLOW

SAE 30 lubricating oil at $T = 38°C$ flows between two parallel plates, one fixed and the other moving at 1.0 m/s. Plates are spaced 0.3 mm apart. What is the shear stress on the plates?

Problem Definition

Situation: SAE 30 lubricating oil between parallel plates.

Find: Shear stress (in N/m²) on top plate.

Properties: From Table A.4, $\mu = 1.0 \times 10^{-1}$ N·s/m².

Sketch:

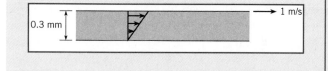

Plan

Calculate shear stress using Eq. (9.5).

Solution

$$\tau = \mu \frac{du}{dy} = \mu \frac{U}{L}$$

$$= (1.0 \times 10^{-1} \text{ N·s/m}^2)(1.0 \text{ m/s})/(3 \times 10^{-4} \text{m})$$

$$\tau = \boxed{333 \text{ N/m}^2}$$

Review

Because the velocity gradient is constant, the shear stress is constant throughout the flow. The magnitude of the shear stress is the same for the bottom plate as the top plate.

Flow Between Stationary Parallel Plates (Hele-Shaw Flow)

Consider the two parallel plates separated by a distance B in Fig. 9.3. In this case, the flow velocity is zero at the surface of both plates, so the boundary conditions for Eq. (9.3) are

$$u = 0 \quad \text{at} \quad y = 0$$
$$u = 0 \quad \text{at} \quad y = B$$

Because the flow is uniform (i.e., there is no change in velocity in the streamwise direction), u is a function of y only. Therefore, d^2u/dy^2 in Eq. (9.3), as well as the gradient in piezometric pressure, must also be equal to a constant in the streamwise direction. Integrating Eq. (9.3) twice gives

$$u = \frac{y^2}{2\mu} \frac{d}{ds}(p + \gamma z) + C_1 y + C_2$$

To satisfy the boundary condition at $y = 0$, set $C_2 = 0$. Applying the boundary condition at $y = B$ requires that C_1 be

$$C_1 = -\frac{B}{2\mu} \frac{d}{ds}(p + \gamma z)$$

so the final equation for the velocity is

$$u = -\frac{1}{2\mu} \frac{d}{ds}(p + \gamma z)(By - y^2) = -\frac{\gamma}{2\mu}(By - y^2)\frac{d(p + \gamma z)}{ds} \tag{9.6}$$

which is a parabolic profile with the maximum velocity occurring on the centerline between the plates, as shown in Fig. 9.3. The maximum velocity is

$$u_{\max} = -\left(\frac{B^2}{8\mu}\right)\frac{d}{ds}(p + \gamma z) \tag{9.7a}$$

Figure 9.3

Uniform flow between two stationary plates (Hele-Shaw flow).

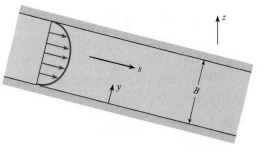

or in terms of piezometric head

$$u_{\max} = -\left(\frac{B^2\gamma}{8\mu}\right)\frac{dh}{ds} \tag{9.7b}$$

The fluid always flows in the direction of decreasing piezometric pressure or piezometric head, so dh/ds is negative, giving a positive value for $u_{\max}$.

The discharge per unit width, q, is obtained by integrating the velocity over the distance between the plates:

$$q = \int_0^B u\,dy = -\left(\frac{B^3}{12\mu}\right)\frac{d}{ds}(p + \gamma z) = -\left(\frac{B^3\gamma}{12\mu}\right)\frac{dh}{ds} \tag{9.8}$$

The average velocity is

$$V = \frac{q}{B} = -\left(\frac{B^2}{12\mu}\right)\frac{d}{ds}(p + \gamma z) = \frac{2}{3}u_{\max} \tag{9.9}$$

Note that flow is the result of a change of the piezometric head, not just a change of p or z alone. Experiments reveal that if the Reynolds number (VB/ν) is less than 1000, the flow is laminar. For a Reynolds number greater than 1000, the flow may be turbulent and the equations developed in this section are invalid.

The flow between parallel plates is often called *Hele-Shaw flow*. It has application in flow visualization studies and in microchannel flows.

A significant difference between Couette flow and Hele-Shaw flow is that the motion of a plate is responsible for Couette flow, whereas a gradient in piezometric pressure provides the force to move a Hele-Shaw flow.

Example 9.2 illustrates how to calculate the pressure gradient required for flow between two parallel plates.

EXAMPLE 9.2   PRESSURE GRADIENT FOR FLOW BETWEEN PARALLEL PLATES

Oil having a specific gravity of 0.8 and a viscosity of 2×10^{-2} N · s/m^2 flows downward between two vertical smooth plates spaced 10 mm apart. If the discharge per meter of width is 0.01 m^2/s, what is the pressure gradient dp/ds for this flow?

Problem Definition

Situation: Oil flows downward between two vertical smooth plates spaced 10 mm apart. The discharge per meter of width is 0.01 (m^2/s).

Find: Pressure gradient dp/ds (in Pa/m) for this flow.

Sketch:

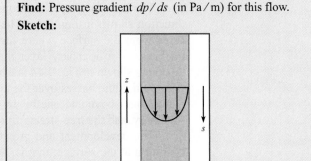

Properties: $S = 0.8$, $\mu = 2 \times 10^{-2}$ N $\cdot$ s/m^2.

Plan

1. Check to see if the flow is laminar $VB/\nu < 1000$. If it is laminar, continue.
2. Calculate piezometric head gradient using Eq. (9.8).
3. Subtract elevation gradient to obtain the pressure gradient.

Solution

1. Laminar flow condition is

$$\text{Re} = \frac{VB}{\nu} = \frac{VB\rho}{\mu} = \frac{q\rho}{\mu}$$

$$= \frac{(0.01 \text{ m}^2/\text{s}) \times 800 \text{ kg/m}^3}{0.02 \text{ N} \cdot \text{s/m}^2} = 400$$

$VB/\nu < 1000$. Flow is laminar, equations apply.

2. Kinematic viscosity:

$$\nu = \mu/\rho = \frac{2 \times 10^{-2} \text{ N} \cdot \text{s/m}^2}{0.8 \times 1000 \text{ kg/m}^3} = 2.5 \times 10^{-5} \text{ m}^2/\text{s}$$

Piezometric head gradient is

$$\frac{dh}{ds} = -\frac{12\mu}{B^3 \gamma}q = -\frac{12\nu}{B^3 g}q$$

$$\frac{dh}{ds} = -\frac{12 \times 2.5 \times 10^{-5} \text{m}^2/\text{s}}{(0.01 \text{ m})^3 \times 9.81 \text{ m}^4/\text{s}^2} \times 0.01 \text{m}^2/\text{s} = -0.306$$

3. Plates are oriented vertically, s *is* positive downward, so $dz/ds = -1$. Thus

$$\frac{dh}{ds} = \frac{d}{ds}\left(\frac{p}{\gamma}\right) + \frac{dz}{ds}$$

$$\frac{d}{ds}\left(\frac{p}{\gamma}\right) = \frac{dh}{ds} - \frac{dz}{ds} = -0.306 + 1 = 0.694$$

or

$$\frac{dp}{ds} = (0.8 \times 9810 \text{ N/m}^3) \times 0.694 = \boxed{5450 \text{ N/m}^2 \text{ per meter}}$$

Review

Note that the pressure increases in the downward direction, which means that the pressure, in part, supports the weight of the fluid.

9.2

Qualitative Description of the Boundary Layer

The purpose of this section is to provide a qualitative description of the *boundary layer*, which is the region adjacent to a surface over which the velocity changes from the free-stream value (with respect to the object) to zero at the surface. This region, which is generally very thin, occurs because of the viscosity of the fluid. The velocity gradient at the surface is responsible for the viscous shear stress and surface resistance.

The boundary-layer development for flow past a thin plate oriented parallel to the flow direction shown in Fig. 9.4*a*. The thickness of the boundary layer, δ, is defined as the distance from the surface where the velocity is 99% of the free-stream velocity. The actual thickness of a boundary layer may be 2%–3% of the plate length, so the boundary-layer thickness shown in Fig. 9.4*a* is exaggerated at least by a factor of five to show details of the flow field. Fluid passes over the top and underneath the plate, so two boundary layers are depicted (one above and one below the plate). For convenience, the surface is assumed to be stationary, and the free-stream fluid is moving at a velocity U_o.

The development and growth of the boundary layer occurs because of the "no-slip" condition at the surface; that is, the fluid velocity at the surface must be zero. As the fluid particles next to the plate pass close to the leading edge of the plate, a retarding force (from the shear stress) begins to act on the particles to slow them down. As these particles progress farther downstream, they continue to be subjected to shear stress from the plate, so they

Figure 9.4

*Development of
boundary layer and
shear stress along a thin,
flat plate.
(a) Flow pattern above
and below the plate.
(b) Shear-stress
distribution on either
side of plate.*

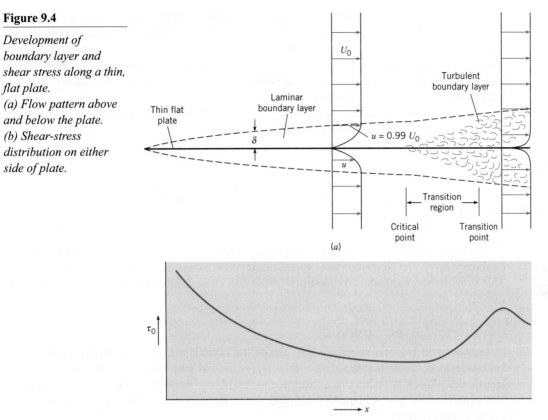

continue to decelerate. In addition, these particles (because of their lower velocity) retard other particles adjacent to them but farther out from the plate. Thus the boundary layer becomes thicker, or "grows," in the downstream direction. The broken line in Fig. 9.4a identifies the outer limit of the boundary layer. As the boundary layer becomes thicker, the velocity gradient at the wall becomes smaller and the local shear stress is reduced.

The initial section of the boundary layer is the laminar boundary layer. In this region the flow is smooth and steady. Thickening of the laminar boundary layer continues smoothly in the downstream direction until a point is reached where the boundary layer becomes unstable. Beyond this point, the critical point, small disturbances in the flow will grow and spread, leading to turbulence. The boundary becomes fully turbulent at the transition point. The region between the critical point and the transition point is called the transition region.

In most problems of practical interest, the extent of the laminar boundary layer is small and contributes little to the total drag force on a body. Still it is important for flow of very viscous liquids and for flow problems with small length scales.

The turbulent boundary layer is characterized by intense cross-stream mixing as turbulent eddies transport high-velocity fluid from the boundary layer edge to the region close to the wall. This cross-stream mixing gives rise to a high effective viscosity, which can be three orders of magnitude higher than the actual viscosity of the fluid itself. The effective viscosity, due to turbulent mixing is not a property of the fluid but rather a property of the flow, namely, the mixing process. Because of this intense mixing, the velocity profile is much "fuller" than the laminar-flow velocity profile as shown in Fig. 9.4a. This situation leads to an increased velocity gradient at the surface and a larger shear stress.

The shear-stress distribution along the plate is shown in Fig. 9.4b. It is easy to visualize that the shear stress must be relatively large near the leading edge of the plate where the velocity gradient is steep, and that it becomes progressively smaller as the boundary layer thickens in the downstream direction. At the point where the boundary layer becomes turbulent, the shear stress at the boundary increases because the velocity profile changes producing a steeper gradient at the surface.

These qualitative aspects of the boundary layer serve as a foundation for the quantitative relations presented in the next section.

9.3

Laminar Boundary Layer

This section summarizes the equations for the velocity profile and shear stress in a laminar boundary layer and describes how to calculate shear stress and shear forces on a surface. This information can be used to estimate drag forces on surfaces in low Reynolds-number flows.

Boundary-Layer Equations

In 1904 Prandtl (1) first stated the essence of the boundary-layer hypothesis, which is that viscous effects are concentrated in a thin layer of fluid (the boundary layer) next to solid boundaries. Along with his discussion of the qualitative aspects of the boundary layer, he also simplified the general equations of motion of a fluid (Navier-Stokes equations) for application to the boundary layer.

In 1908, Blasius, one of Prandtl's students, obtained a solution for the flow in a laminar boundary layer (2) on a flat plate with a constant free-stream velocity. One of Blasius's key assumptions was that the shape of the nondimensional velocity distribution did not vary from section to section along the plate. That is, he assumed that a plot of the relative velocity, u/U_0, versus the relative distance from the boundary, y/δ, would be the same at each section. With this assumption and with Prandtl's equations of motion for boundary layers, Blasius obtained a numerical solution for the relative velocity distribution, shown in Fig. 9.5.* In this plot, x is the distance from the leading edge of the plate, and Re_x is the Reynolds number based on the free-stream velocity and the length along the plate ($\mathrm{Re}_x = U_0 x / v$). In Fig. 9.5 the outer limit of the boundary layer ($u/U_0 = 0.99$) occurs at approximately $y\mathrm{Re}_x^{1/2}/x = 5$. Since $y = \delta$ at this point, the following relationship is derived for the *boundary-layer thickness* in laminar flow on a flat plate:

$$\frac{\delta}{x}\mathrm{Re}_x^{1/2} = 5 \quad \text{or} \quad \delta = \frac{5x}{\mathrm{Re}_x^{1/2}} \tag{9.10}$$

The Blasius solution also showed that

$$\left.\frac{d(u/U_0)}{d[(y/x)\mathrm{Re}_x^{1/2}]}\right|_{y=0} = 0.332$$

---

* Experimental evidence indicates that the Blasius solution is valid except very near the leading edge of the plate. In the vicinity of the leading edge, an error results because of certain simplifying assumptions. However, the discrepancy is not significant for most engineering problems.

Figure 9.5

Velocity distribution in laminar boundary layer. [After Blasius (2).]

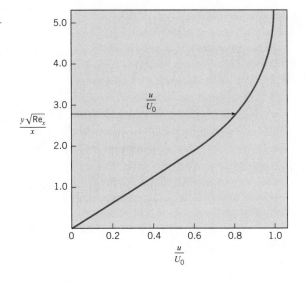

which can be used to find the shear stress at the surface. The velocity gradient at the boundary becomes

$$\left.\frac{du}{dy}\right|_{y=0} = 0.332\frac{U_0}{x}\mathrm{Re}_x^{1/2}$$

$$\left.\frac{du}{dy}\right|_{y=0} = 0.332\frac{U_0^{3/2}}{x^{1/2}\nu^{1/2}} \tag{9.11}$$

Equation (9.11) shows that the velocity gradient (and shear stress) decreases with increasing distance x along the plate.

Shear Stress

The shear stress at the boundary is obtained from

$$\tau_0 = \mu\left.\frac{du}{dy}\right|_{y=0} = 0.332\mu\frac{U_0}{x}\mathrm{Re}_x^{1/2} \tag{9.12}$$

Equation (9.12) is used to obtain the local shear stress at any given section (any given value of x) for the laminar boundary layer.

Example 9.3 illustrates the application of the laminar boundary layer equations for calculating boundary layer thickness and shear stress.

Surface Resistance

Consider one side of a flat plate with width B and length L. Because the shear stress at the boundary, τ_0, varies along the plate, it is necessary to integrate this stress over the entire surface to obtain the total surface resistance, F_s.

$$F_s = \int_0^L \tau_0 B\, dx \tag{9.13}$$

EXAMPLE 9.3 LAMINAR BOUNDARY-LAYER THICKNESS AND SHEAR STRESS

Crude oil at 70°F ($\nu = 10^{-4}$ ft²/s, S = 0.86) with a free-stream velocity of 1 ft/s flows past a thin, flat plate that is 4 ft wide and 6 ft long in a direction parallel to the flow. The flow is laminar. Determine and plot the boundary-layer thickness and the shear stress distribution along the plate.

Problem Definition

Situation: Crude oil flows past a thin, flat plate. Free-stream velocity is 1 ft/s. plate.

Find:

1. Surface shear stress, τ_o, as function of x.
2. Boundary-layer thickness, δ, as function of x.

Properties: For oil, $\nu = 10^{-4}$ ft²/s,   S = 0.86

Sketch:

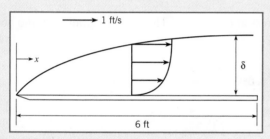

1 ft/s

x

δ

6 ft

Assumptions:

1. Plate is smooth, flat with sharp leading edge.
2. Boundary layer is laminar.

Plan

1. Calculate boundary-layer thickness with Eq. (9.10).
2. Calculate shear-stress distribution with Eq. (9.12).

Solution

1. Reynolds-number variation with distance

$$\text{Re}_x = \frac{U_0 x}{\nu} = \frac{1 \times x}{10^{-4}} = 10^4 x$$

Boundary-layer thickness

$$\delta = \frac{5x}{\text{Re}_x^{1/2}} = \frac{5x}{10^2 x^{1/2}} = 5 \times 10^{-2} x^{1/2} \text{ ft}$$

2. Shear-stress distribution

$$\tau_0 = 0.332 \mu \frac{U_0}{x} \text{Re}_x^{1/2}$$

$$\mu = \rho\nu = 1.94 \text{ slugs/ft}^3 \times 0.86 \times 10^{-4} \text{ ft}^3/s$$
$$= 1.67 \times 10^{-4} \text{ lbf-s/ft}^2$$

$$\tau_0 = 0.332(1.67 \times 10^{-4})\frac{1}{x}(10^2 x^{1/2}) = \frac{5.54 \times 10^{-3}}{x^{1/2}} \text{ psf}$$

The results for Example 9.3 are plotted in the accompanying figure and listed in Table 9.1.

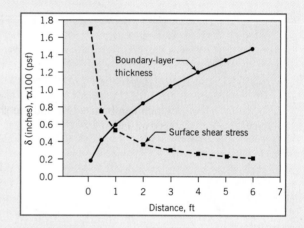

Table 9.1   RESULTS—δ AND τ_0 FOR DIFFERENT VALUES OF x

	x = 0.1 ft	x = 1.0 ft	x = 2 ft	x = 4 ft	x = 6 ft
$x^{1/2}$	0.316	1.00	1.414	2.00	2.45
τ_0, psf	0.018	0.0055	0.0037	0.0028	0.0023
δ, ft	0.016	0.050	0.071	0.10	0.122
δ, in	0.190	0.600	0.848	1.200	1.470

Review

Notice that the boundary-layer thickness increases with distance and that it is very thin. At the end of the plate $\delta/x = 0.02$, or the boundary-layer thickness is 2% of the distance from leading edge.

Notice also that shear stress decreases with distance from leading edge of the plate.

Substituting in Eq. (9.12) for τ_0 and integrating gives

$$F_s = \int_0^L 0.332 B\mu \frac{U_0 U_0^{1/2} x^{1/2}}{x\nu^{1/2}} \, dx$$

$$= 0.664 B\mu U_0 \frac{U_0^{1/2} L^{1/2}}{\nu^{1/2}} \qquad\qquad (9.14)$$

$$= 0.664 B\mu U_0 \mathrm{Re}_L^{1/2}$$

where Re_L is the Reynolds number based on the approach velocity and the length of the plate.

Shear-Stress Coefficients

It is convenient to express the shear stress at the boundary, τ_0, and the total shearing force F_s in terms of π-groups involving the kinetic pressure of the free stream, $\rho U_0^2/2$. The *local shear-stress coefficient*, c_f, is defined as

$$c_f = \frac{\tau_0}{\rho U_0^2/2} \qquad\qquad (9.15)$$

Substituting Eq. (9.12) into Eq. (9.15) gives c_f as a function of Reynolds number based on the distance from the leading edge.

$$c_f = \frac{0.664}{\mathrm{Re}_x^{1/2}} \qquad \text{where} \qquad Re_x = \frac{Ux}{\nu} \qquad\qquad (9.16)$$

The total shearing force, as given by Eq. (9.13), can also be expressed as a π-group

$$C_f = \frac{F_s}{(\rho U_0^2/2)A} \qquad\qquad (9.17)$$

where A is the plate area. This π-group is called the average shear-stress coefficient. Substituting Eq. (9.14) into Eq. (9.17) gives C_f:

$$C_f = \frac{1.33}{\mathrm{Re}_L^{1/2}} \qquad \text{where} \qquad Re_L = \frac{UL}{\nu} \qquad\qquad (9.18)$$

Example 9.4 shows how to calculate the total surface resistance for a laminar boundary layer on a flat plate.

EXAMPLE 9.4   RESISTANCE CALCULATION FOR LAMINAR BOUNDARY LAYER ON A FLAT PLATE

Crude oil at 70°F ($\nu = 10^{-4}$ ft^2/s, S $= 0.86$.) with a free-stream velocity of 1 ft/s flows past a thin, flat plate that is 4 ft wide and 6 ft long in a direction parallel to the flow. The flow is laminar. Determine the resistance on one side of the plate.

Problem Definition

Situation: Crude oil flows past a thin, flat plate. Free-stream velocity is 1 ft/s.

Find: Shear resistance (in lbf) on one side of plate.

Properties: For oil, $\nu = 10^{-4}$ ft^2/s, S $= 0.86$.

Assumptions: Flow is laminar.

Sketch:

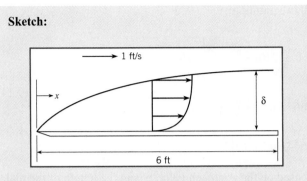

Plan

1. Calculate the Reynolds number based on plate length.
2. Evaluate C_f using Eq. (9.18).
3. Calculate total shear force using Eq. (9.17).

Solution

1. Reynolds number.

$$\mathrm{Re} = \frac{U_0 L}{\nu} = \frac{1\ \mathrm{ft/s} \times 6\ \mathrm{ft}}{10^{-4}\ \mathrm{ft^2/s}} = 6 \times 10^4$$

2. Value for C_f:

$$C_f = \frac{1.33}{\mathrm{Re}_L^{1/2}} = \frac{1.33}{\left(6 \times 10^4\right)^{1/2}} = 0.0054$$

3. Total shear force.

$$F_s = \frac{C_f B L \rho U_0^2}{2}$$

$$= 0.0054 \times 4\ \mathrm{ft} \times 6\ \mathrm{ft} \times 0.86$$

$$\times 1.94\ \mathrm{slugs/ft^3} \times \frac{1^2\,\mathrm{ft/s^2}}{2} = \boxed{0.108\ \mathrm{lbf}}$$

Boundary Layer Transition

Transition is the zone where the laminar boundary layer changes into a turbulent boundary layer as shown in Fig. 9.4a. As the laminar boundary layer continues to grow, the viscous stresses are less capable of damping disturbances in the flow. A point is then reached where disturbances occurring in the flow are amplified, leading to turbulence. The critical point occurs at a Reynolds number of about 10^5 ($\mathrm{Re}_{cr} \cong 10^5$) based on the distance from the leading edge. Vortices created near the wall grow and mutually interact, ultimately leading to a fully turbulent boundary layer at the transition point, which nominally occurs at a Reynolds number of 3×10^6 ($\mathrm{Re}_{tr} \cong 3 \times 10^6$). For purposes of simplicity in this text, it will be assumed that the boundary layer changes from laminar to turbulent flow at a Reynolds number 500,000. The details of the transition region can be found in White (3).

Transition to a turbulent boundary layer can be influenced by several other flow conditions, such as free-stream turbulence, pressure gradient, wall roughness, wall heating, and wall cooling. With appropriate roughness elements at the leading edge, the boundary layer can become turbulent at the very beginning of the plate. In this case it is said that the boundary layer is "tripped" at the leading edge.

Turbulent Boundary Layer

Understanding the mechanics of the turbulent boundary layer is important because in the majority of practical problems it is the turbulent boundary layer that is primarily responsible for surface shear force, or surface resistance. In this section the velocity distribution in the turbulent boundary layer on a flat plate oriented parallel to the flow is presented. The correlations for boundary-layer thickness and shear stress are also included.

Velocity Distribution

The velocity distribution in the turbulent boundary layer is more complicated than the laminar boundary layer. The turbulent boundary has three zones of flow that require different equations for the velocity distribution in each zone, as opposed to the single relationship of the laminar boundary layer. Figure 9.6 shows a portion of a turbulent boundary layer in which the three different zones of flow are identified. The zone adjacent to the wall is the viscous sublayer; the zone immediately above the viscous sublayer is the logarithmic region; and, finally, beyond that region is the velocity defect region. Each of these velocity zones will be discussed separately.

Viscous Sublayer

The zone immediately adjacent to the wall is a layer of fluid that is essentially laminar because the presence of the wall dampens the cross-stream mixing and turbulent fluctuations. This very thin layer is called the *viscous sublayer*. This thin layer behaves as a Couette flow introduced in Section 9.1. In the viscous sublayer, τ is virtually constant and equal to the shear stress at the wall, τ_0. Thus $du/dy = \tau_0/\mu$, which on integration yields

$$u = \frac{\tau_0 y}{\mu} \tag{9.19}$$

Dividing the numerator and denominator by ρ gives

$$u = \frac{\tau_0/\rho}{\mu/\rho}y$$

$$\frac{u}{\sqrt{\tau_0/\rho}} = \frac{\sqrt{\tau_0/\rho}}{\nu}y \tag{9.20}$$

The combination of variables $\sqrt{\tau_0/\rho}$ has the dimensions of velocity and recurs again and again in derivations involving boundary-layer theory. It has been given the special name *shear velocity*. The shear velocity (which is also sometimes called *friction velocity*) is symbolized as u_*. Thus, by definition,

$$u_* = \sqrt{\frac{\tau_0}{\rho}} \tag{9.21}$$

Figure 9.6

Sketch of zones in turbulent boundary layer.

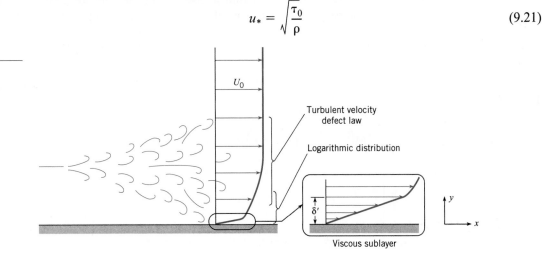

Now, substituting u_* for $\sqrt{\tau_0/\rho}$ in Eq. (9.20), yields the nondimensional velocity distribution in the viscous sublayer:

$$\frac{u}{u_*} = \frac{y}{\nu/u_*} \tag{9.22}$$

Experimental results show that the limit of viscous sublayer occurs when yu_*/ν is approximately 5. Consequently, the thickness of the viscous sublayer, identified by δ', is given as

$$\delta' = \frac{5\nu}{u_*} \tag{9.23}$$

The thickness of the viscous sublayer is very small (typically less than one-tenth the thickness of a dime). The thickness of the viscous sublayer increases as the wall shear stress decreases in the downstream direction.

The Logarithmic Velocity Distribution

The flow zone outside the viscous sublayer is turbulent; therefore, a completely different type of flow is involved. The mixing action of turbulence causes small fluid masses to be swept back and forth in a direction transverse to the mean flow direction. A small mass of fluid swept from a low-velocity zone next to the viscous sublayer into a higher-velocity zone farther out in the stream has a retarding effect on the higher-velocity stream. Similarly, a small mass of fluid that originates farther out in the boundary layer in a high-velocity flow zone and is swept into a region of low velocity has the effect of accelerating the lower-velocity fluid. Although the process just described is primarily a momentum exchange phenomenon, it has the same effect as applying a shear stress to the fluid; thus in turbulent flow these "stresses" are termed *apparent shear stresses*, or *Reynolds stresses* after the British scientist-engineer who first did extensive research in turbulent flow in the late 1800s.

The mixing action of turbulence causes the velocities at a given point in a flow to fluctuate with time. If one places a velocity-sensing device, such as a hot-wire anemometer, in a turbulent flow, one can measure a fluctuating velocity, as illustrated in Fig. 9.7. It is convenient to think of the velocity as composed of two parts: a mean value, $\bar{u}$, plus a fluctuating part, u'. The fluctuating part of the velocity is responsible for the mixing action and the momentum exchange, which manifests itself as an apparent shear stress as noted previously. In fact, the apparent shear stress is related to the fluctuating part of the velocity by

$$\tau_{app} = -\rho\overline{u'v'} \tag{9.24}$$

Figure 9.7

Velocity fluctuations in turbulent flow.

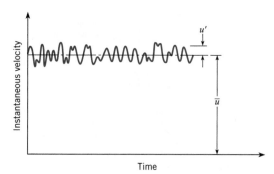

Figure 9.8

Concept of mixing length.

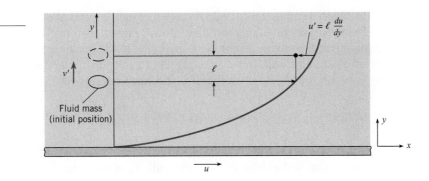

where u' and v' refer to the x and y components of the velocity fluctuations, respectively, and the bar over these terms denotes the product of $u'v'$ averaged over a period of time.* The expression for apparent shear stress is not very useful in this form, so Prandtl developed a theory to relate the apparent shear stress to the temporal mean velocity distribution.

The theory developed by Prandtl is analogous to the idea of molecular transport creating shear stress presented in Chapter 2. In the turbulent boundary layer, the principal flow is parallel to the boundary. However, because of turbulent eddies, there are fluctuating components transverse to the principal flow direction. These fluctuating velocity components are associated with small masses of fluid, as shown in Fig. 9.8, that move across the boundary layer. As the mass moves from the lower-velocity region to the higher-velocity region it tends to retain its original velocity. The difference in velocity between the surrounding fluid and the transported mass is identified as the fluctuating velocity component u'. For the mass shown in Fig. 9.8, u' would be negative and approximated by†

$$u' \approx \ell \frac{du}{dy}$$

where du/dy is the mean velocity gradient and ℓ is the distance the small fluid mass travels in the transverse direction. Prandtl identified this distance as the "mixing length." Prandtl assumed that the magnitude of the transverse fluctuating velocity component is proportional to the magnitude of the fluctuating component in the principal flow direction: $|v'| \cong |u'|$ which seems to be a reasonable assumption because both components arise from the same set of eddies. Also, it should be noted that a positive v' will be associated with a negative u', so the product $\overline{u'v'}$ will be negative. Thus the apparent shear stress can be expressed as

$$\tau_{app} = -\rho\overline{u'v'} = \rho\ell^2\left(\frac{du}{dy}\right)^2 \tag{9.25}$$

A more general form of Eq. (9.25) is

$$\tau_{app} = \rho\ell^2\left|\frac{du}{dy}\right|\frac{du}{dy}$$

which ensures that the sign for the apparent shear stress is correct.

---

* Equation (9.24) can be derived by considering the momentum exchange that results when the transverse component of turbulent flow passes through an area parallel to the x-z plane. Or, by including the fluctuating velocity components in the Navier-Stokes equations, one can obtain the apparent shear stress terms, one of which is Eq. (9.24). Details of these derivations appear in Chapter 18 of Schlichting (4).
† For convenience, the bar used to denote time-averaged velocity is deleted.

The theory leading to Eq. (9.25) is called Prandtl's mixing-length theory and is used extensively in analyses involving turbulent flow.* Prandtl also made the important and clever assumption that the mixing length is proportional to the distance from the wall ($\ell = \kappa y$) for the region close to the wall. If one considers the velocity distribution in a boundary layer where du/dy is positive, as is shown in Fig. 9.8, and substitutes κy for ℓ, then Eq. (9.25) reduces to

$$\tau_{\text{app}} = \rho \kappa^2 y^2 \left(\frac{du}{dy} \right)^2$$

For the zone of flow near the boundary, it is assumed that the shear stress is uniform and approximately equal to the shear stress at the wall. Thus the foregoing equation becomes

$$\tau_0 = \rho \kappa^2 y^2 \left(\frac{du}{dy} \right)^2 \tag{9.26}$$

Taking the square root of each side of Eq. (9.26) and rearranging yields

$$du = \frac{\sqrt{\tau_0/\rho}}{\kappa} \frac{dy}{y}$$

Integrating the above equation and substituting u_* for $\sqrt{\tau_0/\rho}$ gives

$$\frac{u}{u_*} = \frac{1}{\kappa} \ln y + C \tag{9.27}$$

Experiments on smooth boundaries indicate that the constant of integration C can be given in terms of u_*, ν, and a pure number as

$$C = 5.56 - \frac{1}{\kappa} \ln \frac{\nu}{u_*}$$

When this expression for C is substituted into Eq. (9.27), the result is

$$\frac{u}{u_*} = \frac{1}{\kappa} \ln \frac{y u_*}{\nu} + 5.56 \tag{9.28}$$

In Eq. (9.28), κ has sometimes been called the universal turbulence constant, or Karman's constant. Experiments show that this constant is approximately 0.41 (3) for the turbulent zone next to the viscous sublayer. Introducing this value for κ into Eq. (9.28) gives the *logarithmic velocity distribution*

$$\frac{u}{u_*} = 2.44 \ln \frac{y u_*}{\nu} + 5.56 \tag{9.29}$$

Obviously the region where this model is valid is limited because the mixing length cannot continuously increase to the boundary layer edge. This distribution is valid for values of $y u_*/\nu$ ranging from approximately 30 to 500.

The region between the viscous sublayer and the logarithmic velocity distribution is the buffer zone. There is no equation for the velocity distribution in this zone, although various empirical expressions have been developed (6). However, it is common practice to extrapolate the velocity profile for the viscous sublayer to larger values of $y u_*/\nu$ and the logarithmic velocity profile to smaller values of $y u_*/\nu$ until the velocity profiles intersect as shown in Fig. 9.9. The

---

* Prandtl published an account of his mixing-length concept in 1925. G. I. Taylor (5) published a similar concept in 1915, but the idea has been traditionally attributed to Prandtl.

Figure 9.9

Velocity distribution in a turbulent boundary layer.

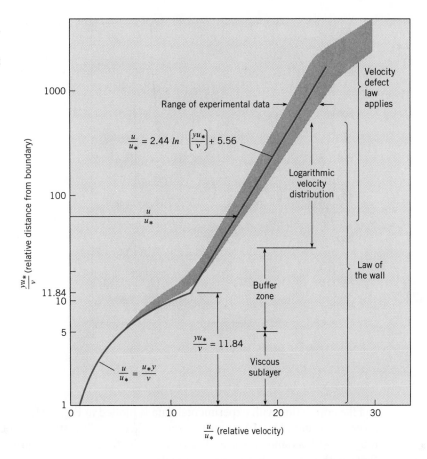

intersection occurs at $yu_*/v = 11.84$ and is regarded as the demarcation between the viscous sublayer and the logarithmic profile. The "nominal" thickness of the viscous sublayer is

$$\delta'_N = 11.84\frac{v}{u_*} \tag{9.30}$$

The combination of the viscous and logarithmic velocity profile for the range of yu_*/v from 0 to approximately 500 is called the *law of the wall*.

Making a semilogarithmic plot of the velocity distribution in a turbulent boundary layer, as shown in Fig. 9.9, makes it straightforward to identify the velocity distribution in the viscous sublayer and in the region where the logarithmic equation applies. However, the logarithmic nature of this plot accentuates the nondimensional distance yu_*/v near the wall. A better perspective of the relative extent of the regions is obtained by plotting the graph on a linear scale, as shown in Fig. 9.10. From this plot one notes that the laminar sublayer and buffer zone are a very small part of the thickness of the turbulent boundary layer.

Velocity Defect Region

For $y/\delta > 0.15$ and $yu_*/v > 500$ the velocity profile corresponding to the law of the wall becomes increasingly inadequate to match experimental data, so a third zone, called the velocity defect region, is identified. The velocity in this region is represented by the *velocity defect law,* which for a flat plate with zero pressure gradient is simply expressed as

$$\frac{U_0 - u}{u_*} = f\left(\frac{y}{\delta}\right) \tag{9.31}$$

Figure 9.10

Figure 9.10

Velocity distribution in a turbulent boundary layer—linear scales.

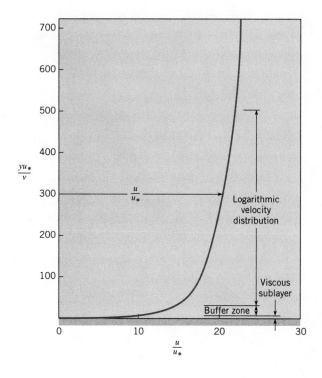

and the correlation with experimental data is plotted in Fig. 9.11. At the edge of the boundary layer $y = \delta$ and $(U_0 - u)/u_* = 0$, so $u = U_0$, or the free-stream velocity. This law applies to rough as well as smooth surfaces. However, the functional relationship has to be modified for flows with free-stream pressure gradients.

Figure 9.11

Velocity defect law for boundary layers on flat plate (zero pressure gradient).
[After Rouse (6).]

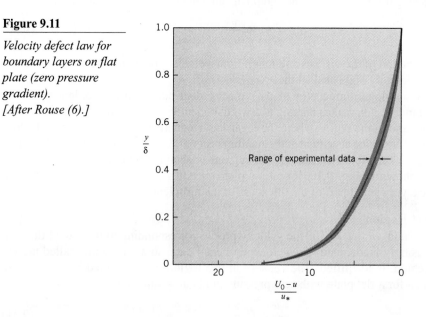

Table 9.2   ZONES FOR TURBULENT BOUNDARY LAYER ON FLAT PLATE

Zone	Velocity Distribution	Range
Viscous Sublayer	$\dfrac{u}{u_*} = \dfrac{yu_*}{\nu}$	$0 < \dfrac{yu_*}{\nu} < 11.84$
Logarithmic Velocity Distribution	$\dfrac{u}{u_*} = 2.44\ln\dfrac{yu_*}{\nu} + 5.56$	$11.84 \leq \dfrac{yu_*}{\nu} < 500$
Velocity Defect Law	$\dfrac{U_0 - u}{u_*} = f\left(\dfrac{y}{\delta}\right)$	$500 \leq \dfrac{yu_*}{\nu}, \dfrac{y}{\delta} > 0.15$

As shown in Fig. 9.9, the demarcation between the law of the wall and the velocity defect regions is somewhat arbitrary, so there is considerable overlap between the two regions. The three zones of the turbulent boundary layer and their range of applicability are summarized in Table 9.2.

Power-Law Formula for Velocity Distribution

Analyses have shown that for a wide range of Reynolds numbers ($10^5 < \text{Re} < 10^7$), the velocity profile in the turbulent boundary layer on a flat plate is approximated reasonably by the *power-law* equation

$$\frac{u}{U_0} = \left(\frac{y}{\delta}\right)^{1/7} \tag{9.32}$$

Comparisons with experimental results show that this formula conforms to those results very closely over about 90% of the boundary layer ($0.1 < y/\delta < 1$). Obviously it is not valid at the surface because $(du)/(dy)|_{y=0} \rightarrow \infty$, which implies infinite surface shear stress. For the inner 10% of the boundary layer, one must resort to equations for the law of the wall (see Fig. 9.9) to obtain a more precise prediction of velocity. Because Eq. (9.32) is valid over the major portion of the boundary layer, it is used to advantage in deriving the overall thickness of the boundary layer as well as other relations for the turbulent boundary layer.

Example 9.5 illustrates the application of various equations to calculate the velocity in the turbulent boundary layer.

EXAMPLE 9.5   TURBULENT BOUNDARY-LAYER PROPERTIES

Water (60°F) flows with a velocity of 20 ft/s past a flat plate. The plate is oriented parallel to the flow. At a particular section downstream of the leading edge of the plate, the boundary layer is turbulent, the shear stress on the plate is 0.896 lbf/ft², and the boundary-layer thickness is 0.0880 ft. Find the velocity of the water at a distance of 0.0088 ft from the plate as determined by
a. The logarithmic velocity distribution
b. The velocity defect law
c. The power-law formula
Also, what is the nominal thickness of the viscous sublayer?

Problem Definition

Situation: Water flows past a flat plate oriented parallel to the flow. At a point downstream of the leading edge of the plate, shear stress on the plate is 0.896 lbf/ft² and boundary layer thickness is 0.0880 ft.

Find:

1. Velocity at 0.0088 ft from plate using
   a. Logarithmic velocity distribution
   b. Velocity defect law
   c. Power-law formula
2. Nominal thickness of viscous sublayer

Properties:
From Table A.5, $\rho = 1.94$ slugs/ft³, $\nu = 1.22 \times 10^{-5}$ ft²/s.

Sketch:

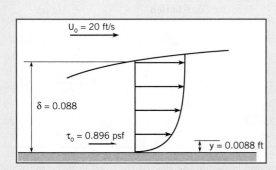

$U_0 = 20$ ft/s

$\delta = 0.088$

$\tau_0 = 0.896$ psf

$y = 0.0088$ ft

Plan

1. Calculate shear velocity, u_*, from Eq. (9.21).
2. Calculate u using Eq. (9.29) for logarithmic profile.
3. Calculate y/δ and find $(U_0 - u)/u_*$ from Fig. 9.11.
4. Calculate u from $(U_0 - u)/u_*$ for velocity defect law.
5. Calculate u from Eq.(9.32) for power law.
6. Calculate δ'_N from Eq. (9.30).

Solution

1. Shear velocity

$$u_* = (\tau_0/\rho)^{1/2}$$

$$= [(0.896 \text{ lbf/ft}^2)/(1.94 \text{slugs/ft}^3)]^{1/2} = 0.680 \text{ ft/s}$$

2. Logarithmic velocity distribution

$$yu_*/\nu = (0.0088 \text{ ft})(0.680 \text{ ft/s})/(1.22 \times 10^{-5} \text{ ft}^2/\text{s}) = 490$$

$$u/u_* = 2.44 \ln(yu_*/\nu) + 5.56$$

$$= 2.44 \times \ln(490) + 5.56 = 20.7$$

$$u = 20.7 \times 0.680 \text{ ft/s} = \boxed{14.1 \text{ ft/s}}$$

3. Nondimensional distance

$$y/\delta = 0.0088 \text{ ft}/0.088 \text{ ft} = 0.10$$

From Fig. 9.11

$$\frac{U_0 - u}{u_*} = 8.2$$

4. Velocity from defect law

$$u = U_0 - 8.2u_*$$

$$= 20 \text{ ft/s} - (8.2)(0.68) \text{ ft/s}$$

$$= \boxed{14.4 \text{ ft/s}}$$

5. Power-law formula

$$u/U_0 = (y/\delta)^{1/7}$$

$$u = (U_0)(0.10)^{1/7}$$

$$= (20 \text{ ft/s})(0.7197)$$

$$= \boxed{14.40 \text{ ft/s}}$$

6. Nominal sublayer thickness

$$\delta'_N = 11.84\nu/u_* = (11.84)(1.22 \times 10^{-5} \text{ ft}^2/\text{s})/(0.68 \text{ ft/s})$$

$$= 2.12 \times 10^{-4} \text{ ft} = \boxed{2.54 \times 10^{-3} \text{ in}}$$

Review

Notice that the velocity obtained using logarithmic distribution and defect law are nearly the same, which indicates that the point is in the overlap region.

Boundary-Layer Thickness and Shear-Stress Correlations

Unlike the laminar boundary layer, there is no analytically derived equation for the thickness of the turbulent boundary layer. There is a way to obtain an equation by using momentum principles, empirical data for the local shear stress, and by assuming the 1/7 power velocity profile (3). The result is

$$\delta = \frac{0.16x}{\text{Re}_x^{1/7}} \tag{9.33}$$

where x is the distance from the leading edge of the plate and Re_x is U_0x/ν.

Many empirical expressions have been proposed for the local shear-stress distribution for the turbulent boundary layer on a flat plate. One of the simplest correlations is

$$c_f = \frac{\tau_0}{\rho U_0^2/2} = \frac{0.027}{\text{Re}_x^{1/7}} \tag{9.34a}$$

and the corresponding average shear-stress coefficient is

$$C_f = \frac{0.032}{\mathrm{Re}_L^{1/7}} \qquad (9.34b)$$

where Re_L is the Reynolds number of the plate based on the length of the plate in the stream-wise direction.

Even though the variation of c_f with Reynolds number given by Eq. (9.34a) provides a reasonably good fit with experimental data for Reynolds numbers less than 10^7, it tends to underpredict the skin friction at higher Reynolds numbers. There are several correlations that have been proposed in the literature; see the review by Schlichting (4). A correlation proposed by White (3) that fits the data for turbulent Reynolds numbers up to 10^{10} is

$$c_f = \frac{0.455}{\ln^2(0.06\mathrm{Re}_x)} \qquad (9.35)$$

The corresponding average shear-stress coefficient is

$$C_f = \frac{0.523}{\ln^2(0.06\mathrm{Re}_L)} \qquad (9.36)$$

These are the correlations for shear-stress coefficients recommended here.

The boundary layer on a flat plate is composed of both a laminar and turbulent part. The purpose here is to develop a correlation valid for the combined boundary layer. As noted in Section 9.3, the boundary layer on a flat plate consists first of a laminar boundary layer that grows in thickness, develops instability, and becomes turbulent. A turbulent boundary layer develops over the remainder of the plate. As discussed earlier in Section 9.4, the transition from a laminar to turbulent boundary layer is not immediate but takes place over a transition length. However for the purposes of analysis here it is assumed that transition occurs at a point corresponding to a transition Reynolds number, Re_{tr}, of about 500,000.

The idea here is to take the turbulent shear force for length L, $F_{s,\,turb}(L)$, assuming the boundary layer is turbulent from the leading edge, subtract the portion up to the transition point, $F_{s,\,turb}(L_{tr})$ and replace it with the laminar shear force up to the transition point $F_{s,\,lam}(L_{tr})$. Thus the composite shear force on the plate is

$$F_s = F_{s,\,turb}(L) - F_{s,\,turb}(L_{tr}) + F_{s,\,lam}(L_{tr})$$

Substituting in Eq. (9.18) for laminar flow and Eq. (9.36) for turbulent flow over a plate of width B gives

$$F_s = \left(\frac{0.523}{\ln^2(0.06\mathrm{Re}_L)}BL - \frac{0.523}{\ln^2(0.06\mathrm{Re}_{tr})}BL_{tr} + \frac{1.33}{\mathrm{Re}_{tr}^{1/2}}BL_{tr} \right)\rho\frac{U_0^2}{2} \qquad (9.37)$$

where Re_{tr} is the Reynolds number at the transition, Re_L is the Reynolds number at the end of the plate, and L_{tr} is the distance from the leading edge of the plate to the transition zone.

Expressing the resistance force in terms of the average shear-stress coefficient, $C_f = F_s/(BL\rho U_0^2/2)$, gives

$$C_f = \frac{0.523}{\ln^2(0.06\mathrm{Re}_L)} + \frac{L_{tr}}{L}\left(\frac{1.33}{\mathrm{Re}_{tr}^{1/2}} - \frac{0.523}{\ln^2(0.06\mathrm{Re}_{tr})} \right)$$

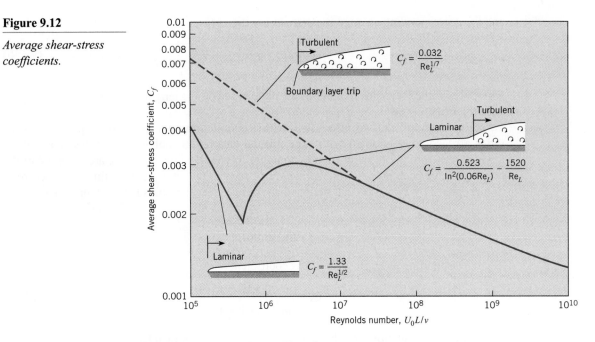

Here $L_{tr}/L = \mathrm{Re}_{tr}/\mathrm{Re}_L$. Therefore,

$$C_f = \frac{0.523}{\ln^2(0.06\mathrm{Re}_L)} + \frac{\mathrm{Re}_{tr}}{\mathrm{Re}_L}\left(\frac{1.33}{\mathrm{Re}_{tr}^{1/2}} - \frac{0.523}{\ln^2(0.06\mathrm{Re}_{tr})}\right)$$

Finally, for $\mathrm{Re}_{tr} = 500{,}000$, the equation for average shear-stress coefficient becomes

$$C_f = \frac{0.523}{\ln^2(0.06\mathrm{Re}_L)} - \frac{1520}{\mathrm{Re}_L} \tag{9.38}$$

The variation of C_f with Reynolds number is shown by the solid line in Fig. 9.12. This curve corresponds to a boundary layer that begins as a laminar boundary layer and then changes to a turbulent boundary layer after the transition Reynolds number. This is the normal condition for a flat-plate boundary layer. Table 9.3 summarizes the equations for boundary-layer-thickness, and for local shear-stress and average shear-stress coefficients for the boundary layer on a flat plate.

Table 9.3 SUMMARY OF EQUATIONS FOR BOUNDARY LAYER ON A FLAT PLATE		
	Laminar Flow $\mathrm{Re}_x, \mathrm{Re}_L < 5\times10^5$	**Turbulent Flow** $\mathrm{Re}_x, \mathrm{Re}_L \geq 5\times10^5$
Boundary-Layer Thickness, δ	$\delta = \dfrac{5x}{\mathrm{Re}_x^{1/2}}$	$\delta = \dfrac{0.16x}{\mathrm{Re}_x^{1/7}}$
Local Shear-Stress Coefficient, c_f	$c_f = \dfrac{0.664}{\mathrm{Re}_x^{1/2}}$	$c_f = \dfrac{0.455}{\ln^2(0.06\mathrm{Re}_x)}$
Average Shear-Stress Coefficient, C_f	$C_f = \dfrac{1.33}{\mathrm{Re}_L^{1/2}}$	$C_f = \dfrac{0.523}{\ln^2(0.06\mathrm{Re}_L)} - \dfrac{1520}{\mathrm{Re}_L}$

Example 9.6 shows the calculation of surface resistance due to a boundary layer on a flat plate.

EXAMPLE 9.6 LAMINAR/TURBULENT BOUNDARY LAYER ON FLAT PLATE

Assume that air 20°C and normal atmospheric pressure flows over a smooth, flat plate with a velocity of 30 m/s. The initial boundary layer is laminar and then becomes turbulent at a transitional Reynolds number of 5×10^5. The plate is 3 m long and 1 m wide. What will be the average resistance coefficient C_f for the plate? Also, what is the total shearing resistance of one side of the plate, and what will be the resistance due to the turbulent part and the laminar part of the boundary layer?

Problem Definition

Situation: Air flows past a flat plate parallel to the flow with a velocity of 30 m/s.

Find:

1. Average shear-stress coefficient, C_f, for the plate.
2. Total shear force (in newtons) on one side of plate.
3. Shear force (in newtons) due to laminar part.
4. Shear force (in newtons) due to turbulent part.

Assumptions: The leading edge of the plate is sharp, and the boundary is not tripped on the leading edge.

Sketch:

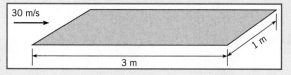

Properties: From Table A.3,

$$\rho = 1.2 \text{ kg/m}^3, \nu = 1.51 \times 10^{-5} \text{ m}^2/\text{s}.$$

Plan

1. Calculate the Reynolds number based on plate length, Re_L.
2. Calculate C_f using Eq. (9.38).
3. Calculate the total shear force on one side of plate using $F_s = (1/2)\rho U_o^2 C_f BL$.
4. Using value for transition Reynolds number, find transition point.

5. Use Eq. (9.18) to find average shear-stress coefficient for laminar portion.
6. Calculate shear force for laminar portion.
7. Subtract laminar portion from total shear force.

Solution

1. Reynolds number based on plate length

$$Re_L = \frac{30 \text{ m/s} \times 3 \text{ m}}{1.51 \times 10^{-5} \text{ m}^2/\text{s}} = 5.96 \times 10^6$$

2. Average shear-stress coefficient

$$C_f = \frac{0.523}{\ln^2(0.06 Re_L)} - \frac{1520}{Re_L} = \boxed{0.00294}$$

3. Total shear force

$$F_s = C_f BL\rho(U_0^2/2)$$

$$= 0.00294 \times 1 \text{ m} \times 3 \text{ m} \times 1.2 \text{ kg/m}^3 \times \frac{(30 \text{ m/s})^2}{2} = \boxed{4.76 \text{ N}}$$

4. Transition point

$$\frac{Ux_{tr}}{\nu} = 500,000$$

$$x_{tr} = \frac{500,000 \times 1.51 \times 10^{-5}}{30} = 0.252 \text{ m}$$

5. Laminar average shear-stress coefficient

$$C_f = \frac{1.33}{Re_{tr}^{1/2}} = 0.00188$$

6. Laminar shear force

$$F_{s, \text{lam}} = 0.00188 \times 1 \text{ m} \times 0.252 \text{ m} \times 1.2 \text{ kg/m}^3 \times \frac{(30 \text{ m/s})^2}{2}$$

$$= \boxed{0.256 \text{ N}}$$

7. Turbulent shear force

$$F_{s, \text{turb}} = 4.76 \text{ N} - 0.26 \text{ N} = \boxed{4.50 \text{ N}}$$

If the boundary layer is "tripped" by some roughness or leading-edge disturbance (such as a wire across the leading edge), the boundary layer is turbulent from the leading edge. This is shown by the dashed line in Fig. 9.12. For this condition the boundary layer thickness, local shear-stress coefficient, and average shear-stress coefficient are fit reasonably well by Eqs. (9.33), (9.34a), and (9.34b).

$$\delta = \frac{0.16x}{\mathrm{Re}_x^{1/7}} \qquad c_f = \frac{0.027}{\mathrm{Re}_x^{1/7}} \qquad C_f = \frac{0.032}{\mathrm{Re}_L^{1/7}} \tag{9.39}$$

which are valid up to a Reynolds number of 10^7. For Reynolds numbers beyond 10^7, the average shear-stress coefficient given by Eq. (9.36) can be used. It is of interest to note that marine engineers incorporate tripping mechanisms for the boundary layer on ship models to produce a boundary layer that can be predicted more precisely than a combination of laminar and turbulent boundary layers.

Example 9.7 illustrates calculating surface resistance with a tripped boundary layer.

EXAMPLE 9.7   RESISTANCE FORCE WITH TRIPPED BOUNDARY LAYER

Air at 20°C flows past a smooth, thin plate with a free-stream velocity of 20 m/s. Plate is 3 m wide and 6 m long in the direction of flow and boundary layer is tripped at the leading edge.

Problem Definition

Situation: Air flows past a smooth, thin plate. Boundary layer is tripped at leading edge.

Find: Total shear resistance (in newtons) on both sides of plate.

Sketch:

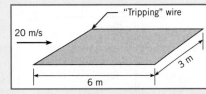

Assumptions: From Table A.3,

$$\rho = 1.2 \text{ kg/m}^3, \quad \mu = 1.81 \times 10^{-5} \text{ N} \cdot \text{s/m}^2.$$

Plan

1. Calculate the Reynolds number based on plate length.
2. Find average shear-stress coefficient from Fig. 9.12.
3. Calculate surface resistance for both sides of plate.

Solution

1. Reynolds number

$$\mathrm{Re}_L = \frac{\rho UL}{\mu} = \frac{1.2 \times 20 \times 6}{1.81 \times 10^{-5}} = 7.96 \times 10^6$$

Reynolds number is less than 10^7.

2. Average shear-stress coefficient

$$C_f = \frac{0.032}{\mathrm{Re}_L^{1/7}}$$

$$= \frac{0.032}{(7.96 \times 10^6)^{1/7}} = 0.0033$$

3. Resistance force

$$F_s = 2 \times C_f A \frac{\rho U_0^2}{2}$$

$$= 0.0033 \times 3 \text{ m} \times 6 \text{ m} \times 1.2 \text{ kg/m}^3 \times (20 \text{ m/s})^2$$

$$= \boxed{28.5 \text{ N}}$$

Even though the equations in this chapter have been developed for flat plates, they are useful for engineering estimates for some surfaces that are not truly flat plates. For example, the skin friction drag of the submerged part of the hull of a ship can be estimated with Eq. (9.38).

9.6

Pressure Gradient Effects on Boundary Layers

In the preceding sections the features of a boundary layer on a flat plate where the external pressure gradient is zero have been presented. The boundary layer begins as laminar, goes through transition, and becomes turbulent with a "fuller" velocity profile and an increase in local shear stress. The purpose of this section is to present some features of the boundary layer over a curved surface where the external pressure gradient is not zero.

The flow over an airfoil section is shown in Fig. 9.13. The variation in static pressure with distance, s, along the surface is also shown on the figure. The point corresponding to $s = 0$ is the forward stagnation point where the pressure is equal to the stagnation pressure. The pressure then decreases toward a minimum value at the midsection. This minimum pressure corresponds to the location of maximum speed as predicted by the Bernoulli equation. The pressure then rises again as the flow accelerates toward the trailing edge. When the pressure decreases with increasing distance ($dp/ds < 0$), the pressure gradient is referred to as a favorable pressure gradient as introduced in Chapter 4. This means that the direction of the force due to the pressure gradient is in the flow direction. In other words, the effect of the pressure gradient is to accelerate the flow. This is the condition between the forward stagnation point and the point of minimum pressure. A rise in pressure with distance ($dp/ds > 0$) is called an adverse pressure gradient and occurs between the point of minimum pressure and the trailing edge. The pressure force due to the adverse pressure gradient acts in the direction opposite to the flow direction and tends to decelerate the flow.

The external pressure gradient effects the properties of the boundary layer. Compared to a flat plate, the laminar boundary layer in a favorable pressure gradient grows more slowly and is more stable. This means that the boundary-layer thickness is less and the local shear stress is increased. Also the transition region is moved downstream, so the boundary layer becomes turbulent somewhat later. Of course, free-stream turbulence and surface roughness will still promote the early transition to a fully turbulent boundary layer.

The effect of external pressure gradient on the boundary layer is most pronounced for the adverse pressure gradient. The development of the velocity profiles for the laminar and turbulent boundary layers in an adverse pressure gradient are shown in Fig. 9.14. The retarding force associated with the adverse pressure gradient decelerates the flow, especially near the surface where the velocities are the lowest. Ultimately there is a reversal of flow at the wall, which gives rise to a recirculatory pattern and the formation of an eddy. This phenomenon is called *boundary-layer separation*. The point of separation is defined where the velocity gradient $\partial u / \partial y$ becomes zero as indicated on the figure. The separation point for the turbulent boundary layer occurs farther downstream because the velocity profile is much fuller (higher velocities persist closer to the wall) than the laminar profile, and it takes longer for the adverse pressure gradient to decelerate the flow. Thus the turbulent boundary layer is less affected by the adverse pressure gradient.

Figure 9.13

Surface pressure distribution on airfoil section.

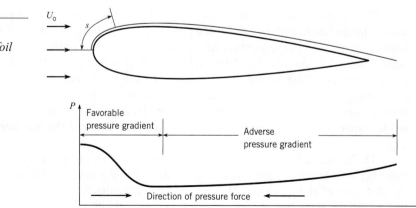

Figure 9.14

Velocity distribution and streamlines for boundary layer separation.
(a) Laminar boundary layer.
(b) Turbulent boundary layer.

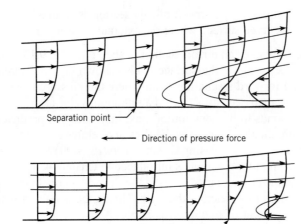

Even though shear stresses on a body in a flow may not contribute significantly to the total drag force, the effect of boundary-layer separation can be very important. When boundary-layer separation takes place on airfoils at a high angle of attack, "stall" occurs, which means the airfoil loses its capability to provide lift. A photograph illustrating boundary-layer separation on an airfoil section is shown in Fig. 4.26. Boundary-layer separation on a cylinder was discussed and illustrated in Section 4.8 on page 111. Understanding and controlling boundary-layer separation is important in the design of fluid dynamic shapes for maximum performance.

9.7

Summary

The variation in velocity for a planar (two-dimensional) steady flow with parallel streamlines is governed by the equation

$$\frac{d^2u}{dy^2} = \frac{1}{\mu}\frac{d}{ds}(p + \gamma z)$$

where the distance y is normal to the streamlines and the distance s is along the streamlines. In this chapter, this equation is used to analyze two flow configurations: Couette flow (flow generated by a moving plate) and Hele-Shaw flow (flow between stationary parallel plates).

The boundary layer is the region where the viscous stresses are responsible for the velocity change between the wall and the free stream. The boundary-layer thickness is the distance from the wall to the location where the velocity is 99% of the free-stream velocity. The laminar boundary layer is characterized by smooth (nonturbulent) flow where the momentum transfer between fluid layers occurs because of the fluid viscosity. As the boundary layer thickness grows, the laminar boundary-layer becomes unstable, and a turbulent boundary layer ensues. The transition point for a boundary layer on a flat plate occurs at a nominal Reynolds number of 5×10^5 based on the free stream velocity and the distance from the leading edge.

The turbulent boundary layer is characterized by an unsteady flow where the momentum exchange between fluid layers occurs because of the mixing of fluid elements normal to the direction of fluid motion. This effect, known as the Reynolds stress, significantly enhances the momentum exchange and leads to a much higher "effective" shear stress.

The local shear-stress coefficient is defined as

$$c_f = \frac{\tau_0}{\frac{1}{2}\rho U_0^2}$$

where τ_0 is the wall shear stress and U_0 is the free-stream velocity. The value for the local shear-stress coefficient on a flat plate depends on the Reynolds number based on the distance from the leading edge. The average shear-stress coefficient is

$$C_f = \frac{F_s}{\frac{1}{2}\rho U_0^2 A}$$

where F_s is the force due to shear-stress, or surface resistance, on a surface with area A. The value for the average shear-stress coefficient for a flat plate depends on the nature of the boundary layer as related to the Reynolds number based on the length of the plate in the flow direction. The laminar boundary layer near the leading edge and the subsequent turbulent boundary layer contribute to the average shear stress on a flat plate. Through leading-edge roughness or other flow disturbance, the boundary layer can be "tripped" at the plate's leading edge, effecting a turbulent boundary layer over the entire plate.

The boundary layer for flow over a curved body is subjected to an external pressure gradient. A favorable pressure gradient produces a force in the flow direction and tends to keep the boundary layer stable. An adverse pressure gradient decelerates the flow and can lead to boundary layer separation.

References

1. Prandtl, L. "Über Flussigkeitsbewegung bei sehr kleiner Reibung." *Verhandlungen des III. Internationalen Mathematiker-Kongresses.* Leipzig, 1905.
2. Blasius, H. "Grenzschichten in Flüssigkeiten mit kleiner Reibung." *Z. Mat. Physik.* (1908), English translation in NACA TM 1256.
3. White, F. M. *Viscous Fluid Flow.* McGraw-Hill, New York, 1991.

4. Schlichting, H. *Boundary Layer Theory.* McGraw-Hill, New York, 1979.
5. Taylor, G. I. "The Transport of Vorticity and Heat through Fluids in Turbulent Motion." *Phil. Trans A,* vol. 215 (1915), p. 1.
6. Rouse, H. and Appel, D. W. *Advanced Mechanics of Fluids.* John Wiley, New York, 1959.

Problems

Uniform Laminar Flow

9.1 ℙℚ◀ The velocity distribution in a Couette flow is linear if the viscosity is constant. If the moving plate is heated and the viscosity of the liquid is decreased near the hot plate, how will the velocity distribution change? Give a qualitative description and the rationale for your argument.

9.2 ℙℚ◀ Consider the flow between two parallel plates. If the viscosity of the fluid is constant along the flow direction, the pressure gradient is linear with distance. How would the pressure gradient differ if the viscosity of the fluid decreased (due to temperature rise) along the flow direction. The density is unchanged. Give a qualitative description of pressure distribution and provide rationale for your answer.

9.3 ℙℚ◀ Consider the flow of a gas between two parallel plates. If there were an increase in temperature due to heat transfer along the flow direction, the gas density would decrase. Assume the viscosity is unaffected. How will the velocity and pressure distribution change from the case with constant density. Sketch the pressure distribution and give the rationale for your result.

9.4 A cube weighing 150 N and measuring 35 cm on a side is allowed to slide down an inclined surface on which there is a film of oil having a viscosity of 10^{-2} N · s/m^2. What is the velocity of the block if the oil has a thickness of 0.1 mm?

9.5 A board 3 ft by 3 ft that weighs 40 lbf slides down an inclined ramp with a velocity of 0.5 fps. The board is separated from the ramp by a layer of oil 0.02 in. thick. Neglecting the edge effects of the board, calculate the approximate dynamic viscosity μ of the oil.

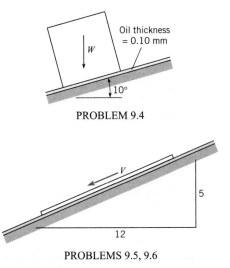

PROBLEM 9.4

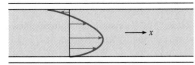

PROBLEMS 9.5, 9.6

9.6 A board 1 m by 1 m that weighs 20 N slides down an inclined ramp with a velocity of 10 cm/s. The board is separated from the ramp by a layer of oil 0.5 mm thick. Neglecting the edge effects of the board, calculate the approximate dynamic viscosity μ of the oil.

9.7 Uniform, steady flow is occurring between horizontal parallel plates as shown.
a. In a few words, tell what other condition must be present to cause the odd velocity distribution.
b. Where is the minimum shear stress located?

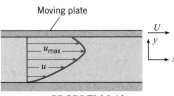

PROBLEM 9.7

9.8 Under certain conditions (pressure decreasing in the x-direction and a moving plate), the laminar velocity distribution will be as shown here. For such a condition, indicate whether each of the following statements is true or false.
a. The greatest shear stress in the liquid occurs next to the fixed plate.
b. The shear stress midway between the plates is zero.
c. The minimum shear stress in the liquid occurs next to the moving plate.
d. The shear stress is greatest where the velocity is the greatest.
e. The minimum shear stress occurs where the velocity is the greatest.

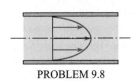

PROBLEM 9.8

9.9 A flat plate is pulled to the right at a speed of 30 cm/s. Oil with a viscosity of 4 N · s/m^2 fills the space between the plate and the solid boundary. The plate is 1 m long ($L = 1$ m) by 30 cm wide, and the spacing between the plate and boundary is 2.0 mm.
a. Express the velocity mathematically in terms of the coordinate system shown.
b. By mathematical means, determine whether this flow is rotational or irrotational.
c. Determine whether continuity is satisfied, using the differential form of the continuity equation.
d. Calculate the force required to produce this plate motion.

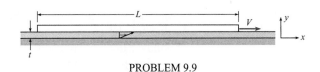

PROBLEM 9.9

9.10 The velocity distribution that is shown represents laminar flow. Indicate which of the following statements are true.
a. The velocity gradient at the boundary is infinitely large.
b. The maximum shear stress in the liquid occurs midway between the walls.
c. The maximum shear stress in the liquid occurs next to the boundary.
d. The flow is irrotational.
e. The flow is rotational.

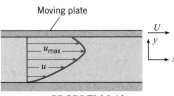

PROBLEM 9.10

9.11 The upper plate shown is moving to the right with a velocity V, and the lower plate is free to move laterally under the action of the viscous forces applied to it. For steady-state conditions, derive an equation for the velocity of the lower plate. Assume that the area of oil contact is the same for the upper plate, each side of the lower plate, and the fixed boundary.

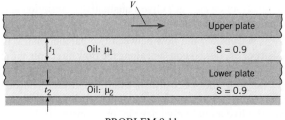

PROBLEM 9.11

9.12 A circular horizontal disk with a 15 cm diameter has a clearance of 2.0 mm from a horizontal plate. What torque is required to rotate the disk about its center at an angular speed of 10 rad/s when the clearance space contains oil ($\mu = 8$ N · s/m²)?

9.13 A plate 2 mm thick and 1 m wide (normal to the page) is pulled between the walls shown in the figure at a speed of 0.40 m/s. Note that the space that is not occupied by the plate is filled with glycerine at a temperature of 20°C. Also, the plate is positioned midway between the walls. Sketch the velocity distribution of the glycerine at section *A-A*. Neglecting the weight of the plate, estimate the force required to pull the plate at the speed given.

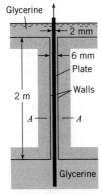

PROBLEM 9.13

9.14 A bearing uses SAE 30 oil with a viscosity of 0.1 N · s/m². The bearing is 30 mm in diameter, and the gap between the shaft and the casing is 1 mm. The bearing has a length of 1 cm. The shaft turns at $\omega = 200$ rad/s. Assuming that the flow between the shaft and the casing is a Couette flow, find the torque required to turn the bearing.

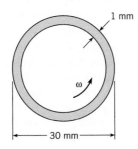

PROBLEM 9.14

9.15 An important application of surface resistance is found in lubrication theory. Consider a shaft that turns inside a stationary cylinder, with a lubricating fluid in the annular region. By considering a system consisting of an annulus of fluid of radius *r* and width Δr, and realizing that under steady-state operation the net torque on this ring is zero, show that $d(r^2\tau)/dr = 0$, where τ is the viscous shear stress. For a flow that has a tangential component of velocity only, the shear stress is related to the velocity by $\tau = \mu r \, d(V/r)/dr$. Show that the torque per unit length acting on

the inner cylinder is given by $T = 4\pi\mu\omega r_s^2 / (1 - r_s^2/r_0^2)$, where ω is the angular velocity of the shaft.

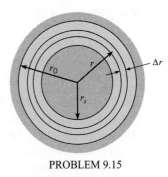

PROBLEM 9.15

9.16 Using the equation developed in Prob. 9.15, find the power necessary to rotate a 2 cm shaft at 60 rad/s if the inside diameter of the casing is 2.2 cm, the bearing is 3 cm long, and SAE 30 oil at 38°C is the lubricating fluid.

9.17 The analysis developed in Prob. 9.15 applies to a device used to measure the viscosity of a fluid. By applying a known torque to the inner cylinder and measuring the angular velocity achieved, one can calculate the viscosity of the fluid. Assume you have a 4 cm inner cylinder and a 4.5 cm outer cylinder. The cylinders are 10 cm long. When a force of 0.6 N is applied to the tangent of the inner cylinder, it rotates at 20 rpm. Calculate the viscosity of the fluid.

9.18 Two horizontal parallel plates are spaced 0.01 ft apart. The pressure decreases at a rate of 12 psf/ft in the horizontal *x*-direction in the fluid between the plates. What is the maximum fluid velocity in the *x* direction? The fluid has a dynamic viscosity of 10^{-3} lbf-s/ft² and a specific gravity of 0.80.

9.19 A viscous fluid fills the space between these two plates, and the pressures at *A* and *B* are 150 psf and 100 psf, respectively. The fluid is not accelerating. If the specific weight of the fluid is 100 lbf/ft³, then one must conclude that (a) flow is downward, (b) flow is upward, or (c) there is no flow.

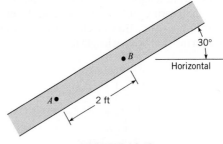

PROBLEM 9.19

9.20 Glycerine at 20°C flows downward between two vertical parallel plates separated by a distance of 0.4 cm. The ends are open, so there is no pressure gradient. Calculate the discharge per unit width, *q*, in m²/s.

9.21 Two vertical parallel plates are spaced 0.01 ft apart. If the pressure decreases at a rate of 60 psf/ft in the vertical z-direction in the fluid between the plates, what is the maximum fluid velocity in the z-direction? The fluid has a viscosity of 10^{-3} lbf-s/ft^2 and a specific gravity of 0.80.

9.22 Two parallel plates are spaced 0.09 in. apart, and motor oil (SAE 30) with a temperature of 100°F flows at a rate of 0.009 cfs per foot of width between the plates. What is the pressure gradient in the direction of flow if the plates are inclined at 60° with the horizontal and if the flow is downward between the plates?

9.23 Glycerine at 20°C flows downward in the annular region between two cylinders. The internal diameter of the outer cylinder is 3 cm, and the external diameter of the inner cylinder is 2.8 cm. The pressure is constant along the flow direction. The flow is laminar. Calculate the discharge. (*Hint:* The flow between the two cylinders can be treated as the flow between two flat plates.)

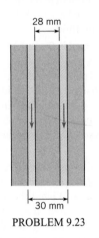

28 mm

30 mm

PROBLEM 9.23

9.24 One type of bearing that can be used to support very large structures is shown in the accompanying figure. Here fluid under pressure is forced from the bearing midpoint (slot A) to the exterior zone B. Thus a pressure distribution occurs as shown. For this bearing, which is 30 cm wide, what discharge of oil from slot A per meter of length of bearing is required? Assume a 50 kN load per meter of bearing length with a clearance space t between the floor and the bearing surface of 0.60 mm. Assume an oil viscosity of 0.20 N · s/m^2. How much oil per hour would have to be pumped per meter of bearing length for the given conditions?

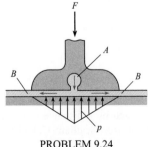

PROBLEM 9.24

9.25 Often in liquid lubrication applications there is a heat generated that is transferred across the lubricating layer. Consider a Couette flow with one wall at a higher temperature than the other. The temperature gradient across the flow affects the fluid viscosity according to the relationship.

$$\mu = \mu_0 \exp\left(-0.1\frac{y}{L}\right)$$

where μ_0 is the viscosity at $y = 0$ and L is the distance between the walls. Incorporate this expression into the Couette flow equation, integrate and express the shear stress in the form

$$\tau = C\frac{U\mu_0}{L}$$

where C is a constant and U is the velocity of the moving wall. Analyze your answer. Should the shear stress be greater or less than that with uniform viscosity?

9.26 Gases form good insulating layers. Consider an application in which there is a Couette flow with the moving plate at a higher temperature than the fixed plate. The viscosity varies between the plates as

$$\mu = \mu_0\left(1 + 0.1\frac{y}{L}\right)^{1/2}$$

where μ_0 is the viscosity at $y = 0$ and L is the distance between the plates. Incorporate this expression into the Couette flow equation, integrate and express the shear stress in the form

$$\tau = C\frac{U\mu_0}{L}$$

where C is a constant and U is the velocity of the moving plate. Analyze your answer. Should the shear stress be greater or less than that with uniform viscosity?

9.27 An engineer is designing a very thin, horizontal channel for cooling electronic circuitry. The channel is 2 cm wide and 5 cm long. The distance between the plates is 0.2 mm. The average velocity is 5 cm/s. The fluid used has a viscosity of 1.2 cp and a density of 800 kg/m^3. Assuming no change in viscosity or density, find the pressure drop in the channel and the power required to move the flow through the channel.

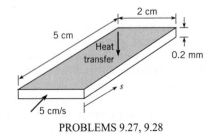

PROBLEMS 9.27, 9.28

9.28 Consider the channel designed for electronic cooling in Prob. 9.27. Because of the heating, the viscosity will change through the channel. Assume the viscosity varies as

$$\mu = \mu_0 \exp\left(-0.1\frac{s}{L}\right)$$

where μ_o is the viscosity at $s = 0$ and L is the length of the channel. Find the percentage change of the pressure drop due to viscosity variation.

Laminar Boundary Layer

9.29 PQ◀ Explain in your own words what is meant by "boundary layer."

9.30 PQ◀ Define "boundary layer thickness."

9.31 PQ◀ List three features of the laminar boundary layer.

9.32 PQ◀ Assume the wall adjacent to a liquid laminar boundary is heated and the viscosity of the fluid is lower near the wall and increases the free-stream value at the edge of the boundary layer. How would this variation in viscosity affect the boundary layer thickness and local shear stress? Give the rationale for your answers.

9.33 A thin plate 6 ft long and 3 ft wide is submerged and held stationary in a stream of water ($T = 60°F$) that has a velocity of 5 ft/s. What is the thickness of the boundary layer on the plate for $\mathrm{Re}_x = 500,000$ (assume the boundary layer is still laminar), and at what distance downstream of the leading edge does this Reynolds number occur? What is the shear stress on the plate at this point?

9.34 What is the ratio of the boundary-layer thickness on a smooth, flat plate to the distance from the leading edge just before transition to turbulent flow?

9.35 A model airplane has a wing span of 3 ft and a chord (leading edge–trailing edge distance) of 5 in. The model flies in air at 60°F and atmospheric pressure. The wing can be regarded as a flat plate so far as drag is concerned. At what speed will a turbulent boundary layer start to develop on the wing? What will be the total drag force on the wing just before turbulence appears?

9.36 Oil ($\mu = 10^{-2}$ N · s/m^2; $\rho = 900$ kg/m^3) flows past a plate in a tangential direction so that a boundary layer develops. If the velocity of approach is 4 m/s, then at a section 30 cm downstream of the leading edge the ratio of τ_δ (shear stress at the edge of the boundary layer) to τ_0 (shear stress at the plate surface) is approximately (a) 0, (b) 0.24, (c) 2.4, or (d) 24.

9.37 A liquid ($\rho = 1000$ kg/m^3; $\mu = 2 \times 10^{-2}$ N · s/m^2; $\nu = 2 \times 10^{-5}$ m^2/s) flows tangentially past a flat plate. If the approach velocity is 1 m/s, what is the liquid velocity 1 m downstream from the leading edge of the plate and 1 mm away from the plate?

9.38 The plate of Prob. 9.37 has a total length of 3 m (parallel to the flow direction), and it is 1 m wide. What is the skin friction drag (shear force) on one side of the plate?

9.39 Oil ($\nu = 10^{-4}$ m^2/s) flows tangentially past a thin plate. If the free-stream velocity is 5 m/s, what is the velocity 1 m downstream from the leading edge and 3 mm away from the plate?

9.40 Oil ($\nu = 10^{-4}$ m^2/s; $S = 0.9$) flows past a plate in a tangential direction so that a boundary layer develops. If the velocity of approach is 1 m/s, what is the oil velocity 1 m downstream from the leading edge and 10 cm away from the plate?

9.41 A thin plate 0.7 m long and 1.5 m wide is submerged and held stationary in a stream of water ($T = 10°C$) that has a velocity of 1.5 m/s. What is the thickness of the boundary layer on the plate for $\mathrm{Re}_x = 500,000$ (assume the boundary layer is still laminar), and at what distance downstream of the leading edge does this Reynolds number occur? What is the shear stress on the plate on this point?

9.42 A flat plate 1.5 m long and 1.0 m wide is towed in water at 20°C in the direction of its length at a speed of 15 cm/s. Determine the resistance of the plate and the boundary layer thickness at its aft end.

9.43 Transition from a laminar to a turbulent boundary-layer occurs between the Reynolds numbers of $\mathrm{Re}_x = 10^5$ and $\mathrm{Re}_x = 3 \times 10^6$. The thickness of the turbulent boundary-layer based on the distance from the leading edge is $\delta = 0.16x/(\mathrm{Re}_x)^{1/7}$. Find the ratio of the thickness of the laminar boundary layer at the beginning of transition to the thickness of the turbulent bounday layer at the end of transition.

Turbulent Boundary Layer

9.44 PQ◀ List three features that describe the difference between the laminar and turbulent boundary layer.

9.45 PQ◀ Assume that a turbulent gas boundary layer was adjacent to a cool wall and the viscosity in the wall region was reduced. How may this affect the features of the boundary layer? Give some rationale for your answers.

9.46 An element for sensing local shear stress is positioned in a flat plate 1 meter from the leading edge. The element simply consists of a small plate, 1 cm × 1 cm, mounted flush with the wall, and the shear force is measured on the plate. The fluid flowing by the plate is air with a free-stream velocity of 30 m/s, a density of 1.2 kg/m^3, and a kinematic viscosity of 1.5×10^{-5} m^2/s. The boundary layer is tripped at the leading edge. What is the magnitude of the force due to shear stress acting on the element?

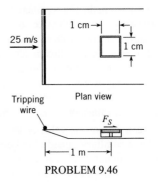

PROBLEM 9.46

9.47 For the conditions of Prob. 9.46, what is the shearing resistance on one side of the plate for the part of the plate that has a Reynolds number, Re_x, less than 500,000? What is the ratio of the laminar shearing force to the total shearing force on the plate?

9.48 An airplane wing of 2 m chord length (leading edge to trailing edge distance) and 11 m span flies at 200 km/hr in air at 30°C. Assume that the resistance of the wing surfaces is like that of a flat plate.
a. What is the friction drag on the wing?
b. What power is required to overcome this?
c. How much of the chord is laminar?
d. What will be the change in drag if a turbulent boundary layer is tripped at the leading edge?

9.49 A turbulent boundary layer exists in the flow of water at 20°C over a flat plate. The local shear stress measured at the surface of the plate is 0.1 N/m². What is the velocity at a point 1 cm from the plate surface?

9.50 A liquid flows tangentially past a flat plate. The fluid properties are $\mu = 10^{-5}$ N·s/m² and $\rho = 1.5$ kg/m³. Find the skin friction drag on the plate per unit width if the plate is 2 m long and the approach velocity is 20 m/s. Also, what is the velocity gradient at a point that is 1 m downstream of the leading edge and just next to the plate ($y = 0$)?

9.51 For the hypothetical boundary layer on the flat plate shown, what is the shear-stress on the plate at the downstream end (point A)? Here $\rho = 1.2$ kg/m³ and $\mu = 1.8 \times 10^{-5}$ N·s/m².

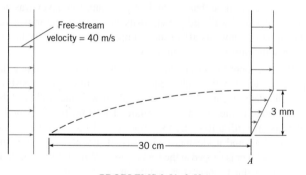

PROBLEMS 9.51, 9.53

9.52 Assume that the velocity profile in a boundary layer is replaced by a step profile, as shown in the figure, where the velocity is zero adjacent to the surface and equal to the free-stream velocity (U) at a distance greater than δ_* from the surface. Assume also that the density is uniform and equal to the free-stream density (ρ_∞). The distance δ_* (displacement thickness) is so chosen that the mass flux corresponding to the step profile is equal to the mass flux through the actual boundary layer. Derive an integral expression for the displacement thickness as a function of u, U, y, and δ.

9.53 Because of the reduction of velocity associated with the boundary layer, the streamlines outside the boundary layer are shifted away from the boundary. This amount of displacement of the streamlines is defined as the displacement thickness δ_*. Using the expression developed in Prob. 9.52, evaluate the displacement

thickness of the boundary layer at the downstream edge of the plate (point A) in Prob. 9.51.

9.54 Use the expression developed in Prob. 9.52 to find the ratio of the displacement thickness to the boundary layer thickness for the turbulent boundary layer profile given by

$$\frac{u}{U_0} = \left(\frac{y}{\delta}\right)^{1/7}$$

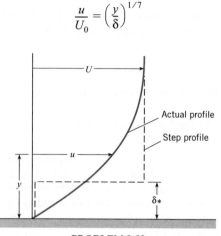

PROBLEM 9.52

9.55 What is the ratio of the skin friction drag of a plate 30 m long and 5 m wide to that of a plate 10 m long and 5 m wide if both plates are towed lengthwise through water ($T = 20$°C) at 10 m/s?

9.56 Estimate the power required to pull the sign shown if it is towed at 40 m/s and if it is assumed that the sign has the same resistance characteristics as a flat plate. Assume standard atmospheric pressure and a temperature of 10°C.

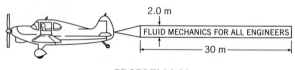

PROBLEM 9.56

9.57 A thin plastic panel (3 mm thick) is lowered from a ship to a construction site on the ocean floor. The plastic panel weighs 300 N in air and is lowered at a rate of 3 m/s. Assuming that the panel remains vertically oriented, calculate the tension in the cable.

9.58 The plate shown in the figure is weighted at the bottom so it will fall stably and steadily in a liquid. The weight of the plate in air is 23.5 N, and the plate has a volume of 0.002 m³. Estimate its falling speed in fresh water at 20°C. The boundary layer is normal; that is, it is not tripped at the leading edge.

In this problem, the final falling speed (terminal velocity) occurs when the weight is equal to the sum of the skin friction and buoyancy.

$$W = B + F_s = \gamma \Psi + \frac{1}{2} C_f \rho U_0^2 S$$

Hint: Find the final falling speed. This problem requires an iterative solution.

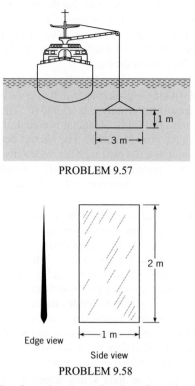

PROBLEM 9.57

PROBLEM 9.58

Edge view / Side view

9.59 A turbulent boundary layer develops from the leading edge of a flat plate with water at 20°C flowing tangentially past the plate with a free-stream velocity of 5 m/s. Determine the thickness of the viscous sublayer, δ', at a distance 1 m downstream from the leading edge.

9.60 A model airplane descends in a vertical dive through air at standard conditions (1 atmosphere and 20°C). The majority of the drag is due to skin friction on the wing (like that on a flat plate). The wing has a span of 1 m (tip to tip) and a chord length (leading edge to trailing edge distance) of 10 cm. The leading edge is rough, so the turbulent boundary layer is "tripped." The model weighs 3 N. Determine the speed (in meters per second) at which the model will fall.

9.61 A flat plate is oriented parallel to a 15 m/s airflow at 20°C and atmospheric pressure. The plate is 1 m long in the flow direction and 0.5 m wide. On one side of the plate, the boundary layer is tripped at the leading edge, and on the other side there is no tripping device. Find the total drag force on the plate.

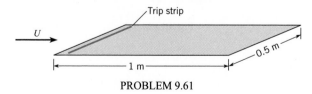

PROBLEM 9.61

9.62 An engineer is designing a horizontal, rectangular conduit that will be part of a system that allows fish to bypass a dam. In-

side the conduit, a flow of water at 40°F will be divided into two streams by a flat, rectangular metal plate. Calculate the viscous drag force on this plate, assuming boundary-layer flow with free-stream velocity of 15 ft/s and plate dimensions of $L = 8$ ft and $W = 4.0$ ft.

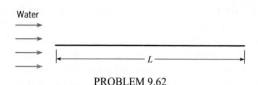

PROBLEM 9.62

9.63 A model is being developed for the entrance region between two flat plates. As shown in the figure, it is assumed that the region is approximated by a turbulent boundary layer originating at the leading edge. The system is designed such that the plates end where the boundary layers merge. The spacing between the plates is 4 mm, and the entrance velocity is 10 m/s. The fluid is water at 20°C. Roughness at the leading edge trips the boundary layers. Find the length L where the boundary layers merge, and find the force per unit depth (into the paper) due to shear stress on both plates.

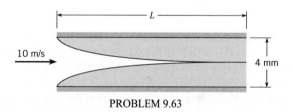

PROBLEM 9.63

9.64 An outboard racing boat "planes" at 70 mph over water at 60°F. The part of the hull in contact with the water has an average width of 3 ft and a length of 8 ft. Estimate the power required to overcome its surface resistance.

9.65 A motor boat pulls a long, smooth, water-soaked log (0.5 m in diameter and 50 m long) at a speed of 1.7 m/s. Assuming total submergence, estimate the force required to overcome the surface resistance of the log. Assume a water temperature of 10°C and that the boundary layer is tripped at the front of the log.

9.66 Modern high-speed passenger trains are streamlined to reduce surface resistance. The cross section of a passenger car of one such train is shown. For a train 150 m long, estimate the surface resistance for a speed of 100 km/hr and for one of 200 km/hr. What power is required for just the surface resistance at these speeds? Assume $T = 10$°C and that the boundary layer is tripped at the front of the train.

9.67 Consider the boundary layer next to the smooth hull of a ship. The ship is cruising at a speed of 45 ft/s in 60°F fresh water. Assuming that the boundary layer on the ship hull develops the same as on a flat plate, determine

a. The thickness of the boundary layer at a distance $x = 100$ ft downstream from the bow.

b. The velocity of the water at a point in the boundary layer at $x = 100$ ft and $y/\delta = 0.50$.

c. The shear stress, τ_0, adjacent to the hull at $x = 100$ ft.

9.68 An Eiffel-type wind tunnel operates by drawing air through a contraction, passing this air through a test section, and then exhausting the air using a large axial fan. Experimental data are recorded in the test section, which is typically a rectangular section of duct that is made of clear plastic (usually acrylic). In the test section, the velocity should have a very uniform distribution; thus, it is important that the boundary layer be very thin at the end of the test section. For the pictured wind tunnel, the test section is square with a dimension of $W = 457$ mm on each side and a length of $L = 914$ mm. Find the ratio of maximum boundary-layer thickness to test section width $[\delta(x = L)/W]$ for two cases: minimum operating velocity (1 m/s) and maximum operating velocity (70 m/s). Assume air properties at 1 atm and 20°C.

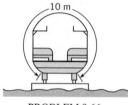

PROBLEM 9.66

9.69 A ship 600 ft long steams at a rate of 25 ft/s through still fresh water ($T = 50°F$). If the submerged area of the ship is 50,000 ft², what is the skin friction drag of this ship?

9.70 A river barge has the dimensions shown. It draws 2 ft of water when empty. Estimate the skin friction drag of the barge when it is being towed at a speed of 10 ft/s through still fresh water at 60°F.

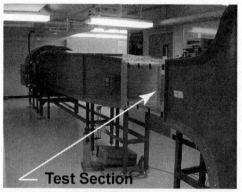

PROBLEM 9.68

9.71 A supertanker has length, breadth, and draught (fully loaded) dimensions of 325 m, 48 m, and 19 m, respectively. In open seas the tanker normally operates at a speed of 18 kt (1 kt = 0.515 m/s).

For these conditions, and assuming that flat-plate boundary-layer conditions are approximated, estimate the skin friction drag of such a ship steaming in 10°C water. What power is required to overcome the skin friction drag? What is the boundary-layer thickness at 300 m from the bow?

9.72 A model test is to be done to predict the wave drag on a ship. The ship is 500 ft long and operates at 30 ft/s in sea water at 10°C. The wetted area of the prototype is 25,000 ft². The model/prototype scale ratio is 1/100. Modeling is done in fresh water at 60°F to match the Froude number. The viscous drag can be calculated by assuming a flat plate with the wetted area of the model and a length corresponding to the length of the model. A total drag of 0.1 lbf is measured in the model tests. Calculate the wave drag on the actual ship.

9.73 A ship is designed so that it is 250 m long, its beam measures 30 m, and its draft is 12 m. The surface area of the ship below the water line is 8800 m². A 1/40 scale model of the ship is tested and is found to have a total drag of 26.0 N when towed at a speed of 1.45 m/s. Using the methods outlined in Section 8.9, answer the following questions, assuming that model tests are made in fresh water (20°C) and that prototype conditions are sea water (10°C).

a. To what speed in the prototype does the 1.45 m/s correspond?

b. What are the model skin friction drag and wave drag?

c. What would the ship drag be in salt water corresponding to the model test conditions in fresh water?

9.74 A hydroplane 3 m long skims across a very calm lake ($T = 20°C$) at a speed of 15 m/s. For this condition, what will be the minimum shear stress along the smooth bottom?

9.75 Estimate the power required to overcome the surface resistance of a water skier if he or she is towed at 30 mph and each ski is 4 ft by 6 in. Assume the water temperature is 60°F.

9.76 If the wetted area of an 80 m ship is 1500 m², approximately how great is the surface drag when the ship is traveling at a speed of 15 m/s. What is the thickness of the boundary layer at the stern? Assume seawater at $T = 10°C$.

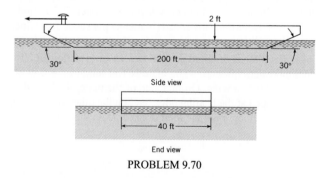

PROBLEM 9.70

CHAPTER 10

Flow in Conduits

The Alaskan pipeline, a significant accomplishment of the engineering profession, transports oil 1286 km across the state of Alaska. The pipe diameter is 1.2 m, and the 44 pumps are used to drive the flow. This chapter presents information for designing systems involving pipes, pumps, and turbines.

SIGNIFICANT LEARNING OUTCOMES

Conceptual Knowledge
- Describe laminar flow, turbulent flow, developing flow, and fully developed flow in a conduit.
- Describe how to characterize total head loss by using component and pipe head loss.
- List the steps used to derive the (a) Darcy-Weisbach equation and (b) Poiseuille flow solution.
- Describe the main features of the Moody diagram.

Procedural Knowledge
- Classify flow as (a) laminar or turbulent and (b) developing or fully developed.
- Using equations or the Moody diagram, find values of the friction factor f.
- Calculate pipe head loss, component head loss, and total head loss.

Typical Applications
- For flow in a pipe, find the pressure drop or head loss.
- For a specified system, find the flow rate.
- For a specified flow rate and pressure drop, determine the size of pipe required.
- For a system with a pump, find the pump specifications (power, head, flow rate).
- For a specified elevation change and flow rate, find the power that can be produced by a turbine.

The fundamentals of the energy equation were presented in Chapter 7, and the fundamentals of shear stress, velocity profiles, and boundary layer in Chapter 9. This chapter combines these ideas to describe flow in conduits. A *conduit* is any pipe, tube, or duct that is completely filled with a flowing fluid. Examples include a pipeline transporting liquefied natural gas, a microchannel transporting hydrogen in a fuel cell, and a duct transporting air for heating of a building. A pipe that is partially filled with a flowing fluid, for example a drainage pipe, is classified as an open-channel flow and will be analyzed in Chapter 15.

The main goal of this chapter is to describe how to predict head loss. Predicting head loss involves classifying flow as laminar or turbulent and then using equations to calculate head losses in pipes and components. This chapter also describes how to use data from a pump manufacturer to select the right size of pump for a given application and how to model a network of pipes.

Classifying Flow

This section describes how to classify flow in a conduit by considering (a) whether the flow is laminar or turbulent, and (b) whether the flow is developing or fully developed. Classifying flow is essential for selecting the proper equation for calculating head loss.

Laminar Flow and Turbulent Flow

Flow in a conduit is classified as being either laminar or turbulent, depending on the magnitude of the Reynolds number. The original research involved visualizing flow in a glass tube as shown in Fig. 10.1a. Reynolds (1) in the 1880s injected dye into the center of the tube and observed the following:

• When the velocity was low, the streak of dye flowed down the tube with little expansion, as shown in Fig. 10.1b. However, if the water in the tank was disturbed, the streak would shift about in the tube.

• If velocity was increased, at some point in the tube, the dye would all at once mix with the water as shown in Fig. 10.1c.

• When the dye exhibited rapid mixing (Fig. 10.1c), illumination with an electric spark revealed eddies in the mixed fluid as shown in Fig. 10.1d.

The flow regimes shown in Fig. 10.1 are laminar flow (Fig. 10.1b) and turbulent flow (Figs. 10.1c and 10.1d). Reynolds showed that the onset of turbulence was related to a π-group that is now called the Reynolds number (Re $= \rho VD/\mu$) in honor of Reynolds' pioneering work. Reynolds discovered that if the fluid in the upstream reservoir was not completely still or if the pipe had some vibrations, then the change from laminar to turbulent flow occurred at Re ~ 2100. However, if conditions were ideal, it was possible to reach a much higher Reynolds number before the flow became turbulent. Reynolds also found that, when going from high velocity to low velocity, the change back to laminar flow occurred at Re ~ 2000. Based

Figure 10.1

Reynolds' experiment.
(a) Apparatus.
(b) Laminar flow of dye in tube.
(c) Turbulent flow of dye in tube.
(d) Eddies in turbulent flow.

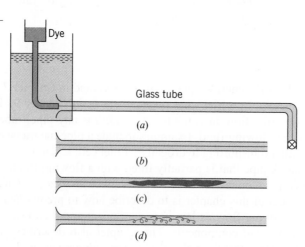

on Reynolds' experiments, engineers use guidelines to establish whether or not flow in a conduit will be laminar or turbulent. The guidelines used in this text are as follows:

$$\begin{array}{ll} \text{Re} \le 2000 & \text{laminar flow} \\ 2000 \le \text{Re} \le 3000 & \text{unpredictable} \\ \text{Re} \ge 3000 & \text{turbulent flow} \end{array} \qquad (10.1)$$

In Eq. (10.1), the middle range $(2000 \le \text{Re} \le 3000)$ corresponds to a the type of flow that is unpredictable because it can changes back and forth between laminar and turbulent states. Recognize that precise values of Reynolds number versus flow regime do not exist. Thus, the guidelines given in Eq. (10.1) are approximate and other references may give slightly different values. For example, some references use $\text{Re} = 2300$ as the criteria for turbulence.

There are several equations for calculating Reynolds number in a pipe.

$$\text{Re} = \frac{VD}{\nu} = \frac{\rho VD}{\mu} = \frac{4Q}{\pi D \nu} = \frac{4\dot{m}}{\pi D \mu} \qquad (10.2)$$

These equations are derived by using the definition of Re, the definition of kinematic viscosity from Eq. (2.8), and the flow rate equations from Eqs. (5.8 and 5.9).

Developing Flow and Fully Developed Flow

Flow in a conduit is classified as being developing flow or fully developed flow. For example, consider laminar fluid entering a pipe from a reservoir as shown in Fig. 10.2. As the fluid moves down the pipe, the velocity distribution changes in the streamwise direction as viscous effects cause the plug-type profile to gradually change into a parabolic profile. This region of changing velocity profile is called *developing flow*. After the parabolic distribution is achieved, the flow profile remains unchanged in the streamwise direction, and flow is called *fully developed flow*.

The distance required for flow to develop is called the *entrance length* (L_e). This length depends on the shear stress that acts on the pipe wall. For laminar flow, the wall shear-stress distribution is shown in Fig. 10.2. Near the pipe entrance, the radial velocity gradient (change in velocity with distance from the wall) is high, so the shear stress is large. As the velocity profile progresses to a parabolic shape, the velocity gradient and the wall shear stress decrease until a constant value is achieved. The entry length is defined as the distance at which the shear stress reaches to within 2% of the fully developed value. Correlations for entry length are

$$\frac{L_e}{D} = 0.05\,\text{Re} \qquad (\text{laminar flow: Re} \le 2000) \qquad (10.3a)$$

$$\frac{L_e}{D} = 50 \qquad (\text{turbulent flow: Re} \ge 3000) \qquad (10.3b)$$

Eq. (10.3) is valid for flow entering a circular pipe from a reservoir under quiescent conditions. Other upstream components such as valves, elbows, and pumps produce complex flow fields that require different lengths to achieve fully developing flow.

In summary, flow in a conduit is classified into four categories: laminar developing, laminar fully developed, turbulent developing, or turbulent fully developed. The key to classification is to calculate the Reynolds number as shown by Example 10.1.

Figure 10.2

In developing flow, the wall shear stress is changing. In fully developed flow, the wall shear stress is constant.

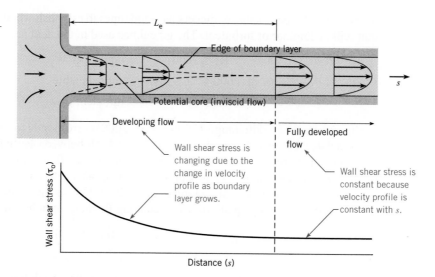

EXAMPLE 10.1 CLASSIFYING FLOW IN CONDUITS

Consider fluid flowing in a round tube of length 1 m and diameter 5 mm. Classify the flow as laminar or turbulent and calculate the entrance length for (a) air (50°C) with a speed of 12 m/s and (b) water (15°C) with a mass flow rate of $\dot{m} = 8$ g/s.

Problem Definition

Situation: Fluid is flowing in a round tube (two cases given).

Find:

1. Whether each flow is laminar or turbulent.
2. Entrance length (in meters) for each case.

Properties:

1. Air (50°C), Table A.3, $\nu = 1.79 \times 10^{-5}$ m²/s.

2. Water (15°C), Table A.5, $\mu = 1.14 \times 10^{-3}$ N·s/m².

Sketch:

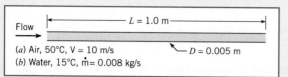

$L = 1.0$ m
Flow
(a) Air, 50°C, V = 10 m/s $D = 0.005$ m
(b) Water, 15°C, $\dot{m}$= 0.008 kg/s

Assumptions:

1. The pipe is connected to a reservoir.
2. The entrance is smooth and tapered.

Plan

1. Calculate the Reynolds number using Eq. (10.2).
2. Establish whether the flow is laminar or turbulent using Eq. (10.1).
3. Calculate the entrance length using Eq. (10.3).

Solution

a. Air

$$\text{Re} = \frac{VD}{\nu} = \frac{(12 \text{ m/s})(0.005\text{m})}{1.79 \times 10^{-5} \text{ m}^2/\text{s}} = 3350$$

Since Re > 3000, the $\boxed{\text{flow is turbulent.}}$

$$L_e = 50D = 50(0.005 \text{ m}) = \boxed{0.25 \text{ m}}$$

b. Water

$$\text{Re} = \frac{4\dot{m}}{\pi D \mu} = \frac{4(0.008 \text{ kg/s})}{\pi(0.005\text{m})(1.14 \times 10^{-3} \text{ N·s/m}^2)}$$

$$= 1787$$

Since Re < 2000, the $\boxed{\text{flow is laminar.}}$

$$L_e = 0.05\text{Re}D = 0.05(1787)(0.005\text{m}) = \boxed{0.447 \text{ m}}$$

Specifying Pipe Sizes

This section describes how to specify pipes using the Nominal Pipe Size (NPS) standard. This information is useful for specifying a size of pipe that is available commercially.

Standard Sizes for Pipes (NPS)

One of the most common standards for pipe sizes is called the Nominal Pipe Size (NPS) system. The terms used in the NPS system are introduced in Fig. 10.3. The ID (pronounced "eye dee") indicates the inner pipe diameter, and the OD ("oh dee") indicates the outer pipe diameter. As shown in Table 10.1, an NPS pipe is specified using two values: a Nominal Pipe Size (NPS) and a schedule. The nominal pipe size determines the outside diameter or OD. For example, pipes with a nominal size of 2 inches have an OD of 2.375 inches. Once the nominal size reaches 14 inches, the nominal size and the OD are equal. That is, a pipe with a nominal size of 24 inches will have an OD of 24 in.

Pipe schedule is related to the thickness of the wall. The original meaning of schedule was the ability of a pipe to withstand pressure, thus pipe schedule correlates with wall thickness. Each nominal pipe size has many possible schedules that range from schedule 5 to schedule 160. The data in Table 10.1 show representative ODs and schedules; more pipe sizes are specified in engineering handbooks and on the Internet.

A larger schedule indicates thicker walls. A schedule 40 pipe has thicker walls than a schedule 10 pipe.

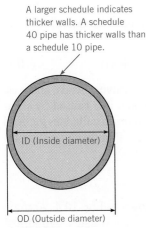

ID (Inside diameter)

OD (Outside diameter)

Figure 10.3

Section view of a pipe.

Table 10.1 NOMINAL PIPE SIZES				
NPS (in)	OD (in)	Schedule	Wall Thickness (in)	ID (in)
1/2	0.840	40	0.109	0.622
		80	0.147	0.546
1	1.315	40	0.133	1.049
		80	0.179	0.957
2	2.375	40	0.154	2.067
		80	0.218	1.939
4	4.500	40	0.237	4.026
		80	0.337	3.826
8	8.625	40	0.322	7.981
		80	0.500	7.625
14	14.000	10	0.250	13.500
		40	0.437	13.126
		80	0.750	12.500
		120	1.093	11.814
24	24.000	10	0.250	23.500
		40	0.687	22.626
		80	1.218	21.564
		120	1.812	20.376

Pipe Head Loss

This section presents the Darcy-Weisbach equation, which is used for calculating head loss in a straight run of pipe. This equation is one of the most useful equations in fluid mechanics.

Combined (Total) Head Loss

Pipe head loss is one type of head loss; the other type is called component head loss. All head loss is classified using these two categories:

$$(\text{Total head loss}) = (\text{Pipe head loss}) + (\text{Component head loss}) \tag{10.4}$$

Component head loss is associated with flow through devices such as valves, bends, and tees. *Pipe head loss* is associated with fully developed flow in conduits, and it is caused by shear stresses that act on the flowing fluid. Note that pipe head loss is sometimes called major head loss, and component head loss is sometimes called minor head loss. Pipe head loss is predicted with the Darcy-Weisbach equation.

Derivation of the Darcy-Weisbach Equation

To derive the Darcy-Weisbach equation, start with the situation shown in Fig. 10.4. Assume fully developed and steady flow in a round tube of constant diameter D. Situate a cylindrical control volume of diameter D and length ΔL inside the pipe. Define a coordinate system with an axial coordinate in the streamwise direction (s direction) and a radial coordinate in the r direction.

Apply the momentum equation (6.5) to the control volume shown in Fig. 10.4.

$$\sum \mathbf{F} = \frac{d}{dt} \int_{cv} \mathbf{v} \rho \, d\mathcal{V} + \int_{cs} \mathbf{v} \rho \mathbf{V} \cdot d\mathbf{A} \tag{10.5}$$

$$(\text{Net forces}) = (\text{Momentum accumulation rate}) + (\text{Net efflux of momentum})$$

Select the streamwise direction and analyze each of the three terms in Eq. (10.5). The net efflux of momentum is zero because the velocity distribution at section 2 is identical to the velocity distribution at section 1. The momentum accumulation term is also zero because

Figure 10.4

Initial situation for the derivation of the Darcy-Weisbach equation.

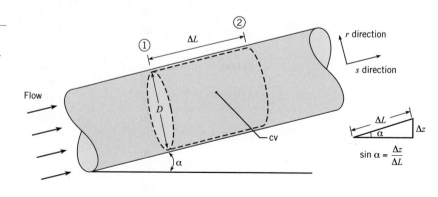

(a) (b)

the flow is steady. Thus, Eq. (10.5) simplifies to $\Sigma F = \mathbf{0}$. Forces are shown in Fig. 10.5. Summing of forces in the streamwise direction gives

$$F_{\text{pressure}} + F_{\text{shear}} + F_{\text{weight}} = 0$$

$$(p_1 - p_2)\left(\frac{\pi D^2}{4}\right) - \tau_0(\pi D \Delta L) - \gamma\left[\left(\frac{\pi D^2}{4}\right)\Delta L\right]\sin\alpha = 0 \qquad (10.6)$$

Figure 10.4b shows that $\sin\alpha = (\Delta z/\Delta L)$. Equation (10.6) becomes

$$(p_1 + \gamma z_1) - (p_2 + \gamma z_2) = \frac{4\Delta L\tau_0}{D} \qquad (10.7)$$

Next, apply the energy equation (7.29) to the control volume shown in Fig. 10.4. Recognize that $h_p = h_t = 0$, $V_1 = V_2$, and $\alpha_1 = \alpha_2$. Thus, the energy equation reduces to

$$\frac{p_1}{\gamma} + z_1 = \frac{p_2}{\gamma} + z_2 + h_L \qquad (10.8)$$

$$(p_1 + \gamma z_1) - (p_2 + \gamma z_2) = \gamma h_L$$

Combine Eqs. (10.7) and (10.8) and replace ΔL by L. Also, introduce a new symbol h_f to represent head loss in pipe.

$$h_f = \left(\begin{array}{c}\text{head loss}\\ \text{in a pipe}\end{array}\right) = \frac{4L\tau_0}{D\gamma} \qquad (10.9)$$

Rearrange the right side of Eq. (10.9).

$$h_f = \left(\frac{L}{D}\right)\left\{\frac{4\tau_0}{\rho V^2/2}\right\}\left\{\frac{\rho V^2/2}{\gamma}\right\} = \left\{\frac{4\tau_0}{\rho V^2/2}\right\}\left(\frac{L}{D}\right)\left\{\frac{V^2}{2g}\right\} \qquad (10.10)$$

Define a new π-group called the *friction factor* f that gives the ratio of wall shear stress (τ_o) to kinetic pressure ($\rho V^2/2$):

$$f \equiv \frac{(4\cdot\tau_0)}{(\rho V^2/2)} \approx \frac{\text{shear stress acting at the wall}}{\text{kinetic pressure}} \qquad (10.11)$$

In the technical literature, the friction factor is identified by several different labels that are synonymous: friction factor, Darcy friction factor, Darcy-Weisbach friction factor, and the resistance coefficient. There is also another coefficient called the Fanning friction factor,

Figure 10.5

Force diagram.

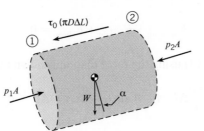

often used by chemical engineers, which is related to the Darcy-Weisbach friction factor by a factor of 4.

$$f_{Darcy} = 4f_{Fanning}$$

This text uses only the Darcy-Weisbach friction factor. Combining Eqs. (10.10) and (10.11) gives the Darcey-Weisbach equation:

$$h_f = f\frac{L}{D}\frac{V^2}{2g} \tag{10.12}$$

To use the Darcy-Weisbach equation, the flow should be fully developed and steady. The Darcy-Weisbach equation is used for either laminar flow or turbulent flow and for either round pipes or nonround conduits such as a rectangular duct.

The Darcy-Weisbach equation shows that head loss depends on the friction factor, the pipe-length-to-diameter ratio, and the mean velocity squared. The key to using the Darcy-Weisbach equation is calculating a value of the friction factor f. This topic is addressed in the next sections of this text.

Stress Distributions in Pipe Flow

This section derives equations for the stress distributions on a plane that is oriented normal to stream lines. These equations, which apply to both laminar and turbulent flow, provide insights about the nature of the flow. Also, these equations are used for subsequent derivations.

In pipe flow the pressure acting on a plane that is normal to the direction of flow is hydrostatic. This means that the pressure distribution varies linearly as shown in Fig. 10.6. The reason that the pressure distribution is hydrostatic can be explained by using Euler's equation (see p. 87).

To derive an equation for the shear-stress variation, consider flow of a Newtonian fluid in a round tube that is inclined at an angle α with respect to the horizontal as shown in Fig. 10.7. Assume that the flow is fully developed, steady, and laminar. Define a cylindrical control volume of length ΔL and radius r.

Apply the momentum equation in the s direction. The net momentum efflux is zero because the flow is fully developed; that is, the velocity distribution at the inlet is the same as the velocity distribution at the exit. The momentum accumulation is also zero because the flow is steady. The momentum equation (6.5) simplifies to force equilibrium.

$$\sum F_s = F_{pressure} + F_{weight} + F_{shear} = 0 \tag{10.13}$$

Analyze each term in Eq. (10.13) using the force diagram shown in Fig. 10.8:

$$pA - \left(p + \frac{dp}{ds}\Delta L\right)A - W\sin\alpha - \tau(2\pi r)\Delta L = 0 \tag{10.14}$$

Figure 10.6

For fully developed flow in a pipe, the pressure distribution on an area normal to streamlines is hydrostatic.

Hydrostatic pressure distribution

Plane normal to streamlines

Flow

Figure 10.7

Sketch for derivation of an equation for shear stress.

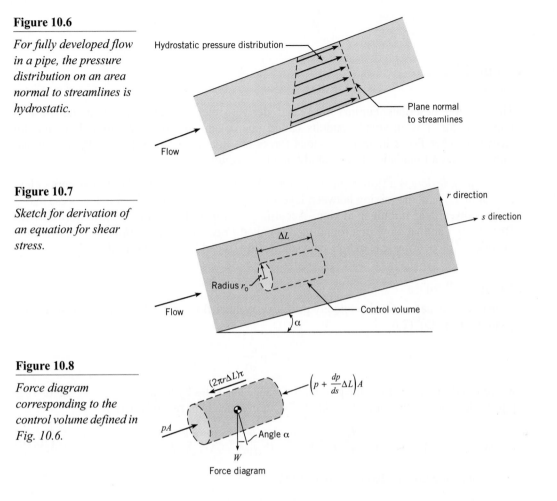

r direction

s direction

ΔL

Radius r_0

Flow

Control volume

α

Figure 10.8

Force diagram corresponding to the control volume defined in Fig. 10.6.

$(2\pi r \Delta L)\tau$

$\left(p + \dfrac{dp}{ds}\Delta L\right)A$

pA

Angle α

W

Force diagram

Let $W = \gamma A \Delta L$, and let $\sin \alpha = \Delta z / \Delta L$ as shown in Fig. 10.4b. Next, divide Eq. (10.14) by $A\Delta L$:

$$\tau = \frac{r}{2}\left[-\frac{d}{ds}(p + \gamma z)\right] \tag{10.15}$$

Equation (10.15) shows that the shear-stress distribution varies linearly with r as shown in Fig. 10.9. Notice that the shear stress is zero at the centerline, it reaches a maximum value of τ_0 at the wall, and the variation is linear in between. This linear shear stress variation applies to both laminar and turbulent flow.

Figure 10.9

In fully developed flow (laminar or turbulent), the shear-stress distribution on an area that is normal to streamlines is linear.

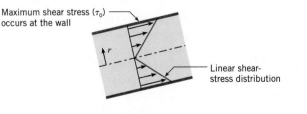

Maximum shear stress (τ_0) occurs at the wall

r

Linear shear-stress distribution

Laminar Flow in a Round Tube

This section describes laminar flow and derives relevant equations. Laminar flow is important for flow in small conduits called microchannels, for lubrication flow, and for analyzing other flows in which viscous forces are dominant. Also, knowledge of laminar flow provides a foundation for the study of advanced topics.

Laminar flow is a flow regime in which fluid motion is smooth, the flow occurs in layers (laminae), and the mixing between layers occurs by molecular diffusion, a process that is much slower than turbulent mixing. According to Eq. (10.1), laminar flow occurs when $\text{Re} \le 2000$. Laminar flow in a round tube is called *Poiseuille flow* or *Hagen-Poiseuille flow* in honor of pioneering researchers who studied low-speed flows in the 1840s.

Velocity Profile

To derive an equation for the velocity profile in laminar flow, begin by relating stress to rate-of-strain using Eq. (2.6)

$$\tau = \mu \frac{dV}{dy}$$

where y is the distance from the pipe wall. Change variables by letting $y = r_0 - r$, where r_0 is pipe radius and r is the radial coordinate. Next, use the chain rule of calculus:

$$\tau = \mu\left(\frac{dV}{dy}\right) = \mu\left(\frac{dV}{dr}\right)\left(\frac{dr}{dy}\right) = -\left(\mu\frac{dV}{dr}\right) \tag{10.16}$$

Substitute Eq. (10.16) into Eq. (10.15).

$$-\left(\frac{2\mu}{r}\right)\left(\frac{dV}{dr}\right) = \frac{d}{ds}(p + \gamma z) \tag{10.17}$$

In Eq. (10.17), the left side of the equation is a function of radius r, and the right side is a function of axial location s. This can be true if and only if each side of Eq. (10.17) is equal to a constant. Thus,

$$\text{constant} = \frac{d}{ds}(p + \gamma z) = \left(\frac{\Delta(p + \gamma z)}{\Delta L}\right) = \left(\frac{\gamma\Delta h}{\Delta L}\right) \tag{10.18}$$

where Δh is the change in piezometric head over a length ΔL of conduit. Combine Eqs. (10.17) and (10.18):

$$\frac{dV}{dr} = -\left(\frac{r}{2\mu}\right)\left(\frac{\gamma\Delta h}{\Delta L}\right) \tag{10.19}$$

Integrate Eq. (10.19):

$$V = -\left(\frac{r^2}{4\mu}\right)\left(\frac{\gamma\Delta h}{\Delta L}\right) + C \tag{10.20}$$

To evaluate the constant of integration C in Eq. (10.20), apply the no-slip condition, which states that the velocity of the fluid at the wall is zero. Thus,

$$V(r = r_0) = 0$$

$$0 = -\frac{r_0^2}{4\mu}\left(\frac{\gamma\Delta h}{\Delta L}\right) + C$$

Solve for C and substitute the result into Eq. (10.20):

$$V = \frac{r_0^2 - r^2}{4\mu}\left[-\frac{d}{ds}(p + \gamma z)\right] = -\left(\frac{r_0^2 - r^2}{4\mu}\right)\left(\frac{\gamma\Delta h}{\Delta L}\right) \tag{10.21}$$

The maximum velocity occurs at $r = r_0$:

$$V_{max} = -\left(\frac{r_0^2}{4\mu}\right)\left(\frac{\gamma\Delta h}{\Delta L}\right) \tag{10.22}$$

Combine Eqs. (10.21) and (10.22):

$$V(r) = -\left(\frac{r_0^2 - r^2}{4\mu}\right)\left(\frac{\gamma\Delta h}{\Delta L}\right) = V_{max}\left(1 - \left(\frac{r}{r_0}\right)^2\right) \tag{10.23}$$

Equation (10.23) shows that velocity varies as radius squared ($V \sim r^2$), meaning that the velocity distribution in laminar flow is parabolic as plotted in Fig. 10.10.

Discharge and Mean Velocity V

To derive an equation for discharge Q, introduce the velocity profile from Eq. (10.23) into the flow rate equation (5.8).

$$Q = \int V\, dA$$

$$= -\int_0^{r_0} \frac{(r_0^2 - r^2)}{4\mu}\left(\frac{\gamma\Delta h}{\Delta L}\right)(2\pi r\, dr) \tag{10.24}$$

Integrate Eq. (10.24):

$$Q = -\left(\frac{\pi}{4\mu}\right)\left(\frac{\gamma\Delta h}{\Delta L}\right)\frac{(r^2 - r_0^2)^2}{2}\Bigg|_0^{r_0} = -\left(\frac{\pi r_0^4}{8\mu}\right)\left(\frac{\gamma\Delta h}{\Delta L}\right) \tag{10.25}$$

To derive an equation for mean velocity, apply $Q = \bar{V}A$ and use Eq. (10.25).

$$\bar{V} = -\left(\frac{r_0^2}{8\mu}\right)\left(\frac{\gamma\Delta h}{\Delta L}\right) \tag{10.26}$$

Figure 10.10

The velocity profile in Poiseuille flow is parabolic.

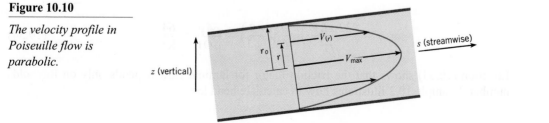

Comparing Eqs. (10.26) and (10.22) reveals that $\overline{V} = V_{max}/2$. Next, substitute $D/2$ for r_0 in Eq. (10.26). The final result is an equation for mean velocity in a round tube.

$$\overline{V} = -\left(\frac{D^2}{32\mu}\right)\left(\frac{\gamma\Delta h}{\Delta L}\right) = \frac{V_{max}}{2} \tag{10.27}$$

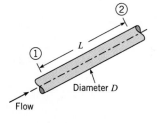

Head Loss and Friction Factor f

To derive an equation for head loss in a round tube, assume fully developed flow in the pipe shown in Fig. 10.11. Apply the energy equation (7.29) from section 1 to 2 and simplify to give

$$\left(\frac{p_1}{\gamma} + z_1\right) = \left(\frac{p_2}{\gamma} + z_2\right) + h_L \tag{10.28}$$

Figure 10.11

Flow in a pipe.

Let $h_L = h_f$ and then Eq. (10.28) becomes

$$\left(\frac{p_1}{\gamma} + z_1\right) = \left(\frac{p_2}{\gamma} + z_2\right) + h_f \tag{10.29}$$

Expand Eq. (10.27).

$$\overline{V} = -\left(\frac{D^2}{32\mu}\right)\left(\frac{1}{\gamma}\right)\left(\frac{\Delta h}{\Delta L}\right) = -\left(\frac{\gamma D^2}{32\mu}\right)\frac{\left(\frac{p_2}{\gamma} + z_2\right) - \left(\frac{p_1}{\gamma} + z_1\right)}{\Delta L} \tag{10.30}$$

Reorganize Eq. (10.30) and replace ΔL with L.

$$\left(\frac{p_1}{\gamma} + z_1\right) = \left(\frac{p_2}{\gamma} + z_2\right) + \frac{32\mu\overline{V}L}{D^2} \tag{10.31}$$

Comparing Eqs. (10.29) and (10.31) gives an equation for head loss in a pipe.

$$h_f = \frac{32\mu L\overline{V}}{\gamma D^2} \tag{10.32}$$

Key assumptions on Eq. (10.32) are (a) laminar flow, (b) fully developed flow, (c) steady flow, and (d) Newtonian fluid.

Equation (10.32) shows that head loss in laminar flow varies linearly with velocity. Also, head loss is influenced by viscosity, pipe length, specific weight, and pipe diameter squared.

To derive an equation for the friction factor f, combine Eq. (10.32) with the Darcy-Weisbach equation (10.12).

$$h_f = \frac{32\mu L V}{\gamma D^2} = f\frac{L}{D}\frac{V^2}{2g} \tag{10.33}$$

or $f = \left(\frac{32\mu L V}{\gamma D^2}\right)\left(\frac{D}{L}\right)\left(\frac{2g}{V^2}\right) = \frac{64\mu}{\rho D V} = \frac{64}{Re} \tag{10.34}$

Equation (10.34) shows that the friction factor for laminar flow depends only on Reynolds number. Example 10.2 illustrates how to calculate head loss.

EXAMPLE 10.2   HEAD LOSS FOR LAMINAR FLOW

Oil (S = 0.85) with a kinematic viscosity of 6×10^{-4} m^2/s flows in a 15 cm pipe at a rate of 0.020 m^3/s. What is the head loss per 100 m length of pipe?

Problem Definition

Situation:

1. Oil is flowing in a pipe at a flow rate of $Q = 0.02$ m^3/s.

2. Pipe diameter is $D = 0.15$ m.

Find: Head loss (in meters) for a pipe length of 100 m.

Assumptions: Fully developed, steady flow.

Properties: Oil: S = 0.85, $\nu = 6 \times 10^{-4}$ m^2/s.

Plan

1. Calculate the mean velocity using the flow rate equation (5.8).

2. Calculate the Reynolds number using Eq. (10.2).

3. Check whether the flow is laminar or turbulent using Eq. (10.1).

4. Calculate head loss using Eq. (10.32).

Solution

1. Mean velocity

$$V = \frac{Q}{A} = \frac{0.020 \text{ m}^3/\text{s}}{(\pi D^2)/4} = \frac{0.020 \text{ m}^3/\text{s}}{\pi((0.15 \text{ m})^2/4)} = 1.13 \text{ m/s}$$

2. Reynolds number

$$\text{Re} = \frac{VD}{\nu} = \frac{(1.13 \text{ m/s})(0.15 \text{ m})}{6 \times 10^{-4} \text{ m}^2/\text{s}} = 283$$

3. Since Re < 2000, the flow is laminar.

4. Head loss (laminar flow).

$$h_f = \frac{32 \mu L V}{\gamma D^2} = \frac{32 \rho \nu L V}{\rho g D^2} = \frac{32 \nu L V}{g D^2}$$

$$= \frac{32(6 \times 10^{-4} \text{ m}^2/\text{s})(100 \text{ m})(1.13 \text{ m/s})}{(9.81 \text{ m/s}^2)(0.15 \text{ m})^2} = \boxed{9.83 \text{ m}}$$

Review

Tip! An alternative way to calculate head loss for laminar flow is to use the Darcy-Weisbach equation (10.12) as follows:

$$f = \frac{64}{\text{Re}} = \frac{64}{283} = 0.226$$

$$h_f = f\left(\frac{L}{D}\right)\left(\frac{V^2}{2g}\right) = 0.226\left(\frac{100 \text{ m}}{0.15 \text{ m}}\right)\left(\frac{(1.13 \text{ m/s})}{2 \times 9.81 \text{ m/s}^2}\right)^2$$

$$= 9.83 \text{ m}$$

Turbulent Flow and the Moody Diagram

This section describes the characteristics of turbulent flow, presents equations for calculating the friction factor f, and presents a famous graph called the Moody diagram. This information is important because most flows in conduits are turbulent.

Qualitative Description of Turbulent Flow

Turbulent flow is a flow regime in which the movement of fluid particles is chaotic, eddying, and unsteady, with significant movement of particles in directions transverse to the flow direction. Because of the chaotic motion of fluid particles, turbulent flow produces high levels of mixing and has a velocity profile that is more uniform or flatter than the corresponding laminar velocity profile. According to Eq. (10.1), turbulent flow occurs when Re ≥ 3000.

Engineers and scientists model turbulent flow by using an empirical approach. This is because the complex nature of turbulent flow has prevented researchers from establishing a mathematical solution of general utility. Still, the empirical information has been used successfully and extensively in system design. Over the years, researchers have proposed many equations for shear stress and head loss in turbulent pipe flow. The empirical equations that have proven to be the most reliable and accurate for engineering use are presented in the next section.

Equations for the Velocity Distribution

The time-average velocity distribution is often described using an equation called the power-law formula.

$$\frac{u(r)}{u_{max}} = \left(\frac{r_0 - r}{r_0}\right)^m \tag{10.35}$$

where u_{max} is velocity in the center of the pipe, r_0 is the pipe radius, and m is an empirically determined variable that depends on Re as shown in Table 10.2. Notice in Table 10.2 that the velocity in the center of the pipe is typically about 20% higher than the mean velocity V. While Eq. (10.35) provides an accurate representation of the velocity profile, it does not predict an accurate value of wall shear stress.

An alternative approach to Eq. (10.35) is to use the turbulent boundary-layer equations presented in Chapter 9. The most significant of these equations, called the logarithmic velocity distribution, is given by Eq. (9.29) and repeated here:

$$\frac{u(r)}{u_*} = 2.44 \ln \frac{u_*(r_0 - r)}{\nu} + 5.56 \tag{10.36}$$

where u_*, the shear velocity, is given by $u_* = \sqrt{\tau_0/\rho}$.

Equations for the Friction Factor, f

To derive an equation for f in turbulent flow, substitute the log law in Eq. (10.36) into the definition of mean velocity given by Eq. (5.8):

$$V = \frac{Q}{A} = \left(\frac{1}{\pi r_0^2}\right) \int_0^{r_0} u(r) 2\pi r\,dr = \left(\frac{1}{\pi r_0^2}\right) \int_0^{r_0} u_* \left[2.44 \ln \frac{u_*(r_0 - r)}{\nu} + 5.56\right] 2\pi r\,dr$$

After integration, algebra, and tweaking the constants to better fit experimental data, the result is

$$\frac{1}{\sqrt{f}} = 2.0\log_{10}(\mathrm{Re}\sqrt{f}) - 0.8 \tag{10.37}$$

Table 10.2   EXPONENTS FOR POWER-LAW EQUATION AND RATIO OF MEAN TO MAXIMUM VELOCITY

Re	4×10^3	2.3×10^4	1.1×10^5	1.1×10^6	3.2×10^6
m	$\frac{1}{6.0}$	$\frac{1}{6.6}$	$\frac{1}{7.0}$	$\frac{1}{8.8}$	$\frac{1}{10.0}$
u_{max}/V	1.26	1.24	1.22	1.18	1.16

SOURCE: Schlichting (2). Used with permission of the McGraw-Hill Companies.

Equation (10.37), first derived by Prandtl in 1935, gives the resistance coefficient for turbulent flow in tubes that have smooth walls. The details of the derivation of Eq. (10.37) are presented by White (21). To determine the influence of roughness on the walls, Nikuradse (4), one of Prandtl's graduate students, glued uniform-sized grains of sand to the inner walls of a tube and then measured pressure drops and flow rates.

Nikuradse's data, Fig. 10.12, shows the friction factor f plotted as function of Reynolds number for various sizes of sand grains. To characterize the size of sand grains, Nikuradse used a variable called the *sand roughness height* with the symbol k_s. The π-group, k_s/D, is given the name *relative roughness*.

In laminar flow, the data in Fig. 10.12 show that wall roughness does not influence f. In particular, notice how the data corresponding to various values of k_s/D collapse into a single blue line that is labeled "laminar flow."

In turbulent flow, the data in Fig. 10.12 show that wall roughness has a major impact on f. When $k_s/D = 0.033$, then values of f are about 0.04. As the relative roughness drops to 0.002, values of f decrease by a factor of about 3. Eventually wall roughness does not matter, and the value of f can be predicted by assuming that the tube has a smooth wall. This latter case corresponds to the blue curve in Fig. 10.12 that is labeled "smooth wall tube." The effects of roughness are summarized by White (5) and presented in Table 10.3. These regions are also labeled in Fig. 10.12.

Moody Diagram

Colebrook (6) advanced Nikarudse's work by acquiring data for commercial pipes and then developing an empirical equation, called the Colebrook-White formula, for the friction factor. Moody (3) used the Colebrook-White formula to generate a design chart similar to that shown in Fig. 10.13. This chart is now known as the *Moody diagram* for commercial pipes.

Figure 10.12

Resistance coefficient f versus Re *for sand-roughened pipe. [After Nikuradse (4)].*

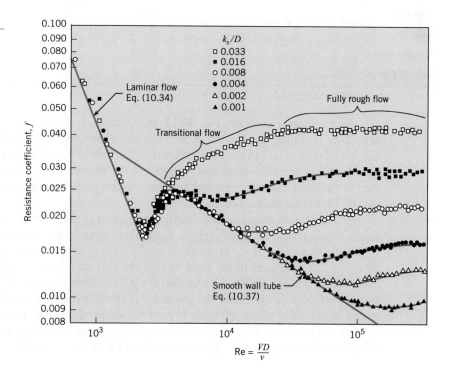

Table 10.3 EFFECTS OF WALL ROUGHNESS		
Type of Flow	**Parameter Ranges**	**Influence of Parameters on f**
Laminar Flow	Re $<$ 2000 \qquad NA	f depends on Reynolds number f is independent of wall roughness (k_s/D)
Turbulent Flow, Smooth Tube	Re $>$ 3000 \qquad $\left(\dfrac{k_s}{D}\right)$Re $<$ 10	f depends on Reynolds number f is independent of wall roughness (k_s/D)
Transitional Turbulent Flow	Re $>$ 3000 \qquad $10 < \left(\dfrac{k_s}{D}\right)$Re < 1000	f depends on Reynolds number f depends on wall roughness (k_s/D)
Fully Rough Turbulent Flow	Re $>$ 3000 \qquad $\left(\dfrac{k_s}{D}\right)$Re > 1000	f is independent of Reynolds number f depends on wall roughness (k_s/D)

Table 10.4 EQUIVALENT SAND-GRAIN ROUGHNESS, (k_S), FOR VARIOUS PIPE MATERIALS		
Boundary Material	**k_s, Millimeters**	**k_s, Inches**
Glass, plastic	Smooth	Smooth
Copper or brass tubing	0.0015	6×10^{-5}
Wrought iron, steel	0.046	0.002
Asphalted cast iron	0.12	0.005
Galvanized iron	0.15	0.006
Cast iron	0.26	0.010
Concrete	0.3 to 3.0	0.012–0.12
Riveted steel	0.9–9	0.035–0.35
Rubber pipe (straight)	0.025	0.001

In the Moody diagram, Fig. 10.13, the variable k_s denotes the *equivalent sand roughness*. That is, a pipe that has the same resistance characteristics at high Re values as a sand-roughened pipe is said to have a roughness equivalent to that of the sand-roughened pipe. Table 10.4 gives the equivalent sand roughness for various kinds of pipes. This table can be used to calculate the relative roughness for a given pipe diameter, which, in turn, is used in Fig. 10.13, to find the friction factor.

In the Moody diagram, Fig. 10.13, the abscissa is the Reynolds number Re, and the ordinate is the resistance coefficient f. Each blue curve is for a constant relative roughness k_s/D, and the values of k_s/D are given on the right at the end of each curve. To find f, given Re and k_s/D, one goes to the right to find the correct relative roughness curve. Then one looks at the bottom of the chart to find the given value of Re and, with this value of Re, moves vertically upward until the given k_s/D curve is reached. Finally, from this point one moves horizontally to the left scale to read the value of f. If the curve for the given value of k_s/D is not plotted in Fig. 10.13, then one simply finds the proper position on the graph by interpolation between the k_s/D curves that bracket the given k_s/D.

Figure 10.13

Resistance coefficient f versus Re. Reprinted with minor variations. [After Moody (3). Reprinted with permission from the ASME.]

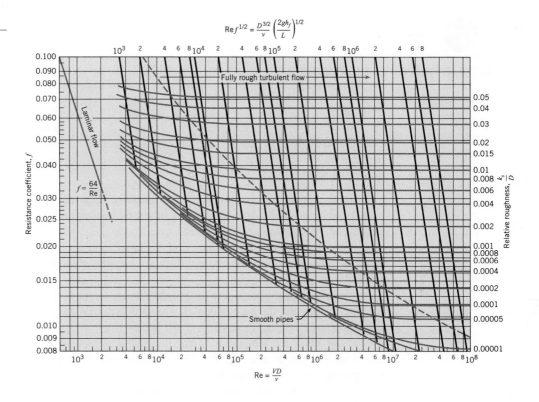

To provide a more convenient solution to some types of problems, the top of the Moody diagram presents a scale based on the parameter $\text{Re}\,f^{1/2}$. This parameter is useful when h_f and k_s/D are known but the velocity V is not. Using the Darcy-Weisbach equation given in Eq. (10.12) and the definition of Reynolds number, one can show that

$$\text{Re}\,f^{1/2} = \frac{D^{3/2}}{\nu}(2gh_f/L)^{1/2} \tag{10.38}$$

In the Moody diagram, Fig. 10.13, curves of constant $\text{Re}\,f^{1/2}$ are plotted using heavy black lines that slant from the left to right. For example, when $\text{Re}\,f^{1/2} = 10^5$ and $k_s/D = 0.004$, then $f = 0.029$. When using computers to carry out pipe-flow calculations, it is much more convenient to have an equation for the friction factor as a function of Reynolds number and relative roughness. By using the Colebrook-White formula, Swamee and Jain (7) developed an explicit equation for friction factor, namely

$$f = \frac{0.25}{\left[\log_{10}\left(\dfrac{k_s}{3.7D} + \dfrac{5.74}{\text{Re}^{0.9}}\right)\right]^2} \tag{10.39}$$

It is reported that this equation predicts friction factors that differ by less than 3% from those on the Moody diagram for $4 \times 10^3 < \text{Re} < 10^8$ and $10^{-5} < k_s/D < 2 \times 10^{-2}$.

Solving Turbulent Flow Problems

This section describes how to solve problems that involve turbulent flow* in a pipe, emphasizing how to classify problems as case 1, 2, or 3. Classification is important because cases 2 and 3 usually require either an iterative approach or they require computer programs that can solve coupled nonlinear equations. Some useful computer programs include TK Solver, EES, MathCAD, and MatLab.

To recognize problems that require iterative approaches or computer solutions, engineers classify problems into three cases based on the goal of the problem and based on what information is known.

Case 1 is when the goal is to find the *head loss*, given the pipe length, pipe diameter, and flow rate. This problem is straightforward because it can be solved using algebra; see Example 10.3.

Case 2 is when the goal is to find the *flow rate*, given the head loss (or pressure drop), the pipe length, and the pipe diameter. This problem usually requires an iterative approach or solver program; see Examples 10.4 and 10.5.

Case 3 is when the goal is to find the *pipe diameter*, given the flow rate, length of pipe, and head loss (or pressure drop). This problem usually requires an iterative approach or a solver program; see Example 10.6.

There are several approaches that sometimes eliminate the need for an iterative approach. For case 2, an iterative approach can sometimes be eliminated by using an explicit equation developed by Swamee and Jain (7):

$$Q = -2.22D^{5/2}\sqrt{gh_f/L}\,\log\left(\frac{k_s}{3.7D} + \frac{1.78\nu}{D^{3/2}\sqrt{gh_f/L}}\right) \qquad (10.40)$$

Using Eq. (10.40) is equivalent to using the top of the Moody diagram, which presents a scale for $\mathrm{Re}\,f^{1/2}$. For case 3, one can sometimes use an explicit equation developed by Swamee and Jain (7) and modified by Streeter and Wylie (8):

$$D = 0.66\left[k_s^{1.25}\left(\frac{LQ^2}{gh_f}\right)^{4.75} + \nu Q^{9.4}\left(\frac{L}{gh_f}\right)^{5.2}\right]^{0.04} \qquad (10.41)$$

Example 10.3 shows an example of a case 1 problem.

---

* Note that laminar flow problems can be solved using algebra because Eq. (10.32) is linear in V. Thus, there is no need to use the iterative approaches described in this section.

EXAMPLE 10.3   HEAD LOSS IN A PIPE (CASE 1)

Water ($T = 20°C$) flows at a rate of $0.05 \text{ m}^3/\text{s}$ in a 20 cm asphalted cast-iron pipe. What is the head loss per kilometer of pipe?

Problem Definition

Situation: Water is flowing in a pipe.

Find: Head loss (in meters) for $L = 1000$ m.

Assumptions: Fully developed flow.

Properties: Water (20°C), Table A.5: $\nu = 1 \times 10^{-6} \text{ m}^2/\text{s}$.

Sketch:

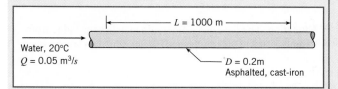

Water, 20°C
$Q = 0.05 \text{ m}^3/s$
$L = 1000$ m
$D = 0.2$m
Asphalted, cast-iron

Plan

Since this is a case 1 problem (head loss is the goal), the solution is straightforward.

1. Calculate the mean velocity using the flow rate equation (5.8).
2. Calculate the Reynolds number using Eq. (10.2).
3. Calculate the relative roughness and then look up f on the Moody diagram.

4. Find head loss by applying the Darcy-Weisbach equation (10.12).

Solution

1. Mean velocity
$$V = \frac{Q}{A} = \frac{0.05 \text{ m}^3/\text{s}}{(\pi/4)(0.02 \text{ m})^2} = 1.59 \text{ m}/\text{s}$$

2. Reynolds number
$$\text{Re} = \frac{VD}{\nu} = \frac{(1.59 \text{ m}/\text{s})(0.20 \text{ m})}{10^{-6} \text{ m}^2/\text{s}} = 3.18 \times 10^5$$

3. Resistance coefficient
   - Equivalent sand roughness (Table 10.4): $k_s = 0.12$ mm
   - Relative roughness:
     $k_s/D = (0.00012 \text{ m})/(0.2 \text{ m}) = 0.0006$
   - Look up f on the Moody diagram for $\text{Re} = 3.18 \times 10^5$ and $k_s/D = 0.0006$:
     $$f = 0.019$$

4. Darcy-Weisbach equation
$$h_f = f\left(\frac{L}{D}\right)\left(\frac{V^2}{2g}\right) = 0.019\left(\frac{1000 \text{ m}}{0.20 \text{ m}}\right)\left(\frac{1.59^2 \text{ m}^2/\text{s}^2}{2(9.81 \text{ m}/\text{s}^2)}\right)$$
$$= \boxed{12.2 \text{ m}}$$

Example 10.4 shows an example of a case 2 problem. Notice that the solution involved application of the scale on the top of the Moody diagram; thereby avoiding an iterative solution.

EXAMPLE 10.4   FLOW RATE IN A PIPE (CASE 2)

The head loss per kilometer of 20 cm asphalted cast-iron pipe is 12.2 m. What is the flow rate of water through the pipe?

Problem Definition

Situation: This is the same situation as Example 10.3 except that the head loss is now specified and the discharge is unknown.

Find: Discharge (m^3/s) in the pipe.

Plan

This is a case 2 problem because flow rate is the goal. However, a direct (i.e., noniterative) solution is possible because head loss is specified. The strategy will be to use the horizontal scale on the top of the Moody diagram.

1. Calculate the parameter on the top of the Moody diagram.
2. Using the Moody diagram, find the friction factor f.
3. Calculate mean velocity using the Darcy-Weisbach equation (10.12).
4. Find discharge using the flow rate equation (5.8).

Solution

1. Compute the parameter $D^{3/2}\sqrt{2gh_f/L}/\nu$.

$$D^{3/2}\frac{\sqrt{2gh_f/L}}{\nu} = (0.20 \text{ m})^{3/2}$$

$$\times \frac{[2(9.81 \text{ m/s}^2)(12.2 \text{ m}/1000 \text{ m})]^{1/2}}{1.0 \times 10^{-6} \text{ m}^2/\text{s}}$$

$$= 4.38 \times 10^4$$

2. Determine resistance coefficient.
   - Relative roughness:
     $$k_s/D = (0.00012 \text{ m})/(0.2 \text{ m}) = 0.0006$$
   - Look up f on the Moody diagram for
     $D^{3/2}\sqrt{2gh_f/L}/\nu = 4.4 \times 10^4$ and $k_s/D = 0.0006$:

     $$f = 0.019$$

3. Find V using the Darcy-Weisbach equation.

$$h_f = f\left(\frac{L}{D}\right)\left(\frac{V^2}{2g}\right)$$

$$12.2 \text{ m} = 0.019\left(\frac{1000 \text{ m}}{0.20 \text{ m}}\right)\left(\frac{V^2}{2(9.81 \text{ m/s}^2)}\right)$$

$$V = 1.59 \text{ m/s}$$

4. Use flow rate equation to find discharge.

$$Q = VA = (1.59 \text{ m/s})(\pi/4)(0.2 \text{ m})^2 = \boxed{0.05 \text{ m}^3/\text{s}}$$

Review

Validation. The calculated flow rate matches the value from Example 10.3. This is expected because the data are the same.

When case 2 problems require iteration, there are several methods that can be used to find a solution. One of the easiest ways is a method called "successive substitution," which is illustrated by Example 10.5.

EXAMPLE 10.5 FLOW RATE IN A PIPE (CASE 2)

Water ($T = 20°C$) flows from a tank through a 50 cm diameter steel pipe. Determine the discharge of water.

Problem Definition

Situation: Water is draining from a tank through a steel pipe.

Find: Discharge (m^3/s) for the system.

Assumptions:

1. Flow is fully developed.
2. Include only the head loss in the pipe.

Properties:

1. Water (20°C), Table A.5: $\nu = 1 \times 10^{-6}$ m^2/s.
2. Steel pipe, Table 10.4, equivalent sand roughness:
   $k_s = 0.046$ mm. Relative roughness (k_s/D) is
   9.2×10^{-5}.

Plan

This is a case 2 problem because flow rate is the goal. An iterative solution is used because V is unknown, so there is no direct way to use the Moody diagram.

1. Apply the energy equation from section 1 to section 2.
2. First trial. Guess a value of f and then solve for V.

Sketch:

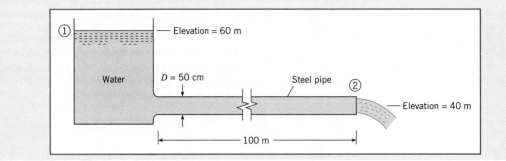

3. Second trial. Using V from the first trial, calculate a new value of f.

4. Convergence. If the value of f is constant within a few percent between trials, then stop. Otherwise, continue with more iterations.

5. Calculate flow rate using the flow rate equation (5.8).

Solution

1. Energy equation (reservoir surface to outlet)

$$\frac{p_1}{\gamma} + \frac{V_1^2}{2g} + z_1 = \frac{p_2}{\gamma} + \frac{V_2^2}{2g} + z_2 + h_L$$

$$0 + 0 + 60 = 0 + \frac{V_2^2}{2g} + 40 + f\frac{L}{D}\frac{V_2^2}{2g}$$

or

$$V = \left(\frac{2g \times 20}{1 + 200f}\right)^{1/2} \qquad (1)$$

2. First trial (iteration 1)

   • Guess a value of $f = 0.020$.

   • Use eq. (1) to calculate $V = 8.86$ m/s.

   • Use $V = 8.86$ m/s to calculate Re $= 4.43 \times 10^6$.

   • Use Re $= 4.43 \times 10^6$ and $k_s/D = 9.2 \times 10^{-5}$ on the Moody diagram to find that $f = 0.012$.

   • Use eq. (1) with $f = 0.012$ to calculate $V = 10.7$ m/s.

3. Second trial (iteration 2)

   • Use $V = 10.7$ m/s to calculate Re $= 5.35 \times 10^6$.

   • Use Re $= 5.35 \times 10^6$ and $k_s/D = 9.2 \times 10^{-5}$ on the Moody diagram to find that $f = 0.012$.

4. Convergence. The value of $f = 0.012$ is unchanged between the first and second trials. Therefore, there is no need for more iterations.

5. Flow rate

$$Q = VA = (10.7 \text{ m/s}) \times (\pi/4) \times (0.50)^2 \text{ m}^2 = \boxed{2.10 \text{ m}^3/\text{s}}$$

In a case 3 problem, derive an equation for diameter D and then use the method of successive substitution to find a solution. Iterative approaches, as illustrated in Example 10.6, can employ a spreadsheet program to perform the calculations.

EXAMPLE 10.6   FINDING PIPE DIAMETER (CASE 3)

What size of asphalted cast-iron pipe is required to carry water (60°F) at a discharge of 3 cfs and with a head loss of 4 ft per 1000 ft of pipe?

Problem Definition

Situation: Water is flowing in a asphalted cast-iron pipe.

$Q = 3$ ft^3/s.

Find: Pipe diameter (in ft) so that head loss is 4 ft per 1000 ft of pipe length.

Assumptions: Fully developed flow.

Properties:

1. Water (60°F), Table A.5: $\nu = 1.22 \times 10^{-5}$ ft^2/s.

2. Asphalted cast-iron pipe, Table 10.4, equivalent sand roughness: $k_s = 0.005$ in.

Plan

Since this is a case 3 problem (pipe diameter is the goal), use an iterative approach.

1. Derive an equation for pipe diameter by using the Darcy-Weisbach equation.

2. For iteration 1, guess f, solve for pipe diameter, and then recalculate f.

3. To complete the problem, build a table in a spreadsheet program.

Solution

1. Develop an equation to use for iteration.

   • Darcy-Weisbach equation

$$h_f = f\left(\frac{L}{D}\right)\left(\frac{V^2}{2g}\right) = f\left(\frac{L}{D}\right)\left(\frac{Q^2/A^2}{2g}\right) = \frac{fLQ^2}{2g(\pi/4)^2 D^5}$$

- Solve for pipe diameter

$$D^5 = \frac{fLQ^2}{0.785^2(2gh_f)} \qquad (1)$$

2. Iteration 1
   - Guess $f = 0.015$.
   - Solve for diameter using eq. (1):

$$D^5 = \frac{0.015(1000\ \text{ft})(3\ \text{ft}^3/\text{s})^2}{0.785^2(64.4\ \text{ft/s}^2)(4\ \text{ft})} = 0.852\ \text{ft}^5$$

$$D = 0.968\ \text{ft}$$

   - Find parameters needed for calculating f:

$$V = \frac{Q}{A} = \frac{3\ \text{ft}^3/\text{s}}{(\pi/4)(0.968^2\ \text{ft}^2)} = 4.08\ \text{ft/s}$$

$$\text{Re} = \frac{VD}{\nu} = \frac{(4.08\ \text{ft/s})(0.968\ \text{ft})}{1.22 \times 10^{-5}\ \text{ft}^2/\text{s}} = 3.26 \times 10^5$$

$$k_s/D = 0.005/(0.97 \times 12) = 0.00043$$

- Calculate f using Eq. (10.39): $f = 0.0178$.

3. In the table below, the first row contains the values from iteration 1. The value of $f = 0.0178$ from iteration 1 is used for the initial value for iteration 2. Notice how the solution has converged by iteration 3.

Iteration #	Initial f	D	V	Re	k_s/D	New f
		(ft)	(ft/s)			
1	0.0150	0.968	4.08	3.26E+05	4.3E–04	0.0178
2	0.0178	1.002	3.81	3.15E+05	4.2E–04	0.0178
3	0.0178	1.001	3.81	3.15E+05	4.2E–04	0.0178
4	0.0178	1.001	3.81	3.15E+05	4.2E–04	0.0178

Specify a pipe with a 12-inch inside diameter.

10.8

Combined Head Loss

Previous sections have described how to calculate head loss in pipes. This section completes the story by describing how to calculate head loss in components. This knowledge is essential for modeling and design of systems.

The Minor Loss Coefficient, K

When fluid flows through a component such as a partially open value or a bend in a pipe, viscous effects cause the flowing fluid to lose mechanical energy. For example, Fig. 10.14 shows flow through a "generic component." At section 2, the head of the flow will be less than at section 1. To characterize component head loss, engineers use a π-group called the *minor loss coefficient K*

$$K \equiv \frac{(\Delta h)}{(V^2/2g)} = \frac{(\Delta p)}{(\rho V^2/2)} \qquad (10.1)$$

where Δh is drop in piezometric head that is caused by a component, Δp is the pressure drop that is caused by the component, and V is mean velocity. As shown in Eq. (10.1), the minor loss coefficient has two useful interpretations:

$$K = \frac{\text{drop in piezometric head across component}}{\text{velocity head}} = \frac{\text{pressure drop due to component}}{\text{kinetic pressure}}$$

Figure 10.14

Flow through a generic component.

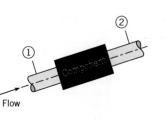

Figure 10.15

Flow at a sharp-edged inlet.

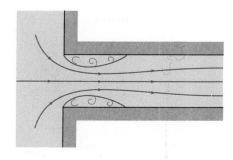

Thus, the head loss across a single component or transition is $h_L = K(V^2/(2g))$, where K is the minor loss coefficient for that component or transition.

Most values of K are found by experiment. For example, consider the setup shown in Fig. 10.14. To find K, flow rate is measured and mean velocity is calculated using $V = (Q/A)$. Pressure and elevation measurements are used to calculate the change in piezometric head.

$$\Delta h = h_2 - h_1 = \left(\frac{p_2}{\gamma} + z_2\right) - \left(\frac{p_1}{\gamma} + z_1\right) \tag{10.2}$$

Then, values of V and Δh are used in Eq. (10.1) to calculate K. The next section presets typical data for K.

Data for the Minor Loss Coefficient

This section presents K data and relates these data to flow separation and wall shear stress. This information is used for nearly all system modeling.

Pipe inlet. Near the entrance to a pipe when the entrance is rounded, flow is developing as shown in Fig. 10.1 and the wall shear stress is higher than that found in fully developed flow. Alternatively, if the pipe inlet is abrupt, or sharp-edged, as in Fig. 10.15, separation occurs just downstream of the entrance. Hence the streamlines converge and then diverge with consequent turbulence and relatively high head loss. The loss coefficient for the abrupt inlet is approximately 0.5. Other values of head loss are summarized in Table 10.5.

Table 10.5   LOSS COEFFICIENTS FOR VARIOUS TRANSITIONS AND FITTINGS

Description	Sketch	Additional Data			K	Source
Pipe entrance $h_L = K_e V^2/2g$		r/d 0.0 0.1 >0.2			K_e 0.50 0.12 0.03	$(10)^\dagger$
Contraction $h_L = K_C V_2^2/2g$		D_2/D_1 0.00 0.20 0.40 0.60 0.80 0.90	K_C $\theta = 60°$ 0.08 0.08 0.07 0.06 0.06 0.06	K_C $\theta = 180°$ 0.50 0.49 0.42 0.27 0.20 0.10		(10)

(Continued)

Description	Sketch	Additional Data			K	Source
Expansion $h_L = K_E V_1^2/2g$		D_1/D_2 0.00 0.20 0.40 0.60 0.80	K_E $\theta = 20°$ 0.30 0.25 0.15 0.10	K_E $\theta = 180°$ 1.00 0.87 0.70 0.41 0.15		(9)
90° miter bend		Without vanes			$K_b = 1.1$	(15)
		With vanes			$K_b = 0.2$	(15)
90° smooth bend		r/d 1 2 4 6 8 10	$K_b = 0.35$ 0.19 0.16 0.21 0.28 0.32			(16) and (9)
Threaded pipe fittings	Globe valve—wide open				$K_v = 10.0$	(15)
	Angle valve—wide open				$K_v = 5.0$	
	Gate valve—wide open				$K_v = 0.2$	
	Gate valve—half open				$K_v = 5.6$	
	Return bend				$K_b = 2.2$	
	Tee					
	Straight-through flow				$K_t = 0.4$	
	Side-outlet flow				$K_t = 1.8$	
	90° elbow				$K_b = 0.9$	
	45° elbow				$K_b = 0.4$	

†Reprinted by permission of the American Society of Heating, Refrigerating and Air Conditioning Engineers, Atlanta, Georgia, from the 1981 *ASHRAE Handbook—Fundamentals*.

Flow in an elbow. In an elbow (90° smooth bend), considerable head loss is produced by secondary flows and by separation that occurs near the inside of the bend and downstream of the midsection as shown in Fig. 10.16.

The loss coefficient for an elbow at high Reynolds numbers depends primarily on the shape of the elbow. If it is a very short-radius elbow, the loss coefficient is quite high. For larger-radius elbows, the coefficient decreases until a minimum value is found at an r/d value of about 4 (see Table 10.3). However, for still larger values of r/d, an increase in loss coefficient occurs because, for larger r/d values, the elbow itself is significantly longer than elbows with small r/d values. The greater length creates an additional head loss.

Figure 10.16

Flow pattern in an elbow.

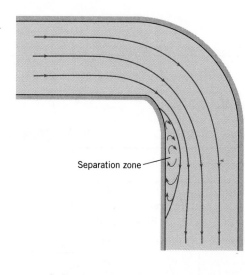

Separation zone

Other components. The loss coefficients for a number of other fittings and flow transitions are given in Table 10.5. This table is representative of engineering practice. For more extensive tables, see references (10–15).

In Table 10.5, the K was found by experiment, so one must be careful to ensure that Re values in the application correspond to Re values of the data.

Combined Head Loss Equation

The total head loss is given by Eq. (10.4), which is repeated here:

$$\{\text{Total head loss}\} = \{\text{Pipe head loss}\} + \{\text{Component head loss}\} \qquad (10.3)$$

To develop an equation for the combined head loss, substitute Eqs. (10.12) and (10.1) in Eq. (10.3):

$$h_L = \underbrace{\sum f \frac{L}{D} \frac{V^2}{2g}}_{\text{pipes}} + \underbrace{\sum K \frac{V^2}{2g}}_{\text{components}} = \frac{V^2}{2g} \left[\underbrace{\sum f \frac{L}{D}}_{\text{pipes}} + \underbrace{\sum K}_{\text{components}} \right] \qquad (10.4)$$

Equation (10.4) is called the *combined head loss equation.* To apply this equation, follow the same approaches that were used for solving pipe problems. That is, classify the flow as case 1, 2, or 3 and apply the usual equations: the energy, Darcy-Weisbach, and flow rate equations. Example 10.7 illustrates this approach for a case 1 problem.

EXAMPLE 10.7   PIPE SYSTEM WITH COMBINED HEAD LOSS

If oil ($\nu = 4 \times 10^{-5}$ m^2/s; $S = 0.9$) flows from the upper to the lower reservoir at a rate of 0.028 m^3/s in the 15 cm smooth pipe, what is the elevation of the oil surface in the upper reservoir?

Problem Definition

Situation: Oil is flowing from a upper reservoir to a lower reservoir.

Find: Elevation (in meters) of the free surface of the upper reservoir.

Properties:
1. Oil: $\nu = 4 \times 10^{-5}$ m^2/s, $S = 0.9$.
2. Minor head loss coefficients, Table 10.5;
   entrance = $K_e = 0.5$; bend = $K_b = 0.19$;
   outlet = $K_E = 1.0$.

Sketch:

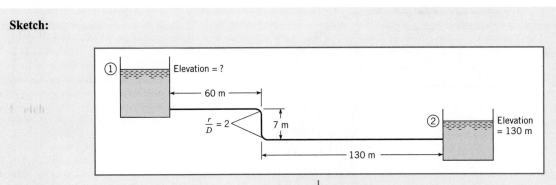

Plan

This is a case 1 problem because flow rate and pipe dimensions are known. Thus, the solution is straightforward.

1. Apply the energy equation [Eq. (7.29)] from 1 to 2.
2. Apply the combined head loss equation (10.45).
3. Develop an equation for z_1 by combining results from steps 1 and 2.
4. Calculate the resistance coefficient f.
5. Solve for z_1 using the equation from step 3.

Solution

1. Energy equation and term-by-term analysis

$$\frac{p_1}{\gamma} + \alpha_1\frac{\overline{V}_1^2}{2g} + z_1 + h_p = \frac{p_2}{\gamma} + \alpha_2\frac{\overline{V}_2^2}{2g} + z_2 + h_t + h_L$$

$$0 + 0 + z_1 + 0 = 0 + 0 + z_2 + 0 + h_L$$

$$z_1 = z_2 + h_L$$

Interpretation: Change in elevation head is balanced by the total head loss.

2. Combined head loss equation

$$h_L = \sum_{\text{pipes}} f\frac{L}{D}\frac{V^2}{2g} + \sum_{\text{components}} K\frac{V^2}{2g}$$

$$h_L = f\frac{L}{D}\frac{V^2}{2g} + \left(2K_b\frac{V^2}{2g} + K_e\frac{V^2}{2g} + K_E\frac{V^2}{2g}\right)$$

$$= \frac{V^2}{2g}\left(f\frac{L}{D} + 2K_b + K_e + K_E\right)$$

3. Combine eqs. (1) and (2).

$$z_1 = z_2 + \frac{V^2}{2g}\left(f\frac{L}{D} + 2K_b + K_e + K_E\right)$$

4. Resistance coefficient
   - Flow rate equation (5.8)

$$V = \frac{Q}{A} = \frac{(0.028 \text{ m}^3/\text{s})}{(\pi/4)(0.15 \text{ m})^2} = 1.58 \text{ m/s}$$

   - Reynolds number

$$\text{Re} = \frac{VD}{\nu} = \frac{1.58 \text{ m/s}(0.15 \text{ m})}{4 \times 10^{-5} \text{ m}^2/\text{s}} = 5.93 \times 10^3$$

   Thus, flow is turbulent.
   - Swamee-Jain equation (10.39)

$$f = \frac{0.25}{\left[\log_{10}\left(\frac{k_s}{3.7D} + \frac{5.74}{\text{Re}^{0.9}}\right)\right]^2} = \frac{0.25}{\left[\log_{10}\left(0 + \frac{5.74}{5930^{0.9}}\right)\right]^2} = 0.036$$

5. Calculate z_1 using (3):

$$z_1 = (130 \text{ m}) + \frac{(1.58 \text{ m/s})^2}{2(9.81)\text{m/s}^2}$$

$$\left(0.036\frac{(197 \text{ m})}{(0.15 \text{ m})} + 2(0.19) + 0.5 + 1.0\right)$$

$$\boxed{z_1 = 136 \text{ m}}$$

Review

1. Notice: There is a big difference between pipe and component head loss:

$$\text{Pipe head loss} \sim \Sigma f\frac{L}{D} = 0.036\frac{(197\text{m})}{(0.15 \text{ m})} = 47.2$$

   Component head loss $\sim \Sigma K = 2(0.19) + 0.5 + 1.0 = 1.88$

   Thus pipe losses $\gg$ component losses for this problem.

2. Tip! When pipe head loss is dominant, make simple estimates of K because these estimates will not impact the prediction very much.

Nonround Conduits

Previous sections have considered round pipes. This section extends this information by describing how to account for conduits that are square, triangular, or any nonround shape. This information is important for applications such as sizing ventilation ducts in buildings and for modeling flow in open channels.

When a conduit has a section area that is noncircular, then engineers modify the Darcy-Weisbach equation, Eq. (10.12), to use hydraulic diameter D_h in place of diameter.

$$h_L = f \frac{L}{D_h} \frac{V^2}{2g} \tag{10.5}$$

Equation (10.5) is derived using the same approach as Eq. (10.12), and the hydraulic diameter that emerges from this derivation is

$$D_h \equiv \frac{4 \times \text{cross-section area}}{\text{wetted perimeter}} \tag{10.6}$$

where the "wetted perimeter" is that portion of the perimeter that is physically touching the fluid. The wetted perimeter of a rectangular duct of dimension $L \times w$ is $2L + 2w$. Thus, the hydraulic diameter of this duct is:

$$D_h \equiv \frac{4 \times Lw}{2L + 2w} = \frac{2Lw}{L + w}$$

Using Eq. (10.6), the hydraulic diameter of a round pipe is the pipe's diameter D. When Eq. (10.5) is used to calculate head loss, the resistance coefficient f is found using a Reynolds number based on hydraulic diameter. Use of hydraulic diameter is an approximation. According to White (21), this approximation introduces an uncertainty of 40% for laminar flow and 15% for turbulent flow.

$$f = \left(\frac{64}{\text{Re}_{D_h}} \right) \pm 40\% \text{ (laminar flow)}$$

$$f_{\text{Moody}} \left(\text{Re}_{D_h}, \frac{k}{D_h} \right) \pm 15\% \text{ (turbulent flow)} \tag{10.7}$$

In addition to hydraulic diameter, engineers also use hydraulic radius, which is defined as

$$R_h \equiv \frac{\text{section area}}{\text{wetted perimeter}} = \frac{D_h}{4} \tag{10.8}$$

Notice that the ratio of R_h to D_h is $1/4$ instead of $1/2$. While this ratio is not logical, it is the convention used in the literature and is useful to remember. Chapter 15, which focuses on open-channel flow, will present examples of hydraulic radius.

To model flow in a nonround conduit, the approaches of the previous sections are followed with the only difference being the use of hydraulic diameter in place of diameter. This is illustrated by Example 10.8.

EXAMPLE 10.8   PRESSURE DROP IN AN HVAC DUCT

Air ($T = 20°C$ and $p = 101$ kPa absolute) flows at a rate of 2.5 m³/s in a horizontal, commercial steel, HVAC duct. (Note that HVAC is an acronym for heating, ventilating, and air conditioning.) What is the pressure drop in inches of water per 50 m of duct?

Problem Definition

Situation: Air is flowing through a duct.

Find: Pressure drop (inch H_2O) in a length of 50 m.

Sketch:

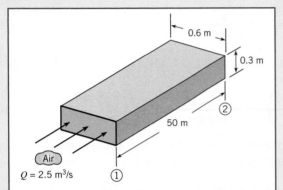

Assumptions:

1. Fully developed flow, meaning that $V_1 = V_2$ and the velocity head terms in the energy equation cancel out.
2. No sources of component head loss.

Properties:

1. Air (20°C, 1 atm), Table A.2: $\rho = 1.2$ kg/m³, $\nu = 15.1 \times 10^{-6}$ m²/s.
2. Steel pipe, Table 10.4: $k_s = 0.046$ mm.

Plan

This is a case 1 problem because flow rate and duct dimensions are known. Thus, the solution is straightforward.

1. Derive an equation for pressure drop by using the energy equation (7.29).

2. Calculate parameters needed to find head loss.
3. Calculate head loss by using the Darcy-Weisbach equation (10.12).
4. Calculate pressure drop Δp by combining steps 1, 2, and 3.

Solution

1. Energy equation (after term-by-term analysis)
$$p_1 - p_2 = \rho g h_L$$

2. Intermediate calculations
   - Flow rate equation
   $$V = \frac{Q}{A} = \frac{2.5 \text{ m}^3/\text{s}}{(0.3 \text{ m})(0.6 \text{ m})} = 13.9 \text{ m/s}$$
   - Hydraulic diameter
   $$D_h \equiv \frac{4 \times \text{section area}}{\text{wetted perimeter}} = \frac{4(0.3 \text{ m})(0.6 \text{ m})}{(2 \times 0.3 \text{ m}) + (2 \times 0.6 \text{ m})} = 0.4 \text{ m}$$
   - Reynolds number
   $$\text{Re} = \frac{V D_h}{\nu} = \frac{(13.9 \text{ m/s})(0.4 \text{ m})}{(15.1 \times 10^{-6} \text{ m}^2/\text{s})} = 368,000$$
   Thus, flow is turbulent.
   - Relative roughness
   $$k_s/D_h = (0.000046 \text{ m})/(0.4 \text{ m}) = 0.000115$$
   - Resistance coefficient (Moody diagram): $f = 0.015$

3. Darcy-Weisbach equation
$$h_f = f\left(\frac{L}{D_h}\right)\left(\frac{V^2}{2g}\right) = 0.015\left(\frac{50 \text{ m}}{0.4 \text{ m}}\right)\left\{\frac{(13.9 \text{ m/s})^2}{2(9.81 \text{ m/s}^2)}\right\}$$
$$= 18.6 \text{ m}$$

4. Pressure drop (from step 1)
$$p_1 - p_2 = \rho g h_L = (1.2 \text{ kg/m}^3)(9.81 \text{ m/s}^2)(18.6 \text{ m}) = 220 \text{ Pa}$$
$$\boxed{p_1 - p_2 = 0.883 \text{ inch } H_2O}$$

10.10

Pumps and Systems of Pipes

Previous sections have presented information for modeling flow in a single round pipe. This section extends this information by describing how to model flow in a network of pipes and how to incorporate performance data for a centrifugal pump. These topics are important because pumps and pipe networks are common.

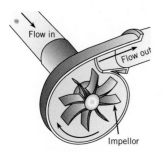

Figure 10.17

A centrifugal pump drives flow with a rotating impellor.

Modeling a Centrifugal Pump

As shown in Fig. 10.17, a *centrifugal pump* is a machine that uses a rotating set of blades situated within a housing to add energy to a flowing fluid. The amount of energy that is added is represented by the head of the pump h_p, and the rate at which work is done on the flowing fluid is $P = \dot{m}gh_p$.

To model a pump in a system, engineers commonly use a graphical solution involving the energy equation given in Eq. (7.29) and data from the pump manufacturer. To illustrate this approach, consider flow of water in the system of Fig. 10.18a. The energy equation applied from the reservoir water surface to the outlet stream is:

$$\frac{p_1}{g} + \frac{V_1^2}{2g} + z_1 + h_p = \frac{p_2}{g} + \frac{V_2^2}{2g} + z_2 + \sum K_L \frac{V^2}{2g} + \sum \frac{fL}{D}\frac{V^2}{2g}$$

For a system with one size of pipe, this simplifies to

$$h_p = (z_2 - z_1) + \frac{V^2}{2g}\left(1 + \sum K_L + \frac{fL}{D}\right) \tag{10.9}$$

Hence, for any given discharge, a certain head h_p must be supplied to maintain that flow. Thus, construct a head-versus-discharge curve, as shown in Fig. 10.18b. Such a curve is called the *system curve*. Now, a given centrifugal pump has a head-versus-discharge curve that is characteristic of that pump at a given pump speed. This curve is called a *pump curve*. A pump curve can be acquired from a pump manufacturer, or it can be measured. A typical pump curve is shown in Fig. 10.18b.

Figure 10.18b reveals that, as the discharge increases in a pipe, the head required for flow also increases. However, the head that is produced by the pump decreases as the discharge increases. Consequently, the two curves intersect, and the operating point is at the point of intersection—that point where the head produced by the pump is just the amount needed to overcome the head loss in the pipe.

To incorporate performance data for a pump, use the energy equation to derive a system curve. Then acquire a pump curve from a manufacturer or other source and plot the two curves together. The point of intersection shows where the pump will operate. This process is illustrated in Example 10.9.

Figure 10.18

(a) Pump and pipe combination.
(b) Pump and system curves.

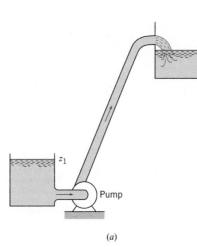

(a)

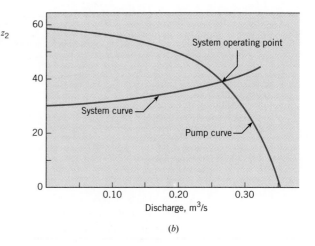

(b)

EXAMPLE 10.9 FINDING A SYSTEM OPERATING POINT

Problem Definition

Situation:

1. A pump has the head-versus-discharge curve shown in Fig. 10.18b.

2. The friction factor is $f = 0.015$.

Find: Discharge (m^3/s) in the system.

Plan

1. Develop an equation for the system curve by applying the energy equation.

2. Plot the given pump curve and the system curve on the same graph.

3. Find discharge Q by finding the intersection of the system and pump curve.

Solution

Energy equation

$$\frac{p_1}{g} + \frac{V_1^2}{2g} + z_1 + h_p = \frac{p_2}{g} + \frac{V_2^2}{2g} + z_2 + \sum h_L$$

$$0 + 0 + 200 + h_p = 0 + 0 + 230 + \left(\frac{fL}{D} + K_e + K_b + K_E\right)\frac{V^2}{2g}$$

Here $K_e = 0.5$, $K_b = 0.35$, and $K_E = 1.0$. Hence

$$h_p = 30 + \frac{Q^2}{2gA^2}\left[\frac{0.015(1000)}{0.40} + 0.5 + 0.35 + 1\right]$$

$$= 30 + \frac{Q^2}{2 \times 9.81 \times [(\pi/4) \times 0.4^2]^2}(39.3)$$

$$= 30 \text{ m} + 127Q^2 \text{ m}$$

Now make a table of Q versus h_p (see below) to give values to produce a system curve that will be plotted with the pump curve. When the system curve is plotted on the same graph as the pump curve, it is seen (Fig. 10.19b) that the operating condition occurs at $Q = 0.27 \text{ m}^3/\text{s}$.

Sketch:

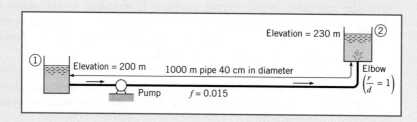

$Q(m^3/s)$	$h_p = 30 \text{ m} + 127Q^2 \text{ m}$
0	30
0.1	31.3
0.2	35.1
0.3	41.4

Pipes in Parallel

Consider a pipe that branches into two parallel pipes and then rejoins, as shown in Fig. 10.19. A problem involving this configuration might be to determine the division of flow in each pipe, given the total flow rate.

No matter which pipe is involved, the pressure difference between the two junction points is the same. Also, the elevation difference between the two junction points is the same. Because $h_L = (p_1/\gamma + z_1) - (p_2/\gamma + z_2)$, it follows that h_L between the two junction points is the same in both of the pipes of the parallel pipe system. Thus,

$$h_{L_1} = h_{L_2}$$

$$f_1 \frac{L_1}{D_1} \frac{V_1^2}{2g} = f_2 \frac{L_2}{D_2} \frac{V_2^2}{2g}$$

Then
$$\left(\frac{V_1}{V_2}\right)^2 = \frac{f_2 L_2 D_1}{f_1 L_1 D_2} \quad \text{or} \quad \frac{V_1}{V_2} = \left(\frac{f_2 L_2 D_1}{f_1 L_1 D_2}\right)^{1/2}$$

If f_1 and f_2 are known, the division of flow can be easily determined. However, some trial-and-error analysis may be required if f_1 and f_2 are in the range where they are functions of the Reynolds number.

Pipe Networks

The most common pipe networks are the water distribution systems for municipalities. These systems have one or more sources (discharges of water into the system) and numerous loads: one for each household and commercial establishment. For purposes of simplification, the loads are usually lumped throughout the system. Figure 10.20 shows a simplified distribution system with two sources and seven loads.

Figure 10.19

Flow in parallel pipes.

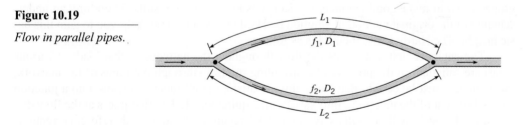

Figure 10.20

Pipe network.

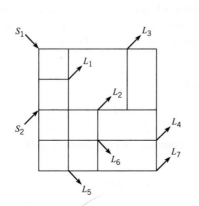

The engineer is often engaged to design the original system or to recommend an economical expansion to the network. An expansion may involve additional housing or commercial developments, or it may be to handle increased loads within the existing area.

In the design of such a system, the engineer will have to estimate the future loads for the system and will need to have sources (wells or direct pumping from streams or lakes) to satisfy the loads. Also, the layout of the pipe network must be made (usually parallel to streets), and pipe sizes will have to be determined. The object of the design is to arrive at a network of pipes that will deliver the design flow at the design pressure for minimum cost. The cost will include first costs (materials and construction) as well as maintenance and operating costs. The design process usually involves a number of iterations on pipe sizes and layouts before the optimum design (minimum cost) is achieved.

So far as the fluid mechanics of the problem are concerned, the engineer must determine pressures throughout the network for various conditions—that is, for various combinations of pipe sizes, sources, and loads. The solution of a problem for a given layout and a given set of sources and loads requires that two conditions be satisfied:

1. The continuity equation must be satisfied. That is, the flow into a junction of the network must equal the flow out of the junction. This must be satisfied for all junctions.
2. The head loss between any two junctions must be the same regardless of the path in the series of pipes taken to get from one junction point to the other. This requirement results because pressure must be continuous throughout the network (pressure cannot have two values at a given point). This condition leads to the conclusion that the algebraic sum of head losses around a given loop must be equal to zero. Here the sign (positive or negative) for the head loss in a given pipe is given by the sense of the flow with respect to the loop, that is, whether the flow has a clockwise or counterclockwise direction.

Only a few years ago, these solutions were made by trial-and-error hand computation, but modern applications using digital computers have made the older methods obsolete. Even with these advances, however, the engineer charged with the design or analysis of such a system must understand the basic fluid mechanics of the system to be able to interpret the results properly and to make good engineering decisions based on the results. Therefore, an understanding of the original method of solution by Hardy Cross (17) may help you to gain this basic insight. The Hardy Cross method is as follows.

The engineer first distributes the flow throughout the network so that loads at various nodes are satisfied. In the process of distributing the flow through the pipes of the network, the engineer must be certain that continuity is satisfied at all junctions (flow into a junction equals flow out of the junction), thus satisfying requirement 1. The first guess at the flow distribution obviously will not satisfy requirement 2 regarding head loss; therefore, corrections are applied. For each loop of the network, a discharge correction is applied to yield a zero net head loss around the loop. For example, consider the isolated loop in Fig. 10.21. In this loop, the loss of head in the clockwise direction will be given by

$$\sum h_{L_c} = h_{L_{AB}} + h_{L_{BC}}$$
$$= \sum kQ_c^n \tag{10.10}$$

The loss of head for the loop in the counterclockwise direction is

$$\sum h_{L_{cc}} = \sum_{cc} kQ_{cc}^n \tag{10.11}$$

Figure 10.21

A typical loop of a pipe network.

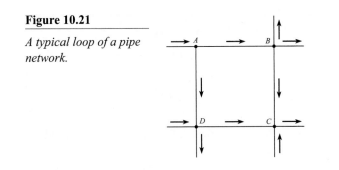

For a solution, the clockwise and counterclockwise head losses have to be equal, or

$$\sum h_{L_c} = \sum h_{L_{cc}}$$

$$\sum k Q_c^n = \sum k Q_{cc}^n$$

As noted, the first guess for flow in the network will undoubtedly be in error; therefore, a correction in discharge, ΔQ, will have to be applied to satisfy the head loss requirement. If the clockwise head loss is greater than the counterclockwise head loss, ΔQ will have to be applied in the counterclockwise direction. That is, subtract ΔQ from the clockwise flows and add it to the counterclockwise flows:

$$\sum k (Q_c - \Delta Q)^n = \sum k (Q_{cc} + \Delta Q)^n \tag{10.12}$$

Expand the summation on either side of Eq. (10.12) and include only two terms of the expansion:

$$\sum k (Q_c^n - n Q_c^{n-1} \Delta Q) = \sum k (Q_{cc}^n + n Q_{cc}^{n-1} \Delta Q)$$

Solve for ΔQ:

$$\Delta Q = \frac{\sum k Q_c^n - \sum k Q_{cc}^n}{\sum n k Q_c^{n-1} + \sum n k Q_{cc}^{n-1}} \tag{10.13}$$

Thus if ΔQ as computed from Eq. (10.13) is positive, the correction is applied in a counterclockwise sense (add ΔQ to counterclockwise flows and subtract it from clockwise flows).

A different ΔQ is computed for each loop of the network and applied to the pipes. Some pipes will have two ΔQs applied because they will be common to two loops. The first set of corrections usually will not yield the final desired result because the solution is approached only by successive approximations. Thus the corrections are applied successively until the corrections are negligible. Experience has shown that for most loop configurations, applying ΔQ as computed by Eq. (10.13) produces too large a correction. Fewer trials are required to solve for Qs if approximately 0.6 of the computed ΔQ is used.

More information on methods of solution of pipe networks is available in references (18) and (19). A search of the Internet under "pipe networks" yields information on software available from various sources.

EXAMPLE 10.10 DISCHARGE IN A PIPING NETWORK

A simple pipe network with water flow consists of three valves and a junction as shown in the figure. The piezometric head at points 1 and 2 is 1 ft and reduces to zero at point 4. There is a wide-open globe valve in line A, a gate valve half open in line B, and a wide-open angle valve in line C. The pipe diameter in all lines is 2 inches. Find the flow rate in each line. Assume that the head loss in each line is due only to the valves.

Problem Definition

Situation:

1. Water flows through a network of pipes.

2. $h_1 = h_2 = 1$ ft.

3. $h_4 = 0$ ft.

4. Pipe diameter (all pipes) is $2/12$ ft.

Find: Flow rate (in cfs) in each pipe.

Assumptions: Head loss is due to valves only.

Sketch:

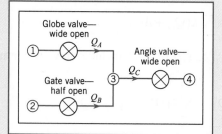

Plan

1. Let $h_{L,1\to3} = h_{L,2\to3}$.

2. Let $h_{L,2\to4} = 1$ ft.

3. Solve equations using the Hardy Cross approach.

Solution

The piezometric heads at points 1 and 2 are equal, so

$$h_{L,1\to3} + h_{L,3\to2} = 0$$

The head loss between points 2 and 4 is 1 ft, so

$$h_{L,2\to3} + h_{L,3\to4} = 0$$

Continuity must be satisfied at point 3, so

$$Q_A + Q_B = Q_C$$

The head loss through a valve is given by

$$h_L = K_V \frac{V^2}{2g}$$

$$= K_V \frac{1}{2g}\left(\frac{Q}{A}\right)^2$$

where K_V is the loss coefficient. For a 2-inch pipe, the head loss becomes

$$h_L = 32.6 K_V Q^2$$

where h_L is in feet and Q is in cfs.
The head loss equation between points 1 and 2 expressed in term of discharge is

$$32.6 K_A Q_A^2 - 32.6 K_B Q_B^2 = 0$$

or

$$K_A Q_A^2 - K_B Q_B^2 = 0$$

where K_A is the loss coefficient for the wide-open globe valve ($K_A = 10$) and K_B is the loss coefficient for the half-open gate valve ($K_B = 5.6$). The head loss equation between points 2 and 4 is

$$32.6 K_B Q_B^2 + 32.6 K_C Q_C^2 = 1$$

or

$$K_B Q_B^2 + K_C Q_C^2 = 0.0307$$

where K_C is the loss coefficient for the wide-open angle valve ($K_C = 5$). The two head loss equations and the continuity equation comprise three equations for Q_A, Q_B, and Q_C. However, the equations are nonlinear and require linearization and solution by iteration (Hardy Cross approach). The discharge is written as

$$Q = Q_0 + \Delta Q$$

where Q_0 is the starting value and ΔQ is the change. Then

$$Q^2 \cdot Q_0^2 + 2Q_0 \Delta Q$$

where the $(\Delta Q)^2$ term is neglected. The equations in terms of ΔQ become

$$2K_A Q_{0,A} \Delta Q_A - 2K_B Q_{0,B} \Delta Q_B = K_B Q_{0,B}^2 - K_A Q_{0,A}^2$$

$$2K_C Q_{0,C} \Delta Q_C - 2K_B Q_{0,B} \Delta Q_B = 0.0307 - K_B Q_{0,B}^2 - K_C Q_{0,C}^2$$

$$\Delta Q_A + \Delta Q_B = \Delta Q_C$$

which can be expressed in matrix form as

$$\begin{bmatrix} 2K_A Q_{0,A} & -2K_B Q_{0,B} & 0 \\ 0 & 2K_B Q_{0,B} & 2K_C Q_{0,C} \\ 1 & 1 & -1 \end{bmatrix} \begin{Bmatrix} \Delta Q_A \\ \Delta Q_B \\ \Delta Q_C \end{Bmatrix}$$

$$= \begin{bmatrix} K_B Q_{0,B}^2 - K_A Q_{0,A}^2 \\ 0.0307 - K_B Q_{0,B}^2 - K_C Q_{0,C}^2 \\ 0 \end{bmatrix}$$

The procedure begins by selecting values for $Q_{0,A}$, $Q_{0,B}$, and $Q_{0,C}$. Assume $Q_{0,A} = Q_{0,B}$ and $Q_{0,C} = 2Q_{0,A}$. Then from the head loss equation from points 2 to 4

$$K_B Q_{0,B}^2 + K_C Q_{0,C}^2 = 0.0307$$

$$(K_B + 4K_C)Q_{0,B}^2 = 0.0307$$

$$(5.6 + 4 \times 5)Q_{0,B}^2 = 0.0307$$

$$Q_{0,B} = 0.0346$$

and $Q_{0,A} = 0.0346$ and $Q_{0,C} = 0.0693$. These values are substituted into the matrix equation to solve for the ΔQ's. The discharges are corrected by $Q_0^{\text{new}} = Q_0^{\text{old}} + \Delta Q$ and substituted into the matrix equation again to yield new ΔQ's. The iterations are continued until sufficient accuracy is obtained. The accuracy is judged by how close the column matrix on the right approaches zero. A table with the results of iterations for this example is shown below.

		Iteration			
	Initial	1	2	3	4
Q_A	0.0346	0.0328	0.0305	0.0293	0.0287
Q_B	0.0346	0.0393	0.0384	0.0394	0.0384
Q_C	0.0693	0.0721	0.0689	0.0687	0.0671

This solution technique is called the Newton-Raphson method for nonlinear systems of algebraic equations. It can be implemented easily on a computer. The solution procedure for more complex systems is the same.

Summary

A conduit *is any* pipe, tube, or duct that is filled with a flowing fluid. This flow can be laminar or turbulent depending on Reynolds number (Re). The guideline used in this text is

$$\text{Re} \leq 2000 \qquad \text{laminar flow}$$
$$\text{Re} \geq 3000 \qquad \text{turbulent flow}$$

Near an entrance, flow is developing, which means that the velocity profile and wall shear stress are changing with distance from the entrance. Once the flow becomes fully developed, the velocity distribution is constant (uniform flow) and the wall shear stress has a constant value.

In fully developed flow, head loss (h_f or h_L) is given by a famous equation called the Darcy-Weisbach equation:

$$h_L = h_f = f \frac{L}{D} \frac{V^2}{2g}$$

where f is the resistance coefficient, L is the pipe length, D is the diameter, and V is the mean velocity. For laminar flow, the Hagen-Poiseuille theory is used to develop an equation for head loss

$$h_f = \frac{32\mu L V}{\gamma D^2}$$

and an equation for the resistance coefficient

$$f = \frac{64}{\text{Re}}$$

For turbulent flow, the resistance coefficient depends on the Reynolds number and the relative roughness:

$$f = f\left(\text{Re}, \frac{k_s}{D}\right)$$

where k_s is the equivalent sand grain roughness. Values for f can be obtained from the Moody diagram or from empirical equations.

The total head loss in a conduit is given by

$$\text{Overall (total) head loss} = \sum(\text{Pipe head loss}) + \sum(\text{Component head loss})$$

$$h_L = \underbrace{\sum f \frac{L}{D}\frac{V^2}{2g}}_{\text{pipes}} + \underbrace{\sum K \frac{V^2}{2g}}_{\text{components}}$$

and K is the minor loss coefficient. Pipe head loss (first group of terms) represents the head loss associated with fully developed flow in straight lengths of conduit. Component head loss (second group of terms) represents the head loss associated with components such as valves, elbows, bends, and transition sections. Values of K are tabulated in this text and in other engineering references.

Noncircular pipes can be analyzed using the hydraulic diameter D_h or the hydraulic radius (R_h), which are defined as

$$D_h = 4R_h = \frac{4 \times \text{section area}}{\text{wetted perimeter}}$$

To find the operating point of a centrifugal pump in a system, the standard approach is a graphical solution. One plots a system curve that is derived using the energy equation, and one plots the head versus flow rate curve of the centrifugal pump. The intersection of these two curves gives the operating point of the system.

The analysis of pipe networks is based on the continuity equation being satisfied at each junction and the head loss between any two junctions being independent of pipe path between the two junctions. A series of equations based on these principles are solved iteratively to obtain the flow rate in each pipe and the pressure at each junction in the network.

References

1. Reynolds, O. "An Experimental Investigation of the Circumstances Which Determine Whether the Motion of Water Shall Be Direct or Sinuous and of the Law of Resistance in Parallel Channels." *Phil. Trans. Roy. Soc. London,* 174, part III (1883).

2. Schlichting, Hermann. *Boundary Layer Theory,* 7th ed. McGraw-Hill, New York, 1979.

3. Moody, Lewis F. "Friction Factors for Pipe Flow." *Trans. ASME,* 671 (November 1944).

4. Nikuradse, J. "Strömungsgesetze in rauhen Rohren." *VDI-Forschungsh.,* no. 361 (1933). Also translated in *NACA Tech. Memo,* 1292.

5. White, F. M. *Viscous Fluid Flow.* McGraw-Hill, New York, 1991.

6. Colebrook, F. "Turbulent Flow in Pipes with Particular Reference to the Transition Region between the Smooth and Rough Pipe Laws." *J. Inst. Civ. Eng.*, vol. 11, 133–156 (1939).

7. Swamee, P. K., and A. K. Jain. "Explicit Equations for Pipe-Flow Problems." *J. Hydraulic Division of the ASCE*, vol. 102, no. HY5 (May 1976).

8. Streeter, V. L., and E. B. Wylie. *Fluid Mechanics,* 7th ed. McGraw-Hill, New York, 1979.

9. Barbin, A. R., and J. B. Jones. "Turbulent Flow in the Inlet Region of a Smooth Pipe." *Trans. ASME, Ser. D: J. Basic Eng.,* vol. 85, no. 1 (March 1963).

10. ASHRAE. *ASHRAE Handbook—1977 Fundamentals.* Am. Soc. of Heating, Refrigerating and Air Conditioning Engineers, Inc., New York, 1977.

11. Crane Co. "Flow of Fluids Through Valves, Fittings and Pipe." Technical Paper No. 410, Crane Co. (1988), 104 N. Chicago St., Joliet, IL 60434.

12. Fried, Irwin, and I. E. Idelchik. *Flow Resistance: A Design Guide for Engineers.* Hemisphere, New York, 1989.

13. Hydraulic Institute. *Engineering Data Book,* 2nd ed., Hydraulic Institute, 30200 Detroit Road, Cleveland, OH 44145.

14. Miller, D. S. *Internal Flow—A Guide to Losses in Pipe and Duct Systems.* British Hydrodynamic and Research Association (BHRA), Cranfield, England (1971).

15. Streeter, V. L. (ed.) *Handbook of Fluid Dynamics.* McGraw-Hill, New York, 1961.

16. Beij, K. H. "Pressure Losses for Fluid Flow in 90% Pipe Bends." *J. Res. Nat. Bur. Std.,* 21 (1938). Information cited in Streeter (20).

17. Cross, Hardy. "Analysis of Flow in Networks of Conduits or Conductors." *Univ. Illinois Bull.,* 286 (November 1936).

18. Hoag, Lyle N., and Gerald Weinberg. "Pipeline Network Analysis by Digital Computers." *J. Am. Water Works Assoc.,* 49 (1957).

19. Jeppson, Roland W. *Analysis of Flow in Pipe Networks.* Ann Arbor Science Publishers, Ann Arbor, MI, 1976.

20. Limerinos, J. T. "Determination of the Manning Coefficient from Measured Bed Roughness in Natural Channels." Water Supply Paper 1898-B, U.S. Geological Survey, Washington, D.C., 1970.

21. White, F. M. *Fluid Mechanics,* 5th Ed. McGraw-Hill, New York, 2003.

Problems

Notes on Pipe Diameter for Chapter 10 Problems

When a pipe diameter is given using the label "NPS" or "nominal," find the dimensions using Table 10.1 on p. 319. Otherwise, assume the specified diameter is an inside diameter (ID).

Classifying Flow

10.1 Kerosene (20°C) flows at a rate of 0.04 m³/s in a 25 cm diameter pipe. Would you expect the flow to be laminar or turbulent? Calculate the entrance length.

10.2 A compressor draws 0.3 m³/s of ambient air (20°C) in from the outside through a round duct that is 10 m long and 150 mm in diameter. Determine the entrance length and establish whether the flow is laminar or turbulent.

10.3 Specify the diameter and length for a tube that carries SAE 10W-30 oil at 38°C. The design requires laminar flow, fully developed flow, and a discharge of $Q = 0.2$ L/s.

Darcy-Weisbach Equation

10.4 P Q ◄ Using Section 10.3 and other resources, answer the following questions. Strive for depth, clarity, and accuracy while also combining sketches, words, and equations in ways that enhance the effectiveness of your communication.
a. What is pipe head loss? How is pipe head loss related to total head loss?
b. What is the friction factor f? How is f related to wall shear stress?
c. What assumptions need to be satisfied to apply the Darcy-Weisbach equation?

10.5 P Q ◄ For each case that follows, apply the Darcy-Weisbach equation from Eq. (10.12) to calculate the head loss in a pipe. Apply the grid method to carry and cancel units.
a. Water flows at a rate of 20 gpm and a mean velocity of 180 ft/min in a pipe of length 200 feet. For a resistance coefficient of $f = 0.02$, find the head loss in feet.
b. The head loss in a section of PVC pipe is 0.8 m, the resistance coefficient is $f = 0.012$, the length is 15 m and the flow rate is 1 cfs. Find the pipe diameter in meters.

10.6 As shown, air (20°C) is flowing from a large tank, through a horizontal pipe, and then discharging to ambient. The pipe length is $L = 50$m, and the pipe is schedule 40 PVC with a nominal diameter of 1 inch. The mean velocity in the pipe is 10 m/s, and $f = 0.015$. Determine the pressure (in Pa) that needs to be maintained in the tank.

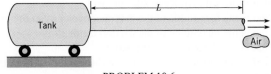

PROBLEM 10.6

10.7 Water (15°C) flows through a garden hose (ID = 18 mm) with a mean velocity of 1.5 m/s. Find the pressure drop for a section of hose that is 20 meters long and situated horizontally. Assume that $f = 0.012$.

10.8 As shown, water (15°C) is flowing from a tank through a tube and then discharging to ambient. The tube has an ID of 8 mm, a length of $L = 6$ m, and the resistance coefficient is $f = 0.015$. The water level is $H = 3$ m. Find the exit velocity in m/s. Find the discharge in $L/$s. Sketch the HGL and the EGL. Assume that the only head loss occurs in the tube.

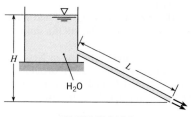

PROBLEM 10.8

10.9 Water flows in the pipe shown, and the manometer deflects 90 cm. What is f for the pipe if $V = 3$ m/s?

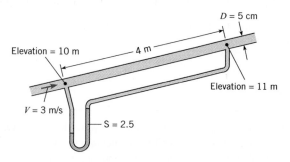

PROBLEM 10.9

Flow in Pipes (Laminar Flow)

10.10 P Q ◀ Using Section 10.5 and other resources, answer the questions that follow. Strive for depth, clarity, and accuracy while also combining sketches, words, and equations in ways that enhance the effectiveness of your communication.
a. What are the main characteristics of laminar flow?
b. What is the meaning of each variable that appears in Eq. (10.27)?
c. In Eq. (10.33), what is the meaning of h_f?

10.11 A fluid ($\mu = 10^{-2}$ N · s/m^2; $\rho = 800$ kg/m^3) flows with a mean velocity of 4 cm/s in a 10 cm smooth pipe.
a. What is the value of Reynolds number?
b. What is the magnitude of the maximum velocity in the pipe?
c. What is the magnitude of the friction factor f?
d. What is the shear stress at the wall?
e. What is the shear stress at a radial distance of 25 mm from the center of the pipe?

10.12 Water (15°C) flows in a horizontal schedule 40 pipe that has a nominal diameter of 0.5 in. The Reynolds number is Re $= 1000$. Work in SI units.
a. What is mass flow rate?
b. What is the magnitude of the friction factor f?
c. What is the head loss per meter of pipe length?
d. What is the pressure drop per meter of pipe length?

10.13 Flow of a liquid in a smooth 3 cm pipe yields a head loss of 2 m per meter of pipe length when the mean velocity is 1 m/s. Calculate f and the Reynolds number. Prove that doubling the flow rate will double the head loss. Assume fully developed flow.

10.14 As shown, a round tube of diameter 0.5 mm and length 750 mm is connected to plenum. A fan produces a negative gage pressure of −1.5 inch H$_2$O in the plenum and draws air (20°C) into the microchannel. What is the mean velocity of air in the microchannel? Assume that the only head loss is in the tube.

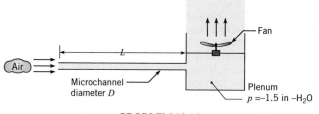

PROBLEM 10.14

10.15 Liquid ($\gamma = 10$ kN/m^3) is flowing in a pipe at a steady rate, but the direction of flow is unknown. Is the liquid moving upward or moving downward in the pipe? If the pipe diameter is 8 mm and the liquid viscosity is 3.0×10^{-3} N · s/m^2, what is the magnitude of the mean velocity in the pipe?

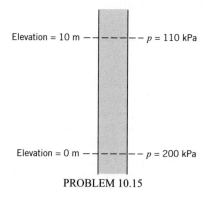

PROBLEM 10.15

10.16 Oil (S $= 0.97$, $\mu = 10^{-2}$ lbf-s/ft^2) is pumped through a nominal 1 in., schedule 80 pipe at the rate of 0.004 cfs. What is the head loss per 100 ft of level pipe?

10.17 A liquid ($\rho = 1000$ kg/m^3; $\mu = 10^{-1}$ N · s/2 m^2; $\nu = 10^{-4}$ m^2/s) flows uniformly with a mean velocity of 1.5 m/s in a pipe with a diameter of 100 mm. Show that the flow is laminar. Also, find the friction factor f and the head loss per meter of pipe length.

10.18 Kerosene (S $= 0.80$ and $T = 68$°F) flows from the tank shown and through the 1/4 in.–diameter (ID) tube. Determine the mean velocity in the tube and the discharge. Assume the only head loss is in the tube.

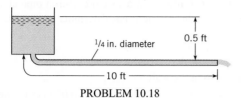

PROBLEM 10.18

10.19 Oil (S = 0.94; μ = 0.048 N · s/m^2) is pumped through a horizontal 5 cm pipe. Mean velocity is 0.5 m/s. What is the pressure drop per 10 m of pipe?

10.20 As shown, SAE 10W-30 oil is pumped through an 8 m length of 1 cm–diameter drawn tubing at a discharge of 7.85×10^{-4} m^3/s. The pipe is horizontal, and the pressures at points 1 and 2 are equal. Find the power necessary to operate the pump, assuming the pump has an efficiency of 100%. Properties of SAE 10W-30 oil: kinematic viscosity = 7.6×10^{-5} m^2/s; specific weight = 8630 N/m^3.

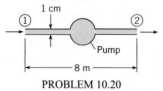

PROBLEM 10.20

10.21 Oil (S = 0.9; μ = 10^{-2} lbf-s/ft^2; ν = 0.0057 ft^2/s) flows downward in the pipe, which is 0.10 ft in diameter and has a slope of 30° with the horizontal. Mean velocity is 2 ft/s. What is the pressure gradient (dp/ds) along the pipe?

PROBLEM 10.21

10.22 In the pipe system for a given discharge, the ratio of the head loss in a given length of the 1 m pipe to the head loss in the same length of the 2 m pipe is (a) 2, (b) 4, (c) 16, or (d) 32.

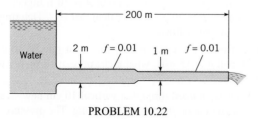

PROBLEM 10.22

10.23 Glycerine (T = 68°F) flows in a pipe with a 1/2 ft diameter at a mean velocity of 2 ft/s. Is the flow laminar or turbulent? Plot the velocity distribution across the flow section.

10.24 Glycerine (T = 20°C) flows through a funnel as shown. Calculate the mean velocity of the glycerine exiting the tube. Assume the only head loss is due to friction in the tube.

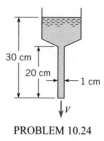

PROBLEM 10.24

10.25 What nominal size of steel pipe should be used to carry 0.2 cfs of castor oil at 90°F a distance of 0.5 mi with an allowable pressure drop of 10 psi (μ = 0.085 lbf-s/ft^2)? Assume S = 0.85.

10.26 Velocity measurements are made in a 30 cm pipe. The velocity at the center is found to be 1.5 m/s, and the velocity distribution is observed to be parabolic. If the pressure drop is found to be 1.9 kPa per 100 m of pipe, what is the kinematic viscosity ν of the fluid? Assume that the fluid's specific gravity is 0.80.

10.27 The velocity of oil (S = 0.8) through the 5 cm smooth pipe is 1.2 m/s. Here L = 12 m, z_1 = 1 m, z_2 = 2 m, and the manometer deflection is 10 cm. Determine the flow direction, the resistance coefficient f, whether the flow is laminar or turbulent, and the viscosity of the oil.

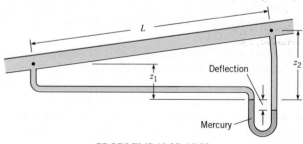

PROBLEMS 10.27, 10.28

10.28 The velocity of oil (S = 0.8) through the 2 in. smooth pipe is 5 ft/s. Here L = 30 ft, z_1 = 2 ft, z_2 = 4 ft, and the manometer deflection is 4 in. Determine the flow direction, the resistance coefficient f, whether the flow is laminar or turbulent, and the viscosity of the oil.

10.29 Glycerine at 20°C flows at 0.6 m/s in the 2 cm commercial steel pipe. Two piezometers are used as shown to measure the piezometric head. The distance along the pipe between the standpipes is 1 m. The inclination of the pipe is 20°. What is the height difference Δh between the glycerine in the two standpipes?

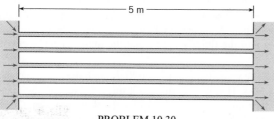

PROBLEM 10.29

10.30 Water is pumped through a heat exchanger consisting of tubes 5 mm in diameter and 5 m long. The velocity in each tube is 12 cm/s. The water temperature increases from 20°C at the entrance to 30°C at the exit. Calculate the pressure difference across the heat exchanger, neglecting entrance losses but accounting for the effect of temperature change by using properties at average temperatures.

PROBLEM 10.30

Flow in Pipes (Turbulent Flow)

10.31 Water (70°F) flows through a nominal 4 in., schedule 40, PVC pipe at the rate of 2 cfs. What is the resistance coefficient f?

10.32 Water at 20°C flows through a 3 cm ID smooth brass tube at a rate of 0.002 m³/s. What is f for this flow?

10.33 Water (10°C) flows through a 25 cm smooth pipe at a rate of 0.05 m³/s. What is the resistance coefficient f?

10.34 What is f for the flow of water at 10°C through a 10 cm cast-iron pipe with a mean velocity of 4 m/s? Also, apply Eq. (10.36) to plot the velocity distribution for this flow.

10.35 A fluid ($\mu = 10^{-2}$ N · s/m²; $\rho = 800$ kg/m³) flows with a mean velocity of 500 mm/s in a 100 mm diameter smooth pipe. Answer the following questions relating to the given flow conditions.
a. What is the magnitude of the maximum velocity in the pipe?
b. What is the magnitude of the resistance coefficient f?
c. What is the shear velocity?
d. What is the shear stress at a radial distance of 25 mm from the center of the pipe?
e. If the discharge is doubled, will the head loss per length of pipe also be doubled?

10.36 Water (20°C) flows in a 15 cm cast-iron pipe at a rate of 0.075 m³/s. For these conditions, determine or estimate the following:
a. Reynolds number
b. Friction factor f
c. Shear stress at the wall, τ_0

10.37 In a 4 in. uncoated cast-iron pipe, 0.02 cfs of water flows at 60°F. Determine f from Fig. 10.13.

10.38 Determine the head loss in 900 ft of a concrete pipe with a 6 in. diameter ($k_s = 0.0002$ ft) carrying 3.0 cfs of fluid. The properties of the fluid are $\nu = 3.33 \times 10^{-3}$ ft²/s and $\rho = 1.5$ slug/ft³.

10.39 Points A and B are 1 km apart along a 15 cm new steel pipe. Point B is 20 m higher than A. With a flow from A to B of 0.03 m³/s of crude oil (S = 0.82) at 10°C ($\mu = 10^{-2}$ N · s/m²), what pressure must be maintained at A if the pressure at B is to be 300 kPa?

10.40 A pipe can be used to measure the viscosity of a fluid. A liquid flows in a 1 cm smooth pipe 1 m long with an average velocity of 3 m/s. A head loss of 50 cm is measured. Estimate the kinematic viscosity.

10.41 For a 40 cm pipe, the resistance coefficient f was found to be 0.06 when the mean velocity was 3 m/s and the kinematic viscosity was 10^{-5} m²/s. If the velocity were doubled, would you expect the head loss per meter of length of pipe to double, triple, or quadruple?

10.42 Water (50°F) flows with a speed of 5 ft/s through a horizontal run of PVC pipe. The length of the pipe is 100 ft, and the pipe is schedule 40 with a nominal diameter of 2.5 inches. Calculate (a) the pressure drop in psi, (b) the head loss in feet, and (c) the power in horsepower needed to overcome the head loss.

10.43 Water (10°C) flows with a speed of 2 m/s through a horizontal run of PVC pipe. The length of the pipe is 50 m, and the pipe is schedule 40 with a nominal diameter of 2.5 inches. Calculate (a) the pressure drop in kilopascals, (b) the head loss in meters, and (c) the power in watts needed to overcome the head loss.

10.44 Air flows in a 3 cm smooth tube at a rate of 0.015 m³/s. If $T = 20$°C and $p = 110$ kPa absolute, what is the pressure drop per meter of length of tube?

10.45 Points A and B are 3 mi apart along a 24 in. new cast-iron pipe carrying water ($T = 50$°F). Point A is 30 ft higher than B. The pressure at B is 20 psi greater than that at A. Determine the direction and rate of flow.

10.46 Air flows in a 1 in. smooth tube at a rate of 30 cfm. If $T = 80$°F and $p = 15$ psia, what is the pressure drop per foot of length of tube?

10.47 Water is pumped through a vertical 10 cm new steel pipe to an elevated tank on the roof of a building. The pressure on the discharge side of the pump is 1.6 MPa. What pressure can be expected at a point in the pipe 80 m above the pump when the flow is 0.02 m³/s? Assume $T = 20$°C.

10.48 A train travels through a tunnel as shown. The train and tunnel are circular in cross section. Clearance is small, causing all air (60°F) to be pushed from the front of the train and discharged from the tunnel. The tunnel is 10 ft in diameter and is concrete. The train speed is 50 fps. Assume the concrete is very rough ($k_s = 0.05$ ft).

a. Determine the change in pressure between the front and rear of the train that is due to *pipe friction* effects.

b. Sketch the energy and hydraulic grade lines for the train position shown.

c. What power is required to produce the air flow in the tunnel?

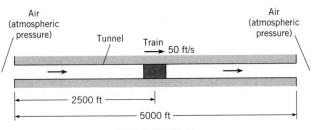

PROBLEM 10.48

10.49 Water (60°F) is pumped from a reservoir to a large, pressurized tank as shown. The steel pipe is 4 in. in diameter and 300 ft long. The discharge is 1 cfs. The initial water levels in the tanks are the same, but the pressure in tank B is 10 psig and tank A is open to the atmosphere. The pump efficiency is 90%. Find the power necessary to operate the pump for the given conditions.

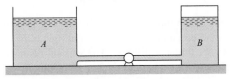

PROBLEM 10.49

Flow in Pipes (Iterative Solutions)

10.50 PQ◀ Using the information on page 332, classify each problem given below as case 1, case 2, or case 3. For each of your choices, state your rationale.

a. Problem 10.49.

b. Problem 10.52.

c. Problem 10.55.

10.51 A plastic siphon hose with $D = 1.2$ cm and $L = 5.5$ m is used to drain water (15°C) out a tank. Calculate the velocity in the tube for the two situations given below. Use $H = 3$ m and $h = 1$ m.

a. Assume the Bernoulli equation applies (neglect all head loss).

b. Assume the component head loss is zero, and the pipe head loss is nonzero.

10.52 A plastic siphon hose of length 7 m is used to drain water (15°C) out of a tank. For a flow rate of 1.5 L/s, what hose diameter is needed? Use $H = 5$ m and $h = 0.5$ m. Assume all head loss occurs in the tube.

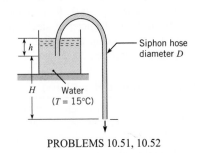

PROBLEMS 10.51, 10.52

10.53 As shown, water (70°F) is draining from a tank through a galvanized iron pipe. The pipe length is $L = 10$ ft, the tank depth is $H = 4$ ft, and the pipe is 1 inch NPS schedule 40. Calculate the velocity in the pipe and the flow rate. Neglect component head loss.

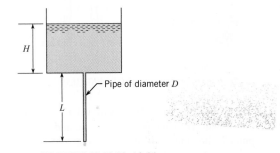

PROBLEMS 10.53, 10.54

10.54 As shown, water (15°C) is draining from a tank through a galvanized iron pipe. The pipe length is $L = 2$ m, the tank depth is $H = 1$ m, and the pipe is a 0.5 inch NPS schedule 40. Calculate the velocity in the pipe. Neglect component head loss.

10.55 Air (40°C, 1 atm) will be transported in a straight horizontal copper tube over a distance of 150 m at a rate of 0.1 m³/s. If the pressure drop in the tube should not exceed 6 in H_2O, what is the minimum pipe diameter?

10.56 A fluid with $v = 10^{-6}$ m²/s and $\rho = 800$ kg/m³ flows through the 8 cm galvanized iron pipe. Estimate the flow rate for the conditions shown in the figure.

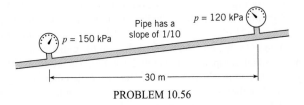

PROBLEM 10.56

10.57 Determine the diameter of commercial steel pipe required to convey 300 cfs of water at 60°F with a head loss of 1 ft per 1000 ft of pipe. Assume pipes are available in the even sizes when the diameters are expressed in inches (that is, 10 in., 12 in., etc.).

10.58 A pipeline is to be designed to carry crude oil (S = 0.93, $v = 10^{-5}$ m²/s) with a discharge of 0.10 m³/s and a head loss per kilometer of 50 m. What diameter of steel pipe is needed? What power output from a pump is required to maintain this flow? Available pipe diameters are 20, 22, and 24 cm.

Flow in Systems (Combined Head Loss)

10.59 The sketch shows a test of an electrostatic air filter. The pressure drop for the filter is 3 inches of water when the air-speed is 10 m/s. What is the minor loss coefficient for the filter? Assume air properties at 20°C.

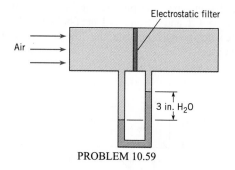

PROBLEM 10.59

10.60 If the flow of 0.10 m³/s of water is to be maintained in the system shown, what power must be added to the water by the pump? The pipe is made of steel and is 15 cm in diameter.

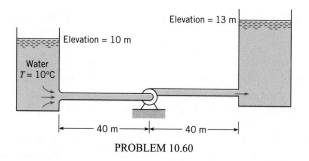

PROBLEM 10.60

10.61 Water will be siphoned through a 3/16 in.–diameter, 50 in.–long Tygon tube from a jug on an upside-down wastebasket into a graduated cylinder as shown. The initial level of the water in the jug is 21 in. above the table top. The graduated cylinder is a 500 ml cylinder, and the water surface in the cylinder is 12 in. above the table top when the cylinder is full. The bottom of the cylinder is 1/2 in. above the table. The inside diameter of the jug is 7 in. Calculate the time it will take to fill the cylinder from an initial depth of 2 in. of water in the cylinder.

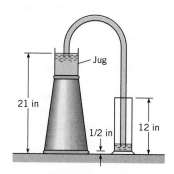

PROBLEM 10.61

10.62 Water flows from a tank through a 2.6 m length of galvanized iron pipe 26 mm in diameter. At the end of the pipe is an angle valve that is wide open. The tank is 2 m in diameter. Calculate the time required for the level in the tank to change from 10 m to 2 m. *Hint:* Develop an equation for dh/dt where h is the level and t if time. Then solve this equation numerically.

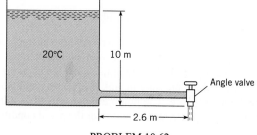

PROBLEM 10.62

10.63 A tank and piping system is shown. The galvanized pipe diameter is 1.5 cm, and the total length of pipe is 10 m. The two 90° elbows are threaded fittings. The vertical distance from the water surface to the pipe outlet is 5 m. The velocity of the water in the tank is negligible. Find (a) the exit velocity of the water and (b) the height (h) the water jet would rise on exiting the pipe. The water temperature is 20°C.

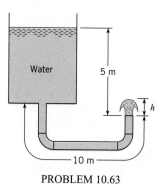

PROBLEM 10.63

10.64 A pump is used to fill a tank from a reservoir as shown. The head provided by the pump is given by $h_p = h_0(1 - (Q^2/Q^2_{max}))$ where h_0 is 50 meters, Q is the discharge through the pump, and

Q_{max} is 2 m³/s. Assume $f = 0.018$ and the pipe diameter is 90 cm. Initially the water level in the tank is the same as the level in the reservoir. The cross-sectional area of the tank is 100 m². How long will it take to fill the tank to a height, h, of 40 m?

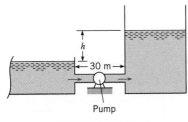

Pump

PROBLEM 10.64

10.65 A water turbine is connected to a reservoir as shown. The flow rate in this system is 5 cfs. What power can be delivered by the turbine if its efficiency is 80%? Assume a temperature of 70°F.

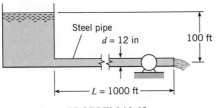

PROBLEM 10.65

10.66 What power must the pump supply to the system to pump the oil from the lower reservoir to the upper reservoir at a rate of 0.20 m³/s? Sketch the HGL and the EGL for the system.

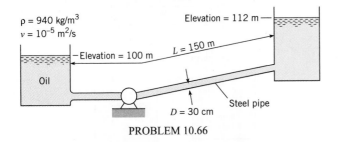

PROBLEM 10.66

10.67 A cast-iron pipe 1.0 ft in diameter and 200 ft long joins two water (60°F) reservoirs. The upper reservoir has a water-surface elevation of 100 ft, and the lower on has a water-surface elevation of 40 ft. The pipe exits from the side of the upper reservoir at an elevation of 70 ft and enters the lower reservoir at an elevation of 30 ft. There are two wide-open gate valves in the pipe. Draw the EGL and the HGL for the system, and determine the discharge in the pipe.

10.68 An engineer is making an estimate of hydroelectric power for a home owner. This owner has a small stream ($Q = 2$ cfs, $T = 40°F$) that is located at an elevation $H = 34$ ft above the owner's residence. The owner is proposing to divert the stream and operate a water turbine connected to an electric generator to supply electrical power to the residence. The maximum acceptable head loss in the penstock (a penstock is a conduit that supplies a turbine) is 3 ft. The penstock has a length of 87 ft. If the penstock is going to be fabricated from commercial-grade, plastic pipe, find the minimum diameter that can be used. Neglect component head losses. Assume that pipes are available in even sizes—that is, 2 in., 4 in., 6 in., etc.

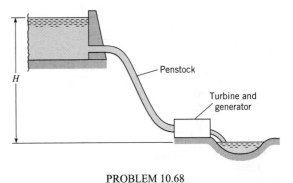

PROBLEM 10.68

10.69 The water-surface elevation in a reservoir is 120 ft. A straight pipe 100 ft long and 6 in. in diameter conveys water from the reservoir to an open drain. The pipe entrance (it is abrupt) is at elevation 100 ft, and the pipe outlet is at elevation 70 ft. At the outlet the water discharges freely into the air. The water temperature is 50°F. If the pipe is asphalted cast iron, what will be the discharge rate in the pipe? Consider all head losses. Also draw the HGL and the EGL for this system.

10.70 A heat exchanger is being designed as a component of a geothermal power system in which heat is transferred from the geothermal brine to a "clean" fluid in a closed-loop power cycle. The heat exchanger, a shell-and-tube type, consists of 100 galvanized-iron tubes 2 cm in diameter and 5 m long, as shown. The temperature of the fluid is 200°C, the density is 860 kg/m³, and the viscosity is 1.35×10^{-4} N · s/m². The total mass flow rate through the exchanger is 50 kg/s.

a. Calculate the power required to operate the heat exchanger, neglecting entrance and outlet losses.

b. After continued use, 2 mm of scale develops on the inside surfaces of the tubes. This scale has an equivalent roughness of 0.5 mm. Calculate the power required under these conditions.

10.71 The heat exchanger shown consists of 20 m of drawn tubing 2 cm in diameter with 19 return bends. The flow rate is 3×10^{-4} m³/s. Water enters at 20°C and exits at 80°C. The elevation difference between the entrance and the exit is 0.8 m. Calculate the pump power required to operate the heat exchanger if the pressure at 1 equals the pressure at 2. Use the viscosity corresponding to the average temperature in the heat exchanger.

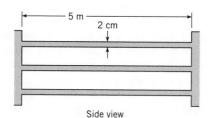

Side view

PROBLEM 10.70

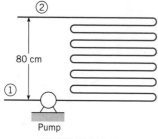

PROBLEM 10.71

10.72 A heat exchanger consists of a closed system with a series of parallel tubes connected by 180° elbows as shown in the figure. There are a total of 14 return elbows. The pipe diameter is 2 cm, and the total pipe length is 10 m. The head loss coefficient for each return elbow is 2.2. The tube is copper. Water with an average temperature of 40°C flows through the system with a mean velocity of 10 m/s. Find the power required to operate the pump if the pump is 80% efficient.

10.73 A heat exchanger consists of 15 m of copper tubing with an internal diameter of 15 mm. There are 14 return elbows in the system with a loss coefficient of 2.2 for each elbow. The pump in the system has a pump curve given by

$$h_p = h_{p0}\left[1 - \left(\frac{Q}{Q_{max}}\right)^3\right]$$

where h_{p0} is head provided by the pump at zero discharge and Q_{max} is 10^{-3} m^3/s. Water at 40°C flows through the system. Find the system operating point for values of h_{p0} of 2 m, 10 m, and 20 m.

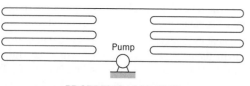

PROBLEMS 10.72, 10.73

10.74 Gasoline ($T = 50$°F) is pumped from the gas tank of an automobile to the carburetor through a 1/4 in. fuel line of drawn tubing 10 ft long. The line has five 90° smooth bends with an r/d

of 6. The gasoline discharges through a 1/32 in. jet in the carburetor to a pressure of 14 psia. The pressure in the tank is 14.7 psia. The pump is 80% efficient. What power must be supplied to the pump if the automobile is consuming fuel at the rate of 0.12 gpm? Obtain gasoline properties from Figs. A.2 and A.3.

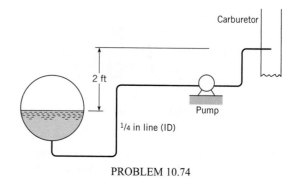

PROBLEM 10.74

10.75 Find the loss coefficient K_v of the partially closed valve that is required to reduce the discharge to 50% of the flow with the valve wide open as shown.

10.76 The pressure at a water main is 300 kPa gage. What size of pipe is needed to carry water from the main at a rate of 0.025 m^3/s to a factory that is 160 m from the main? Assume that galvanized-steel pipe is to be used and that the pressure required at the factory is 60 kPa gage at a point 10 m above the main connection.

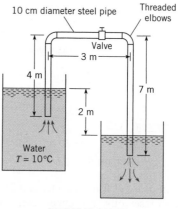

PROBLEM 10.75

10.77 The 10 cm galvanized-steel pipe is 1000 m long and discharges water into the atmosphere. The pipeline has an open globe valve and four threaded elbows; $h_1 = 3$ m and $h_2 = 15$ m. What is the discharge, and what is the pressure at A, the midpoint of the line?

10.78 Water is pumped at a rate of 25 m^3/s from the reservoir and out through the pipe, which has a diameter of 1.50 m. What power must be supplied to the water to effect this discharge?

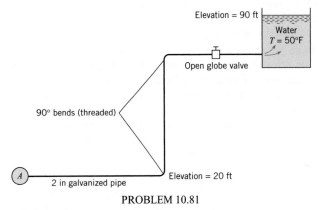

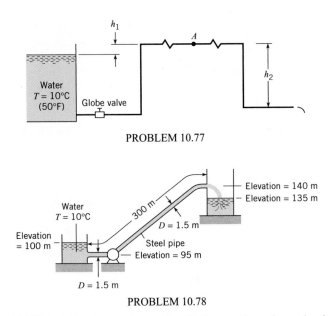

PROBLEM 10.77

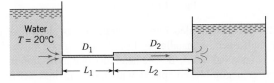

PROBLEM 10.78

10.79 Both pipes shown have an equivalent sand roughness k_s of 0.10 mm and a discharge of 0.1 m³/s. Also, D_1 = 15 cm, L_1 = 50 m, D_2 = 30 cm, and L_2 = 160 m. Determine the difference in the water-surface elevation between the two reservoirs.

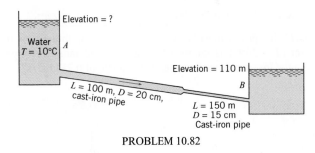

PROBLEM 10.79

10.80 Liquid discharges from a tank through the piping system shown. There is a venturi section at A and a sudden contraction at B. The liquid discharges to the atmosphere. Sketch the energy and hydraulic gradelines. Where might cavitation occur?

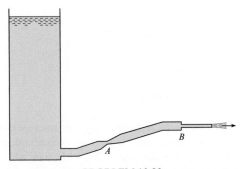

PROBLEM 10.80

10.81 The steel pipe shown carries water from the main pipe A to the reservoir and is 2 in. in diameter and 240 ft long. What must be the pressure in pipe A to provide a flow of 50 gpm?

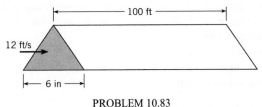

PROBLEM 10.81

10.82 If the water surface elevation in reservoir B is 110 m, what must be the water surface elevation in reservoir A if a flow of 0.03 m³/s is to occur in the cast-iron pipe? Draw the HGL and the EGL, including relative slopes and changes in slope.

PROBLEM 10.82

Nonround Conduits

10.83 Air at 60°F and atmospheric pressure flows in a horizontal duct with a cross section corresponding to an equilateral triangle (all sides equal). The duct is 100 ft long, and the dimension of a side is 6 in. The duct is constructed of galvanized iron (k_s = 0.0005 ft). The mean velocity in the duct is 12 ft/s. What is the pressure drop over the 100 ft length?

PROBLEM 10.83

10.84 A cold-air duct 100 cm by 15 cm in cross section is 100 m long and made of galvanized iron. This duct is to carry air at a rate of 6 m³/s at a temperature of 15°C and atmospheric pressure. What is the power loss in the duct?

10.85 Air (20°C) flows with a speed of 10 m/s through a horizontal rectangular air-conditioning duct. The duct is 20 m long and has a cross section of 4 by 10 in. (102 by 254 mm). Calculate

(a) the pressure drop in inches of water and (b) the power in watts needed to overcome head loss. Assume the roughness of the duct is $k_s = 0.004$ mm. Neglect component head losses.

10.86 An air-conditioning system is designed to have a duct with a rectangular cross section 1 ft by 2 ft, as shown. During construction, a truck driver backed into the duct and made it a trapezoidal section, as shown. The contractor, behind schedule, installed it anyway. For the same pressure drop along the pipe, what will be the ratio of the velocity in the trapezoidal duct to that in the rectangular duct? Assume the Darcy-Weisbach resistance coefficient is the same for both ducts.

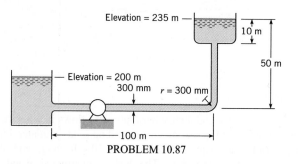

PROBLEM 10.86

Modeling Pumps in Systems

10.87 What power must be supplied by the pump to the flow if water ($T = 20°C$) is pumped through the 300 mm steel pipe from the lower tank to the upper one at a rate of 0.314 m^3/s?

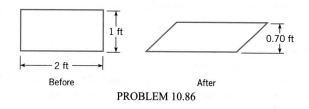

PROBLEM 10.87

10.88 If the pump for Fig. 10.18*b* is installed in the system of Prob. 10.87, what will be the rate of discharge of water from the lower tank to the upper one?

10.89 A pump that has the characteristic curve shown in the accompanying graph is to be installed as shown. What will be the discharge of water in the system?

10.90 If the liquid of Prob. 10.89 is a superliquid (zero head loss occurs with the flow of this liquid), then what will be the pumping rate, assuming that the pump curve is the same?

Pipes in Parallel and in Networks

10.91 A pipe system consists of a gate valve, wide open ($K_v = 0.2$), in line *A* and a globe valve, wide open ($K_v = 10$), in line *B*. The cross-sectional area of pipe *A* is half of the cross-sectional area of pipe *B*. The head loss due to the junction, elbows, and pipe friction are negligible compared with the head loss through the valves. Find the ratio of the discharge in line *B* to that in line *A*.

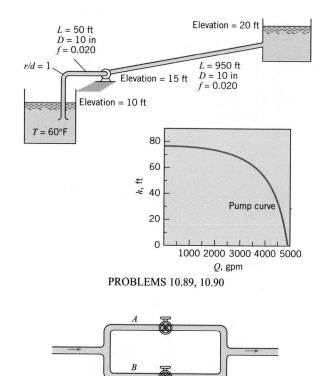

PROBLEMS 10.89, 10.90

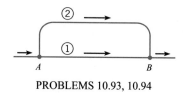

PROBLEMS 10.91, 10.92

10.92 A flow is divided into two branches as shown. A gate valve, half open, is installed in line *A*, and a globe valve, fully open, is installed in line *B*. The head loss due to friction in each branch is negligible compared with the head loss across the valves. Find the ratio of the velocity in line *A* to that in line *B* (include elbow losses for threaded pipe fittings).

10.93 In the parallel system shown, pipe 1 has a length of 1000 m and is 50 cm in diameter. Pipe 2 is 1500 m long and 40 cm in diameter. The pipe is commercial steel. What is the division of the flow of water at 10°C if the total discharge is to be 1.2 m^3/s?

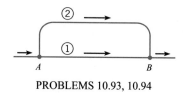

PROBLEMS 10.93, 10.94

10.94 Pipes 1 and 2 are the same kind (cast-iron pipe), but pipe 2 is four times as long as pipe 1. They are the same diameter (1 ft). If the discharge of water in pipe 2 is 1 cfs, then what will be the discharge in pipe 1? Assume the same value of f in both pipes.

10.95 Water flows from left to right in this parallel pipe system. The pipe having the greatest velocity is (a) pipe A, (b) pipe B, or (c) pipe C.

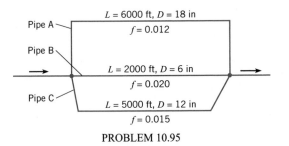

PROBLEM 10.95

10.96 Two pipes are connected in parallel. One pipe is twice the diameter of the other and three times as long. Assume that f in the larger pipe is 0.010 and f in the smaller one is 0.014. Determine the ratio of the discharges in the two pipes.

10.97 With a total flow of 14 cfs, determine the division of flow and the head loss from A to B.

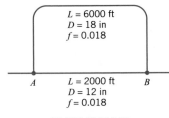

PROBLEM 10.97

10.98 The pipes shown in the system are all concrete. With a flow of 25 cfs of water, find the head loss and the division of flow in the pipes from A to B. Assume $f = 0.030$ for all pipes.

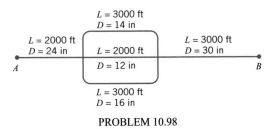

PROBLEM 10.98

10.99 A parallel pipe system is set up as shown. Flow occurs from A to B. To augment the flow, a pump having the characteristics shown in Fig. 10.15 is installed at point C. For a total discharge of 0.60 m³/s, what will be the division of flow between the pipes and what will be the head loss between A and B? Assume commercial steel pipe.

10.100 For the given source and loads shown, how will the flow be distributed in the simple network, and what will be the pressures at the load points if the pressure at the source is 60 psi? Assume horizontal pipes and $f = 0.012$ for all pipes.

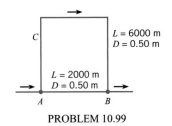

PROBLEM 10.99

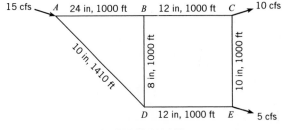

PROBLEM 10.100

10.101 Frequently in the design of pump systems, a bypass line will be installed in parallel to the pump so that some of the fluid can recirculate as shown. The bypass valve then controls the flow rate in the system. Assume that the head-versus-discharge curve for the pump is given by $h_p = 100 - 100Q$, where h_p is in meters and Q is in m³/s. The bypass line is 10 cm in diameter. Assume the only head loss is that due to the valve, which has a head-loss coefficient of 0.2. The discharge leaving the system is 0.2 m³/s. Find the discharge through the pump and bypass line.

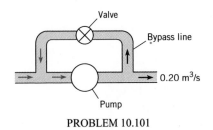

PROBLEM 10.101

C H A P T E R

11

Drag and Lift

SIGNIFICANT LEARNING OUTCOMES

Conceptual Knowledge

- Describe lift and drag.
- Describe form drag and friction drag.
- Relate lift and drag to pressure and shear-stress distributions.
- Explain the significance and meaning of vortex shedding, streamlining, and terminal velocity.
- Relate drag to flow separation.

Procedural Knowledge

- Find the coefficient of drag and calculate the drag force.
- Find the coefficient of lift and calculate the lift force.

Applications (Typical Examples)

- For objects moving through a fluid (e.g., automobile, bird), determine power requirements.
- For structures (e.g., bridge, sign, tower), determine wind loads.
- For falling objects (e.g., pollen spore, parachute), calculate terminal velocity.
- For a wing, determine the lift force.

Previous chapters have described various forces that are caused by fluids. This list includes hydrostatic force on a panel, buoyant force on a submerged object, and shear force on a flat plate. This chapter expands this list by describing lift and drag forces.

When a body moves through a stationary fluid or when a fluid flows past a body, the fluid exerts a resultant force. The component of this resultant force that is parallel to the free-stream velocity is called the *drag force*. Similarly, the *lift force* is the component of the resultant force that is perpendicular to the free stream. For example, as air flows over a kite it creates a resultant force that can be resolved in lift and drag components as shown in Fig. 11.1. By definition, lift and drag forces are limited to those forces produced by a flowing fluid.

Figure 11.1

(a) A kite.
(b) Forces acting on the kite due to the air flowing over the kite.

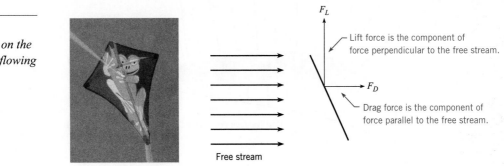

Lift force is the component of force perpendicular to the free stream.

F_L

F_D

Drag force is the component of force parallel to the free stream.

Free stream

(a)   *(b)*

11.1

Relating Lift and Drag to Stress Distributions

This section explains how lift and drag forces are related to stress distributions. This section also introduces the concepts of form and friction drag. These ideas are fundamental to understanding of lift and drag.

Integrating a Stress Distribution to Yield Force

Lift and drag forces are related to the stress distribution on a body through integration. For example, consider the stress acting on the airfoil shown in Fig. 11.2. As shown, there is a pressure distribution and a shear-stress distribution. To relate stress to force, select a differential area as shown in Fig. 11.3. The magnitude of the pressure force is $dF_p = p\,dA$, and the magnitude of the viscous force is $dF_v = \tau\,dA$.* The differential lift force is normal to the free-stream direction

$$dF_L = -p\,dA\sin\theta - \tau\,dA\cos\theta$$

and the differential drag is parallel to the free-stream direction

$$dF_D = -p\,dA\cos\theta + \tau\,dA\sin\theta$$

Integration over the surface of the airfoil gives lift force (F_L) and drag force (F_D):

$$F_L = \int (-p\sin\theta - \tau\cos\theta)\,dA \tag{11.1}$$

$$F_D = \int (-p\cos\theta + \tau\sin\theta)\,dA \tag{11.2}$$

Equations (11.1) and (11.2) show that the lift and drag forces are related to the stress distributions through integration.

Figure 11.2

Pressure and shear stress acting on an airfoil.

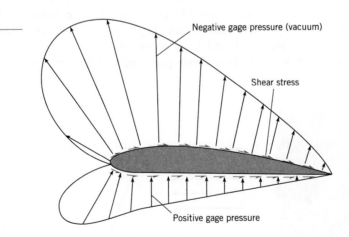

Negative gage pressure (vacuum)

Shear stress

Positive gage pressure

---

* The sign convention on τ is such that a clockwise sense of $\tau\,dA$ on the surface of the foil signifies a positive sign for τ.

Figure 11.3

Pressure and viscous forces acting on a differential element of area.

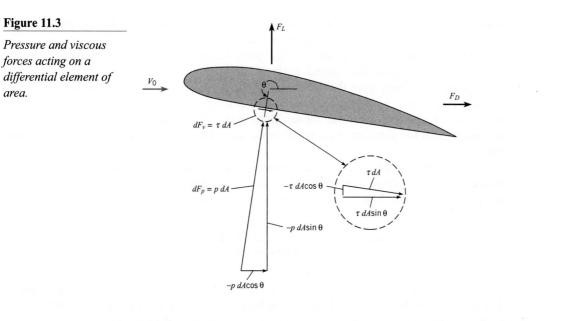

Form Drag and Friction Drag

Notice that Eq. (11.2) can be written as the sum of two integrals.

$$F_D = \underbrace{\int (-p\cos\theta)\, dA}_{\text{form drag}} + \underbrace{\int (\tau\sin\theta)\, dA}_{\text{friction drag}} \tag{11.3}$$

Form drag is the portion of the total drag force that is associated with the pressure distribution. *Friction drag* is the portion of the total drag force that is associated with the viscous shear-stress distribution. The drag force on any body is the sum of form drag and friction drag. In words, Eq. (11.3) can be written as

$$(\text{total drag force}) = (\text{form drag}) + (\text{friction drag}) \tag{11.4}$$

Calculating Drag Force

This section introduces the drag force equation, the coefficient of drag, and presents data for two-dimensional bodies. This information is used to calculate drag force on objects.

Drag Force Equation

The drag force F_D on a body is found by using the drag force equation:

$$F_D = C_D\, A\left(\frac{\rho V_0^2}{2}\right) \tag{11.5}$$

where C_D is called the coefficient of drag, A is a reference area of the body, ρ is the fluid density, and V_0 is the free-stream velocity measured relative to the body.

The reference area A depends on the type of body. One common reference area, called *projected area* and given the symbol A_p, is the silhouetted area that would be seen by a person looking at the body from the direction of flow. For example, the projected area of a plate normal to the flow is $b\ell$, and the projected area of a cylinder with its axis normal to the flow is $d\ell$. Other geometries use different reference areas; for example, the reference area for an airplane wing is the planform area, which is the area observed when the wing is viewed from above.

The *coefficient of drag* C_D is a parameter that characterizes the drag force associated with a given body shape. For example, an airplane might have $C_D = 0.03$, and a baseball might have $C_D = 0.4$. The coefficient of drag is a π-group that is defined by

$$C_D \equiv \frac{F_D}{A(\rho V_0^2/2)} = \frac{(\text{drag force})}{(\text{reference area})(\text{kinetic pressure})} \tag{11.6}$$

Values of the coefficient of drag C_D are usually found by experiment. For example, drag force F_D can be measured using a force balance in a wind tunnel. Then C_D can be calculated using Eq. (11.6). For this calculation, speed of the air in the wind tunnel V_0 can be measured using a Pitot-static tube or similar device, and air density can be calculated by applying the ideal gas law using measured values of temperature and pressure.

Equation (11.5) shows that drag force is related to four variables. Drag is related to the shape of an object because shape is characterized by the value of C_D. Drag is related to the size of the object because size is characterized by the reference area. Drag is related to the density of ambient fluid. Finally, drag is related to the speed of the fluid squared. This means that if the wind velocity doubles and C_D is constant, then the wind load on a building goes up by a factor of four.

Coefficient of Drag (Two-Dimensional Bodies)

This section presents C_D data and describes how C_D varies with the Reynolds number for objects that can be classified as two-dimensional. A *two-dimensional body* is a body with a uniform section area and a flow pattern that is independent of the ends of the body. Examples of two-dimensional bodies are shown in Fig. 11.4. In the aerodynamics literature, C_D values for two-dimensional bodies are called *sectional drag coefficients*. Two-dimensional bodies can be visualized as objects that are infinitely long in the direction normal to the flow.

The sectional drag coefficient can be used to estimate C_D for real objects. For example, C_D for a cylinder with a length to diameter ratio of 20 (e.g., $L/D \geq 20$) approaches the sectional drag coefficient because the end effects have an insignificant contribution to the total drag force. Alternatively, the sectional drag coefficient would be inaccurate for a cylinder with a small L/D ratio (e.g., $L/D \approx 1$) because the end effects would be important.

As shown in Fig. 11.4, the Reynolds number sometimes, but not always, influences the sectional drag coefficient. The value of C_D for the flat plate and square rod are independent of Re. The sharp edges of these bodies produces flow separation, and the drag force is due to the pressure distribution (form drag) and not on the shear-stress distribution (friction drag, which depends on Re). Alternatively, C_D for the cylinder and the streamlined strut show strong Re dependence because both form and friction drag are significant.

Figure 11.4

Coefficient of drag versus Reynolds number for two-dimensional bodies. [Data sources: Bullivant (1), DeFoe (2), Goett and Bullivant (3), Jacobs (4), Jones (5), and Lindsey (6).]

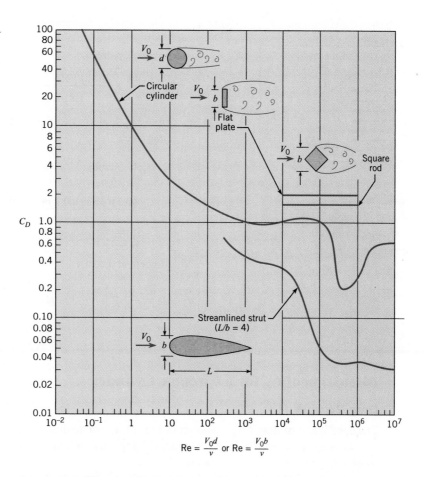

To calculate drag force on an object, find a suitable coefficient of drag and then apply the drag force equation. This approach is illustrated by Example 11.1.

EXAMPLE 11.1   DRAG FORCE ON A CYLINDER

A vertical cylinder that is 30 m high and 30 cm in diameter is being used to support a television transmitting antenna. Find the drag force acting on the cylinder and the bending moment at its base. The wind speed is 35 m/s, the air pressure is 1 atm, and temperature is 20°C.

Problem Definition

Situation: Wind is blowing across a tall cylinder.

Find:

1. Drag force (in N) on the cylinder.
2. Bending moment (in N · m) acting at the base of the cylinder.

Assumptions:

1. Wind speed is steady.
2. Effects associated with the ends of the cylinder are negligible because $L/D = 100$.
3. Neglect drag force on the antenna because the frontal area is much less than the frontal area of the cylinder.
4. The line of action of the drag force is at an elevation of 15 m, halfway up to the cylinder.

Properties: Air (20°C), Table A.5: $\rho = 1.2$ kg/m^3, and $\mu = 1.81 \times 10^{-5}$ N · s/m^2.

Sketch:

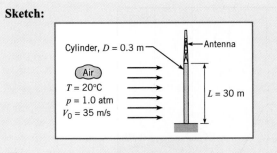

Cylinder, $D = 0.3$ m — Antenna

Air

$T = 20°C$
$p = 1.0$ atm
$V_0 = 35$ m/s

$L = 30$ m

Plan

1. Calculate the Reynolds number.
2. Find coefficient of drag using Fig. 11.4.
3. Calculate drag force using Eq. (11.5).
4. Calculate bending moment using $M = F_D \cdot L/2$.

Solution

1. Reynolds number

$$\mathrm{Re} = \frac{V_0 d\rho}{\mu} = \frac{35 \text{ m/s} \times 0.30 \text{ m} \times 1.20 \text{ kg/m}^3}{1.81 \times 10^{-5} \text{ N} \cdot \text{s/m}^2} = 7.0 \times 10^5$$

2. From Fig. 11.4, the coefficient of drag is $C_D = 0.20$.
3. Drag force

$$F_D = \frac{C_D A_p \rho V_0^2}{2}$$

$$= \frac{(0.2)(30 \text{ m})(0.3 \text{ m})(1.20 \text{ kg/m}^3)(35^2 \text{m}^2/\text{s}^2)}{2}$$

$$= \boxed{1323 \text{ N}}$$

4. Moment at the base

$$M = F_D\left(\frac{L}{2}\right) = (1323 \text{ N})\left(\frac{30}{2} \text{ m}\right) = \boxed{19{,}800 \text{ N} \cdot \text{m}}$$

Discussion of C_D for a Circular Cylinder

Drag Regimes

The coefficient of drag C_D, as shown in Fig. 11.4, can be described in terms of three regimes.

Regime I ($\mathrm{Re} < 10^3$). In this regime, C_D depends on both form drag and friction drag. As shown, C_D decreases with increasing Re.

Regime II ($10^3 < \mathrm{Re} < 10^5$). In this regime, C_D has a nearly constant value. The reason is that form drag, which is associated with the pressure distribution, is the dominant cause of drag. Over this range of Reynolds numbers, the flow pattern around the cylinder remains virtually unchanged, thereby producing very similar pressure distributions. This characteristic, the constancy of C_D at high values of Re, is representative of most bodies that have angular form.

Regime III ($10^5 < \mathrm{Re} < 5 \times 10^5$). In this regime, C_D decreases by about 500%, a remarkable change! This change occurs because the boundary layer on the circular cylinder changes. For Reynolds numbers less than 10^5, the boundary layer is laminar, and separation occurs about midway between the upstream side and downstream side of the cylinder (Fig. 11.5). Hence the entire downstream half of the cylinder is exposed to a relatively low pressure, which in turn produces a relatively high value for C_D. When the Reynolds number is increased to about 10^5, the boundary layer becomes turbulent, which causes higher-velocity fluid to be mixed into the region close to the wall of the cylinder. As a consequence of the presence of this high-velocity, high-momentum fluid in the boundary layer, the flow proceeds farther downstream along the surface of the cylinder against the adverse pressure before separation occurs (Fig. 11.6). This change in separation produces a much smaller zone of low pressure and the lower value of C_D.

Surface Roughness

Surface roughness has a major influence on drag. For example, if the surface of the cylinder is slightly roughened upstream of the midsection, the boundary layer will be forced to become turbulent at lower Reynolds numbers than those for a smooth cylinder surface. The same

Figure 11.5

Flow pattern around a cylinder for $10^3 < \mathrm{Re} < 10^5$.

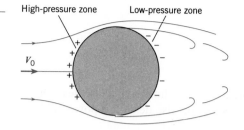

Figure 11.6

Flow pattern around a cylinder for $\mathrm{Re} > 5 \times 10^5$.

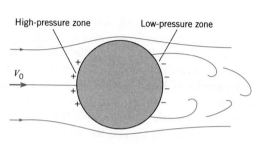

trend can also be produced by creating abnormal turbulence in the approach flow. The effects of roughness are shown in Fig. 11.7 for cylinders that were roughened with sand grains of size k. A small to medium size of roughness ($10^{-3} < k/d < 10^{-2}$) on a cylinder triggers an early onset of reduction of C_D. However, when the relative roughness is quite large ($10^{-2} < k/d$), the characteristic dip in C_D is absent.

Figure 11.7

Effects of roughness on C_D for a cylinder. [After Miller, et al. (7).]

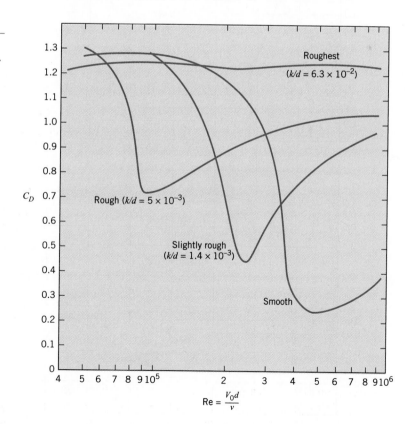

Drag of Axisymmetric and 3D Bodies

Section 11.2 described drag for two-dimensional bodies. Drag on other body shapes is presented in this section. This section also describes power and rolling resistance.

Drag Data

An object is classified as an *axisymmetric body* when the flow direction is parallel to an axis of symmetry of the body and the resulting flow is also symmetric about his axis. Examples of axisymmetric bodies include a sphere, bullet, and javelin. When flow is not aligned with an axis of symmetry, the flow field is three-dimensional (3D), and the body is classified as a three-dimensional or *3D body*. Examples of 3D bodies include a tree, a building, and an automobile.

The principles that apply to two-dimensional flow over a body also apply to axisymmetric flows. For example, at very low values of the Reynolds number, the coefficient of drag is given by exact equations relating C_D and Re. At high values of Re, the coefficient of drag becomes constant for angular bodies, whereas rather abrupt changes in C_D occur for rounded bodies. All of these characteristics can be seen in Fig. 11.8, where C_D is plotted against Re for several axisymmetric bodies.

Figure 11.8

Coefficient of drag versus Reynolds number for axisymmetric bodies. [Data sources: Abbott (9), Brevoort and Joyner (10), Freeman (11), and Rouse (12).]

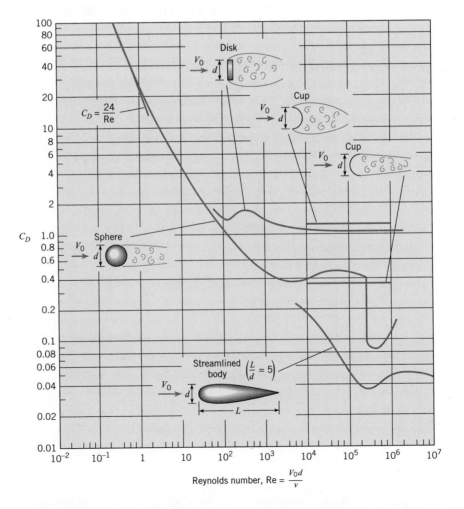

The drag coefficient of a sphere is of special interest because many applications involve the drag of spherical or near-spherical objects, such as particles and droplets. Also, the drag of a sphere is often used as a standard of comparison for other shapes. For Reynolds numbers less than 0.5, the flow around the sphere is laminar and amenable to analytical solutions. An exact solution by Stokes yielded the following equation, which is called Stokes's equation, for the drag of a sphere:

$$F_D = 3\pi\mu V_0 d \tag{11.7}$$

Note that the drag for this laminar flow condition varies directly with the first power of V_0. This is characteristic of all laminar flow processes. For completely turbulent flow, the drag is a function of the velocity to the second power. When the drag force given by Eq. (11.7) is substituted into Eq. (11.6), the result is the drag coefficient corresponding to Stokes's equation:

$$C_D = \frac{24}{\mathrm{Re}} \tag{11.8}$$

Thus for flow past a sphere, when $\mathrm{Re} \leq 0.5$, one may use the direct relation for C_D given in Eq. (11.8).

Several correlations for the drag coefficient of a sphere are available (13). One such correlation has been proposed by Clift and Gauvin (14):

$$C_D = \frac{24}{\mathrm{Re}}(1 + 0.15\mathrm{Re}^{0.687}) + \frac{0.42}{1 + 4.25 \times 10^4 \mathrm{Re}^{-1.16}} \tag{11.9}$$

which deviates from the *standard drag curve** by –4% to 6% for Reynolds numbers up to 3×10^5. Note that as the Reynolds number approaches zero, this correlation reduces to the equation for Stokes flow.

Values for C_D for other axisymmetric and 3D bodies at high Reynolds numbers ($\mathrm{Re} > 10^4$) are given in Table 11.1. Extensive data on the drag of various shapes is available in Hoerner (15).

Table 11.1 APPROXIMATE C_D VALUES FOR VARIOUS BODIES				
	Type of Body	**Length Ratio**	**Re**	**C_D**
	Rectangular plate	$l/b = 1$	$>10^4$	1.18
		$l/b = 5$	$>10^4$	1.20
		$l/b = 10$	$>10^4$	1.30
		$l/b = 20$	$>10^4$	1.50
		$l/b = \infty$	$>10^4$	1.98
	Circular cylinder with axis parallel to flow	$l/d = 0$ (disk)	$>10^4$	1.17
		$l/d = 0.5$	$>10^4$	1.15
		$l/d = 1$	$>10^4$	0.90
		$l/d = 2$	$>10^4$	0.85
		$l/d = 4$	$>10^4$	0.87
		$l/d = 8$	$>10^4$	0.99
	Square rod	∞	$>10^4$	2.00

(Continued)

* The *standard drag curve* represents the best fit of the cumulative data that have been obtained for drag coefficient of a sphere.

	Type of Body	Length Ratio	Re	C_D
	Square rod	∞	$>10^4$	1.50
	Triangular cylinder	∞	$>10^4$	1.39
	Semicircular shell	∞	$>10^4$	1.20
	Semicircular shell	∞	$>10^4$	2.30
	Hemispherical shell		$>10^4$	0.39
	Hemispherical shell		$>10^4$	1.40
	Cube		$>10^4$	1.10
	Cube		$>10^4$	0.81
	Cone—60° vertex		$>10^4$	0.49
	Parachute		$\approx 3 \times 10^7$	1.20

SOURCES: Brevoort and Joyner (10), Lindsey (6), Morrison (16), Roberson et al. (17), Rouse (12), and Scher and Gale (18).

To find the drag force on an object, find or estimate the coefficient of drag and then apply the drag force equation. This approach is illustrated by Example 11.2.

EXAMPLE 11.2 DRAG ON A SPHERE

What is the drag of a 12 mm sphere that drops at a rate of 8 cm/s in oil ($\mu = 10^{-1}$ N · s/m², S = 0.85)?

Problem Definition

Situation:

1. A sphere ($d = 0.012$ m) is falling in oil.
2. Speed of the sphere is $V = 0.08$ m/s.

Find: Drag force (in newtons) on the sphere.

Assumptions: Sphere is moving at a steady speed (terminal velocity).

Properties:

Oil: $\mu = 10^{-1}$ N · s/m², S = 0.85, $\rho = 850$ kg/m³.

Plan

1. Calculate the Reynolds number.
2. Find the coefficient of drag using Fig. 11.8.
3. Calculate drag force using Eq. (11.5).

Solution

1. Reynolds number

$$\text{Re} = \frac{Vd\rho}{\mu} = \frac{(0.08 \text{ m/s})(0.012 \text{ m})(850 \text{ kg/m}^3)}{10^{-1} \text{ N} \cdot \text{s/m}^2} = 8.16$$

2. Coefficient of drag (from Fig. 11.8) is $C_D = 5.3$.

3. Drag force

$$F_D = \frac{C_D A_p \rho V_0^2}{2}$$

$$F_D = \frac{(5.3)(\pi/4)(0.012^2 \text{ m}^2)(850 \text{ kg/m}^3)(0.08 \text{ m/s})^2}{2}$$

$$= \boxed{1.63 \times 10^{-3} \text{ N}}$$

Power and Rolling Resistance

When power is involved in a problem, the power equation from Chapter 7 is applied. For example, consider a car moving at a steady speed on a level road. Because the car in not accelerating, the horizontal forces are balanced as shown in Fig. 11.9. Force equilibrium gives

$$F_{\text{Drive}} = F_{\text{Drag}} + F_{\text{Rolling resistance}}$$

The driving force (F_{Drive}) is the frictional force between the driving wheels and the road. The drag force is the resistance of the air on the car. The rolling resistance is the frictional force that occurs when an object such as a ball or tire rolls. It is related to the deformation and types of the materials that are in contact. For example, a rubber tire on asphalt will have a larger rolling resistance than a steel train wheel on a steel rail. The rolling resistance is calculated using

$$F_{\text{Rolling resistance}} = F_r = C_r N \tag{11.10}$$

where C_r is the coefficient of rolling resistance and N is the normal force.

The power required to move the car shown in Fig. 11.9 at a constant speed is given by Eq. (7.2a)

$$P = FV = F_{\text{Drive}} V_{\text{Car}} = (F_{\text{Drag}} + F_{\text{Rolling resistance}}) V_{\text{Car}} \tag{11.11}$$

Thus, when power is involved in a problem, apply the equation $P = FV$ while concurrently using a free-body diagram to determine the appropriate force. This approach is illustrated in Example 11.3.

Figure 11.9

Horizontal forces acting on car that is moving at a steady speed.

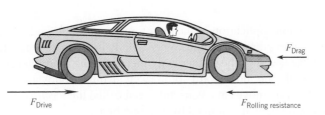

F_{Drive} $F_{\text{Rolling resistance}}$

EXAMPLE 11.3   SPEED OF A BICYCLE RIDER

A bicyclist of mass 70 km supplies 300 watts of power while riding into a 3 m/s head wind. The frontal area of the cyclist and bicycle together is 3.9 ft^2 = 0.362 m^2, the drag coefficient is 0.88, and the coefficient of rolling resistance is 0.007. Determine the speed V_c of the cyclist. Express your answer in mph and in m/s.

Problem Definition

Situation: A bicycle rider is cycling into a head wind of magnitude V_w = 3 m/s.

Find: Speed (m/s and mph) of the rider.

Sketch:

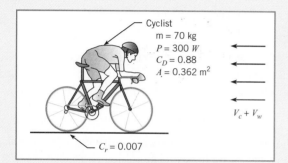

Cyclist
m = 70 kg
P = 300 W
C_D = 0.88
A = 0.362 m^2

$V_c + V_w$

C_r = 0.007

Assumptions:

1. The path is level, with no hills.
2. Mechanical losses in the bike gear train are zero.

Properties: Air (20°C, 1 atm), Table A.2: ρ = 1.2 kg/m^3.

Plan

1. Relate bike speed (V_c) to power using Eq. (11.11).
2. Calculate rolling resistance.
3. Develop an equation for drag force using Eq. (11.5).
4. Combine steps 1 to 3.
5. Solve for V_c.

Solution

1. Power equation
   - The power from the bike rider is being used to overcome drag and rolling resistance. Thus,
$$P = (F_D + F_r)V_c$$

2. Rolling resistance
$$F_r = C_r N = C_r mg = 0.007(70 \text{ kg})(9.81 \text{ m/s}^2) = 4.81 \text{ N}$$

3. Drag force
   - V_0 = speed of the air relative to the bike rider
$$V_0 = V_c + 3 \text{ m/s}$$

   - Drag force
$$F_D = C_D A\left(\frac{\rho V_0^2}{2}\right) = \frac{0.88(0.362 \text{ m}^2)(1.2 \text{ kg/m}^3)}{2}$$
$$\times (V_c + 3 \text{ m/s})^2$$
$$= 0.1911(V_c + 3 \text{ m/s})^2$$

4. Combine results:
$$P = (F_D + F_r)V_c$$
$$300 \text{ W} = (0.1911(V_c + 3)^2 + 4.81)V_c$$

5. Since the equation is cubic, use a spreadsheet program as shown. In this spreadsheet, let V_c vary and then search for the value of V_c that causes the right side of the equation to equal 300. The result is

$$V_c = \boxed{9.12 \text{ m/s} = 20.4 \text{ mph}}$$

V_c	RHS
(m/s)	(W)
0	0.0
5	85.2
8	223.5
9	291.0
9.1	298.4
9.11	299.1
9.12	299.9
9.13	300.6

11.4

Terminal Velocity

Another common application of the drag force equation is finding the steady-state speed of a body that is falling through a fluid. When a body is dropped, it accelerates under the action of gravity. As the speed of the falling body increases, the drag increases until the upward force (drag) equals the net downward force (weight minus buoyant force). Once the forces are balanced, the body moves at a constant speed called the *terminal velocity,* which is identified as the maximum velocity attained by a falling body.

To find terminal velocity, balance the forces acting on the object, and then solve the resulting equation. In general this process is iterative as illustrated by Example 11.4.

EXAMPLE 11.4   TERMINAL VELOCITY OF A SPHERE IN WATER

A 20 mm plastic sphere ($S = 1.3$) is dropped in water. Determine its terminal velocity. Assume $T = 20°C$.

Problem Definition

Situation: A smooth sphere ($D = 0.02$ m, $S = 1.3$) is falling in water.

Find: Terminal velocity (m/s) of the sphere.

Properties: Water (20°C), Table A.5, $\nu = 1 \times 10^{-6}$ m^2/s, $\rho = 998$ kg/m^3, and $\gamma = 9790$ N/m^3.

Plan

This problem requires an iterative solution because the terminal velocity equation is implicit.

1. Apply force equilibrium.
2. Develop an equation for terminal velocity.
3. To solve the terminal velocity equation, set up a procedure for iteration.
4. To implement the iterative solution, build a table in a spreadsheet program.

Solution

1. Force equilibrium
   - Sketch a free-body diagram.

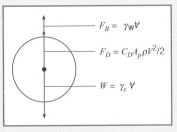

   - Apply force equilibrium (vertical direction):
   $$F_{\text{Drag}} + F_{\text{Buoyancy}} = W$$

2. Terminal velocity equation
   - Analyze terms in the equilibrium equation:

   $$C_D A\left(\frac{\rho V_0^2}{2}\right) + \gamma_w \forall = \gamma_s \forall$$

   $$C_D\left(\frac{\pi d^2}{4}\right)\left(\frac{\rho V_0^2}{2}\right) + \gamma_w\left(\frac{\pi d^3}{6}\right) = \gamma_s\left(\frac{\pi d^3}{6}\right)$$

- Solve for V_0

$$V_0 = \left[\frac{(\gamma_s - \gamma_w)(4/3)d}{C_D \rho_w}\right]^{1/2}$$

$$= \left[\frac{(12.7 - 9.79)(10^3 \text{ N/m}^3)(4/3)(0.02 \text{ m})}{C_D \times 998 \text{ kg/m}^3}\right]^{1/2}$$

$$V_0 = \left(\frac{0.0778}{C_D}\right)^{1/2} = \frac{0.279}{C_D^{1/2}} \text{ m/s}$$

3. Iteration 1
   - Initial guess: $V_0 = 1.0$ m/s
   - Calculate Re:

   $$\text{Re} = \frac{Vd}{\nu} = \frac{(1.0 \text{ m/s})(0.02 \text{ m})}{1 \times 10^{-6} \text{ m}^2/\text{s}} = 20000$$

   - Calculate C_D using Eq. (11.9):

   $$C_D = \frac{24}{20000}(1 + 0.15(20000^{0.687}))$$

   $$+ \frac{0.42}{1 + 4.25 \times 10^4(20000)^{-1.16}} = 0.456$$

   - Find new value of V_0 (use equation from step 2):

   $$V_0 = \left(\frac{0.0778}{C_D}\right)^{1/2} = \frac{0.279}{0.456^{0.5}} = 0.413 \text{ m/s}$$

4. Iterative solution
   - As shown, use a spread sheet program to build a table. The first row shows the results of iteration 1.
   - The terminal velocity from iteration 1 $V_0 = 0.413$ m/s is used as the initial velocity for iteration 2.
   - The iteration process is repeated until the terminal velocity reaches a constant value of $V_0 = 0.44$ m/s. Notice that convergence is reached in two iterations.

Iteration #	Initial V_0	Re	C_D	New V_0
	(m/s)			(m/s)
1	1.000	20000	0.456	0.413
2	0.413	8264	0.406	0.438
3	0.438	8752	0.409	0.436
4	0.436	8721	0.409	0.436
5	0.436	8723	0.409	0.436
6	0.436	8722	0.409	0.436

$$\boxed{V_0 = 0.44 \text{ m/s}}$$

11.5

Vortex Shedding

This section introduces vortex shedding, which is important for two reasons: It can be used to enhance heat transfer and mixing, and it can cause unwanted vibrations and failures of structures.

Flow past a bluff body generally produces a series of vortices that are shed alternatively from each side, thereby producing a series of alternating vortices in the wake. This phenomenon is call *vortex shedding*. Vortex shedding for a cylinder occurs for Re $\geqslant$ 50 and gives the flow pattern sketched in Fig. 11.10. In this figure, a vortex is in the process of formation near the top of the cylinder. Below and to the right of the first vortex is another vortex, which was formed and shed a short time before. Thus the flow process in the wake of a cylinder involves the formation and shedding of vortices alternately from one side and then the other. This alternate formation and shedding of vortices creates a cyclic change in pressure with consequent periodicity in side thrust on the cylinder. Vortex shedding was the primary cause of failure of the Tacoma Narrows suspension bridge in the state of Washington in 1940.

Experiments reveal that the frequency of shedding can be represented by plotting Strouhal number (St) as a function of Reynolds number. The Strouhal number is a π-group defined as

$$\text{St} = \frac{nd}{V_0} \tag{11.12}$$

where n is the frequency of shedding of vortices from one side of the cylinder, in Hz, d is the diameter of the cylinder, and V_0 is the free-stream velocity. The Strouhal number for vortex shedding from a circular cylinder is given in Fig. 11.11. Other cylindrical and two-dimensional bodies also shed vortices. Consequently, the engineer should always be alert to vibration problems when designing structures that are exposed to wind or water flow.

Figure 11.10

Formation of a vortex behind a cylinder.

Figure 11.11

Strouhal number versus Reynolds number for flow past a circular cylinder. [After Jones (5) and Roshko (8)]

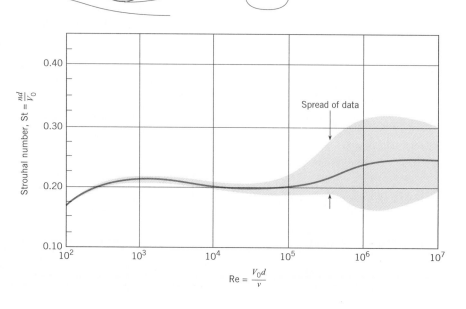

Reducing Drag by Streamlining

An engineer can design a body shape to minimize the drag force. This process is called *streamlining* and is often focused on reducing form drag. The reason for focusing on form drag is that drag on most bluff objects (e.g., a cylindrical body at Re $>$ 1000) is predominantly due to the pressure variation associated with flow separation. In this case, streamlining involves modifying the body shape to reduce or eliminate separation. The impacts of streamlining can be dramatic. For example, Fig. 11.4 shows that C_D for the streamlined shape is about 1/6 of C_D for the circular cylinder when Re $\approx 5 \times 10^5$.

While streamlining reduces form drag, friction drag is typically increased. This is because there is more surface area on a streamlined body as compared to a nonstreamlined body. Consequently, when a body is streamlined the optimum condition results when the sum of form drag and friction drag is minimum.

Streamlining to produce minimum drag at high Reynolds numbers will probably not produce minimum drag at very low Reynolds numbers. For example, at Re $<$ 1, the majority of the drag of a cylinder is friction drag. Hence, if the cylinder is streamlined, the friction drag will likely be magnified, and C_D will increase.

Another advantage of streamlining at high Reynolds numbers is that vortex shedding is eliminated. Example 11.5 shows how to estimate the impact of streamlining by using a ratio of C_D values.

EXAMPLE 11.5   COMPARING DRAG ON BLUFF AND STREAMLINED SHAPES

Compare the drag of the cylinder of Example 11.1 with the drag of the streamlined shape shown in Fig. 11.4. Assume that both shapes have the same projected area.

Problem Definition

Situation: The cylinder from Example 11.1 is being compared to a streamlined shape.

Find: Ratio of drag force on the streamlined body to drag force on the cylinder.

Assumptions:

1. The cylinder and the streamlined body have the same projected area.
2. Both objects are two-dimensional bodies (neglect end effects).

Plan

1. Retrieve Re and C_D from Example 11.1.
2. Find the coefficient of drag for the streamlined shape using Fig. 11.4.
3. Calculate the ratio of drag forces using Eq. (11.4).

Solution

1. From Example 11.1, Re $= 7 \times 10^5$ and C_D(cylinder) = 0.2.
2. Using this Re and Fig. 11.4 gives C_D(streamlined shape) = 0.034.
3. Drag force ratio (derived from Eq. 11.4) is

$$\frac{F_D(\text{streamlined shape})}{F_D(\text{cylinder})} = \frac{C_D(\text{streamlined shape})}{C_D(\text{cylinder})}$$

$$\times \left(\frac{A_p(\rho V_0^2/2)}{A_p(\rho V_0^2/2)} \right)$$

$$\frac{F_D(\text{streamlined shape})}{F_D(\text{cylinder})} = \frac{0.034}{0.2} = \boxed{0.17}$$

Review

Notice that streamlining provided nearly a sixfold reduction in drag!

11.7

Drag in Compressible Flow

So far, this chapter has described drag for flows with constant density. This section describes drag when the density of a gas is changing due to pressure variations. These types of flow are called *compressible flows*. This information is important for modeling of projectiles such as bullets and rockets.

In steady flow, the influence of compressibility depends on the ratio of fluid velocity to the speed of sound. This ratio is a π-group called the Mach number.

The variation of drag coefficient with Mach number for three axisymmetric bodies is shown in Fig. 11.12. In each case, the drag coefficient increases only slightly with the Mach number at low Mach numbers and then increases sharply as transonic flow ($M \approx 1$) is approached. Note that the rapid increase in drag coefficient occurs at a higher Mach number (closer to unity) if the body is slender with a pointed nose. The drag coefficient reaches a maximum at a Mach number somewhat larger than unity and then decreases as the Mach number is further increased.

The slight increase in drag coefficient with low Mach numbers is attributed to an increase in form drag due to compressibility effects on the pressure distribution. However, as the flow velocity is increased, the maximum velocity on the body finally becomes sonic. The Mach number of the free-stream flow at which sonic flow first appears on the body is called the *critical Mach number*. Further increases in flow velocity result in local regions of supersonic flow ($M > 1$), which lead to wave drag due to shock wave formation and an appreciable increase in drag coefficient.

The critical Mach number for a sphere is approximately 0.6. Note in Fig. 11.12 that the drag coefficient begins to rise sharply at about this Mach number. The critical Mach number for the pointed body is larger, and, correspondingly, the rise in drag coefficient occurs at a Mach number closer to unity.

The drag coefficient data for the sphere shown in Fig. 11.12 are for a Reynolds number of the order of 10^4. The data for the sphere shown in Fig. 11.8, on the other hand, are for very low Mach numbers. The question then arises about the general variation of the drag coefficient

Figure 11.12

Drag characteristics of projectile, sphere, and cylinder with compressibility effects. [After Rouse (12)]

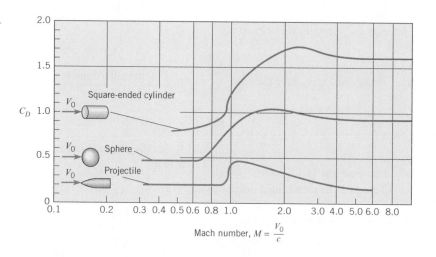

Figure 11.13

Contour plot of the drag coefficient of the sphere versus Reynolds and Mach numbers.

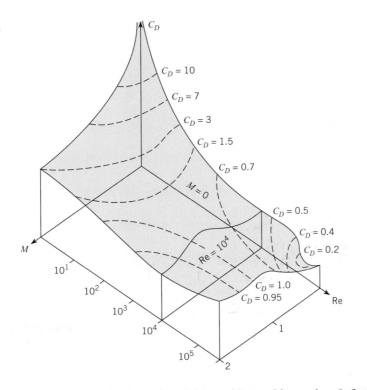

of a sphere with both Mach number and Reynolds number. Information of this nature is often needed to predict the trajectory of a body through the upper atmosphere or to model the motion of a nanoparticle.

A contour plot of the drag coefficient of a sphere versus both Reynolds and Mach numbers based on available data (19) is shown in Fig. 11.13. Notice the C_D-versus-Re curve from Fig. 11.8 in the M = 0 plane. Correspondingly, notice the C_D-versus-M curve from Fig. 11.12 in the Re = 10^4 plane. At low Reynolds numbers C_D decreases with increasing Mach number, whereas at high Reynolds numbers the opposite trend is observed. Using this figure, the engineer can determine the drag coefficient of a sphere at any combination of Re and M. Of course, corresponding C_D contour plots can be generated for any body, provided the data are available.

11.8

Theory of Lift

This section introduces circulation, the basic cause of lift, as well as the coefficient of lift.

Circulation

Circulation, a characteristic of a flow field, gives a measure of the average rate of rotation of fluid particles that are situated in an area that is bounded by a closed curve. Circulation is defined by the path integral as shown in Fig. 11.14. Along any differential segment of the path, the velocity can be resolved into components that are tangent and normal to the path. Signify the tangential component of velocity as V_L. Integrate $V_L \, dL$ around the curve; the resulting

Figure 11.14

Concept of circulation.

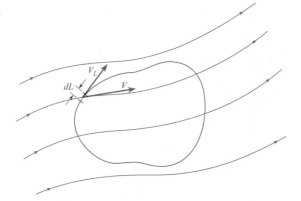

quantity is called circulation, which is represented by the Greek letter Γ (capital gamma). Hence

$$\Gamma = \oint V_L \, dL \qquad (11.13)$$

Sign convention dictates that in applying Eq. (11.13), one uses tangential velocity vectors that have a counterclockwise sense around the curve as negative and take those that have a clockwise direction as having a positive contribution.* For example, consider finding the circulation for an irrotational vortex. The tangential velocity at any radius is C/r, where a positive C means a clockwise rotation. Therefore, if circulation is evaluated about a curve with radius r, the differential circulation is

$$d\Gamma = V_L \, dL = \frac{C}{r_1} r_1 \, d\theta = C \, d\theta \qquad (11.14)$$

Integrate this around the entire circle:

$$\Gamma = \int_0^{2\pi} C \, d\theta = 2\pi C \qquad (11.15)$$

One way to induce circulation physically is to rotate a cylinder about its axis. Fig. 11.15a shows the flow pattern produced by such action. The velocity of the fluid next to the surface of the cylinder is equal to the velocity of the cylinder surface itself because of the no-slip condition that must prevail between the fluid and solid. At some distance from the cylinder, however, the velocity decreases with r, much like it does for the irrotational vortex. The next section shows how circulation produces lift.

Combination of Circulation and Uniform Flow around a Cylinder

Superpose the velocity field produced for uniform flow around a cylinder, Fig. 11.15b, onto a velocity field with circulation around a cylinder, Fig. 11.15a. Observe that the velocity is reinforced on the top side of the cylinder and reduced on the other side (Fig. 11.15c). Also observe that the stagnation points have both moved toward the low-velocity side of the cylinder. Consistent with the Bernoulli equation (assuming irrotational flow throughout), the pressure on the high-velocity side is lower than the pressure on the low-velocity side. Hence a pressure differential exists that causes a side thrust, or lift, on the cylinder. According to ideal flow theory, the lift per unit length of an infinitely long cylinder is given by $F_L/\ell = \rho V_0 \Gamma$, where F_L is the lift on the segment of length ℓ. For this ideal irrotational flow there is

---

* The sign convention is the opposite of that for the mathematical definition of a line integral.

Figure 11.15

Ideal flow around a cylinder.
(a) Circulation.
(b) Uniform flow.
(c) Combination of circulation and uniform flow.

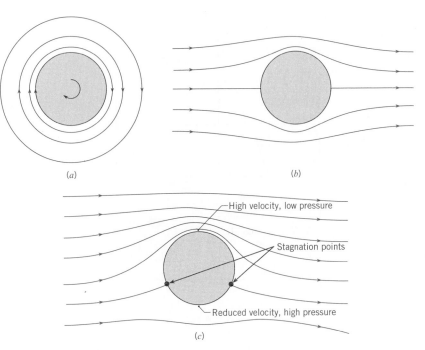

no drag on the cylinder. For the real-flow case, separation and viscous stresses do produce drag, and the same viscous effects will reduce the lift somewhat. Even so, the lift is significant when flow occurs past a rotating body or when a body is translating and rotating through a fluid. Hence the reason for the "curve" on a pitched baseball or the "drop" on a Ping-Pong ball is a fore spin. This phenomenon of lift produced by rotation of a solid body is called the *Magnus effect* after a nineteenth-century German scientist who made early studies of the lift on rotating bodies. A paper by Mehta (28) offers an interesting account of the motion of rotating sports balls.

Coefficients of lift and drag for the rotating cylinder with end plates are shown in Fig. 11.16. In this figure, the parameter $r\omega/V_0$ is the ratio of cylinder surface speed to the free-stream velocity, where r is the radius of the cylinder and ω is the angular speed in radians per second. The corresponding curves for the rotating sphere are given in Fig. 11.17.

Coefficient of Lift

The *coefficient of lift* is a parameter that characterizes the lift that is associated with a body. For example, a wing at a high angle of attack will have a high coefficient of lift, and a wing that has a zero angle of attack will have a low or zero coefficient of lift. The coefficient of lift is defined using a π-group:

$$C_L \equiv \frac{F_L}{A(\rho V_o^2/2)} = \frac{\text{lift force}}{(\text{reference area})(\text{dynamic pressure})} \tag{11.16}$$

To calculate lift force, engineers use the lift equation:

$$F_L = C_L A\left(\frac{\rho V_o^2}{2}\right) \tag{11.17}$$

where the reference area for a rotating cylinder or sphere is the projected area A_p.

Figure 11.16

Coefficients of lift and drag as functions of $r\omega/V_0$ for a rotating cylinder. [After Rouse (12).]

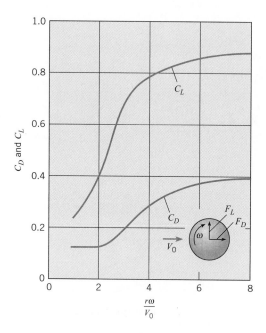

Figure 11.17

Coefficients of lift and drag for a rotating sphere. [After Barkla et al. (20). Reprinted with the permission of Cambridge University Press.]

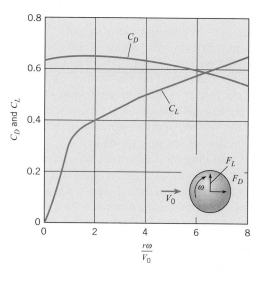

EXAMPLE 11.6   LIFT ON A ROTATING SPHERE

A Ping-Pong ball is moving at 10 m/s in air and is spinning at 100 revolutions per second in the clockwise direction. The diameter of the ball is 3 cm. Calculate the lift and drag force and indicate the direction of the lift (up or down). The density of air is 1.2 kg/m^3.

Problem Definition

Situation: A Ping-Pong ball is moving horizontally and rotating.

Sketch:

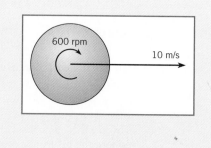

Find:

1. Drag force (in newtons) on the ball.
2. Lift force (in newtons) on the ball.
3. The direction of lift (up or down?).

Properties: Air: $\rho = 1.2 \text{ kg/m}^3$.

Plan

1. Calculate the value of $r\omega/V_0$.
2. Use the value of $r\omega/V_0$ to look up the coefficients of lift and drag on Fig. 11.7.
3. Calculate lift force using Eq. (11.17).
4. Calculate drag force using Eq. (11.5).

Solution

The rotation rate in rad/s is

$$\omega = (100 \text{ rev/s})(2\pi \text{ rad/rev}) = 628 \text{ rad/s}$$

The rotational parameter is

$$\frac{\omega r}{V_0} = \frac{(628 \text{ rad/s})(0.015 \text{ m})}{10 \text{ m/s}} = 0.942$$

From Fig. 11.17, the lift coefficient is approximately 0.26, and the drag coefficient is 0.64. The lift force is

$$F_L = \frac{1}{2}\rho V_0^2 C_L A_p$$

$$= \frac{1}{2}(1.2 \text{ kg/m}^3)(10 \text{ m/s})^2(0.26)\frac{\pi}{4}(0.03 \text{ m})^2$$

$$= \boxed{1.10 \times 10^{-2} \text{ N}}$$

$\boxed{\text{The lift force is downward.}}$ The drag force is

$$F_D = \frac{1}{2}\rho V_0^2 C_D A_p$$

$$= \boxed{27.1 \times 10^{-3} \text{ N}}$$

11.9

Lift and Drag on Airfoils

This section presents information on how to calculate lift and drag on wing-like objects. Some typical applications include calculating the takeoff weight of an airplane, determining the size of wings needed, and estimating power requirements to overcome drag force.

Lift of an Airfoil

An *airfoil* is a body designed to produce lift from the movement of fluid around it. Specifically, lift is a result of circulation in the flow produced by the airfoil. To see this, consider flow of an ideal flow (nonviscous and incompressible) past an airfoil as shown in Fig. 11.18a. Here, as for irrotational flow past a cylinder, the lift and drag are zero. There is a stagnation point on the bottom side near the leading edge, and another on the top side near the trailing edge of the foil. In the real flow (viscous fluid) case, the flow pattern around the upstream half of the foil is plausible. However, the flow pattern in the region of the trailing edge, as shown in Fig. 11.18a, cannot occur. A stagnation point on the upper side of the foil indicates that fluid must flow from the lower side around the trailing edge and then toward the stagnation point. Such a flow pattern implies an infinite acceleration of the fluid particles as they turn the corner around the trailing edge of the wing. This is a physical impossibility, and as we have seen in previous sections of the text, separation occurs at the sharp edge. As a consequence of the separation, the upstream stagnation point moves to the trailing edge. Flow from both the top and bottom sides of the airfoil in the vicinity of the trailing edge then leaves the airfoil smoothly and essentially parallel to these surfaces at the trailing edge (Fig. 11.18b).

Figure 11.18

Patterns of flow around an airfoil.
(a) Ideal flow—no circulation.
(b) Real flow—circulation.

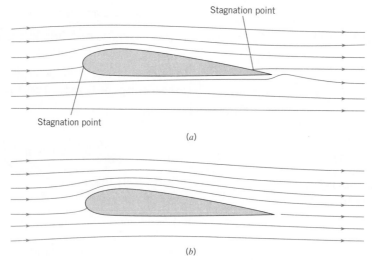

To bring theory into line with the physically observed phenomenon, it was hypothesized that a circulation around the airfoil must be induced in just the right amount so that the downstream stagnation point is moved all the way back to the trailing edge of the airfoil, thus allowing the flow to leave the airfoil smoothly at the trailing edge. This is called the *Kutta condition* (21), named after a pioneer in aerodynamic theory. When analyses are made with this simple assumption concerning the magnitude of the circulation, very good agreement occurs between theory and experiment for the flow pattern and the pressure distribution, as well as for the lift on a two-dimensional airfoil section (no end effects). Ideal flow theory then shows that the magnitude of the circulation required to maintain the rear stagnation point at the trailing edge (the Kutta condition) of a symmetric airfoil with a small angle of attack is given by

$$\Gamma = \pi c V_0 \alpha \tag{11.18}$$

where Γ is the circulation, c is the chord length of the airfoil, and α is the angle of attack of the chord of the airfoil with the free-stream direction (see Fig. 11.19 for a definition sketch).

Like that for the cylinder, the lift per unit length for an infinitely long wing is

$$F_L/\ell = \rho V_0 \Gamma$$

The planform area for the length segment ℓ is ℓc. Hence the lift on segment ℓ is

$$F_L = \rho V_0^2 \pi c \ell \alpha \tag{11.19}$$

For an airfoil the coefficient of lift is

$$C_L = \frac{F_L}{S\rho V_0^2/2} \tag{11.20}$$

Figure 11.19

Definition sketch for an airfoil section.

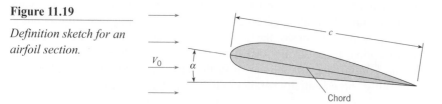

where the reference area S is the planform area of the wing—that is, the area seen from the plan view. On combining Eqs. (11.18) and (11.19) and identifying S as the area associated with length segment ℓ, one finds that C_L for irrotational flow past a two-dimensional airfoil is given by

$$C_L = 2\pi\alpha \qquad (11.21)$$

Equations (11.19) and (11.21) are the theoretical lift equations for an infinitely long airfoil at a small angle of attack. Flow separation near the leading edge of the airfoil produces deviations (high drag and low lift) from the ideal flow predictions at high angles of attack. Hence experimental wind-tunnel tests are always made to evaluate the performance of a given type of airfoil section. For example, the experimentally determined values of lift coefficient versus α for two NACA airfoils are shown in Fig. 11.20. Note in this figure that the coefficient of lift increases with the angle of attack, α, to a maximum value and then decreases with further increase in α. This condition, where C_L starts to decrease with a further increase in α, is called *stall*. Stall occurs because of the onset of separation over the top of the airfoil, which changes the pressure distribution in such a way as not only to decrease lift but also to increase drag. Data for many other airfoil sections are given by Abbott and Von Doenhoff (22).

Airfoils of Finite Length—Effect on Drag and Lift

The drag of a two-dimensional foil at a low angle of attack (no end effects) is primarily viscous drag. However, wings of finite length also have an added drag and a reduced lift associated with vortices generated at the wing tips. These vortices occur because the high pressure below the wing and the low pressure on top cause fluid to circulate around the end of the wing from the high-pressure zone to the low-pressure zone, as shown in Fig. 11.21. This induced flow has the effect of adding a downward component of velocity, w, to the approach velocity V_0. Hence, the "effective" free-stream velocity is now at an angle ($\phi \approx w/V_0$) to the direction of the original free-stream velocity, and the resultant force is tilted back as shown in Fig. 11.22. Thus the effective lift is smaller than the lift for the infinitely long wing because the effective angle of incidence is smaller. This resultant force has a component parallel to

Figure 11.20

Values of C_L for two NACA airfoil sections. [After Abbott and Van Doenhoff (22).]

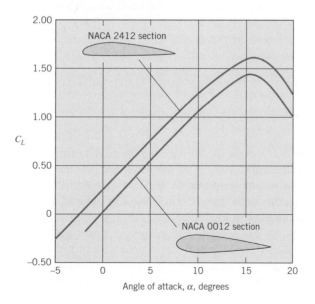

Figure 11.21

Formation of tip vortices.

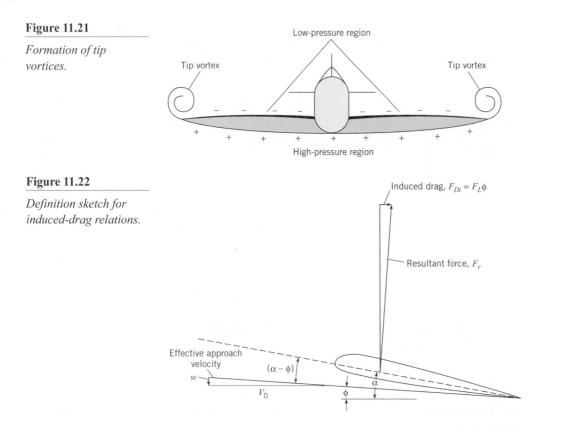

Figure 11.22

Definition sketch for induced-drag relations.

V_0 that is called the *induced drag* and is given by $F_L \phi$. Prandtl (23) showed that the induced velocity w for an elliptical spanwise lift distribution is given by the following equation:

$$w = \frac{2F_L}{\pi \rho V_0 b^2} \qquad (11.22)$$

where b is the total length (or span) of the finite wing. Hence

$$F_{Di} = F_L \phi = \frac{2F_L^2}{\pi \rho V_0^2 b^2} = \frac{C_L^2}{\pi} \frac{S^2 \rho V_0^2}{b^2 \, 2} \qquad (11.23)$$

From Eq. (11.23) it can be easily shown that the coefficient of induced drag, C_{Di}, is given by

$$C_{Di} = \frac{C_L^2}{\pi (b^2/S)} \qquad (11.24)$$

which happens to represent the minimum induced drag for any wing planform. Here the ratio b^2/S is called the aspect ratio Λ of the wing, and S is the planform area of the wing. Thus, for a given wing section (constant C_L and constant chord c), longer wings (larger aspect ratios) have smaller induced-drag coefficients. The induced drag is a significant portion of the total drag of an airplane at low velocities and must be given careful consideration in airplane design. Aircraft (such as gliders) and even birds (such as the albatross and gull) that are required to be airborne for long periods of time with minimum energy expenditure are noted for their long, slender wings. Such a wing is more efficient because the induced drag is small. To illustrate the effect of finite span, look at Fig. 11.23, which shows C_L and C_D versus α for wings with several aspect ratios.

Figure 11.23

Coefficients of lift and drag for three wings with aspect ratios of 3, 5, and 7. [After Prandtl (23).]

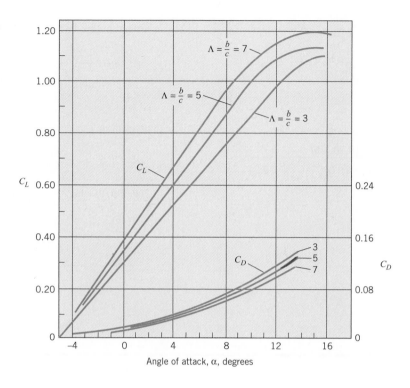

The total drag of a rectangular wing is computed by

$$F_D = (C_{D0} + C_{Di})\frac{bc\rho V_0^2}{2} \qquad (11.25)$$

where C_{D0} is the coefficient of form drag of the wing section and C_{Di} is the coefficient of induced drag.

EXAMPLE 11.7   WING AREA FOR AN AIRPLANE

An airplane with a weight of 10,000 lbf is flying at 600 ft/s at 36,000 ft, where the pressure is 3.3 psia and the temperature is –67°F. The lift coefficient is 0.2. The span of the wing is 54 ft. Calculate the wing area (in ft²) and the minimum induced drag.

Problem Definition

Situation:

1. An airplane ($W = 10{,}000$ lbf) is traveling at $V_0 = 600$ ft/s.
2. Coefficient of lift is $C_L = 0.2$.
3. Wing span is $b = 54$ ft.

Find:

1. Required wing area (in ft²).
2. Minimum value of induced drag (in N).

Properties: Atmosphere (36,000 ft): $T = -67°F$, $p = 3.3$ psia.

Plan

1. Apply the idea gas law to calculate density of air.
2. Apply force equilibrium to derive an equation for the the required wing area.
3. Calculate induced drag using Eq. (11.24) with $C_{D0} = 0$.

Solution

Ideal gas law

$$\rho = \frac{p}{RT}$$

$$= \frac{(3.3\ \text{lbf/in}^2)(144\ \text{in}^2/\text{ft}^2)}{(1716\ \text{ft-lbf/slug-°R})(-67 + 460°R)}$$

$$= 0.000705\ \text{slug/ft}^3$$

For steady flight, the lift force is equal to the weight,

$$W = F_L = \frac{1}{2}\rho V_0^2 C_L S$$

so

$$S = \frac{2W}{\rho V_0^2 C_L}$$

$$= \frac{2 \times 10{,}000 \text{ lbf}}{(0.000705 \text{ slug/ft}^3)(600^2 \text{ ft}^2/\text{s}^2)(0.2)}$$

$$= \boxed{394 \text{ ft}^2}$$

The minimum induced drag coefficient is

$$C_{Di} = \frac{C_L^2}{\pi\left(\dfrac{b^2}{S}\right)} = \frac{0.2^2}{\pi\left(\dfrac{54^2}{394}\right)} = 0.00172$$

The induced drag is

$$D_i = \frac{1}{2}\rho V_0^2 C_{Di} S$$

$$= \frac{1}{2}(0.000705 \text{ slug/ft}^3)(600 \text{ ft/s})^2(0.00172)(394 \text{ ft}^2)$$

$$= \boxed{86.0 \text{ lbf}}$$

A graph showing C_L and C_D versus α is given in Fig. 11.24. Note in this graph that C_D is separated into the induced-drag coefficient C_{Di} and the form drag coefficient C_{D0}.

Figure 11.24

Coefficients of lift and drag for a wing with an aspect ratio of 5. [After Prandtl (23).]

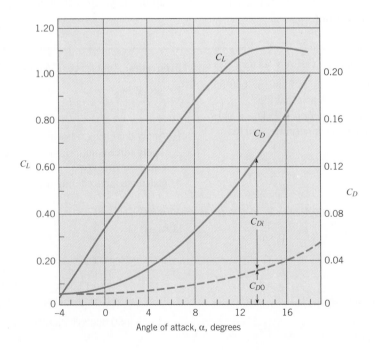

EXAMPLE 11.8 TAKEOFF CHARACTERISTICS OF AN AIRPLANE

A light plane (weight = 10 kN) has a wingspan of 10 m and a chord length of 1.5 m. If the lift characteristics of the wing are like those given in Fig. 11.23, what must be the angle of attack for a takeoff speed of 140 km/h? What is the stall speed? Assume two passengers at 800 N each and standard atmospheric conditions.

Problem Definition

Situation:

1. An airplane (W = 10 kN) with two passengers W = 1.6 kN is taking off.
2. Wing span is b = 10 m, and chord length is c = 1.5 m.
3. Lift coefficient information is given by Fig. 11.23.
4. Takeoff speed is V_0 = 140 km/h.

Find:

1. Angle of attack (in degrees).
2. Stall speed (in km/h).

Assumptions:

1. Ground effects can be neglected.
2. Standard atmospheric conditions prevail.

Properties: Air: ρ = 1.2 kg/m^3.

Plan

1. Find the lift by applying force equilibrium.
2. Calculate the coefficient of lift using Eq. (11.20).
3. Find the angle of attack α from Fig. 11.23.
4. Read the maximum angle of attack from Fig. 11.23, and then calculate the corresponding stall speed using the lift force equation (11.17).

Solution

Force equilibrium (y direction), so lift = weight = 11.6 kN. Thus,

$$C_L = \frac{F_L}{S\rho V_0^2/2}$$

$$= \frac{11,600 \text{ N}}{(15 \text{ m}^2)(1.2 \text{ kg/m}^3)[(140,000/3600)^2 \text{ m}^2/\text{s}^2]/2}$$

$$= 0.852$$

The aspect ratio is

$$\Lambda = \frac{b}{c} = \frac{10}{1.5} = 6.67$$

From Fig. 11.23, the angle of attack is

$$\boxed{\alpha = 7°}$$

From Fig. 11.23, stall will occur when

$$C_L = 1.18$$

Applying the lift force force equation gives

$$F_L = C_L A\left(\frac{\rho V_0^2}{2}\right)$$

$$11,600 = 1.18(15)\left(\frac{1.2}{2}\right)(V_{\text{stall}})^2$$

$$V_{\text{stall}} = 33.0 \text{ m/s} = \boxed{119 \text{ km/h}}$$

Review

Notice that the stall speed (119 km/h) is less than the takeoff speed (140 km/h).

11.10

Lift and Drag on Road Vehicles

Early in the development of cars, aerodynamic drag was a minor factor in performance because normal highway speeds were quite low. Thus in the 1920s, coefficients of drag for cars were around 0.80. As highway speeds increased and the science of metal forming became more advanced, cars took on a less angular shape, so that by the 1940s drag coefficients were 0.70 and lower. In the 1970s the average C_D for U.S. cars was approximately 0.55. In the early 1980s the average C_D for American cars dropped to 0.45, and currently auto manufacturers are giving even more attention to reducing drag in designing their cars. All major U.S., Japanese, and European automobile companies now have models with C_Ds of about 0.33, and some companies even report C_Ds as low as 0.29 on new models. European manufacturers were the leaders in the streamlining of cars because European gasoline prices (including tax) have been, for a number of years, about three times those in the United States. Table 11.2 shows the C_D for a 1932 Fiat and for other, more contemporary car models.

Great strides have been made in reducing the drag coefficients for passenger cars. However, significant future progress will be very hard to achieve. One of the most streamlined cars was the "Bluebird," which set a world land-speed record in 1938. Its C_D was 0.16. The minimum C_D of well-streamlined racing cars is about 0.20. Thus, lowering the C_D for passenger cars below 0.30 will require exceptional design and workmanship. For example, the underside of most cars is aerodynamically very rough (axles, wheels, muffler, fuel tank, shock absorbers, and so on). One way to smooth the underside is to add a panel to the bottom of the car. But then clearance may become a problem, and adequate dissipation of heat from the muffler may be hard to achieve. Other basic features of the automobile that contribute to drag but are not very amenable to drag-reduction modifications are interior airflow systems for engine cooling, wheels, exterior features such as rear-view mirrors and antennas, and other surface protrusions. The reader is directed to two books on road-vehicle aerodynamics, (24) and (25), which address all aspects of the drag and lift of road vehicles in considerably more detail than is possible here.

Table 11.2 COEFFICIENTS OF DRAG FOR CARS		
Make and Model	**Profile**	C_D
1932 Fiat Balillo		0.60
Volkswagen "Bug"		0.46
Plymouth Voyager		0.36
Toyota Paseo		0.31
Dodge Intrepid		0.31
Ford Taurus		0.30
Mercedes-Benz E320		0.29
Ford Probe V (concept car)		0.14
GM Sunraycer (experimental solar vehicle)		0.12

To produce low-drag vehicles, the basic teardrop shape is an idealized starting point. This shape can be altered to accommodate the necessary functional features of the vehicle. For example, the rear end of the teardrop shape must be lopped off to yield an overall vehicle length that will be manageable in traffic and will fit in our garages. Also, the shape should be wider than its height. Wind-tunnel tests are always helpful in producing the most efficient design. One such test was done on a 3/8 scale model of a typical notchback sedan. Wind-tunnel test results for such a sedan are shown in Fig. 11.25. Here the centerline pressure distribution (distribution of C_P) for the conventional sedan is shown by a solid line and that for a sedan with a 68 mm rear-deck lip is shown by a dashed line. Clearly the rear-deck lip causes the pressure on the rear of the car to increase (C_P is less negative), thereby reducing the drag on the car itself. It also decreases the lift, thereby improving traction. Of course, the lip itself produces some drag, and these tests show that the optimum lip height for greatest overall drag reduction is about 20 mm.

Research and development programs to reduce the drag of automobiles continue. As an entry in the PNGV (Partnership for a New Generation of Vehicles), General Motors (26) has exhibited a vehicle with a drag coefficient as low as 0.163, which is approximately one-half that of the typical midsize sedan. These automobiles will have a rear engine to eliminate the exhaust system underneath the vehicle, and allow a flat underbody. Cooling air for the engine is drawn in through inlets on the rear fenders and exhausted out the rear, reducing the drag due to the wake. The protruding rear-view mirrors are also removed to reduce the drag. The cumulative effect of these design modifications is a sizable reduction in aerodynamic drag.

The drag of trucks can be reduced by installing vanes near the corners of the truck body to deflect the flow of air more sharply around the corner, thereby reducing the degree of separation. This in turn creates a higher pressure on the rear surfaces of the truck, which reduces the drag of the truck.

One of the desired features in racing cars is the generation of negative lift to improve the stability and traction at high speeds. One idea (27) is to generate negative gage pressure underneath the car by installing a *ground-effect pod*. This is an airfoil section mounted across the bottom of the car that produces a venturi effect in the channel between the airfoil section and the road surface. The design of ground-effect vehicles involves optimizing design parameters to avoid separation and possible increase in drag. Another scheme to generate negative lift is the use of vanes as shown in Fig. 11.26. Sometimes "gurneys" are mounted on these vanes to reduce separation effects. Gurneys are small ribs mounted on the upper surface of the vanes near the trailing edge to induce local separation, reduce the separation on the lower surface of the vane, and increase the magnitude of the negative lift. As the speed of racing cars continues to increase, automobile aerodynamics will play an ever-increasing role in traction, stability, and control.

Figure 11.25

Effect of rear-deck lip on model surface. Pressure coefficients are plotted normal to the surface. [After Schenkel (25). Reprinted with permission from SAE Paper No. 770389. ©1977 Society of Automotive Engineers, Inc.]

Figure 11.26

Racing car with negative-lift devices.

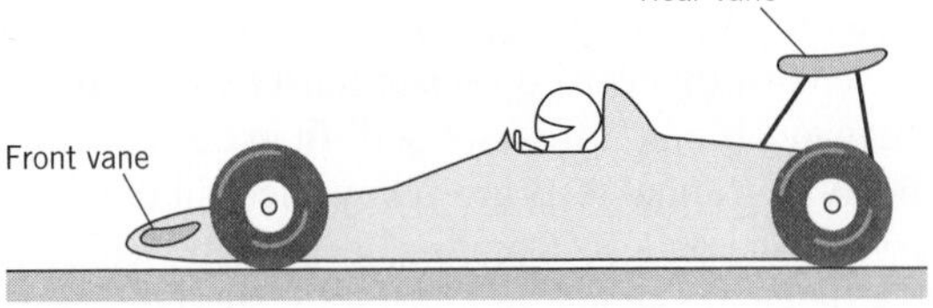

EXAMPLE 11.9 NEGATIVE LIFT ON A RACE CAR

The rear vane installed on the racing car of Fig. 11.26 is at an angle of attack of 8° and has characteristics like those given in Fig. 11.23. Estimate the downward thrust (negative lift) and drag from the vane that is 1.5 m long and has a chord length of 250 mm. Assume the racing car travels at a speed of 270 km/h on a track where normal atmospheric pressure and a temperature of 30°C prevail.

Problem Definition

Situation:

1. A racing car experiences downward lift from a rear mounted vane.
2. Vane overall length is $\ell = 1.5$ m, and chord length is $c = 0.25$ m.
3. Car speed is $V_0 = 270$ km/h $= 75$ m/s.

Find:

1. Downward lift force from vane (in newtons).
2. Drag force from vane (in newtons).

Properties: Air: $\rho = 1.17$ kg/m^3.

Plan

1. Find the coefficient of lift C_L and the coefficient of drag C_D from Fig. 11.23.
2. Calculate the downward thrust using the lift force equation (11.17).
3. Calculate the drag using the drag force equation (11.5).

Solution

1. The aspect ratio is

$$\Lambda = \frac{\ell}{c} = \frac{1.5}{0.25} = 6$$

From Fig. 11.23, the lift and drag coefficients are

$$C_L = 0.93 \quad \text{and} \quad C_D = 0.070$$

2. Lift force equation

$$F_L = C_L A \left(\frac{\rho V_o^2}{2} \right)$$

$$F_L = 0.93 \times 1.5 \times 0.25 \times 1.17 \times (75)^2/2 = \boxed{1148 \text{ N}}$$

3. Drag force equation

$$F_D = C_D A \left(\frac{\rho V_0^2}{2} \right) = \left(\frac{C_D}{C_L} \right) F_L$$

$$F_D = (0.070/0.93) \times 1148 = \boxed{86.4 \text{ N}}$$

11.11

Summary

A body immersed in a flowing fluid is subjected to pressure and shear-stress distributions. When the stress distributions are integrated, the resulant force is resolved into lift and drag. By definition, the drag force is parallel to the free-stream velocity and the lift force is perpendicular.

The drag force on a body is evaluated using

$$F_D = C_D A \left(\frac{\rho V_0^2}{2} \right)$$

where C_D is a the drag coefficient, V_0 is the relative speed between the body and free-stream velocity, and A is the reference area. The drag coefficient is typically a function of Reynolds number based on the relative speed and a characteristic dimension of the body. Values of the drag coefficient for various shapes are determined analytically or experimentally and are published as equations or in tabular or graphical format. The drag force is the combination of two effects, form drag and skin friction drag. Form drag is due to pressure forces acting on the body, whereas skin friction is due to shear stress on the body surface.

The drag coefficient of a sphere for Reynolds numbers less than 0.5 is

$$C_D = \frac{24}{\text{Re}}$$

For bluff bodies in high–Reynolds number flows, the drag force is primarily form drag and results from the reduced pressure in the body's wake. For streamlined bodies the form drag is reduced, and skin friction drag plays a more important role. The drag coefficients of cylinders and spheres show a marked decrease at Reynolds numbers near 10^5. This effect is attributed to the flow in the boundary layer changing from laminar to turbulent, moving the separation point downstream, reducing the wake region, and decreasing the form drag. The Reynolds number where the drag coefficient decreases is the critical Reynolds number, and the phenomenon is known as the critical Reynolds number effect.

Cylinders and bluff bodies in a cross-flow produce vortex shedding in which vortices are released alternately from each side of the body. The frequency of vortex shedding is given by the Strouhal number

$$\text{St} = \frac{nd}{V_0}$$

where n is the rate at which vortices are shed (Hz) from one side and d is the cross-stream dimension of the body.

Increasing the Mach number of flow past a body increases the drag coefficient. The free-stream Mach number where sonic flow first occurs on a body is the critical Mach number.

The lift force on a body is quantified by

$$F_L = C_L A \left(\frac{\rho V_0^2}{2} \right)$$

where C_L is the lift coefficient and A is the reference area. The values for the lift coefficient for various bodies are obtained by analysis or experiment.

The lift on an airfoil is due to the circulation produced by the airfoil on the surrounding fluid. This circulatory motion causes a change in the momentum of the fluid and a lift on the airfoil. The lift coefficient for a symmetric two-dimensional wing (no tip effect) is

$$C_L = 2\pi\alpha$$

where α is the angle of attack (expressed in radians) and the reference angle is the product of the chord and a unit length of wing. As the angle of attack increases, the airfoil stalls and the lift coefficient decreases. A wing of finite span produces trailing vortices that reduce the angle of attack and produce an induced drag. The drag coefficient corresponding to the minimum induced drag is

$$C_{Di} = \frac{C_L^2}{\pi(b^2/S)}$$

where b is the wing span and S is the planform area of the wing.

References

1. Bullivant, W. K. "Tests of the NACA 0025 and 0035 Airfoils in the Full Scale Wind Tunnel." *NACA Rept.,* 708 (1941).

2. DeFoe, G. L. "Resistance of Streamline Wires." *NACA Tech. Note,* 279 (March 1928).

3. Goett, H. J., and W. K. Bullivant. "Tests of NACA 0009, 0012, and 0018 Airfoils in the Full Scale Tunnel." *NACA Rept.,* 647 (1938).

4. Jacobs, E. N. "The Drag of Streamline Wires." *NACA Tech. Note,* 480 (December 1933).

5. Jones, G. W., Jr. "Unsteady Lift Forces Generated by Vortex Shedding about a Large, Stationary, and Oscillating Cylinder at High Reynolds Numbers." *Symp. Unsteady Flow, ASME* (1968).

6. Lindsey, W. F. "Drag of Cylinders of Simple Shapes." *NACA Rept.,* 619 (1938).

7. Miller, B. L., J. F. Mayberry, and I. J. Salter. "The Drag of Roughened Cylinders at High Reynolds Numbers." *NPL Rept. MAR Sci.,* R132 (April 1975).

8. Roshko, A. "Turbulent Wakes from Vortex Streets." *NACA Rept.,* 1191 (1954).

9. Abbott, I. H. "The Drag of Two Streamline Bodies as Affected by Protuberances and Appendages." *NACA Rept.,* 451 (1932).

10. Brevoort, M. J., and U. T. Joyner. "Experimental Investigation of the RobinsonType Cup Anemometer." *NACA Rept.,* 513 (1935).

11. Freeman, H. B. "Force Measurements on a 1/40-Scale Model of the U.S. Airship 'Akron.'" *NACA Rept.,* 432 (1932).

12. Rouse, H. *Elementary Mechanics of Fluids.* John Wiley, New York, 1946.

13. Clift, R., J. R. Grace, and M. E. Weber. *Bubbles, Drops and Particles.* Academic Press, San Diego, CA, 1978.

14. Clift, R., and W. H. Gauvin. "The Motion of Particles in Turbulent Gas Streams." *Proc. Chemeca '70,* vol. 1, pp. 14–28 (1970).

15. Hoerner, S. F. *Fluid Dynamic Drag.* Published by the author, 1958.

16. Morrison, R. B. (ed). *Design Data for Aeronautics and Astronautics.* John Wiley, New York, 1962.

17. Roberson, J. A., et al. "Turbulence Effects on Drag of Sharp-Edged Bodies." *J. Hydraulics Div., Am. Soc. Civil Eng.* (July 1972).

18. Scher, S. H., and L. J. Gale. "Wind Tunnel Investigation of the Opening Characteristics, Drag, and Stability of Several Hemispherical Parachutes." *NACA Tech. Note,* 1869 (1949).

19. Crowe, C. T., et al. "Drag Coefficient for Particles in Rarefied, Low Mach-Number Flows." In *Progress in Heat and Mass Transfer,* vol. 6, pp. 419–431. Pergamon Press, New York, 1972.

20. Barkla, H. M., et al. "The Magnus or Robins Effect on Rotating Spheres." *J. Fluid Mech.,* vol. 47, part 3 (1971).

21. Kuethe, A. M., and J. D. Schetzer. *Foundations of Aerodynamics.* John Wiley, New York, 1967.

22. Abbott, H., and A. E. Von Doenhoff. *Theory of Wing Sections.* Dover, New York, 1949.

23. Prandtl, L. "Applications of Modern Hydrodynamics to Aeronautics." *NACA Rept.,* 116 (1921).

24. Hucho, Wolf-Heinrich, ed. *Aerodynamics of Road Vehicles.* Butterworth, London, 1987.

25. Schenkel, Franz K. "The Origins of Drag and Lift Reductions on Automobiles with Front and Rear Spoilers." *SAE Paper,* no. 770389 (February 1977).

26. Sharke, P. "Smooth Body." *Mechanical Engineer,* vol. 121, pp. 74–77 (1999).

27. Smith, C. *Engineer to Win.* Motorbooks International, Osceola, WI (1984).

28. Mehta, R. D. "Aerodynamics of Sports Balls." *Annual Review of Fluid Mechanics,* 17, p. 151 (March 1985).

Problems

Relating Pressure Distribution and C_D

11.1 A hypothetical pressure coefficient variation over a long (length normal to the page) plate is shown. What is the coefficient of drag for the plate in this orientation and with the given pressure distribution? Assume that the reference area is the surface area (one side) of the plate.

11.2 Flow is occurring past the square rod. The pressure coefficient values are as shown. From which direction do you think the flow is coming? (a) SW direction, (b) SE direction, (c) NW direction, or (d) NE direction.

11.3 The hypothetical pressure distribution on a rod of triangular (equilateral) cross section is shown, where flow is from left to right. That is, C_p is maximum and equal to +1.0 at the leading edge and decreases linearly to zero at the trailing edges. The pressure coefficient on the downstream face is constant with a value of –0.5. Neglecting skin friction drag, find C_D for the rod.

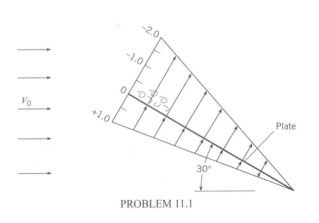

PROBLEM 11.1

11.4 The pressure distribution on a rod having a triangular (equilateral) cross section is shown, where flow is from left to right. What is C_D for the rod?

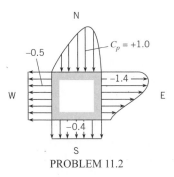

PROBLEM 11.2

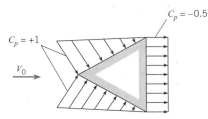

PROBLEM 11.3

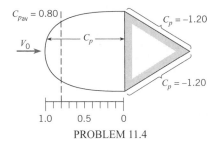

PROBLEM 11.4

Drag Calculations

11.5 $\mathbb{P}\mathbb{Q}\blacktriangleleft$ Apply the grid method to each situation that follows.

a. Use Eq. (11.5) to predict the drag force in newtons for a automobile that is traveling at $V = 60$ mph on a summer day. Assume that the frontal area is 2 m², and the coefficient of drag is $C_D = 0.4$.

b. Apply Eq. (11.5) to predict the the speed in mph of a bicycle rider that is subject to a drag force of 5 lbf on a summer's day. Assume the frontal area of the rider is $A = 0.5$ m², and the coefficient of drag is $C_D = 0.3$.

11.6 $\mathbb{P}\mathbb{Q}\blacktriangleleft$ Using the first two sections in this chapter and using other resources, answer the questions that follow. Strive for depth, clarity, and accuracy. Also, strive for effective use of sketches, words, and equations.

a. What are the four most important factors that influence the drag force?

b. How are stress and drag related?

c. What is form drag? What is friction drag?

11.7 $\mathbb{P}\mathbb{Q}\blacktriangleleft$ Use information in sections 11.2 and 11.3 to find the the coefficient of drag for each case described here.

a. A sphere is falling through water, Re = 10,000.

b. Air is blowing normal to a very long circular cylinder, and Re = 7,000.

c. Wind is blowing normal to a billboard that is 20 ft wide by 10 ft high.

11.8 Estimate the wind force on a billboard 10 ft high and 30 ft wide when a 50 mph wind ($T = 60°F$) is blowing normal to it.

11.9 If Stokes's law is considered valid below a Reynolds number of 0.5, what is the largest raindrop that will fall in accordance with Stokes's law?

11.10 Determine the drag of a 4 ft × 8 ft sheet of plywood held at a right angle to a stream of air (60°F, 1 atm) having a velocity of 45 mph.

11.11 Estimate the drag of a thin square plate (3 m by 2 m) when it is towed through water (10°C). Assume a towing speed of about 2 m/s.

a. The plate is oriented for minimum drag.

b. The plate is oriented for maximum drag.

11.12 A cooling tower, used for cooling recirculating water in a modern steam power plant, is 350 ft high and 250 ft average diameter. Estimate the drag on the cooling tower in a 200 mph wind ($T = 60°F$).

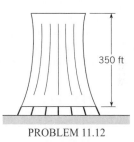

PROBLEM 11.12

11.13 Estimate the wind force that would act on *you* if you were standing on top of a tower in a 30 m/s (115 ft/s) wind on a day when the temperature was 20°C (68°F) and the atmospheric pressure was 96 kPa (14 psia).

11.14 As shown, wind is blowing on a 55-gallon drum. Estimate the wind speed needed to tip the drum over. Work in SI units. The drum weighs 48 lbm, the diameter is 22.5 in., and the height is 34.5 in.

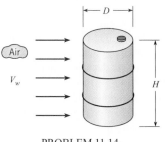

PROBLEM 11.14

11.15 What drag is produced when a disk 0.75 m in diameter is submerged in water at 10°C and towed behind a boat at a speed of 4 m/s? Assume orientation of the disk so that maximum drag is produced.

11.16 A circular billboard having a diameter of 6 m is mounted so as to be freely exposed to the wind. Estimate the total force exerted on the structure by a wind that has a direction normal to the structure and a speed of 30 m/s. Assume $T = 10°C$ and $p = 101$ kPa absolute.

11.17 Consider a large rock situated at the bottom of a river and acted on by a strong current. Estimate a typical speed of the current that will cause the rock to move downstream along the bottom of the river. List and justify all your major assumptions. Shown all calculations and work in SI units.

11.18 Compute the overturning moment exerted by a 35 m/s wind on a smokestack that has a diameter of 2.5 m and a height of 75 m. Assume that the air temperature is 20°C and that p_a is 99 kPa absolute.

11.19 What is the moment in SI units at the bottom of a flagpole 35 m high and 10 cm in diameter in a 30 m/s wind? The atmospheric pressure is 100 kPa, and the temperature is 20°C.

11.20 A cylindrical anchor (vertical axis) made of concrete ($\gamma = 15$ kN/m^3) is reeled in at a rate of 1.0 m/s by a man in a boat. If the anchor is 30 cm in diameter and 30 cm long, what tension must be applied to the rope to pull it up at this rate? Neglect the weight of the rope.

11.21 A Ping-Pong ball of mass 2.6 g and diameter 38 mm is supported by an air jet. The air is at a temperature of 18°C and a pressure of 27 in-Hg. What is the minimum speed of the air jet?

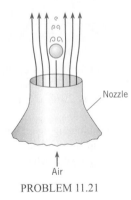

Nozzle

Air

PROBLEM 11.21

11.22 Estimate the moment at ground level on a signpost supporting a sign measuring 3 m by 2 m if the wind is normal to the surface and has a speed of 40 m/s and the center of the sign is 3 m above the ground. Neglect the wind load on the post itself. Assume $T = 10°C$ and $p = 1$ atm.

11.23 Windstorms sometimes blow empty boxcars off their tracks. The dimensions of one type of boxcar are shown. What minimum wind velocity normal to the side of the car would be required to blow the car over?

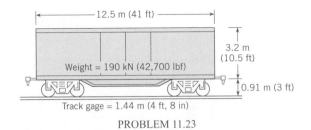

12.5 m (41 ft)

3.2 m (10.5 ft)

Weight = 190 kN (42,700 lbf)

0.91 m (3 ft)

Track gage = 1.44 m (4 ft, 8 in)

PROBLEM 11.23

11.24 A semiautomatic popcorn popper is shown. After the unpopped corn is placed in screen S, the fan F blows air past the heating coils C and then past the popcorn. When the corn pops, its projected area increases; thus it is blown up and into a container. Unpopped corn has a mass of about 0.15 g per kernel and an average diameter of approximately 6 mm. When the corn pops, its average diameter is about 18 mm. Within what range of airspeeds in the chamber will the device operate properly?

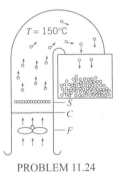

$T = 150°C$

S

C

F

PROBLEM 11.24

11.25 Hoerner (15) presents data that show that fluttering flags of moderate-weight fabric have a drag coefficient (based on the flag area) of about 0.14. Thus the total drag is about 14 times the skin friction drag alone. Design a flagpole that is 100 ft high and is to fly an American flag 6 ft high. Make your own assumptions regarding other required data.

Power, Energy, and Rolling Resistance

11.26 How much power is required to move a spherical-shaped submarine of diameter 1.5 m through sea water at a speed of 10 knots? Assume the submarine is fully submerged.

11.27 A dirigible flies at 25 ft/s at an altitude where the specific weight of the air is 0.07 lbf/ft^3 and the kinematic viscosity is 1.3×10^{-4} ft^2/s. The dirigible has a length-to-diameter ratio of 5 and has a drag coefficient corresponding to the streamlined body in Fig. 11.8. The diameter of the dirigible is 100 ft. What is the power required to propel the dirigible at this speed?

11.28 Estimate the energy in joules and kcal (food calories) that a runner supplies to overcome aerodynamic drag during a 10 km race. The runner runs a 6:30 pace (i.e., each mile takes 6 minutes and 30 seconds). The product of frontal area and coefficient of drag is $C_D A = 8$ ft^2. (One "food calorie" is equivalent to 4186 J.) Assume an air density of 1.22 kg/m^3.

11.29 A cylindrical rod of diameter d and length L is rotated in still air about its midpoint in a horizontal plane. Assume the drag force at each section of the rod can be calculated assuming a two-dimensional flow with an oncoming velocity equal to the relative velocity component normal to the rod. Assume C_D is constant along the rod.

a. Derive an expression for the average power needed to rotate the rod.
b. Calculate the power for $\omega = 50$ rad/s, $d = 2$ cm, $L = 1.5$ m, $\rho = 1.2$ kg/m^3, and $C_D = 1.2$.

PROBLEM 11.29

11.30 Estimate the additional power (in hp) required for the truck when it is carrying the rectangular sign at a speed of 30 m/s over that required when it is traveling at the same speed but is not carrying the sign.

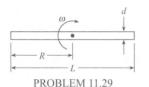

PROBLEM 11.30

11.31 Estimate the added power (in hp) required for the car when the cartop carrier is used and the car is driven at 100 km/h in a 25 km/h head wind over that required when the carrier is not used in the same conditions.

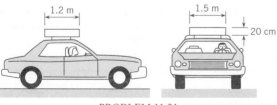

PROBLEM 11.31

11.32 The resistance to motion of an automobile consists of rolling resistance and aerodynamic drag. The weight of an automobile is 3000 lbf, and it has a frontal area of 20 ft^2. The drag coefficient is 0.30, and the coefficient of rolling friction is 0.02. Determine the percentage savings in gas mileage that one achieves when one drives at 55 mph instead of 65 mph on a level road. Assume an air temperature of 60°F.

11.33 A car coasts down a very long hill. The weight of the car is 2500 lbf, and the slope of the grade is 6%. The rolling friction coefficient is 0.01. The frontal area of the car is 20 ft^2, and the drag coefficient is 0.32. The density of the air is 0.002 slugs/ft^3. Find the maximum coasting speed of the car in mph.

11.34 An automobile with a mass of 1000 kg is driven up a hill where the slope is 3° (5.2% grade). The automobile is moving at 30 m/s. The coefficient of rolling friction is 0.02, the drag coefficient is 0.4, and the cross-sectional area is 4 m^2. Find the power (in kW) needed for this condition. The air density is 1.2 kg/m^3.

11.35 A bicyclist is coasting down a hill with a slope of 8° into a head wind (measured with respect to the ground) of 5 m/s. The mass of the cyclist and bicycle is 80 kg, and the coefficient of rolling friction is 0.02. The drag coefficient is 0.5, and the projected area is 0.5 m^2. The air density is 1.2 kg/m^3. Find the speed of the bicycle in meters per second.

11.36 A bicyclist is capable of delivering 275 W of power to the wheels. How fast can the bicyclist travel in a 3 m/s head wind if his or her projected area is 0.5 m^2, the drag coefficient is 0.3, and the air density is 1.2 kg/m^3? Assume the rolling resistance is negligible.

11.37 Assume that the horsepower of the engine in the original 1932 Fiat Balillo (see Table 11.2) was 40 bhp (brake horsepower) and that the maximum speed at sea level was 60 mph. Also assume that the projected area of the automobile is 30 ft^2. Assume that the automobile is now fitted with a modern 220 bhp motor with a weight equal to the weight of the original motor; thus the rolling resistance is unchanged. What is the maximum speed of the "souped up" Balillo at sea level?

11.38 One way to reduce the drag of a blunt object is to install vanes to suppress the amount of separation. Such a procedure was used on model trucks in a wind-tunnel study by Kirsch and Bettes. For tests on a van-type truck, they noted that without vanes the C_D was 0.78. However, when vanes were installed around the top and side leading edges of the truck body (see the figure), a 25% reduction in C_D was achieved. For a truck with a projected area of 8.36 m^2, what reduction in drag force will be effected by installation of the vanes when the truck travels at 100 km/h? Assume standard atmospheric pressure and a temperature of 20°C.

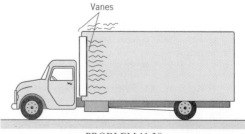

PROBLEM 11.38

11.39 For the truck of Prob. 11.38, assume that the total resistance is given by $R = F_D + C$, where F_D is the air drag and C is the resistance due to bearing friction. If C is constant at 350 N for the given truck, what fuel-savings percentage will be effected by the installation of the vanes when the truck travels at 100 km/h?

Terminal Velocity

11.40 PQ◄ Suppose you are designing an object to fall through seawater with a terminal velocity of exactly 1 m/s. What variables will have the most influence on the terminal velocity? List these variables and justify your decisions.

11.41 As shown a 20 cm diameter parachute supporting a mass of 10 g is falling through air (20°C). Assume a coefficient of drag of $C_D = 1.4$ and estimate the terminal velocity V_0. Use a projected area of $(\pi D^2)/4$.

PROBLEM 11.41

11.42 Consider a small air bubble (approximately 4 mm diameter) rising in a very tall column of liquid. Will the bubble accelerate or decelerate as it moves upward in the liquid? Will the drag of the bubble be largely skin friction or form drag? Explain.

11.43 Determine the terminal velocity in water ($T = 10$°C) of a 10 cm ball that weighs 20 N in air.

11.44 This cube is weighted so that it will fall with one edge down as shown. The cube weighs 19.8 N in air. What will be its terminal velocity in water?

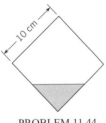

PROBLEM 11.44

11.45 A spherical rock weighs 30 N in air and 5 N in water. Estimate its terminal velocity as it falls in water (20°C).

11.46 A spherical balloon 2 m in diameter that is used for meteorological observations is filled with helium at standard conditions. The empty weight of the balloon is 3 N. What velocity of ascent will it attain under standard atmospheric conditions?

11.47 A sphere 2 cm in diameter rises in oil at a velocity of 1.5 cm/s. What is the specific weight of the sphere if the oil density is 900 kg/m³ and the dynamic viscosity is 0.096 N · s/m²?

11.48 Estimate the terminal velocity of a 1.5 mm plastic sphere in oil. The oil has a specific gravity of 0.95 and a kinematic viscosity of 10^{-4} m²/s. The plastic has a specific gravity of 1.07. The volume of a sphere is given by $\pi D^3/6$.

11.49 A 120 lbf (534 N) skydiver is free-falling at an altitude of 6500 ft (1980 m). Estimate the terminal velocity in mph for minimum and maximum drag conditions. At maximum drag conditions, the product of frontal area and coefficient of drag is $C_D A = 8$ ft² (0.743 m²). At minimum drag conditions, $C_D A = 1$ ft² (0.0929 m²). Assume the pressure and temperature at sea level are 14.7 psia (101 kPa) and 60°F (15°C). To calculate air properties, use the lapse rate for the U.S. standard atmosphere (see Chapter 3).

PROBLEM 11.49

11.50 What is the terminal velocity of a 0.5 cm hailstone in air that has an atmospheric pressure of 96 kPa absolute and a temperature of 0°C? Assume that the hailstone has a specific weight of 6 kN/m³.

11.51 A drag chute is used to decelerate an airplane after touchdown. The chute has a diameter of 12 ft and is deployed when the aircraft is moving at 200 ft/s. The mass of the aircraft is 20,000 lbm, and the density of the air is 0.075 lbm/ft³. Find the initial deceleration of the aircraft due to the chute.

11.52 A paratrooper and parachute weigh 900 N. What rate of descent will they have if the parachute is 7 m in diameter and the air has a density of 1.20 kg/m³?

11.53 If a balloon weighs 0.15 N (empty) and is inflated with helium to a diameter of 50 cm, what will be its terminal velocity in air (standard atmospheric conditions)? The helium is at standard conditions.

11.54 A 2 cm plastic ball with a specific gravity of 1.2 is released from rest in water at 20°C. Find the time and distance needed to achieve 99% of the terminal velocity. Write out the equation of motion by equating the mass times acceleration to the buoyant force, weight, and drag force and solve by developing a computer program or using available software. Use Eq. (11.9) for the drag coefficient. [*Hint:* The equation of motion can be expressed in the form

$$\frac{dv}{dt} = -\left(\frac{C_D \, \mathrm{Re}}{24}\right)\frac{18\mu}{\rho_b d^2}v + \frac{\rho_b - \rho_w}{\rho_b}g$$

where ρ_b is the density of the ball and ρ_w is the density of the water. This form avoids the problem of the drag coefficient approaching infinity when the velocity approaches zero because $C_D \, \mathrm{Re}/24$ approaches unity as the Reynolds number approaches zero. An "if-statement" is needed to avoid a singularity in Eq. (11.9) when the Reynolds number is zero.]

Lift Force

11.55 $\mathbb{PQ}\blacktriangleleft$ From the following list, select one topic that is interesting to you. Then, use references such as the Internet to research your topic and prepare one page of written documentation that you could use to present your topic to your peers.
a. Explain how an airplane works.
b. Describe the aerodynamics of a flying bird.
c. Explain how a propellor produces thrust.
d. Explain how a kite flies.

11.56 $\mathbb{PQ}\blacktriangleleft$ Apply the grid method to each situation below.
a. Use Eq. (11.17) to predict the lift force in newtons for a spinning baseball. Use a coefficient of lift of $C_L = 1.2$. The speed of the baseball is 90 mph. Calculate area using $A = \pi r^2$, where the radius of a baseball is $r = 1.45$ in. Assume a hot summer day.
b. Use Eq. (11.17) to predict the size of wing in mm^2 needed for a model aircraft that has a mass of 570 g. Wing size is specified by giving the wing area (A) as veiwed by an observer looking down on the wing. Assume the airplane is traveling at 80 mph on a hot summer day. Use a coefficient of lift of $C_L = 1.2$. Assume straight and level flight so lift force balances weight.

11.57 $\mathbb{PQ}\blacktriangleleft$ Using Section 11.8 and other resources, answer the following questions. Strive for depth, clarity, and accuracy. Also, strive for effective use of sketches, words, and equations.
a. What is circulation? Why is it important?
b. What is lift force?
c. What variables influence the magnitude of the lift force?

11.58 As shown, a glider traveling at a constant velocity will move along a straight glide path that has an angle θ with respect to the horizontal. The angle θ, also called the glide ratio, is given by $\theta = (C_D/C_L)$. Use basic principles to prove the above statement.

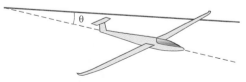

PROBLEM 11.58

11.59 The baseball is thrown from west to east with a spin about its vertical axis as shown. Under these conditions it will "break" toward the (a) north, (b) south, or (c) neither.

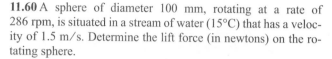

Plan view

PROBLEM 11.59

11.60 A sphere of diameter 100 mm, rotating at a rate of 286 rpm, is situated in a stream of water (15°C) that has a velocity of 1.5 m/s. Determine the lift force (in newtons) on the rotating sphere.

11.61 Analyses of pitched baseballs indicate that C_L of a rotating baseball is approximately three times that shown in Fig. 11.17. This greater C_L is due to the added circulation caused by the seams of the ball. What is the lift of a ball pitched at a speed of 85 mph and with a spin rate of 35 rps? Also, how much will the ball be deflected from its original path by the time it gets to the plate as a result of the lift force? *Note:* The mound-to-plate distance is 60 ft, the weight of the baseball is 5 oz, and the circumference is 9 in. Assume standard atmospheric conditions, and assume that the axis of rotation is vertical.

11.62 An airplane wing having the characteristics shown in Fig. 11.23 is to be designed to lift 2000 lbf when the airplane is cruising at 200 ft/s with an angle of attack of 3°. If the chord length is to be 4 ft, what span of wing is required? Assume $\rho = 0.0024$ slugs/ft^3.

11.63 A boat of the hydrofoil type has a lifting vane with an aspect ratio of 4 that has the characteristics shown in Fig. 11.23. If the angle of attack is 4° and the weight of the boat is 5 tons, what foil dimensions are needed to support the boat at a velocity of 60 fps?

11.64 One wing (wing A) is identical (same cross section) to another wing (wing B) except that wing B is twice as long as wing A. Then for a given wind speed past both wings and with the same angle of attack, one would expect the total lift of wing B to be (a) the same as that of wing A, (b) less than that of wing A, (c) double that of wing A, or (d) more than double that of wing A.

11.65 What happens to the value of the induced drag coefficient for an aircraft that increases speed in level flight? (a) it increases, (b) it decreases, (c) it does not change.

11.66 The total drag coefficient for an airplane wing is $C_D = C_{D0} + C_L^2/\pi\Lambda$, where C_{D0} is the form drag coefficient, C_L is the lift coefficient and Λ is the aspect ratio of the wing. The power is given by $P = F_D V = 1/2C_D\rho V^3 S$. For level flight the lift is equal to the weight, so $W/S = 1/2\rho C_L V^2$, where W/S is called the "wing loading." Find an expression for V for which the power is a minimum in terms of $V_{\mathrm{MinPower}} = f(\rho,\Lambda,W/S,C_{D0})$, and find the V for minimum power when $\rho = 1$ kg/m^3, $\Lambda = 10$, $W/S = 600$ N/m^2, and $C_D = 0.02$.

11.67 The airstream affected by the wing of an airplane can be considered to be a cylinder (stream tube) with a diameter equal to the wing span, b. Far downstream from the wing, the tube is-deflected through an angle θ from the original direction. Apply the momentum equation to the stream tube between sections 1 and 2 and find the lift of the wing as a function of b, ρ, V, and θ. Relating the lift to the lift coefficient, find θ as a function of b, C_L, and wing area, S. Using the relation for induced drag, $F_{Di} = F_L\theta/2$, show that $C_{Di} = C_L^2/\pi\Lambda$, where Λ is the wing aspect ratio.

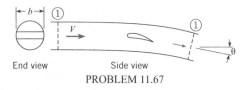

End view   Side view
PROBLEM 11.67

11.68 The landing speed of an airplane is 8 m/s faster than its stalling speed. The lift coefficient at landing speed is 1.2, and the maximum lift coefficient (stall condition) is 1.4. Calculate both the landing speed and the stalling speed.

11.69 An airplane has a rectangular-planform wing that has an elliptical spanwise lift distribution. The airplane has a mass of 1200 kg, a wing area of 20 m², and a wingspan of 14 m, and it is flying at 60 m/s at 3000 m altitude in a standard atmosphere. If the form drag coefficient is 0.01, calculate the total drag on the wing and the power ($P = F_D V$) necessary to overcome the drag.

11.70 The figure shows the relative pressure distribution for a Göttingen 387-FB lifting vane (19) when the angle of attack is 8°. If such a vane with a 20 cm chord were used as a hydrofoil at a depth of 70 cm, at what speed in 10°C fresh water would cavitation begin? Also, estimate the lift per unit of length of foil at this speed.

11.71 Consider the distribution of C_p as given for the wing section in Prob. 11.70. For this distribution of C_p, the lift coefficient C_L will fall within which range of values: (a) $0 < C_L < 1.0$; (b) $1.01 < C_L < 2.0$; (c) $2.01 < C_L < 3.0$; (d) $3.0 < C_L$?

11.72 The total drag coefficient for a wing with an elliptical lift distribution is $C_D = C_{D_0} + C_L^2/\pi\Lambda$, where Λ is the aspect ratio. Derive an expression for C_L that corresponds to minimum C_D/C_L (maximum C_L/C_D) and the corresponding C_L/C_D.

11.73 A glider at 1,000 m altitude has a mass of 200 kg and a wing area of 20 m². The glide angle is 1.7°, and the air density is 1.2 kg/m³. If the lift coefficient of the glider is 0.8, how many minutes will it take to reach sea level on a calm day?

11.74 The wing loading on an airplane is defined as the aircraft weight divided by the wing area. An airplane with a wing loading of 2000 N/m² has the aerodynamic characteristics given by Fig. 11.24. Under cruise conditions the lift coefficient is 0.3. If the wing area is 10 m², find the drag force.

11.75 An ultralight airplane has a wing with an aspect ratio of 5 and with lift and drag coefficients corresponding to Fig. 11.23. The planform area of the wing is 200 ft². The weight of the airplane and pilot is 400 lbf. The airplane flies at 50 ft/s in air with a density of 0.002 slugs/ft³. Find the angle of attack and the drag force on the wing.

11.76 Your objective is to design a human-powered aircraft using the characteristics of the wing in Fig. 11.23. The pilot weighs 130 pounds and is capable of outputting 1/2 horsepower (225 ft-lbf/s) of continuous power. The aircraft without the wing has a weight of 40 lbf, and the wing can be designed with a weight of 0.12 lbf per square foot of wing area. The drag consists of the drag of the structure plus the drag of the wing. The drag coefficient of the structure, C_{D0} is 0.05, so that the total drag on the craft will be

$$F_D = (C_{D_0} + C_D)\frac{1}{2}\rho V_0^2 S$$

where C_D is the drag coefficient from Fig. 11.23. The power required is equal to $F_D V_0$. The air density is 0.00238 slugs/ft³. Assess whether the airfoil is adequate and, if it is, find the optimum design (wing area and aspect ratio).

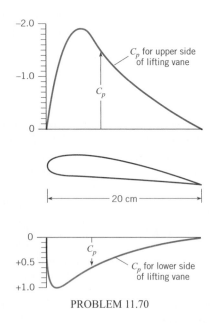

PROBLEM 11.70

12

Compressible Flow

The compressibility effects in gas flows become significant when the Mach number exceeds 0.3. The performance of high-speed aircraft, the flow in rocket nozzles, and the re-entry mechanics of spacecraft require inclusion of compressible flow effects. The purpose of this chapter is to introduce the basic concepts for compressible flow and demonstrate their applications.

12.1

Wave Propagation in Compressible Fluids

Wave propagation in a fluid is the mechanism through which the presence of boundaries is communicated to the flowing fluid. In a liquid the propagation speed of the pressure wave is much higher than the flow velocities, so the flow has adequate time to adjust to a change in boundary shape. Gas flows, on the other hand, can achieve speeds that are comparable to and even exceed the speed at which pressure disturbances are propagated. In this situation, with compressible fluids, the propagation speed is an important parameter and must be incorporated into the flow analysis. In this section it will be shown how the speed of an infinitesimal pressure disturbance can be evaluated and what its significance is to flow of a compressible fluid.

Speed of Sound

Everyone has had the experience during a thunderstorm of seeing lightning flash and hearing the accompanying thunder an instant later. Obviously, the sound was produced by the lightning, so the sound wave must have traveled at a finite speed. If the air were totally incompressible

Figure 12.1

Section view of a sound wave.

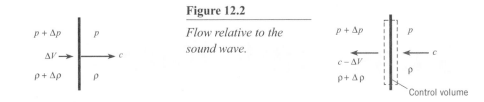

Figure 12.2

Flow relative to the sound wave.

(if that were possible), the sound of thunder and the lightening flash would be simultaneous, because all disturbances propagate at infinite speed through incompressible media.* It is analogous to striking one end of a bar of incompressible material and recording instantaneously the response at the other end. Actually, all materials are compressible to some degree and propagate disturbances at finite speeds.

The *speed of sound* is defined as the rate at which an infinitesimal disturbance (pressure pulse) propagates in a medium with respect to the frame of reference of that medium. Actual sound waves, comprised of pressure disturbances of finite amplitude, such that the ear can detect them, travel only slightly faster than the "speed of sound." In Chapter 6 it was found that the speed at which a pressure wave travels through a fluid depends on the bulk modulus of the fluid and its density. The analysis was performed by considering the unsteady flow within a control volume as the wave passed through the control volume. Here the equation for the speed of sound is developed assuming the control volume moves with the wave, thereby analyzing a steady-flow problem.

Consider a small section of a pressure wave as it propagates at velocity c through a medium, as depicted in Fig. 12.1. As the wave travels through the gas at pressure p and density ρ, it produces infinitesimal changes of Δp, $\Delta \rho$, and ΔV. These changes must be related through the laws of conservation of mass and momentum. Select a control surface around the wave and let the control volume travel with the wave. The velocities, pressures, and densities relative to the control volume (which is assumed to be very thin) are shown in Fig. 12.2. Conservation of mass in a steady flow requires that the net mass flux across the control surface be zero. Thus

$$- \rho c A + (\rho + \Delta \rho)(c - \Delta V)A = 0 \tag{12.1}$$

where A is the cross-sectional area of the control volume. Neglecting products of higher-order terms ($\Delta \rho \Delta V$) and dividing by the area reduces the conservation-of-mass equation to

$$- \rho \Delta V + c \Delta \rho = 0 \tag{12.2}$$

The momentum equation for a non-accelerating steady flow,

$$\sum \mathbf{F} = \dot{m}_o V_o - \dot{m} V_i \tag{12.3}$$

applied to the control volume containing the pressure wave gives

$$(p + \Delta p)A - pA = (-c)(-\rho A c) + (-c + \Delta V)\rho A c \tag{12.4}$$

where the direction to the right is defined as positive. The momentum equation reduces to

$$\Delta p = \rho c \Delta V \tag{12.5}$$

---

* Actually, the thunder would be heard before the lightning was seen, because light also travels at a finite, though very high, speed! However, this would violate one of the basic tenets of relativity theory. No medium can be completely incompressible and propagate disturbances exceeding the speed of light.

Substituting the expression for ΔV obtained from Eq. (12.2) into Eq. (12.5) gives

$$c^2 = \frac{\Delta p}{\Delta \rho} \tag{12.6}$$

which shows how the speed of propagation is related to the pressure and density change across the wave. It is immediately obvious from this equation that if the flow were ideally incompressible, $\Delta \rho = 0$, the propagation speed would be infinite, which confirms the argument presented earlier.

Equation (12.6) provides an expression for the speed of a general pressure wave. The sound wave is a special type of pressure wave. By definition, a sound wave produces only infinitesimal changes in pressure and density, so it can be regarded as a reversible process. There is also negligibly small heat transfer, so one can assume the process is *adiabatic*. A reversible, adiabatic process is an *isentropic* process; thus the resulting expression for the speed of sound is

$$c^2 = \frac{\partial p}{\partial \rho}\bigg|_s \tag{12.7}$$

This equation is valid for the speed of sound in any substance. However, for many substances the relationship between p and ρ at constant entropy is not very well known.

To reiterate, the speed of sound is the speed at which an infinitesimal pressure disturbance travels through a fluid. Waves of finite strength (finite pressure change across the wave) travel faster than sound waves. Sound speed is the *minimum* speed at which a pressure wave can propagate through a fluid.

For an isentropic process in an ideal gas, the following relationship exists between pressure and density (1)

$$\frac{p}{\rho^k} = \text{constant} \tag{12.8}$$

where k is the ratio of specific heats; that is, the ratio of specific heat at constant pressure to that at constant volume.

$$k = \frac{c_p}{c_v} \tag{12.9}$$

The values of k for some commonly used gases are given in Table A.2. Taking the derivative of Eq. (12.8) to obtain $\partial p/\partial \rho|_s$ results in

$$\frac{\partial p}{\partial \rho}\bigg|_s = \frac{kp}{\rho} \tag{12.10}$$

However, from the ideal gas law,

$$\frac{p}{\rho} = RT$$

so the *speed of sound* is given by

$$c = \sqrt{kRT} \tag{12.11}$$

Thus the speed of sound in an ideal gas varies with the square root of the temperature. Using this equation to predict sound speeds in real gases at standard conditions gives results very near the measured values. Of course, if the state of the gas is far removed from ideal conditions (high pressures, low temperatures), then using Eq. (12.11) is not valid.

Example 12.1 illustrates the calculation of sound speed for a given temperature.

EXAMPLE 12.1 SPEED OF SOUND CALCULATION

Problem Definition

Find: The speed of sound in air at 15°C

Assumptions: Air is an ideal gas.

Properties: Table A.2: $R = 287$ J/kg K, and $k = 1.4$.

Plan

Apply the speed of sound equation, Eq. (12.11), with $T = 288$ K.

Solution

$$c = \sqrt{kRT}$$

$$c = [(1.4)(287 \text{ J/kg K})(288 \text{ K})]^{1/2} = \boxed{340 \text{ m/s}}$$

Review

Hint: The absolute temperature must always be used in speed of sound equation.

It is possible to demonstrate, in a very simple way, the significance of sound in a compressible flow. Consider the airfoil traveling at speed V in Fig. 12.3. As this airfoil travels through the fluid, the pressure disturbance generated by the airfoil's motion propagates as a wave at sonic speed ahead of the airfoil. These pressure disturbances travel a considerable distance ahead of the airfoil before being attenuated by the viscosity of the fluid, and they "warn" the upstream fluid that the airfoil is coming. In turn, the fluid particles begin to move apart in such a way that there is a smooth flow over the airfoil by the time it arrives. If a pressure disturbance created by the airfoil is essentially attenuated in time Δt, then the fluid at a distance $\Delta t(c - V)$ ahead is alerted to prepare for the airfoil's impending arrival.

What happens as the speed of the airfoil is increased? Obviously, the relative velocity $c - V$ is reduced, and the upstream fluid has less time to prepare for the airfoil's arrival. The flow field is modified by smaller streamline curvatures, and the form drag on the airfoil is increased. If the airfoil speed increases to the speed of sound or greater, the fluid has no warning whatsoever that the airfoil is coming and cannot prepare for its arrival. Nature, at this point, resolves the problem by creating a shock wave that stands off the leading edge, as shown in Fig. 12.4. As the fluid passes through the shock wave near the leading edge, it is decelerated to a speed less than sonic speed and therefore has time to divide and flow around the airfoil. Shock waves will be treated in more detail in Section 12.3.

Figure 12.3

Propagation of a sound wave by an airfoil.

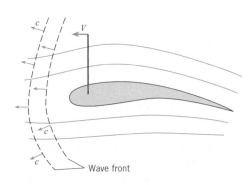

Figure 12.4

Standing shock wave in front of an airfoil.

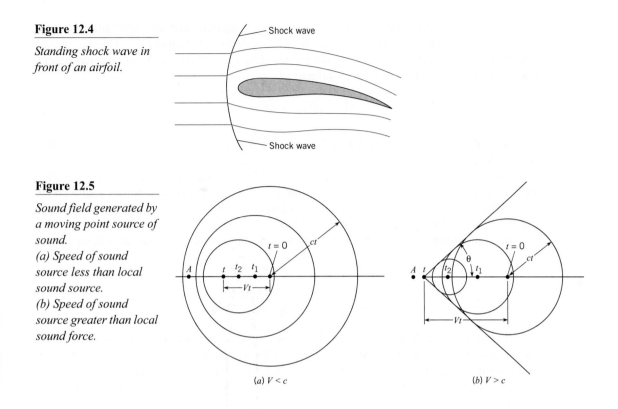

Figure 12.5

Sound field generated by a moving point source of sound.
(a) Speed of sound source less than local sound source.
(b) Speed of sound source greater than local sound force.

Another approach to appreciating the significance of sound propagation in a compressible fluid is to consider a point source of sound moving in a quiescent fluid, as shown in Fig. 12.5. The sound source is moving at a speed less than the local sound speed in Fig. 12.5*a* and faster than the local sound speed in Fig. 12.5*b*. At time $t = 0$ a sound pulse is generated and propagates radially outward at the local speed of sound. At time t_1 the sound source has moved a distance Vt_1, and the circle representing the sound wave emitted at $t = 0$ has a radius of ct_1. The sound source emits a new sound wave at t_1 that propagates radially outward. At time t_2 the sound source has moved to Vt_2, and the sound waves have moved outward as shown.

When the sound source moves at a speed less than the speed of sound, the sound waves form a family of nonintersecting eccentric circles, as shown in Fig. 12.5*a*. For an observer stationed at A the frequency of the sound pulses would appear higher than the emitted frequency because the sound source is moving toward the observer. In fact, the observer at A will detect a frequency of

$$f = f_0 / (1 - V/c)$$

where f_0 is the emitting frequency of the moving sound source. This change in frequency is known as the *Doppler effect*.

When the sound source moves faster than the local sound speed, the sound waves intersect and form the locus of a cone with a half-angle of

$$\theta = \sin^{-1}(c/V)$$

The observer at A will not detect the sound source until it has passed. In fact, only an observer within the cone is aware of the moving sound source.

In view of the physical arguments given, it is apparent that an important parameter relating to sound propagation and compressibility effects is the ratio V/c. This parameter, already introduced in Chapter 1, was first proposed by Ernst Mach, an Austrian scientist, and bears his name. The Mach number is defined as

$$M = \frac{V}{c} \tag{12.12}$$

The conical wave surface depicted in Fig. 12.5*b* is known as a *Mach wave* and the conical half-angle as the *Mach angle.*

Besides the qualitative argument presented for the Mach number, it is also recalled from Chapter 8 that the Mach number is the ratio of the inertial to elastic forces acting on the fluid. If the Mach number is small, the inertial forces are ineffective in compressing the fluid, and the fluid can be regarded as incompressible.

Compressible flows are characterized by their Mach number regimes as follows:

$$
\begin{array}{ll}
M < 1 & \text{subsonic flow} \\
M \approx 1 & \text{transonic flow} \\
M > 1 & \text{supersonic flow}
\end{array}
$$

Flows with Mach numbers exceeding 5 are sometimes referred to as *hypersonic.* Airplanes designed to travel near sonic speeds and faster are equipped with Mach meters because of the significance of the Mach number with respect to aircraft performance.

Evaluation of the Mach number of an airplane flying at altitude is demonstrated in Example 12.2.

EXAMPLE 12.2   MACH-NUMBER CALCULATION

An F-16 fighter is flying at an altitude of 13 km with a speed of 470 m/s. Assume a U.S. standard atmosphere and calculate the Mach number of the aircraft.

Problem Definition

Situation: Aircraft flying at 470 m/s at an altitude of 13 km.

Find: The Mach number of the aircraft.

Assumptions: Temperature varies as U.S. standard atmosphere.

Properties: From Table A.2: $R_{\text{air}} = 287 \text{ J/kg K}$, and $k = 1.4$.

Plan

1. Find temperature at 13 km using Eq. (3.13).
2. Calculate the speed of sound.
3. Calculate the Mach number.

Solution

1. Temperature at 13 km

$$T = T_0 - \alpha z$$

$$T = 296 - 5.87 \text{ K/km} \times 13 \text{ km} = 220 \text{ K}$$

2. Speed of sound

$$c = \sqrt{kRT} = \sqrt{1.4 \times 287 \times 220} = 297 \text{ m/s}$$

3. Mach number

$$M = \frac{V}{c} = \frac{470 \text{ m/s}}{297 \text{ m/s}} = \boxed{1.58}$$

Review

The aircraft is flying at supersonic speed.

Mach Number Relationships

In this section it will be shown how fluid properties vary the Mach number in compressible flows. Consider a control volume bounded by two streamlines in a steady compressible flow, as shown in Fig. 12.6. Applying the energy equation, Eq. (7.20), to this control volume, realizing that the shaft work is zero, gives

$$-\dot{m}_1\left(h_1 + \frac{V_1^2}{2} + gz_1\right) + \dot{m}_2\left(h_2 + \frac{V_2^2}{2} + gz_2\right) = \dot{Q} \qquad (12.13)$$

As pointed out in Chapter 4, the elevation terms (z_1 and z_2) can usually be neglected for gaseous flows. If the flow is adiabatic ($\dot{Q} = 0$), the energy equation reduces to

$$\dot{m}_1\left(h_1 + \frac{V_1^2}{2}\right) = \dot{m}_2\left(h_2 + \frac{V_2^2}{2}\right) \qquad (12.14)$$

From the principle of continuity, the mass flow rate is constant, $\dot{m}_1 = \dot{m}_2$, so

$$h_1 + \frac{V_1^2}{2} = h_2 + \frac{V_2^2}{2} \qquad (12.15)$$

Since positions 1 and 2 are arbitrary points on the same streamline, one can say that

$$h + \frac{V^2}{2} = \text{constant along a streamline in an adiabatic flow} \qquad (12.16)$$

The constant in this expression is called the *total enthalpy, h_t.* It is the enthalpy that would arise if the flow velocity were brought to zero in a adiabatic process. Thus the energy equation along a streamline under adiabatic conditions is

$$h + \frac{V^2}{2} = h_t \qquad (12.17)$$

If h_t is the same for all streamlines, the flow is *homenergic.*

It is instructive at this point to compare Eq. (12.17) with the Bernoulli equation. Expressing the specific enthalpy as the sum of the specific internal energy and p/ρ, Eq. (12.17) becomes

$$u + \frac{p}{\rho} + \frac{V^2}{2} = \text{constant}$$

Figure 12.6

Control volume enclosed by streamlines.

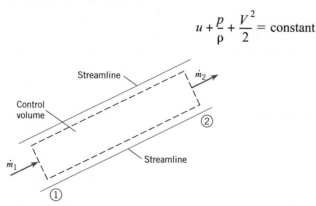

If the fluid is incompressible and there is no heat transfer, the specific internal energy is constant and the equation reduces to the Bernoulli equation (excluding the pressure due to elevation change).

Temperature

The enthalpy of an ideal gas can be written as

$$h = c_p T \tag{12.18}$$

where c_p is the specific heat at constant pressure. Substituting this relation into Eq. (12.17) and dividing by $c_p T$, results in

$$1 + \frac{V^2}{2 c_p T} = \frac{T_t}{T} \tag{12.19}$$

where T_t is the *total temperature*. From thermodynamics (1) it is known for an ideal gas that

$$c_p - c_v = R \tag{12.20}$$

or

$$k - 1 = \frac{R}{c_v} = \frac{kR}{c_p}$$

Therefore

$$c_p = \frac{kR}{k-1} \tag{12.21}$$

Substituting this expression for c_p back into Eq. (12.19) and realizing that kRT is the speed of sound squared results in the *total temperature* equation

$$T_t = T\left(1 + \frac{k-1}{2}M^2\right) \tag{12.22}$$

The temperature T is called the *static temperature*—the temperature that would be registered by a thermometer moving with the flowing fluid. Total temperature is analogous to total enthalpy in that it is the temperature that would arise if the velocity were brought to zero adiabatically. If the flow is adiabatic, the total temperature is constant along a streamline. If not, the total temperature varies according to the amount of thermal energy transferred.

Example 12.3 illustrates the evaluation of the total temperature on an aircraft's surface.

EXAMPLE 12.3 TOTAL TEMPERATURE CALCULATION

An aircraft is flying at M = 1.6 at an altitude where the atmospheric temperature is –50°C. The temperature on the aircraft's surface is approximately the total temperature. Estimate the surface temperature, taking $k = 1.4$.

Problem Definition

Situation: Aircraft flying at M = 1.6 with static temperature of –50°C.

Find: Total temperature.

Plan

This problem can be visualized as the aircraft being stationary and an airstream with a static temperature of –50°C flowing past the aircraft at a Mach number of 1.6.

1. Convert the local static temperature to degrees K.
2. Use total temperature equation, Eq. (12.22).

Solution

1. Static temperature in absolute temperature units

$$T = 273 - 50 = 223 \text{ K}$$

2. Total temperature

$$T_t = 223[1 + 0.2(1.6)^2] = \boxed{337 \text{ K or } 64°C}$$

If the flow is isentropic, thermodynamics shows that the following relationship for pressure and temperature of an ideal gas between two points on a streamline is valid (1):

$$\frac{p_1}{p_2} = \left(\frac{T_1}{T_2}\right)^{k/(k-1)} \tag{12.23}$$

Isentropic flow means that there is no heat transfer, so the total temperature is constant along the streamline. Therefore

$$T_t = T_1\left(1 + \frac{k-1}{2}M_1^2\right) = T_2\left(1 + \frac{k-1}{2}M_2^2\right) \tag{12.24}$$

Solving for the ratio T_1/T_2 and substituting into Eq. (12.23) shows that the pressure variation with the Mach number is given by

$$\frac{p_1}{p_2} = \left\{\frac{1 + [(k-1)/2]M_2^2}{1 + [(k-1)/2]M_1^2}\right\}^{k/(k-1)} \tag{12.25}$$

In the ideal gas law used to derive Eq. (12.23), absolute pressures must always be used in calculations with these equations.

The *total pressure* in a compressible flow is given by

$$p_t = p\left(1 + \frac{k-1}{2}M^2\right)^{k/(k-1)} \tag{12.26}$$

which is the pressure that would result if the flow were decelerated to zero speed reversibly and adiabatically. Unlike total temperature, total pressure may not be constant along streamlines in adiabatic flows. For example, it will be shown that flow through a shock wave, although adiabatic, is not reversible and, therefore, not isentropic. The total pressure variation along a streamline in an adiabatic flow can be obtained by substituting Eqs. (12.26) and (12.24) into Eq. (12.25) to give

$$\frac{p_{t_1}}{p_{t_2}} = \frac{p_1}{p_2}\left\{\frac{1 + [(k-1)/2]M_1^2}{1 + [(k-1)/2]M_2^2}\right\}^{k/(k-1)} = \frac{p_1}{p_2}\left(\frac{T_2}{T_1}\right)^{k/(k-1)} \tag{12.27}$$

Unless the flow is also reversible and Eq. (12.23) is applicable, the total pressures at points 1 and 2 will not be equal. However, if the flow is isentropic, total pressure is constant along streamlines.

Density

Analogous to the total pressure, the *total density* in a compressible flow is given by

$$\rho_t = \rho\left(1 + \frac{k-1}{2}M^2\right)^{1/(k-1)} \tag{12.28}$$

where ρ is the local or static density. If the flow is isentropic, then ρ_t is a constant along streamlines and Eq. (12.28) can be used to determine the variation of gas density with the Mach number.

In literature dealing with compressible flows, one often finds reference to "stagnation" conditions—that is, "stagnation temperature" and "stagnation pressure." By definition, *stagnation* refers to the conditions that exist at a point in the flow where the velocity is zero, regardless of whether or not the zero velocity has been achieved by an adiabatic, or reversible, process. For example, if one were to insert a Pitot-static tube into a compressible flow, strictly speaking one would measure stagnation pressure, not total pressure, since the deceleration of the flow would not be reversible. In practice, however, the difference between stagnation and total pressure is insignificant.

Kinetic Pressure

The kinetic pressure, $q = \rho V^2/2$, is often used, as seen in Chapter 11, to calculate aerodynamic forces with the use of appropriate coefficients. It can also be related to the Mach number. Using the ideal gas law to replace ρ gives

$$q = \frac{1}{2}\frac{pV^2}{RT} \tag{12.29}$$

Then using the equation for the speed of sound, Eq. (12.11), results in

$$q = \frac{k}{2}p\mathrm{M}^2 \tag{12.30}$$

where p must always be an absolute pressure since it derives from the ideal gas law.

    The use of the equation for kinetic pressure to evaluate the drag force is shown in Example 12.4.

    The Bernoulli equation is not valid for compressible flows. Consider what would happen if one decided to measure the Mach number of a high-speed air flow with a Pitot-static tube, assuming that the Bernoulli equation was valid. Assume a total pressure of 180 kPa and a static pressure of 100 kPa were measured. By the Bernoulli equation, the kinetic pressure is equal to the difference between the total and static pressures, so

$$\frac{1}{2}\rho V^2 = p_t - p \quad\quad \text{or} \quad\quad \frac{k}{2}p\mathrm{M}^2 = p_t - p$$

Solving for the Mach number,

$$\mathrm{M} = \sqrt{\frac{2}{k}\left(\frac{p_t}{p} - 1\right)}$$

EXAMPLE 12.4 DRAG FORCE ON A SPHERE

The drag coefficient for a sphere at a Mach number of 0.7 is 0.95. Determine the drag force on a sphere 10 mm in diameter in air if p = 101 kPa.

Problem Definition

Situation: A sphere is moving at a Mach number of 0.7 in air.

Find: The drag force (in newtons) on the sphere.

Properties: From Table A.2, k_{air} = 1.4.

Plan

The drag force on a sphere is $F_D = qC_DA$.

1. Calculate the kinetic pressure q from Eq. (12.30).
2. Calculate the drag force.

Solution

1. Kinetic pressure

$$q = \frac{k}{2}p\mathrm{M}^2 = \frac{1.4}{2}(101\ \text{kPa})(0.7)^2 = 34.6\ \text{kPa}$$

2. Drag force:

$$F_D = C_D q\left(\frac{\pi}{4}\right)D^2 = 0.95\left(34.6 \times 10^3\ \frac{\text{N}}{\text{m}^2}\right)\left(\frac{\pi}{4}\right)(0.01\ \text{m})^2$$

$$= \boxed{2.58\ \text{N}}$$

and substituting in the measured values, one obtains

$$M = 1.07$$

The correct approach is to relate the total and static pressures in a compressible flow using Eq. (12.26). Solving that equation for the Mach number gives

$$M = \left\{ \frac{2}{k-1} \left[\left(\frac{p_t}{p} \right)^{(k-1)/k} - 1 \right] \right\}^{1/2} \tag{12.31}$$

and substituting in the measured values yields

$$M = 0.96$$

Thus applying the Bernoulli equation would have led one to say that the flow was supersonic, whereas the flow was actually subsonic. In the limit of low velocities ($p_t/p \rightarrow 1$), Eq. (12.31) reduces to the expression derived using the Bernoulli equation, which is indeed valid for very low ($M \ll 1$) Mach numbers.

It is instructive to see how the pressure coefficient at the stagnation (total pressure) condition varies with Mach number. The pressure coefficient is defined by

$$C_p = \frac{p_t - p}{\frac{1}{2}\rho V^2}$$

Using Eq. (12.30) for the kinetic pressure enables one to express C_p as a function of the Mach number and the ratio of specific heats.

$$C_p = \frac{2}{kM^2} \left[\left(1 + \frac{k-1}{2} M^2 \right)^{k/(k-1)} - 1 \right]$$

The variation of C_p with Mach number is shown in Fig. 12.7. At a Mach number of zero, the pressure coefficient is unity, which corresponds to incompressible flow. The pressure coefficient begins to depart significantly from unity at a number of about 0.3. From this observation it is inferred that compressibility effects in the flow field are unimportant for Mach numbers less than 0.3.

Figure 12.7

Variation of the pressure coefficient with Mach number.

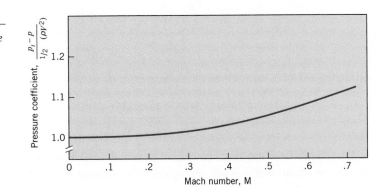

12.3

Normal Shock Waves

Normal shock waves are wave fronts normal to the flow across which a supersonic flow is decelerated to a subsonic flow with an attendant increase in static temperature, pressure, and density. The normal shock wave is analogous to the water hammer introduced in Chapter 6. The purpose of this section is to develop relations for property changes across normal shock waves.

Change in Flow Properties Across a Normal Shock Wave

The most straightforward way to analyze a normal shock wave is to draw a control surface around the wave, as shown in Fig. 12.8, and write down the continuity, momentum, and energy equations.

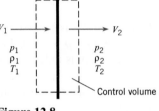

Figure 12.8

Control volume enclosing a normal shock wave.

The net mass flux into the control volume is zero because the flow is steady. Therefore

$$-\rho_1 V_1 A + \rho_2 V_2 A = 0 \tag{12.32}$$

where A is the cross-sectional area of the control volume. Equating the net pressure forces acting on the control surface to the net efflux of momentum from the control volume gives

$$\rho_1 V_1 A(-V_1 + V_2) = (p_1 - p_2)A \tag{12.33}$$

The energy equation can be expressed simply as

$$T_{t_1} = T_{t_2} \tag{12.34}$$

because the temperature gradients on the control surface are assumed negligible and thus heat transfer is neglected (adiabatic).

Using the equation for the speed of sound, Eq. (12.11), and the ideal gas law, the continuity equation can be rewritten to include the Mach number as follows:

$$\frac{p_1}{RT_1}M_1\sqrt{kRT_1} = \frac{p_2}{RT_2}M_2\sqrt{kRT_2} \tag{12.35}$$

The Mach number can be introduced into the momentum equation in the following way:

$$\rho_2 V_2^2 - \rho_1 V_1^2 = p_1 - p_2$$

$$p_1 + \frac{p_1}{RT_1}V_1^2 = p_2 + \frac{p_2}{RT_2}V_2^2 \tag{12.36}$$

$$p_1(1 + kM_1^2) = p_2(1 + kM_2^2)$$

Rearranging Eq. (12.36) for the *static-pressure ratio* across the shock wave results in

$$\frac{p_2}{p_1} = \frac{(1 + kM_1^2)}{(1 + kM_2^2)} \tag{12.37}$$

As will be shown later, the Mach number of a normal shock wave is always greater than unity upstream and less than unity downstream, so the static pressure always increases across a shock wave.

Rewriting the energy equation in terms of the temperature and Mach number, as done in Eq. (12.22), by utilizing the fact that $T_{t_2}/T_{t_1} = 1$, yields the static *temperature ratio* across the shock wave.

$$\frac{T_2}{T_1} = \frac{\{1 + [(k-1)/2]M_1^2\}}{\{1 + [(k-1)/2]M_2^2\}} \tag{12.38}$$

Substituting Eqs. (12.37) and (12.38) into Eq. (12.35) gives the following relationship for the Mach numbers upstream and downstream of a normal shock wave:

$$\frac{M_1}{1 + kM_1^2}\left(1 + \frac{k-1}{2}M_1^2\right)^{1/2} = \frac{M_2}{1 + kM_2^2}\left(1 + \frac{k-1}{2}M_2^2\right)^{1/2} \tag{12.39}$$

Solving this equation for M_2 as a function of M_1, results in two solutions. One solution is trivial, $M_1 = M_2$ which corresponds to no shock wave in the control volume. The other solution gives the Mach number downstream of the shock wave:

$$M_2^2 = \frac{(k-1)M_1^2 + 2}{2kM_1^2 - (k-1)} \tag{12.40}$$

Note: Because of the symmetry of Eq. (12.39), one can also use Eq. (12.40) to solve for M_1 given M_2 by simply interchanging the subscripts on the Mach numbers.

Setting $M_1 = 1$ in Eq. (12.40) results in M_2 also being equal to unity. Equations (12.38) and (12.39) also show that there would be no pressure or temperature increase across such a wave. In fact, the wave corresponding to $M_1 = 1$ is the sound wave across which, by definition, pressure and temperature changes are infinitesimal. Thus the sound wave represents a degenerate normal shock wave.

Example 12.5 demonstrates how to calculate properties downstream of a normal shock wave given the upstream Mach number.

The changes in flow properties across a shock wave are presented in Table A.1 for a gas, such as air, for which $k = 1.4$.

A shock wave is an adiabatic process in which no shaft work is done. Thus for ideal gases the total temperature (and total enthalpy) is unchanged across the wave. The total pressure, however, does change across a shock wave. The total pressure upstream of the wave in Example 12.5 is

$$p_{t_1} = p_1\left(1 + \frac{k-1}{2}M_1^2\right)^{k/(k-1)}$$

$$= 100 \text{ kPa}[1 + (0.2)(1.6^2)]^{3.5} = 425 \text{ kPa}$$

The total pressure downstream of the same wave is

$$p_{t_2} = p_2\left(1 + \frac{k-1}{2}M_2^2\right)^{k/(k-1)}$$

$$= 282 \text{ kPa}[1 + (0.2)(0.668^2)]^{3.5} = 380 \text{ kPa}$$

Thus the total pressure decreases through the wave, which occurs because the flow through the shock wave is not an isentropic process. Total pressure remains constant along streamlines only in isentropic flow. Values for the ratio of total pressure across a normal shock wave are also provided in Table A.1.

EXAMPLE 12.5 PROPERTY CHANGES ACROSS NORMAL SHOCK WAVE

A normal shock wave occurs in air flowing at a Mach number of 1.6. The static pressure and temperature of the air upstream of the shock wave are 100 kPa absolute and 15°C. Determine the Mach number, pressure, and temperature downstream of the shock wave.

Problem Definition

Situation: The Mach number upstream of a normal shock wave in air is 1.6.

Find: The downstream Mach number, pressure, and temperature.

Sketch:

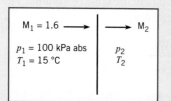

Properties: From table A.2, $k = 1.4$.

Plan

1. Use Eq. (12.40) to calculate M_2.
2. Use Eq. (12.37) to calculate p_2.
Convert upstream temperature to degrees Kelvin and use Eq. (12.38) to find T_2.

Solution

1. Downstream Mach number

$$M_2^2 = \frac{(k-1)M_1^2 + 2}{2kM_1^2 - (k-1)} = \frac{(0.4)(1.6)^2 + 2}{(2.8)(1.6)^2 - 0.4} = 0.447$$

$$M_2 = \boxed{0.668}$$

2. Downstream pressure

$$p_2 = p_1\left(\frac{1 + kM_1^2}{1 + kM_2^2}\right)$$

$$= (100 \text{ kPa})\left[\frac{1 + (1.4)(1.6)^2}{1 + (1.4)(0.668)^2}\right] = \boxed{282 \text{ kPa, absolute}}$$

3. Downstream temperature

$$T_2 = T_1\left\{\frac{1 + [(k-1)/2]M_1^2}{1 + [(k-1)/2]M_2^2}\right\}$$

$$= (288 \text{ K})\left[\frac{1 + (0.2)(2.56)}{1 + (0.2)(0.447)}\right] = \boxed{400 \text{ K or } 127°C}$$

Review

Note that absolute values for the pressure and temperature have to be used in the equations for property changes across shock waves.

Existence of Shock Waves Only in Supersonic Flows

Refer back to Eq. (12.40), which gives the Mach number downstream of a normal shock wave. If one were to substitute a value for M_1 less than unity, it is easy to see that a value for M_2 would be larger than unity. For example, if $M_1 = 0.5$ in air, then

$$M_2^2 = \frac{(0.4)(0.5)^2 + 2}{(2.8)(0.5)^2 - 0.4}$$

$$M_2 = 2.65$$

Is it possible to have a shock wave in a subsonic flow across which the Mach number becomes supersonic? In this case the total pressure would also increase across the wave; that is,

$$\frac{p_{t_2}}{p_{t_1}} > 1$$

The only way to determine whether such a solution is possible is to invoke the second law of thermodynamics, which states that for any process the entropy of the universe must remain unchanged or increase.

$$\Delta s_{univ} \geq 0 \tag{12.41}$$

Because the shock wave is an adiabatic process, there is no change in the entropy of the surroundings; thus the entropy of the system must remain unchanged or increase.

$$\Delta s_{sys} \geq 0 \tag{12.42}$$

The entropy change of an ideal gas between pressures p_1 and p_2 and temperatures T_1 and T_2 is given by (1)

$$\Delta s_{1 \rightarrow 2} = c_p \ln \frac{T_2}{T_1} - R \ln \frac{p_2}{p_1} \tag{12.43}$$

Using the relationship between c_p and R, Eq. (12.21), one can express the entropy change as

$$\Delta s_{1 \rightarrow 2} = R \ln \left[\frac{p_1}{p_2} \left(\frac{T_2}{T_1} \right)^{k/(k-1)} \right] \tag{12.44}$$

Note that the quantity in the square brackets is simply the total pressure ratio as given by Eq. (12.27). Therefore the entropy change across a shock wave can be rewritten as

$$\Delta s = R \ln \frac{p_{t_1}}{p_{t_2}} \tag{12.45}$$

A shock wave across which the Mach number changes from subsonic to supersonic would give rise to a total pressure ratio less than unity and a corresponding decrease in entropy,

$$\Delta s_{sys} < 0$$

which violates the second law of thermodynamics. Therefore shock waves can exist only in supersonic flow.

The total pressure ratio approaches unity for $M_1 \longrightarrow 1$, which conforms with the definition that sound waves are isentropic ($\ln 1 = 0$). Example 12.6 demonstrates the increase in entropy across a normal shock wave.

More examples of shock waves will be given in the next section. This section is concluded by qualitatively discussing other features of shock waves.

Besides the normal shock waves studied here, there are oblique shock waves that are inclined with respect to the flow direction. Look once again at the shock wave structure in front of a blunt body, as depicted qualitatively in Fig. 12.9. The portion of the shock wave immediately in front of the body behaves like a normal shock wave. As the shock wave bends in the free-stream direction, oblique shock waves result. The same relationships derived earlier for the normal shock waves are valid for the velocity components normal to oblique waves. The oblique shock waves continue to bend in the downstream direction until the Mach number of the velocity component normal to the wave is unity. Then the oblique shock has degenerated into a so-called Mach wave across which changes in flow properties are infinitesimal.

The familiar sonic booms are the result of weak oblique shock waves that reach ground level. One can appreciate the damage that would ensue from stronger oblique shock waves if aircraft were permitted to travel at supersonic speeds near ground level.

EXAMPLE 12.6   ENTROPY INCREASE ACROSS SHOCK WAVE

A normal shock wave occurs in air flowing at a Mach number of 1.5. Find the change in entropy across the wave.

Problem Definition

Situation: A normal shock wave in air with upstream Mach number of 1.5.

Find: The change in entropy (in J/kg K) across the wave.

Properties: From Table A.2, $R_{air} = 287$ J/kg K, and $k = 1.4$.

Plan

1. Calculate downstream Mach number using Eq. (12.40).
2. Calculate pressure ratio across wave using Eq. (12.37).
3. Calculate temperature across the wave using Eq. (12.38).
4. Calculate entropy change using Eq. (12.44).

Solution

1. Downstream Mach number

$$M_2^2 = \frac{(k-1)M_1^2 + 2}{2kM_1^2 - (k-1)} = \frac{(0.4)(1.5)^2 + 2}{(2.8)(1.5)^2 - 0.4} = 0.492$$

$$M_2 = 0.701$$

2. Pressure ratio

$$\frac{p_2}{p_1} = \left(\frac{1 + kM_1^2}{1 + kM_2^2}\right)$$

$$= \left[\frac{1 + (1.4)(1.5)^2}{1 + (1.4)(0.701)^2}\right] = 2.46$$

3. Temperature ratio

$$\frac{T_2}{T_1} = \left\{\frac{1 + [(k-1)/2]M_1^2}{1 + [(k-1)/2]M_2^2}\right\}$$

$$= \left[\frac{1 + (0.2)(2.25)}{1 + (0.2)(0.492)}\right] = 1.32$$

4. Entropy change

$$\Delta s = R\ln\left[\left(\frac{p_1}{p_2}\right)\left(\frac{T_2}{T_1}\right)^{k/(k-1)}\right]$$

$$= 287(\text{J/kg K})\ln\left[\left(\frac{1}{2.46}\right)(1.32)^{3.5}\right]$$

$$= \boxed{20.5 \text{ J/kg K}}$$

Figure 12.9

Shock wave structure in front of a blunt body.

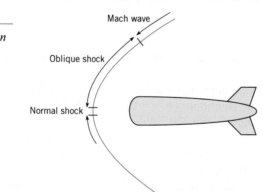

Mach wave

Oblique shock

Normal shock

12.4

Isentropic Compressible Flow Through a Duct with Varying Area

With the flow of incompressible fluids through a venturi configuration, as the flow approaches the throat (smallest area), the velocity increases and the pressure decreases; then as the area again increases, the velocity decreases. The same velocity-area relationship is not always found for compressible flows. The purpose of this section is to show the dependence of flow properties on changes in cross-sectional area with compressible flow in variable area ducts.

Figure 12.10

Duct with variable area.

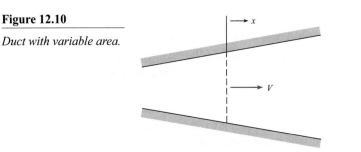

Dependence of the Mach Number on Area Variation

Consider the duct of varying area shown in Fig. 12.10. It is assumed that the flow is isentropic and that the flow properties at each section are uniform. This type of analysis, in which the flow properties are assumed to be uniform at each section yet in which the cross-sectional area is allowed to vary (nonuniform), is classified as "quasi one-dimensional."

The mass flow through the duct is given by

$$\dot{m} = \rho A V \tag{12.46}$$

where A is the duct's cross-sectional area. Since the mass flow is constant along the duct,

$$\frac{d\dot{m}}{dx} = \frac{d(\rho A V)}{dx} = 0 \tag{12.47}$$

which can be written as*

$$\frac{1}{\rho}\frac{d\rho}{dx} + \frac{1}{A}\frac{dA}{dx} + \frac{1}{V}\frac{dV}{dx} = 0 \tag{12.48}$$

The flow is assumed to be inviscid, so Euler's equation, Eq. (4.8), is valid. For steady flow

$$\rho V\frac{dV}{dx} + \frac{dp}{dx} = 0$$

Making use of Eq. (12.7), which relates $dp / d\rho$ to the speed of sound in an isentropic flow, gives

$$\frac{-V}{c^2}\frac{dV}{dx} = \frac{1}{\rho}\frac{d\rho}{dx} \tag{12.49}$$

Using this relationship to eliminate ρ in Eq. (12.48) results in

$$\frac{1}{V}\frac{dV}{dx} = \frac{1}{M^2 - 1}\frac{1}{A}\frac{dA}{dx} \tag{12.50a}$$

which can be written in an alternate form as

$$\frac{dV}{dA} = \frac{V}{A}\frac{1}{M^2 - 1} \tag{12.50b}$$

This equation, although simple, leads to the following important, far-reaching conclusions.

---

* This step can easily be seen by first taking the logarithm of Eq. (12.46):

$$\ln(\rho A V) = \ln\rho + \ln A + \ln V$$

and then taking the derivative of each term:

$$\frac{d}{dx}[\ln(\rho A V)] = 0 = \frac{1}{\rho}\frac{d\rho}{dx} + \frac{1}{A}\frac{dA}{dx} + \frac{1}{V}\frac{dV}{dx}$$

Subsonic Flow

For subsonic flow, $M^2 - 1$ is negative so $dV/dA < 0$, which means that a decreasing area leads to an increasing velocity, and, correspondingly, an increasing area leads to a decreasing velocity. This velocity area relationship parallels the trend for incompressible flows.

Supersonic Flow

For supersonic flow, $M^2 - 1$ is positive so $dV/dA > 0$, which means that a decreasing area leads to a decreasing velocity, and an increasing area leads to an increasing velocity. Thus the velocity at the minimum area of a duct with supersonic compressible flow is a minimum. This is the principle underlying the operation of diffusers on jet engines for supersonic aircraft, as shown in Fig. 12.11. The purpose of the diffuser is to decelerate the flow so that there is sufficient time for combustion in the chamber. Then the diverging nozzle accelerates the flow again to achieve a larger kinetic energy of the exhaust gases and an increased engine thrust.

Transonic Flow ($M \approx 1$)

Stations along a duct corresponding to $dA/dx = 0$ represent either a local minimum or a local maximum in the duct's cross-sectional area, as illustrated in Fig. 12.12. If at these stations the flow were either subsonic ($M < 1$) or supersonic ($M > 1$), then by Eq. (12.50a) $dV/dx = 0$, so the flow velocity would have either a maximum or a minimum value. In particular, if the flow were supersonic through the duct of Fig. 12.12a, then the velocity would be a minimum at the throat; if subsonic, a maximum.

Now, what happens if the Mach number is unity? Equation (12.50a) states that if the Mach number is unity and dA/dx is not equal to zero, the velocity gradient dV/dx is infinite—a physically impossible situation. Therefore, dA/dx must be zero where the Mach number is unity in order for a finite, physically reasonable velocity gradient to exist.*

Figure 12.11

Engine for supersonic aircraft.

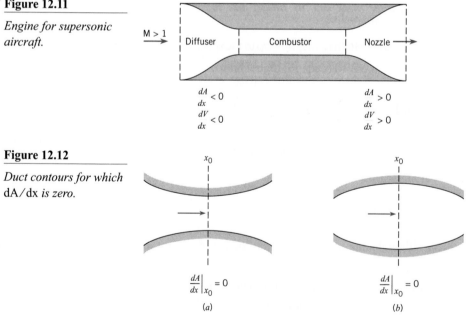

Figure 12.12

Duct contours for which dA/dx is zero.

* Actually, the velocity gradient is indeterminate because the numerator and denominator are both zero. It can be shown by application of L'Hôpital's rule, however, that the velocity gradient is finite.

The argument can be taken one step further here to show that sonic flow can occur only at a minimum area. Consider Fig. 12.12a. If the flow is initially subsonic, the converging duct accelerates the flow toward a sonic velocity. If the flow is initially supersonic, the converging duct decelerates the flow toward a sonic velocity. Using this same reasoning, one can prove that sonic flow is impossible in the duct depicted in Fig. 12.12b. If the flow is initially supersonic, the diverging duct increases the Mach number even more. If the flow is initially subsonic, the diverging duct decreases the Mach number; thus sonic flow cannot be achieved at a maximum area. Hence the Mach number in a duct of varying cross-sectional area can be unity only at a local area minimum (throat). This does not imply, however, that the Mach number must always be unity at a local area minimum.

Laval Nozzle

The Laval nozzle is a duct of varying area that produces supersonic flow. The nozzle is named after its inventor, de Laval (1845–1913), a Swedish engineer. According to the foregoing discussion, the nozzle must consist of a converging section to accelerate the subsonic flow, a throat section for transonic flow, and a diverging section to further accelerate the supersonic flow. Thus the shape of the Laval nozzle is as shown in Fig. 12.13.

One very important application of the Laval nozzle is the supersonic wind tunnel, which has been an indispensable tool in the development of supersonic aircraft. Basically, the supersonic wind tunnel, as illustrated in Fig. 12.14, consists of a high-pressure source of gas, a Laval nozzle to produce supersonic flow, and a test section. The high-pressure source may be from a large pressure tank, which is connected to the Laval nozzle through a regulator valve to maintain a constant upstream pressure, or from a pumping system that provides a continuous high-pressure supply of gas.

The equations relating to the compressible flow through a Laval nozzle have already been developed. Since the mass flow rate is the same at every cross section,

$$\rho V A = \text{constant}$$

and the constant is usually evaluated corresponding to those conditions that exist when the Mach number is unity. Thus

$$\rho V A = \rho_* V_* A_* \tag{12.51}$$

where the asterisk signifies conditions wherein the Mach number is equal to unity. Rearranging Eq. (12.51) gives

$$\frac{A}{A_*} = \frac{\rho_* V_*}{\rho V}$$

However, the velocity is the product of the Mach number and the local speed of sound. Therefore

$$\frac{A}{A_*} = \frac{\rho_*}{\rho} \frac{M_* \sqrt{kRT_*}}{M \sqrt{kRT}} \tag{12.52}$$

By definition $M_* = 1$, so

$$\frac{A}{A_*} = \frac{\rho_*}{\rho} \left(\frac{T_*}{T}\right)^{1/2} \frac{1}{M} \tag{12.53}$$

Figure 12.13

Laval nozzle.

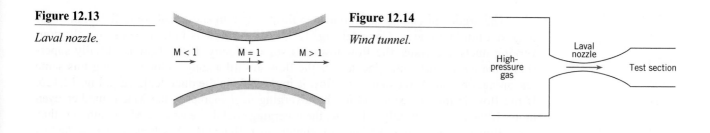

M < 1   M = 1   M > 1

Figure 12.14

Wind tunnel.

High-pressure gas   Laval nozzle   Test section

Because the flow in a Laval nozzle is assumed to be isentropic, the total temperature and total pressure (and total density) are constant throughout the nozzle. From Eq. (12.28),

$$\frac{\rho_*}{\rho} = \left\{ \frac{1 + [(k-1)/2]M^2}{(k+1)/2} \right\}^{1/(k-1)}$$

and from Eq. (12.24)

$$\frac{T_*}{T} = \frac{1 + [(k-1)/2]M^2}{(k+1)/2}$$

Substituting these expressions into Eq. (12.53) yields the following relationship for *area ratio* as a function of Mach number in a variable area duct:

$$\frac{A}{A_*} = \frac{1}{M} \left\{ \frac{1 + [(k-1)/2]M^2}{(k+1)/2} \right\}^{(k+1)/2(k-1)} \tag{12.54}$$

This equation is valid, of course, for all Mach numbers—subsonic, transonic, and supersonic. The area ratio A/A_* is the ratio of the area at the station where the Mach number is M to the area where M is equal to unity. Many supersonic wind tunnels are designed to maintain the same test-section area and to vary the Mach number by varying the throat area.

Example 12.7 illustrates the use of the Mach-number–area ratio expression to size the test section of a supersonic wind tunnel.

Example 12.7 demonstrates that it is a straightforward task to calculate the area ratio given the Mach number and ratio of specific heats. However, in practice, one usually knows the area ratio and wishes to determine the Mach number. It is not possible to solve Eq. (12.54) for the Mach number as an explicit function of the area ratio. For this reason, compressible-flow tables have been developed that allow one to obtain the Mach number easily given the area ratio (as shown in Table A.1).

EXAMPLE 12.7   TEST SECTION SIZE IN SUPERSONIC WIND TUNNEL

Suppose a supersonic wind tunnel is being designed to operate with air at a Mach number of 3. If the throat area is 10 cm^2, what must the cross-sectional area of the test section be?

Problem Definition

Situation: Design of supersonic wind tunnel with air for Mach number 3 in test section.

Find: The cross-sectional area (in cm^2) of test section.

Sketch:

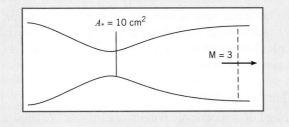

$A_* = 10$ cm^2

$M = 3$

Properties: From Table A.2, $k_{air} = 1.4$.

Plan

1. Use Eq. (12.54), which gives area ratio with respect to the throat section.
2. Calculate area of test section.

Solution

1. Area ratio

$$\frac{A}{A_*} = \frac{1}{M}\left\{\frac{1 + [(k-1)/2]M^2}{(k+1)/2}\right\}^{(k+1)/2(k-1)}$$

$$= \frac{1}{3}\left[\frac{1 + (0.2)3^2}{1.2}\right]^3 = 4.23$$

2. Cross-sectional area of test section

$$A = 4.23 \times 10 \text{ cm}^2 = \boxed{42.3 \text{ cm}^2}$$

EXAMPLE 12.8   FLOW PROPERTIES IN SUPERSONIC WIND TUNNEL

The test section of a supersonic wind tunnel using air has an area ratio of 10. The absolute total pressure and temperature are 4 MPa and 350 K. Find the Mach number, pressure,

Problem Definition

Situation: Supersonic wind tunnel has an area ratio of 10 at test section.

Find: The Mach number, pressure, temperature, and velocity at test section.

Sketch:

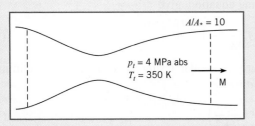

$A/A_* = 10$

$p_t = 4$ MPa abs
$T_t = 350$ K

M

Properties: From Table A.2, $k_{air} = 1.4$, $R_{air} = 287$ J/kg K.

Plan

1. Use Table A.1 and interpolate to find the Mach number at test section.
2. Use Table A.1 to find the pressure and temperature ratios at test section.
3. Evaluate the pressure and temperature in test section.
4. Calculate the speed of sound using Eq. (12.11).
5. Find the velocity using $V = Mc$.

Solution

1. From Table A.1

M	A/A_*
3.5	6.79
4.0	10.72

Interpolating between the two points gives M = 3.91 at $A/A_* = 10.0$.

2. Interpolation using Table A.1 to find the pressure and temperature ratios:

$$\frac{p}{p_t} = 0.00743 \quad \text{and} \quad \frac{T}{T_t} = 0.246$$

Solution	*Review*

Solution

3. In the test section

$$p = 0.00743 \times 4 \text{ MPa} = 29.7 \text{ kPa}$$

$$T = 0.246 \times 350 \text{ K} = 86 \text{ K}$$

4. Speed of sound

$$c = \sqrt{kRT} = \sqrt{1.4 \times 287 \times 86} = 186 \text{ m/s}$$

5. Velocity

$$V = 3.91 \times 186 \text{ m/s} = 727 \text{ m/s}$$

Review

Notice that the temperature of air in the test section is only 86 K, or −187°C. At this temperature, the water vapor in the air can condense out, creating fog in the tunnel and compromising tunnel utility.

Consider again Table A.1. This table has been developed for a gas, such as air, for which $k = 1.4$. The symbols that head each column are defined at the beginning of the table. Tables for both subsonic and supersonic flow are provided. Example 12.8 shows how to use the tables to find flow properties at a given area ratio.

Mass Flow Rate Through a Laval Nozzle

An important consideration in the design of a supersonic wind tunnel is size. A large wind tunnel requires a large mass flow rate, which, in turn, requires a large pumping system for a continuous-flow tunnel or a large tank for sufficient run time in an intermittent tunnel. The purpose of this section is to develop an equation for the mass flow rate.

The easiest station at which to calculate the mass flow rate is the throat, because there the Mach number is unity.

$$\dot{m} = \rho_* A_* V_* = \rho_* A_* \sqrt{kRT_*}$$

It is more convenient, however, to express the mass flow in terms of total conditions. The local density and static temperature at sonic velocity are related to the total density and temperature by

$$\frac{T_*}{T_t} = \left(\frac{2}{k+1}\right)$$

$$\frac{\rho_*}{\rho_t} = \left(\frac{2}{k+1}\right)^{1/(k-1)}$$

which, when substituted into the foregoing equation, give

$$\dot{m} = \rho_t \sqrt{kRT_t} A_* \left(\frac{2}{k+1}\right)^{(k+1)/2(k-1)} \tag{12.55}$$

Usually, the total pressure and temperature are known. Using the ideal gas law to eliminate ρ_t, yields the expression for *critical mass flow rate*

$$\dot{m} = \frac{p_t A_*}{\sqrt{RT_t}} k^{1/2} \left(\frac{2}{k+1}\right)^{(k+1)/2(k-1)} \tag{12.56}$$

For gases with a ratio of specific heats of 1.4,

$$\dot{m} = 0.685 \frac{p_t A_*}{\sqrt{RT_t}} \tag{12.57}$$

For gases with $k = 1.67$,

$$\dot{m} = 0.727 \frac{p_t A_*}{\sqrt{R T_t}} \tag{12.58}$$

Example 12.9 illustrates how to calculate mass flow rate in a supersonic wind tunnel given the conditions in the test section.

EXAMPLE 12.9 MASS FLOW RATE IN SUPERSONIC WIND TUNNEL

A supersonic wind tunnel with a square test section 15 cm by 15 cm is being designed to operate at a Mach number of 3 using air. The static temperature and pressure in the test section are –20°C and 50 kPa abs, respectively. Calculate the mass flow rate.

Problem Definition

Situation: A Mach-3 supersonic wind tunnel has 15 cm by 15 cm test section.

Find: Mass flow rate (kg/s) in tunnel.

Properties: From Table A.2, $k_{air} = 1.4$ and $R_{air} = 287$ J/kg K.

Sketch:

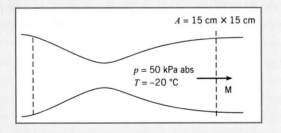

$A = 15$ cm × 15 cm
$p = 50$ kPa abs
$T = -20\ °C$
M

Plan

1. Use Eq. (12.54) to find area ratio and calculate throat area.
2. Use Eq. (12.22) to find total temperature.
3. Use Eq. (12.26) to find total pressure.
4. Use Eq. (12.56) to find the mass flow rate.

Solution

1. Area ratio

$$\frac{A}{A_*} = \frac{1}{M} \left\{ \frac{1 + [(k-1)/2]M^2}{(k+1)/2} \right\}^{(k+1)/2(k-1)}$$

$$= \frac{1}{3} \left[\frac{1 + 0.2 \times 3^2}{1.2} \right]^3 = 4.23$$

Throat area

$$A_* = \frac{225\ \text{cm}^2}{4.23} = 53.2\ \text{cm}^2 = 0.00532\ \text{m}^2$$

2. Total temperature

$$T_t = T\left(1 + \frac{k-1}{2}M^2\right) = 253\ \text{K}(2.8) = 708\ \text{K}$$

3. Total pressure

$$p_t = p\left(1 + \frac{k-1}{2}M^2\right)^{k/(k-1)} = 50\ \text{kPa}(36.7)$$

$$= 1840\ \text{kPa} = 1.84\ \text{MPa}$$

4. Mass flow rate

$$\dot{m} = 0.685 \frac{p_t A_*}{\sqrt{R T_t}} = \frac{(0.685)[1.840(10^6\ \text{N/m}^2)](0.00532\ \text{m}^2)}{[(287\ \text{J/kg K})(708\ \text{K})]^{1/2}}$$

$$= \boxed{14.9\ \text{kg/s}}$$

Review

1. An alternate way to do this problem is to calculate the density in the test section using the ideal gas law, calculate the speed of sound with the speed of sound equation, find the air speed using the Mach number, and finally determine the mass flow rate with $\dot{m} = \rho V A$.

2. A pump capable of moving air at this rate against a 1.8 MPa pressure would require over 6000 kW of power input. Such a system would be large and costly to build and to operate.

Classification of Nozzle Flow by Exit Conditions

Nozzles are classified by the conditions at the nozzle exit. Consider the Laval nozzle depicted in Fig. 12.15 with the corresponding pressure and Mach-number distributions plotted beneath it. The pressure at the nozzle entrance is very near the total pressure, because the Mach number is small. As the area decreases toward the throat, the Mach number increases and the pressure decreases. The static-to-total-pressure ratio at the throat, where conditions are sonic, is called the *critical pressure ratio*. It has a value of

$$\frac{p_*}{p_t} = \left(\frac{2}{k+1}\right)^{k/(k-1)}$$

which for air with $k = 1.4$ is

$$\frac{p_*}{p_t} = 0.528$$

It is called a critical pressure ratio because to achieve sonic flow with air in a nozzle, it is necessary that the exit pressure be equal to or less than 0.528 times the total pressure. The pressure continues to decrease until it reaches the exit pressure corresponding to the nozzle-exit area ratio. Similarly, the Mach number monotonically increases with distance down the nozzle.

The nature of the exit flow from the nozzle depends on the difference between the exit pressure, p_e, and the back pressure (the pressure to which the nozzle exhausts). If the exit pressure is higher than the back pressure, an expansion wave exists at the nozzle exit, as shown in Fig. 12.16a. These waves, which will not be studied here, effect a turning and further acceleration of the flow to achieve the back pressure. As one watches the exhaust of a rocket motor as it rises through the ever-decreasing pressure of higher altitudes, one can see the plume fan out as the flow turns more in response to the lower back pressure. A nozzle for which the exit pressure is larger than the back pressure is called an *underexpanded nozzle* because the flow could have expanded further.

If the exit pressure is less than the back pressure, shock waves occur. If the exit pressure is only slightly less than the back pressure, then pressure equalization can be obtained by oblique shock waves at the nozzle exit, as shown in Fig. 12.16b.

Figure 12.15

Distribution of static pressure and Mach number in a Laval nozzle.

Figure 12.16

Conditions at a nozzle exit. (a) Expansion waves. (b) Oblique shock waves. (c) Normal shock wave.

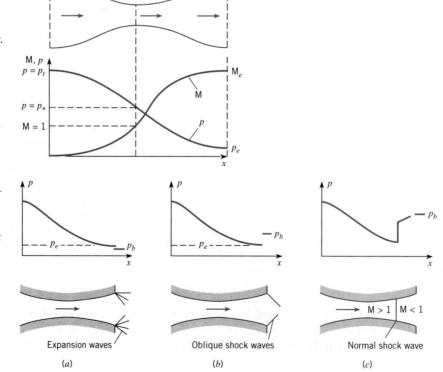

If, however, the difference between back pressure and exit pressure is larger than can be accommodated by oblique shock waves, a normal shock wave will occur in the nozzle, as shown in Fig. 12.16*c*. A pressure jump occurs across the normal shock wave. The flow becomes subsonic and decelerates in the remaining portion of the diverging section in such a way that the exit pressure is equal to the back pressure. As the back pressure is further increased, the shock wave moves toward the throat region until, finally, there is no region of supersonic flow. A nozzle in which the exit pressure corresponding to the exit area ratio of the nozzle is less than the back pressure is called an *overexpanded nozzle.* Any flow that exits from a duct (or pipe) subsonically must always exit at the local back pressure.

A nozzle with supersonic flow in which the exit pressure is equal to the back pressure is *ideally expanded.*

The assessment of the nozzle exit conditions is provided by Example 12.10.

EXAMPLE 12.10   NOZZLE EXIT CONDITION

The total pressure in a nozzle with an area ratio (A/A_*) of 4 is 1.3 MPa. Air is flowing through the nozzle. If the back pressure is 100 kPa, is the nozzle overexpanded, ideally expanded, or underexpanded?

Problem Definition

Situation: Air flows through a nozzle with exit area ratio of 4.

Find: The state of the exit condition (ideally expanded, overexpanded or underexpanded).

Sketch:

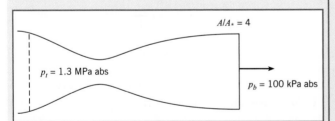

Plan

1. Interpolate Table A.1 to find Mach number corresponding to exit area ratio.
2. Calculate exit pressure using Eq. (12.26).
3. Compare exit pressure with back pressure to determine exit condition.

Solution

1. Interpolation for Mach number from Table A.1.

M	A/A_*
2.90	3.850
3.00	4.235

$$M = 2.94 \text{ at } A/A_* = 4.0.$$

2. Exit pressure

$$\frac{p_t}{p_e} = \left(1 + \frac{k-1}{2}M^2\right)^{k/(k-1)}$$

$$p_e = \frac{1300 \text{ kPa}}{\left(1 + 0.2 \times 2.94^2\right)^{3.5}} = 38.7 \text{ kPa}.$$

3. Because $p_e < p_b$, the nozzle is overexpanded.

Review

Because the nozzle is overexpanded, there will be a shock wave structure inside the nozzle to achieve pressure equilibration at the nozzle exit.

Example 12.11 illustrates how to calculate the static pressure at the exit of a Laval nozzle with overexpanded flow.

EXAMPLE 12.11   SHOCK WAVE IN LAVAL NOZZLE

The Laval nozzle shown in the figure has an expansion ratio of 4 (exit area / throat area). Air flows through the nozzle, and a normal shock wave occurs where the area ratio is 2. The total pressure upstream of the shock is 1 MPa. Determine the static pressure at the exit.

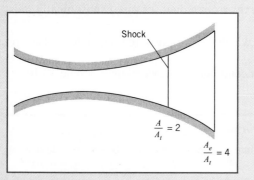

Situation: Air flows in Laval nozzle with an exit area ratio (A_e/A_*) of 4 and normal shock at $A/A_* = 2$.

Find: Static pressure (in kPa) at exit.

Properties: $k_{air} = 1.4$.

Sketch:

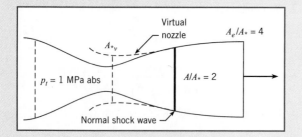

Plan

This problem will require the identification of a "virtual nozzle" shown in the sketch. The virtual nozzle is an expanding nozzle with subsonic flow and with a Mach number equal to the downstream Mach number behind the normal shock wave.

1. From Table A.1, interpolate to find the Mach number for $A/A_* = 2$.
2. Using the same table, find the Mach number downstream of shock and total pressure ratio across shock.
3. Calculate total pressure downstream of shock wave.
4. Treat the problem as flow in virtual subsonic nozzle with Mach number equal to the Mach number behind the wave with new total pressure. Calculate exit area ratio of virtual nozzle.
5. Use subsonic flow table to find subsonic Mach number at exit.
6. Use total pressure equation to calculate static pressure at exit.

Solution

1. From interpolation of the supersonic-flow part of Table A.1,

$$\text{at } A/A_* = 2, \text{ and } M = 2.2 .$$

2. From the same entry in the table,

$$M_2 = 0.547$$

$$\frac{p_{t_2}}{p_{t_1}} = 0.6281$$

3. Total pressure downstream of the shock wave

$$p_{t_2} = 0.6281 \times 1 \text{ MPa} = 6.28 \text{ kPa}$$

4. From the subsonic part of Table A.1,

$$\text{at } M = 0.547, \text{ and } A/A_{*_v} = 1.26.$$

5. Exit area ratio of virtual nozzle

$$\frac{A_e}{A_{*_v}} = \frac{A_e}{A_*} \times \frac{A_*}{A_s} \times \frac{A_s}{A_{*_v}}$$

$$= 4 \times \frac{1}{2} \times 1.26 = 2.52$$

where A_s is cross-sectional area at the shock wave.

6. From subsonic part of Table A.1,

$$\text{at } A/A_* = 2.52, M = 0.24$$

Exit pressure from Eq. (12.26)

$$\frac{p_t}{p_e} = \left(1 + \frac{k-1}{2}M^2\right)^{k/(k-1)}$$

$$p_e = \frac{628 \text{ kPa}}{[1 + (0.2)(0.24)^2]^{3.5}} = \boxed{603 \text{ kPa}}$$

Mass Flow Through a Truncated Nozzle

The *truncated nozzle* is a Laval nozzle cut off at the throat, as shown in Fig. 12.17. The nozzle exits to a back pressure p_b. This type of nozzle is important to engineers because of its frequent use as a flow-metering device for compressible flows. The purpose of this section is to develop an equation for mass flow through a truncated nozzle.

To calculate the mass flow, one must first determine whether the flow at the exit is sonic or subsonic. Of course, the flow at the exit could never be supersonic, since the nozzle area does not diverge. First calculate the value of the critical pressure ratio:

$$\frac{p_*}{p_t} = \left(\frac{2}{k+1}\right)^{k/(k-1)}$$

which, for air, is 0.528. Then evaluate the ratio of back pressure to total pressure, p_b/p_t, and compare it with the critical pressure ratio:

1. If $p_b/p_t \le p_*/p_t$, the exit pressure is higher than or equal to the back pressure, so the exit flow must be sonic. Pressure equilibration is achieved after exit by a series of expansion waves. The mass flow is calculated using Eq. (12.56), where A_* is the area at the truncated station.
2. If $p_b/p_t > p_*/p_t$, the flow exits subsonically. In this case the exit pressure is equal to the back pressure. One must first determine the Mach number at the exit by using Eq. (12.31):

$$M_e = \sqrt{\frac{2}{k-1}\left[\left(\frac{p_t}{p_b}\right)^{(k-1)/k} - 1\right]}$$

Then, using this value for Mach number, one calculates the static temperature and speed of sound at the exit:

$$T_e = \frac{T_t}{\{1 + [(k-1)/2]M_e^2\}}$$

$$c_e = \sqrt{kRT_e}$$

The gas density at the nozzle exit is determined by using the ideal gas law with the exit temperature and back pressure:

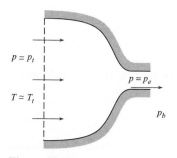

$p = p_t$

$T = T_t$

$p = p_e$

p_b

$$\rho_e = \frac{p_b}{RT_e}$$

Finally, the mass flow is given by

$$\dot{m} = \rho_e A_e M_e c_e$$

where A_e is the area at the truncated section.

Example 12.12 shows how to calculate mass flow in a truncated nozzle.

Figure 12.17

Truncated nozzle.

EXAMPLE 12.12 MASS FLOW IN TRUNCATED NOZZLE

Air exhausts through a truncated nozzle 3 cm in diameter from a reservoir at a pressure of 160 kPa and a temperature of 80°C. Calculate the mass flow rate if the back pressure is 100 kPa.

Problem Definition

Situation: Air flows through 3 cm diameter truncated nozzle.

Find: Mass flow rate (in kg/s) through nozzle.

Properties: From Table A.2, $k_{air} = 1.4$.

Sketch:

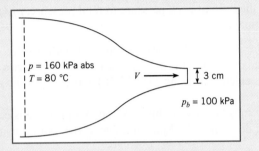

$p = 160$ kPa abs
$T = 80$ °C
$V \longrightarrow$ 3 cm
$p_b = 100$ kPa

Plan

1. Determine exit condition by comparing exit pressure with back pressure. If $p_b/p_t < p_*/p_t$, exit flow is sonic. If $p_b/p_t > p_*/p_t$, exit flow is subsonic.
2. Calculate mass flow according to exit condition.

Solution

1. Ratio of exit pressure to total pressure

$$p_b/p_t = 100/160 = 0.625$$

Because 0.625 is larger than the critical pressure ratio for air (0.528), the flow at the nozzle exit must be subsonic.

2. Mach number at exit. From total pressure equation, Eq. (12.26),

$$M_e^2 = \frac{2}{k-1}\left[\left(\frac{p_t}{p_b}\right)^{(k-1)/k} - 1\right]$$

$$M_e = 0.85$$

Static temperature at exit. From total temperature equation, Eq. (12.22),

$$T_e = \frac{T_t}{\{1 + [(k-1)/2]M_e^2\}} = 308 \text{ K}$$

Static density at exit. From ideal gas law,

$$\rho_e = \frac{p_b}{RT_e} = \frac{100 \times 10^3 \text{ N/m}^2}{(287 \text{ J/kg K})(309 \text{ K})} = 1.13 \text{ kg/m}^3$$

Speed of sound at the exit from speed-of-sound equation, Eq. (12.11),

$$c_e = \sqrt{kRT_e} = [(1.4)(287 \text{ J/kg K})(309 \text{ K})]^{1/2}$$

$$= 352 \text{ m/s}$$

Mass flow rate.

$$\dot{m} = \rho_e A_e M_e c_e$$

$$\dot{m} = (1.13 \text{ kg/m}^3)(\pi/4)(0.03^2 \text{ m}^2)(0.85)(352 \text{ m/s})$$

$$= \boxed{0.239 \text{ kg/s}}$$

Review

Had p_b/p_t been less than 0.528, then Eq. (12.56) would have been used to calculate the mass flow rate.

Further information and other topic areas in compressible flow can be found in other sources, such as Anderson (2) and Shapiro (3).

Summary

The speed of sound is the speed at which an infinitesimal pressure disturbance travels through a fluid. The speed of sound in an ideal gas is

$$c = \sqrt{kRT}$$

where k is the ratio of specific heats, R is the gas constant, and T is the absolute temperature. The Mach number is defined as

$$M = \frac{V}{c}$$

Compressible flows are classified as

$$M < 1 \quad \text{subsonic}$$
$$M \approx 1 \quad \text{transonic}$$
$$M > 1 \quad \text{supersonic}$$

In general, if the Mach number is less than 0.3, a steady flow can be regarded as incompressible.

For an adiabatic flow (no heat transfer), the temperature varies along a streamline according to

$$T = T_t\left(1 + \frac{k-1}{2}M^2\right)^{-1}$$

where T_t, the total temperature, is the temperature attained if the flow is decelerated to zero velocity. If the flow is isentropic, the pressure varies along a streamline as

$$p = p_t\left(1 + \frac{k-1}{2}M^2\right)^{k/(k-1)}$$

where p_t is the total pressure, the pressure achieved if the flow is decelerated to zero velocity isentropically.

A normal shock wave is a narrow region where a supersonic flow is decelerated to a subsonic flow with an attendant rise in pressure, temperature, and density. The total temperature does not change through a shock wave, but the total pressure decreases. The shock wave is a nonisentropic process and can only occur in supersonic flows.

A Laval nozzle is a duct with a converging and expanding area that is used to accelerate a compressible fluid to supersonic speeds. Sonic flow can occur only at the nozzle throat (minimum area). The ratio of the area at a location in the nozzle to the throat area, A/A_*, is a function of the local Mach number and the ratio of specific heats. The flow rate through a Laval nozzle is given by

$$\dot{m} = 0.685\frac{p_t A_*}{\sqrt{RT_t}}$$

A Laval nozzle is classified by comparing the pressure at the exit, p_e, for supersonic flow in the nozzle with the back (ambient) pressure, p_b.

$$p_e/p_b > 1 \quad \text{underexpanded}$$
$$p_e/p_b = 1 \quad \text{ideally expanded}$$
$$p_e/p_b < 1 \quad \text{overexpanded}$$

Shock waves occur in overexpanded nozzles, yielding a subsonic flow at the exit.

A truncated nozzle is a Laval nozzle terminated at the throat typically used for mass flow measurement.

References

1. Cengel, Y. A., and M. A. Boles, *Thermodynamics*. McGraw-Hill, New York, 1994.

2. Anderson, J. D., Jr. *Modern Compressible Flow with Historical Perspective*. McGraw-Hill, New York, 1991.

3. Shapiro, A. H. *The Dynamics and Thermodynamics of Compressible Fluid Flow*. Ronald Press, New York, 1953.

Problems

Speed of Sound and Mach Number

12.1 $\mathbb{PQ}$◄ Find from available sources the Mach number at which modern day airliners fly at altitude. Discuss whether it is possible to have regions of supersonic flow on the aircraft.

12.2 $\mathbb{PQ}$◄ The speed of sound in air is 340 m/s. What is this speed in miles per hour?

12.3 $\mathbb{PQ}$◄ It takes 3 seconds between seeing lightning and hearing the thunder, how far away (miles) is the storm (T = 50°F)?

12.4 $\mathbb{PQ}$◄ From available sources, find the orbital velocity of a satellite circling the earth. If the satellite entered the earth's atmosphere at this speed and the air temperature were –60°C, what would the Mach number be? Classify the flow.

12.5 How fast (in meters per second) will a sound wave travel in methane at 0°C?

12.6 Calculate the speed of sound in helium at 60°C.

12.7 Calculate the speed of sound in hydrogen at 68°F.

12.8 How much faster will a sound wave propagate in helium than in nitrogen if the temperature of both gases is 20°C?

12.9 Determine what the equation for the speed of sound in an ideal gas would be if the sound wave were an isothermal process.

12.10 The relationship between pressure and density for the propagation of a sound wave through a fluid is

$$p - p_0 = E_v \ln(\rho/\rho_0)$$

where p_0 and ρ_0 are the reference pressure and density (constants) and E_v is the bulk modulus of elasticity. Determine the equation for the speed of a sound wave in terms of E_v and ρ. Calculate the sound speed for water with $\rho = 1000$ kg/m³ and $E_v = 2.20$ GN/m².

12.11 A supersonic aircraft is flying at Mach 1.5 through air at –30°C. What temperature could be expected on exposed aircraft surfaces?

12.12 What is the temperature on the nose of a supersonic fighter flying at Mach 2 through air at 273 K?

12.13 A high-performance aircraft is flying at a Mach number of 1.8 at an altitude of 10,000 m, where the temperature is –44°C and the pressure is 30.5 kPa. How fast is the aircraft traveling in kilometers per hour?

a. The total temperature is an estimate of surface temperature on the aircraft. What is the total temperature under these conditions?

b. If the aircraft slows down, at what speed (kilometers per hour) will the Mach number be unity?

12.14 An airplane travels at 800 km/h at sea level where the temperature is 15°C. How fast would the airplane be flying at the same Mach number at an altitude where the temperature was –40°C?

12.15 An airplane flies at a Mach number of 0.95 at a l0,000 m altitude, where the static temperature is –44°C and the pressure is 30 kPa absolute. The lift coefficient of the wing is 0.05. Determine the wing loading (lift force/wing area).

Mach-Number Relationships

12.16 $\mathbb{PQ}$◄ Early passenger aircraft used to fly at a cruising speed of 250 mph at 15,000 ft altitude. Did the designers of these aircraft have to be concerned about compressible flow effects? Explain.

12.17 $\mathbb{PQ}$◄ A total heat tube inserted in the flow of a compressible fluid measures the stagnation pressure. Explain the difference between the total and stagnation pressure.

12.18 An object is immersed in an airflow with a static pressure of 200 kPa absolute, a static temperature of 20°C, and a velocity of 250 m/s. What are the pressure and temperature at the stagnation point?

12.19 An airflow at M = 0.75 passes through a conduit with a cross-sectional area of 50 cm². The total absolute pressure is 360 kPa, and the total temperature is 10°C. Calculate the mass flow rate through the conduit.

12.20 Oxygen flows from a reservoir in which the temperature is 200°C and the pressure is 300 kPa absolute. Assuming isentropic flow, calculate the velocity, pressure, and temperature when the Mach number is 0.9.

12.21 One problem in creating high-Mach-number flows is condensation of the oxygen component in the air when the temperature reaches 50 K. If the temperature of the reservoir is 300 K and the flow is isentropic, at what Mach number will condensation of oxygen occur?

12.22 Hydrogen flows from a reservoir where the temperature is 20°C and the pressure is 500 kPa absolute to a section 2 cm in diameter where the velocity is 250 m/s. Assuming isentropic flow, calculate the temperature, pressure, Mach number, and mass flow rate at the 2 cm section.

12.23 The total pressure in a Mach-2.5 wind tunnel operating with air is 600 kPa absolute. A sphere 2 cm in diameter, positioned in the wind tunnel, has a drag coefficient of 0.95. Calculate the drag of the sphere.

12.24 Using Eq. (12.26), develop an expression for the pressure coefficient at stagnation conditions—that is, $C_p = (p_t - p)/[(1/2)\rho V^2]$—in terms of Mach number and ratio of specific heats, $C_p = f(k, M)$. Evaluate C_p at M = 0, 2, and 4 for $k = 1.4$. What would its value be for incompressible flow?

12.25 For low velocities, the total pressure is only slightly larger than the static pressure. Thus one can write $p_t/p = 1 + \epsilon$, where ϵ is a small positive number ($\epsilon \ll 1$). Using this approximation, show that as $\epsilon \to 0$ (M $\to$ 0), Eq. (12.31) reduces to

$$M = \left[\frac{2(p_t/p - 1)}{k}\right]^{1/2}$$

Normal Shock Waves

12.26 $\mathbb{PQ}$◀ Which of the following statements are true? (a) Shock waves only occur in supersonic flows. (b) The static pressure increases across a normal shock wave. (c) The Mach number downstream of a normal shock wave can be supersonic.

12.27 $\mathbb{PQ}$◀ Can normal shock waves occur in subsonic flows? Explain your answer.

12.28 A normal shock wave exists in a 500 m/s stream of nitrogen having a static temperature of –50°C and a static pressure of 70 kPa. Calculate the Mach number, pressure, and temperature downstream of the wave and the entropy increase across the wave.

12.29 A normal shock wave exists in a Mach 2 stream of air having a static temperature and pressure of 45°F and 30 psia. Calculate the Mach number, pressure, and temperature downstream of the shock wave.

12.30 A Pitot-static tube is used to measure the Mach number on a supersonic aircraft. The tube, because of its bluntness, creates a normal shock wave as shown. The absolute total pressure downstream of the shock wave (p_{t_2}) is 150 kPa. The static pressure of the free stream ahead of the shock wave (p_1) is 40 kPa and is sensed by the static pressure tap on the probe. Determine the Mach number (M_1) graphically.

PROBLEM 12.30

12.31 A shock wave occurs in a methane stream in which the Mach number is 3, the static pressure is 100 kPa absolute, and the static temperature is 20°C. Determine the downstream Mach number, static pressure, static temperature, and density.

12.32 The Mach number downstream of a shock wave in helium is 0.9, and the static temperature is 100°C. Calculate the velocity upstream of the wave.

12.33 Show that the lowest Mach number possible downstream of a normal shock wave is

$$M_2 = \sqrt{\frac{k-1}{2k}}$$

and that the largest density ratio possible is

$$\frac{\rho_2}{\rho_1} = \frac{k+1}{k-1}$$

What are the limiting values of M_2 and ρ_2/ρ_1 for air?

12.34 Show that the Mach number downstream of a weak wave ($M \approx 1$) is approximated by

$$M_2^2 = 2 - M_1^2$$

[*Hint:* Let $M_1^2 = 1 + \epsilon$, where $\epsilon \ll 1$, and expand Eq. (12.40) in terms of ϵ.] Compare values for M_2 obtained using this equation with values for M_2 from Table A.1 for $M_1 = 1, 1.05, 1.1,$ and 1.2.

Mass Flow in Truncated Nozzle

12.35 $\mathbb{PQ}$◀ What is meant by "back pressure"?

12.36 Develop a computer program for calculating the mass flow in a truncated nozzle. The input to the program would be total pressure, total temperature, back pressure, ratio of specific heats, gas constant, and nozzle diameter. Run the program and compare the results with Example 12.12. Run the program for back pressures of 80, 90, 110, 120, and 130 kPa and make a table for the variation of mass flow rate with back pressure. What trends do you observe?

This program will be useful for Probs. 12.37, 12.38, 12.40, and 12.41.

12.37 The truncated nozzle shown in the figure is used to meter the mass flow of air in a pipe. The area of the nozzle is 3 cm². The total pressure and total temperature measured upstream of the nozzle in the pipe are 300 kPa absolute and 20°C. The pressure downstream of the nozzle (back pressure) is 90 kPa absolute. Calculate the mass flow rate.

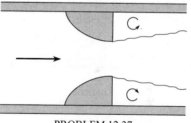

PROBLEM 12.37

12.38 The truncated nozzle shown in Prob. 12.37 is used to monitor the mass flow rate of methane. The area of the nozzle is 3 cm², and the area of the pipe is 12 cm². The upstream total pressure and total temperature are 150 kPa absolute and 30°C. The back pressure is 100 kPa.

a. Calculate the mass flow rate of methane.
b. Calculate the mass flow rate assuming the Bernoulli equation is valid, with the density being the density of the gas at the nozzle exit.

12.39 A truncated nozzle with an exit area of 5 cm² is used to measure a mass flow of air of 0.30 kg/s. The static temperature of the air at the exit is 10°C, and the back pressure is 100 kPa. Determine the total pressure.

12.40 A truncated nozzle with a 10 cm² exit area is supplied from a helium reservoir in which the absolute pressure is first 130 kPa and then 350 kPa. The temperature in the reservoir is 28°C, and the back pressure is 100 kPa. Calculate the mass flow rate of helium for the two reservoir pressures.

12.41 A sampling probe is used to draw gas samples from a gas stream for analysis. In sampling, it is important that the velocity entering the probe equal the velocity of the gas stream (isokinetic condition). Consider the sampling probe shown, which has a truncated nozzle inside it to control the mass flow rate. The probe has an inlet diameter of 4 mm and a truncated nozzle diameter of 2 mm. The probe is in a hot-air stream with a static temperature of 600°C, a static pressure of 100 kPa absolute, and a velocity of 60 m/s. Calculate the pressure required in the probe (back pressure) to maintain the isokinetic sampling condition.

12.42 Truncated nozzles are often used for flow-metering devices. Assume that you have to design a truncated nozzle, or a series of truncated nozzles, to measure the performance of an air compressor. The compressor is rated at 100 scfm (standard cubic feet per minute) at 120 psig. (A standard cubic foot is the volume the air would occupy at atmospheric conditions.) A performance curve for the compressor would be a plot of flow rate versus supply pressure. Explain how you would carry out the test program.

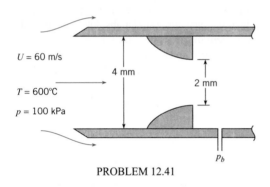

PROBLEM 12.41

Flow in Laval Nozzles

12.43 $\mathbb{P}\mathbb{Q}$◀ Sketch how the Mach number and velocity vary through a Laval nozzle from the entrance to the exit? How is the velocity variation different from a venturi configuration?

12.44 $\mathbb{P}\mathbb{Q}$◀ When a Laval nozzle has expansion ratio of 4, what does that mean?

12.45 Develop a computer program that requires the Mach number and ratio of specific heats as input and prints out the area ratio, the ratio of static to total pressure, the ratio of static to total temperature, the ratio of density to total density, and the ratio of pressure before and after a shock wave. Run the program for a Mach number of 2 and a ratio of specific heats of 1.4, and compare with results in Table A.1. Then run the program for the same Mach number with ratios of specific heats of 1.3 (carbon dioxide) and 1.67 (helium).

This program will be useful for Probs. 12.47, 12.48, 12.51, 12.55, 12.56, 12.58, and 12.59.

12.46 Develop a computer program that, given the area ratio, ratio of specific heats, and flow condition (subsonic or supersonic) as input, provides the Mach number. This will require some numerical root-finding scheme. Run the program for an area ratio of 5 and ratio of specific heats of 1.4. Compare the results with those in Table A.1. Then run the program for the same area ratio but with the ratios of specific heats of 1.67 (helium) and 1.31 (methane).

This program will be useful for Probs. 12.49, 12.50, and 12.53–12.59.

12.47 A wind tunnel using air is designed to have a Mach number of 3, a static pressure of 1.5 psia, and a static temperature of −10°F in the test section. Determine the area ratio of the nozzle required and the reservoir conditions that must be maintained if air is to be used.

12.48 A Laval nozzle is to be designed to operate supersonically and expand ideally to an absolute pressure of 30 kPa. If the stagnation pressure in the nozzle is 1 MPa, calculate the nozzle area ratio required. Determine the nozzle throat area for a mass flow of 5 kg/s and a stagnation temperature of 550 K. Assume that the gas is nitrogen.

12.49 A rocket nozzle with an area ratio of 4 is operating at a total absolute pressure of 1.3 MPa and exhausting to an atmosphere with an absolute pressure of 30 kPa. Determine whether the nozzle is overexpanded, underexpanded, or ideally expanded. Assume $k = 1.4$.

12.50 Repeat Prob. 12.49 for a nozzle with the same area ratio but with a ratio of specific heats of 1.2. Classify the nozzle flow.

12.51 A Laval nozzle with an exit area ratio of 1.688 exhausts air from a large reservoir into ambient conditions at $p = 100$ kPa.

a. Show that the reservoir pressure must be 782.5 kPa to achieve ideally expanded exit conditions at M = 2.

b. What are the static temperature and pressure at the throat if the reservoir temperature is 17°C with the pressure as in (a)?

c. If the reservoir pressure were lowered to 700 kPa, what would be the exit condition (overexpanded, ideally expanded, underexpanded, subsonic flow in entire nozzle)?

d. What reservoir pressure would cause a normal shock to form at the exit?

12.52 Determine the Mach number and area ratio at which the dynamic pressure is maximized in a Laval nozzle with air. [*Hint:* Express q in terms of p and M, and use Eq. (12.26) for p. Differentiate with respect to M and equate to zero.]

12.53 A rocket motor operates at an altitude where the atmospheric pressure is 30 kPa. The expansion ratio of the nozzle is 4 (exit area/throat area). The chamber pressure of the motor (total pressure) is 1.2 MPa, and the chamber temperature (total temperature) is 3000°C. The ratio of the specific heats of the exhaust gas is 1.2, and the gas constant is 400 J/kg K. The throat area of the rocket nozzle is 100 cm².

a. Determine the Mach number, density, pressure, and velocity at the nozzle exit.

b. Determine the mass flow rate.

c. Calculate the thrust of the rocket using

$$T = \dot{m}V_e + (p_e - p_0)A_e$$

d. What would the chamber pressure of the rocket have to be to have an ideally expanded nozzle? Calculate the rocket thrust under this condition.

12.54 A rocket motor is being designed to operate at sea level, where the pressure is 100 kPa absolute. The chamber pressure (total pressure) is 2.0 MPa, and the chamber temperature (total temperature) is 3300 K. The throat area of the nozzle is 10 cm². The ratio of the specific heats (k) of the exhaust gas is 1.2, and the gas constant is 400 J/kg K.

a. Determine the nozzle expansion ratio that is required to achieve an ideally expanded nozzle, and determine the nozzle thrust under these conditions (see Prob. 12.53 for the thrust equation).

b. Determine the thrust that would be obtained if the expansion ratio were reduced by 10% to achieve an underexpanded nozzle.

12.55 Air flows through a Laval nozzle with an expansion ratio of 4. The total pressure of the air entering the nozzle is 200 kPa, and the back pressure is 100 kPa. Determine the area ratio at which the shock wave occurs in the expansion section of the nozzle. (*Hint:* This problem can be solved graphically by calculating the exit pressure corresponding to different shock wave locations and finding the location where the exit pressure is equal to the back pressure.)

12.56 A rocket nozzle has the configuration shown. The diameter of the throat is 4 cm, and the exit diameter is 8 cm. The half-angle of the expansion cone is 15°. Gases with a specific heat ratio of 1.2 flow into the nozzle with a total pressure of 250 kPa. The back pressure is 100 kPa. First, using an iterative or graphical method, determine the area ratio at which the shock occurs. Then determine the shock wave's distance from the throat in centimeters.

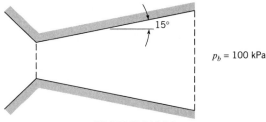

PROBLEM 12.56

12.57 A normal shock wave occurs in a nozzle at an area ratio of 5. Determine the entropy increase if the gas is hydrogen.

12.58 Consider airflow in the variable-area channel shown in the figure. Determine the Mach number, static pressure, and stagnation pressure at station 3. Assume isentropic flow except for normal shock waves.

12.59 Determine the back pressure necessary for the shock wave to position itself as shown in the figure. The fluid is air.

12.60 Design a supersonic wind tunnel that achieves a Mach number of 1.5 in a test section 5 cm by 5 cm. The tunnel is to be attached to a vacuum tank as shown in the figure. After the tank is evacuated, the valve is opened and atmospheric air is drawn through the tunnel into the tank. The tunnel should operate for 30 seconds before the pressure rises to the point in the tank that supersonic flow is no longer achievable. Do a preliminary design of this system including details such as nozzle dimensions, configuration, and tank size.

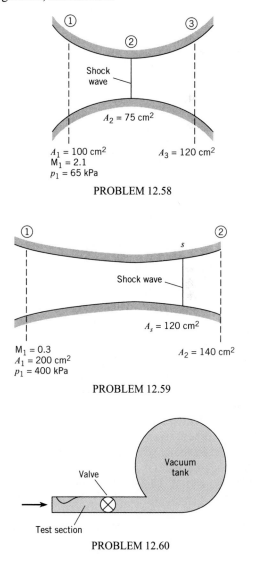

PROBLEM 12.58

PROBLEM 12.59

PROBLEM 12.60

C H A P T E R

Flow Measurements

13

The pattern of flow that occurs when an air jet is discharged into quiescent air is revealed in this photo. (Courtesy Cecilia D. Richards.)

SIGNIFICANT LEARNING OUTCOMES

Contextual Knowledge
- Sketch common measuring instruments.
- Explain the operating principles of common measuring instruments.
- List the advantages and disadvantages of common measuring instruments (common measurment instruments include the hot-wire anemometer, orifice meter, laser-Doppler anemometer venturi meter, rotameter, and the weir).

Procedural Knowledge
- Calculate flow rate for an orifice meter, a venturi meter, or a weir.
- Calculate flow rate by integrating velocity distribution data.
- For a weir or an orifice meter, estimate uncertainty using the RMS method.

Measurement techniques are important because fluid mechanics relies heavily on experiments. Thus, Chapter 3 described instruments for measuring pressure including the piezometer, the manometer, the Bourdon-tube gage, and the pressure transducer. Chapter 4 describes the Pitot-static tube and the Pitot tube. This chapter builds on this knowledge by introducing additional ways to measure flow rate, pressure, and velocity. Also, this chapter describes how to estimate the uncertainty of a measurement.

Measuring Velocity and Pressure

Stagnation (Pitot) Tube

The stagnation tube, also called the Pitot tube, is shown in Fig. 13.1a. A Pitot tube measures stagnation pressure with an open tube that is aligned parallel with the velocity direction and then senses pressure in the tube using a pressure gage or transducer.

When the stagnation tube was introduced in Chapter 4, viscous effects were not discussed. Viscous effects are noteable because they can influence the accuracy of a measurement. The effects of viscosity, from reference (1), are shown in Fig. 13.2. This shows the pressure coefficient C_p plotted as a function of the Reynolds number. Viscous effects are important when $C_p > 1.0$.

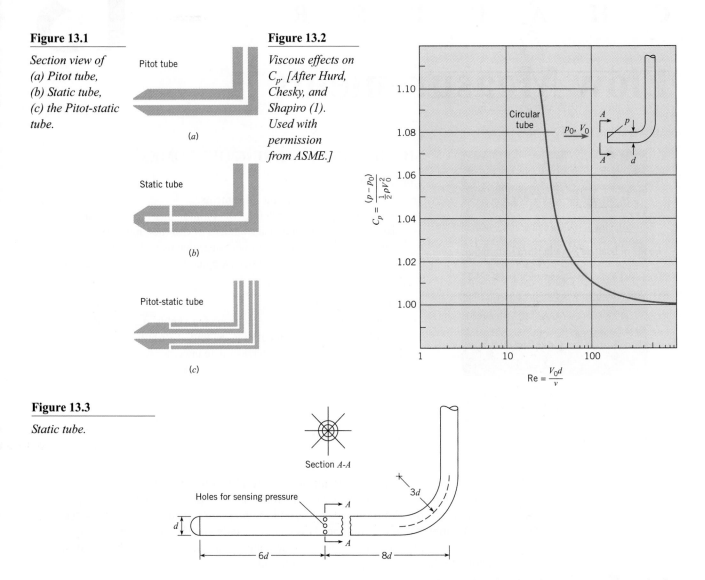

Figure 13.1

Section view of (a) Pitot tube, (b) Static tube, (c) the Pitot-static tube.

Pitot tube

(a)

Static tube

(b)

Pitot-static tube

(c)

Figure 13.2

Viscous effects on C_p. [After Hurd, Chesky, and Shapiro (1). Used with permission from ASME.]

$$C_p = \frac{(p - p_0)}{\frac{1}{2}\rho V_0^2}$$

Circular tube

$$Re = \frac{V_0 d}{\nu}$$

Figure 13.3

Static tube.

Section A-A

Holes for sensing pressure

6d

8d

3d

In Fig. 13.2 it is seen that when the Reynolds number for the circular stagnation tube is greater than 60, the error in measured velocity is less than 1%. For boundary-layer measurements a stagnation tube with a flattened end can be used. By flattening the end of the tube, the velocity measurement can be taken nearer the boundary than if a circular tube were used. For these flattened tubes, the pressure coefficient remains near unity for a Reynolds number as low as 30. See reference (15) for more information on flattened-end stagnation tubes.

Static Tube

A *static tube*, as shown in Fig. 13.1b, is an instrument for measuring static pressure. Static pressure is the pressure in a fluid that is stationary or in a fluid that is flowing. When the fluid is flowing, the static pressure must be measured in a way that does not disturb the pressure. Thus, in the design of the static tube, as shown in Fig. 13.3, the placement of the holes along the probe is critical because the rounded nose on the tube causes some decrease of pressure along the tube and the downstream stem causes an increase in pressure in front of it. Hence the location for sensing the static pressure must be at the point where these two effects cancel each other. Experiments reveal that the optimum location is at a point approximately six diameters downstream of the front of the tube and eight diameters upstream from the stem.

Figure 13.4

Various types of yaw meters.
(a) Cylindrical-tube yaw meter.
(b) Two-tube yaw meter.
(c) Three-dimensional yaw meter.

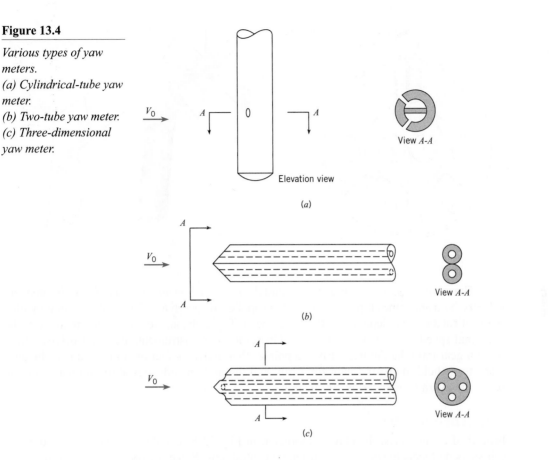

Pitot-Static Tube

The Pitot-static tube, Fig. 13.1c, measures velocity by using concentric tubes to measure static pressure and dynamic pressure. Application of the Pitot-static tube is presented in Chapter 4.

Yaw Meters

A *yaw meter*, Fig. 13.4, is an instrument for measuring velocity by using multiple pressure ports to determine the magnitude and direction of fluid velocity. The first two yaw meters in Fig. 13.4 can be used for two-dimensional flow, where flow direction in only one plane needs to be found. The third yaw meter in Fig. 13.4 is used for determining flow direction in three dimensions. In all these devices, the tube is turned until the pressure on symmetrically opposite openings is equal. This pressure is sensed by a differential pressure gage or manometer connected to the openings in the yaw meter. The flow direction is sensed when a null reading is indicated on the differential gage. The velocity magnitude is found by using equations that depend on the type of yaw meter that is used.

The Vane or Propeller Anemometer

The term *anemometer* originally meant an instrument that was used to measure the velocity of the wind. However, anemometer now means an instrument that is used to measure fluid velocity, because anemometers are used in water, air, nitrogen, blood, and many other fluids. See (18) for an overview of the many types of anemometers.

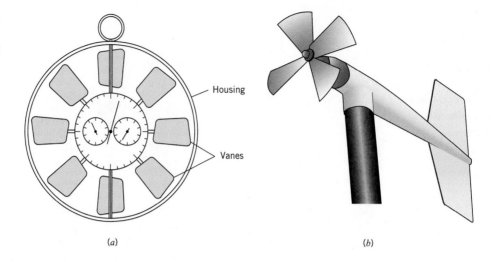

Housing

Vanes

(a)   (b)

The *vane anemometer* (Fig. 13.5*a*) and the *propeller anemometer* (Fig. 13.5*b*) measure velocity by using vanes typical of a fan or propeller, respectively. These blades rotate with a speed of rotation that depends on the wind speed. Typically, an electronic circuit converts the rotational speed into a velocity reading. On some older instruments the rotor drives a low-friction gear train that, in turn, drives a pointer that indicates feet on a dial. Thus if the anemometer is held in an airstream for 1 min and the pointer indicates a 300 ft change on the scale, the average airspeed is 300 ft/min.

Cup Anemometer

Instead of using vanes, the cup anemometer, in Fig. 13.6, is a device that uses the drag on cup-shaped objects to spin a rotor around a central axis. Since the rotational speed of the rotor is related to drag force, the frequency of rotation is related to the fluid velocity by appropriate calibration data. A typical rotor comprises three to five hemispherical or conical cups. In addition to applications in air, engineers use a cup anemometer to measure the velocity in streams and rivers.

Hot-wire and Hot-film Anemometers

The *hot-wire anemometer* (HWA), Fig. 13.7, is an instrument for measuring velocity by sensing the heat transfer from a heated wire. As velocity increases, more energy is needed to keep the wire hot and the corresponding changes in electrical characteristics can be used to determine the velocity of the fluid that is passing by the wire.

The HWA has advantages over other instruments. The HWA is well suited for measuring velocity fluctuations that occur in turbulent flow, whereas instruments such as the Pitot-static tube are only suitable for measuring velocity that either is steady or changes slowly with time. The sensing element of the HWA is quite small, allowing the HWA to be used in locations such as the boundary layer, where the velocity is varying in a region that is small in size. Many other instruments are too large for recording velocity in a region that is geometrically small. Another advantage of the HWA is that it is sensitive to low-velocity flows, a characteristic lacking in the Pitot tube and other instruments. The main disadvantages of the HWA are its delicate nature (the sensor wire is easily broken), its relatively high cost, and its need for an experienced user.

The basic principle of the hot-wire anemometer is described as follows: A wire of very small diameter—the sensing element of the hot-wire anemometer—is welded to supports as

Figure 13.7

Probe for hot-wire anemometer (enlarged).

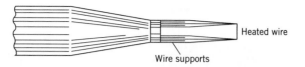

Heated wire

Wire supports

shown in Fig. 13.7. In operation the wire either is heated by a fixed flow of electric current (the constant-current anemometer) or is maintained at a constant temperature by adjusting the current (the constant-temperature anemometer).

A flow of fluid past the hot wire causes the wire to cool because of convective heat transfer. In the constant-current anemometer, the cooling of the wire causes its resistance to change, and a corresponding voltage change occurs across the wire. Because the rate of cooling is a function of the speed of flow past the heated wire, the voltage across the wire is correlated with the flow velocity. The more popular type of anemometer, the constant-temperature anemometer, operates by varying the current in such a manner as to keep the resistance (and temperature) constant. The flow of current is correlated with the speed of the flow: the higher the speed, the greater the current needed to maintain a constant temperature. Typically, the wires are 1 mm to 2 mm in length and heated to 150°C. The wires may be 10 μm or less in diameter; the time response improves with the smaller wire. The lag of the wire's response to a change in velocity (thermal inertia) can be compensated for more easily, using modern electronic circuitry, in constant-temperature anemometers than in constant-current anemometers. The signal from the hot wire is processed electronically to give the desired information, such as mean velocity or the root-mean-square of the velocity fluctuation.

To illustrate the versatility of these instruments, note that the hot-wire anemometer can measure accurately gas flow velocities from 30 cm/s to 150 m/s; it can measure fluctuating velocities with frequencies up to 100,000 Hz; and it has been used satisfactorily for both gases and liquids.

The single hot wire mounted normal to the mean flow direction measures the fluctuating component of velocity in the mean flow direction. Other probe configurations and electronic circuitry can be used to measure other components of velocity.

For velocity measurements in liquids or dusty gases, where wire breakage is a problem, the hot-film anemometer is more suitable. This anemometer consists of a thin conducting metal film (less than 0.1 μm thick) mounted on a ceramic support, which may be 50 μm in diameter. The hot film operates in the same fashion as the hot wire. Recently, the split film has been introduced. It consists of two semicylindrical films mounted on the same cylindrical support and electrically insulated from each other. The split film provides both speed and directional information.

For more detailed information on the hot-wire and hot-film anemometers, see King and Brater (2) and Lomas (3).

Laser-Doppler Anemometer

The *laser-Doppler anemometer* (LDA) is an instrument for measuring velocity by using the Doppler shift that occurs when a particle in a flow scatters light from crossed laser beams. Advantages of the LDA are (a) the flow field is not disturbed by the presence of a probe, and (b) it provides excellent spatial resolution. Disadvantages of the LDA include cost, complexity, the need for a transparent fluid, and requirement for particle seeding.

There are several different configurations for the LDA. The dual-beam mode, Fig. 13.8, splits a laser beam into two parallel beams and then uses a converging lens to cause the two beams to cross. The point where beams cross is called the measuring volume, which might best be described as an ellipsoid that is typically 0.3 mm in diameter and 2 mm long, illustrating the excellent spatial resolution achievable. The interference of the two beams generates a series of light and dark fringes in the measuring volume perpendicular to the plane of the two

Figure 13.8

*Dual-beam laser-Doppler
anemometer.*

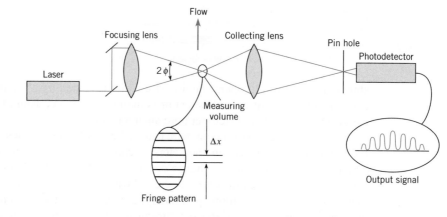

beams. As a particle passes through the fringe pattern, light is scattered and a portion of the scattered light passes through the collecting lens toward the photodetector. A typical signal obtained from the photodetector is shown in the figure.

It can be shown from optics theory that the spacing between the fringes is given by

$$\Delta x = \frac{\lambda}{2 \sin \phi} \tag{13.1}$$

where λ is the wavelength of the laser beam and ϕ is the half-angle between the crossing beams. By suitable electronic circuitry, the frequency of the signal (f) is measured, so the velocity is given by

$$U = \frac{\Delta x}{\Delta t} = \frac{\lambda f}{2 \sin \phi} \tag{13.2}$$

The operation of the laser-Doppler anemometer depends on the presence of particles to scatter the light. These particles need to move at the same velocity as the fluid. Thus the particles need to be small relative to the size of flow patterns, and they need to have a density near that of the ambient fluid. In liquid flows, impurities of the fluid can serve as scattering centers. In water flows, adding a few drops of milk is common. In gaseous flows, it is common to "seed" the flow with small particles. Smoke is often used for this seeding.

Laser-Doppler anemometers that provide two or three velocity components of a particle traveling through the measuring volume are now available. This is accomplished by using laser-beam pairs of different colors (wavelengths). The measuring volumes for each color are positioned at the same physical location but oriented differently to measure a different component. The signal-processing system can discriminate the signals from each color and thereby provide component velocities.

Another recent technological advance in laser-Doppler anemometry is the use of fiber optics. The fiber optics transmit the laser beams from the laser to a probe that contains optical elements to cross the beams and generate a measuring volume. Thus measurements at different locations can be made by moving the probe and without moving the laser. For more applications of the laser-Doppler technique see Durst (4).

Marker Methods

The marker method for determining velocity involves particles that are placed in the stream. By analyzing the motion of these particles, one can deduce the velocity of the flow itself. Of course, this requires that the markers follow virtually the same path as the surrounding fluid elements. It means, then, that the marker must have nearly the same density as the fluid or that it must be so small that its motion relative to the fluid is negligible. Thus for water flow it is common to use colored droplets from a liquid mixture that has nearly the same density as the water. For example, Macagno (6) used a mixture of *n*-butyl phthalate and xylene with a bit of white paint to yield a mixture that had the same density as water and could be photographed effectively. Solid particles, such as plastic beads, that have densities near that of the liquid being studied can also be used as markers.

Hydrogen bubbles have also been used for markers in water flow. Here an electrode placed in flowing water causes small bubbles to be formed and swept downstream, thus revealing the motion of the fluid. The wire must be very small so that the resulting bubbles do not have a significant rise velocity with respect to the water. By pulsing the current through the electrode, it is possible to add a time frame to the visualization technique, thus making it a useful tool for velocity measurements. Figure 13.9 shows patches of tiny hydrogen bubbles that were released with a pulsing action from noninsulated segments of a wire located to the left of the picture. Flow is from left to right, and the necked-down section of the flow passage has higher water velocity. Therefore, the patches are longer in that region. Next to the walls the patches of bubbles are shorter, indicating less distance traveled per unit of time. Other details concerning the marker methods of flow visualization are described by Macagno (6).

A relatively new marker method is particle image velocimetry (PIV), which provides a measurement of the velocity field. In PIV, the marker or seeding particles may be minuscule spheres of aluminum, glass, or polystyrene. Or they may be oil droplets, oxygen bubbles (liquids only), or smoke particles (gases only). The seeding particles are illuminated in order to produce a photographic record of their motion. In particular, a sheet of light passing through a cross section of the flow is pulsed on twice, and the scattered light from the particles is recorded by a camera. The first pulse of light records the position of each particle at time *t*, and the second pulse of light records the position at time $t + \Delta t$. Thus, the displacement $\Delta \mathbf{r}$ of each particle is recorded on the photograph. Dividing $\Delta \mathbf{r}$ by Δt yields the velocity of each particle. Because PIV uses a sheet of light, the method provides a simultaneous measurement of velocity at locations throughout a cross section of the flow. Hence, PIV is identified as a whole-field technique. Other velocity measurements, the LDA method, for example, are limited to measurements at one location.

PIV measurement of the velocity field for flow over a backward-facing step is shown in Fig. 13.10. This experiment was carried out in water using 15 μm–diameter, silver-coated hollow spheres as seeding particles. Notice that the PIV method provided data over the cross section of the flow. Although the data shown in Fig. 13.10 are qualitative, numerical values of the velocity at each location are also available.

The PIV method is typically performed using digital hardware and computers. For example, images may be recorded with a digital camera. Each resulting digital image is evaluated with software that calculates the velocity at points throughout the image. This evaluation

Figure 13.9

Combined time-streak markers (hydrogen bubbles); flow is from left to right. [After Kline (5) Courtesy of Education Development Center, Inc., Newton, MA.]

Figure 13.10

Velocity vectors from PIV measurements. (Courtesy TSI, Inc., and Florida State University.)

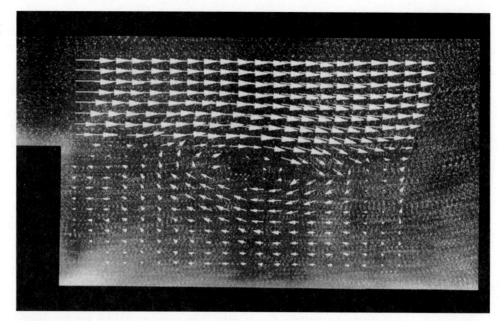

proceeds by dividing the image into small sub-areas called "interrogation areas." Within a given interrogation area, the displacement vector ($\Delta\mathbf{r}$) of each particle is found by using statistical techniques (auto- and cross-correlation). After processing, the PIV data are typically available on a computer screen. Additional information on PIV systems is provided by Raffel et al. (7).

Smoke is often used as a marker in flow measurement. One technique is to suspend a wire vertically across the flow field and allow oil to flow down the wire. The oil tends to accumulate in droplets along the wire. Applying a voltage to the wire vaporizes the oil, creating streaks from the droplets. Figure 13.11 is an example of a flow pattern revealed by such a method. Smoke generators that provide smoke by heating oils are also commercially available. It is also possible to position a thin sheet of laser light through the smoke field to obtain an improved spatial definition of the flow field indicated by the smoke. Another technique is to introduce titanium tetrachloride ($TiCl_4$) in a dried-air flow, which reacts with the water vapor in the ambient air to produce micron-sized titanium oxide particles, which serve as tracers. The flow pattern obtained for an upward-flowing air jet using this technique in conjunction with a laser light sheet

Figure 13.11

Flow pattern in the wake of a flat plate.

is shown in the photograph at the beginning of this chapter (p. 435). This jet is subjected to an acoustic field, which enhances the vortex shedding pattern observed in the jet.

13.2

Measuring Flow Rate (Discharge)

Measuring flow rate is important in research, design, testing, and in many commercial applications.

Direct Measurement of Volume or Weight

For liquids, a simple and accurate method is to collect a sample of the flowing fluid over a given period of time Δt. Then the sample is weighed, and the average weight rate of flow is $\Delta W / \Delta t$, where ΔW is the weight of the sample. The volume of a sample can also be measured (usually in a calibrated tank), and from this the average volume rate of flow is calculated as $\Delta V\!\!\!/ / \Delta t$, where $\Delta V\!\!\!/$ is the volume of the sample. This method has several disadvantages: It cannot be used for an unsteady flow, and it is not always possible to collect a sample.

Integrating a Measured Velocity Distribution

Flow rate can be found by measuring a velocity distribution and then integrating using the volume flow rate equation (5.8):

$$Q = \int_A V dA$$

Measure velocity at center of each sub-area

Sub-area ΔA_i

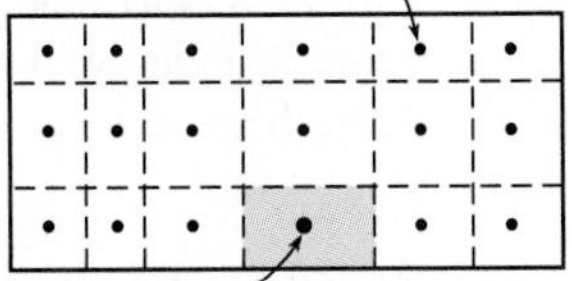

Figure 13.12

Dividing a rectangular conduit into sub-areas for approximating discharge.

For example, one can divide a rectangular conduit into sub-areas and then measure velocity at the center of each sub-area as shown in Fig. 13.12. Then flow rate is determined by

$$Q = \int_A V \, dA \approx \sum_{i=1}^{N} V_i (\Delta A)_i \tag{13.3}$$

where N is the number of sub-areas. When the flow area occurs in a round pipe, then the sub-area is a ring as shown by Example 13.1.

EXAMPLE 13.1   DISCHARGE FROM VELOCITY DATA

The data given in the table are for a velocity traverse of air flow in a pipe 100 cm in diameter. What is the volume rate of flow in cubic meters per second?

r (cm)	V (m/s)
0.00	50.0
5.00	49.5
10.00	49.0
15.00	48.0
20.00	46.5
25.00	45.0
30.00	43.0
35.00	40.5
40.00	37.5
45.00	34.0
47.50	25.0
50.00	0.0

EXAMPLE 13.2

Situation:

1. Air is flowing in a round pipe ($D = 1.0$ m).
2. Velocity in m/s is known as a function of radius (see table).

Find: Volume flow rate (in m^3/s) in the pipe.

Assumptions: The velocity distribution is symmetric around the centerline of the pipe.

Plan

1. Develop an equation for a round pipe by applying Eq. (13.3).
2. Find discharge by using a spreadsheet program.

Solution

The flow rate is given by

$$Q = \sum_{i=1}^{N} V_i (\Delta A)_i$$

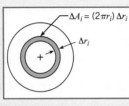

The area ΔA_i is shown in the sketch above. Visualize this area as a strip of length $2\pi r_i$ and width Δr_i. Then $\Delta A_i \approx (2\pi r_i)\Delta r_i$. The flow rate equation becomes

$$Q = \sum_{i=1}^{N} V_i (\Delta A)_i = \sum_{i=1}^{N} V_i (2\pi r_i) \Delta r_i$$

To perform the sum, use a spreadsheet as shown. To see how the table is set up, consider the row $i = 2$. The area is

$$\Delta A_2 = (2\pi r_2)\Delta r_2 = (2\pi(0.05 \text{ m}))(0.05 \text{ m}) = 0.0157 \text{ m}^2$$

which is given in the sixth column. The last column gives

$$V_2(\Delta A)_2 = (49.5 \text{ m/s})(0.0157 \text{ m}^2) = 0.778 \text{ m}^3/\text{s}$$

i	r_i (cm)	V_i (m/s)	$2*\pi*r_i$ (m)	Δr_i (m)	ΔA_i (m²)	$V_i*\Delta A_i$ (m³/s)
1	0.0	50.0	0.0000	0.0250	0.0000	0.000
2	5.0	49.5	0.3142	0.0500	0.0157	0.778
3	10.0	49.0	0.6283	0.0500	0.0314	1.539
4	15.0	48.0	0.9425	0.0500	0.0471	2.262
5	20.0	46.5	1.2566	0.0500	0.0628	2.922
6	25.0	45.0	1.5708	0.0500	0.0785	3.534
7	30.0	43.0	1.8850	0.0500	0.0942	4.053
8	35.0	40.5	2.1991	0.0500	0.1100	4.453
9	40.0	37.5	2.5133	0.0500	0.1257	4.712
10	45.0	34.0	2.8274	0.0375	0.1060	3.605
11	47.5	25.0	2.9845	0.0250	0.0746	1.865
12	50.0	0.0	3.1416	0.0125	0.0393	0.000
			SUM==>	0.50	0.79	29.72

Discharge is found by summing the last column. As shown

$$Q = \sum_{i=1}^{12} V_i(\Delta A)_i = \boxed{29.7 \frac{\text{m}^3}{\text{s}}}$$

To check the validity of the calculation, sum the column labeled Δr_i and check to ensure that this value equals the radius of the pipe. As shown, this sum is 0.5 m. Similarly, the pipe area of

$$A = \pi r^2 = \pi (0.5 \text{ m})^2 = 0.785 \text{m}^2$$

should be produced by summing the column labeled ΔA_i. As shown, this is the case.

Calibrated Orifice Meter

An *orifice meter* is an instrument for measuring flow rate by using a carefully designed plate with a round opening and situating this device in a pipe, as shown in Fig. 13.13. Flow rate is found by measuring the pressure drop across the orifice and then using an equation to calculate the appropriate flow rate. One common application of the orifice meter is metering of natural gas in pipelines. Because large quantities of natural gas are measured and the associated costs are high, accuracy is very important. This section describes the main ideas associated with orifice meters. Details about using orifice meters are presented in standards such as reference (10).

Flow through a sharp-edged orifice is shown in Fig. 13.13. Note that the streamlines continue to converge a short distance downstream of the plane of the orifice. Hence the minimum-flow area is actually smaller than the area of the orifice. To relate the minimum-

Figure 13.13

Flow through a sharp-edged pipe orifice.

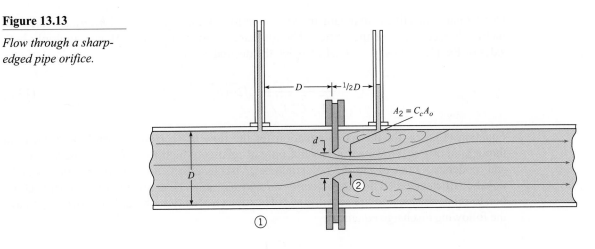

flow area, often called the contracted area of the jet, or *vena contracta*, to the area of the orifice A_o, one uses the contraction coefficient, which is defined as

$$A_j = C_c A_o$$

$$C_c = \frac{A_j}{A_o}$$

Then, for a circular orifice,

$$C_c = \frac{(\pi/4)d_j^2}{(\pi/4)d^2} = \left(\frac{d_j}{d}\right)^2$$

Because d_j and d_2 are identical, $C_c = (d_2/d)^2$. At low values of the Reynolds number, C_c is a function of the Reynolds number. However, at high values of the Reynolds number, C_c is only a function of the geometry of the orifice. For d/D ratios less than 0.3, C_c has a value of approximately 0.62. However, as d/D is increased to 0.8, C_c increases to a value of 0.72.

To derive the orifice equation, consider the situation shown in Fig. 13.13. Apply the Bernoulli equation between section 1 and section 2:

$$\frac{p_1}{\gamma} + \frac{V_1^2}{2g} + z_1 = \frac{p_2}{\gamma} + \frac{V_2^2}{2g} + z_2$$

V_1 is eliminated by means of the continuity equation $V_1 A_1 = V_2 A_2$. Then solving for V_2 gives

$$V_2 = \left\{\frac{2g[(p_1/\gamma + z_1) - (p_2/\gamma + z_2)]}{1 - (A_2/A_1)^2}\right\}^{1/2} \tag{13.4a}$$

However, $A_2 = C_c A_o$ and $h = p/\gamma + z$, so Eq. (13.4a) reduces to

$$V_2 = \sqrt{\frac{2g(h_1 - h_2)}{1 - C_c^2 A_o^2 / A_1^2}} \tag{13.4b}$$

Our primary objective is to obtain an expression for discharge in terms of h_1, h_2, and the geometric characteristics of the orifice. The discharge is given by $V_2 A_2$. Hence, multiply both sides of Eq. (13.4b) by $A_2 = C_c A_o$, to give the desired result:

$$Q = \frac{C_c A_o}{\sqrt{1 - C_c^2 A_o^2 / A_1^2}} \sqrt{2g(h_1 - h_2)} \tag{13.5}$$

Equation (13.5) is the discharge equation for the flow of an incompressible inviscid fluid through an orifice. However, it is valid only at relatively high Reynolds numbers. For low and moderate values of the Reynolds number, viscous effects are significant, and an additional coefficient called the *coefficient of velocity, C_v*, must be applied to the discharge equation to relate the ideal to the actual flow.* Thus for viscous flow through an orifice, we have the following discharge equation:

$$Q = \frac{C_v C_c A_o}{\sqrt{1 - C_c^2 A_o^2 / A_1^2}} \sqrt{2g(h_1 - h_2)}$$

The product $C_v C_c$ is called the *discharge coefficient, C_d*, and the combination $C_v C_c / (1 - C_c^2 A_o^2 / A_1^2)^{1/2}$ is called the *flow coefficient, K*. Thus, $Q = KA_o \sqrt{2g(h_1 - h_2)}$, where

$$K = \frac{C_d}{\sqrt{1 - C_c^2 A_o^2 / A_1^2}} \tag{13.6}$$

If Δh is defined as $h_1 - h_2$, then the final form of the orifice equation reduces to

$$Q = KA_o \sqrt{2g\Delta h} \tag{13.7a}$$

If a differential pressure transducer is connected across the orifice, it will sense a piezometric pressure change that is equivalent to $\gamma \Delta h$, so the orifice equation becomes

$$Q = KA_o \sqrt{2\frac{\Delta p_z}{\rho}} \tag{13.7b}$$

Experimentally determined values of K as a function of d/D and Reynolds number based on orifice size are given in Fig. 13.14. If Q is given, Re_d is equal to $4Q / \pi dv$. Then K is obtained from Fig. 13.14 (using the vertical lines and the bottom scale), and Δh is computed from Eq. (13.7a), or Δp_z can be computed from Eq. (13.7b). However, one is often confronted with the problem of determining the discharge Q when a certain value of Δh or a certain value of Δp_z is given. When Q is to be determined, there is no direct way to obtain K by entering Fig. 13.14 with Re, because Re is a function of the flow rate, which is still unknown. Hence another scale, which does not involve Q, is constructed on the graph of Fig. 13.14. The variables for this scale are obtained in the following manner: Because $Re_d = 4Q / \pi dv$ and $Q = K(\pi d^2 / 4) \sqrt{2g\Delta h}$, write Re_d in terms of Δh:

$$Re_d = K \frac{d}{v} \sqrt{2g\Delta h}$$

* At low Reynolds numbers the coefficient of velocity may be quite small; however, at Reynolds numbers above 10^5, C_v typically has a value close to 0.98. See Lienhard (8) for C_v analyses.

Figure 13.14

Flow coefficient K and
Re$_d$/K versus the
Reynolds number for
orifices, nozzles, and
venturi meters. [After
Tuve and Sprenkle (9)
and ASME (10).
Permission to use Tuve
granted by
Instrumentation &
Control Systems
magazine, formerly
Instruments *magazine.]*

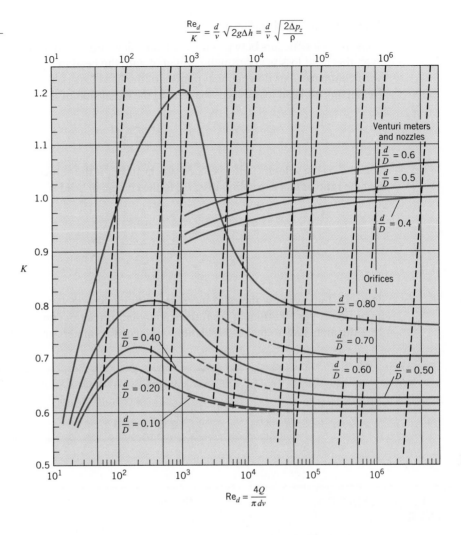

or

$$\frac{\mathrm{Re}_d}{K} = \frac{d}{v}\sqrt{2g\Delta h} = \frac{d}{v}\sqrt{\frac{2\Delta p_z}{\rho}}$$

Thus the slanted dashed lines and the top scale are used in Fig. 13.14 when Δh is known and the flow rate is to be determined. If a certain value of Δp is given, one can apply Fig. 13.14 by using $\Delta p_z/\rho$ in place of $g\Delta h$ in the parameter at the top of Fig. 13.14.

The literature on orifice flow contains numerous discussions concerning the optimum placement of pressure taps on both the upstream side and the downstream side of an orifice. The data given in Fig. 13.14 are for "corner taps." That is, on the upstream side the pressure readings were taken immediately upstream of the orifice plate (at the corner of the orifice plate and the pipe wall), and the downstream tap was at a similar downstream location. However, pressure data from flange taps (1 in. upstream and 1 in. downstream) and from the taps shown in Fig. 13.13 all yield virtually the same values for K—the differences are no greater than the deviations involved in reading Fig. 13.14. For more precise values of K with specific types of taps, see the ASME report on fluid meters (10).

Head Loss for Orifices

Some head loss occurs between the upstream side of the orifice and the vena contracta. However, this head loss is very small compared with the head loss that occurs downstream of the vena contracta. This downstream portion of the head loss is like that for an abrupt expansion. Neglecting all head loss except that due to the expansion of the flow, gives

$$h_L = \frac{(V_2 - V_1)^2}{2g} \tag{13.8}$$

where V_2 is the velocity at the vena contracta and V_1 is the velocity in the pipe. It can be shown that the ratio of this expansion loss, h_L, to the change in head across the orifice, Δh, is given as

$$\frac{h_L}{\Delta h} = \frac{\dfrac{V_2}{V_1} - 1}{\dfrac{V_2}{V_1} + 1} \tag{13.9}$$

Table 13.1 shows how the ratio increases with increasing values of V_2/V_1. It is obvious that an orifice is very inefficient from the standpoint of energy conservation. Examples 13.2 and 13.3 illustrate how to make calculations when orifice meters are used.

TABLE 13.1 RELATIVE HEAD LOSS FOR ORIFICES

$V_2/V_1 \rightarrow$	1	2	4	6	8	10
$h_L/\Delta h \rightarrow$	0	0.33	0.60	0.71	0.78	0.82

EXAMPLE 13.3 ANALYSIS OF AN ORIFICE METER

A 15 cm orifice is located in a horizontal 24 cm water pipe, and a water-mercury manometer is connected to either side of the orifice. When the deflection on the manometer is 25 cm, what is the discharge in the system, and what head loss is produced by the orifice? Assume the water temperature is 20°C.

Problem Definition

Situation:

1. Water flows through an orifice ($d = 0.15$ m) in a pipe ($D = 0.24$ m).
2. A mercury-water manometer is used to measure pressure drop.

Find:

1. Discharge (in m^3/s) in pipe.
2. Head loss (in meters) produced by the orifice.

Properties:

1. Water (20°C), Table A.5: $\nu = 1 \times 10^{-6}$ m^2/s.
2. Mercury (20°C), Table A.4: $S = 13.6$.

Sketch:

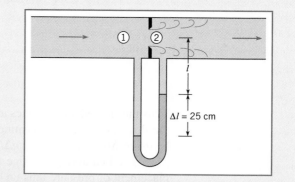

Plan

1. Calculate $\Delta h = h_1 - h_2$ using the manometer equation (3.18).
2. Find the flow coefficient K using Fig. 13.14.
3. Find discharge Q using Eq. (13.7a).
4. Calculate the coefficient of contraction C_c using Eq. (13.6).

5. Solve for the velocity V_2 at the vena contracta.

6. Calculate head loss using Eq. (13.8).

Solution

1. Change in piezometric head
   - Apply manometer equation from 1 to 2.

$$p_1 + \gamma_w(l + \Delta l) - \gamma_{Hg}\Delta l - \gamma_w l = p_2$$

   - Solve for Δh.

$$\Delta h = \frac{p_1 - p_2}{\gamma_w} = \Delta l \frac{\gamma_{Hg} - \gamma_w}{\gamma_w} = \Delta l \left(\frac{\gamma_{Hg}}{\gamma_w} - 1\right)$$

$$\Delta h = (0.25 \text{ m})(13.6 - 1) = 3.15 \text{ m of water}$$

2. Flow coefficient
   - Calculate (Re_d/K).

$$\frac{\text{Re}_d}{K} = \frac{d\sqrt{2g\Delta h}}{\nu} = \frac{0.15 \text{ m}\sqrt{2(9.81 \text{ m/s}^2)(3.15 \text{ m})}}{1.0 \times 10^{-6} \text{ m}^2/\text{s}}$$

$$= 1.2 \times 10^6$$

   - From Fig. 13.14 with $d/D = 0.625$, $K = 0.66$ (interpolated).

3. Discharge

$$Q = 0.66A_o\sqrt{2g\Delta h}$$

$$= 0.66\frac{\pi}{4}d^2\sqrt{2(9.81 \text{ m/s}^2)(3.15 \text{ m})}$$

$$= 0.66(0.785)(0.15^2 \text{ m}^2)(7.86 \text{ m/s}) = \boxed{0.092 \text{ m}^3/\text{s}}$$

4. Coefficient of contraction C_c

$$K = \frac{C_d}{1 - C_c^2 A_o^2/A_1^2}$$

   Let $K = 0.66$. The ratio $(A_o/A_1)^2 = (0.625)^4 = 0.1526$ and $C_d = C_v C_c$. Assuming $C_v = 0.98$ (see the footnote on page 446) and solving for C_c, gives $C_c = 0.633$.

5. Velocity at the vena contracta

$$V_2 = Q/(C_c A_o)$$

$$(0.092 \text{ m}^3/\text{s})/[(0.633)(\pi/4)(0.15^2 \text{ m}^2)] = 8.23 \text{ m/s}$$

$$V_1 = Q/A_{\text{pipe}}$$

$$(0.092 \text{ m}^3/\text{s})/[(\pi/4)(0.24^2 \text{ m}^2)] = 2.03 \text{ m/s}$$

6. Head loss

$$h_L = (V_2 - V_1)^2/2g = (8.23 - 2.03)^2/(2 \times 9.81)$$

$$= \boxed{1.96 \text{ m}}$$

EXAMPLE 13.4   MANOMETER DEFLECTION FOR AN ORIFICE METER

An air-water manometer is connected to either side of an 8 in. orifice in a 12 in. water pipe. If the maximum flow rate is 2 cfs, what is the deflection on the manometer? The water temperature is 60°F.

Problem Definition

Situation:

1. Water flows ($Q = 2$ cfs) through an orifice ($d = 8$ in.) in a pipe ($D = 12$ in.).

2. An air-water manometer is used to measure pressure drop.

Find: Deflection (in ft) of water in the manometer.

Sketch:

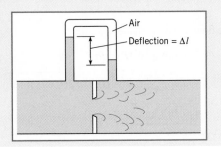

Properties: Water (60°F), Table A.5: $\nu = 1.22 \times 10^{-5}$ ft^2/s.

Plan

1. Calculate Reynolds number.

2. Find the flow coefficient K from Fig. 13.14.

3. Solve for Δh by using Eq. (13.7a).

4. Solve for Δl by using the manometer equation (3.19)

Solution

1. Reynolds number.

$$\text{Re} = \frac{4Q}{\pi d\nu} = \frac{(4)(2 \text{ ft}^3/\text{s})}{\pi((8/12) \text{ ft})(1.22 \times 10^{-5} \text{ ft}^2/\text{s})} = \boxed{3.1 \times 10^5}$$

2. Flow coefficient.
   - Use Fig. 13.14. Interpolate for $d/D = 8/12 = 0.667$ to find $K \approx 0.68$.

3. Change in piezometric head
    • From $Q = KA_o\sqrt{2g\Delta h}$, solve for Δh:

$$\Delta h = \frac{Q^2}{2gK^2A_o^2} = \frac{4}{64.4(0.68^2)[((\pi)/4)(8/12)^2]^2} = 1.1 \text{ ft}$$

4. Manometer deflection
    • The deflection is related to Δh by

$$\Delta h = \Delta l\left(\frac{\gamma_w - \gamma_{air}}{\gamma_w}\right)$$

    • Since $\gamma_w \gg \gamma_{air}$, $\Delta l = \Delta h = 1.1$ ft. $\boxed{\Delta l = 1.1 \text{ ft}}$

The sharp-edged orifice can also be used to measure the mass flow rate of gases. The discharge equation [Eq. (13.7b)] is multiplied by the upstream gas density and an empirical factor to account for compressibility effects (10). The resulting equation is

$$\dot{m} = YA_oK\sqrt{2\rho_1(p_1 - p_2)} \tag{13.10}$$

where K, the flow coefficient, is found using Fig. 13.14 and Y is the compressibility factor given by the empirical equation

$$Y = 1 - \left\{\frac{1}{k}\left(1 - \frac{p_2}{p_1}\right)\left[0.41 + 0.35\left(\frac{A_o}{A_1}\right)^2\right]\right\} \tag{13.11}$$

In this case both the pressure difference across the orifice and the absolute pressure of the gas are needed. One must remember when using the equation for the compressibility factor that the absolute pressure must be used.

EXAMPLE 13.5 MASS FLOW RATE OF NATURAL GAS

The mass flow rate of natural gas is to be measured using a sharp-edged orifice. The upstream pressure of the gas is 101 kPa absolute, and the pressure difference across the orifice is 10 kPa. The upstream temperature of the methane is 15°C. The pipe diameter is 10 cm, and the orifice diameter is 7 cm. What is the mass flow rate?

Problem Definition

Situation:

1. Natural gas (methane) is flowing through a sharp-edged orifice.
2. Pipe diameter is $D = 0.1$ m. Orifice diameter is $d = 0.07$ m.
3. Pressure difference across orifice is 10 kPa.

Find: Mass flow rate (in kg/s).

Properties: Natural gas (15°C, 1 atm), Table A.2:
$\rho = 0.678$ kg/m³, $\nu = 1.59 \times 10^{-5}$ m²/s, $K = 1.31$.

Plan

1. Find the flow coefficient K from Fig. 13.14.
2. Calculate the compressibility factor Y using Eq. (13.11).
3. Calculate the mass flow rate using Eq. (13.10).

Solution

1. Flow coefficient
    • Calculate (Re_d/K):

$$\frac{\text{Re}_d}{K} = \frac{d}{\nu}\sqrt{2\frac{\Delta p}{\rho_1}} = \frac{0.07}{1.59(10^{-5})}\sqrt{2\frac{10^4}{0.678}} = 7.56 \times 10^5$$

    • Using Fig. 13.14, $K = 0.7$.

2. Compressibility factor

$$Y = 1 - \left\{\frac{1}{1.31}\left(1 - \frac{91}{101}\right)(0.41 + 0.35 \times 0.7^4)\right\} = 0.962$$

3. Mass flow rate of methane

$$\dot{m} = YA_oK\sqrt{2\rho_1(p_1 - p_2)}$$

$$= 0.962\left(\frac{\pi}{4}0.07^2\right)(0.7)\sqrt{2(0.678)(10^4)}$$

$$= \boxed{0.302 \text{ kg/s}}$$

Figure 13.15

Typical venturi meter.

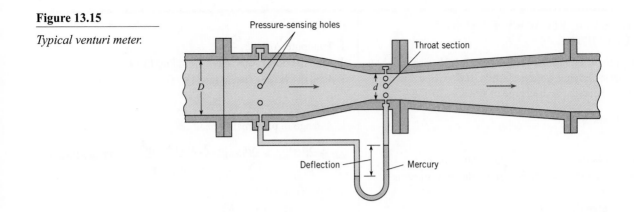

The foregoing examples involved the determination of either Q or Δh for a given size of orifice. Another type of problem is determination of the diameter of the orifice for a given Q and Δh. For this type of problem a trial-and-error procedure is required. Because one knows the approximate value of K, that is guessed first. Then the diameter is solved for, after which a better value of K can be determined, and so on.

Venturi Meter

The *venturi meter*, Fig. 13.15, is an instrument for measuring flow rate by using measurements of pressure across a converging-diverging flow passage. The main advantage of the venturi meter as compared to the orifice meter is that the head loss for a venturi meter is much smaller. The lower head loss results from streamlining the flow passage, as shown in Fig. 13.15. Such streamlining eliminates any jet contraction beyond the smallest flow section. Consequently, the coefficient of contraction has a value of unity, and the venturi equation is

$$Q = \frac{A_t C_d}{\sqrt{1 - (A_t / A_p)^2}} \sqrt{2g(h_p - h_t)} \tag{13.12}$$

$$Q = K A_t \sqrt{2g\Delta h} \tag{13.13}$$

where A_t is the throat area and Δh is the difference in piezometric head between the venturi entrance (pipe) and the throat. Note that the venturi equation is the same as the orifice equation. However, K for the venturi meter approaches unity at high values of the Reynolds number and small d/D ratios. This trend can be seen in Fig. 13.14, where values of K for the venturi meter are plotted along with similar data for the orifice.

Flow Nozzles

The *flow nozzle,* Fig. 13.16, is an instrument for measuring flow rate by using the pressure drop across a nozzle that is typically placed inside a conduit. Similar to an orifice meter, design and application of the flow nozzle is described by engineering standards (10). As compared to an orifice meter, the flow nozzle is better in flows that cause wear (e.g., particle-laden flow). The reason is that erosion of an orifice will produce more change in the pressure-drop versus flow-rate relationship. Both the flow nozzle and orifice meter will produce about the same overall head loss.

EXAMPLE 13.6 FLOW RATE USING A VENTURI METER

The pressure difference between the taps of a horizontal venturi meter carrying water is 35 kPa. If $d = 20$ cm and $D = 40$ cm, what is the discharge of water at 10°C?

Problem Definition

Situation:

1. Water flows through a horizontal venturi meter.
2. Pipe diameter is $D = 0.40$ m. Venturi throat diameter is $d = 0.2$ m.

Find: Discharge (in m^3/s).

Properties: Water (10°C), Table A.5: $\nu = 1.31 \times 10^{-6}$ m^2/s, and $\gamma = 9810$ N/m^3.

Plan

1. Compute $\Delta h = h_1 - h_2$.
2. Find the flow coefficient K from Fig. 13.14.
3. Find discharge Q using Eq. (13.7a).

Solution

1. Change in piezometric head

$$\Delta h = \frac{\Delta p}{\gamma} + \Delta z = \frac{\Delta p}{\gamma} + 0 = \frac{35{,}000 \text{ N}/m^2}{9810 \text{ N}/m^3} = 3.57 \text{ m of water}$$

2. Flow coefficient
   - Calculate (Re_d/K):

$$\frac{\mathrm{Re}_d}{K} = \frac{d\sqrt{2g\Delta h}}{\nu} = \frac{0.20\sqrt{2(9.81)(3.57)}}{1.31(10^{-6})} = 1.28 \times 10^6$$

   - From Fig. 13.14, find that $K = 1.02$.
3. Discharge

$$Q = 1.02 A_2 \sqrt{2g\Delta h}$$

$$= 1.02(0.785)(0.20^2)\sqrt{2(9.81)(3.57)} = \boxed{0.268 \text{ m}^3/s}$$

Figure 13.16

Typical flow nozzle.

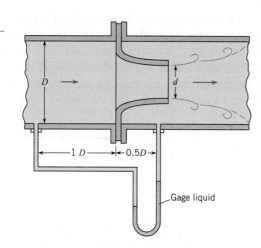

Electromagnetic Flow Meter

All of the flow meters described so far require that some sort of obstruction be placed in the flow. The obstruction may be the rotor of a vane anemometer or the reduced cross-section of an orifice or venturi meter. A meter that neither obstructs the flow nor requires pressure taps, which are subject to clogging, is the *electromagnetic flow meter*. Its basic principle is that a conductor that moves in a magnetic field produces an electromotive force. Hence liquids having a degree of conductivity generate a voltage between the electrodes, as in Fig. 13.17, and this voltage is proportional to the velocity of flow in the conduit. It is interesting to note that the basic principle of the electromagnetic flow meter was investigated by Faraday in 1832. However, practical application of the principle was not made until approximately a century later, when it was used to measure blood flow. Recently, with the need for a meter to

Figure 13.17

Electromagnetic flow meter.

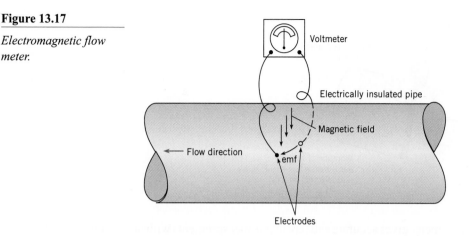

measure the flow of liquid metal in nuclear reactors and with the advent of sophisticated electronic signal detection, this type of meter has found extensive commercial use.

The main advantages of the electromagnetic flow meter are that the output signal varies linearly with the flow rate and that the meter causes no resistance to the flow. The major disadvantages are its high cost and its unsuitability for measuring gas flow.

For a summary of the theory and application of the electromagnetic flow meter, the reader is referred to Shercliff (11). This reference also includes a comprehensive bibliography on the subject.

Ultrasonic Flow Meter

Another form of nonintrusive flow meter that is used in diverse applications ranging from blood flow measurement to open-channel flow is the *ultrasonic flow meter*. Basically, there are two different modes of operation for ultrasonic flow meters. One mode involves measuring the difference in travel time for a sound wave traveling upstream and downstream between two measuring stations. The difference in travel time is proportional to flow velocity. The second mode of operation is based on the Doppler effect. When an ultrasonic beam is projected into an inhomogeneous fluid, some acoustic energy is scattered back to the transmitter at a different frequency (Doppler shift). The measured frequency difference is related directly to the flow velocity.

Turbine Flow Meter

The *turbine flow meter* consists of a wheel with a set of curved vanes (blades) mounted inside a duct. The volume rate of flow through the meter is related to the rotational speed of the wheel. This rotational rate is generally measured by a blade passing an electromagnetic pickup mounted in the casing. The meter must be calibrated for the given flow conditions. The turbine meter is versatile in that it can be used for either liquids or gases. It has an accuracy of better than 1% over a wide range of flow rates, and it operates with small head loss. The turbine flow meter is used extensively in monitoring flow rates in fuel-supply systems.

Vortex Flow Meter

The *vortex flow meter*, shown in Fig. 13.18, measures flow rate by relating vortex shedding frequency to flow rate. The vortices are shed from a sensor tube that is situated in the center of a pipe. These vortices cause vibrations, which are sensed by piezoelectric crystals that are located inside the sensor tube, and are converted to an electronic signal that is directly proportional to flow rate.

Figure 13.18

Vortex flow meter.

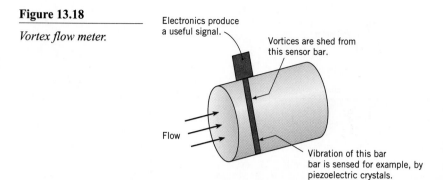

This vortex meter gives accurate and repeatable measurements with no moving parts. However, the corresponding head loss is comparable to that from other obstruction-type meters.

Rotameter

The *rotameter*, Fig. 13.19, is an instrument for measuring flow rate by sensing the position of an active element (weight) that is situated in a tapered tube. The equilibrium position of the weight is related to the flow rate. Because the velocity is lower at the top of the tube (greater flow section there) than at the bottom, the rotor seeks a neutral position where the drag on it just balances its weight. Thus the rotor "rides" higher or lower in the tube depending on the rate of flow. The weight is designed so that it spins, thus it stays in the center of the tube. A calibrated scale on the side of the tube indicates the rate of flow. Although venturi and orifice meters have better accuracy (approximately 1% of full scale) than the rotameter (approximately 5% of full scale), the rotameter offers other advantages, such as ease of use and low cost.

Figure 13.19

Rotameter.

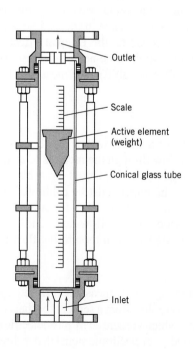

Rectangular Weir

A *weir*, shown in Fig. 13.20, is an instrument for determining flow rate in liquids by measuring the height of water relative to an obstruction in an open channel. The discharge over the weir is a function of the weir geometry and of the head on the weir. Consider flow over the weir in a rectangular channel, shown in Fig. 13.20. The head H on the weir is defined as the vertical distance between the weir crest and the liquid surface taken far enough upstream of the weir to avoid local free-surface curvature (see Fig. 13.20).

The discharge equation for the weir is derived by integrating $V\,dA = VL\,dh$ over the total head on the weir. Here L is the length of the weir and V is the velocity at any given distance h below the free surface. Neglecting streamline curvature and assuming negligible velocity of approach upstream of the weir, one obtains an expression for V by writing the Bernoulli equation between a point upstream of the weir and a point in the plane of the weir (see Fig. 13.21). Assuming the pressure in the plane of the weir is atmospheric, this equation is

$$\frac{p_1}{\gamma} + H = (H - h) + \frac{V^2}{2g} \tag{13.14}$$

Figure 13.20

Definition sketch for sharp-crested weir.
(a) Plan view.
(b) Elevation view.

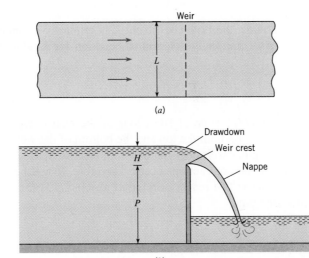

(a)

(b)

Figure 13.21

Theoretical velocity distribution over a weir.

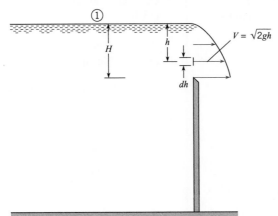

Here the reference elevation is the elevation of the crest of the weir, and the reference pressure is atmospheric pressure. Therefore $p_1 = 0$, and Eq. (13.14) reduces to

$$V = \sqrt{2gh}$$

Then $dQ = \sqrt{2gh}\,L\,dh$, and the discharge equation becomes

$$Q = \int_0^H \sqrt{2gh}\,L\,dh$$
$$= \tfrac{2}{3}L\sqrt{2g}H^{3/2}$$

$$(13.15)$$

In the case of actual flow over a weir, the streamlines converge downstream of the plane of the weir, and viscous effects are not entirely absent. Consequently, a discharge coefficient C_d must be applied to the basic expression on the right-hand side of Eq. (13.15) to bring the theory in line with the actual flow rate. Thus the rectangular weir equation is

$$Q = \tfrac{2}{3}C_d\sqrt{2g}LH^{3/2}$$
$$= K\sqrt{2g}LH^{3/2}$$

$$(13.16)$$

For low-viscosity liquids, the flow coefficient K is primarily a function of the relative head on the weir, H/P. An empirically determined equation for K adapted from Kindsvater and Carter (12) is

$$K = 0.40 + 0.05\frac{H}{P}$$

$$(13.17)$$

This is valid up to an H/P value of 10 as long as the weir is well ventilated so that atmospheric pressure prevails on both the top and the bottom of the weir nappe.

When the rectangular weir does not extend the entire distance across the channel, as in Fig. 13.22, additional end contractions occur. Therefore, K will be smaller than for the weir without end contractions. The reader is referred to King (13) for additional information on flow coefficients for weirs.

EXAMPLE 13.7 FLOW RATE FOR A RECTANGULAR WEIR

The head on a rectangular weir that is 60 cm high in a rectangular channel that is 1.3 m wide is measured to be 21 cm. What is the discharge of water over the weir?

Problem Definition

Situation:

1. Water flows over a rectangular weir.
2. The weir has a height of $P = 0.6$ m and a width of $L = 1.3$ m.
3. Head on the weir is $H = 0.21$ m.

Find: Discharge (in m^3/s).

Plan

1. Calculate the flow coefficient K using Eq. (13.17).
2. Calculate flow rate using the rectangular weir equation (13.16).

Solution

1. Flow coefficient

$$K = 0.40 + 0.05\frac{H}{P} = 0.40 + 0.05\left(\frac{21}{60}\right) = 0.417$$

2. Discharge

$$Q = K\sqrt{2g}LH^{3/2} = 0.417\sqrt{2(9.81)}(1.3)(0.21^{3/2})$$
$$= 0.23 \ m^3/s$$

Figure 13.22

*Rectangular weir with
end contractions.
(a) Plan view.
(b) Elevation view.*

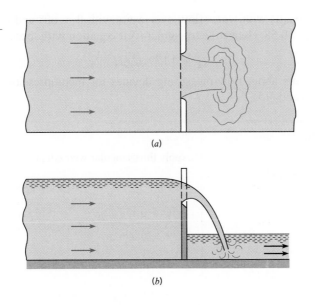

(a)

(b)

Triangular Weir

A definition sketch for the triangular weir is shown in Fig. 13.23. The primary advantage of the triangular weir is that it has a higher degree of accuracy over a much wider range of flow than does the rectangular weir, because the average width of the flow section increases as the head increases.

The discharge equation for the triangular weir is derived in the same manner as that for the rectangular weir. The differential discharge $dQ = V\,dA = VL\,dh$ is integrated over the total head on the weir to give

$$Q = \int_0^H \sqrt{2gh}\,(H-h)2\,\tan\!\left(\frac{\theta}{2}\right)dh$$

which integrates to

$$Q = \frac{8}{15}\sqrt{2g}\,\tan\!\left(\frac{\theta}{2}\right)H^{5/2}$$

However, a coefficient of discharge must still be used with the basic equation. Hence

$$Q = \frac{8}{15}C_d\sqrt{2g}\,\tan\!\left(\frac{\theta}{2}\right)H^{5/2} \tag{13.18}$$

Figure 13.23

*Definition sketch for the
triangular weir.*

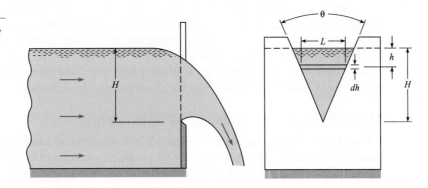

Experimental results with water flow over weirs with $\theta = 60°$ and $H > 2$ cm indicate that C_d has a value of 0.58. Hence the triangular weir equation with these limitations is

$$Q = 0.179\sqrt{2g}H^{5/2} \tag{13.19}$$

More details about flow-measuring devices for incompressible flow can be found in references (14) and (15).

EXAMPLE 13.8 FLOW RATE FOR A TRIANGULAR WEIR	*Plan*
	Apply the triangular weir equation Eq. (13.19).

EXAMPLE 13.8 FLOW RATE FOR A TRIANGULAR WEIR

The head on a 60° triangular weir is measured to be 43 cm. What is the flow of water over the weir?

Problem Definition

Situation:

1. Water flows over a 60° triangular weir.
2. Head on the weir is $H = 0.43$ m.

Find: Discharge (in m^3/s).

Plan

Apply the triangular weir equation Eq. (13.19).

Solution

$$Q = 0.179\sqrt{2g}H^{5/2} = 0.179 \times \sqrt{2 \times 9.81} \times (0.43)^{5/2}$$
$$= \boxed{0.096 \ m^3/s}$$

13.3

Measurement in Compressible Flow

This section describes how to measure velocity, pressure, and flow rate in compressible flows. Since fluid density is changing in these flows, the Bernoulli equation is invalid. Thus, compressible flow theory from Chapter 12 will be applied to develop valid measurement techniques.

Pressure Measurements

Static-pressure measurements can be made using the conventional static-pressure taps of a probe. However, if the boundary layer is disturbed by the presence of a shock wave in the vicinity of the pressure tap, the reading may not give the correct static pressure. The effect of the shock wave on the boundary layer is smaller if the boundary layer is turbulent. Therefore an effort is sometimes made to trip the boundary layer and ensure a turbulent boundary layer in the region of the pressure tap.

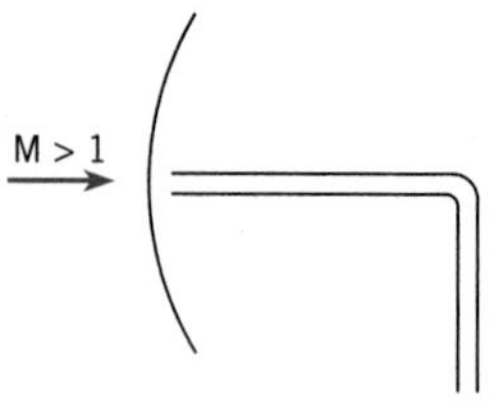

Figure 13.24

Stagnation tube in supersonic flow.

The stagnation pressure can be measured with a stagnation tube aligned with the local velocity vector. If the flow is supersonic, however, a shock wave forms around the tip of the probe, as shown in Fig. 13.24, and the stagnation pressure measured is that downstream of the shock wave and not that of the free stream. The stagnation pressure in the free stream can be calculated using the normal shock relationships, provided the free-stream Mach number is known. See Chapter 12 for more details about normal shock waves.

Mach Number and Velocity Measurements

A Pitot-static tube can be used to measure Mach numbers in compressible flows. Taking the measured stagnation pressure as the total pressure, one can calculate the Mach number in subsonic flows from the total-to-static-pressure ratio according to Eq. (12.31):

$$M = \left\{ \frac{2}{k-1} \left[\left(\frac{p_t}{p} \right)^{(k-1)/k} - 1 \right] \right\}^{1/2}$$

It is interesting to note here that one must measure the stagnation and static pressures separately to determine the pressure ratio, whereas one needs only the pressure difference to calculate the velocity of a flow.

If the flow is supersonic, then the indicated stagnation pressure is the pressure behind the shock wave standing off the tip of the tube. By taking this pressure as the total pressure downstream of a normal shock wave and the measured static pressure as the static pressure upstream of the shock wave, one can determine the Mach number of the free stream (M_1) from the static-to-total-pressure ratio (p_1/p_{t_2}) according to the expression

$$\frac{p_1}{p_{t_2}} = \frac{\{[2k/(k+1)]M_1^2 - [(k-1)/(k+1)]\}^{1/(k-1)}}{\{[(k+1)/2]M_1^2\}^{k/(k-1)}} \tag{13.20}$$

which is called the Rayleigh supersonic Pitot formula. Note, however, that M_1 is an implicit function of the pressure ratio and must be determined graphically or by some numerical procedure. Many normal-shock tables, such as those in reference (16), have p_1/p_{t_2} tabulated versus M_1, which enables one to find M_1 quite easily by interpolation.

Once the Mach number is determined, more information is needed to evaluate the velocity—namely, the local speed of sound. This can be done by inserting a probe into the flow to measure total temperature and then calculating the static temperature using Eq. (12.22):

$$T = \frac{T_t}{1 + [(k-1)/2]M_1^2}$$

The local speed of sound is then determined by Eq. (12.11):

$$c = \sqrt{kRT}$$

and the velocity is calculated from

$$V = M_1 c$$

The hot-wire anemometer can also be used to measure velocity in compressible flows, provided it is calibrated to account for Mach-number effects.

Mass Flow Measurement

Measuring the flow rate of a compressible fluid using a truncated nozzle was discussed in some detail in Chapter 12. Basically, the flow nozzle is a truncated nozzle located in a pipe, so the equations developed in Chapter 12 can be used to determine the flow rate through the flow nozzle. Strictly speaking, the flow rate so calculated should be multiplied by the discharge coefficient. For the high Reynolds numbers characteristic of compressible flows, however, the discharge coefficient can be taken as unity. If the flow at the throat of the flow nozzle is sonic (i.e., Mach number at the throat is 1.0), it is conceivable that the complex flow field existing downstream of the nozzle will make the reading from the downstream pressure tap difficult to interpret. That is, there can be no assurance that the measured pressure is the true back pressure. In such a case, it is advisable to use a venturi meter because the pressure is measured directly at the throat.

The mass flow rate of a compressible fluid through a venturi meter can easily be analyzed using the equations developed in Chapter 12. Consider the venturi meter shown in Fig. 13.25. Writing the energy equation, Eq. (12.15), for the flow of an ideal gas between stations 1 and 2 gives

$$\frac{V_1^2}{2} + \frac{kRT_1}{k-1} = \frac{V_2^2}{2} + \frac{kRT_2}{k-1} \qquad (13.21)$$

By conservation of mass, the velocity V_1 can be expressed as

$$V_1 = \frac{\rho_2 A_2 V_2}{\rho_1 A_1}$$

Substituting this result into Eq. (13.21), using the ideal-gas law to eliminate temperature, and solving for V_2 gives

$$V_2 = \left\{ \frac{[2k/(k-1)][(p_1/\rho_1) - (p_2/\rho_2)]}{1 - (\rho_2 A_2/\rho_1 A_1)^2} \right\}^{1/2} \qquad (13.22)$$

Assuming that the flow is isentropic,

$$\frac{p_1}{p_2} = \left(\frac{\rho_1}{\rho_2}\right)^k$$

the equation for the velocity at the throat can be rewritten as

$$V_2 = \left\{ \frac{[2k/(k-1)](p_1/\rho_1)[1 - (p_2/p_1)^{(k-1)/k}]}{1 - (p_2/p_1)^{2/k}(D_2/D_1)^4} \right\}^{1/2} \qquad (13.23)$$

The mass flow is obtained by multiplying V_2 by $\rho_2 A_2$. This analysis, however, has been based on a one-dimensional flow, and two-dimensional effects can be accounted for by the discharge coefficient C_d. The final result is

$$\dot{m} = C_d \rho_2 A_2 V_2 = C_d A_2 \left(\frac{p_2}{p_1}\right)^{1/k} \left\{ \frac{[2k/(k-1)]p_1 \rho_1 [1 - (p_2/p_1)^{(k-1)/k}]}{1 - (p_2/p_1)^{2/k}(D_2/D_1)^4} \right\}^{1/2} \qquad (13.24)$$

This equation is valid for all flow conditions, subsonic or supersonic, provided no shock waves occur between station 1 and station 2. It is good design practice to avoid supersonic flows in the venturi meter in order to prevent the formation of shock waves and the attendant total pressure losses. Also, the discharge coefficient can generally be taken as unity if no shock waves occur between 1 and 2.

Figure 13.25

Venturi meter.

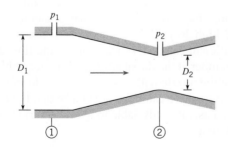

EXAMPLE 13.9   FLOW RATE FOR AIR THROUGH A VENTURI METER (COMPRESSIBLE FLOW)

Calculate the mass flow rate of air (inlet static temperature = 27°C) flowing through a venturi meter. The venturi throat is 1 cm in diameter (D_2), and the pipe is 3 cm in diameter (D_1). Upstream static pressure is 150 kPa, and throat pressure is 100 kPa.

Problem Definition

Situation:

1. Air flows through a venturi meter.
2. Pipe diameter is $D = 0.03$ m. Venturi throat diameter is $d = 0.01$ m.
3. Upstream conditions: Static temperature is 27°C; static pressure is 150 kPa.
4. Pressure in throat = 100 kPa.

Find: Mass flow rate (in kg/s).

Properties: Air (27°C), Table A.2: $k = 1.4$, and $R = 287$ J/kg · K.

Sketch:

Plan

1. Calculate density of air in the pipe (upstream) using the ideal gas law.
2. Calculate mass flow rate using Eq. (13.24).

Solution

1. Ideal gas law

$$\rho_1 = \frac{p_1}{RT_1} = \frac{150 \times 10^3 \text{ N/m}^2}{(287 \text{ J/kg K})(300 \text{ K})} = 1.74 \text{ kg/m}^3$$

2. Mass flow rate

$$\dot{m} = 1 \times 0.785 \times 10^{-4} \text{ m}^2 \left(\frac{1}{1.5}\right)^{0.714}$$

$$\times \left\{ \frac{7 \times 150 \times 10^3 \text{ N/m}^2 \times 1.74 \text{ kg/m}^3 [1 - (1/1.5)^{0.286}]}{[1 - (1/1.5)^{1.43}(1/3)^4]} \right\}^{1/2}$$

$$= \boxed{0.0264 \text{ kg/s}}$$

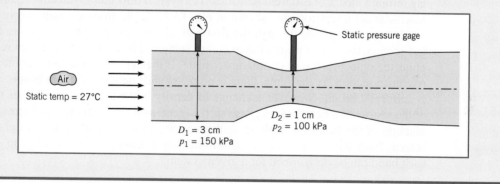

Static pressure gage

Air
Static temp = 27°C

D_1 = 3 cm
p_1 = 150 kPa

D_2 = 1 cm
p_2 = 100 kPa

Shock Wave Visualization

When studying supersonic flow in a wind tunnel, it is important to be able to locate and identify the shock wave pattern. Unfortunately, shock waves cannot be seen with the naked eye, so the application of some type of optical technique is necessary. There are three techniques by which shock waves can be seen: the shadowgraph, the interferometer, and the schlieren system. Each technique has its special application related to the type of information on density variation that is desired. The schlieren technique, however, finds frequent use in shock wave visualization.

An illustration of the essential features of the schlieren system is given in Fig. 13.26. Light from the source s is collimated by lens L_1 to produce a parallel light beam. The light then passes through a second lens L_2 and produces an image of the source at plane f. A third lens L_3 focuses the image on the display screen. A sharp edge, usually called the knife edge, is positioned at plane f so as to block out a portion of the light.

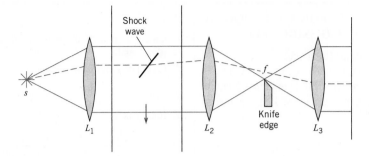

If a shock wave occurs in the test section, the light is refracted by the density change across the wave. As illustrated by the dashed line in Fig. 13.26, the refracted ray escapes the blocking effect of the knife edge, and the shock wave appears as a lighter region on the screen. Of course, if the beam is refracted in the other direction, the knife edge blocks out more light, and the shock wave appears as a darker region. The contrast can be increased by intercepting more light with the knife edge.

Interferometry

The interferometer allows one to map contours of constant density and to measure the density changes in the flow field. The underlying principle is the phase shift of a light beam on passing through media of different densities. The system now employed almost universally is the Mach-Zender interferometer, shown in Fig. 13.27. Light from a common source is split into two beams as it passes through the first half-silvered mirror. One beam passes through the test section, the other through the reference section. The two beams are then recombined and projected onto a screen or photographic plate. If the density in the test section and that in the reference section are the same, there is no phase shift between the two beams, and the screen is uniformly bright. However, a change of density in the test section changes the light speed of the test section beam, and a phase shift is generated between the two beams. Upon recombination of the beams, this phase shift gives rise to a series of dark and light bands on the screen. Each band represents a uniform-density contour, and the change in density across each band can be determined for a given system.

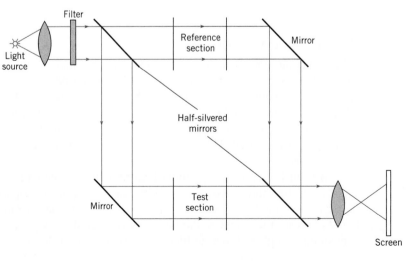

13.4

Accuracy of Measurements

When a parameter is measured, it is important to assess the accuracy of the measurement. The resulting analysis, called an *uncertainty analysis,* provides an estimate of the upper and lower bounds of the parameter. For example, if Q is a measured value of discharge, uncertainty analysis provides an estimate of the uncertainty U_Q in this measurement. The measurement would then be reported as $Q \pm U_Q$.

Commonly, a parameter of interest is not directly measured but is calculated from other variables. For example, discharge for an orifice meter is calculated using Eq. (13.7a). Such an equation is called a data reduction equation. Consider a data reduction equation of the form

$$x = f(y_1, y_2, \ldots, y_n)$$

where x is the parameter of interest and y_1 through y_n are the independent variables. Then, the uncertainty in x, which is written as U_x, is given by

$$U_x = \left[\left(\frac{\partial x}{\partial y_1} U_{y_1} \right)^2 + \left(\frac{\partial x}{\partial y_2} U_{y_2} \right)^2 + \cdots + \left(\frac{\partial x}{\partial y_n} U_{y_n} \right)^2 \right]^{0.5} \tag{13.25}$$

where U_{y_i} is the uncertainty in variable y_i. Equation (13.25), known as the uncertainty equation, is very useful for quantifying the accuracy of an experimental measurement, and for planning experiments. Additional information about uncertainty analysis is provided by Coleman and Steele (17).

EXAMPLE 13.10   UNCERTAINTY ESTIMATE FOR AN ORIFICE METER

For the orifice meter described in Example 13.2, estimate the uncertainty of the calculated discharge. Assume that uncertainty in K is 0.03, the uncertainty in diameter is 0.15 mm, and the uncertainty in measured head is 10 mm-Hg.

Problem Definition

Situation:

1. Water flows through an orifice ($d = 0.15$ m) in a pipe ($D = 0.24$ m).
2. A mercury-water manometer is used to measure pressure drop.

Find: Uncertainty (in m^3/s) for the calculated discharge Q.

Plan

1. Identify the data reduction equation (DRE).
2. Within the DRE, identify each variable that contributes to uncertainty.
3. Develop an equation for uncertainty by applying Eq. (13.25).
4. Calculate uncertainty by using the equation developed in step 3.

Sketch:

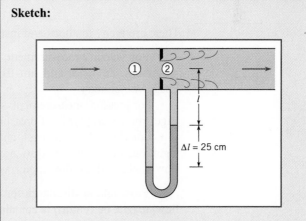

Solution

1. The data reduction equation is the orifice equation, Eq. (13.7a).

$$Q = K(\pi d^2/4)\sqrt{2g\Delta h}$$

2. Variables that cause uncertainty are K, d, g, and h. Neglect the influence of g.

3. Derive an equation for the uncertainty

$$U_Q^2 = \left(\frac{\partial Q}{\partial K}U_K\right)^2 + \left(\frac{\partial Q}{\partial d}U_d\right)^2 + \left(\frac{\partial Q}{\partial h}U_h\right)^2$$

Evaluate each partial derivative and then divide both sides of this equation by Q^2:

$$\left(\frac{U_Q}{Q}\right)^2 = \left(\frac{U_K}{K}\right)^2 + \left(\frac{2U_d}{d}\right)^2 + \left(\frac{U_h}{2h}\right)^2$$

4. Substitute values from Example 13.2:

$$\left(\frac{U_Q}{Q}\right)^2 = \left(\frac{0.03}{0.66}\right)^2 + \left(\frac{2\times0.15}{150}\right)^2 + \left(\frac{10}{2\times250}\right)^2$$

$$\left(\frac{U_Q}{Q}\right)^2 = (20.7\times10^{-4}) + (0.04\times10^{-4}) + (4\times10^{-4})$$

$$U_Q = 0.0497 Q = 0.0497 \times (0.092~\text{m}^3/\text{s})$$

$$= 0.0046~\text{m}^3/\text{s}$$

Thus

$$\boxed{Q = (0.092 \pm 0.046)~\text{m}^3/\text{s}}$$

Review

The primary source of uncertainty in the discharge is due to U_K. The term U_h has a small effect, and U_d has a negligible effect.

13.5

Summary

There are many methods and instruments for measuring velocity, pressure, and flow rate:

• For velocity measurement: stagnation tube, Pitot tube, yaw meter, vane and cup anemometers, hot-wire and hot-film anemometers, laser-Doppler anemometer, and particle image velocimetry

• For pressure measurement: static tube, piezometer, differential manometer, Bourdon-tube gage, and several types of pressure transducers

• For flow rate measurement: direct volume or weight measurement, velocity-area integration, orifice meter, flow nozzle, venturi meter, electromagnetic flow meter, ultrasonic flow meter, turbine flow meter, vortex flow meter, rotameter, and weir

Flow rate or discharge for a flow meter that uses a restricted opening (i.e., an orifice, flow nozzle, or venturi) is calculated using

$$Q = KA_o\sqrt{2g\Delta h} = KA_o\sqrt{2\Delta p_z/\rho}$$

where K is a flow coefficient that depends on Reynolds number and the type of flow meter, A_o is the area of the opening, Δh is the change in piezometric head across the flowmeter, and Δp_z is drop in piezometric pressure across the flowmeter.

Discharge for a rectangular weir of length L is given by

$$Q = K\sqrt{2g}LH^{3/2}$$

where K is the flow coefficient that depends on H/P. The term H is the height of the water above the crest of the weir, as measured upstream of the weir, and P is the height of the weir. Discharge for a 60° triangular weir with $H > 2$ cm is given by

$$Q = 0.179\sqrt{2g}H^{5/2}$$

When flow is compressible, instruments such as the stagnation tube, hot-wire anemometer, Pitot tube, and flow nozzle may be used. However, equations correlating velocity and discharge need to be altered to account for the effects of compressibility. To observe shock waves in compressible flow, a schlieren technique or an interferometer may be used.

Uncertainty analysis provides a way to quantify the accuracy of a measurement. When a parameter of interest x is evaluated using an equation of the form $x = f(y_1, y_2, \ldots, y_n)$, where y_1 through y_n are the independent variables, the uncertainty in x is given by

$$U_x = \left[\left(\frac{\partial x}{\partial y_1}U_{y_1}\right)^2 + \left(\frac{\partial x}{\partial y_2}U_{y_2}\right)^2 + \cdots + \left(\frac{\partial x}{\partial y_n}U_{y_n}\right)^2\right]^{0.5}$$

where U_{y_i} is the uncertainty in variable y_i. This equation, known as the uncertainty equation, is very useful for estimating uncertainty and for planning experiments.

References

1. Hurd, C. W., K. P. Chesky, and A. H. Shapiro. "Influence of Viscous Effects on Impact Tubes." *Trans. ASME J. Applied Mechanics,* vol. 75 (June 1953).

2. King, H. W., and E. F. Brater. *Handbook of Hydraulics.* McGraw-Hill, New York, 1963.

3. Lomas, Charles C. *Fundamentals of Hot Wire Anemometry.* Cambridge University Press, 1986.

4. Durst, Franz. *Principles and Practice of Laser-Doppler Anemometry.* Academic Press, New York, 1981.

5. Kline, J. J. "Flow Visualization." In *Illustrated Experiments in Fluid Mechanics: The NCFMF Book of Film Notes.* Educational Development Center, 1972.

6. Macagno, Enzo O. "Flow Visualization in Liquids." *Iowa Inst. Hydraulic Res. Rept.,* 114 (1969).

7. Raffel, M., C. Wilbert, and J. Kompenhans. *Particle Image Velocimetry.* Springer, New York, 1998.

8. Lienhard, J. H., V, and J. H. Lienhard, IV. "Velocity Coefficients for Free Jets from Sharp-Edged Orifices." *Trans. ASME J. Fluids Engineering,* 106, (March 1984).

9. Tuve, G. L., and R. E. Sprenkle. "Orifice Discharge Coefficients for Viscous Liquids." *Instruments,* vol. 8 (1935).

10. ASME. *Fluid Meters,* 6th ed. ASME, 1971.

11. Shercliff, J. A. *Electromagnetic Flow-Measurement.* Cambridge University Press, New York, 1962.

12. Kindsvater, Carl E., and R. W. Carter. "Discharge Characteristics of Rectangular Thin-Plate Weirs." *Trans. Am. Soc. Civil Eng.,* 124 (1959), 772–822.

13. King, L. V. *Phil. Trans. Roy. Soc. London, Ser. A,* 14 (1914), 214.

14. Miller, R. W. *Flow Measurement Engineering Handbook.* McGraw-Hill, New York, 1983.

15. Scott, R. W. W., ed. *Developments in Flow Measurement–1.* Applied Science, Englewood, NJ, 1982.

16. NACA. "Equations, Tables, and Charts for Compressible Flow." TR 1135 (1953).

17. Coleman, Hugh W., and W. Glenn Steele, *Experimentation and Uncertainty Analysis for Engineers.* Wiley, New York, 1989.

18. Wikipedia contributors, "Anemometer," Wikipedia, The Free Encyclopedia, http://en.wikipedia.org/w/index.php?title=Anemometer&oldid= 156121777 (accessed September 8, 2007).

Problems

Overview

13.1 List five different instruments or approaches that engineers use to measure fluid velocity. For each instrument or approach, list two advantages and two disadvantages.

13.2 List five different instruments or approaches that engineers use to measure flow rate (discharge). For each instrument or approach, list two advantages and two disadvantages.

Stagnation Tubes

13.3 Consider measuring the speed of automobile by building a stagnation tube from a drinking straw and then using this device with a water-filled *U*-tube manometer.

a. Make a sketch that illustrates how you would propose making this measurement.

b. Determine the lowest velocity that could be measured. Assume that the lower limit is based on the resolution of the manometer.

13.4 Without exceeding an error of 2.5%, what is the minimum air velocity that can be obtained using a 1 mm circular stagnation tube if the formula.

$$V = \sqrt{2\Delta p_{stag}/\rho} = \sqrt{2gh_{stag}}$$

is used for computing the velocity? Assume standard atmospheric conditions.

13.5 Without exceeding an error of 1%, what is the minimum water velocity that can be obtained using a 1.5 mm circular stagnation tube if the formula

$$V = \sqrt{2\Delta p_{stag}/\rho} = \sqrt{2gh_{stag}}$$

is used for computing the velocity? Assume the water temperature is 20°C.

13.6 A stagnation tube 2 mm in diameter is used to measure the velocity in a stream of air as shown. What is the air velocity if the deflection on the air-water manometer is 1.0 mm? Air temperature = 10°C, and p = 1 atm.

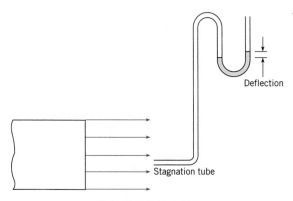

PROBLEMS 13.6, 13.7

13.7 If the velocity in an airstream (p_a = 98 kPa; T = 10°C) is 12 m/s, what deflection will be produced in an air-water manometer if the stagnation tube is 2 mm in diameter?

13.8 What would be the error in velocity determination if one used a C_p value of 1.00 for a circular stagnation tube instead of the true value? Assume the measurement is made with a stagnation tube 2 mm in diameter that is measuring air (T = 25°C, p = 1 atm) velocity for which the stagnation pressure reading is 5.00 Pa.

13.9 A velocity-measuring probe used frequently for measuring stack gas velocities is shown. The probe consists of two tubes bent away from and toward the flow direction and cut off on a plane normal to the flow direction, as shown. Assume the pressure coefficient is 1.0 at A and −0.4 at B. The probe is inserted in a stack where the temperature is 300°C and the pressure is 100 kPa absolute. The gas constant of the stack gases is 410 J/kg K. The probe is connected to a water manometer, and a 1.0 cm deflection is measured. Calculate the stack gas velocity.

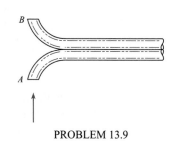

PROBLEM 13.9

Volume Flow Rate (Discharge)

13.10 Water from a pipe is diverted into a tank for 4 min. If the weight of diverted water is measured to be 10 kN, what is the discharge in cubic meters per second? Assume the water temperature is 20°C.

13.11 Water from a test apparatus is diverted into a calibrated volumetric tank for 5 min. If the volume of diverted water is measured to be 80 m³, what is the discharge in cubic meters per second, gallons per minute, and cubic feet per second?

13.12 A velocity traverse in a 24 cm oil pipe yields the data in the table. What are the discharge, mean velocity, and ratio of maximum to mean velocity? Does the flow appear to be laminar or turbulent?

r (cm)	V (m/s)	r (cm)	V (m/s)
0	8.7	7	5.8
1	8.6	8	4.9
2	8.4	9	3.8
3	8.2	10	2.5
4	7.7	10.5	1.9
5	7.2	11.0	1.4
6	6.5	11.5	0.7

13.13 A velocity traverse inside a 16 in. circular air duct yields the data in the table. What is the rate of flow in cubic feet per second and cubic feet per minute? What is the ratio of V_{max} to V_{mean}? Does it appear that the flow is laminar or turbulent? If p = 14.3 psia and T = 70°F, what is the mass flow rate?

$y*$	V (ft/s)	$y*$	V (ft/s)
0.0	0	2.0	110
0.1	72	3.0	117
0.2	79	4.0	122
0.4	88	5.0	126
0.6	93	6.0	129
1.0	100	7.0	132
1.5	106	8.0	135

*Distance from pipe wall, in.

13.14 The asymmetry of the flow in stacks means that flow velocity must be measured at several locations on the cross-flow plane. Consider the cross section of the cylindrical stack shown. The two access holes through which probes can be inserted are separated by 90°. Velocities can be measured at the five points shown (five-point method).

a. Determine the ratio r_m/D such that the areas of the five measuring segments are equal.

b. Determine the ratio r/D (probe location) that corresponds to the centroid of the segment.

c. The data in the table are taken for a stack 2 m in diameter in which the gas temperature is 300°C, the pressure is 110 kPa absolute, and the gas constant is 400 J/kg K. The data represent the deflection on a water manometer connected to a Pitot-static tube located at the measuring stations. Calculate the mass flow rate.

Station	Δh (cm)
1	1.2
2	1.1
3	1.1
4	0.9
5	1.05

PROBLEM 13.14

13.15 Repeat Prob. 13.14 for the case in which three access holes are separated by 60° and seven measuring points are used. The diameter of the stack is 1.5 m, the gas temperature is 250°C, the pressure is 115 kPa absolute, and the gas constant is 420 J/kg K. The data in the following table represent the deflection of a water manometer connected to a Pitot-static tube at the measuring stations. Calculate the mass flow rate.

Station	Δh (mm)
1	8.2
2	8.6
3	8.2
4	8.9
5	8.0
6	8.5
7	8.4

13.16 Theory and experimental verification indicate that the mean velocity along a vertical line in a wide stream is closely approximated by the velocity at 0.6 depth. If the indicated velocities at 0.6 depth in a river cross section are measured, what is the discharge in the river?

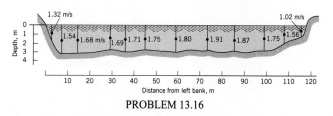

PROBLEM 13.16

Laser-Doppler Anemometers

13.17 $\mathbb{PQ}$◄ Literature Review. On the Internet, locate quality resources relevant to the LDA. Skim these resources, and then

a. Write down five findings that are relevant to engineering practice and interesting to you.

b. Write down two questions about LDAs that are interesting and insightful.

13.18 A laser-Doppler anemometer (LDA) system is being used to measure the velocity of air in a tube. The laser is an argon-ion laser with a wavelength of 4880 angstroms. The angle between the laser beams is 20°. The time interval is determined by measuring the time between five spikes, as shown, on the signal from the photodetector. The time interval between the five spikes is 12 microseconds. Find the velocity.

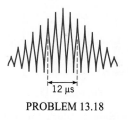

PROBLEM 13.18

Orifice Meters

13.19 $\mathbb{PQ}$◄ On the Internet, locate quality knowledge resources relevant to orifice meters. Skim these resources, and then

a. Write down five findings that are relevant to engineering practice and interesting to you.

b. Write down two questions that are interesting, insightful, and relevant to orifice meters.

13.20 For the jet and orifice shown, determine C_v, C_c, and C_d.

13.21 A fluid jet discharging from a 3 cm orifice has a diameter of 2.7 cm at its vena contracta. What is the coefficient of contraction?

13.22 Figure 13.13 is of a sharp-edged orifice. Note that the metal surface immediately downstream of the leading edge makes an acute angle with the metal of the upstream face of the orifice. Do you think the orifice would operate the same (have the same flow coefficient, K) if that angle were 90°? Explain how you came to your conclusion.

13.23 New orifices such as that shown in Fig. 13.13 will have definite flow coefficients as given in Fig. 13.14. With age,

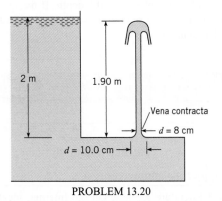

PROBLEM 13.20

values. Is the deflection on manometer C the same as the deflection on manometer F? Determine the deflections.

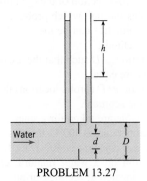

PROBLEM 13.27

however, physical changes could occur to the orifice. Explain what changes these might be and how (if at all) these physical changes might affect the flow coefficients.

13.24 A 6 in. orifice is placed in a 10 in. pipe, and a mercury manometer is connected to either side of the orifice. If the flow rate of water (60°F) through this orifice is 4.5 cfs, what will be the manometer deflection?

13.25 Determine the discharge of water through this 6 in. orifice that is installed in a 12 in. pipe.

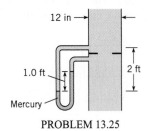

PROBLEM 13.25

13.26 The flow coefficient values for orifices given in Fig. 13.14 were obtained by testing orifices in relatively smooth pipes. If an orifice were used in a pipe that was very rough, do you think you would get a valid indication of discharge by using the flow coefficient of Fig. 13.14? Justify your conclusion.

13.27 Determine the discharge of water ($T = 60°F$) through the orifice shown if $h = 5$ ft, $D = 6$ in., and $d = 3$ in.

13.28 A pressure transducer is connected across an orifice to measure the flow rate of kerosene at 20°C. The pipe diameter is 2 cm, and the ratio of orifice diameter to pipe diameter is 0.6. The pressure differential as indicated by the transducer is 10 kPa. What is the mean velocity of the kerosene in the pipe?

13.29 The 10 cm orifice in the horizontal 30 cm pipe shown is the same size as the orifice in the vertical pipe. The manometers are mercury-water manometers, and water ($T = 20°C$) is flowing in the system. The gages are Bourdon-tube gages. The flow, at a rate of 0.1 m³/s, is to the right in the horizontal pipe and therefore downward in the vertical pipe. Is Δp as indicated by gages A and B the same as Δp as indicated by gages D and E? Determine their

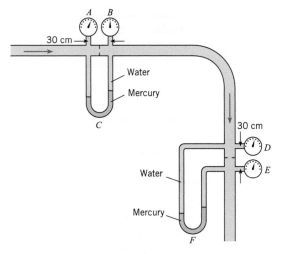

PROBLEM 13.29

13.30 A 15 cm plate orifice at the end of a 30 cm pipe is enlarged to 20 cm. With the same pressure drop across the orifice (approximately 50 kPa), what will be the percentage of increase in discharge?

13.31 If water (20°C) is flowing through this 5 cm orifice, estimate the rate of flow.

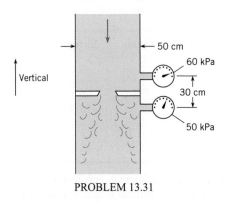

PROBLEM 13.31

13.32 A pressure transducer is connected across an orifice as shown. The pressure at the upstream pressure tap is p_1, and the pressure at the downstream tap is p_2. The pressure at the transducer connected to the upstream tap is $p_{T,1}$ and to the downstream pressure tap, $p_{T,2}$. Show that the difference in piezometric pressure defined as $(p_1 + \gamma z_1) - (p_2 + \gamma z_2)$ is equal to the pressure difference across the transducer, $p_{T,1} - p_{T,2}$.

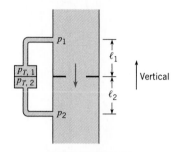

PROBLEM 13.32

13.33 Water ($T = 50°F$) is pumped at a rate of 20 cfs through the system shown in the figure. What differential pressure will occur across the orifice? What power must the pump supply to the flow for the given conditions? Also, draw the HGL and the EGL for the system. Assume $f = 0.015$ for the pipe.

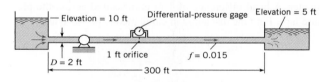

PROBLEM 13.33

13.34 Determine the size of orifice required in a 15 cm pipe to measure 0.03 m³/s of water with a deflection of 1 m on a mercury-water manometer.

13.35 What is the discharge of gasoline (S = 0.68) in a 10 cm horizontal pipe if the differential pressure across a 6 cm orifice in the pipe is 50 kPa?

13.36 What size orifice is required to produce a change in head of 6 m for a discharge of 2 m³/s of water in a pipe 1 m in diameter?

13.37 An orifice is to be designed to have a change in pressure of 50 kPa across it (measured with a differential-pressure transducer) for a discharge of 3.0 m³/s of water in a pipe 1.2 m in diameter. What diameter should the orifice have to yield the desired results?

13.38 Hemicircular orifices such as the one shown are sometimes used to measure the flow rate of liquids that also transport sediments. The opening at the bottom of the pipe allows free passage of the sediment. Derive a formula for Q as a function of Δp, D, and other relevant variables associated with the problem. Then, using that formula and guessing any unknown data, estimate the water discharge through such an orifice when Δp is read as 80 kPa and flow is in a 30 cm pipe.

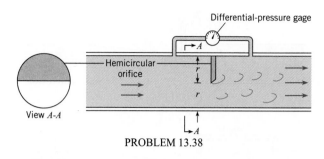

PROBLEM 13.38

Venturi Meters

13.39 PQ◄ What is the main advantage of a venturi meter versus an orifice meter? The main disadvantage?

13.40 Water flows through a venturi meter that has a 30 cm throat. The venturi meter is in a 60 cm pipe. What deflection will occur on a mercury-water manometer connected between the upstream and throat sections if the discharge is 0.75 m³/s? Assume $T = 20°C$.

13.41 What is the throat diameter required for a venturi meter in a 200 cm horizontal pipe carrying water with a discharge of 10 m³/s if the differential pressure between the throat and the upstream section is to be limited to 200 kPa at this discharge?

13.42 Estimate the rate of flow of water through the venturi meter shown.

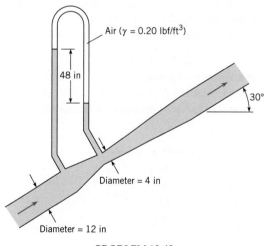

PROBLEM 13.42

13.43 When no flow occurs through the venturi meter, the indicator on the differential-pressure gage is straight up and indicates a Δp of zero. When 5 cfs of water flows to the right, the differential-pressure gage indicates $\Delta p = +10$ psi. If the flow is now reversed and 5 cfs flow to the left through the venturi meter, in which range would Δp fall? (a) $\Delta p < -10$ psi, (b) -10 psi $< \Delta p < 0$, (c) $0 < \Delta p < 10$ psi, or (d) $\Delta p = 10$ psi.

13.44 The pressure differential across this venturi meter is 100 kPa. What is the discharge of water through it?

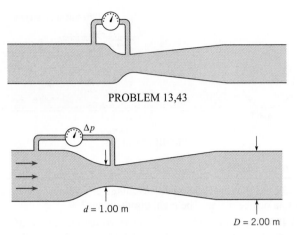

PROBLEM 13,43

PROBLEM 13.44

13.45 Engineers are calibrating a poorly designed venturi meter for the flow of an incompressible liquid by relating the pressure difference between taps 1 and 2 to the discharge. By applying the Bernoulli equation and assuming a quasi–one-dimensional flow (velocity uniform across every cross section), the engineers find that

$$Q_0 = A_2[2(p_1 - p_2)/\rho]^{0.5}[1 - (d/D)^4]^{-0.5}$$

where D and d are the duct diameters at stations 1 and 2. However, they realize that the flow is not quasi–one-dimensional and that the pressure at tap 2 is not equal to the average pressure in the throat because of streamline curvature. Thus the engineers introduce a correction factor K into the foregoing equation to yield

$$Q = KQ_0$$

Use your knowledge of pressure variation across curved streamlines to decide whether K is larger or smaller than unity, and support your conclusion by presenting a rational argument.

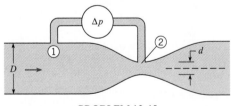

PROBLEM 13.45

13.46 The differential-pressure gage on the venturi meter shown reads 6.2 psi, $h = 25$ in., $d = 6$ in., and $D = 12$ in. What is the discharge of water in the system? Assume $T = 50°F$.

13.47 The differential-pressure gage on the venturi meter reads 45 kPa, $d = 20$ cm, $D = 40$ cm, and $h = 80$ cm. What is the discharge of gasoline (S = 0.69; $\mu = 3 \times 10^{-4}$ N · s/m²) in the system?

13.48 A flow nozzle has a throat diameter of 2 cm and a beta ratio (d/D) of 0.5. Water flows through the nozzle, creating a

pressure difference across the nozzle of 8 kPa. The viscosity of the water is 10^{-6} m²/s, and the density is 1000 kg/m³. Find the discharge.

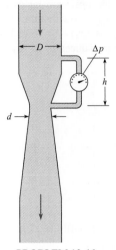

PROBLEM 13.46

13.49 Water flows through an annular venturi consisting of a body of revolution mounted inside a pipe. The pressure is measured at the minimum area and upstream of the body. The pipe is 5 cm in diameter, and the body of revolution is 2.5 cm in diameter. A head difference of 1 m is measured across the pressure taps. Find the discharge in cubic meters per second.

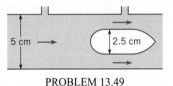

PROBLEM 13.49

Miscellaneous Measurement Techniques

13.50 What is the head loss in terms of $V_0^2/2g$ for the flow nozzle shown?

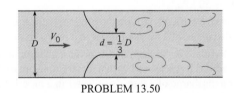

PROBLEM 13.50

13.51 A vortex flow meter is used to measure the discharge in a duct 5 cm in diameter. The diameter of the shedding element is 1 cm. The Strouhal number based on the shedding frequency from one side of the element is 0.2. A signal frequency of 50 Hz is measured by a pressure transducer mounted downstream of the element. What is the discharge in the duct?

13.52 A rotameter operates by aerodynamic suspension of a weight in a tapered tube. The scale on the side of the rotameter is calibrated in scfm of air—that is, cubic feet per minute at standard conditions ($p = 1$ atm and $T = 68°F$). By considering the balance of weight and aerodynamic force on the weight inside the tube, determine how the readings would be corrected for nonstandard conditions. In other words, how would the actual cubic feet per minute be calculated from the reading on the scale, given the pressure, temperature, and gas constant of the gas entering the rotameter?

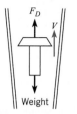

PROBLEM 13.52

13.53 A rotameter is used to measure the flow rate of a gas with a density of 1.0 kg/m^3. The scale on the rotameter indicates 5 liters/s. However, the rotameter is calibrated for a gas with a density of 1.2 kg/m^3. What is the actual flow rate of the gas (in liters per second)?

13.54 One mode of operation of ultrasonic flow meters is to measure the travel times between two stations for a sound wave traveling upstream and then downstream with the flow. The downstream propagation speed with respect to the measuring stations is $c + V$, where c is the sound speed and V is the flow velocity. Correspondingly, the upstream propagation speed is $c - V$.
a. Derive an expression for the flow velocity in terms of the distance between the two stations, L; the difference in travel times, Δt; and the sound speed.
b. The sound speed is typically much larger than V ($c \gg V$). With this approximation, express V in terms of L, c, and Δt.
c. A 10 ms time difference is measured for waves traveling 20 m in a gas where the speed of sound is 300 m/s. Calculate the flow velocity.

Weirs

13.55 $\mathbb{P}\mathbb{Q}$◀What variables influence flow rate through a rectangular weir?

13.56 $\mathbb{P}\mathbb{Q}$◀ On the Internet locate quality resources relevant to weirs, skim these resources, and write down five important findings.

13.57 Water flows over a rectangular weir that is 4 m wide and 30 cm high. If the head on the weir is 20 cm, what is the discharge in cubic meters per second?

13.58 The head on a 60° triangular weir is 25 cm. What is the discharge over the weir in cubic meters per second?

13.59 Water flows over two rectangular weirs. Weir A is 5 ft long in a channel 10 ft wide; weir B is 5 ft long in a channel 5 ft wide. Both weirs are 2 ft high. If the head on both weirs is 1.00 ft, then one can conclude that (a) $Q_A = Q_B$, (b) $Q_A > Q_B$, or (c) $Q_A < Q_B$.

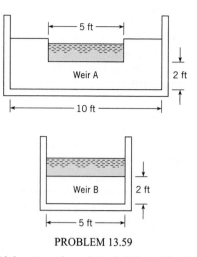

PROBLEM 13.59

13.60 A 1 ft–high rectangular weir (weir 1) is installed in a 2 ft–wide rectangular channel, and the head on the weir is observed for a discharge of 10 cfs. Then the 1 ft weir is replaced by a 2 ft–high rectangular weir (weir 2), and the head on the weir is observed for a discharge of 10 cfs. The ratio H_1/H_2 should be (a) equal to 1.00, (b) less than 1.00, or (c) greater than 1.00.

13.61 A 3 m–long rectangular weir is to be constructed in a 3 m–wide rectangular channel, as shown (a). The maximum flow in the channel will be 4 m^3/s. What should be the height P of the weir to yield a depth of water of 2 m in the channel upstream of the weir?

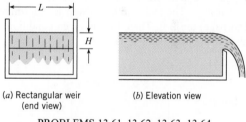

(a) Rectangular weir   (b) Elevation view
(end view)

PROBLEMS 13.61, 13.62, 13.63, 13.64

13.62 Consider the rectangular weir described in Prob. 13.61. When the head is doubled, the discharge is (a) doubled, (b) less than doubled, or (c) more than doubled.

13.63 A basin is 50 ft long, 2 ft wide, and 4 ft deep. A sharp-crested rectangular weir is located at one end of the basin, and it spans the width of the basin (the weir is 2 ft long). The crest of the weir is 2 ft above the bottom of the basin. At a given instant water in the basin is 3 ft deep; thus water is flowing over the weir and out of the basin. Estimate the time it will take for the water in the basin to go from the 3 ft depth to a depth of 2 ft 2 in.

13.64 Water at 50°F is piped from a reservoir to a channel like that shown. The pipe from the reservoir to the channel is a 4 in. steel pipe 100 ft in total length. There are two 90° bends, $r/D = 1$, in the line, and the entrance and exit are sharp-edged. The weir is 2 ft long. The elevation of the water surface in the reservoir is 100 ft, and the elevation of the bottom of the channel is 70 ft. The crest of the weir is 3 ft above the bottom of the channel. For steady flow conditions determine the water surface elevation in the channel and the discharge in the system.

13.65 At one end of a rectangular tank 1 m wide is a sharp-crested rectangular weir 1 m high. In the bottom of the tank is a 10 cm sharp-edged orifice. If 0.10 m^3/s of water flows into the tank and leaves the tank both through the orifice and over the weir, what depth will the water in the tank attain?

13.66 What is the water discharge over a rectangular weir 3 ft high and 10 ft long in a rectangular channel 10 ft wide if the head on the weir is 1.5 ft?

13.67 A reservoir is supplied with water at 60°F by a pipe with a venturi meter as shown. The water leaves the reservoir through a triangular weir with an included angle of 60°. The flow coefficient of the venturi is unity, the area of the venturi throat is 12 in.^{2}, and the measured Δp is 10 psi. Find the head, H, of the triangular weir.

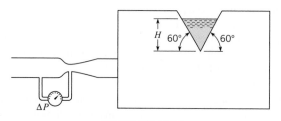

PROBLEM 13.67

13.68 At a particular instant water flows into the tank shown through pipes A and B, and it flows out of the tank over the rectangular weir at C. The tank width and weir length (dimensions normal to page) are 2 ft. Then, for the given conditions, is the water level in the tank rising or falling?

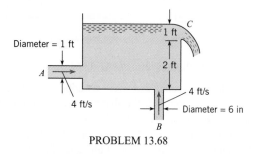

PROBLEM 13.68

13.69 Water flows from the first reservoir to the second over a rectangular weir with a width-to-head ratio of 3. The height P of the weir is twice the head. The water from the second reservoir flows over a 60° triangular weir to a third reservoir. The discharge across both weirs is the same. Find the ratio of the head on the rectangular weir to the head on the triangular weir.

13.70 Given the initial conditions of Prob. 13.69, tell, qualitatively and quantitatively, what will happen if the flow entering the first reservoir is increased 50%.

13.71 A rectangular irrigation canal 3 m wide carries water with a discharge of 6 m^3/s. What height of rectangular weir installed

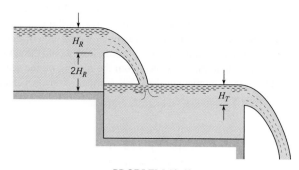

PROBLEM 13.69

across the canal will raise the water surface to a level 2 m above the canal floor?

13.72 The head on a 60° triangular weir is 1.5 ft. What is the discharge of water over the weir?

13.73 An engineer is designing a triangular weir for measuring the flow rate of a stream of water that has a discharge of 10 cfm. The weir has an included angle of 45° and a coefficient of discharge of 0.6. Find the head on the weir.

13.74 A pump is used to deliver water at 10°C from a well to a tank. The bottom of the tank is 2 m above the water surface in the well. The pipe is commercial steel 2.5 m long with a diameter of 5 cm. The pump develops a head of 20 m. A triangular weir with an included angle of 60° is located in a wall of the tank with the bottom of the weir 1 m above the tank floor. Find the level of the water in the tank above the floor of the tank.

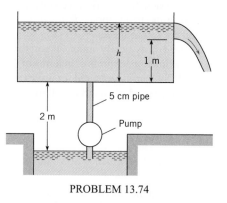

PROBLEM 13.74

Measurements in Compressible Flow

13.75 A Pitot-static tube is used to measure the Mach number in a compressible subsonic flow of air. The stagnation pressure is 140 kPa, and the static pressure is 100 kPa. The total temperature of the flow is 300 K. Determine the Mach number and the flow velocity.

13.76 Use the normal shock wave relationships developed in Chapter 12 to derive the Rayleigh supersonic Pitot formula.

13.77 The static and stagnation pressures measured by a Pitot-static tube in a supersonic air flow are 54 kPa and 200 kPa, respectively.

The total temperature is 350 K. Determine the Mach number and the velocity of the free stream.

13.78 A venturi meter is used to measure the flow of helium in a pipe. The pipe is 1 cm in diameter, and the throat diameter is 0.5 cm. The measured upstream and throat pressures are 120 kPa and 80 kPa, respectively. The static temperature of the helium in the pipe is 17°C. Determine the mass flow rate.

13.79 Hydrogen at atmospheric pressure and 15°C flows through a sharp-edged orifice with a beta ratio, d/D, of 0.5 in a 2 cm pipe. The pipe is horizontal, and the pressure change across the orifice is 1 kPa. The flow coefficient is 0.62. Find the mass flow (in kilograms per second) through the orifice.

13.80 A hole 0.2 in. in diameter is accidentally punctured in a line carrying natural gas (methane). The pressure in the pipe is 50 psig, and the atmospheric pressure is 14 psia. The temperature in the line is 70°F. What is the rate at which the methane leaks through the hole (in lbm/s)? The hole can be treated as a truncated nozzle.

Uncertainty Analysis

13.81 Consider the stagnation tube of Prob. 13.6. If the uncertainty in the manometer measurement is 0.1 mm, calculate the velocity and the uncertainty in the velocity. Assume that $C_p = 1.00$, $\rho_{air} = 1.25 \text{ kg/m}^3$, and the only uncertainty is due to the manometer measurement.

13.82 Consider the orifice meter in Prob. 13.25. Calculate the flow rate and the uncertainty in the flow rate. Assume the following values of uncertainty: 0.03 in flow coefficient, 0.05 in. in orifice diameter, and 0.5 in. in height of mercury.

13.83 Consider the weir in Prob. 13.66. Calculate the discharge and the uncertainty in the discharge. Assume the uncertainty in K is 5%, in H is 3 in., and in L is 1 in.

14

Turbomachinery

Machines to move fluids or to extract power from moving fluids have been designed and used by human beings since the beginning of recorded history. Ancient designs included buckets attached to a rope to transport water from a well or river. In the third century B.C., Archimedes invented the screw pump, which the Romans later used in their water supply systems. Water wheels were used in ancient China for grinding grain.

Fluid machines are used everywhere. They are the essential components of the automobiles we drive, the supply systems for the water we drink, the power generation plants for the electricity we use, and the air-conditioning and heating systems which provide the comfort we enjoy.

This chapter will introduce the concepts underlying various types of turbomachines and show how different designs are best suited to specific applications.

Fluid machines are separated into two broad categories: positive-displacement machines and turbomachines. Positive-displacement machines operate by forcing fluid into or out of a chamber. Examples include the bicycle tire pump, the gear pump, the peristaltic pump, and the human heart. Turbomachines involve the flow of fluid through rotating blades or rotors that remove or add energy to the fluid. Examples include propellers, fans, water pumps, windmills, and compressors. Axial-flow turbomachines operate with the flow entering and leaving the machine in the direction that is parallel to the axis of rotation of blades. A radial-flow machine can have the flow either entering or leaving the machine in the radial direction which is normal to the axis of rotation.

Table 14.1	CATEGORIES OF TURBOMACHINERY	
	Power Absorbing	**Power Producing**
Axial machines	Axial pumps Axial fans Propellers Axial compressors	Axial turbine (Kaplan) Wind turbine Gas turbine
Radial machines	Centrifugal pump Centrifugal fan Centrifugal compressor	Impulse turbine (Pelton wheel) Reaction turbine (Francis turbine)

Table 14.1 provides a classification for turbomachinery. Power-absorbing machines require power to increase head (or pressure). A power-producing machine provides shaft power at the expense of head (or pressure) loss. Pumps are associated with liquids whereas fans (blowers) and compressors are associated with gases. Both gases and liquids produce power through turbines. Oftentimes the gas turbine refers to an engine that has both a compressor and turbine and produces power.

14.1

Propellers

A propeller is a fan that converts rotational motion into thrust. The design of a propeller is based on the fundamental principles of airfoil theory. For example, consider a section of the propeller in Fig. 14.1, and notice the analogy between the lifting vane and the propeller. This propeller is rotating at an angular speed ω, and the speed of advance of the airplane and propeller is V_0. Focusing on an elemental section of the propeller, Fig. 14.1c, it is noted that the given section has a velocity with components V_0 and V_t. Here V_t is tangential velocity, $V_t = r\omega$, resulting from the rotation of the propeller. Reversing and adding the velocity vectors V_0 and V_t yield the velocity of the air relative to the particular propeller section (Fig. 14.1d).

The angle θ is given by

$$\theta = \arctan\left(\frac{V_0}{r\omega}\right) \tag{14.1}$$

For a given forward speed and rotational rate, this angle is a minimum at the propeller tip ($r = r_0$) and increases toward the hub as the radius decreases. The angle β is known as the *pitch angle*. The local angle of attack of the elemental section is

$$\alpha = \beta - \theta \tag{14.2}$$

The propeller can be analyzed as a series of elemental sections (of width dr) producing lift and drag, which provide the propeller thrust and create resistive torque. This torque multiplied by the rotational speed is the power input to the propeller.

The propeller is designed to produce thrust, and since the greatest contribution to thrust comes from the lift force F_L, the goal is to maximize lift and minimize drag, F_D. For a given shape of propeller section, the optimum angle of attack can be determined from data such as

Figure 14.1

Propeller motion.
(a) Airplane motion.
(b) View A-A.
(c) View B-B.
(d) Velocity relative to blade element.

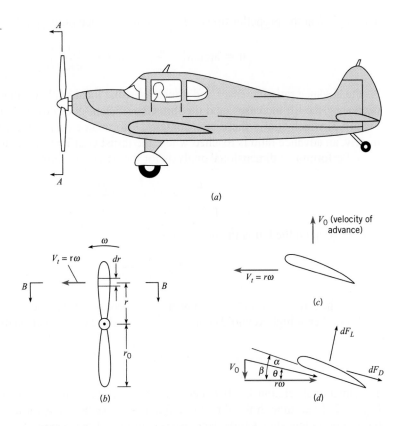

are given in Fig. 11.23. Because the angle θ increases with decreasing radius, the local pitch angle has to change to achieve the optimum angle of attack. This is done by twisting the blade.

A dimensional analysis can be performed to determine the π-groups that characterize the performance of a propeller. For a given propeller shape and pitch distribution, the thrust of a propeller T, will depend on the propeller diameter D, the rotational speed n, the forward speed V_0, the fluid density ρ, and the fluid viscosity μ.

$$T = f(D, \omega, V_0, \rho, \mu) \tag{14.3}$$

Performing a dimensional analysis using the methods presented in Chapter 8, results in

$$\frac{T}{\rho n^2 D^4} = f\left(\frac{V_0}{nD}, \frac{\rho D^2 n}{\mu}\right) \tag{14.4}$$

It is conventional practice to express the rotational rate, n, as revolutions per second (rps). The π-group on the left is called the *thrust coefficient*,

$$C_T = \frac{T}{\rho n^2 D^4} \tag{14.5}$$

The first π-group on the right is the *advance ratio*. The second group is a Reynolds number based on the tip speed and diameter of the propeller. For most applications, the Reynolds number is high, and experience shows that the thrust coefficient is unaffected by the Reynolds number, so

$$C_T = f\left(\frac{V_0}{nD}\right) \tag{14.6}$$

The angle θ at the propeller tip is related to the advance ratio by

$$\theta = \arctan\left(\frac{V_0}{\omega r_0}\right) = \arctan\left(\frac{1}{\pi}\frac{V_0}{nD}\right) \tag{14.7}$$

As the advance ratio increases and θ increases, the local angle of attack at the blade element decreases, the lift increases, and the thrust coefficient goes down. This trend is illustrated in Fig. 14.2, which shows the dimensionless performance curves for a typical propeller. Ultimately, an advance ratio is reached where the thrust coefficient goes to zero.

Performing a dimensional analysis for the power, P, shows

$$\frac{P}{\rho n^3 D^5} = f\left(\frac{V_0}{nD}, \frac{\rho D^2 n}{\mu}\right) \tag{14.8}$$

The π-group on the left is the *power coefficient*,

$$C_P = \frac{P}{\rho n^3 D^5} \tag{14.9}$$

As with the thrust coefficient, the power coefficient is not significantly influenced by the Reynolds number at high Reynolds numbers, so C_P reduces to a function of the advance ratio only

$$C_P = f\left(\frac{V_0}{nD}\right) \tag{14.10}$$

The functional relationship between C_P and V_0/nD for an actual propeller is also shown in Fig. 14.2. Even though the thrust coefficient approaches zero for a given advance ratio, the power coefficient shows little decrease because it still takes power to overcome the torque on the propeller blade.

Figure 14.2

Dimensionless performance curves for a typical propeller; $D = 2.90$ m, $n = 1400$ rpm. [After Weick (1).]

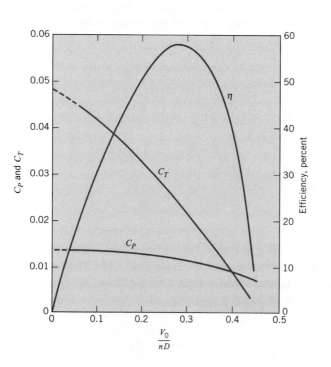

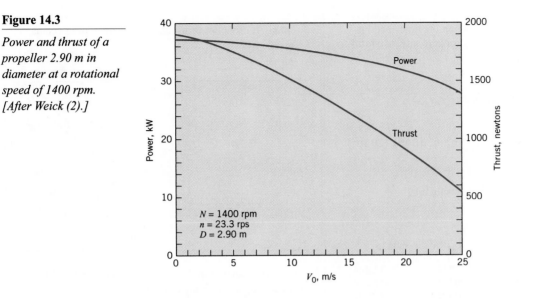

Figure 14.3

Power and thrust of a propeller 2.90 m in diameter at a rotational speed of 1400 rpm. [After Weick (2).]

$N = 1400$ rpm
$n = 23.3$ rps
$D = 2.90$ m

The curves for C_T and C_P are evaluated from performance characteristics of a given propeller operating at different values of V_0 as shown in Fig. 14.3. Although the data for the curves are obtained for a given propeller, the values for C_T and C_P, as a function of advance ratio, can be applied to geometrically similar propellers of different sizes and angular speeds.* Example 14.1 illustrates such an application.

EXAMPLE 14.1 PROPELLER APPLICATION

A propeller having the characteristics shown in Fig. 14.2 is to be used to drive a swamp boat. If the propeller is to have a diameter of 2 m and a rotational speed of $N = 1200$ rpm, what should be the thrust starting from rest? If the boat resistance (air and water) is given by the empirical equation $F_D = 0.003 \rho V_0^2/2$, where V_0 is the boat speed in meters per second, F_D is the drag, and ρ is the mass density of the water, what will be the maximum speed of the boat and what power will be required to drive the propeller? Assume $\rho_{air} = 1.20$ kg/m^3 and $\rho_{water} = 1000$ kg/m^3.

Problem Definition

Situation: Propeller used to drive a swamp boat.

Find:

1. Thrust (in N) starting from rest.
2. Maximum speed (in m/s) of swamp boat.
3. Power required (in kW) to operate propeller.

Sketch:

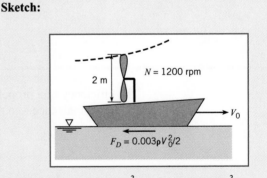

2 m

$N = 1200$ rpm

V_0

$F_D = 0.003 \rho V_0^2/2$

Properties: $\rho = 1.2$ kg/m^3, $\rho_w = 1000$ kg/m^3.

Plan

1. From Fig. 14.2, find thrust coefficient for zero advance ratio.
2. Calculate thrust using Eq. (14.5).
3. To calculate maximum speed, plot propeller thrust versus boat speed and on same graph plot resistance of swamp boat versus boat speed. The maximum speed is where the curves intersect.

* The speed of sound was not included in the dimensional analysis. However, the propeller performance is reduced because the Mach number based on the propeller tip speed leads to shock waves and other compressible-flow effects.

4. The maximum power will be when the boat speed is zero, so use Eq. (14.9) with C_P for zero advance ratio from Fig. 14.2.

Solution

1. From Fig. 14.2, $C_T = 0.048$ for $V_0/nD = 0$.

2. Thrust

$$F_T = C_T \rho_a D^4 n^2 = 0.048(1.20 \text{ kg/m}^3)(2 \text{ m})^4(20 \text{ rps})^2$$

$$= \boxed{369 \text{ N}}$$

3. Table of thrust versus speed of swamp boat

V_0	V_0/nD	C_T	$F_T = C_T\rho_a D^4 n^2$	$F_D = 0.003\rho_w V_0^2/2$
5 m/s	0.125	0.040	307 N	37.5 N
10 m/s	0.250	0.027	207 N	150 N
15 m/s	0.375	0.012	90 N	337 N

Graph of propeller thrust and swamp boat drag versus speed

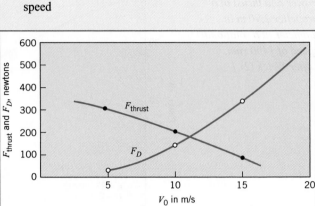

Curves intersect at $V_0 = 11$ m/s. Hence maximum speed of swamp boat is 11 m/s.

4. At $V_0/nD = 0$, $C_P = 0.014$.

$$P = 0.014(1.20 \text{ kg/m}^3)(2 \text{ m})^5(20 \text{ rps})^3$$

$$= 4300 \text{ m} \cdot \text{N/s} = \boxed{4.30 \text{ kW}}$$

Review

In an actual application the calculated maximum power is somewhat misleading. The starting rotational rate of propeller need not be 1200 rpm but can be a lower value. After the boat is gaining speed the rotational rate can be increased to achieve maximum speed.

The efficiency of a propeller is defined as the ratio of the power output—that is, thrust times velocity of advance—to the power input. Hence the efficiency η is given as

$$\eta = \frac{F_T V_0}{P} = \frac{C_T \rho D^4 n^2 V_0}{C_P \rho D^5 n^3}$$

or

$$\eta = \frac{C_T}{C_P}\left(\frac{V_0}{nD}\right) \tag{14.11}$$

The variation of efficiency with advance ratio for a typical propeller is also shown in Fig. 14.2. The efficiency can be calculated directly from C_T and C_P performance curves. Note at low advance ratios, the efficiency increases with advance ratio and then reaches a maximum value before the decreasing thrust coefficient causes the efficiency to drop toward zero. The maximum efficiency represents the best operating point for fuel efficiency.

Many propeller systems are designed to have variable pitch; that is, pitch angles can be changed during propeller operation. Different efficiency curves corresponding to varying pitch angle are shown in Fig. 14.4. The envelope for the maximum efficiency is also shown in the figure. During operation of the aircraft, the pitch angle can be controlled to achieve maximum efficiency corresponding to the propeller rpm and forward speed.

Figure 14.4

Efficiency curves for variable-pitch propeller.

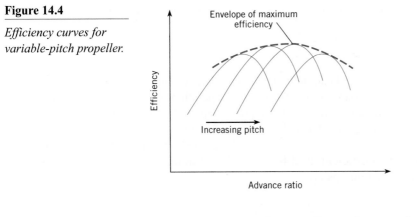

Figure 14.4

Efficiency curves for variable-pitch propeller.

The best source for propeller performance information is from propeller manufacturers. There are many speciality manufacturers from everything from marine to aircraft applications.

Axial-Flow Pumps

The axial flow pump acts much like a propeller enclosed in a housing as shown in Fig. 14.5. The rotating element, the impeller, causes a pressure change between the upstream and downstream sections of the pump. In practical applications, axial-flow machines are best suited to deliver relatively low heads and high flow rates. Hence pumps used for dewatering lowlands, such as those behind dikes, are almost always of the axial-flow type. Water turbines in low-head dams (less than 30 m) where the flow rate and power production are large are also generally of the axial type.

Head and Discharge Coefficients for Pumps

The thrust coefficient is defined as $F_T/\rho D^4 n^2$ for use with propellers, and if the same variables are applied to flow in an axial pump, the thrust can be expressed as $F_T = \Delta p A = \gamma \Delta H A$ or

$$C_T = \frac{\gamma \Delta H A}{\rho D^4 n^2} = \frac{\pi}{4} \frac{\gamma \Delta H D^2}{\rho D^4 n^2} = \frac{\pi}{4} \frac{g \Delta H}{D^2 n^2} \tag{14.12}$$

Figure 14.5

Axial-flow blower in a duct.

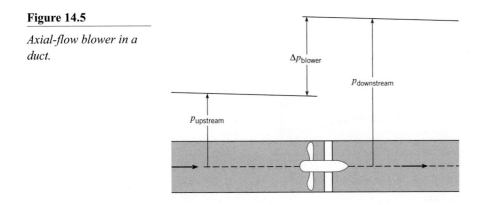

A new parameter, called the *head coefficient* C_H, is defined using the variables of Eq. (14.12), as

$$C_H = \frac{4}{\pi} C_T = \frac{\Delta H}{D^2 n^2 / g} \qquad (14.13)$$

which is a π-group that relates head delivered to fan diameter and rotational speed.

The independent π-group relating to propeller operation is V_0/nD; however, multiplying the numerator and denominator by the diameter squared gives $V_0 D^2/nD^3$, and $V_0 D^2$ is proportional to the discharge, Q. Thus the π-group for pump similarity studies is Q/nD^3 and is identified as the *discharge coefficient* C_Q. The power coefficient used for pumps is the same as the power coefficient used for propellers. Summarizing, the π-groups used in the similarity analyses of pumps are

$$C_H = \frac{\Delta H}{D^2 n^2 / g} \qquad (14.14)$$

$$C_P = \frac{P}{\rho D^5 n^3} \qquad (14.15)$$

$$C_Q = \frac{Q}{nD^3} \qquad (14.16)$$

where C_H and C_P are functions of C_Q for a given type of pump.

Figure 14.6 is a set of curves of C_H and C_P versus C_Q for a typical axial-flow pump. Also plotted on this graph is the efficiency of the pump as a function of C_Q. The dimensional curves (head and power versus Q for a constant speed of rotation) from which Fig. 14.6 was developed are shown in Fig. 14.7. Because curves like those shown in Fig. 14.6 or Fig. 14.7 characterize pump performance, they are often called *characteristic curves* or *performance curves*. These curves are obtained by experiment.

Figure 14.6

Dimensionless performance curves for a typical axial-flow pump. [After Stepanoff (3).]

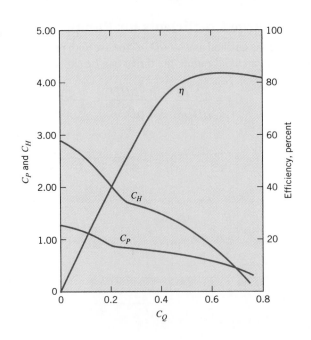

Figure 14.7

Performance curves for a typical axial-flow pump. [After Stepanoff (3).]

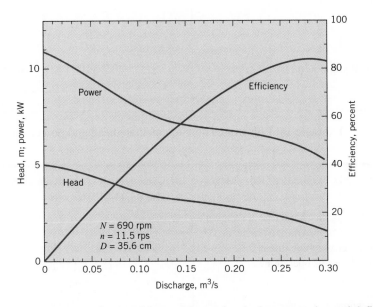

There can be a problem with overload when operating axial-flow pumps. As seen in Fig. 14.6, when the pump flow is throttled below maximum-efficiency conditions, the required power increases with decreasing flow, thus leading to the possibility of overloading at low-flow conditions. For very large installations, special operating procedures are followed in order to avoid such overloading. For instance, the valve in the bypass from the pump discharge back to the pump inlet can be adjusted to maintain a constant flow through the pump. However, for small-scale applications, it is often desirable to have complete flexibility in flow control without the complexity of special operating procedures.

Performance curves are used to predict prototype operation from model tests or the effect of changing the speed of the pump. Example 14.2 shows how to use pump curves to calculate discharge and power.

EXAMPLE 14.2 DISCHARGE AND POWER FOR AXIAL-FLOW PUMP

For the pump represented by Figs. 14.6 and 14.7, what discharge of water in cubic meters per second will occur when the pump is operating against a 2 m head and at a speed of 600 rpm? What power in kilowatts is required for these conditions?

Problem Definition

Situation: Axial flow pump with water.

Find:

1. Discharge (in m^3/s).
2. Power (in kW).

Sketch:

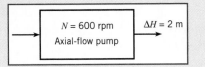

Assumptions: Assume $\rho = 1000 \ kg/m^3$.

Plan

1. Calculate C_H.
2. From Fig. 14.6 find C_Q and C_P.
3. Use C_Q to calculate discharge.
4. Use C_P to calculate power.

Solution

1. Rotational rate is $(600 \ rev/min)/(60 \ s/min) = 10 \ rps$. $D = 35.6 \ cm$.

$$C_H = \frac{2 \ m}{(0.356 \ m)^2 (10^2 \ s^{-2})/(9.81 \ m/s^2)} = 1.55$$

2. From Fig. 14.6, $C_Q = 0.40$ and $C_P = 0.72$.

3. Discharge is

$$Q = C_Q n D^3$$

$$Q = 0.40(10 \text{ s}^{-1})(0.356 \text{ m})^3 = \boxed{0.180 \text{ m}^3/\text{s}}$$

4. Power is

$$P = 0.72\rho D^5 n^3$$

$$= 0.72(10^3 \text{ kg/m}^3)(0.356 \text{ m})^5(10 \text{ s}^{-1})^3$$

$$= 4.12 \text{ km} \cdot \text{N/s} = 4.12 \text{ kJ/s} = \boxed{4.12 \text{ kW}}$$

Example 14.3 illustrates how to calculate head and power for an axial-flow pump.

EXAMPLE 14.3   HEAD AND POWER FOR AXIAL-FLOW PUMP

If a 30 cm axial-flow pump having the characteristics shown in Fig. 14.6 is operated at a speed of 800 rpm, what head ΔH will be developed when the water-pumping rate is 0.127 m^3/s? What power is required for this operation?

Problem Definition
Situation: 30 cm axial flow pump with water.

1. Head (in meters) developed.
2. Power (in kW) required.

Sketch:

$$Q = 0.127 \text{ m}^3/\text{s} \longrightarrow \boxed{\begin{array}{c} N = 800 \text{ rpm} \\ \text{Axial-flow pump} \end{array}} \begin{array}{c} \Delta H = ? \\ \longrightarrow \\ P = ? \end{array}$$

Assumptions: Water with $\rho = 10^3 \text{kg/m}^3$.

Plan

1. Calculate the discharge coefficient, C_Q.
2. From Fig. 14.6, read C_H, and C_P.

3. Use Eq. (14.14) to calculate head produced.
4. Use Eq. (14.15) to calculate power required.

Solution

1. Discharge coefficient is

$$Q = 0.127 \text{ m}^3/\text{s}$$

$$n = \frac{800}{60} = 13.3 \text{ rps}$$

$$D = 30 \text{ cm}$$

$$C_Q = \frac{0.127 \text{ m}^3/\text{s}}{(13.3 \text{ s}^{-1})(0.30 \text{ m})^3} = 0.354$$

2. From Fig. 14.6, $C_H = 1.70$ and $C_P = 0.80$.
3. Head produced is

$$\Delta H = \frac{C_H D^2 n^2}{g} = \frac{1.70(0.30 \text{ m})^2(13.3 \text{ s}^{-1})^2}{(9.81 \text{ m/s}^2)} = \boxed{2.76 \text{ m}}$$

4. Power required is

$$P = C_P \rho D^5 n^3$$

$$= 0.80(10^3 \text{ kg/m}^3)(0.30 \text{ m})^5(13.3 \text{ s}^{-1})^3 = \boxed{4.57 \text{ kW}}$$

Fan Laws

The *fan laws* are used extensively by designers and practitioners involved with axial fans and blowers. The fan laws are equations that provide the discharge, pressure rise, and power requirements for a fan that operates at different speeds. The laws are based on the discharge, head, and power coefficients being the same at any other state as at the reference state, o; namely, $C_Q = C_{Qo}$, $C_H = C_{Ho}$, and $C_P = C_{Po}$. Because the size and design of fan is unchanged, the discharge at speed n is

$$Q = Q_o \frac{n}{n_o} \tag{14.17a}$$

and the pressure rise is

$$\Delta p = \Delta p_o \left(\frac{n}{n_o} \right)^2 \tag{14.17b}$$

and finally the power required is

$$P = P_o \left(\frac{n}{n_o} \right)^3 \tag{14.17c}$$

These fan laws cannot be applied between fans of different size and design. Of course, the fan laws do not provide exact values because of design considerations and manufacturing tolerances, but they are very useful in estimating fan performance.

14.3

Radial-Flow Machines

Radial-flow machines are characterized by the radial flow of the fluid through the machine. Radial-flow pumps and fans are better suited for larger heads at lower flow rates than axial machines.

Centrifugal Pumps

A sketch of the *centrifugal pump* is shown in Fig. 14.8. Fluid from the inlet pipe enters the pump through the eye of the impeller and then travels outward between the vanes of the impeller to its edge, where the fluid enters the casing of the pump and is then conducted to the discharge pipe. The principle of the radial-flow pump is different from that of the axial-flow pump in that the change in pressure results in large part from rotary action (pressure increases outward like that in the rotating tank in Section 4.4 produced by the rotating impeller). Additional pressure increase is produced in the radial-flow pump when the high velocity of the flow leaving the impeller is reduced in the expanding section of the casing.

Although the basic designs are different for radial- and axial-flow pumps, it can be shown that the same similarity parameters (C_Q, C_P, and C_H) apply for both types. Thus the methods that have already been discussed for relating size, speed, and discharge in axial-flow machines also apply to radial-flow machines.

The major practical difference between axial- and radial-flow pumps so far as the user is concerned is the difference in the performance characteristics of the two designs. The dimensional performance curves for a typical radial-flow pump operating at a constant speed

Figure 14.8

Centrifugal pump.

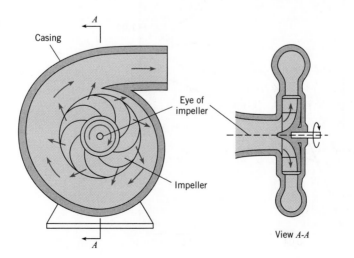

Figure 14.9

Performance curves for a typical centrifugal pump; D = 37.1 cm. [After Daugherty and Franzini (4). Used with the permission of the McGraw-Hill Companies.]

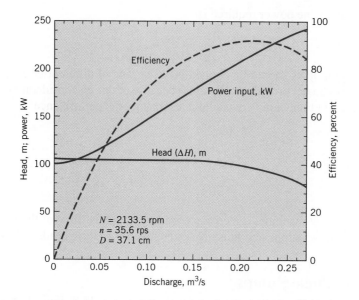

of rotation are shown in Fig. 14.9. The corresponding dimensionless performance curves for the same pump are shown in Fig. 14.10. Note that the power required at shutoff flow is less than that required for flow at maximum efficiency. Normally, the motor used to drive the pump is chosen for conditions corresponding to maximum pump efficiency. Hence the flow can be throttled between the limits of shutoff condition and normal operating conditions with no chance of overloading the pump motor. In this latter case, a radial-flow pump offers a distinct advantage over axial flow pumps.

Radial-flow pumps are manufactured in sizes from 1 hp or less and heads of 50 or 60 ft to thousands of horsepower and heads of several hundred feet. Figure 14.11 shows a cutaway view of a single-suction, single-stage, horizontal-shaft radial pump. Fluid enters in the direction of the rotating shaft and is accelerated outward by the rotating impeller. There are many other configurations designed for specific applications.

Figure 14.10

Dimensionless performance curves for a typical centrifugal pump, from data given in Fig. 14.9. [After Daugherty and Franzini (4).]

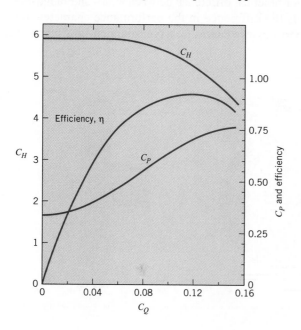

Figure 14.11

Cutaway view of a single-suction, single-stage, horizontal-shaft radial pump. Pump inlet, outlet, and impeller shown on photograph. (Courtesy of Ingersol Rand Co.)

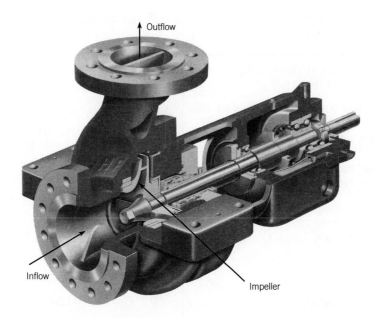

Example 14.4 shows how to find the speed and discharge for a centrifugal pump needed to provide a given head.

EXAMPLE 14.4   SPEED AND DISCHARGE OF CENTRIFUGAL PUMP

A pump that has the characteristics given in Fig. 14.9 when operated at 2133.5 rpm is to be used to pump water at maximum efficiency under a head of 76 m. At what speed should the pump be operated, and what will the discharge be for these conditions?

Problem Definition

Situation: Centrifugal pump operated at 2133.5 rpm pumps water to head of 76 m at maximum efficiency.

Find:

1. Operational speed of pump (in rpm).
2. Discharge (in m³/s).

Assumptions: Assume pump is the same size as that corresponding to Fig. 14.9 and water properties are the same.

Plan

The C_H, C_P, C_Q, and η are the same for any pump with the same characteristics operating at maximum efficiency. Thus

$$(C_H)_N = (C_H)_{2133.5 \text{ rpm}}$$

where N represents the unknown speed. Also
$(C_Q)_N = (C_Q)_{2133.5 \text{ rpm}}.$

1. Calculate speed using same head coefficient.
2. Calculate discharge using same discharge coefficient.

Solution

1. Speed calculation: From Fig. 14.9, at maximum efficiency $\Delta H = 90$ m.

$$\left(\frac{g\Delta H}{n^2 D^2}\right)_N = \left(\frac{g\Delta H}{n^2 D^2}\right)_{2133.5}$$

$$\frac{76 \text{ m}}{N^2} = \frac{90 \text{ m}}{2133.5^2 \text{ rpm}^2}$$

$$N = 2133.5 \times \left(\frac{76}{90}\right)^{1/2} = \boxed{1960 \text{ rpm}}$$

2. Discharge calculation: From Fig. 14.9, at maximum efficiency $Q = 0.255$ m³/s.

$$\left(\frac{Q}{nD^3}\right)_N = \left(\frac{Q}{nD^3}\right)_{2133.5}$$

$$\frac{Q_{1960}}{Q_{2133.5}} = \frac{1960}{2133.5} = 0.919$$

$$Q_{1960} = \boxed{0.234 \text{ m}^3/\text{s}}$$

Example 14.5 shows how to scale up data for a specific centrifugal pump to predict performance.

EXAMPLE 14.5 HEAD, DISCHARGE, AND POWER OF CENTRIFUGAL PUMP

The pump having the characteristics shown in Figs. 14.9 and 14.10 is a model of a pump that was actually used in one of the pumping plants of the Colorado River Aqueduct [see Daugherty and Franzini (4)]. For a prototype that is 5.33 times larger than the model and operates at a speed of 400 rpm, what head, discharge, and power are to be expected at maximum efficiency?

Problem Definition

Situation: Pump 5.33 times larger than model and operates at 400 rpm.

Find: At maximum efficiency,

1. Head (in meters).
2. Discharge (in m³/s).
3. Power (in kW).

Assumptions: Pumping water with $\rho = 10^3$ kg/m³.

Plan

1. Find C_Q, C_H, and C_P at maximum efficiency from Fig. 14.10.
2. Evaluate speed in rps and calculate new diameter.

3. Use Eqs. (14.14) through (14.16) to calculate head, discharge, and power.

Solution

1. From Fig. 14.10 at maximum efficiency, $C_Q = 0.12$, $C_H = 5.2$ and $C_P = 0.69$.
2. Speed in rps: $n = (400/60)$ rps = 6.67 rps
   $D = 0.371 \times 5.33 = 1.98$ m.
3. Pump performance
   - Head

   $$\Delta H = \frac{C_H D^2 n^2}{g} = \frac{5.2(1.98 \text{ m})^2 (6.67 \text{ s}^{-1})^2}{(9.81 \text{ m/s}^2)} = \boxed{92.4 \text{ m}}$$

   - Discharge

   $$Q = C_Q n D^3 = 0.12(6.67 \text{ s}^{-1})(1.98 \text{ m})^3 = \boxed{6.21 \text{ m}^3/\text{s}}$$

   - Power

   $$P = C_P \rho D^5 n^3 = 0.69((10^3 \text{ kg})/\text{m}^3)(1.98 \text{ m})^5 (6.67 \text{ s}^{-1})^3$$
   $$= \boxed{6230 \text{ kW}}$$

14.4

Specific Speed

From the discussion in preceding sections it was pointed out that axial-flow pumps are best suited for high discharge and low head whereas radial machines perform better for low discharge and high head. A tool for selecting the best pump is the value of a π-group called the specific speed, n_s. The *specific speed* is obtained by combining both C_H and C_Q in such a manner that the diameter D is eliminated:

$$n_s = \frac{C_Q^{1/2}}{C_H^{3/4}} = \frac{(Q/nD^3)^{1/2}}{[\Delta H/(D^2 n^2/g)]^{3/4}} = \frac{nQ^{1/2}}{g^{3/4}\Delta H^{3/4}}$$

Thus specific speed relates different types of pumps without reference to their sizes.

As shown in Fig. 14.12, when efficiencies of different types of pumps are plotted against n_s, it is seen that certain types of pumps have higher efficiencies for certain ranges of n_s. For low specific speeds, the radial-flow pump is more efficient whereas high specific speeds favor axial flow machines. In the range between the completely axial-flow machine and the completely radial-flow machine, there is a gradual change in impeller shape to accommodate the particular flow conditions with maximum efficiency. The boundaries between axial, mixed, and radial machines are somewhat vague, but the value of the specific speed provides some guidance as to which machine would be most suitable. The final choice

Figure 14.12

Optimum efficiency and impeller design versus specific speed.

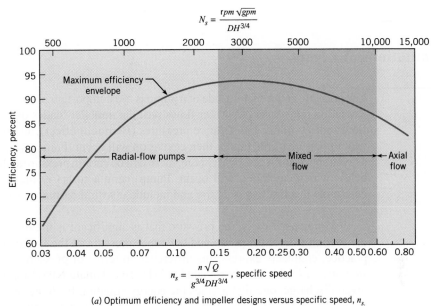

$$N_s = \frac{rpm \sqrt{gpm}}{DH^{3/4}}$$

$$n_s = \frac{n\sqrt{Q}}{g^{3/4}DH^{3/4}}, \text{ specific speed}$$

(a) Optimum efficiency and impeller designs versus specific speed, n_s.

(b) Radial-flow impellers.   (c) Mixed-flow impellers.   (d) Axial flow.

would depend on which pumps were commercially available as well as their purchase price, operating cost, and reliability.

It should be noted that the specific speed traditionally used for pumps in the United States is defined as $N_s = NQ^{1/2}/\Delta H^{3/4}$. Here the speed N is in revolutions per minute, Q is in gallons per minute, and ΔH is in feet. This form is not dimensionless. Therefore its values are much larger than those found for n_s (the conversion factor is 17,200). Most texts and references published before the introduction of the SI system of units use this traditional definition for specific speed.

Example 14.6 illustrates the use of specific speed to select a pump type.

EXAMPLE 14.6   PUMP SELECTION USING SPECIFIC SPEED

What type of pump should be used to pump water at the rate of 10 cfs and under a head of 600 ft? Assume $N = 1100$ rpm.

Problem Definition

Situation: Select pump to pump water at 10 cfs under head of 600 ft.

Find: Type of pump for application.

Plan

1. Calculate specific speed.
2. Use Fig. 14.12 to select pump type.

Solution

1. Rotational rate in rps

$$n = \frac{1100}{60} = 18.33 \text{ rps}$$

Specific speed

$$n_s = \frac{n\sqrt{Q}}{(g\Delta H)^{3/4}}$$

$$= \frac{18.33 \text{ rps} \times (10 \text{ cfs})^{1/2}}{(32.2 \text{ ft/s}^2 \times 600 \text{ ft})^{3/4}} = 0.035$$

2. From Fig. 14.12 radial-flow pump is the best choice.

14.5

Suction Limitations of Pumps

The pressure at the suction side of a pump is most important because there is the possibility that cavitation may occur. As water flows past the impeller blades of a pump, local high-velocity flow zones produce low relative pressures (Bernoulli effect), and if these pressures reach the vapor pressure of the liquid, then cavitation will occur. For a given type of pump operating at a given speed and a given discharge, there will be certain pressure at the suction side of the pump below which cavitation will occur. Pump manufacturers in their testing procedures always determine this limiting pressure and include it with their pump performance curves.

More specifically, the pressure that is significant is the difference in pressure between the suction side of the pump and the vapor pressure of the liquid being pumped. Actually, in practice, engineers express this difference in terms of pressure head, called the *net positive suction head,* which is abbreviated NPSH. To calculate NPSH for a pump that is delivering a given discharge, one first applies the energy equation from the reservoir from which water is being pumped to the section of the intake pipe at the suction side of the pump. Then one subtracts the vapor pressure head of the water to determine NPSH.

In Fig. 14.13, points 1 and 2 are the points between which the energy equation would be written to evaluate NPSH.

A more general parameter for indicating susceptibility to cavitation is specific speed. However, instead of using head produced (ΔH), one uses NPSH for the variable to the 3/4 power. This is

$$n_{ss} = \frac{nQ^{1/2}}{g^{3/4}(\text{NPSH})^{3/4}}$$

Here n_{ss} is called the suction specific speed. The more traditional suction specific speed used in the United States is $N_{ss} = NQ^{1/2}/(\text{NPSH})^{3/4}$, where N is in rpm, Q is in gallons per minute (gpm), and NPSH is in feet. Analyses of data from pump tests show that the value of the suction specific speed is a good indicator of whether cavitation may be expected. For example, the Hydraulic Institute (5) indicates that the critical value of N_{ss} is 8500. The reader is directed to manufacturer's data or the Hydraulic Institute for more details about critical NPSH or N_{ss}.

An analysis to find NPSH for a pump system is illustrated in Example 14.7.

Figure 14.13

Locations used to evaluate NPSH for a pump.

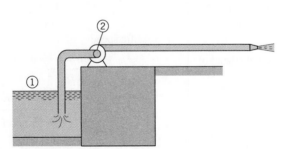

EXAMPLE 14.7   NET POSITIVE SUCTION HEAD

In Fig. 14.13 the pump delivers 2 cfs flow of 80°F water, and the intake pipe diameter is 8 in. The pump intake is located 6 ft above the water surface level in the reservoir. The pump operates at 1750 rpm. What is the net positive suction head and the traditional suction specific speed for these conditions?

Problem Definition

Situation: Pump delivers 2 cfs flow of 80°F water.

Find:

1. Net positive suction head (NPSH).
2. Traditional suction specific speed (N_{ss}).

Assumptions:

1. Atmospheric pressure is 14.7 psi.
2. Entrance loss coefficient = 0.10.
3. Bend loss coefficient = 0.20.

Properties: Table A.5, (Water at 80°F) $\gamma = 62.2$ lbf/ft^3, and $p_{vap} = 0.506$ psi.

Plan

The net positive suction head is the difference between pressure at pump inlet and the vapor pressure.

1. Determine the atmospheric pressure in head of water for reservoir surface.
2. Determine velocity in 8 in. pipe.
3. Apply the energy equation [Eq. (7.29)] between the reservoir and pump entrance.
4. Calculate NPSH.
5. Calculate N_{ss} with $N_{ss} = (NQ^{1/2})/(\text{NPSH})^{3/4}$.

Solution

1. Pressure head at reservoir

$$\frac{p_1}{\gamma} = \frac{14.7 \text{ lbf/in}^2 \times 144 \ (\text{in}^2/\text{ft}^2)}{62.2 \text{ lbf/ft}^3} = 34 \text{ ft}$$

2. Velocity in pipe

$$V_2 = \frac{Q}{A} = \frac{2 \text{ cfs}}{\pi \times ((4 \text{ in})/12)^2} = 5.73 \text{ ft/s}$$

3. Energy equation between points 1 and 2:

$$\frac{p_1}{\gamma} + \frac{V_1^2}{2g} + z_1 = \frac{p_2}{\gamma} + \frac{V_2^2}{2g} + z_2 + \sum h_L$$

- Input values

$$V_1 = 0, \quad z_1 = 0, \quad z_2 = 6$$

- Head loss

$$\sum h_L = (0.1 + 0.2)\frac{V_2^2}{2g}$$

- Head at pump entrance

$$\frac{p_2}{\gamma} = \frac{p_1}{\gamma} - z_2 - \frac{V_2^2}{2g}(1 + 0.3)$$

$$= 34 - 6 - 1.3 \times \frac{5.73^2}{2 \times 32.2} = 27.3 \text{ ft}$$

4. Vapor pressure in feet of head

$$0.506 \times 144/62.2 = 1.17 \text{ ft.}$$

Net positive suction head

$$\text{NPSH} = 27.3 - 1.17 = 26.1 \text{ ft}$$

5. Traditional suction specific speed

$$Q = 2 \text{ cfs} = 898 \text{ gpm}$$

$$N_{ss} = (1750)(898)^{1/2}/(26.1)^{3/4} = \boxed{4540}$$

Review

1. For a typical single-stage centrifugal pump with an intake diameter of 8 in. and pumping 2 cfs, the critical NPSH is normally about 10 ft; therefore, the pump of this example is operating well within the safe range with respect to cavitation susceptibility.

2. This value of N_{ss} is much below the critical limit of 8500; therefore, it is in a safe operating range so far as cavitation is concerned.

A typical pump performance curve for a centrifugal pump that would be supplied by a pump manufacturer is shown in Fig. 14.14. The solid lines labeled from 5 in. to 7 in. represent different impeller sizes that can be accommodated by the pump housing. These curves give the head delivered as a function of discharge. The dashed lines represent the power

Figure 14.14

*Centrifugal pump
performance curve.
[After McQuiston and
Parker (6). Used with
permission of John Wiley
and Sons.]*

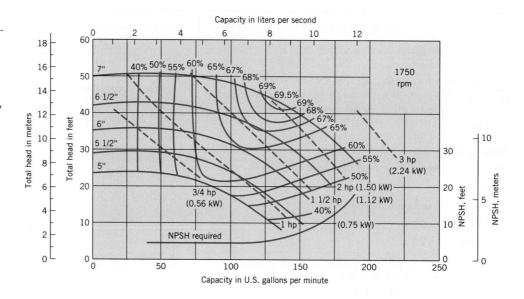

required by the pump for a given head and discharge. Lines of constant efficiency are also shown. Obviously, when selecting an impeller one would like to have the operating point as close as possible to the point of maximum efficiency. The NPSH value gives the minimum head (absolute head) at the pump intake for which the pump will operate without cavitation.

Viscous Effects

In the foregoing sections, similarity parameters were developed to predict prototype results from model tests, neglecting viscous effects. The latter assumption is not necessarily valid, especially if the model is quite small. To minimize the viscous effects in modeling pumps, the Hydraulic Institute standards (5) recommend that the size of the model be such that the model impeller is not less than 30 cm in diameter. These same standards state that "the model should have complete geometric similarity with the prototype, not only in the pump proper, but also in the intake and discharge conduits."

Even with complete geometric similarity, one can expect the model to be less efficient than the prototype. An empirical formula proposed by Moody (7) is used for estimating prototype efficiencies of radial- and mixed-flow pumps and turbines from model efficiencies. That formula is

$$\frac{1 - e_1}{1 - e} = \left(\frac{D}{D_1}\right)^{1/5} \tag{14.18}$$

Here e_1 is the efficiency of the model and e is the efficiency of the prototype.

Example 14.8 shows how to estimate the efficiency due to viscous effects.

EXAMPLE 14.8   VISCOUS EFFECTS ON PUMP EFFICIENCY

A model having an impeller diameter of 45 cm is tested and found to have an efficiency of 85%. If a geometrically similar prototype has an impeller diameter of 1.80 m, estimate its efficiency when it is operating under conditions that are dynamically similar to those in the model test $(C_{Q,\text{model}} = C_{Q,\text{prototype}})$.

Problem Definition

Situation: Model with 45 cm impeller has 85% efficiency.

Find: Efficiency of pump with 1.6 m impeller.

Assumptions: Efficiency difference due to viscous effects.

Plan

Use Eq. (14.18) to determine viscous effects.

Solution

Efficiency

$$e = 1 - \frac{1 - e_1}{(D/D_1)^{1/5}} = 1 - \frac{0.15}{1.32} = 1 - 0.11 = 0.89$$

or

$$e = 89\%$$

Centrifugal Compressors

Centrifugal compressors are similar in design to centrifugal pumps. Because the density of the air or gases used is much less than the density of a liquid, the compressor must rotate at much higher speeds than the pump to effect a sizable pressure increase. If the compression process were isentropic and the gases ideal, the power necessary to compress the gas from p_1 to p_2 would be

$$P_{\text{theo}} = \frac{k}{k-1} Q_1 p_1 \left[\left(\frac{p_2}{p_1} \right)^{(k-1)/k} - 1 \right] \tag{14.19}$$

where Q_1 is the volume flow rate into the compressor and k is the ratio of specific heats. The power calculated using Eq. (14.19) is referred to as the *theoretical adiabatic power*. The efficiency of a compressor with no water cooling is defined as the ratio of the theoretical adiabatic power to the actual power required at the shaft. Ordinarily the efficiency improves with higher inlet-volume flow rates, increasing from a typical value of 0.60 at 0.6 m³/s to 0.74 at 40 m³/s. Higher efficiencies are obtainable with more expensive design refinements.

Example 14.9 shows how to calculate shaft power required to operate a compressor.

EXAMPLE 14.9   CENTRIFUGAL COMPRESSOR

Determine the shaft power required to operate a compressor that compresses air at the rate of 1 m³/s from 100 kPa to 200 kPa. The efficiency of the compressor is 65%.

Problem Definition

Situation: Compressor compresses air at 1 m³/s from 100 kPa to 200 kPa.

Find: Required shaft power (in kW).

Sketch:

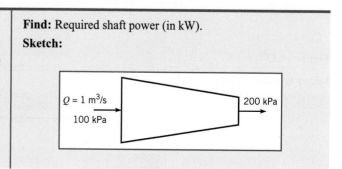

Properties: From Table A.2, $k = 1.4$.

Plan

1. Use Eq. (14.19) to calculate theoretical power.
2. Divide theoretical power by efficiency to find shaft (required) power.

Solution

1. Theoretical power

$$P_{theo} = \frac{k}{k-1} Q_1 P_1 \left[\left(\frac{p_2}{p_1} \right)^{(k-1)/k} - 1 \right]$$

$$= (3.5)(1 \text{ m}^3/\text{s})(10^5 \text{ N/m}^2)[(2)^{0.286} - 1]$$

$$= 0.767 \times 10^5 \text{ N} \cdot \text{m/s} = 76.7 \text{ kW}$$

2. Shaft power

$$P_{shaft} = \frac{76.7}{0.65} \text{ kW} = \boxed{118 \text{ kW}}$$

Cooling is necessary for high-pressure compressors because of the high gas temperatures resulting from the compression process. Cooling can be achieved through the use of water jackets or intercoolers that cool the gases between stages. The efficiency of water-cooled compressors is based on the power required to compress ideal gases isothermally, or

$$P_{theo} = p_1 Q_1 \ln \frac{p_2}{p_1} \tag{14.20}$$

which is usually called the *theoretical isothermal power*. The efficiencies of water-cooled compressors are generally lower than those of noncooled compressors. If a compressor is cooled by water jackets, its efficiency characteristically ranges between 55% and 60%. The use of intercoolers results in efficiencies from 60% to 65%.

Application to Fluid Systems

The selection of a pump, fan, or compressor for a specific application depends on the desired flow rate. This process requires the acquisition or generation of a system curve for the flow system of interest and a performance curve for the fluid machine. The intersection of these two curves provides the operating point as discussed in Chapter 10.

As an example, consider using the centrifugal pump with the characteristics shown in Fig. 14.14 to pump water at 60°F from a wall into a tank as shown in Fig. 14.15. A pumping capacity of at least 80 gpm is required. Two hundred feet of standard schedule-40 2 inch galvanized iron pipe are to be used. There is a check valve in the system as well as an open gate valve. There is a 20 ft elevation between the well and the top of the fluid in the tank. Applying the energy equation developed in Chapter 7, the head required by the pump is

$$h_p = \Delta z + \frac{V^2}{2g} \left(\frac{fL}{D} + \sum K_L \right)$$

Figure 14.15

System for pumping water from a well into a tank.

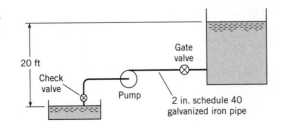

Figure 14.16

*System and pump
performance curves for
pumping application.*

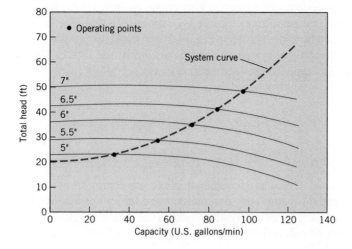

where K_L represents the head loss coefficients for the entrance, check valve, gate valve, and sudden-expansion loss entering the tank. Using representative values for the loss coefficients and evaluating the friction factor from the Moody diagram in Chapter 10 leads to

$$h_p = 20 + 0.00305Q^2$$

where Q is the flow rate in gpm. This is the system curve.

The result of plotting the system curve on the pump performance curves is shown in Fig. 14.16. The locations where the lines cross are the operating points. One notes that a discharge of just over 80 gpm is achieved with the 6.5 in. impeller. Also, referring back to Fig. 14.14, the efficiency at this point is about 62%. To ensure that the design requirements are satisfied, the engineer may select the larger impeller, which has an operating point of 95 gpm. If the pump is to be used in continuous operation and the efficiency is important to operating costs, the engineer may choose to consider another pump that would have a higher efficiency at the operation point. An engineer experienced in the design of pump systems would be very familiar with the trade-offs for economy and performance and could make a design decision relatively quickly.

In some systems it may be advantageous to use two pumps in series or in parallel. If two pumps are used in series, the performance curve is the sum of the pump heads of the two machines at the same flow rate, as shown in Fig. 14.17a. This configuration would be desirable

Figure 14.17

*Pump performance
curves for pumps
connected in series (a)
and in parallel (b).*

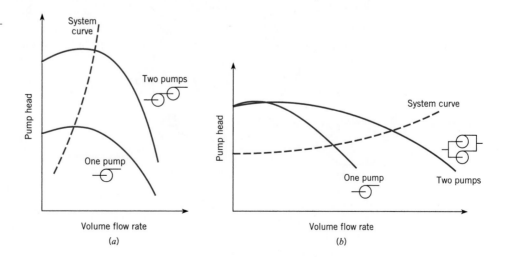

for a flow system with a steep system curve, as shown in the figure. If two pumps are connected in parallel, the performance curve is obtained by adding the flow rates of the two pumps at the same pump heads, as shown in Fig. 14.17*b*. This configuration would be advisable for flow systems with shallow system curves, as shown in the figure. The concepts presented here for pumps also apply to fans and compressors.

Turbines

A *turbine* is defined as a machine that extracts energy from a moving fluid. Much of the basic theory and most similarity parameters used for pumps also apply to turbines. However, there are some differences in physical features and terminology. The details of the flow through the impellers of radial-flow machines have not yet been considered. These topics will now be addressed.

The two main categories of hydraulic machines are the *impulse* and *reaction* turbines. In a reaction turbine, the water flow is used to rotate a turbine wheel or runner through the action of vanes or blades attached to the wheel. When the blades are oriented like a propeller, the flow is axial and the machine is called a *Kaplan turbine*. When the vanes are oriented like an impeller in a centrifugal pump, the flow is radial and the machine is called a *Francis turbine*. In an impulse turbine, the water accelerates through a nozzle and impinges on vanes attached to the rim of the wheel. This machine is called a *Pelton wheel*.

Impulse Turbine

In the impulse turbine a jet of fluid issuing from a nozzle impinges on vanes of the turbine wheel, or *runner*, thus producing power as the runner rotates (see Fig. 14.18). Figure 14.19 shows a runner for the Henry Borden hydroelectric plant in Brazil. The primary feature of the impulse turbine with respect to fluid mechanics is the power production as the jet is deflected

Figure 14.18

Impulse turbine.

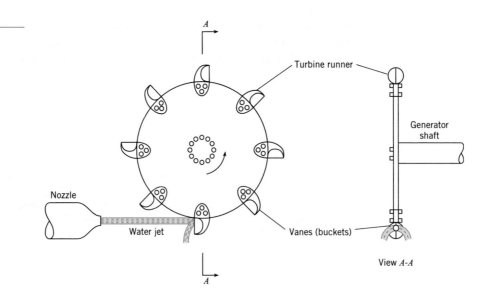

by the moving vanes. When the momentum equation is applied to this deflected jet, it can be shown [see Daugherty and Franzini (4)] for idealized conditions that the maximum power will be developed when the vane speed is one-half of the initial jet speed. Under such conditions the exiting jet speed will be zero—all of the kinetic energy of the jet will have been expended in driving the vane. Thus if one applies the energy equation, Eq. (7.29), between the incoming jet and the exiting fluid (assuming negligible head loss and negligible kinetic energy at exit), it is found that the head given up to the turbine is $h_t = (V_j^2/2g)$, and the power thus developed is

$$P = Q\gamma h_t \tag{14.21}$$

where Q is the discharge of the incoming jet, γ is the specific weight of jet fluid, and $h_t = V_j^2/2g$, or the velocity head of the jet. Thus Eq. (14.21) reduces to

$$P = \rho Q \frac{V_j^2}{2} \tag{14.22}$$

To obtain the torque on the turbine shaft, the angular-momentum equation (6.27) is applied to a control volume, as shown in Fig. 14.20. For steady flow

$$\sum M = \sum_{cs} \mathbf{r}_o \times (\dot{m}_o \mathbf{v}_o) - \sum_{cs} \mathbf{r}_i \times (\dot{m}_i \mathbf{v}_i)$$

Figure 14.20

Control-volume approach for the impulse turbine using the angular-momentum principle.

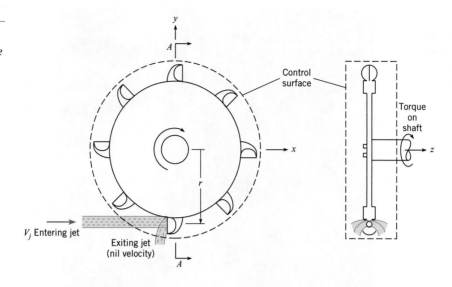

Generally it is assumed that the exiting fluid has negligible angular momentum. The moment acting on the system is the torque T acting on the shaft. Thus the angular-momentum equation reduces to

$$T = -\dot{m}rV_j \qquad (14.23)$$

The mass flow rate across the control surface is ρQ, so the torque is

$$T = -\rho Q V_j r$$

The minus sign indicates that the torque applied to the system (to keep it rotating at constant angular velocity) is in the clockwise direction. However, the torque applied by the system to the shaft is in the counterclockwise direction, which is the direction of wheel rotation, so

$$T = \rho Q V_j r \qquad (14.24)$$

The power developed by the turbine is $T\omega$, or

$$P = \rho Q V_j r \omega \qquad (14.25)$$

Furthermore, if the velocity of the turbine vanes is $(1/2)V_j$ for maximum power, as noted earlier, then $P = \rho Q V_j^2/2$, which is the same as Eq. (14.22).

The calculation of torque for an impulse turbine is illustrated in Example 14.10.

EXAMPLE 14.10 IMPULSE TURBINE

What power in kilowatts can be developed by the impulse turbine shown if the turbine efficiency is 85%? Assume that the resistance coefficient f of the penstock is 0.015 and the head loss in the nozzle itself is negligible. What will be the angular speed of the wheel, assuming ideal conditions $(V_j = 2V_{bucket})$, and what torque will be exerted on the turbine shaft?

Problem Definition

Situation: Impulse turbine with 85% efficiency.

Find:

1. Power (in kW) developed by turbine.
2. Angular (in rpm) speed of wheel for maximum efficiency.
3. Torque (in kN · m) on turbine shaft.

Assumptions:

1. There is no entrance loss.
2. Head loss in nozzle is negligible.
3. Water density is 1000 kg / m³.

Sketch:

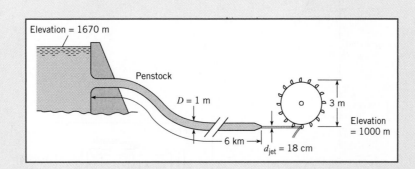

Elevation = 1670 m

Penstock

$D = 1$ m

3 m

Elevation = 1000 m

6 km

$d_{jet} = 18$ cm

Plan

1. Apply energy equation, Eq. (7.29), to find nozzle velocity.
2. Use Eq. (14.22) for power.
3. For maximum efficiency, $\omega r = (V_j/2)$.
4. Calculate torque from $P = T\omega$.

Solution

1. Energy equation

$$\frac{p_1}{\gamma} + \frac{V_1^2}{2g} + z_1 = \frac{p_j}{\gamma} + \frac{V_j^2}{2g} + z_j + h_L$$

- Values in energy equation

$p_1 = 0$, $z_1 = 1670$ m, $V_1 = 0$, $p_j = 0$, $z_j = 1000$ m

- Penstock-supply pipe velocity ratio

$$V_{penstock} = \frac{V_j A_j}{A_{penstock}} = V_j \left(\frac{0.18 \text{ m}}{1 \text{ m}}\right)^2 = 0.0324 V_j$$

- Head loss

$$h_L = f \frac{L}{D} \frac{1}{2g} V_{penstock}^2$$

$$= \frac{0.015 \times 6000}{1}(0.0324)^2 \frac{V_j^2}{2g} = 0.094 \frac{V_j^2}{2g}$$

- Jet velocity

$$z_1 - z_2 = 1.094 \frac{V_j^2}{2g}$$

$$V_j = \left(\frac{2 \times 9.81 \text{ m/s}^2 \times 670 \text{ m}}{1.094}\right)^{1/2} = 109.6 \text{ m/s}$$

2. Gross power

$$P = Q\gamma \frac{V_j^2}{2g} = \frac{\gamma A_j V_j^3}{2g}$$

$$= \frac{9810(\pi/4)(0.18)^2(109.6)^3}{2 \times 9.81} = 16{,}750 \text{ kW}$$

Power delivered

$$P = 16{,}750 \times \text{efficiency} = \boxed{14{,}240 \text{ kW}}$$

3. Angular speed of wheel

$$V_{bucket} = \tfrac{1}{2}(109.6 \text{ m/s}) = 54.8 \text{ m/s}$$

$$r\omega = 54.8 \text{ m/s}$$

$$\omega = \frac{54.8 \text{ m/s}}{1.5 \text{ m}} = 36.5 \text{ rad/s}$$

Wheel speed

$$N = (36.5 \text{ rad/s})\frac{1 \text{ rev}}{2\pi \text{ rad}}(60 \text{ s/min}) = \boxed{349 \text{ rpm}}$$

4. Torque

$$T = \frac{\text{power}}{\omega} = \frac{14{,}240 \text{ kW}}{36.5 \text{ rad/s}} = \boxed{390 \text{ kN} \cdot \text{m}}$$

Reaction Turbine

In contrast to the impulse turbine, where a jet under atmospheric pressure impinges on only one or two vanes at a time, flow in a reaction turbine is under pressure and reacts on all vanes of the impeller turbine simultaneously. Also, this flow completely fills the chamber in which the impeller is located (see Fig. 14.21). There is a drop in pressure from the outer radius of the impeller, r_1, to the inner radius, r_2. This is another point of difference with the impulse turbine,

Figure 14.21

Schematic view of a reaction-turbine installation. (a) Elevation view. (b) Plan view, section A-A.

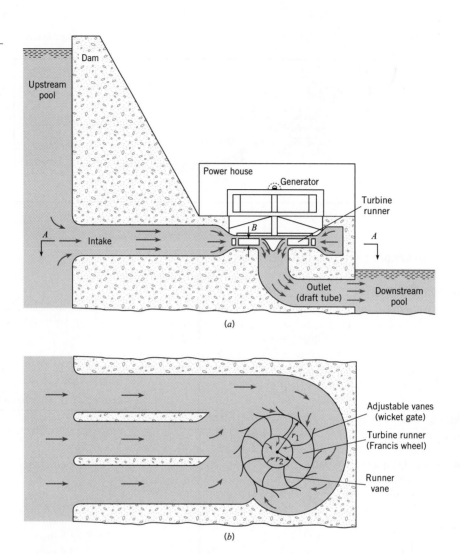

Figure 14.21

Schematic view of a reaction-turbine installation. (a) Elevation view. (b) Plan view, section A-A.

in which the pressure is the same for the entering and exiting flows. The original form of the reaction turbine, first extensively tested by J. B. Francis, had a completely radial-flow impeller (Fig. 14.22). That is, the flow passing through the impeller had velocity components only in a plane normal to the axis of the runner. However, more recent impeller designs, such as the mixed-flow and axial-flow types, are still called reaction turbines.

Torque and Power Relations for the Reaction Turbine

As for the impulse turbine, the angular-momentum equation is used to develop formulas for the torque and power for the reaction turbine. The segment of turbine runner shown in Fig. 14.22 depicts the flow conditions that occur for the entire runner. The guide vanes outside the runner itself cause the fluid to have a tangential component of velocity around the entire circumference of the runner. Thus the fluid has an initial amount of angular momentum with respect to the turbine axis when it approaches the turbine runner. As the fluid passes through the passages of the runner, the runner vanes effect a change in the magnitude and direction of its velocity. Thus the angular momentum of the fluid is changed, which produces a torque on the runner. This torque drives the runner, which, in turn, generates power.

Figure 14.22

Velocity diagrams for the impeller for a Francis turbine.

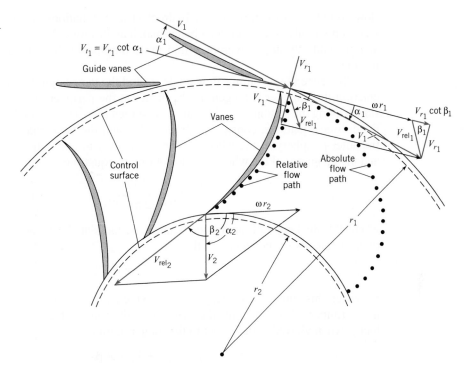

To quantify the above, let V_1 and α_1 represent the incoming velocity and the angle of the velocity vector with respect to a tangent to the runner, respectively. Similar terms at the inner-runner radius are V_2 and α_2. Applying the angular-momentum equation for steady flow, Eq. (6.27), to the control volume shown in Fig. 14.22 yields

$$T = \dot{m}(-r_2 V_2 \cos\alpha_2) - \dot{m}(-r_1 V_1 \cos\alpha_1)$$
$$= \dot{m}(r_1 V_1 \cos\alpha_1 - r_2 V_2 \cos\alpha_2) \tag{14.26}$$

The power from this turbine will be $T\omega$, or

$$P = \rho Q \omega (r_1 V_1 \cos\alpha_1 - r_2 V_2 \cos\alpha_2) \tag{14.27}$$

Equation (14.27) shows that the power production is a function of the directions of the flow velocities entering and leaving the impeller—that is, α_1 and α_2.

It is interesting to note that even though the pressure varies within the flow in a reaction turbine, it does not enter into the expressions derived using the angular-momentum equation. The reason it does not appear is that the chosen outer and inner control surfaces are concentric with the axis about which the moments and angular momentum are evaluated. The pressure forces acting on these surfaces all pass through the given axis; therefore they do not produce moments about the given axis.

Vane Angles

It should be apparent that the head loss in a turbine will be less if the flow enters the runner with a direction tangent to the runner vanes than if the flow approaches the vane with an angle of attack. In the latter case, separation will occur with consequent head loss. Thus vanes of an impeller designed for a given speed and discharge and with fixed guide vanes will have a particular optimum blade angle β_1. However, if the discharge is changed from the condition of the original design, the guide vanes and impeller vane angles will not "match" the new flow condition. Most turbines for hydroelectric installations are made with movable guide vanes on the inlet side to effect a better match at all flows. Thus α_1 is increased or decreased automatically through governor action to accommodate fluctuating power demands on the turbine.

To relate the incoming-flow angle α_1 and the vane angle β_1, first assume that the flow entering the impeller is tangent to the blades at the periphery of the impeller. Likewise, the flow leaving the stationary guide vane is assumed to be tangent to the guide vane. To develop the desired equations, consider both the radial and the tangential components of velocity at the outer periphery of the wheel ($r = r_1$). It is easy to compute the radial velocity, given Q and the geometry of the wheel, by the continuity equation:

$$V_{r_1} = \frac{Q}{2\pi r_1 B} \tag{14.28}$$

where B is the height of the turbine blades. The tangential (tangent to the outer surface of the runner) velocity of the incoming flow is

$$V_{t_1} = V_{r_1} \cot \alpha_1 \tag{14.29}$$

However, this tangential velocity is equal to the tangential component of the relative velocity in the runner, $V_{r_1} \cot \beta_1$, plus the velocity of the runner itself, ωr_1. Thus the tangential velocity, when viewed with respect to the runner motion, is

$$V_{t_1} = r_1 \omega + V_{r_1} \cot \beta_1 \tag{14.30}$$

Now, eliminating V_{t_1} between Eqs. (14.29) and (14.30) results in

$$V_{r_1} \cot \alpha_1 = r_1 \omega + V_{r_1} \cot \beta_1 \tag{14.31}$$

Equation (14.31) can be rearranged to yield

$$\alpha_1 = \text{arccot}\left(\frac{r_1 \omega}{V_{r_1}} + \cot \beta_1\right) \tag{14.32}$$

Example 14.11 illustrates how to calculate the inlet blade angle to avoid separation.

EXAMPLE 14.11 FRANCIS TURBINE

A Francis turbine is to be operated at a speed of 600 rpm and with a discharge of 4.0 m³/s. If $r_1 = 0.60$ m, $\beta_1 = 110°$, and the blade height B is 10 cm, what should be the guide vane angle α_1 for a nonseparating flow condition at the runner entrance?

Problem Definition

Situation: Francis turbine with speed of 600 rpm and discharge of 4.0 m³/s.

Find: Inlet guide vane angle, α_1.

Plan

Use Eq. (14.32) for inlet guide angle.

Solution

Inlet guide vane angle

$$\alpha_1 = \text{arccot}\left(\frac{r_1 \omega}{V_{r_1}} + \cot \beta_1\right)$$

$$r_1 \omega = 0.6 \times 600 \text{ rpm} \times 2\pi \text{ rad/rev} \times 1/60 \text{ min/s}$$

$$= 37.7 \text{ m/s}$$

Radial velocity at inlet

$$V_{r_1} = \frac{Q}{2\pi r_1 B} = \frac{4.00 \text{ m}^3/\text{s}}{2\pi \times 0.6 \text{ m} \times 0.10 \text{ m}} = 10.61 \text{ m/s}$$

$$\cot \beta_1 = \cot(110°) = -0.364$$

$$\alpha_1 = \text{arccot}\left(\frac{37.7}{10.61} - 0.364\right) = \boxed{17.4°}$$

Specific Speed for Turbines

Because of the attention focused on the production of power by turbines, the specific speed for turbines is defined in terms of power:

$$n_s = \frac{nP^{1/2}}{g^{3/4}\gamma^{1/2}h_t^{5/4}}$$

It should also be noted that large water turbines are innately more efficient than pumps. The reason for this is that as the fluid leaves the impeller of a pump, it decelerates appreciably over a relatively short distance. Also, because guide vanes are generally not used in the flow passages with pumps, large local velocity gradients develop, which in turn cause intense mixing and turbulence, thereby producing large head losses. In most turbine installations, the flow that exits the turbine runner is gradually reduced in velocity through a gradually expanding *draft tube,* thus producing a much smoother flow situation and less head loss than for the pump. For additional details of hydropower turbines, see Daugherty and Franzini (4).

Gas Turbines

The conventional gas turbine consists of a compressor that pressurizes the air entering the turbine and delivers it to a combustion chamber. The high-temperature, high-pressure gases resulting from combustion in the combustion chamber expand through a turbine, which both drives the compressor and delivers power. The theoretical efficiency (power delivered/rate of energy input) of a gas turbine depends on the pressure ratio between the combustion chamber and the intake; the higher the pressure ratio, the higher the efficiency. The reader is directed to Cohen et al. (8) for more detail.

Wind Turbines

Wind energy is discussed frequently as an alternative energy source. The application of wind turbines* as potential sources for power becomes more attractive as utility power rates increase and the concern over greenhouse gases grows. In many European countries, especially northern Europe, the wind turbine is playing an ever-increasing role in power generation.

In essence, the wind turbine is just a reverse application of the process of introducing energy into an airstream to derive a propulsive force. The wind turbine extracts energy from the wind to produce power. There is one significant difference, however. The theoretical upper limit of efficiency of a propeller supplying energy to an airstream is 100%; that is, it is theoretically possible, neglecting viscous and other effects, to convert all the energy supplied to a propeller into energy of the airstream. This is not the case for a wind turbine.

A sketch of a horizontal-axis wind turbine is shown in Fig. 14.23. The wind blows along the axis of the turbine. The area of the circle traced out by the rotating blades is the *capture area.* The power associated with the wind passing through the capture area is

$$P = \rho Q \frac{V^2}{2} = \rho A \frac{V^3}{2} \tag{14.33}$$

where ρ is the air density and V is the wind speed. In an analysis attributed to [Glauert/Betz (9)], the theoretical maximum power attainable from a wind turbine is 16/27 or 59.3% of this power or

$$P_{\text{max}} = \frac{16}{27}\left(\frac{1}{2}\rho V^3 A\right) \tag{14.34}$$

* The word "wind turbine" is used to convey the idea of conversion of wind to electrical energy. A windmill converts wind energy to mechanical energy.

Figure 14.23

Horizontal-axis wind turbine showing capture area.

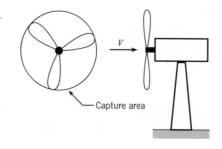

Other factors, such as swirl of the airstream and viscous effects, further reduce the power achievable from a wind turbine.

The power output of any wind turbine is related to the wind speed through the wind-turbine power curve. A typical curve is shown in Fig. 14.24. This curve can usually be obtained from the manufacturer. The wind turbine is inoperative below the cut-in speed. After cut-in, the power increases with wind speed reaching a maximum value, which is the rated power output for the turbine. Engineering design and safety constraints impose an upper limit on the rotational velocity and establish the cut-out speed. A braking system is used to prevent operation of the wind turbine beyond this velocity.

The conventional horizontal-axis wind turbine has been the focus of most research and design. Considerable effort has also been devoted to assessment of the Savonius rotor and the Darrieus turbine, both of which are vertical-axis turbines, as shown in Fig. 14.25. The Savonius rotor consists of two curved blades forming an S-shaped passage for the air flow. The Darrieus turbine consists of two or three airfoils attached to a vertical shaft; the unit resembles an egg beater. The advantage of vertical-axis turbines is that their operation is independent of wind direction. The Darrieus wind turbine is considered superior in performance but has a disadvantage in that it is not self-starting. Frequently, a Savonius rotor is mounted on the axis of a Darrieus turbine to provide the starting torque.

For more information on wind turbines and wind turbine systems, refer to *Wind Energy Explained* (10). Considerable information on wind turbines is also available from the Internet.

Figure 14.24

Typical wind turbine power curve.

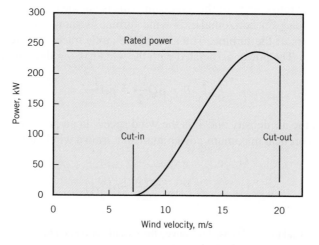

Figure 14.25

*Wind trubine
configurations.
(a) Savonius rotor.
(b) Darrieus turbine.*

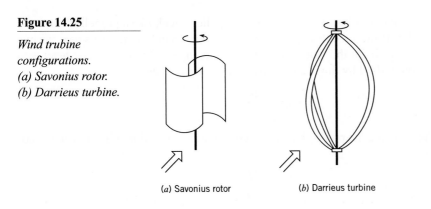

(*a*) Savonius rotor   (*b*) Darrieus turbine

**EXAMPLE 14.12 CAPTURE AREA OF
WIND TURBINE**

Calculate the minimum capture area necessary for a windmill that has to operate five 100-watt bulbs if the wind velocity is 20 km/h and the air density is 1.2 kg/m³.

Problem Definition

Situation: A wind turbine produces of 500 watts.

Find: Capture area of windmill.

Plan

Use equation for maximum power of windmill.

Solution

Capture area for maximum power

$$A = P_{max} \frac{54}{16} \frac{1}{\rho V^3}$$

Wind velocity in m/s

$$20 \text{ km/h} = \frac{20 \times 1000}{3600} = 5.56 \text{ m/s}$$

Minimum capture area

$$A = 500 \text{ W} \times \frac{54}{16} \times \frac{1}{1.2 \text{ kg/m}^3 \times (5.56 \text{ m/s})^3}$$

$$= \boxed{8.18 \text{ m}^2}$$

Review

This area corresponds to a windmill diameter of 3.23 m or about 10.6 ft.

Summary

The thrust of a propeller is calculated using

$$F_T = C_T \rho n^2 D^4$$

where ρ is the fluid density, n is the rotational rate of the propeller, and D is the propeller diameter. The thrust coefficient C_T is a function of the advance ratio V_0/nD. The efficiency of a propeller is the ratio of the power delivered by the propeller to the power provided to the propeller.

$$\eta = \frac{F_T V_0}{P}$$

An axial-flow pump, or blower, consists of an impeller, much like a propeller, mounted in a housing. In a radial-flow, or centrifugal pump, on the other hand, fluid enters near the eye of the impeller, passes through the vanes, and exits at the edge of the vanes. The head provided by a pump is quantified by the head coefficient, C_H, defined as

$$C_H = \frac{g \Delta H}{n^2 D^2}$$

where ΔH is the head across the pump. The head coefficient is a function of the discharge coefficient, which is

$$C_Q = \frac{Q}{nD^3}$$

where Q is the discharge. Pump performance curves show head delivered, power required, and efficiency as a function of discharge. The specific speed of a pump can be used to select an appropriate pump for a given application. Axial-flow pumps are best suited for high-discharge, low-head applications, whereas radial-flow pumps are best suited for low-discharge, high-head applications.

Turbines convert the energy associated with a moving fluid to shaft work. The impulse turbine consists of a liquid jet impinging on vanes of a turbine wheel or runner. A reaction turbine consists of a series of rotating vanes where liquid enters from the outside and exits at the center. The pressure on the vanes provides the torque for the power. Wind turbines consist of the conventional horizontal-axis design, the Darrieus turbine, or Savonius rotor. The maximum power derivable from a wind turbine is

$$P_{max} = \frac{16}{27} \left(\frac{1}{2} \rho V_0^3 A \right)$$

where A is the capture area of the wind turbine (projected area from direction of wind) and V_0 is the wind speed.

References

1. Weick, F. E. *Aircraft Propeller Design.* McGraw-Hill, New York, 1930.

2. Weick, Fred E. "Full Scale Tests on a Thin Metal Propeller at Various Pit Speeds." *NACA Report*, 302 (January 1929).

3. Stepanoff, A. J. *Centrifugal and Axial Flow Pumps*, 2nd ed. John Wiley, New York, 1957.

4. Daugherty, Robert L., and Joseph B. Franzini. *Fluid Mechanics with Engineering Applications.* McGraw-Hill, New York, 1957.

5. Hydraulic Institute. *Centrifugal Pumps.* Hydraulic Institute, Parsippany, NJ, 1994.

6. McQuiston, F. C., and J. D. Parker. *Heating, Ventilating and Air Conditioning.* John Wiley, New York, 1994.

7. Moody, L. F. "Hydraulic Machinery." In *Handbook of Applied Hydraulics*, ed. C. V. Davis. McGraw-Hill, New York, 1942.

8. Cohen, H., G. F. C. Rogers, and H. I. H. Saravanamuttoo. *Gas Turbine Theory.* John Wiley, New York, 1972.

9. Glauert, H. "Airplane Propellers." *Aerodynamic Theory*, vol. IV, ed. W. F. Durand. Dover, New York, 1963.

10. Manwell, J. F., J. G. McGowan, and A. L. Rogers. *Wind Energy Explained: Theory, Design and Application.* John Wiley, Chichester, UK, 2002.

Problems

Propellers

14.1 $\mathbb{P}\mathbb{Q}$ ◀ Explain why the thrust of a fixed-pitch propeller decreases with increasing forward speed.

14.2 $\mathbb{P}\mathbb{Q}$ ◀ What limits the rotational speed of a propeller?

14.3 What thrust is obtained from a propeller 3 m in diameter that has the characteristics given in Fig. 14.2 when the propeller is operated at an angular speed of 1400 rpm and an advance velocity of zero? Assume $\rho = 1.05 \ kg/m^3$.

14.4 What thrust is obtained from a propeller 3 m in diameter that has the characteristics given in Fig. 14.2 when the propeller is operated at an angular speed of 1400 rpm and an advance velocity of 80 km/h? What power is required to operate the propeller under these conditions? Assume $\rho = 1.05 \ kg/m^3$.

14.5 A propeller 8 ft in diameter has the characteristics shown in Fig. 14.2. What thrust is produced by the propeller when it is operating at an angular speed of 1000 rpm and a forward speed of 25 mph? What power input is required under these operating conditions? If the forward speed is reduced to zero, what is the thrust? Assume $\rho = 0.0024 \ slugs/ft^3$.

14.6 A propeller 8 ft in diameter, like the one for which characteristics are given in Fig. 14.2, is to be used on a swamp boat and is to operate at maximum efficiency when cruising. If the cruising speed is to be 30 mph, what should the angular speed of the propeller be? Assume $\rho = 0.0024 \ slugs/ft^3$.

14.7 For the propeller and conditions described in Prob. 14.6 determine the thrust and the power input.

14.8 A propeller is being selected for an airplane that will cruise at 2000 m altitude, where the pressure is 60 kPa absolute and the temperature is 10°C. The mass of the airplane is 1200 kg, and the planform area of the wing is 10 m². The lift-to-drag ratio is 30:1. The lift coefficient is 0.4. The engine speed at cruise conditions is 3000 rpm. The propeller is to operate at maximum efficiency, which corresponds to a thrust coefficient of 0.025. Calculate the diameter of the propeller and the speed of the aircraft.

14.9 If the tip speed of a propeller is to be kept below $0.9c$, where c is the speed of sound, what is the maximum allowable angular speed of propellers having diameters of 2 m (6.56 ft), 3 m (9.84 ft), and 4 m (13.12 ft)? Take the speed of sound as 335 m/s (1099 ft/s).

14.10 A propeller 2 m in diameter, like the one for which characteristics are given in Fig. 14.2, is to be used on a swamp boat and is to operate at maximum efficiency when cruising. If the cruising speed is to be 40 km/h, what should the angular speed of the propeller be?

14.11 For the propeller and conditions described in Prob. 14.10, determine the thrust and the power input. Assume $\rho = 1.2 \ kg/m^3$.

14.12 A propeller 2 m in diameter and like the one for which characteristics are given in Fig. 14.2 is used on a swamp boat. If the angular speed is 1000 rpm and if the boat and passengers have a combined mass of 300 kg, estimate the initial acceleration of the boat when starting from rest. Assume $\rho = 1.1 \ kg/m^3$.

Axial Flow Pumps and Fans

14.13 PQ◀ Axial-flow pumps are best suited for what conditions of head produced and discharge?

14.14 PQ◀ For an axial pump how does the head produced by the pump and the power required to operate a pump vary with flow rate through the pump?

14.15 If a pump having the characteristics shown in Fig. 14.6 has a diameter of 40 cm and is operated at a speed of 1000 rpm, what will be the discharge when the head is 3 m?

14.16 The pump used in the system shown has the characteristics given in Fig. 14.7. What discharge will occur under the conditions shown, and what power is required?

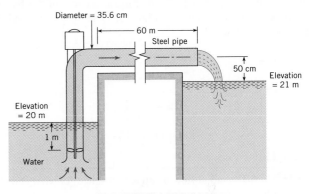

PROBLEMS 14.16, 14.17

14.17 If the conditions are the same as in Prob. 14.16 except that the speed is increased to 900 rpm, what discharge will occur, and what power is required for the operation?

14.18 For a pump having the characteristics given in Fig. 14.6 or 14.7, what water discharge and head will be produced at maximum efficiency if the pump diameter is 20 in. and the angular speed is 1100 rpm? What power is required under these conditions?

14.19 A pump has the characteristics given by Fig. 14.6. What discharge and head will be produced at maximum efficiency if the pump size is 50 cm and the angular speed is 45 rps? What power is required when pumping water at 10°C under these conditions?

14.20 For a pump having the characteristics of Fig. 14.6, plot the head-discharge curve if the pump is 14 in. in diameter and is operated at a speed of 1000 rpm.

14.21 For a pump having the characteristics of Fig. 14.6, plot the head-discharge curve if the pump diameter is 60 cm and the speed is 690 rpm.

14.22 An axial-flow blower is used for a wind tunnel that has a test section measuring 60 cm by 60 cm and is capable of airspeeds up to 40 m/s. If the blower is to operate at maximum efficiency at the highest speed and if the rotational speed of the blower is 2000 rpm at this condition, what are the diameter of the blower and the power required? Assume that the blower has the characteristics shown in Fig. 14.6. Assume $\rho = 1.2 \ kg/m^3$.

14.23 An axial-flow blower is used to air-condition an office building that has a volume of $10^5 \ m^3$. It is decided that the air at 60°F in the building must be completely changed every 15 min. Assume that the blower operates at 600 rpm at maximum efficiency and has the characteristics shown in Fig. 14.6.

Calculate the diameter and power requirements for two blowers operating in parallel.

14.24 An axial fan 2 m in diameter is used in a wind tunnel as shown (test section 1.2 m in diameter; test section velocity of 60 m/s). The rotational speed of the fan is 1800 rpm. Assume the density of the air is constant at 1.2 kg/m³. There are negligible losses in the tunnel. The performance curve of the fan is identical to that shown in Fig. 14.6. Calculate the power needed to operate the fan.

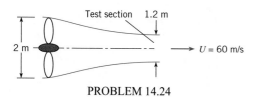

PROBLEM 14.24

Radial Flow Pumps

14.25 $\mathbb{PQ}$◀ The radial flow pump is best suited for what conditions of head produced and discharge?

14.26 $\mathbb{PQ}$◀ A pump is used to pump water out of a reservoir. What limits the depth for which the pump can draw water?

14.27 If a pump having the characteristics given in Fig. 14.9 is doubled in size but halved in speed, what will be the head and discharge at maximum efficiency?

14.28 A pump having the characteristics given in Fig. 14.9 pumps water at 20°C from a reservoir at an elevation of 366 m to a reservoir at an elevation of 450 m through a 36 cm steel pipe. If the pipe is 610 m long, what will be the discharge through the pipe?

14.29 If a pump having the characteristics given in Fig. 14.9 or 14.10 is operated at a speed of 1600 rpm, what will be the discharge when the head is 150 ft?

14.30 If a pump having the performance curve shown is operated at a speed of 1600 rpm, what will be the maximum possible head developed?

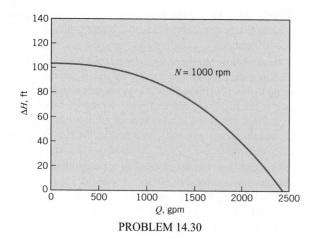

PROBLEM 14.30

14.31 If a pump having the characteristics given in Fig. 14.9 is operated at a speed of 30 rps, what will be the shutoff head?

14.32 If a pump having the characteristics given in Fig. 14.10 is 40 cm in diameter and is operated at a speed of 25 rps, what will be the discharge when the head is 50 m?

14.33 A centrifugal pump 20 cm in diameter is used to pump kerosene at a speed of 5000 rpm. Assume that the pump has the characteristics shown in Fig. 14.10. Calculate the flow rate, the pressure rise across the pump, and the power required if the pump operates at maximum efficiency.

14.34 Plot the five performance curves from Fig. 14.14 for the different impeller diameters in terms of the head and discharge coefficients. Use impeller diameter for D.

Pump Selection

14.35 $\mathbb{PQ}$◀ What is the difference between a system curve and a pump curve. Explain.

14.36 $\mathbb{PQ}$◀ The operating point for a pump system is established by what condition?

14.37 $\mathbb{PQ}$◀ The value of the specific speed suggests the type of pump to be used for a given application. A high specific speed suggests the use of what kind of pump?

14.38 The pump curve for a given pump is represented by

$$h_{p,\text{pump}} = 20\left[1 - \left(\frac{Q}{100}\right)^2\right]$$

where $h_{p,\text{pump}}$ is the head provided by the pump in feet and Q is the discharge in gpm. The system curve for a pumping application is

$$h_{p,\text{sys}} = 5 + 0.002Q^2$$

where $h_{p,\text{sys}}$ is the head in feet required to operate the system and Q is the discharge in gpm. Find the operating point (Q) for (a) one pump, (b) two identical pumps connected in series, and (c) two identical pumps connected in parallel.

14.39 What is the suction specific speed for the pump that is operating under the conditions given in Prob. 14.16? Is this a safe operation with respect to susceptibility to cavitation?

14.40 What type of pump should be used to pump water at a rate of 10 cfs and under a head of 30 ft? Assume $N = 1500$ rpm.

14.41 For the most efficient operation, what type of pump should be used to pump water at a rate of 0.30 m³/s and under a head of 8 m? Assume $n = 25$ rps.

14.42 What type of pump should be used to pump water at a rate of 0.40 m³/s and under a head of 70 m? Assume $N = 1100$ rpm.

14.43 An axial-flow pump is to be used to lift water against a head (friction and static) of 15 ft. If the discharge is to be 5000 gpm, what maximum speed in revolutions per minute is allowed if the suction head is 5 ft?

14.44 A pump is beeded to pump water at a rate of $1.0 \text{ m}^3/\text{s}$ from the lower to the upper reservoir shown in the figure. What type of pump would be best for this operation if the impeller speed is to be 600 rpm?

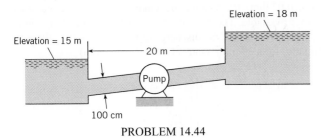

PROBLEM 14.44

Compressors

14.45 $\mathbb{PQ}\blacktriangleleft$ The pressure rise associated with gases in a compressor causes the gas temperature to increase as well. The ratio of final temperature to initial temperature is less than the ratio of final pressure to initial pressure. Will the final density be less or greater than the initial density?

14.46 Methane flowing at the rate of 1 kg/s is to be compressed by a noncooled centrifugal compressor from 100 kPa to 150 kPa. The temperature of the methane entering the compressor is 27°C. The efficiency of the compressor is 70%. Calculate the shaft power necessary to run the compressor.

14.47 A 12 kW (shaft output) motor is available to run a noncooled compressor for carbon dioxide. The pressure is to be increased from 90 kPa to 140 kPa. If the compressor is 60% efficient, calculate the volume flow rate into the compressor.

14.48 A water-cooled centrifugal compressor is used to compress air from 100 kPa to 400 kPa at the rate of 1 kg/s. The temperature of the inlet air is 15°C. The efficiency of the compressor is 50%. Calculate the necessary shaft power.

Impulse Turbine

14.49 $\mathbb{PQ}\blacktriangleleft$ An impulse turbine will produce no power if the velocity of the jet striking the bucket is the same as the bucket velocity. Explain.

14.50 A penstock 1 m in diameter and 10 km long carries water at 10°C from a reservoir to an impulse turbine. If the turbine is 85% efficient, what power can be produced by the system if the upstream reservoir elevation is 650 m above the turbine jet and the jet diameter is 16.0 cm? Assume that $f = 0.016$ and neglect head losses in the nozzle. What should the diameter of the turbine wheel be if it is to have an angular speed of 360 rpm? Assume ideal conditions for the bucket design $[V_{\text{bucket}} = (1/2)V_j]$.

14.51 Consider an idealized bucket on an impulse turbine that turns the water through 180°. Prove that the bucket speed should be one-half the incoming jet speed for a maximum power production. (*Hint:* Set up the momentum equation to solve for the force on the bucket in terms of V_j and V_{bucket}; then

the power will be given by this force times V_{bucket}. You can use your mathematical talent to complete the problem.)

14.52 Consider a single jet of water striking the buckets of the impulse wheel as shown. Assume ideal conditions for power generation [$V_{\text{bucket}} = (1/2)V_j$ and the jet is turned through 180° of arc]. With the foregoing conditions, solve for the jet force on the bucket and then solve for the power developed. Note that this power is not the same as that given by Eq. (14.22)! Study the figure to resolve the discrepancy.

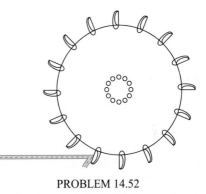

PROBLEM 14.52

Reaction Turbine

14.53 $\mathbb{PQ}\blacktriangleleft$ How does a reaction turbine differ from a centrifugal pump?

14.54 $\mathbb{PQ}\blacktriangleleft$ What is meant by the "runner" in a reaction turbine?

14.55 For a given Francis turbine, $\beta_1 = 60°$, $\beta_2 = 90°$, $r_1 = 5$ m, $r_2 = 3$ m, and $B = 1$ m. The discharge is $126 \text{ m}^3/\text{s}$ and the rotational speed is 60 rpm. Assume $T = 10°C$.
a. What should α_1 be for a nonseparating flow condition at the entrance to the runner.
b. What is the maximum attainable power with the conditions noted?
c. If you were to redesign the turbine blades of the runner, what changes would you suggest to increase the power production if the discharge and overall dimensions are to be kept the same?

14.56 A Francis turbine is to be operated at a speed of 60 rpm and with a discharge of $3.0 \text{ m}^3/\text{s}$. If $r_1 = 1.5$ m, $r_2 = 1.20$ m, $B = 30$ cm, $\beta_1 = 85°$, and $\beta_2 = 165°$, what should α_1 be for nonseparating flow to occur through the runner? What power and torque should result with this operation? Assume $T = 10°C$.

14.57 A Francis turbine is to be operated at a speed of 120 rpm and with a discharge of $113 \text{ m}^3/\text{s}$. If $r_1 = 2.5$ m, $B = 0.90$ m, and $\beta_1 = 45°$, what should α_1 be for nonseparating flow at the runner inlet?

14.58 Shown is a preliminary layout for a proposed small hydroelectric project. The initial design calls for a discharge of 8 cfs through the penstock and turbine. Assume 80% turbine efficiency. For this setup, what power output could be expected from the power plant? Draw the HGL and EGL for the system.

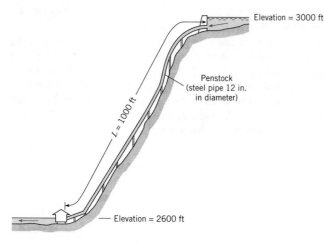

PROBLEM 14.58

Wind Turbine

14.59 $\mathbb{PQ}\blacktriangleleft$ What determines the minimum and maximum wind speeds at which a wind turbine can operate?

14.60 Calculate the maximum power derivable from a conventional horizontal-axis wind turbine with a propeller 2.5 m in diameter in a 50 km/h wind whose density is 1.2 kg/m³.

14.61 A wind "farm" consists of 20 Darrieus turbines, each 15 m high. The total output from the turbines is to be 2 MW in a wind of 20 m/s and an air density of 1.2 kg/m³. The Darrieus turbine shown has the shape of an arc of a circle. Find the minimum width, W, of the turbine needed to provide this power output.

14.62 A windmill is connected directly to a mechanical pump that is to pump water from a well 10 ft deep as shown. The windmill is a conventional horizontal-axis type with a fan diameter of 10 ft. The efficiency of the mechanical pump is 80%. The density of the air is 0.07 lbm/ft³. Assume the windmill delivers the maximum power available. There is 20 ft of 2 inch galvanized pipe in the system. What would the discharge of the pump be (in gallons per minute) for a 30 mph wind? (1 cfm = 7.48 gpm)

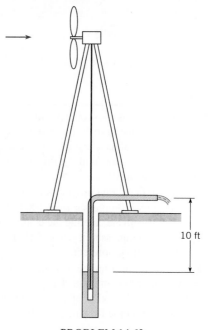

PROBLEM 14.62

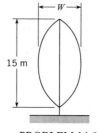

PROBLEM 14.61

Flow in Open Channels

SIGNIFICANT LEARNING OUTCOMES

Conceptual Knowledge

- Describe differences between uniform flow, gradually varied flow, and rapidly varied flow.
- Describe critical depth, specific energy, supercritical flow, and subcritical flow.
- Describe what causes head loss in open-channel flow.
- Describe the factors used to classify surface profiles in gradually varied flow.
- Explain the conditions leading to a hydraulic jump.

Procedural Knowledge

- Apply Darcy-Weisbach and Manning's equations to uniform flow.
- Find the best hydraulic section.
- Calculate the depth, velocity, and head loss in a hydraulic jump.
- Apply the Froude number to classify flow as critical, subcritical, or supercritical.

Typical Applications

- For a pipe that is half full, calculate the water depth and head loss.
- For a concrete channel, calculate the dimensions necessary to carry a desired flow rate.
- To dissipate energy in water exiting a hydroelectric dam, design a stilling basin with hydraulic jump.

An *open channel* is one in which a liquid flows with a free surface. A *free surface* means that the liquid surface is exposed to the atmosphere. Examples of open channels are natural creeks and rivers, artificial channels such as irrigation ditches and canals, and pipelines or sewers flowing less than full. In most cases, water or waste-water is the flowing liquid.

Description of Open-Channel Flow

Flow in an open channel is described as uniform or nonuniform, as distinguished in Fig. 15.1. As defined in Chapter 4, *uniform flow* means that the velocity is constant along a streamline, which in open-channel flow means that depth and cross section are constant along the length of a channel. The depth for uniform-flow conditions is called *normal depth* and is designated by y_n. For *nonuniform flow*, the velocity changes from section to section along the channel, thus one observes changes in depth. The velocity change may be due to a change in channel configuration, such as a bend, change in cross-sectional shape, or change in channel slope. For example, Fig. 15.1 shows steady flow over a spillway of constant width, where the water

Figure 15.1

Distingishing uniform and nonuniform flow. This example shows steady flow over a spillway, such as the emergency overflow channel of a dam.

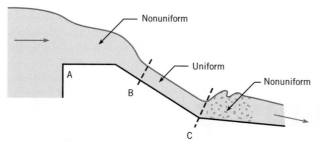

must flow progressively faster as it goes over the brink of the spillway (from A to B), caused by the suddenly steeper slope. The faster velocity requires a smaller depth, in accordance with conservation of mass (continuity). From reach B to C, the flow is uniform because the velocity, and thus depth, are constant. After reach C the abrupt flattening of channel slope requires the velocity to suddenly, and turbulently, slow down. Thus there is a deeper depth downstream of C than in reach B to C.

The most complicated open-channel flow is unsteady nonuniform flow. An example of this is a breaking wave on a sloping beach. Theory and analysis of unsteady nonuniform flow are reserved for more advanced courses.

Dimensional Analysis in Open-Channel Flow

Open-channel flow results from gravity moving water from higher to lower elevations, and is impeded by friction forces caused by the roughness of the channel. Thus the functional equation $Q = f(\mu, \rho, g, V, L)$ and dimensional anaysis as presented in Chapter 8 lead to two important independent π-groups to characterize open-channel flow: the Froude number and the Reynolds number. The Froude number is the ratio of inertial force to gravity force:

$$\text{Fr}^2 = \frac{\text{inertial force}}{\text{gravity force}} = \frac{\rho L^2 V^2}{\Delta \gamma L^3} = \frac{V^2}{L \Delta \gamma / \rho} \tag{15.1}$$

$$\text{Fr} = \frac{V}{\sqrt{gL}} \tag{15.2}$$

The Froude number is important if the gravitational force influences the direction of flow, such as in flow over a spillway, or the formation of surface waves. However, it is unimportant when gravity causes only a hydrostatic pressure distribution, such as in a closed conduit.

The use of Reynolds number for determining whether the flow in open channels will be laminar or turbulent depends upon the *hydraulic radius*, given by

$$R_h = \frac{A}{P} \tag{15.3}$$

where A is the cross-sectional area of flow and P is the wetted perimeter. The characteristic length R_h is analogous to diameter D in pipe flow. Recall that for pipe flow (Chapter 10), if the Reynolds number ($VD\rho/\mu = VD/\nu$) is less than 2000, the flow will be laminar, and if it is greater than about 3000, one can expect the flow to be turbulent. The Reynolds number criterion for open-channel flow would be 2000 if one replaced D in the Reynolds number by $4R_h$, where R_h is the hydraulic radius. For this definition of Reyholds number, laminar flow would occur in open channels if $V(4R_h)/\nu < 2000$.

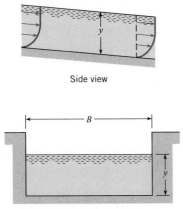

Side view

End view

Figure 15.2

Open-channel relations.

However, the standard convention in open-channel flow analysis is to define the Reynolds number as

$$\text{Re} = \frac{VR_h}{\nu} \qquad (15.4)$$

Therefore, in open channels, if the Reynolds number is less than 500, the flow is laminar, and if Re is greater than about 750, one can expect to have turbulent flow. A brief analysis of this turbulent criterion (see Example 15.1) will show that water flow in channels will usually be turbulent unless the velocity and/or the depth is very small.

It should be noted that for rectangular channels (see Fig. 15.2), the hydraulic radius is

$$R_h = \frac{A}{P} = \frac{By}{B + 2y} \qquad (15.5)$$

Example 15.1 shows that for very wide, shallow channels the hydraulic radius approaches the depth y.

Most open-channel flow problems involve turbulent flow. If one calculates the conditions needed to maintain laminar flow, as in Example 15.1, one sees that laminar flow is uncommon.

EXAMPLE 15.1   CONDITIONS FOR LAMINAR OPEN-CHANNEL FLOW

Water (60°F) flows in a 10 ft–wide rectangular channel at a depth of 6 ft. What is the Reynolds number if the mean velocity is 0.1 ft/s? With this velocity, at what maximum depth can one be assured of having laminar flow?

Problem Definition

Situation: Constant velocity in rectangular channel, so uniform flow.

Find:

1. Reynolds number for given mean velocity.
2. Maximum depth for which flow is laminar.

Properties: $\nu = 1.22 \times 10^{-5} \text{ ft}^2$, from Table A.5.

Plan

1. Calculate Reynolds number using Eq. (15.4)
2. Find the depth for which Re = 500 using Eq. (15.5).

Solution

1. Reynolds number

$$\text{Re} = VR_h/\nu$$

where

$$V = 0.1 \text{ ft/s}$$
$$R_h = A/P = By/(B + 2y)$$
$$= (10 \times 6)/(10 + 2 \times 6)$$
$$= 2.73 \text{ ft}$$

$$\text{Re} = (0.1 \text{ ft/s})(2.73 \text{ ft})/(1.22 \times 10^{-5} \text{ ft}^2/\text{s}) = \boxed{22{,}377}$$

Since Re > 500, flow is turbulent.

2. Depth for which Re = 500.

$$\text{Re} = VR_h/\nu = (0.10 \text{ ft/s})R_h/(1.22 \times 10^{-5} \text{ ft}^2/\text{s}) = 500$$

$$R_h = (500)(1.22 \times 10^{-5} \text{ ft}^2/\text{s})/(0.10 \text{ ft/s}) = 0.061 \text{ ft}$$

For a rectangular channel,

$$R_h = (By)/(B + 2y)$$

$$(By)/(B + 2y) = (10y)/(10 + 2y) = 0.061 \text{ ft}$$

$$y = \boxed{0.062 \text{ ft}}$$

Review

1. Note: Velocity or depth must be very small to yield laminar flow of water in an open channel.
2. Note: Depth and hydraulic radius are virtually the same when depth is very small relative to width.

Figure 15.3

Definition sketch for flow in open channels.

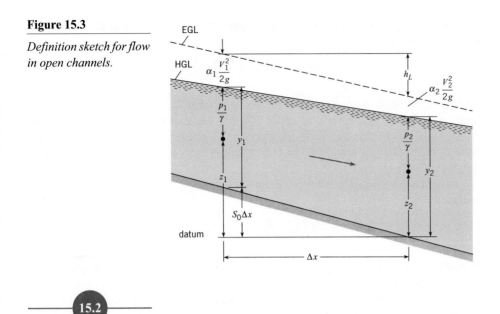

Energy Equation for Steady Open-Channel Flow

To derive the energy equation for flow in an open channel, begin with Eq. (7.29) and let the pump head and turbine head equal zero: $h_p = h_t = 0$. Equation (7.29) becomes

$$\frac{p_1}{\gamma} + \alpha_1 \frac{V_1^2}{2g} + z_1 = \frac{p_2}{\gamma} + \alpha_2 \frac{V_2^2}{2g} + z_2 + h_L \tag{15.6}$$

Use Fig. 15.3 to show that

$$\frac{p_1}{\gamma} + z_1 = y_1 + S_0 \Delta x \quad \text{and} \quad \frac{p_2}{\gamma} + z_2 = y_2$$

where S_0 is the slope of the channel bottom, and y is the depth of flow. Assume the flow in the channel is turbulent, so $\alpha_1 = \alpha_2 \approx 1.0$. Equation (15.6) becomes

$$y_1 + \frac{V_1^2}{2g} + S_0 \Delta x = y_2 + \frac{V_2^2}{2g} + h_L \tag{15.7}$$

In addition to the foregoing assumptions, Eq. (15.7) also requires that the channel have a uniform cross section, and the flow be steady.

Steady Uniform Flow

Uniform flow requires that velocity be constant in the flow direction, so the shape of the channel and the depth of fluid is the same from section to section. Consideration of the foregoing slope equations shows that for uniform flow, the slope of the HGL will be the same as the channel slope, because the velocity and depth are the same in both sections. The HGL, and thus the slope of the water surface, is controlled by head loss. If one restates the Darcy-Weisbach equation introduced in Chapter 10 with D replaced by $4R_h$, the head loss is

$$h_f = \frac{fL}{4R_h}\frac{V^2}{2g} \quad \text{or} \quad \frac{h_f}{L} = \frac{f}{4R_h}\frac{V^2}{2g} \tag{15.8}$$

From Fig. 15.3, $S_0 = $ [slope of HGL], which is a function of the head loss, so $S_0 = (h_f/L)$, yielding the following equation for velocity:

$$V = \sqrt{\frac{8g}{f}R_hS_0} \tag{15.9}$$

To solve Eq. (15.9) for velocity, the friction factor f can be found from the Moody diagram (Fig. 10.14) and can then be used to solve iteratively for the velocity for a given uniform-flow condition. This is demonstrated in Example 15.2.

EXAMPLE 15.2   ESTIMATING Q FOR UNIFORM FLOW USING DARCY-WEISBACH EQUATION

Estimate the discharge of water that a concrete channel 10 ft wide can carry if the depth of flow is 6 ft and the slope of the channel is 0.0016.

Problem Definition

Situation: Uniform flow, concrete surface.

Find: Discharge in ft^3/s.

Properties: Concrete, Table 10.4: $k_s = 0.012$–0.12 inches, or 0.001–0.01 ft.

Plan

1. Find channel velocity by relating channel slope to h_f/L with Eq. (15.9).
    *   Use the Moody diagram to find f.
    *   Assume a roughness for first estimate of $k_s/4R_h$ to use with Reynolds number.
    *   Select a first estimate of f, which is opposite $k_s/4R_h$ on the Moody diagram
    *   Solve for V, first iteration.
    *   Calculate new Reynolds number with this value of V; check f against reasonable convergence criterion.
2. Calculate $Q = VA$.

Solution

1. For Eq. (15.9), $V = \sqrt{\frac{8g}{f}R_hS_0}$; need to get a value for f.

2a. Assume $k_s = 0.005$ ft. Relative roughness is

$$\frac{k_s}{4R_h} = \frac{0.005 \text{ ft}}{4(60 \text{ ft}^2/22 \text{ ft})} = \frac{0.005 \text{ ft}}{4(2.73 \text{ ft})} = 0.00046$$

2b. Use value of $k_s/4R_h = 0.00046$ as a guide to estimate $f = 0.016$.

2c. First iteration for V gives

$$V = \sqrt{\frac{8(32.2 \text{ ft/s}^2)(2.73 \text{ ft})(0.0016)}{0.016}}$$

$$= \sqrt{70.6 \text{ ft}^2/\text{s}^2} = 8.39 \text{ ft/s}$$

2d. Recalculate Reynolds number.

$$\text{Re} = V\frac{4R_h}{\nu} = \frac{8.39 \text{ ft/s}(10.9 \text{ ft})}{1.2(10^{-5} \text{ ft}^2/\text{s})} = 7.62 \times 10^6$$

Using this new value of Re, and with $k_s/4R_h = 0.00046$, read f as 0.016. This value of f is the same as previous estimate—meets reasonable convergence criterion.

$$V = \boxed{8.39 \text{ ft/s}}$$

3. Compute Q.

$$Q = VA = 8.39 \text{ ft/s}(60 \text{ ft}^2) = \boxed{503 \text{ cfs}}$$

Rock-Bedded Channels

For rock-bedded channels such as those in some natural streams or unlined canals, the larger rocks produce most of the resistance to flow, and essentially none of this resistance is due to viscous effects. Thus, the friction factor is independent of the Reynolds number. This is analogous to the fully rough region of the Moody diagram for pipe flow. For a rock-bedded channel, Limerinos (1) has shown that the resistance coefficient f can be given in terms of the size of rock in the stream bed as

$$f = \frac{1}{\left[1.2 + 2.03\log\left(\dfrac{R_h}{d_{84}}\right)\right]^2} \tag{15.10}$$

where d_{84} is a measure of the rock size.* See Example 15.3

EXAMPLE 15.3 RESISTANCE COEFFICIENT FOR BOULDERS

Determine the value of the resistance coefficient, f, for a natural rock-bedded channel that is 100 ft wide and has an average depth of 4.3 ft. The d_{84} size of boulders in the stream bed is 0.72 ft.

Problem Definition

Situation: Boulders in natural channel bottom will control magnitude of f.

Find: Friction factor, f.

Plan

1. Simplify calculation of R_h for wide channel; take R_h to be depth.
2. Use Eq. (15.10) to find f on the basis of the d_{84} boulder size.

Solution

1. R_h is 4.3 ft.
2. Evaluate f.

$$f = \frac{1}{\left[1.2 + 2.03\log\left(\dfrac{4.3}{0.72}\right)\right]^2} = \boxed{0.130}$$

The Chezy Equation

Leaders in open-channel research have recommended the use of the methods already presented (involving the Reynolds number and relative roughness k_s) for channel design (2). However, many engineers continue to use two traditional methods, the Chezy equation and the Manning equation.

As noted earlier, the depth in uniform flow, called *normal depth, y_n,* is constant. Consequently, h_f/L is the slope S_0 of the channel, and Eq. (15.8) can be written as

$$R_h S_0 = \frac{f}{8g}V^2$$

or

$$V = C\sqrt{R_h S_0} \tag{15.11}$$

where

$$C = \sqrt{8g/f} \tag{15.12}$$

Since $Q = VA$, the discharge in a channel is given by

$$Q = CA\sqrt{R_h S_0} \tag{15.13}$$

This equation is known as the *Chezy equation* after a French engineer of that name. For practical application, the coefficient C must be determined. One way to determine C is by knowing an acceptable value of the friction factor f and using Eq. (15.2).

---

* Most river-worn rocks are somewhat elliptical in shape. Limerinos (1) showed that the intermediate dimension d_{84} correlates best with f. The d_{84} refers to the size of rock (intermediate dimension) for which 84% of the rocks in the random sample are smaller than the d_{84} size. Details for choosing the sample are given by Wolman (3).

The Manning Equation

The second, and more common, way to determine C in the SI system of units is given as:

$$C = \frac{R_h^{1/6}}{n} \tag{15.14}$$

where n is a resistance coefficient called *Manning's n*, which has different values for different types of boundary roughness. When this expression for C is inserted into Eq. (15.3), the result is a common form of the discharge equation for uniform flow in open channels for SI units, referred to as the Manning equation:

$$Q = \frac{1.0}{n} A R_h^{2/3} S_0^{1/2} \tag{15.15}$$

Table 15.1 gives values of n for various types of boundary surfaces. The major limitation of this approach is that the viscous or relative-roughness effects are not present in the design formula. Hence, application outside the range of normal-sized channels carrying water is not recommended.

Manning Equation—Traditional System of Units

The form of the Manning equation depends on the system of units because Manning's equation is not dimensionally homogeneous. In Eq. (15.15), notice that the primary dimensions on the left side of the equation are L^3/T and the primary dimensions on the right side are $L^{8/3}$.

Table 15.1 TYPICAL VALUES OF ROUGHNESS COEFFICIENT, MANNING'S n	
Lined Canals	**n**
Cement plaster	0.011
Untreated gunite	0.016
Wood, planed	0.012
Wood, unplaned	0.013
Concrete, troweled	0.012
Concrete, wood forms, unfinished	0.015
Rubble in cement	0.020
Asphalt, smooth	0.013
Asphalt, rough	0.016
Corrugated metal	0.024
Unlined Canals	
Earth, straight and uniform	0.023
Earth, winding and weedy banks	0.035
Cut in rock, straight and uniform	0.030
Cut in rock, jagged and irregular	0.045
Natural Channels	
Gravel beds, straight	0.025
Gravel beds plus large boulders	0.040
Earth, straight, with some grass	0.026
Earth, winding, no vegetation	0.030
Earth, winding, weedy banks	0.050
Earth, very weedy and overgrown	0.080

To convert the Manning equation from SI to traditional units, one must apply a factor equal to 1.49 if the same value of n is used in the two systems. Thus in the traditional system the discharge equation using Manning's n is

$$Q = \frac{1.49}{n} A R_h^{2/3} S_0^{1/2} \tag{15.16}$$

In Example 15.4, a value for Manning's n is calculated from known information about a channel and compared to tabulated values for n in Table 15.1.

EXAMPLE 15.4 CALCULATING DISCHARGE AND MANNING'S n USING CHEZY EQUATION

If a channel with boulders has a slope of 0.0030, is 100 ft wide, has an average depth of 4.3 ft, and is known to have a friction factor of 0.130, what is the discharge in the channel and what is the numerical value of Manning's n for this channel?

Problem Definition

Situation: Uniform flow, channel with known f.

Find:

1. Discharge, Q, in cfs.
2. Numerical value of Manning's n for this channel with boulders.

Plan

1. Find velocity using Eq. (15.9) with known f. Estimate R_h to be y, which is 4.3 ft.
2. Calculate discharge from $Q = VA$.
3. Solve for Manning's n using Chezy equation for traditional units (Eq. 15.16).

Solution

1. Velocity

$$V = \left[\sqrt{\frac{(8)(32.2 \text{ ft/s}^2)}{0.130}} \right] \left[\sqrt{(4.3 \text{ ft})(0.0030)} \right] = 5.06 \text{ ft/s}$$

2. Discharge

$$Q = VA = (5.06)(100 \times 4.3) = \boxed{2176 \text{ cfs}}$$

3. Manning's n, using the traditional unit equation (Eq. 15.16)

$$n = \frac{1.49}{Q} A R_h^{2/3} S_0^{1/2}$$

$$n = \left(\frac{1.49}{2176 \text{ ft}^3/\text{s}} \right) (100 \times 4.3 \text{ ft}^2)(4.3 \text{ ft})^{2/3} (0.003)^{1/2}$$

$$n = \boxed{0.0426}$$

Review

Note: This calculated value of n is within the range of typical values in Table 15.1 under the category of "Unlined Canals, Cut in rock."

Note: This example and Example 15.3 show that f in the Darcy-Weisbach equation can be related to Manning's n for uniform-flow conditions.

In Example 15.5 the Chezy equation for traditional units is used to compute discharge.

EXAMPLE 15.5 DISCHARGE USING CHEZY EQUATION

Using the Chezy equation with Manning's n, compute the discharge in a concrete channel 10 ft wide if the depth of flow is 6 ft and the slope of the channel is 0.0016.

Problem Definition

Situation: Uniform flow, concrete channel, known geometry and depth.

Find: Discharge, Q.

Properties: $n = 0.015$ for concrete channel (Table 15.1).

Plan

Use the Chezy equation for traditional units, Eq. (15.16).

Solution

$$Q = \frac{1.49}{n} A R_h^{2/3} S_0^{1/2}$$

$$R_h = \frac{60}{22} = 2.73 \text{ ft} \quad \text{and} \quad R_h^{2/3} = 1.95$$

$$S_0^{1/2} = 0.04 \quad \text{and} \quad A = 60 \text{ ft}^2$$

$$Q = \frac{1.49}{0.015} (60)(1.96)(0.04) = \boxed{467 \text{ cfs}}$$

The two results (Examples 15.4 and 15.5) are within expected engineering accuracy for this type of problem. For a more complete discussion of the historical development of Manning's equation and the choice of n values for use in design or analysis, refer to Yen (4) and Chow (5).

Best Hydraulic Section for Uniform Flow

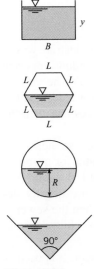

Figure 15.4

Best hydraulic sections for different geometries.

The *best hydraulic section* is the channel geometry that yields a minimum wetted perimeter for a given cross-sectional area. Therefore, it yields the least viscous energy loss for a given area. Consider the quantity $AR_h^{2/3}$ in Manning's equation given in Eqs. (15.15 and 15.16), which is referred to as the section factor. Because $R_h = A/P$, the section factor relating to uniform flow is given by $A(A/P)^{2/3}$. Thus, for a channel of given resistance and slope, the discharge will increase with increasing cross-sectional area but decrease with increasing wetted perimeter P. For a given area, A, and a given shape of channel—for example, rectangular cross section—there will be a certain ratio of depth to width (y/B) for which the section factor will be maximum. This ratio is the best hydraulic section.

Example 15.6 shows that the best hydraulic section for a rectangular channel occurs when $y = \frac{1}{2}B$. It can be shown that the best hydraulic section for a trapezoidal channel is half a hexagon as shown; for the circular section, it is the half circle with depth equal to radius; and for the triangular section, it is a triangle with a vertex of 90° (Fig 15.4). Of all the various shapes, the half circle has the best hydraulic section because it has the smallest peri-meter for a given area.

The best hydraulic section can be relevant to the cost of the channel. For example, if a trapezoidal channel were to be excavated and if the water surface were to be at adjacent ground level, the minimum amount of excavation (and excavation cost) would result if the channel of best hydraulic section were used.

EXAMPLE 15.6   BEST HYDRAULIC SECTION FOR A RECTANGULAR CHANNEL

Determine the best hydraulic section for a rectangular channel with depth y and width B.

Problem Definition

Situation: Rectangular channel with depth y and width B.

Find: Best hydraulic section.

Plan

1. Set $A = By$ and $P = B + 2y$ so that both are a function of y.
2. Let A be constant, and minimize P.
   - Differentiate P with respect to y and set the derivative equal to zero.
   - Express the result of minimizing P as a relation between y and B.

Solution

1. Relate A and P in terms of y.

$$P = \frac{A}{y} + 2y$$

2a. Minimize P.

$$\frac{dP}{dy} = \frac{-A}{y^2} + 2 = 0$$

$$\frac{A}{y^2} = 2$$

2b. Express result in terms of y and B.

$A = By$, so

$$\frac{By}{y^2} = 2 \quad \text{or} \quad \boxed{y = \frac{1}{2}B}$$

The best hydraulic section for a rectangular channel occurs when the depth is one-half the width of the channel, see Fig. 15.4.

Figure 15.5

Culvert under a highway embankment.

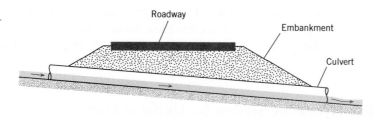

Uniform Flow in Culverts and Sewers

Sewers are conduits that carry sewage (liquid domestic, commercial, or industrial waste) from households, businesses, and factories to sewage disposal sites. These conduits are often circular in cross section, but elliptical and rectangular conduits are also used. The volume rate of sewage varies throughout the day and season, but of course sewers are designed to carry the maximum design discharge flowing full or nearly full. At discharges less than the maximum, the sewers will operate as open channels.

Sewage usually consists of about 99% water and 1% solid waste. Because most sewage is so dilute, it is assumed that it has the same physical properties as water for purposes of discharge computations. However, if the velocity in the sewer is too small, the solid particles may settle out and cause blockage of the flow. Therefore, sewers are usually designed to have a minimum velocity of about 2 ft/s (0.60 m/s) at times when the sewer is flowing full. This condition is met by choosing a slope on the sewer line to achieve the desired velocity.

A culvert is a conduit placed under a fill such as a highway embankment. It is used to convey stream-flow from the uphill side of the fill to the downhill side. Figure 15.5 shows the essential features of a culvert. A culvert should be able to convey runoff from a *design storm* without overtopping the fill and without erosion of the fill at either the upstream or downstream end of the culvert. The design storm, for example, might be the maximum storm that could be expected to occur once in 50 years at the particular site.

The flow in a culvert is a function of many variables, including cross-sectional shape (circular or rectangular), slope, length, roughness, entrance design, and exit design. Flow in a culvert may occur as an open channel throughout its length, it may occur as a completely full pipe, or it may occur as a combination of both. The complete design and analysis of culverts are beyond the scope of this text; therefore, only simple examples are included here (Examples 15.7 and 15.8). For more extensive treatment of culverts, please refer to Chow (5), Henderson (6), and American Concrete Pipe Assoc. (7).

EXAMPLE 15.7 SIZING A ROUND CONCRETE SEWER LINE

A sewer line is to be constructed of concrete pipe to be laid on a slope of 0.006. If $n = 0.013$ and if the design discharge is 110 cfs, what size pipe (commercially available) should be selected for a full-flow condition? What will be the mean velocity in the sewer pipe for these conditions? (It should be noted that concrete pipe is readily available in commercial sizes of 8 in., 10 in., and 12 in. diameter and then in 3 in. increments up to 36 in. diameter. From 36 in. diameter up to 144 in. the sizes are available in 6 in. increments.)

Problem Definition

Situation: Given slope, find Manning's n, design discharge, traditional unit system.

Find: Pipe diameter large enough to carry design discharge and that allows $V \geq 2$ ft/s at full-flow condition.

Assumptions: Can only use a standard pipe size.

Plan

1. Use Chezy equation for traditional units, Eq. (15.16).
2. Solve for $AR^{2/3}$.
3. For pipe flowing full, relate A and P to diameter through R_h.
4. Solve for diameter, and use the next commerical size larger.
5. Check that velocity for full flow is greater than 2 ft/s.

Solution

1. Chezy equation for traditional units is

$$Q = \frac{1.49}{n} AR^{2/3} S_0^{1/2}$$

$$Q = 110 \ \text{ft}^3/\text{s}$$

$$n = 0.013$$

$$S_0 = 0.006 \ \text{(assume atmospheric pressure in the pipe)}$$

2. Solve for $AR^{2/3}$. Note that units of $AR^{2/3}$ are $\text{ft}^{8/3}$ because A is in ft^2 and R_h is in $\text{ft}^{2/3}$.

$$AR^{2/3} = \frac{(110 \ \text{ft}^3/\text{s})(0.013)}{(1.49)(0.006)^{1/2}} = 12.39 \ \text{ft}^{8/3}$$

3. Relate A and P to diameter by relating to R_h.

$$R_h = \frac{A}{P} \quad \text{and} \quad R_h^{2/3} = \left(\frac{A}{P}\right)^{2/3}$$

$$AR_h^{2/3} = \frac{A^{5/3}}{P^{2/3}} = 12.39 \ \text{ft}^{8/3}$$

For a pipe flowing full, $A = \pi D^2/4$ and $P = \pi D$, or

$$\frac{(\pi D^2/4)^{5/3}}{(\pi D)^{2/3}} = 12.39 \ \text{ft}^{8/3}$$

4. Solving for diameter yields $D = 3.98 \ \text{ft} = 47.8 \ \text{in}$. Use the next commercial size larger, which is $\boxed{D = 48 \ \text{in}}$

$$A = \frac{\pi D^2}{4} = 50.3 \ \text{ft}^2 \ \text{(for pipe flowing full)}$$

5. Verify that velocity of full flow is greater than $2 \ \text{ft}/\text{s}$.

$$V = \frac{Q}{A} = \frac{(110 \ \text{ft}^3/\text{s})}{(50.3 \ \text{ft}^2)} = \boxed{2.19 \ \text{ft}/\text{s}}$$

Example 15.8 demonstrates the calculation of necessary slope given all sources of head loss, and a required discharge.

EXAMPLE 15.8   CULVERT DESIGN

A 54 in. diameter culvert laid under a highway embankment has a length of 200 ft and a slope of 0.01. This was designed to pass a 50-year flood flow of 225 cfs under full-flow conditions (figure below). For these conditions, what head H is required? When the discharge is only 50 cfs, what will be the uniform flow depth in the culvert? Assume $n = 0.012$.

Problem Definition

Situation: Culvert has been designed to carry 225 cfs with given dimensions.

Find:

1. The height H required between the two free surfaces when flowing full.

2. The uniform flow depth in the culvert when $Q = 50$ cfs.

Assumptions: Uniform flow, so that pipe head loss h_f can be related to S_0.

Sketch:

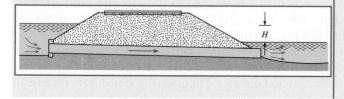

Plan

1. Use energy equation between the two end sections, accounting for head loss.

2. Document all sources of head loss.

3. Find pipe head loss h_f using Eq. (15.17) and the fact that

$$S_0 = \frac{h_f}{L}.$$

4. Use continuity equation to find V, the uniform flow velocity, needed to calculate head loss.

5. Solve for H.

6. Solve for depth of flow, for $Q = 50$, using Eq. (15.17) and pipe geometry relations for pipe flowing partly full.

Solution

1. Energy equation

$$\frac{p_1}{g} + \frac{V_1^2}{2g} + z_1 = \frac{p_2}{g} + \frac{V_2^2}{2g} + z_2 + \sum h_L$$

Let points 1 and 2 be at the upstream and downstream water surfaces, respectively.

Thus, $(p_1 = p_2 = 0 \ \text{gage} \quad \text{and} \quad V_1 = V_2 = 0)$

Also, $(z_1 - z_2 = H)$

Therefore, $(H = \sum h_L)$

2. Head losses occur at culvert entrance and exit, as well as over the length of pipe.

H = pipe head loss + entrance head loss + exit head loss

$$H = \frac{V^2}{2g}(K_e + K_E) + \text{pipe head loss}$$

$$K_e = 0.50 \text{ (from Table 10.5)}$$
$$K_E = 1.00 \text{ (from Table 10.5)}$$

3. Pipe head loss is

$$Q = \frac{1.49}{n}AR_h^{2/3}S_0^{1/2}$$

$$Q = 225 \text{ ft}^3/\text{s}$$

$$A = \frac{\pi D^2}{4} = 15.90 \text{ ft}^2$$

$$R_h = \frac{A}{P} = \frac{\pi D^2/4}{\pi D} = \frac{D}{4} = 1.125 \text{ ft}$$

$$R_h^{2/3} = (1.125 \text{ ft})^{2/3} = 1.0817 \text{ ft}^{2/3}$$

$$S_0 = \frac{h_f}{L}$$

$$225 = \frac{1.49}{0.012}(15.90 \text{ ft}^2)(1.0817 \text{ ft}^{2/3})\left(\frac{h_f}{200}\right)^{1/2}$$

$$h_f = 2.22 \text{ ft}$$

4. Continuity equation yields

$$V = \frac{Q}{A} = \frac{225 \text{ ft}^3/\text{s}}{15.90 \text{ ft}^2} = 14.15 \text{ ft/s}$$

5. Solve for H.

$$H = \frac{14.15^2}{64.4}(0.50 + 1.0) + 2.22$$

$$H = 4.66 \text{ ft} + 2.22 \text{ ft} = \boxed{6.88 \text{ ft}}$$

6. Depth of flow for $Q = 50$ cfs is

$$50 = \frac{1.49}{0.012}AR_h^{2/3}(0.01)^{1/2}$$

Values of A and R_h will depend upon geometry of partly full pipe, as shown:

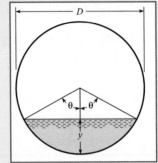

Area A if angle θ is given in degrees

$$A = \left[\left(\frac{\pi D^2}{4}\right)\left(\frac{2\theta}{360°}\right)\right] - \left(\frac{D}{2}\right)^2(\sin\theta\ \cos\theta)$$

Wetted perimeter will be $P = \pi D(\pi/180°)$, so

$$R_h = \frac{A}{P} = \left(\frac{D}{4}\right)\left[1 - \left(\frac{\sin\theta\cos\theta}{(\pi\theta/180°)}\right)\right]$$

Substituting these relations for A and R_h into the discharge equation and solving for θ yields $\theta = 70°$. Therefore, y is

$$y = \frac{D}{2} - \frac{D}{2}\cos\theta = \left(\frac{54 \text{ in}}{2}\right)(1 - 0.342) = \boxed{17.8 \text{ in}}$$

Steady Nonuniform Flow

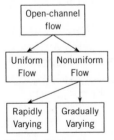

Figure 15.6

Classifying nonuniform flow.

As stated in the beginning of this chapter, and shown in Fig. 15.1, all open-channel flows are classified as either uniform or nonuniform. Recall that uniform flow has constant velocity along a streamline, and thus has constant depth for a constant cross section. In steady nonuniform flow, the depth and velocity change over distance (but not with time). For all such cases, the energy equation as generally introduced in Section 15.2 is invoked to compare two cross sections. However, for analysis of nonuniform flow, it is useful to distinguish whether the depth and velocity change occurs over a short distance, referred to as *rapidly varied flow*, or over a long reach of the channel, referred to as *gradually varied flow* (Fig.15.6). The head loss term is different for these two cases. For rapidly varied flow, one can neglect the resistance of the channel walls and bottom because it occurs over a short distance. For gradually varied flow, because of the long distances involved, the surface resistance is a signficant variable in the energy balance.

Rapidly Varied Flow

Rapidly varied flow is analyzed with the energy equation presented previously for open-channel flow, Eq. (15.7), with the additional assumptions that the channel bottom is horizontal ($S_0 = 0$) and the head loss is zero ($h_L = 0$). Therefore, Eq. (15.7) becomes

$$y_1 + \frac{V_1^2}{2g} = y_2 + \frac{V_2^2}{2g} \tag{15.17}$$

Specific Energy

The sum of the depth of flow and the velocity head is defined as the *specific energy*:

$$E = y + \frac{V^2}{2g} \tag{15.18}$$

Note that specific energy has dimensions [L]; that is, it is an energy head. Equation (15.17) states that the specific energy at section 1 is equal to the specific energy at section 2, or $E_1 = E_2$. The continuity equation between sections 1 and 2 is

$$A_1 V_1 = A_2 V_2 = Q \tag{15.19}$$

Therefore, Eq. (15.17) can be expressed as

$$y_1 + \frac{Q^2}{2gA_1^2} = y_2 + \frac{Q^2}{2gA_2^2} \tag{15.20}$$

Because A_1 and A_2 are functions of the depths y_1 and y_2, respectively, the magnitude of the specific energy at section 1 or 2 is solely a function of the depth at each section. If, for a given channel and given discharge, one plots depth versus specific energy, a relationship such as that shown in Fig. 15.7 is obtained. By studying Fig. 15.7 for a given value of spe-

Figure 15.7

Relation between depth and specific energy.

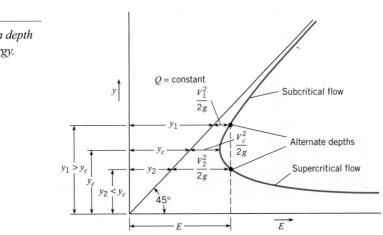

cific energy, one can see that the depth may be either large or small. This means that for the small depth, the bulk of the energy of flow is in the form of kinetic energy—that is, $Q^2/(2gA^2) \gg y$—whereas for a larger depth, most of the energy is in the form of potential energy. Flow under a *sluice gate* (Fig. 15.8) is an example of flow in which two depths occur for a given value of specific energy. The large depth and low kinetic energy occur upstream of the gate; the low depth and large kinetic energy occur downstream. The depths as used here are called *alternate depths*. That is, for a given value of E, the large depth is alternate to the low depth, or vice versa. Returning to the flow under the sluice gate, one finds that if the same rate of flow is maintained, but the gate is set with a larger opening, as in Fig. 15.8b, the upstream depth will drop, and the downstream depth will rise. This results in different alternate depths and a smaller value of specific energy than before. This is consistent with the diagram in Fig. 15.7.

Finally, it can be seen in Fig. 15.7 that a point will be reached where the specific energy is minimum and only a single depth occurs. At this point, the flow is termed *critical*. Thus one definition of critical flow is the flow that occurs when the specific energy is minimum for a given discharge. The flow for which the depth is less than critical (velocity is greater than critical) is termed *supercritical flow*, and the flow for which the depth is greater than critical (velocity is less than critical) is termed *subcritical flow*. Therefore, subcritical flow occurs upstream and supercritical flow occurs downstream of the sluice gate in Fig. 15.8. It should be noted that some engineers refer to subcritical and supercritical flow as *tranquil* and *rapid* flow, respectively. Other aspects of critical flow are shown in the next section.

Figure 15.8

Flow under a sluice gate.
(a) Smaller gate opening.
(b) Larger gate opening.

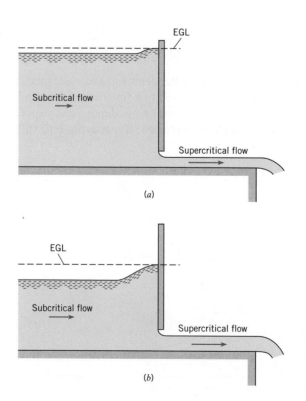

Characteristics of Critical Flow

Critical flow occurs when the specific energy is minimum for a given discharge. The depth for this condition may be determined by solving for dE/dy from $E = y + Q^2/2gA^2$ and setting dE/dy equal to zero:

$$\frac{dE}{dy} = 1 - \frac{Q^2}{gA^3} \cdot \frac{dA}{dy} \tag{15.21}$$

However, $dA = T\,dy$, where T is the width of the channel at the water surface, as shown in Fig. 15.9. Then Eq. (15.21), with $dE/dy = 0$, will reduce to

$$\frac{Q^2 T_c}{gA_c^3} = 1 \tag{15.22}$$

or

$$\frac{A_c}{T_c} = \frac{Q^2}{gA_c^2} \tag{15.23}$$

If the *hydraulic depth, D,* is defined as

$$D = \frac{A}{T} \tag{15.24}$$

then Eq. (15.23) will yield a critical hydraulic depth D_c, given by

$$D_c = \frac{Q^2}{gA_c^2} = \frac{V^2}{g} \tag{15.25}$$

Dividing Eq. (15.25) by D_c and taking the square root yields

$$1 = \frac{V}{\sqrt{gD_c}} \tag{15.26}$$

Note: $V/\sqrt{gD_c}$ is the Froude number. Therefore, it has been shown that the Froude number is equal to unity when critical flow prevails.

If a channel is of rectangular cross section, then A/T is the actual depth, and $Q^2/A^2 = q^2/y^2$, so the condition for *critical depth* (Eq. 15.23) for a rectangular channel becomes

$$y_c = \left(\frac{q^2}{g}\right)^{1/3} \tag{15.27}$$

where q is the discharge per unit width of channel.

Figure 15.9

Open-channel relations.

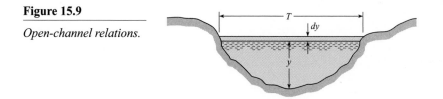

EXAMPLE 15.9   CRITICAL DEPTH IN A CHANNEL

Determine the critical depth in this trapezoidal channel for a discharge of 500 cfs. The width of the channel bottom is $B = 20$ ft, and the sides slope upward at an angle of 45°.

Problem Definition

Situation: Trapezoidal channel with known geometry.

Find: Critical depth.

Sketch:

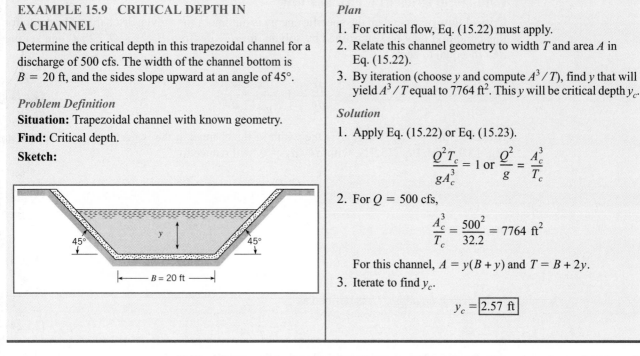

Plan

1. For critical flow, Eq. (15.22) must apply.
2. Relate this channel geometry to width T and area A in Eq. (15.22).
3. By iteration (choose y and compute A^3/T), find y that will yield A^3/T equal to 7764 ft². This y will be critical depth y_c.

Solution

1. Apply Eq. (15.22) or Eq. (15.23).

$$\frac{Q^2 T_c}{g A_c^3} = 1 \quad \text{or} \quad \frac{Q^2}{g} = \frac{A_c^3}{T_c}$$

2. For $Q = 500$ cfs,

$$\frac{A_c^3}{T_c} = \frac{500^2}{32.2} = 7764 \text{ ft}^2$$

For this channel, $A = y(B + y)$ and $T = B + 2y$.

3. Iterate to find y_c.

$$y_c = \boxed{2.57 \text{ ft}}$$

Critical flow may also be examined in terms of how the discharge in a channel varies with depth for a given specific energy. For example, consider flow in a rectangular channel where

$$E = y + \frac{Q^2}{2gA^2}$$

or

$$E = y + \frac{Q^2}{2gy^2 B^2}$$

If one considers a unit width of the channel and lets $q = Q/B$, then the foregoing equation becomes

$$E = y + \frac{q^2}{2gy^2}$$

If one determines how q varies with y for a constant value of specific energy, one sees that critical flow occurs when the discharge is maximum (see Fig. 15.10).

Figure 15.10

Variation of q and y with constant specific energy.

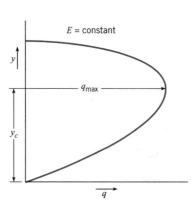

Originally, the term *critical flow* probably related to the unstable character of the flow for this condition. Referring to Fig. 15.7, one can see that only a slight change in specific energy will cause the depth to increase or decrease a significant amount; this is a very unstable condition. In fact, observations of critical flow in open channels show that the water surface consists of a series of standing waves. Because of the unstable nature of the depth in critical flow, designing canals so that normal depth is either well above or well below critical depth is usually best. The flow in canals and rivers is usually subcritical; however, the flow in steep chutes or over spillways is supercritical.

In this section, various characteristics of critical flow have been explored. The main ones can be summarized as follows:

1. Critical flow occurs when specific energy is minimum for a given discharge (Fig. 15.7).
2. Critical flow occurs when the discharge is maximum for a given specific energy.
3. Critical flow occurs when

$$\frac{A^3}{T} = \frac{Q^2}{g}$$

4. Critical flow occurs when Fr $= 1$.
5. For rectangular channels, critical depth is given as $y_c = (q^2/g)^{1/3}$.

Common Occurrence of Critical Flow

Critical flow occurs when a liquid passes over a broad-crested weir (Fig. 15.11). The principle of the broad-crested weir is illustrated by first considering a closed sluice gate that prevents water from being discharged from the reservoir, as shown in Fig. 15.11*a*. If the gate is opened a small amount (gate position a'-a'), the flow upstream of the gate will be subcritical and the flow downstream will be supercritical (as in the condition shown in Fig. 15.8). As the gate is opened further, a point is finally reached where the depths immediately upstream and downstream of the gate are the same. This is the critical condition. At this gate opening and beyond, the gate has no influence on the flow; this is the condition shown in Fig. 15.11*b*, the broad-crested weir. If the depth of flow over the weir is measured, the rate of flow can easily be computed from Eq. (15.27):

$$q = \sqrt{gy_c^3}$$

or
$$Q = L\sqrt{gy_c^3} \tag{15.28}$$

where L is the length of the weir crest normal to the flow direction.

Figure 15.11

Flow over a broad-crested weir. (a) Depth of flow controlled by sluice gate. (b) Depth of flow is controlled by weir, and is y_c.

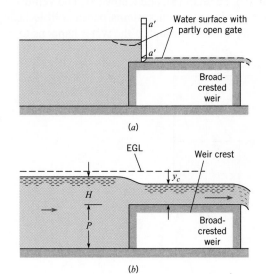

Because $y_c/2 = (V_c^2/2g)$, from Eq. (15.25), it can be shown that $y_c = (2/3E)$, where E is the total head above the crest $(H + V_{approach}^2/2g)$; hence Eq. (15.28) can be rewritten as

$$Q = L\sqrt{g}\left(\frac{2}{3}\right)^{3/2}E^{3/2}$$

or
$$Q = 0.385L\sqrt{2g}E_c^{3/2} \qquad (15.29)$$

For high weirs, the upstream velocity of approach is almost zero. Hence Eq. (15.29) can be expressed as

$$Q_{theor} = 0.385L\sqrt{2g}H^{3/2} \qquad (15.30)$$

If the height P of the broad-crested weir is relatively small, then the velocity of approach may be significant, and the discharge produced will be greater than that given by Eq. (15.30). Also, head loss will have some effect. To account for these effects, a discharge coefficient C is defined as

$$C = Q/Q_{theor} \qquad (15.31)$$

Then
$$Q = 0.385CL\sqrt{2g}H^{3/2} \qquad (15.32)$$

where Q is the actual discharge over the weir. An analysis of experimental data by Raju (15) shows that C varies with $H/(H + P)$ as shown in Fig. 15.12. The curve in Fig. 15.12 is for a weir with a vertical upstream face and a sharp corner at the intersection of the upstream face and the weir crest. If the upstream face is sloping at a 45° angle, the discharge coefficient should be increased 10% over that given in Fig. 15.12. Rounding of the upstream corner will also produce a coefficient of discharge as much as 3% greater.

Equation (15.32) reveals a definite relationship for Q as a function of the head, H. This type of discharge-measuring device is in the broad class of discharge meters called *critical-flow flumes*. Another very common critical-flow flume is the *venturi flume,* which was developed and calibrated by Parshall (8). Figure 15.13 shows the essential features of the venturi flume. The discharge equation for the venturi flume is in the same form as Eq. (15.32), the only difference being that the experimentally determined coefficient C will have a different value from the C for the broad-crested weir. For more details on the venturi flume, you may refer to Roberson et al. (9), Parshall (8), and Chow (5). The venturi flume is especially useful for discharge measurement in irrigation systems because little head loss is required for its use, and sediment is easily flushed through if the water happens to be silty.

Figure 15.12

Discharge coefficient for a broad-crested weir for 0.1 < H/L < 0.8.

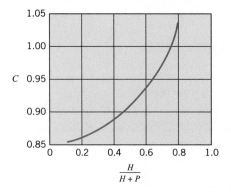

Figure 15.13

Flow through a venturi flume.

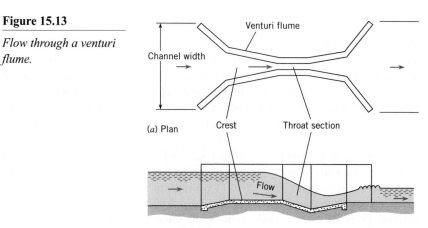

(a) Plan

(b) Profile

The depth also passes through a critical stage in channel flow where the slope changes from a mild one to a steep one. A *mild slope* is defined as a slope for which the normal depth y_n is greater than y_c. Likewise, a *steep slope* is one for which $y_n < y_c$. This condition is shown in Fig. 15.14. Note that y_c is the same for both slopes in the figure because y_c is a function of the discharge only. However, normal depth (uniform-flow depth) for the mild upstream channel is greater than critical, whereas the normal depth for the steep downstream channel is less than critical; hence it is obvious that the depth must pass through a critical stage. Experiments show that critical depth occurs a very short distance upstream of the intersection of the two channels.

Another place where critical depth occurs is upstream of a free overfall at the end of a channel with a mild slope Fig. 15.15. Critical depth will occur at a distance of $3y_c$ to $4y_c$ upstream of the brink. Such occurrences of critical depth (at a break in grade or at a brink) are useful in computing surface profiles because they provide a point for starting surface-profile calculations.*

Figure 15.14

Critical depth at a break in grade.

Figure 15.15

Critical depth at a free overfall.

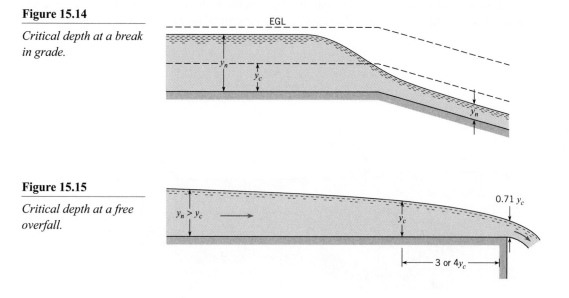

* The procedure for making these computations starts on p. 541 (water surface profiles).

Channel Transitions

Whenever a channel's cross-sectional configuration (shape or dimension) changes along its length, the change is termed a *transition*. Concepts previously presented are used to show how the flow depth changes when the floor of a rectangular channel is increased in elevation or when the width of the channel is decreased. In these developments negligible energy losses are assumed. First, the case where the floor of the channel is raised (an upstep) is considered. Later in this section, configurations of transitions used for subcritical flow from a rectangular to a trapezoidal channel are presented.

Consider the rectangular channel shown in Fig. 15.16, where the floor rises an amount Δz. To help in evaluating depth changes, one can use a diagram of specific energy versus depth, which is similar to Fig. 15.7. This diagram is applied both at the section upstream of the transition and at the section just downstream of the transition. Because the discharge, Q, is the same at both sections, the same diagram is valid at both sections. As noted in Fig. 15.16, the depth of flow at section 1 can be either large (subcritical) or small (supercritical) if the specific energy E_1 is greater than that required for critical flow. It can also be seen in Fig. 15.16 that when the upstream flow is subcritical, a decrease in depth occurs in the region of the elevated channel bottom. This occurs because the specific energy at this section, E_2, is less than that at section 1 by the amount Δz. Therefore, the specific-energy diagram indicates that y_2 will be less than y_1. In a similar manner it can be seen that when the upstream flow is supercritical, the depth as well as the actual water-surface elevation increases from section 1 to section 2. A further note should be made about the effect on flow depth of a change in bottom-surface elevation. If the channel bottom at section 2 is at an elevation greater than that just sufficient to establish critical flow at section 2, then there is not enough head at section 1 to cause flow to occur over the rise under steady-flow conditions. Instead, the water level upstream will rise until it is just sufficient to reestablish steady flow.

When the channel bottom is kept at the same elevation but the channel is decreased in width, then the discharge per unit of width between sections 1 and 2 increases, but the specific energy E remains constant. Thus when utilizing the diagram of q versus depth for the given specific energy E, one notes that the depth in the restricted section increases if the upstream flow is supercritical and decreases if it is subcritical (see Fig. 15.17).

Figure 15.16

Change in depth with change in bottom elevation of a rectangular channel.

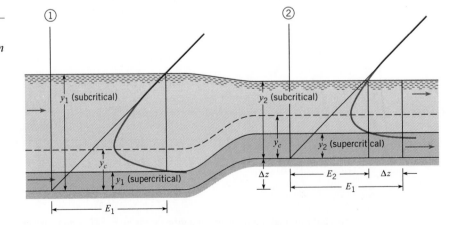

Figure 15.17

Change in depth with change in channel width.

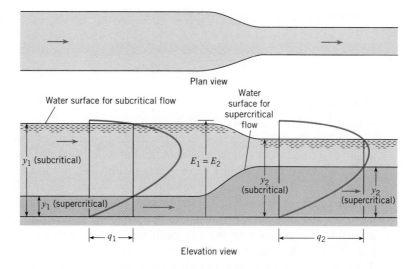

Plan view

Elevation view

The foregoing paragraphs describe gross effects for the simplest transitions. In practice, it is more common to find transitions between a channel of one shape (rectangular cross section, for example) and a channel having a different cross section (trapezoidal, for example). A very simple transition between two such channels consists of two straight vertical walls joining the two channels, as shown by half section in Fig. 15.18.

This type of transition can work, but it will produce excessive head loss because of the abrupt change in cross section, and the ensuing separation that will occur. To reduce the head losses, a more gradual type of transition is used. Figure 15.19 is a half section of a transition similar to that of Fig. 15.18, but with the angle θ much greater than 90°. This is called a *wedge transition*.

The *warped-wall* transition shown in Fig. 15.20 will yield even smoother flow than either of the other two, and it will thus have less head loss. In the practical design and analysis of transitions, engineers usually use the complete energy equation, including the kinetic energy factors α_1 and α_2 as well as a head loss term h_L, to define velocity and water-surface elevation through the transition. Analyses of transitions utilizing the one-dimensional form of the energy equation are applicable only if the flow is subcritical. If the flow is supercritical, then a much more involved analysis is required. For more details on the design and analysis of transitions, you are referred to Hinds (10), Chow (5), U.S. Bureau of Reclamation (11), and Rouse (12).

Figure 15.18

Simplest type of transition between a rectangular channel and a trapezoidal channel.

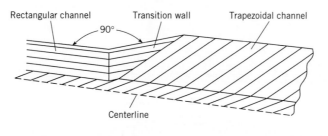

Figure 15.19

Half section of a wedge transition.

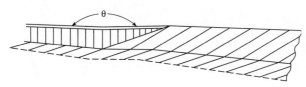

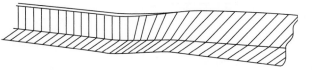

Wave Celerity

Wave celerity is the velocity at which an infinitesimally small wave travels relative to the velocity of the fluid. It can be used to characterize the velocity of waves in the ocean, or propagation of a flood wave following a dam failure. A derivation of wave celerity, c, follows.

Consider a small solitary wave moving with velocity c in a calm body of liquid of small depth (Fig. 15.21a). Because the velocity in the liquid changes with time, this is a condition of unsteady flow. However, if one referred all velocities to a reference frame moving with the wave, the shape of the wave would be fixed, and the flow would be steady. Then the flow is amenable to analysis with the Bernoulli equation. The steady-flow condition is shown in Fig. 15.21b. When the Bernoulli equation is written between a point on the surface of the undisturbed fluid and a point at the wave crest, the following equation results:

$$\frac{c^2}{2g} + y = \frac{V^2}{2g} + y + \Delta y \tag{15.33}$$

In Eq. (15.33), V is the velocity of the liquid in the section where the crest of the wave is located. From the continuity equation, $cy = V(y + \Delta y)$. Hence

$$V = \frac{cy}{y + \Delta y}$$

and

$$V^2 = \frac{c^2 y^2}{(y + \Delta y)^2} \tag{15.34}$$

When Eq. (15.34) is substituted into Eq. (15.33), it yields

$$\frac{c^2}{2g} + y = \frac{c^2 y^2}{2g[y^2 + 2y\Delta y + (\Delta y)^2]} + y + \Delta y \tag{15.35}$$

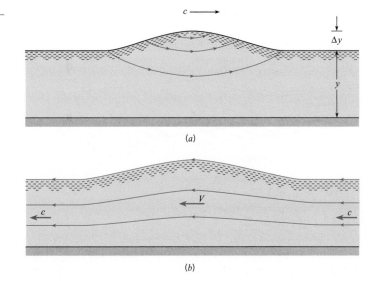

(a)

(b)

Solving Eq. (15.35) for c after discarding terms with $(\Delta y)^2$, assuming an infinitesimally small wave, yields the *wave celerity equation*

$$c = \sqrt{gy} \qquad (15.36)$$

It has thus been shown that the speed of a small solitary wave is equal to the square root of the product of the depth and g.

15.6

Hydraulic Jump

Occurrence of the Hydraulic Jump

An interesting and important case of rapidly varied flow is the hydraulic jump. A *hydraulic jump* occurs when the flow is supercritical in an upstream section of a channel and is then forced to become subcritical in a downstream section (the change in depth can be forced by a sill in the downstream part of the channel or just by the prevailing depth in the stream further downstream), resulting in an abrupt increase in depth, and considerable energy loss. Hydraulic jumps (Fig. 15.22) are often considered in the design of open channels and spillways of dams. If a channel is designed to carry water at supercritical velocities, the designer must be certain that the flow will not become subcritical prematurely. If it did, overtopping of the channel walls would undoubtedly occur, with consequent failure of the structure. Because the energy loss in the hydraulic jump is initially not known, the energy equation is not a suitable tool for analysis of the velocity-depth relationships. Because there is a significant difference in hydrostatatic head on both sides of the equation causing opposing pressure forces, the momentum equation can be applied to the problem, as developed in the following sections.

Derivation of Depth Relationships in Hydraulic Jumps

Consider flow as shown in Fig. 15.22. Here it is assumed that uniform flow occurs both upstream and downstream of the jump and that the resistance of the channel bottom over the relatively short distance L is negligible. The derivation is for a horizontal channel, but experiments show that the results of the derivation will apply to all channels of moderate slope ($S_0 < 0.02$). The derivation is started by applying the momentum equation in the x-direction to the control volume shown in Fig. 15.23:

$$\sum F_x = \dot{m}_2 V_2 - \dot{m}_1 V_1$$

Figure 15.22

Definition sketch for the hydraulic jump.

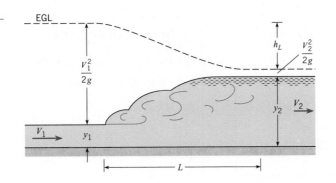

Figure 15.23

Control-volume analysis for the hydraulic jump.

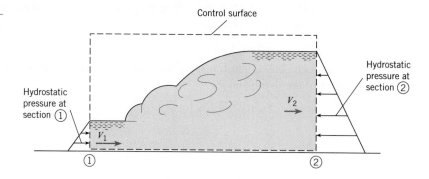

The forces are the hydrostatic forces on each end of the system; thus the following is obtained:

$$\overline{p}_1 A_1 - \overline{p}_2 A_2 = \rho V_2 A_2 V_2 - \rho V_1 A_1 V_1$$

or

$$\overline{p}_1 A_1 + \rho Q V_1 = \overline{p}_2 A_2 + \rho Q V_2 \tag{15.37}$$

In Eq. (15.37), $\overline{p}_1$ and $\overline{p}_2$ are the pressures at the centroids of the respective areas A_1 and A_2.

A representative problem (e.g., Example 15.10) is to determine the downstream depth y_2 given the discharge and upstream depth. The left-hand side of Eq. (15.37) would be known because V, A, and p are all functions of y and Q, and the right-hand side is a function of y_2; therefore, y_2 can be determined.

EXAMPLE 15.10 DOWNSTREAM DEPTH IN HYDRAULIC JUMP

Water flows in a trapezoidal channel at a rate of 300 cfs. The channel has a bottom width of 10 ft and side slopes of 1 vertical to 1 horizontal. If a hydraulic jump is forced to occur where the upstream depth is 1.0 ft, what will be the downstream depth and velocity? What are the values of Fr_1 and Fr_2?

Problem Definition

Situation: Known upstream conditions, hydraulic jump is forced to occur (details not described).

Find:

1. Downstream depth and velocity.
2. Values of Fr_1 and Fr_2.

Sketch:

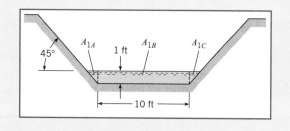

Properties: Water (50°F), Table A.5:

$\gamma = 62.4 \ \mathrm{lbf/ft}^3$, and $\rho = 1.94 \ \mathrm{slugs/ft}^3$.

Plan

1. Find cross section, velocity, and hydraulic depth in the upstream section.
2. Find pressure in the upstream section to use for left-hand side of Eq. (15.37).
3. Use channel geometry information to solve for y_2 in right-hand side of Eq. (15.37).
4. Use Eq. (15.2) to solve for the Froude number at both sections.

Solution

1. By inspection, for the upstream section, the cross-sectional flow area is 11 ft².
   Therefore, the mean velocity is $V_1 = Q/A_1 = 27.3 \ \mathrm{ft/s}$.
   The hydraulic depth is $D_1 = A_1/T_1 = 11 \ \mathrm{ft}^2/12 \ \mathrm{ft} = 0.9167$ ft.

2. The location of the centroid ($\bar{y}$) of the area A_1 can be obtained by taking moments of the sub-areas about the water surface (see example sketch).

$$A_1\bar{y}_1 = A_{1A} \times 0.333 \text{ ft} + A_{1B} \times 0.500 \text{ ft} + A_{1C} \times 0.333 \text{ ft}$$

$$(11 \text{ ft}^2)\bar{y}_1 = (0.333 \text{ ft})(0.500 \text{ ft}^2 \times 2) + (0.50 \text{ ft})(10.00 \text{ ft}^2)$$

$$\bar{y} = 0.485 \text{ ft}$$

Pressure $p_1 = 62.4 \text{ lbf/ft}^3 \times 0.485 \text{ ft} = 30.26 \text{ lbf/ft}^2$. Therefore,

$$30.26 \times 11 + 1.94 \times 300 \times 27.3 = \bar{p}_2 A_2 + \rho Q V_2$$

3. Using right-hand side of Eq. (15.37), solve for y_2.

$$\bar{p}_2 A_2 + \rho Q V_2 = 16{,}221 \text{ lbf}$$

$$\gamma \bar{y}_2 A_2 + \frac{\rho Q^2}{A_2} = 16{,}221$$

$$\bar{y}_2 = \frac{\sum A_i y_i}{A_2} = \frac{B y_2^2/2 + y_2^3/3}{A_2}$$

Using $B = 10$ ft, $Q = 300 \text{ ft}^2/\text{s}$, and material properties assumed earlier,

$$y_2 = \boxed{5.75 \text{ ft}}$$

4. Froude numbers at both sections are

$$\text{Fr}_1 = \frac{V_1}{\sqrt{gD_1}} = \frac{27.3 \text{ ft/s}}{\sqrt{32.2 \text{ ft/s}^2 \times 0.9167 \text{ ft}}} = \boxed{5.02}$$

$$V_2 = \frac{Q}{A_2} = \frac{300}{57.5 + 5.75^2} = 3.31 \text{ ft/s}$$

$$D_2 = \frac{A_2}{T_2} = \frac{90.56}{21.5} = 4.21 \text{ ft}$$

$$\text{Fr}_2 = \frac{V}{\sqrt{gD}} = \frac{3.31}{\sqrt{32.2 \times 4.21}} = \boxed{0.284}$$

Hydraulic Jump in Rectangular Channels

If one writes Eq. (15.37) for a unit width of a rectangular channel where $\bar{p}_1 = \gamma y_1/2$, $\bar{p}_2 = \gamma y_2/2$, $Q = q$, $A_1 = y_1$, and $A_2 = y_2$, this will yield

$$\gamma \frac{y_1^2}{2} + \rho q V_1 = \gamma \frac{y_2^2}{2} + \rho q V_2 \tag{15.38a}$$

but $q = Vy$, so Eq. (15.38a) can be rewritten as

$$\frac{\gamma}{2}(y_1^2 - y_2^2) = \frac{\gamma}{g}(V_2^2 y_2 - V_1^2 y_1) \tag{15.39b}$$

The preceding equation can be further manipulated to yield

$$\frac{2V_1^2}{gy_1} = \left(\frac{y_2}{y_1}\right)^2 + \frac{y_2}{y_1} \tag{15.40}$$

The term on the left-hand side of Eq. (15.40) will be recognized as twice Fr_1^2. Hence Eq. (15.40) is written as

$$\left(\frac{y_2}{y_1}\right)^2 + \frac{y_2}{y_1} - 2\text{Fr}_1^2 = 0 \tag{15.41}$$

By use of the quadratic formula, it is easy to solve for y_2/y_1 in terms of the upstream Froude number. This yields an equation for *depth ratio* across a hydraulic jump:

$$\frac{y_2}{y_1} = \frac{1}{2}(\sqrt{1 + 8\text{Fr}_1^2} - 1) \tag{15.42}$$

or

$$y_2 = \frac{y_1}{2}(\sqrt{1 + 8\text{Fr}_1^2} - 1) \tag{15.43}$$

The other solution of Eq. (15.41) gives a negative downstream depth, which is not physically possible. Hence the downstream depth is expressed in terms of the upstream depth and the upstream Froude number. In Eqs. (15.42) and (15.43), the depths y_1 and y_2 are said to be *conjugate* or *sequent* (both terms are in common use) to each other, in contrast to the alternate depths obtained from the energy equation. Numerous experiments show that the relation represented by Eqs. (15.42) and (15.43) is valid over a wide range of Froude numbers.

Although no theory has been developed to predict the length of a hydraulic jump, experiments [see Chow (5)] show that the relative length of the jump, L/y_2, is approximately 6 for Fr_1 ranging from 4 to 18.

Head Loss in a Hydraulic Jump

In addition to determining the geometric characteristics of the hydraulic jump, it is often desirable to determine the head loss produced by it. This is obtained by comparing the specific energy before the jump to that after the jump, the head loss being the difference between the two specific energies. It can be shown that the head loss for a jump in a rectangular channel is

$$h_L = \frac{(y_2 - y_1)^3}{4y_1 y_2} \tag{15.44}$$

For more information on the hydraulic jump, see Chow (5). The following example shows that Eq. (15.44) yields a magnitude that equals the difference between the specific energies at the two ends of the hydraulic jump.

EXAMPLE 15.11 HEAD LOSS IN HYDRAULIC JUMP

Water flows in a rectangular channel at a depth of 30 cm with a velocity of 16 m/s, as shown in the sketch that follows. If a downstream sill (not shown) forces a hydraulic jump, what will be the depth and velocity downstream of the jump? What head loss is produced by the jump?

Problem Definition

Situation: Channel is rectangular; upstream depth and velocity known.

Find:

1. Downstream depth and velocity.
2. Head loss produced by the jump.

Sketch:

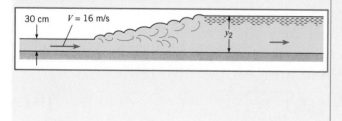

Plan

1. In order to calculate h_L using Eq. (15.44), must calculate y_2 from the depth ratio equation (Eq. 15.43). This requires Fr_1.
2. Check validity of head loss by comparing to $E_1 - E_2$.

Solution

1. Calculate Fr_1, y_2, V_2, and h_L from Eqs. (Eq. 15.43) and (15.44).

$$Fr_1 = \frac{V}{\sqrt{gy_1}} = \frac{16}{\sqrt{9.81(0.30)}} = 9.33$$

$$y_2 = \frac{0.30}{2}[\sqrt{1 + 8(9.33)^2} - 1] = \boxed{3.81 \text{ m}}$$

$$V_2 = \frac{q}{y_2} = \frac{(16 \text{ m/s})(0.30 \text{ m})}{3.81 \text{m}} = \boxed{1.26 \text{ m/s}}$$

$$h_L = \frac{(3.81 - 0.30)^3}{4(0.30)(3.81)} = \boxed{9.46 \text{ m}}$$

2. Compare the head loss to $E_1 - E_2$.

$$h_L = \left(0.30 + \frac{16^2}{2 \times 9.81}\right) - \left(3.81 + \frac{1.26^2}{2 \times 9.81}\right) = 9.46 \text{ m}$$

The value is the same, so

$\boxed{\text{validity of } h_L \text{ equation is verified.}}$

Use of Hydraulic Jump on Downstream End of Dam Spillway

Previously it was shown that the transition from supercritical to subcritical flow produces a hydraulic jump, and that the relative height of the jump (y_2/y_1) is a function of Fr_1. Because flow over the spillway of a dam invariably results in supercritical flow at the lower end of the spillway, and because flow in the channel downstream of a spillway is usually subcritical, it is obvious that a hydraulic jump must form near the base of the spillway (see Fig. 15.26). The downstream portion of the spillway, called the spillway *apron*, must be designed so that the hydraulic jump always forms on the concrete structure itself. If the hydraulic jump were allowed to form beyond the concrete structure, as in Fig. 15.25, severe erosion of the foundation material as a result of the high-velocity supercritical flow could undermine the dam and cause its complete failure. One way to solve this problem might be to incorporate a long, sloping apron into the design of the spillway, as shown in Fig. 15.26. A design like this would work

Figure 15.24

Spillway of dam and hydraulic jump.

Figure 15.25

Hydraulic jump occurring downstream of spillway apron.

Figure 15.26

Long sloping apron.

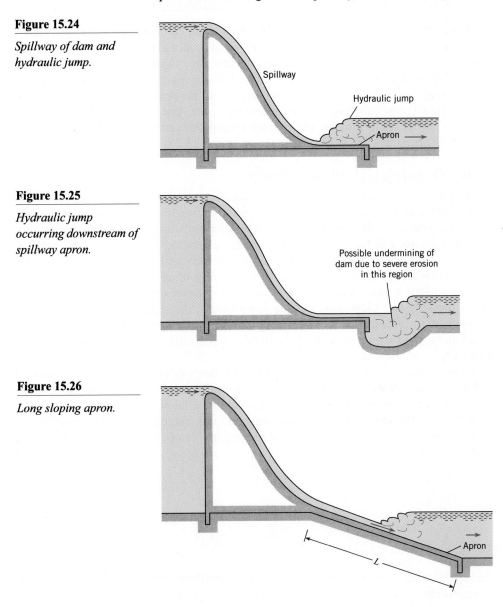

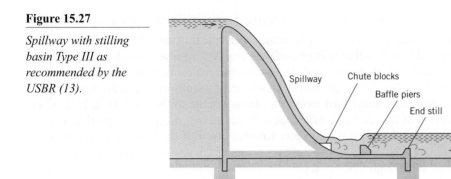

very satisfactorily from the hydraulics point of view. For all combinations of Fr_1 and water-surface elevation in the downstream channel, the jump would always form on the sloping apron. However, its main drawback is cost of construction. Construction costs will be reduced as the length, L, of the stilling basin is reduced. Much research has been devoted to the design of stilling basins that will operate properly for all upstream and downstream conditions and yet be relatively short to reduce construction cost. Research by the U.S. Bureau of Reclamation (13) has resulted in sets of standard designs that can be used. These designs include sills, baffle piers, and chute blocks, as shown in Fig. 15.27.

Naturally Occurring Hydraulic Jumps

Hydraulic jumps can occur naturally in creeks and rivers, providing spectacular standing waves, called rollers. Kayakers and white-water rafters must exercise considerable skill when navigating hydraulic jumps because the significant energy loss that occurs over a short distance can be dangerous, potentially engulfing the boat in turbulence. A special case of hydraulic jump, referred to as a submerged hydraulic jump, can be deadly to whitewater enthusiasts because it is not easy to see. A *submerged hydraulic jump* occurs when the downstream depth predicted by conservation of momentum is exceeded by the tailwater elevation, and the jump cannot move upstream in response to this disequilibrium because of a buried obstacle [see Valle and Pasternak (14)]. Thus, the visual markers of a hydraulic jump, particularly the rolling waves depicted in Figs. 15.22 and 15.23, are hidden.

A *surge*, or *tidal bore*, is a moving hydraulic jump that may occur for a high tide entering a bay or river mouth. Tides are generally low enough that the waves they produce are smooth and nondestructive. However, in some parts of the world the tides are so high that their entry into shallow bays or mouths of rivers causes a surge to form. Surges may be very hazardous to small boats. The same analytical methods used for the jump can be used to solve for the speed of the surge.

Gradually Varied Flow

For gradually varied flow, channel resistance is a significant factor in the flow process. Therefore, the energy equation is invoked by comparing S_0 and S_f.

Basic Differential Equation for Gradually Varied Flow

There are a number of cases of open-channel flow in which the change in water-surface profile is so gradual that it is possible to integrate the relevant differential equation from one section to another to obtain the desired change in depth. This may be either an analytical integration or, more commonly, a numerical integration. In Section 15.2, the energy equation was written between two sections of a channel Δx distance apart. Because the only head loss here is the channel resistance, the h_L is given by Δh_f, and Eq. (15.7) becomes

$$y_1 + \frac{V_1^2}{2g} + S_0 \Delta x = y_2 + \frac{V_2^2}{2g} + \Delta h_f \tag{15.45}$$

The friction slope S_f is defined as the slope of the EGL, or $\Delta h_f / \Delta x$. Thus $\Delta h_f = S_f \Delta x$, and defining $\Delta y = y_2 - y_1$, then

$$\frac{V_2^2}{2g} - \frac{V_1^2}{2g} = \frac{d}{dx}\left(\frac{V^2}{2g}\right)\Delta x \tag{15.46}$$

Therefore, Eq. (15.45) becomes

$$\Delta y = S_0 \Delta x - S_f \Delta x - \frac{d}{dx}\left(\frac{V^2}{2g}\right)\Delta x$$

Dividing through by Δx and taking the limit as Δx approaches zero gives us

$$\frac{dy}{dx} + \frac{d}{dx}\left(\frac{V^2}{2g}\right) = S_0 - S_f \tag{15.47}$$

The second term is rewritten as $[d(V^2/2g)/dy]\,dy/dx$, so that Eq. (15.47) simplifies to

$$\frac{dy}{dx} = \frac{S_0 - S_f}{1 + d(V^2/2g)/dy} \tag{15.48}$$

To put Eq. (15.48) in a more usable form, the denominator is expressed in terms of the Froude number. This is accomplished by observing that

$$\frac{d}{dy}\left(\frac{V^2}{2g}\right) = \frac{d}{dy}\left(\frac{Q^2}{2gA^2}\right) \tag{15.49}$$

After differentiating the right side of Eq. (15.35), the equation becomes

$$\frac{d}{dy}\left(\frac{V^2}{2g}\right) = \frac{-2Q^2}{2gA^3} \cdot \frac{dA}{dy}$$

But $dA/dy = T$ (top width), and $A/T = D$ (hydraulic depth); therefore,

$$\frac{d}{dy}\left(\frac{V^2}{2g}\right) = \frac{-Q^2}{gA^2 D}$$

or

$$\frac{d}{dy}\left(\frac{V^2}{2g}\right) = -\mathrm{Fr}^2$$

Hence, when the expression for $d(V^2/2g)/dy$ is substituted into Eq. (15.48), the result is

$$\frac{dy}{dx} = \frac{S_0 - S_f}{1 - \mathrm{Fr}^2} \tag{15.50}$$

This is the general differential equation for gradually varied flow. It is used to describe the various types of water-surface profiles that occur in open channels. Note that, in the derivation of the equation, S_0 and S_f were taken as positive when the channel and energy grade lines, respectively, were sloping downward in the direction of flow. Also note that y is measured from the bottom of the channel. Therefore, $dy/dx = 0$ if the slope of the water surface is equal to the slope of the channel bottom, and dy/dx is positive if the slope of the water surface is less than the channel slope.

Introduction to Water-Surface Profiles

In the design of projects involving the flow in channels (rivers or irrigation canals, for example), the engineer must often estimate the *water-surface profile* (elevation of the water surface along the channel) for a given discharge. For example, when a dam is being designed for a river project, the water-surface profile in the river upstream must be defined so that the project planners will know how much land to acquire to accommodate the upstream pool. The first step in defining a water-surface profile is to locate a point or points along the channel where the depth can be computed for a given discharge. For example, at a change in slope from mild to steep, critical depth will occur just upstream of the break in grade (see Fig. 15.31). At that point one can solve for y_c with Eq. (15.22) or (15.27). Also, for flow over the spillway of a dam, there will be a discharge equation for the spillway from which one can calculate the water-surface elevation in the reservoir at the face of the dam. Such points where there is a unique relationship between discharge and water-surface elevation are called *controls*. Once the water-surface elevations at these controls are determined, then the water-surface profile can be extended upstream or downstream from the control points to define the water-surface profile for the entire channel. The completion of the profile is done by numerical integration. However, before this integration is performed, it is usually helpful for the engineer to sketch in the profiles. To assist in the process of sketching the possible profiles, the engineer can refer to different categories of profiles (water-surface profiles have unique characteristics depending upon the relationship between normal depth, critical depth, and the actual depth of flow in the channel). This initial sketching of the profiles helps the engineer to scope the problem and to obtain a solution, or solutions, in a minimum amount of time. The next section describes the various types of water-surface profiles.

Types of Water-Surface Profiles

There are 12 different types of water-surface profiles for gradually varied flow in channels, and these are shown schematically in Fig. 15.28. Each profile is identified by a letter and number designator. For example, the first water-surface profile in column 1 of Fig. 15.28 is identified as an M1 profile. The letter indicates the type of slope of the channel—that is, whether the slope is mild (M), critical (C), steep (S), horizontal (H), or adverse (A). The slope is defined as mild if the uniform flow depth, y_n, is greater than the critical flow depth, y_c. Conversely, if y_n is less than y_c, the channel would be termed steep. Or if $y_n = y_c$, this would be a channel with critical slope. The designation M, S, or C is determined by computing y_n and y_c for the given channel for a given discharge. Equations (15.11) through (15.15) are used to compute y_n, and Eq. (15.22) or (15.27) is used to compute y_c. Figure 15.29 shows the relationship between y_n and y_c for the H, M, S, C, and A designations. As the name implies, a horizontal slope is one where the channel actually has a zero slope, and an adverse slope is one where the slope of the channel is upward in the direction of flow. Normal depth does not exist for these two cases (for example, water cannot flow at uniform depth in either a horizontal channel or one with adverse slope); therefore, they are given the special designations H and A, respectively.

Figure 15.28

Classification of water-surface profiles of gradually varied flow. [Adapted from Open Channel Hydraulics *by Chow (5). Copyright © 1959, McGraw-Hill Book Company, New York; used with permission of McGraw-Hill Book Company.]*

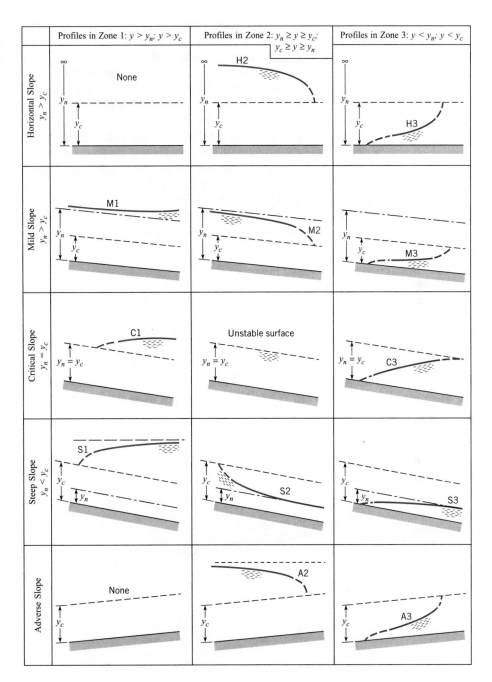

Figure 15.29

Letter designators as a function of the relationship between y_n and y_c.

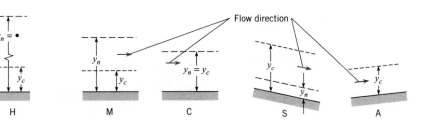

The number designator for the type of profile relates to the position of the *actual* water surface in relation to the position of the water surface for uniform and critical flow in the channel. If the actual water surface is above that for uniform and critical flow ($y > y_n$; $y > y_c$), then that condition is given a 1 designation; if the actual water surface is between those for uniform and critical flow, then it is given a 2 designation; and if the actual water surface lies below those for uniform and critical flow, then it is given a 3 designation. Figure 15.30 depicts these conditions for mild and steep slopes.

Figure 15.31 shows how different water-surface profiles can develop in certain field situations. More specifically, if one considers in detail the flow downstream of the sluice gate (see Fig. 15.32), one can see that the discharge and slope are such that the normal depth is greater than the critical depth; therefore the slope is termed mild. The actual depth of flow shown in Fig. 15.32 is less than both y_c and y_n. Hence a type 3 water-surface profile exists. The complete classification of the profile in Fig. 15.32, therefore, is a mild type 3 profile, or simply an M3 profile. Using these designations, one would categorize the profile upstream of the sluice gate as type M1.

Figure 15.30

Number designator as a function of the location of the actual water surface in relation to y_n and y_c.

Figure 15.31

Water-surface profiles associated with flow behind a dam, flow under a sluice gate, and flow in a channel with a change in grade.

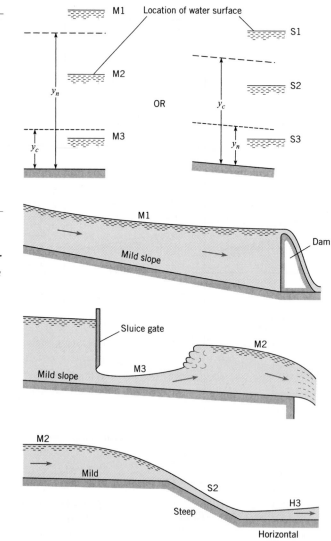

Figure 15.32

Water-surface profile,
M3 type.

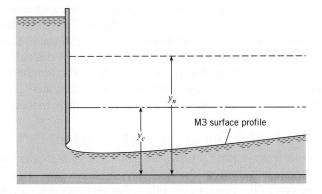

y_n

M3 surface profile

y_c

EXAMPLE 15.12   CLASSIFICATION OF WATER-SURFACE PROFILES

Classify the water-surface profile for the flow downstream of the sluice gate in Fig. 15.8 if the slope is horizontal, and that for the flow immediately downstream of the break in grade in Fig. 15.14.

Problem Definition

Situation: Nonuniform flow. Figures from which channel steepness and water surface steepness can be determined, and from which one can infer whether depth is more or less than critical depth.

Find: The water-surface profile classification for two different flow situations.

Plan

1. Select a number designator based upon the location of the actual water surface relative to y_n and y_c (see Fig. 15.30).

2. Select a letter designator to describe the steepness of the slopes, which can also be characterized by the relative size of y_n and y_c (see Fig. 15.29).

Solution

For Fig. 15.8

1. The actual depth is less than critical; thus the profile is type 3.
2. The channel is horizontal; hence the profile is designated

   type H3.

For Fig. 15.14

1. The actual depth is greater than normal but less than critical, so the profile is type 2.
2. The uniform-flow depth (normal depth y_n) is less than the critical depth; hence the slope is steep. Therefore the water-surface profile is designated   type S2.

With the previous introduction to the classification of water-surface profiles, one can refer to Eq. (15.50) to describe the shapes of the profiles. Again, for example, if one considers the M3 profile, it is known that Fr > 1 because the flow is supercritical ($y < y_c$), and that $S_f > S_0$ because the velocity is greater than normal velocity. Hence a head loss greater than that for normal flow must exist. Inserting these relative values into Eq. (15.50) reveals that both the numerator and the denominator are negative. Thus dy/dx must be positive (the depth increases in the direction of flow), and as critical depth is approached, the Froude number approaches unity. Hence the denominator of Eq. (15.50) approaches zero. Therefore, as the depth approaches critical depth, $dy/dx \to \infty$. What actually occurs in cases where the critical depth is approached in supercritical flow is that a hydraulic jump forms and a discontinuity in profile is thereby produced.

Certain general features of profiles, as shown in Fig. 15.28, are evident. First, as the depth becomes very great, the velocity of flow approaches zero. Hence Fr $\to 0$ and $S_f \to 0$ and dy/dx approaches S_0 because $dy/dx = (S_0 - S_f)(1 - \text{Fr}^2)$. In other words, the depth increases at the same rate at which the channel bottom drops away from the horizontal. Thus the water surface approaches the horizontal. The profiles that show this tendency are types M1, S1, and C1. A physical example of the M1 type is the water-surface profile upstream of a dam, as shown in Fig. 15.31. The second general feature of several of the profiles is that those that approach normal depth do so asymptotically. This is shown in the S2, S3, M1, and M2 profiles. Also note in Fig. 15.28 that profiles that approach critical depth are shown by dashed lines. This is done because near critical depth either discontinuities develop (hydraulic jump), or the streamlines are very curved (such as near a brink). These profiles cannot be accurately predicted by Eq. (15.50) because this equation is based on one-dimensional flow, which, in these regions, is invalid.

Quantitative Evaluation of the Water-Surface Profile

In practice, most water-surface profiles are generated by numerical integration, that is, by dividing the channel into short reaches and carrying the computation for water-surface elevation from one end of the reach to the other. For one method, called the *direct step method*, the depth and velocity are known at a given section of the channel (one end of the reach), and one arbitrarily chooses the depth at the other end of the reach. Then the length of the reach is solved for. The applicable equation for quantitative evaluation of the water-surface profile is the energy equation written for a finite reach of channel, Δx:

$$y_1 + \frac{V_1^2}{2g} + S_0 \Delta x = y_2 + \frac{V_2^2}{2g} + S_f \Delta x$$

or

$$\Delta x (S_f - S_0) = \left(y_1 + \frac{V_1^2}{2g} \right) - \left(y_2 + \frac{V_2^2}{2g} \right)$$

or

$$\Delta x = \frac{(y_1 + V_1^2/2g) - (y_2 + V_2^2/2g)}{S_f - S_0} = \frac{(y_1 - y_2) + (V_1^2 - V_2^2)/2g}{S_f - S_0} \qquad (15.51)$$

The procedure for evaluation of a profile starts by ascertaining which type applies to the given reach of channel (using the methods of the preceding subsection). Then, starting from a known depth, one computes a finite value of Δx for an arbitrarily chosen change in depth. The process of computing Δx, step by step, up (negative Δx) or down (positive Δx) the channel is repeated until the full reach of channel has been covered. Usually small changes of y are taken, so that the friction slope is approximated by the following equation:

$$S_f = \frac{h_f}{\Delta x} = \frac{f V^2}{8 g R_h} \qquad (15.52)$$

Here V is the mean velocity in the reach, and R_h is the mean hydraulic radius. That is, $V = (V_1 + V_2)/2$, and $R_h = (R_{h1} + R_{h2})/2$. It is obvious that a numerical approach of this type is ideally suited for solution by computer.

EXAMPLE 15.13 CLASSIFICATION AND NUMERICAL ANALYSIS

1. Water discharges from under a sluice gate into a horizontal rectangular channel at a rate of 1 m³/s per meter of width, as shown in the following sketch. What is the classification of the water-surface profile? Quantitatively evaluate the profile downstream of the gate and determine whether it will extend all the way to the abrupt drop 80 m downstream. Make the simplifying assumptions that the resistance factor f is equal to 0.02 and that the hydraulic radius R_h is equal to the depth y.

Problem Definition

Situation: Sluice gate with gradually increasing depth after water exits gate.

Find:

1. Classification of the downstream profile.
2. Whether increasing slope will increase all the way to a point of interest 80 m downstream.

Assumptions:

1. Resistance factor f is equal to 0.02.
2. Hydraulic radius R_h is equal to the depth y.

Sketch:

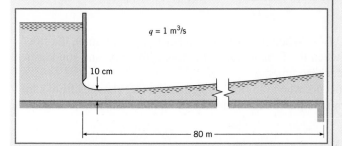

Plan

1. Determine the letter designation of channel using Fig. 15.29.
2. For flow leaving sluice gate, determine critical depth y_c, and compare to actual depth of flow. Use this information to refine the classification.
3. Solve for depth versus distance using Eqs. (15.51) and (15.52).

Solution

1. Channel is horizontal, so letter designation is H.
2. Determine critical depth y_c using Eq. (15.27).

$$y_c = (q^2/g)^{1/3} = [(1^2 \text{ m}^4/\text{s}^2)/(9.81 \text{ m/s}^2)]^{1/3}$$
$$= \boxed{0.467 \text{ m}}$$

Thus, the depth of flow from sluice gate is less than the critical depth. Therefore the water-surface profile is classified as

$$\boxed{\text{type H3.}}$$

3. To determine depth versus distance along the channel, apply Eqs. (15.51) and (15.52), using a numerical approach. The results of the computation are given in the table shown on p. 547. From the numerical results one plots the profile shown in the accompanying figure, which shows that the

$$\boxed{\text{profile extends to the abrupt drop.}}$$

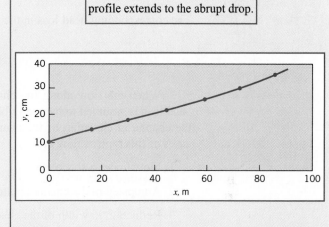

Summary

An open channel is one in which a liquid flows with a free surface. Steady open-channel flow is classified as either uniform (constant velocity with distance) or nonuniform (varying velocity with distance). For uniform open-channel flow, the head loss corresponds to the potential energy change associated with the slope of the channel. The discharge in an open channel is given by the Manning equation:

$$Q = \frac{1}{n}AR_h^{2/3}S_0^{1/2}$$

where A is the flow area, S_0 is the slope of the channel, and n is the resistance coefficient (Manning's n), which has been tabulated for different surfaces.

Nonuniform flow in open channels is characterized as either rapidly varied flow or gradually varied flow. In rapidly varied flow the channel resistance is negligible, and flow changes (depth and velocity change) occur over relatively short distances.

The significant π-group is Froude number

$$\frac{V}{\sqrt{gD_c}}$$

where D_c is the hydraulic depth, A/T. When the Froude number is equal to unity, the flow is critical. Subcritical flow occurs when the Froude number is less than unity, and supercritical when the Froude number is greater than unity.

A hydraulic jump usually occurs when the flow along the channel changes from supercritical to subcritical. The governing equation for hydraulic jump in a horizontal, rectangular channel is

$$y_2 = \frac{y_1}{2}(\sqrt{1 + 8Fr_1^2} - 1)$$

The corresponding head loss in the hydraulic jump is

$$h_L = \frac{(y_2 - y_1)^3}{4y_1 y_2}$$

When the flow along the channel changes from subcritical to supercritical flow, the head loss is assumed to be negligible, and the depth and velocity relationship is governed by the change in elevation of the channel bottom and the specific energy, $y + V^2/2g$. Typical cases of this type of flow are

1. Flow under a sluice gate

2. An upstep in the channel bottom

3. Reduction in width of the channel

For gradually varied flow the governing differential equation is

$$\frac{dy}{dx} = \frac{S_0 - S_f}{1 - Fr^2}$$

When this equation is integrated along the length of the channel the depth, y, is determined as a function of distance, x, along the channel. This yields the water surface profile for the reach of the channel.

SOLUTION TO EXAMPLE 15.13

Section Number Downstream of Gate	Depth y, m	Velocity at Section V, m/s	Mean Velocity in Reach, $(V_1 + V_2)/2$	V^2	Mean Hydraulic Radius, $R_m = (y_1 + y_2)/2$	$S_f = \dfrac{fV_{mean}^2}{8gR_m}$	$\Delta x = \dfrac{(y_1 - y_2) + \dfrac{(V_1^2 - V_2^2)}{2g}}{(S_f - S_0)}$	Distance from Gate x, m
1 (at gate)	0.1	10	...	100	...	...	...	0
	...	...	8.57	73.4	0.12	0.156	15.7	
2	0.14	7.14	...	51.0	...	...	...	15.7
	...	...	6.35	40.3	0.16	0.064	15.3	
3	0.18	5.56	...	30.9	...	...	...	31.0
	...	...	5.05	25.5	0.20	0.032	15.1	
4	0.22	4.54	...	20.6	...	...	...	46.1
	...	...	4.19	17.6	0.24	0.019	13.4	
5	0.26	3.85	...	14.8	...	...	...	59.5
	...	...	3.59	12.9	0.28	0.012	12.4	
6	0.30	3.33	...	11.1	...	...	...	71.9
	...	...	3.13	9.8	0.32	0.008	10.9	
7	0.34	2.94	...	8.6	...	...	...	82.8

References

1. Limerinos, J. T. "Determination of the Manning Coefficient from Measured Bed Roughness in Natural Channels." Water Supply Paper 1898-B, U.S. Geological Survey, Washington, D.C., 1970.

2. Committee on Hydromechanics of the Hydraulics Division of American Society of Civil Engineers. "Friction Factors in Open Channels." *J. Hydraulics Div., Am. Soc. Civil Eng.* (March 1963).

3. Wolman, M. G. "The Natural Channel of Brandywine Creek, Pennsylvania." Prof. Paper 271, U.S. Geological Survey, Washington D.C., 1954.

4. Yen, B. C. (ed.) *Channel Flow Resistance: Centennial of Manning's Formula.* Water Resources Publications, Littleton, CO, 1992.

5. Chow, Ven Te. *Open Channel Hydraulics.* McGraw-Hill, New York, 1959.

6. Henderson, F. M. *Open Channel Flow.* Macmillan, New York, 1966.

7. American Concrete Pipe Assoc. *Concrete Pipe Design Manual.* American Concrete Pipe Assoc., Vienna, Va., 1980.

8. Parshall, R. L. "The Improved Venturi Flume." *Trans. ASCE,* 89 (1926), 841–851.

9. Roberson, J. A., J. J. Cassidy, and M. H. Chaudhry. *Hydraulic Engineering.* John Wiley, New York, 1988.

10. Hinds, J. "The Hydraulic Design of Flume and Siphon Transitions." *Trans. ASCE,* 92 (1928), pp. 1423–1459.

11. U.S. Bureau of Reclamation. *Design of Small Canal Structures.* U.S. Dept. of Interior, U.S. Govt. Printing Office, Washington, D.C., 1978.

12. Rouse, H. (ed.). *Engineering Hydraulics.* John Wiley, New York, 1950.

13. U.S. Bureau of Reclamation. *Hydraulic Design of Stilling Basin and Bucket Energy Dissipators.* Engr. Monograph no. 25, U.S. Supt. of Doc., 1958.

14. Valle, B. L., and G. B. Pasternak. "Submerged and Unsubmerged Natural Hydraulic Jumps in a Bedrock Step-Pool Mountain Channel." *Geomorphology,* 82 (2006), pp. 146-159.

15. Raju, K. G. R. *Flow Through Open Channels.* Tata McGraw-Hill, New Delhi, 1981.

Problems

Describing Open-Channel Flow

15.1 $\mathbb{PQ}$◄ Why is the Reynolds number for onset of turbulence given by Re > 2000 in fully flowing pipes, and Re > 500 in partly flowing pipes and other open channels?

15.2 $\mathbb{PQ}$◄ A rectangular open channel has a base of length $2b$, and the water is flowing with a depth of b.
a. Sketch this channel.
a. What is the hydraulic radius of this channel?

15.3 $\mathbb{PQ}$◄ Two channels have the same cross-sectional area, but different geometry, as shown.
a. Which channel has the largest wetted perimeter?
b. Which channel has more contact between water and channelwall?
c. Which channel will have more energy loss to friction?

Uniform Open-Channel Flow

15.4 Consider uniform flow of water in the two channels shown. They both have the same slope, the same wall roughness, and the same cross-sectional area. Then one can conclude that (a) $Q_A = Q_B$, (b) $Q_A < Q_B$, or (c) $Q_A > Q_B$.

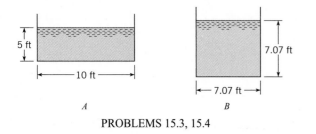

PROBLEMS 15.3, 15.4

15.5 This wood flume has a slope of 0.0015. What will be the discharge of water in it for a depth of 1 m?

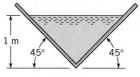

PROBLEM 15.5

15.6 Estimate the discharge in a rock-bedded stream ($d_{84} = 30$ cm) that has an average depth of 2.21 m, a slope of 0.0037, and a width of 48 m. Assume $k_s = d_{84}$.

15.7 Estimate the discharge of water $(T = 10°C)$ that flows 1.5 m deep in a long rectangular concrete channel that is 3 m wide and is on a slope of 0.001.

15.8 A rectangular concrete channel is 12 ft wide and has uniform water flow. If the channel drops 5 ft in a length of 8000 ft, what is the discharge? Assume $T = 60°F$. The depth of flow is 4 ft.

15.9 Consider channels of rectangular cross section carrying 100 cfs of water flow. The channels have a slope of 0.001. Determine the cross sectional areas required for widths of 2 ft, 4 ft, 6 ft, 8 ft, 10 ft, and 15 ft. Plot A versus y/b, and see how the results compare with the accepted result for the best hydraulic section.

15.10 A concrete sewer pipe 3 ft in diameter is laid so it has a drop in elevation of 1.0 ft per 1000 ft of length. If sewage (assume the properties are the same as those of water) flows at a depth of 1.5 ft in the pipe, what will be the discharge?

15.11 Determine the discharge in a 5 ft–diameter concrete sewer pipe on a slope of 0.001 that is carrying water at a depth of 4 ft.

15.12 Water flows at a depth of 6 ft in the trapezoidal, concrete-lined channel shown. If the channel slope is 1 ft in 2000 ft, what is the average velocity and what is the discharge?

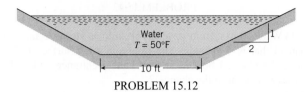

PROBLEM 15.12

15.13 What will be the depth of flow in a trapezoidal concrete-lined channel that has a water discharge of 1000 cfs? The channel has a slope of 1 ft in 500 ft. The bottom width of the channel is 10 ft, and the side slopes are 1 vertical to 1 horizontal.

15.14 What discharge of water will occur in a trapezoidal channel that has a bottom width of 10 ft and side slopes of 1 vertical to 1 horizontal if the slope of the channel is 4 ft/mi and the depth is to be 5 ft? The channel will be lined with concrete.

15.15 A rectangular concrete channel 4 m wide on a slope of 0.004 is designed to carry a water ($T = 10°C$) discharge of 25 m³/s. Estimate the uniform flow depth for these conditions. The channel has a rectangular cross section.

15.16 A rectangular troweled concrete channel 12 ft wide with a slope of 10 ft in 8000 ft is designed for a discharge of 500 cfs. For a water temperature of 40°F, estimate the depth of flow.

15.17 A concrete-lined trapezoidal channel having a bottom width of 10 ft and side slopes of 1 vertical to 2 horizontal is designed to carry a flow of 3000 cfs. If the slope of the channel is 0.001, what will be the depth of flow in the channel?

15.18 Design a canal having a trapezoidal cross section to carry a design discharge of irrigation water of 900 cfs. The slope of the canal is to be 0.002. The canal is to be lined with concrete, and it is to have the best hydraulic section for the design flow.

Nonuniform Open-Channel Flow

15.19 ℙℚ◄ How are head loss and slope related for nonuniform flow, as compared to uniform flow?

15.20 ℙℚ◄ Is critical flow a desirable or undesirable flow condition? Why?

15.21 Water flows at a depth of 4 in. with a velocity of 28 ft/s in a rectangular channel. Is the flow subcritical or supercritical? What is the alternate depth?

15.22 The water discharge in a rectangular channel 16 ft wide is 900 cfs. If the depth of water is 3 ft, is the flow subcritical or supercritical?

15.23 The discharge in a rectangular channel 18 ft wide is 420 cfs. If the water velocity is 9 ft/s, is the flow subcritical or supercritical?

15.24 Water flows at a rate of 12 m³/s in a rectangular channel 3 m wide. Determine the Froude number and the type of flow

(subcritical, critical, or supercritical) for depths of 30 cm, 1.0 m, and 2.0 m. What is the critical depth?

15.25 For the discharge and channel of Prob. 15.24, what is the alternate depth to the 30 cm depth? What is the specific energy for these conditions?

15.26 Water flows at the critical depth with a velocity of 5 m/s. What is the depth of flow?

15.27 Water flows uniformly at a rate of 320 cfs in a rectangular channel that is 12 ft wide and has a bottom slope of 0.005. If n is 0.014, is the flow subcritical or supercritical?

15.28 The discharge in a trapezoidal channel is 10 m³/s. The bottom width of the channel is 3.0 m, and the side slopes are 1 vertical to 1 horizontal. If the depth of flow is 1.0 m, is the flow supercritical or subcritical?

15.29 For the channel of Prob. 15.28, determine the critical depth for a discharge of 20 m³/s.

15.30 A rectangular channel is 6 m wide, and the discharge of water in it is 18 m³/s. Plot depth versus specific energy for these conditions. Let specific energy range from E_{min} to $E = 7$ m. What are the alternate and sequent depths to the 30 cm depth?

15.31 A long rectangular channel that is 4 m wide and has a mild slope ends in a free outfall. If the water depth at the brink is 0.35 m, what is the discharge in the channel?

15.32 A rectangular channel that is 15 ft wide and has a mild slope ends in a free outfall. If the water depth at the brink is 1.20 ft, what is the discharge in the channel?

15.33 A horizontal rectangular channel 14 ft wide carries a discharge of water of 500 cfs. If the channel ends with a free outfall, what is the depth at the brink?

15.34 What discharge of water will occur over a 2 ft–high, broad-crested weir that is 10 ft long if the head on the weir is 1.5 ft?

15.35 What discharge of water will occur over a 2 m–high, broad-crested weir that is 5 m long if the head on the weir is 60 cm?

15.36 The crest of a high, broad-crested weir has an elevation of 100 m. If the weir is 10 m long and the discharge of water over the weir is 25 m³/s, what is the water-surface elevation in the reservoir upstream?

15.37 The crest of a high, broad-crested weir has an elevation of 300 ft. If the weir is 40 ft long and the discharge of water over the weir is 1200 cfs, what is the water-surface elevation in the reservoir upstream?

15.38 Water flows with a velocity of 3 m/s and at a depth of 3 m in a rectangular channel. What is the change in depth and in water-surface elevation produced by a gradual upward change in bottom elevation (upstep) of 30 cm? What would be the depth and elevation changes if there were a gradual downstep of 30 cm? What is the maximum size of upstep that could exist before upstream depth changes would result?

15.39 Water flows with a velocity of 2 m/s and at a depth of 3 m in a rectangular channel. What is the change in depth and in water-surface elevation produced by a gradual upward change in bottom elevation (upstep) of 60 cm? What would be the depth and elevation changes if there were a gradual downstep of 15 cm? What is the maximum size of upstep that could exist before upstream depth changes would result?

15.40 Assuming no energy loss, what is the maximum value of Δz that will permit the unit flow rate of 6 m^2/s to pass over the hump without increasing the upstream depth? Sketch carefully the water-surface shape from section 1 to section 2. On the sketch give values for Δz, the depth, and the amount of rise or fall in the water surface from section 1 to section 2.

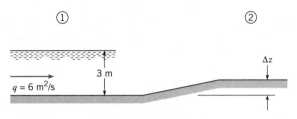

PROBLEM 15.40

15.41 Water flows with a velocity of 3 m/s in a rectangular channel 3 m wide at a depth of 3 m. What is the change in depth and in water-surface elevation produced when a gradual contraction in the channel to a width of 2.6 m takes place? Determine the greatest contraction allowable without altering the specified upstream conditions.

15.42 Because of the increased size of ships, the phenomenon called "ship squat" has produced serious problems in harbors where the draft of vessels approaches the depth of the ship channel. When a ship steams up a channel, the resulting flow situation is analogous to open-channel flow in which a constricting flow section exists (the ship reduces the cross-sectional area of the channel). The problem may be analyzed by referencing the water velocity to the ship and applying the energy equation. Thus, at the section of the channel where the ship is located, the relative water velocity in the channel will be greatest, and the water level in the channel will be reduced as dictated by the energy equation. Consequently, the ship itself will be at a lower elevation than if it were stationary; this lowering is referred to as "ship squat." Estimate the squat of the fully loaded supertanker *Bellamya* when it is steaming at 5 kt (1 kt = 0.515 m/s) in a channel that is 35 m deep and 200 m wide. The draft of the *Bellamya* when fully loaded is 29 m. Its width and length are 63 m and 414 m, respectively.

15.43 A rectangular channel that is 10 ft wide is very smooth except for a small reach that is roughened with angle irons attached to the bottom. Water flows in the channel at a rate of 200 cfs and at a depth of 1.0 ft upstream of the rough section. Assume frictionless flow except over the roughened part, where the total drag of all roughness (all of the angle irons) is assumed to be 2000 lbf. Determine the depth downstream of the roughness for the assumed conditions.

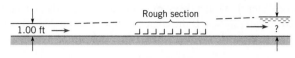

PROBLEM 15.43

15.44 Water flows from a reservoir into a steep rectangular channel that is 4 m wide. The reservoir water surface is 3 m above the channel bottom at the channel entrance. What discharge will occur in the channel?

15.45 A small wave is produced in a pond that is 8 in. deep. What is the speed of the wave in the pond?

15.46 A small wave in a pool of water having constant depth travels at a speed of 1.5 m/s. How deep is the water?

15.47 As waves in the ocean approach a sloping beach, they curve so that they are nearly parallel to the beach when they finally break (see accompanying figure). Explain why the waves curve like this.

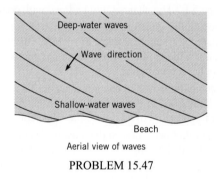

Aerial view of waves

PROBLEM 15.47

Hydraulic Jumps

15.48 The baffled ramp shown is used as an energy dissipator in a two-dimensional open channel. For a discharge of 18 cfs per foot of width, calculate the head lost, the power dissipated, and the horizontal component of force exerted by the ramp on the water.

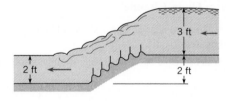

PROBLEM 15.48

15.49 The spillway shown has a discharge of 2.5 m^3/s per meter of width occurring over it. What depth y_2 will exist downstream of the hydraulic jump? Assume negligible energy loss over the spillway.

15.50 The flow of water downstream from a sluice gate in a horizontal channel has a depth of 30 cm and a flow rate of 3.60 m^3/m of width. The sluice gate is 2 m wide. Could a hydraulic jump be caused to form downstream of this section? If so, what would be the depth downstream of the jump?

PROBLEM 15.49

15.51 It is known that the discharge per unit width is 65 cfs/ft and that the height (H) of the hydraulic jump is 14 ft. What is the depth y_1?

PROBLEM 15.51

15.52 Water flows in a channel at a depth of 40 cm and with a velocity of 8 m/s. An obstruction causes a hydraulic jump to be formed. What is the depth of flow downstream of the jump?

15.53 Water flows in a trapezoidal channel at a depth of 40 cm and with a velocity of 10 m/s. An obstruction causes a hydraulic jump to be formed. What is the depth of flow downstream of the jump? The bottom width of the channel is 5 m, and the side slopes are 1 vertical to 1 horizontal.

15.54 A hydraulic jump occurs in a wide rectangular channel. If the depths upstream and downstream are 0.50 ft and 10 ft, respectively, what is the discharge per foot of width of channel?

15.55 The 20 ft–wide rectangular channel shown has three different reaches. $S_{0,1} = 0.01$; $S_{0,2} = 0.0004$; $S_{0,3} = 0.00317$; $Q = 500$ cfs; $n_1 = 0.015$; normal depth for reach 2 is 5.4 ft and that for reach 3 is 2.7 ft. Determine the critical depth and normal depth for reach 1 (use Manning's equation). Then classify the flow in each reach (supercritical, subcritical, critical), and determine whether a hydraulic jump could occur. In which reach(es) might it occur if it does occur?

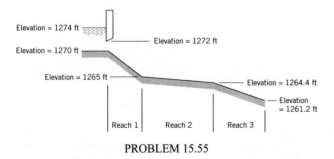

PROBLEM 15.55

15.56 Water flows from under the sluice gate as shown and continues on to a free overfall (also shown). Upstream from the overfall the flow soon reaches a normal depth of 1.1 m. The profile immediately downstream of the sluice gate is as it would be if there were no influence from the part nearer the overfall. Will a hydraulic jump form for these conditions? If so, locate its position. If not, sketch the full profile and label each part. Draw the energy grade line for the system.

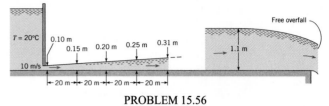

PROBLEM 15.56

15.57 Water is flowing as shown under the sluice gate in a horizontal rectangular channel that is 5 ft wide. The depths of y_0 and y_1 are 65 ft and 1 ft, respectively. What will be the horsepower lost in the hydraulic jump?

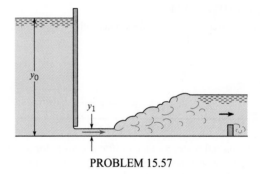

PROBLEM 15.57

15.58 Water flows uniformly at a depth $y_1 = 40$ cm in the concrete channel shown, which is 10 m wide. Estimate the height of the hydraulic jump that will form when a sill is installed to force it to form. Assume Manning's n value is $n = 0.012$.

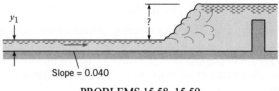

PROBLEMS 15.58, 15.59

15.59 For the derivation of Eq. (15.28) it is assumed that the bottom shearing force is negligible. For water flowing uniformly at a depth $y_1 = 40$ cm in the concrete channel shown, which is 10 m wide, a sill is installed to force a hydraulic jump to form. Estimate the magnitude of the shearing force F_s associated with the hydraulic jump and then determine F_s/F_H, where F_H is the net hydrostatic force on the hydraulic jump. Assume Manning's n value is $n = 0.012$.

15.60 The normal depth in the channel downstream of the sluice gate shown is 1 m. What type of water-surface profile occurs downstream of the sluice gate? Also, estimate the shear stress on the smooth bottom at a distance 0.5 m downstream of the sluice gate.

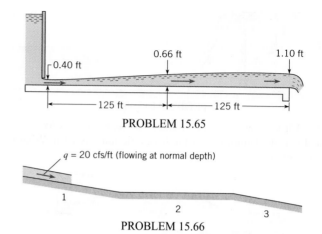

PROBLEM 15.60

15.61 Water flows at a rate of 100 ft³/s in a rectangular channel 10 ft wide. The normal depth in that channel is 2 ft. The actual depth of flow in the channel is 4 ft. The water-surface profile in the channel for these conditions would be classified as (a) S1, (b) S2, (c) M1, or (d) M2.

15.62 The water-surface profile labeled with a question mark is (a) M2, (b) S2, (c) H2, or (d) A2.

PROBLEM 15.62

15.63 The partial water-surface profile shown is for a rectangular channel that is 3 m wide and has water flowing in it at a rate of 5 m³/s. Sketch in the missing part of the water-surface profile and identify the type(s).

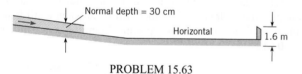

PROBLEM 15.63

15.64 A very long 10 ft–wide concrete rectangular channel with a slope of 0.0001 ends with a free overfall. The discharge in the channel is 120 cfs. One mile upstream the flow is uniform. What kind (classification) of water surface occurs upstream of the brink?

15.65 The horizontal rectangular channel downstream of the sluice gate is 10 ft wide, and the water discharge therein is 108 cfs. The water-surface profile was computed by the direct step method. If a 2 ft–high sharp-crested weir is installed at the end of the channel, do you think a hydraulic jump would develop in the channel? If so, approximately where would it be located? Justify your answers by appropriate calculations. Label any water-surface profiles that can be classified.

15.66 The discharge per foot of width in this rectangular channel is 20 cfs. The normal depths for parts 1 and 3 are 0.5 ft and 1.00 ft, respectively. The slope for part 2 is 0.001 (sloping upward in the direction of flow). Sketch all possible water-surface profiles for flow in this channel, and label each part with its classification.

PROBLEM 15.65

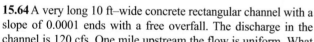

PROBLEM 15.66

15.67 Water flows from under a sluice gate into a horizontal rectangular channel at a rate of 3 m³/s per meter of width. The channel is concrete, and the initial depth is 20 cm. Apply Eq. (15.37) to construct the water-surface profile up to a depth of 60 cm. In your solution, compute reaches for adjacent pairs of depths given in the following sequence: $d = 20$ cm, 30 cm, 40 cm, 50 cm, and 60 cm. Assume that f is constant with a value of 0.02. Plot your results.

15.68 A horizontal rectangular concrete channel terminates in a free outfall. The channel is 4 m wide and carries a discharge of water of 12 m³/s. What is the water depth 300 m upstream from the outfall?

15.69 Consider the hydraulic jump shown for the long horizontal rectangular channel. What kind of water-surface profile (classification) is located upstream of the jump? What kind of water-surface profile is located downstream of the jump? If baffle blocks are put on the bottom of the channel in the vicinity of A to increase the bottom resistance, what changes are likely to occur given the same gate opening? Explain and/or sketch the changes.

PROBLEM 15.69

15.70 The steep rectangular concrete spillway shown is 4 m wide and 500 m long. It conveys water from a reservoir and delivers it to a free outfall. The channel entrance is rounded and smooth (negligible head loss at the entrance). If the water-surface elevation in the reservoir is 2 m above the channel bottom, what will the discharge in the channel be?

15.71 The concrete rectangular channel shown is 3.5 m wide and has a bottom slope of 0.001. The channel entrance is rounded and smooth (negligible head loss at the entrance), and the reservoir water surface is 2.5 m above the bed of the channel at the entrance.

a. Estimate the discharge in it if the channel is 3000 m long.
b. Tell how you would solve for the discharge in it if the channel were only 100 m long.

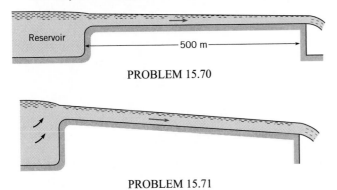

PROBLEM 15.70

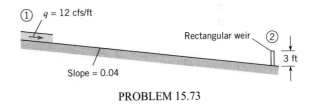

PROBLEM 15.71

15.72 A dam 50 m high backs up water in a river valley as shown. During flood flow, the discharge per meter of width, q, is equal to 10 m³/s. Making the simplifying assumptions that $R = y$ and $f = 0.030$, determine the water-surface profile upstream from the dam to a depth of 6 m. In your numerical calculation, let the first increment of depth change be y_c; use increments of depth change of 10 m until a depth of 10 m is reached; and then use 2 m increments until the desired limit is reached.

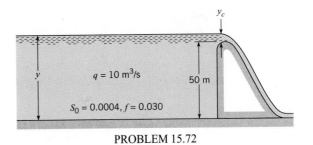

PROBLEM 15.72

15.73 Water flows at a steady rate of 12 cfs per foot of width ($q = 12$ cfs) in the wide rectangular concrete channel shown. Determine the water-surface profile from section 1 to section 2.

PROBLEM 15.73

A P P E N D I X

Figure A.1

Centroids and moments of inertia of plane areas.

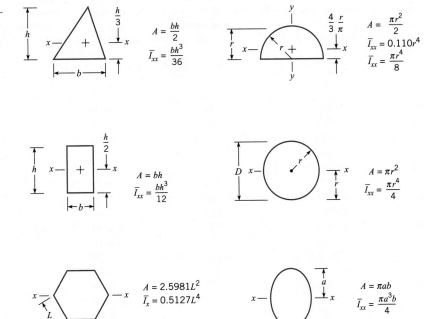

Volume and Area Formulas:

$$A_{circle} = \pi r^2 = \pi D^2 / 4$$

$$A_{sphere\ surface} = \pi D^2$$

$$V_{sphere} = \tfrac{1}{6}\pi D^3 = \tfrac{4}{3}\pi r^3$$

Table A.1 COMPRESSIBLE FLOW TABLES FOR AN IDEAL GAS WITH $k = 1.4$

M or M_1 = local number or Mach number upstream of a normal shock wave; p/p_t = ratio of static pressure to total pressure; ρ/ρ_t = ratio of static density to total density; T/T_t = ratio of static temperature to total temperature; A/A_* = ratio of local cross-sectional area of an isentropic stream tube to cross-sectional area at the point where M = 1; M_2 = Mach number downstream of a normal shock wave; p_2/p_1 = static pressure ratio across a normal shock wave; T_2/T_1 = static pressure ratio across a normal shock wave; p_{t_2}/p_{t_1} = total pressure ratio across normal shock wave.

Subsonic Flow				
M	p/p_t	ρ/ρ_t	T/T_t	A/A_*
0.00	1.0000	1.0000	1.0000	∞
0.05	0.9983	0.9988	0.9995	11.5914
0.10	0.9930	0.9950	0.9980	5.8218
0.15	0.9844	0.9888	0.9955	3.9103
0.20	0.9725	0.9803	0.9921	2.9630
0.25	0.9575	0.9694	0.9877	2.4027
0.30	0.9395	0.9564	0.9823	2.0351
0.35	0.9188	0.9413	0.9761	1.7780
0.40	0.8956	0.9243	0.9690	1.5901
0.45	0.8703	0.9055	0.9611	1.4487
0.50	0.8430	0.8852	0.9524	1.3398
0.52	0.8317	0.8766	0.9487	1.3034
0.54	0.8201	0.8679	0.9449	1.2703
0.56	0.8082	0.8589	0.9410	1.2403
0.58	0.7962	0.8498	0.9370	1.2130
0.60	0.7840	0.8405	0.9328	1.1882
0.62	0.7716	0.8310	0.9286	1.1657
0.64	0.7591	0.8213	0.9243	1.1452
0.66	0.7465	0.8115	0.9199	1.1265
0.68	0.7338	0.8016	0.9153	1.1097
0.70	0.7209	0.7916	0.9107	1.0944
0.72	0.7080	0.7814	0.9061	1.0806
0.74	0.6951	0.7712	0.9013	1.0681
0.76	0.6821	0.7609	0.8964	1.0570
0.78	0.6691	0.7505	0.8915	1.0471
0.80	0.6560	0.7400	0.8865	1.0382
0.82	0.6430	0.7295	0.8815	1.0305
0.84	0.6300	0.7189	0.8763	1.0237
0.86	0.6170	0.7083	0.8711	1.0179
0.88	0.6041	0.6977	0.8659	1.0129
0.90	0.5913	0.6870	0.8606	1.0089
0.92	0.5785	0.6764	0.8552	1.0056
0.94	0.5658	0.6658	0.8498	1.0031
0.96	0.5532	0.6551	0.8444	1.0014
0.98	0.5407	0.6445	0.8389	1.0003
1.00	0.5283	0.6339	0.8333	1.0000

(*Continued*)

Table A.1 COMPRESSIBLE FLOW TABLES FOR AN IDEAL GAS WITH $k = 1.4$ (*Continued*)

	Supersonic Flow				Normal Shock Wave			
M_1	p/p_t	ρ/ρ_t	T/T_t	A/A_*	M_2	p_2/p_1	T_2/T_1	p_{t_2}/p_{t_1}
1.00	0.5283	0.6339	0.8333	1.000	1.0000	1.000	1.000	1.0000
1.01	0.5221	0.6287	0.8306	1.000	0.9901	1.023	1.007	0.9999
1.02	0.5160	0.6234	0.8278	1.000	0.9805	1.047	1.013	0.9999
1.03	0.5099	0.6181	0.8250	1.001	0.9712	1.071	1.020	0.9999
1.04	0.5039	0.6129	0.8222	1.001	0.9620	1.095	1.026	0.9999
1.05	0.4979	0.6077	0.8193	1.002	0.9531	1.120	1.033	0.9998
1.06	0.4919	0.6024	0.8165	1.003	0.9444	1.144	1.039	0.9997
1.07	0.4860	0.5972	0.8137	1.004	0.9360	1.169	1.046	0.9996
1.08	0.4800	0.5920	0.8108	1.005	0.9277	1.194	1.052	0.9994
1.09	0.4742	0.5869	0.8080	1.006	0.9196	1.219	1.059	0.9992
1.10	0.4684	0.5817	0.8052	1.008	0.9118	1.245	1.065	0.9989
1.11	0.4626	0.5766	0.8023	1.010	0.9041	1.271	1.071	0.9986
1.12	0.4568	0.5714	0.7994	1.011	0.8966	1.297	1.078	0.9982
1.13	0.4511	0.5663	0.7966	1.013	0.8892	1.323	1.084	0.9978
1.14	0.4455	0.5612	0.7937	1.015	0.8820	1.350	1.090	0.9973
1.15	0.4398	0.5562	0.7908	1.017	0.8750	1.376	1.097	0.9967
1.16	0.4343	0.5511	0.7879	1.020	0.8682	1.403	1.103	0.9961
1.17	0.4287	0.5461	0.7851	1.022	0.8615	1.430	1.109	0.9953
1.18	0.4232	0.5411	0.7822	1.025	0.8549	1.458	1.115	0.9946
1.19	0.4178	0.5361	0.7793	1.026	0.8485	1.485	1.122	0.9937
1.20	0.4124	0.5311	0.7764	1.030	0.8422	1.513	1.128	0.9928
1.21	0.4070	0.5262	0.7735	1.033	0.8360	1.541	1.134	0.9918
1.22	0.4017	0.5213	0.7706	1.037	0.8300	1.570	1.141	0.9907
1.23	0.3964	0.5164	0.7677	1.040	0.8241	1.598	1.147	0.9896
1.24	0.3912	0.5115	0.7648	1.043	0.8183	1.627	1.153	0.9884
1.25	0.3861	0.5067	0.7619	1.047	0.8126	1.656	1.159	0.9871
1.30	0.3609	0.4829	0.7474	1.066	0.7860	1.805	1.191	0.9794
1.35	0.3370	0.4598	0.7329	1.089	0.7618	1.960	1.223	0.9697
1.40	0.3142	0.4374	0.7184	1.115	0.7397	2.120	1.255	0.9582
1.45	0.2927	0.4158	0.7040	1.144	0.7196	2.286	1.287	0.9448
1.50	0.2724	0.3950	0.6897	1.176	0.7011	2.458	1.320	0.9278
1.55	0.2533	0.3750	0.6754	1.212	0.6841	2.636	1.354	0.9132
1.60	0.2353	0.3557	0.6614	1.250	0.6684	2.820	1.388	0.8952
1.65	0.2184	0.3373	0.6475	1.292	0.6540	3.010	1.423	0.8760
1.70	0.2026	0.3197	0.6337	1.338	0.6405	3.205	1.458	0.8557
1.75	0.1878	0.3029	0.6202	1.386	0.6281	3.406	1.495	0.8346
1.80	0.1740	0.2868	0.6068	1.439	0.6165	3.613	1.532	0.8127
1.85	0.1612	0.2715	0.5936	1.495	0.6057	3.826	1.569	0.7902
1.90	0.1492	0.2570	0.5807	1.555	0.5956	4.045	1.608	0.7674
1.95	0.1381	0.2432	0.5680	1.619	0.5862	4.270	1.647	0.7442
2.00	0.1278	0.2300	0.5556	1.688	0.5774	4.500	1.688	0.7209
2.10	0.1094	0.2058	0.5313	1.837	0.5613	4.978	1.770	0.6742
2.20	$0.9352^{-1\dagger}$	0.1841	0.5081	2.005	0.5471	5.480	1.857	0.6281
2.30	0.7997^{-1}	0.1646	0.4859	2.193	0.5344	6.005	1.947	0.5833
2.50	0.5853^{-1}	0.1317	0.4444	2.637	0.5130	7.125	2.138	0.4990
2.60	0.5012^{-1}	0.1179	0.4252	2.896	0.5039	7.720	2.238	0.4601
2.70	0.4295^{-1}	0.1056	0.4068	3.183	0.4956	8.338	2.343	0.4236
2.80	0.3685^{-1}	0.9463^{-1}	0.3894	3.500	0.4882	8.980	2.451	0.3895
2.90	0.3165^{-1}	0.8489^{-1}	0.3729	3.850	0.4814	9.645	2.563	0.3577
3.00	0.2722^{-1}	0.7623^{-1}	0.3571	4.235	0.4752	10.330	2.679	0.3283
3.50	0.1311^{-1}	0.4523^{-1}	0.2899	6.790	0.4512	14.130	3.315	0.2129
4.00	0.6586^{-2}	0.2766^{-1}	0.2381	10.72	0.4350	18.500	4.047	0.1388

(*Continued*)

Table A.1 COMPRESSIBLE FLOW TABLES FOR AN IDEAL GAS WITH $k = 1.4$ *(Continued)*

	Supersonic Flow				Normal Shock Wave			
M_1	p/p_t	ρ/ρ_t	T/T_t	A/A_*	M_2	p_2/p_1	T_2/T_1	p_{t_2}/p_{t_1}
4.50	0.3455^{-2}	0.1745^{-1}	0.1980	16.56	0.4236	23.460	4.875	0.9170^{-1}
5.00	0.1890^{-2}	0.1134^{-1}	0.1667	25.00	0.4152	29.000	5.800	0.6172^{-1}
5.50	0.1075^{-2}	0.7578^{-2}	0.1418	36.87	0.4090	35.130	6.822	0.4236^{-1}
6.00	0.6334^{-2}	0.5194^{-2}	0.1220	53.18	0.4042	41.830	7.941	0.2965^{-1}
6.50	0.3855^{-2}	0.3643^{-2}	0.1058	75.13	0.4004	49.130	9.156	0.2115^{-1}
7.00	0.2416^{-3}	0.2609^{-2}	0.9259^{-1}	104.1	0.3974	57.000	10.47	0.1535^{-1}
7.50	0.1554^{-3}	0.1904^{-2}	0.8163^{-1}	141.8	0.3949	65.460	11.88	0.1133^{-1}
8.00	0.1024^{-3}	0.1414^{-2}	0.7246^{-1}	190.1	0.3929	74.500	13.39	0.8488^{-2}
8.50	0.6898^{-4}	0.1066^{-2}	0.6472^{-1}	251.1	0.3912	84.130	14.99	0.6449^{-2}
9.00	0.4739^{-4}	0.8150^{-3}	0.5814^{-1}	327.2	0.3898	94.330	16.69	0.4964^{-2}
9.50	0.3314^{-4}	0.6313^{-3}	0.5249^{-1}	421.1	0.3886	105.100	18.49	0.3866^{-2}
10.00	0.2356^{-4}	0.4948^{-3}	0.4762^{-1}	535.9	0.3876	116.500	20.39	0.3045^{-2}

$\dagger x^{-n}$ means $x \cdot 10^{-n}$.

SOURCE: Abridged with permission from R. E. Bolz and G. L. Tuve, *The Handbook of Tables for Applied Engineering Sciences,* CRC Press, Inc., Cleveland, 1973. Copyright © 1973 by The Chemical Rubber Co., CRC Press, Inc.

Figure A.2

Absolute viscosities of certain gases and liquids [Adapted from Fluid Mechanics, *5th ed., by V. L. Streeter. Copyright © 1971, McGraw-Hill Book Company, New York. Used with permission of the McGraw-Hill Book Company.]*

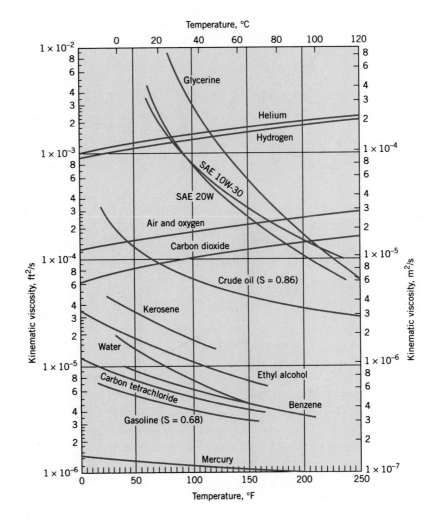

Figure A.3

Kinematic viscosities of certain gases and liquids. The gases are at standard pressure. [Adapted from Fluid Mechanics, *5th ed., by V. L. Streeter. Copyright © 1971, McGraw-Hill Book Company, New York. Used with permission of the McGraw-Hill Book Company.]*

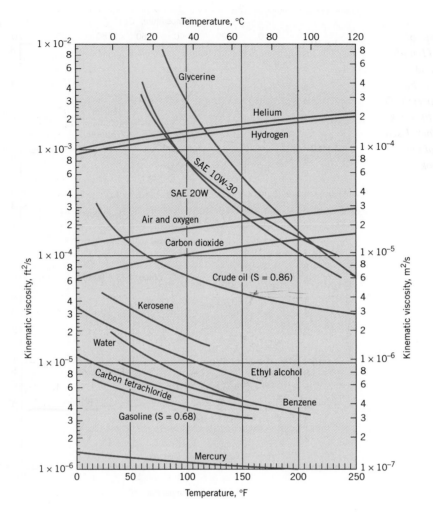

				c_p		
Gas	**Density** kg/m^3 **(slugs/ft^3)**	**Kinematic Viscosity** m^2/s **(ft^2/s)**	**R, Gas Constant** J/kg K **(ft-lbf/slug-°R)**	**$\dfrac{J}{kg\ K}$** $\left(\dfrac{Btu}{lbm\text{-}°R}\right)$	**$k = \dfrac{c_p}{c_v}$**	**S Sutherland's Constant** K **(°R)**
Air	1.22 (0.00237)	1.46×10^{-5} (1.58×10^{-4})	287 (1716)	1004 (0.240)	1.40	111 (199)
Carbon dioxide	1.85 (0.0036)	7.84×10^{-6} (8.48×10^{-5})	189 (1130)	841 (0.201)	1.30	222 (400)
Helium	0.169 (0.00033)	1.14×10^{-4} (1.22×10^{-3})	2077 (12,419)	5187 (1.24)	1.66	79.4 (143)
Hydrogen	0.0851 (0.00017)	1.01×10^{-4} (1.09×10^{-3})	4127 (24,677)	14,223 (3.40)	1.41	96.7 (174)
Methane (natural gas)	0.678 (0.0013)	1.59×10^{-5} (1.72×10^{-4})	518 (3098)	2208 (0.528)	1.31	198 (356)
Nitrogen	1.18 (0.0023)	1.45×10^{-5} (1.56×10^{-4})	297 (1776)	1041 (0.249)	1.40	107 (192)
Oxygen	1.35 (0.0026)	1.50×10^{-5} (1.61×10^{-4})	260 (1555)	916 (0.219)	1.40	

Table A.2   PHYSICAL PROPERTIES OF GASES [$T = 15°C$ (59 °F), $p = 1$ atm]

SOURCE: V. L. Streeter (ed.), *Handbook of Fluid Dynamics,* McGraw-Hill Book Company, New York, 1961; also R. E. Bolz and G. L. Tuve, *Handbook of Tables for Applied Engineering Science,* CRC Press, Inc. Cleveland, 1973; and *Handbook of Chemistry and Physics,* Chemical Rubber Company, 1951.

Table A.3   MECHANICAL PROPERTIES OF AIR AT STANDARD ATMOSPHERIC PRESSURE

Temperature	Density	Specific Weight	Dynamic Viscosity	Kinematic Viscosity
	kg/m^3	N/m^3	$N \cdot s/m^2$	m^2/s
−20°C	1.40	13.70	1.61×10^{-5}	1.16×10^{-5}
−10°C	1.34	13.20	1.67×10^{-5}	1.24×10^{-5}
0°C	1.29	12.70	1.72×10^{-5}	1.33×10^{-5}
10°C	1.25	12.20	1.76×10^{-5}	1.41×10^{-5}
20°C	1.20	11.80	1.81×10^{-5}	1.51×10^{-5}
30°C	1.17	11.40	1.86×10^{-5}	1.60×10^{-5}
40°C	1.13	11.10	1.91×10^{-5}	1.69×10^{-5}
50°C	1.09	10.70	1.95×10^{-5}	1.79×10^{-5}
60°C	1.06	10.40	2.00×10^{-5}	1.89×10^{-5}
70°C	1.03	10.10	2.04×10^{-5}	1.99×10^{-5}
80°C	1.00	9.81	2.09×10^{-5}	2.09×10^{-5}
90°C	0.97	9.54	2.13×10^{-5}	2.19×10^{-5}
100°C	0.95	9.28	2.17×10^{-5}	2.29×10^{-5}
120°C	0.90	8.82	2.26×10^{-5}	2.51×10^{-5}
140°C	0.85	8.38	2.34×10^{-5}	2.74×10^{-5}
160°C	0.81	7.99	2.42×10^{-5}	2.97×10^{-5}
180°C	0.78	7.65	2.50×10^{-5}	3.20×10^{-5}
200°C	0.75	7.32	2.57×10^{-5}	3.44×10^{-5}
	$slugs/ft^3$	lbf/ft^3	$lbf\text{-}s/ft^2$	ft^2/s
0°F	0.00269	0.0866	3.39×10^{-7}	1.26×10^{-4}
20°F	0.00257	0.0828	3.51×10^{-7}	1.37×10^{-4}
40°F	0.00247	0.0794	3.63×10^{-7}	1.47×10^{-4}
60°F	0.00237	0.0764	3.74×10^{-7}	1.58×10^{-4}
80°F	0.00228	0.0735	3.85×10^{-7}	1.69×10^{-4}
100°F	0.00220	0.0709	3.96×10^{-7}	1.80×10^{-4}
120°F	0.00213	0.0685	4.07×10^{-7}	1.91×10^{-4}
150°F	0.00202	0.0651	4.23×10^{-7}	2.09×10^{-4}
200°F	0.00187	0.0601	4.48×10^{-7}	2.40×10^{-4}
300°F	0.00162	0.0522	4.96×10^{-7}	3.05×10^{-4}
400°F	0.00143	0.0462	5.40×10^{-7}	3.77×10^{-4}

SOURCE: Reprinted with permission from R. E. Bolz and G. L. Tuve, *Handbook of Tables for Applied Engineering Science*, CRC Press, Inc., Cleveland, 1973. Copyright © 1973 by The Chemical Rubber Co., CRC Press, Inc.

Table A.4 APPROXIMATE PHYSICAL PROPERTIES OF COMMON LIQUIDS AT ATMOSPHERIC PRESSURE

Liquid and Temperature	Density kg/m³ (slugs/ft³)	Specific Gravity	Specific Weight N/m³ (lbf/ft³)	Dynamic Viscosity N·s/m² (lbf-s/ft²)	Kinematic Viscosity m²/s (ft²/s)	Surface Tension N/m* (lbf/ft)
Ethyl alcohol[1][3] 20°C (68°F)	799 (1.55)	0.79	7,850 (50.0)	1.2×10^{-3} (2.5×10^{-5})	1.5×10^{-6} (1.6×10^{-5})	2.2×10^{-2} (1.5×10^{-3})
Carbon tetrachloride[3] 20°C (68°F)	1,590 (3.09)	1.59	15,600 (99.5)	9.6×10^{-4} (2.0×10^{-5})	6.0×10^{-7} (6.5×10^{-6})	2.6×10^{-2} (1.8×10^{-3})
Glycerine[3] 20°C (68°F)	1,260 (2.45)	1.26	12,300 (78.5)	1.41 (2.95×10^{-2})	1.12×10^{-3} (1.22×10^{-2})	6.3×10^{-2} (4.3×10^{-3})
Kerosene[1][2] 20°C (68°F)	814 (1.58)	0.81	8,010 (51)	1.9×10^{-3} (4.0×10^{-5})	2.37×10^{-6} (2.55×10^{-5})	2.9×10^{-2} (2.0×10^{-3})
Mercury[1][3] 20°C (68°F)	13,550 (26.3)	13.55	133,000 (847)	1.5×10^{-3} (3.1×10^{-5})	1.2×10^{-7} (1.3×10^{-6})	4.8×10^{-1} (3.3×10^{-2})
Sea water 10°C at 3.3% salinity	1,026 (1.99)	1.03	10,070 (64.1)	1.4×10^{-3} (2.9×10^{-5})	1.4×10^{-6} (1.5×10^{-5})	
Oils—38°C (100°F) SAE 10W[4]	870 (1.69)	0.87	8,530 (54.4)	3.6×10^{-2} (7.5×10^{-4})	4.1×10^{-5} (4.4×10^{-4})	
SAE 10W-30[4]	880 (1.71)	0.88	8,630 (55.1)	6.7×10^{-2} (1.4×10^{-3})	7.6×10^{-5} (8.2×10^{-4})	
SAE 30[4]	880 (1.71)	0.88	8,630 (55.1)	1.0×10^{-1} (2.1×10^{-3})	1.1×10^{-4} (1.2×10^{-3})	

*Liquid–air surface tension values.
SOURCES: (1) V. L. Streeter, *Handbook of Fluid Dynamics*, McGraw-Hill, New York, 1961; (2) V. L. Streeter, *Fluid Mechanics*, 4th ed., McGraw-Hill, New York, 1966; (3) A. A. Newman, *Glycerol*, CRC Press, Cleveland, 1968; (4) R. E. Bolz and G. L. Tuve, *Handbook of Tables for Applied Engineering Sciences*, CRC Press, Cleveland, 1973.

Table A.5   APPROXIMATE PHYSICAL PROPERTIES OF WATER* AT ATMOSPHERIC PRESSURE

Temperature	Density	Specific Weight	Dynamic Viscosity	Kinematic Viscosity	Vapor Pressure
	kg/m^3	N/m^3	$N \cdot s/m^2$	m^2/s	N/m^2 abs
0°C	1000	9810	1.79×10^{-3}	1.79×10^{-6}	611
5°C	1000	9810	1.51×10^{-3}	1.51×10^{-6}	872
10°C	1000	9810	1.31×10^{-3}	1.31×10^{-6}	1,230
15°C	999	9800	1.14×10^{-3}	1.14×10^{-6}	1,700
20°C	998	9790	1.00×10^{-3}	1.00×10^{-6}	2,340
25°C	997	9781	8.91×10^{-4}	8.94×10^{-7}	3,170
30°C	996	9771	7.97×10^{-4}	8.00×10^{-7}	4,250
35°C	994	9751	7.20×10^{-4}	7.24×10^{-7}	5,630
40°C	992	9732	6.53×10^{-4}	6.58×10^{-7}	7,380
50°C	988	9693	5.47×10^{-4}	5.53×10^{-7}	12,300
60°C	983	9643	4.66×10^{-4}	4.74×10^{-7}	20,000
70°C	978	9594	4.04×10^{-4}	4.13×10^{-7}	31,200
80°C	972	9535	3.54×10^{-4}	3.64×10^{-7}	47,400
90°C	965	9467	3.15×10^{-4}	3.26×10^{-7}	70,100
100°C	958	9398	2.82×10^{-4}	2.94×10^{-7}	101,300
	$slugs/ft^3$	lbf/ft^3	$lbf\text{-}s/ft^2$	ft^2/s	psia
40°F	1.94	62.43	3.23×10^{-5}	1.66×10^{-5}	0.122
50°F	1.94	62.40	2.73×10^{-5}	1.41×10^{-5}	0.178
60°F	1.94	62.37	2.36×10^{-5}	1.22×10^{-5}	0.256
70°F	1.94	62.30	2.05×10^{-5}	1.06×10^{-5}	0.363
80°F	1.93	62.22	1.80×10^{-5}	0.930×10^{-5}	0.506
100°F	1.93	62.00	1.42×10^{-5}	0.739×10^{-5}	0.949
120°F	1.92	61.72	1.17×10^{-5}	0.609×10^{-5}	1.69
140°F	1.91	61.38	0.981×10^{-5}	0.514×10^{-5}	2.89
160°F	1.90	61.00	0.838×10^{-5}	0.442×10^{-5}	4.74
180°F	1.88	60.58	0.726×10^{-5}	0.385×10^{-5}	7.51
200°F	1.87	60.12	0.637×10^{-5}	0.341×10^{-5}	11.53
212°F	1.86	59.83	0.593×10^{-5}	0.319×10^{-5}	14.70

* Notes: (1) Bulk modulus E_v of water is approximately 2.2 GPa (3.2×10^5 psi); (2) water–air surface tension is approximately 7.3×10^{-2} N/m (5×10^{-3} lbf/ft) from 10°C to 50°C.

SOURCE: Reprinted with permission from R. E. Bolz and G. L. Tuve, *Handbook of Tables for Applied Engineering Science*, CRC Press, Inc., Cleveland, 1973. Copyright © 1973 by The Chemical Rubber Co., CRC Press, Inc.

A N S W E R S

Answers to Even Problems

Chapter 1
1.10 $\Delta p = 432$ Pa $= 0.0626$ psi $= 1.74$ in.-H_2O

1.12 $P = 239$ W, $\Delta E = 205$ calories

Chapter 2
2.4   Local conditions: $\rho = 1.09$ kg/m^3;
       table value: $\rho = 1.22$ kg/m^3

2.6 $\rho_{methane} = 1.74$ kg/m^3,
       $\gamma_{methane} = 17.1$ N/m^3

2.8 $\rho_{water}/\rho_{air} = 203$

2.10   Mass released is 26.7 kg.

2.12 $M = 3.49 \times 10^8$ slugs,
       $M = 5.09 \times 10^9$ kg

2.14 $\forall = 2.54$ m^3, $M = 5.66$ kg

2.16 $S_f = 0.972$, percent alcohol by volume $= 13.7\%$

2.18   For water, $\Delta\mu = -9.06 \times 10^{-4}$ N·s/m^2
       and $\Delta\rho = -22$ kg/m^3; for air,
       $\Delta\mu = 2.8 \times 10^{-6}$ N·s/m^2 and
       $\Delta\rho = -0.22$ kg/m^3.

2.20   For oil, $\mu = 6.7 \times 10^{-2}$ N·s/m^2 and
       $\nu = 7.6 \times 10^{-5}$ m^2/s; for kerosene,
       $\mu = 1.4 \times 10^{-3}$ N·s/m^2 and
       $\nu = 1.7 \times 10^{-6}$ m^2/s; for water,
       $\mu = 6.8 \times 10^{-4}$ N·s/m^2 and
       $\nu = 6.8 \times 10^{-7}$ m^2/s

2.22 $\nu/\nu_0 = (p_0/p)(T/T_0)^{5/2}(T_0 + S)/$
       $(T + S)$

2.24 $\nu = 1.99 \times 10^{-5}$ m^2/s

2.26 $\nu = 1.66 \times 10^{-3}$ ft^2/s

2.28 $S = 903°R$

2.30 $\mu = 1.32 \times 10^{-3}$ lbf-s/ft^2

2.32   For air, $\mu_{air} = 1.91 \times 10^{-5}$ N·s/m^2 and
       $\nu_{air} = 10.1 \times 10^{-6}$ m^2/s;
       for water, $\mu_{water} = 6.53 \times 10^{-5}$ N·s/m^2 and
       $\nu_{water} = 6.58 \times 10^{-7}$ m^2/s.

2.34 $\tau(y = 1$ mm$) = 1.49$ Pa

2.36   (a) $\tau_{max} = 1.0$ N/m^2; (b) minimum shear stress will occur
       midway between the two walls

2.38   (a) Maximum shear stress will occur at $y = H$;
       (b) $y = (H/2) - \mu u_t/(Hdp/ds)$;
       (c) $u_t = (1/2\mu)(dp/ds)H^2$

2.40 $V_{fall} = 0.17$ m/s

2.42 $T = \frac{1}{16}\pi\mu\omega D^4/s$

2.44   Ethyl alcohol is easier to compress.

2.46 $\Delta p = 44$ MPa

2.50 $\Delta p = 4\sigma/R$; $\Delta p_{4mm} = 73.0$ N/m^2

2.52   For $d = 1/4$ in., $\Delta h_{ST} = 0.185$ in.;
       for $d = 1/8$ in., $\Delta h_{ST} = 0.369$ in.;
       for $d = 1/32$ in., $\Delta h_{ST} = 1.48$ in.

2.54 $p = 292$ N/m^2

2.56   (a)

2.58 $\sigma = 0.0961$ N/m

2.60   Vapor pressure increases.

2.62 $P = 2340$ Pa abs

2.64   Boiling temperature (3000m) $= 89.7°C$

Chapter 3
3.2   (a) $\rho = 0.175$ kg/m^3, (b) $\rho = 0.531$ kg/m^3

3.4   % error $= 1.01\%$

3.10   The height decreases by 2.55 m.

3.12   Selection (a) is correct.

3.14 $p = 490$, $p_{50}/p_{atm} = 5.83$

3.18 $\Delta h = 5.00$ cm

3.20 $h_2 = 0.812$ m

3.22 $F_B = 2850$ lbf

3.24 $F_d = 2950$ lbf (acts down; metal is in tension)

3.26 $\rho_{34}/\rho_8 = 1.62$

3.30 p(center of pipe) $= 0.0$ lbf/ft^2

3.32 $p_A = 395$ Pa gage

3.34 $p_{container} = 891$ Pa gage

3.36 $p_A = 39.5$ kPa gage $= 5.72$ psig

3.38   The surface of the water is located 468 mm above the centerline
       of the horizontal leg. The surface of the mercury is located 121
       mm above the centerline of the horizontal leg. $p_{max} = 16.1$ kPa
       gage.

3.40 $p_A - p_B = 4.17$ kPa, $h_A - h_B = -0.50$ m

3.42 $p_A - p_B = 108$ psf, $h_A - h_B = 3.32$ ft

3.44 $p_A - p_B = 1.57$ kPa, $p_{zA} - p_{zB} = 0.589$ kPa

3.46 $T_{boiling, 2000\,m} \approx 93.2°C$, $T_{boiling, 4000\,m} \approx 86.7°C$

3.48   28.4 breaths per minute

3.50 $T = 287\,\text{K} = 516°\text{R}, p_a = 86.0\,\text{kPa} = 12.2\,\text{psia}, \rho = 1.04\,\text{kg}/\text{m}^3$
       $= 0.00199\,\text{slugs}/\text{ft}^3$

3.52 $p(z = 8\,\text{km}) = 3.31\,\text{mbar}, p(z = 30\,\text{km}) = 0.383\,\text{mbar}$

3.56   Valid statements are a, b, and e.

3.58 $F = 7.85\,\text{kN}, y_{cp} - \bar{y} = 22.4\,\text{mm}$

3.60 $F_{\text{hydrostatic}} = 6075\,\text{lbf}/\text{ft}, F_{\text{bottom tie}} = 8100\,\text{lbf}$ (tension)

3.62 $R_A = 557\,\text{kN}$ (acting normal to the gate)

3.64 $h = \ell/3$

3.66   Gate will stay in position

3.68 $F = 5\gamma Wh^2/3\sqrt{3}$, $R_T/F = 3/10$

3.70   Unstable

3.72 $n = 5$ bolts

3.74   Resultant passes through the pin.

3.76 $F = (3,120,000\mathbf{i} + 565,344\mathbf{j})\,\text{lbf}$

3.78 $F_{\text{horizontal}} = 61.6\,\text{kN}$ (applied to the left to hold dome in place)
       Line of action is 0.125 m below the center of the dome
       $F_{\text{vertical}} = 20.6\,\text{kN}$ (applied downward to hold dome in place)

3.80   (a) $F_B \approx 5.9\,\text{N}$, (b) $F_B = 4.19 \times 10^{-6}\,\text{N}$, (c) $F_B = 0.175\,\text{N}$

3.82   Selection (c) is correct

3.84 $W_{\text{scrap}} = 3420\,\text{N}$

3.86 $V\!\!\!\!/ = 39.9\,\text{m}^3$

3.88 $V\!\!\!\!/ = 40.8\,\text{L}, \gamma_{\text{block}} = 17.2\,\text{kN}/\text{m}^3$

3.90 $\Delta h = 0.368\,\text{cm}$

3.92 $d = 2.17L$

3.94   Due to the addition of ice, the water level will increase by $\Delta h = 0.306\,\text{in}$. Melting of the ice will not cause any additional change.

3.96 $\ell = 8.59\,\text{m}$

3.98 $z = 22.8\,\text{km}$

3.100 $S = 0.938$

3.102 $1.11 \leq S \leq 1.39$

3.104 $\ell/w = 0.211, S = 0.211$

3.106 The block will not float in a stable manner with its ends horizontal.

Chapter 4

4.4   (a) Unsteady, uniform; (b) nonuniform, steady

4.6   Nonuniform, steady or unsteady

4.8   (a) 2-d, (b) 1-d, (c) 1-d, (d) 2-d, (e) 3-d, (f) 3-d, (g) 2-d

4.16   (a) Steady; (b) Two-dimensional; (c) No; (d) Yes

4.18 $a_x = -(3\,U_0^2 r_0^3/x^4)(1 - r_0^3/x^3)$

4.20 $a_c = 5\,\text{ft}/\text{s}^2$

4.22 $a_l = 4q_0/(Bt_0)$

4.24 $a_l = 3.56\,\text{ft}/\text{s}^2, a_c = 37.9\,\text{ft}/\text{s}^2$

4.28 $\partial p/\partial z = -65.7\,\text{lbf}/\text{ft}^3$

4.30 $p_2 = 187\,\text{psfg}$

4.32 $\partial p/\partial s = -6000\,\text{N}/\text{m}^3$

4.34 $a_z = -141\,\text{ft}/\text{s}^2$

4.36 $dp/dx = -5330\,\text{psf}/\text{ft}$

4.38 $p_B - p_A = 12.7\,\text{kPa}, p_C - p_A = 44.6\,\text{kPa}$

4.42 $p_B = 527\,\text{psf}$

4.44 $p_B - p_A = 48.1\,\text{kPa}$

4.46 $\ell = 0$

4.48 $a_n = 4g$

4.50 $\omega = \sqrt{7.5g/\ell}$

4.52 $\omega = 17.7\,\text{rad}/\text{s}$

4.54 $\partial p/\partial z\,(-1) = -34.8\,\text{kPa}/\text{m}$,
       $\partial p/\partial z\,(0) = -9.81\,\text{kPa}/\text{m}$,
       $\partial p/\partial z\,(+1) = 15.2\,\text{kPa}/\text{m}$

4.56 $\Delta p_{\text{max}} = 737\,\text{lbf}/\text{ft}^2, z(p_{\text{min}}) = 0.206\,\text{ft}$ above axis

4.60 $V_2 = 5.46\,\text{m}/\text{s}$

4.62 $V_1 = 7.13\,\text{m}/\text{s}$

4.64 $h = 0.815\,\text{m}$

4.66 $V = 92.7\,\text{m}/\text{s}$

4.68 $V_0 = 1.33\,\text{m}/\text{s}$

4.72 $V = 18.3\,\text{fps}$

4.74 $V = 96.3\,\text{fps}$

4.76   (c)

4.78   Same reading

4.80 $V_0 = 8.82\,\text{m}/\text{s}$

4.82 $V_{\text{true}} = 80.6\,\text{m}/\text{s}$

4.88   Irrotational

4.90   Irrotational

4.92 $\Delta\theta = 8y/(1 - 4y^2)$

4.94 $z_2 - z_1 = 0.045\,\text{m}$

4.96 $p_A = 129\,\text{kPa, gage}$

4.98   (c)

4.100 $p_2 - p_1 = 0.96\,\text{kPa}$

4.102 $p = 914\,\text{mbar}$

1.106 Toward vortex center

4.108 $\theta_{\text{sep}} = 81.1°$

Chapter 5

5.4   (c)

5.6 $Q = 4.19\,\text{cfs}, 1880\,\text{gpm}$

5.8 $\dot{m} = 0.239 \text{ kg/s}$

5.10 $D = 1.25 \text{ m}$

5.12 $\bar{V}/V_0 = 1/3/$

5.14 $Q = 163 \text{ cfs, } 73,400 \text{ gpm}$

5.16 $Q = 5 \text{ m}^3/\text{s, } V = 5 \text{ m/s, } \dot{m} = 6.0 \text{ kg/s}$

5.18 $Q = 0.93 \text{ m}^3/\text{s}$

5.20 $Q = 1.70 \times 10^{-3} \text{ m}^3/\text{s}$

5.22 $q = 3.09 \text{ m}^2/\text{s, } V = 2.57 \text{ m/s}$

5.24 $V = [1/(n + 1)]V_c$

5.26 $V = 0.979 \text{ fps}$

5.28 $Q = 0.0849 \text{ cfs, } 37.9 \text{ gpm}$

5.30 $Q = 0.0276 \text{ m}^3/\text{s}$

5.38 (a) T, (b) T, (c) T, (d) T, (e) T

5.44 Rising

5.46 $V_{\text{exit}} = 2.8 \text{ m/s, } a_{\text{exit}} = 3.6 \text{ m/s}^3$

5.48 $a_{2s} = -5060 \text{ m/s}^2, a_{3s} = -11,400 \text{ m/s}^2$

5.50 $t = 14.4 \text{ s}$

5.52 $V_R = 2/3 \text{ fps}$

5.54 $V_1 = 12 \text{ m/s, } V_2 = 36 \text{ m/s}$

5.56 $V_{15} = 5.66 \text{ m/s, } V_{20} = 6.37 \text{ m/s}$

5.58 $V_B = 5.00 \text{ m/s}$

5.60 $Q_B = 3.33 \text{ cfm, leaving tank}$

5.62 Rising, $dh/dt = 1/8 \text{ fps}$

5.64 $Q_p = 7.5 \text{ cfs}$

5.66 $\dot{m} = 7.18 \text{ slug/s, } V_C = 20.4 \text{ ft/s,}$
      $S = 0.925$

5.68 $V = 10.8 \text{ m/s}$

5.70 $Q = 0.658 \, A_0\sqrt{2(p_1 - p_2)/\rho}$
      $Q = 5.20 \times 10^{-4} \text{ m}^3/\text{s}$

5.72 $A = 1.25 \times 10^{-6} \text{ in.}^2$

5.74 $t = 185 \text{ s}$

5.76 $\Delta t = 5.48 \text{ min, } 10.6 \text{ min (for } p = 0)$

5.78 $\Delta t = 24.8 \text{ min}$

5.80 $V_e = 2800 \text{ m/s}$

5.82 $p_c = (a\rho_p/0.65)^{1/(1-n)}(A_g/A_t)^{1/(1-n)}(RT_c)^{1/[2(1-n)]},$
      $\Delta p_c = 4.54 \text{ MPa}$

5.84 $p_B = 19.3 \text{ psi}$

5.86 $Q = 231 \text{ cfm}$

5.88 $\Delta h = 0.160 \text{ ft}$

5.90 $V_{e,\max} = 3.62 \text{ m/s, } Q_{\max} = 0.00362 \text{ m}^3/\text{s, } L_{\max} = 8850 \text{ N}$

5.92 $V_2 = 24 \text{ ft/s, } F = 45.2 \text{ lbf}$

5.96 $Q = 0.165 \text{ m}^3/\text{s}$

5.98 $V_0 = 12.4 \text{ m/s}$

5.100 $V_0 = 49.0 \text{ fps}$

5.102 $V_0 = 46.0 \text{ fps}$

5.104 Continuity not satisfied

5.106 $v = -Ay^2/2 + C(x)$

Chapter 6

6.2 $a_n = 0.112 \text{ ft/s}^2, a_n/g = 0.0035$

6.6 $F = 0.31 \text{ N, } v = 57.6 \text{ m/s}$

6.8 $\mu = 0.218$

6.10 $v_1 = 51.5 \text{ ft/s}$

6.12 $F_1 = 182 \text{ N, } F_2 = 169 \text{ N}$

6.14 $\dot{m} = 200 \text{ kg/s, } d = 7.14 \text{ cm}$

6.16 $F_A = 643 \text{ lbf, } F_B = 88.9 \text{ lbf}$

6.18 $F_y = 2\rho v^2 tr$

6.20 $F_x = -4.42 \text{ kN (to left)}$
      $F_y = -1.15 \text{ kN (downward)}$

6.22 $F_x = 890 \text{ lbf/ft(to left),}$
      $F_y = 382 \text{ lbf/ft (upward)}$

6.24 $F_f = 2.01 \text{ N, } N = 17.3 \text{ N, block moves}$

6.26 $F = 1.48 \text{ N (to left)}$

6.28 $h = 3.21 \text{ m}$

6.30 $\mathbf{F}\text{(on vane)} = (1470\mathbf{i} + 332\mathbf{j}) \text{ N}$

6.32 $F_x = 4.05 \text{ kN (to left)}$

6.34 $a_s = -80 \text{ m/s}^2$

6.36 $D = 91.8 \text{ lbf, } L = 5260 \text{ lbf}$

6.38 $F_x = -4.15 \text{ lbf}$

6.40 $\mathbf{F} = (-5370\mathbf{i} + 387\mathbf{k}) \text{ lbf}$

6.42 $\mathbf{F} = (808\mathbf{i} - 356\mathbf{j} + 350\mathbf{k}) \text{ lbf}$

6.44 $F_a = 3310 \text{ lbf}$

6.46 $F_y = 12,200 \text{ lbf (downward)}$

6.48 $F_x = 1840 \text{ lbf (opposite flow direction)}$

6.50 $F_x = 1140 \text{ N}$

6.52 $F_x = -167 \text{ lbf (to left)}$

6.54 $F_x = -9.96 \text{ kN (to left)}$
      $F_y = -1.8 \text{ kN (downward)}$

6.58 $v_e = 51.1 \text{ m/s, } F = 98.3 \text{ lbf (to left)}$

6.62 $F_x = -40.7 \text{ N (to left)}$

6.64 $F_x = -18.3$ kN

6.66 F(per bolt) $= 1410$ N

6.68 $p_g = 13.3$ kPa, $F_s = -1.38$ kN/m

6.70 $\mathbf{F} = (-524\mathbf{i} - 58.9\mathbf{j})$ lbf

6.72 $\mathbf{F} = (-36.8\mathbf{i} + 119\mathbf{j})$ N

6.74 $F_x = -474$ lbf/ft

6.76 F(stationary) $= 161$ lbf,
       F(moving) $= 113$ lbf

6.78 $(p_2 - p_1) = \rho v_1^2 (D_0^2 - D_j^2)/D_0^2$
       $+ \rho v_j^2 D_j^2/D_0^2 - \rho v^2$
       $(p_2 - p_1) = 32$ kPa

6.80 $D = 1.3$ N, $L = 3.33$ N

6.82 $T = 15.3$ kN

6.86 $F_r = 60$ N

6.88 $t = 10$ s

6.90 $v_{max} = 15.2$ m/s

6.92 $\Delta p = 5.93$ MPa

6.94 $\Delta p_{max} = 525$ psi

6.96 $\Delta p = \rho vc$

6.98 $Q = 1.22$ m^3/s, $L = 2160$ m

6.100 $P = 1.11$ kW

6.102 $\mathbf{F} = (1240\mathbf{i} - 420\mathbf{j})$ lbf
       $\mathbf{M} = 1260\mathbf{k}$ ft-lbf

6.104 $\mathbf{F}_A = -908\mathbf{i}$ lbf
       $\mathbf{M}_A = (-1820\mathbf{j} + 299\mathbf{k})$ ft-lbf

Chapter 7

7.4   (a) cost $= \$0.67$, (b) $P = 1010$ W, (c) $P = 13.6$ hp

7.6 $P = 55.2$ W

7.10   (a) $\alpha = 1.0$, (b) $\alpha > 1.0$, (c) $\alpha > 1.0$, (d) $\alpha > 1.0$

7.12 $\alpha = 27/20$

7.14 $\alpha = (n + 1)^3/(3n + 1)$, $\alpha = 1.05$

7.20   (b)

7.22 $p_A = -250$ psf, $V_2 = 30.0$ ft/s

7.24 $p_2/\gamma = 13.16$ m

7.26 $p_2 = 16.9$ psig

7.28 $p_1 = 9.23$ psig

7.30 $p_1 = 152$ Pa

7.32 $p_B = -2.38$ psig

7.34 $z_A = $ depth $= 9.07$ m

7.36 $Q = 0.302$ m^3/s

7.38 $t = 6.63$ h

7.40 $P_{electrical} = 1.49$ W

7.42 $P = 1.76$ MW

7.44 $h = 120$ ft

7.46 $P = 95.9$ kW

7.48 $P = 14.9$ kW

7.50 $P = 61.6$ kW

7.52 $P = 1590$ hp $= 1.18$ MW

7.54 $t = 9260$ s $= 2.57$ h

7.56 $h_L = 2.52$ ft

7.58 $h_L = 0.104$ m

7.60 $Q = 0.0149$ m^3/s

7.62 $F_j = 11.7$ lbf acting to the left

7.64 $F_{wall} = 198$ lbf acting upward

7.66 $F_d = \rho U^2 \pi D^2/8[1/(D^2/d^2 - 1)^2]$,
       $F_d = 0.372$ N

7.70   Possible if the fluid is being accelerated to the left

7.78 $Q = 6.23$ ft^3/s

7.80 $P = 49.0$ hp

7.82 $z_L = 129$ ft

7.84 $Q = 0.320$ m^3/s, $p_m = -78.7$ kPa

Chapter 8

8.4   (a) homo, (b) nonhomo, (c) homo, (d) homo

8.6 $\Delta h/d = f(D/d, \gamma t^2/\rho d, h_1/d)$

8.8 $(F_D/\mu V d) = C$

8.10 $(\Delta p D^2)/(\Delta \ell \mu V) = C$

8.12 $\Delta p/(n^2 \rho D^2) = f(Q/nD^3)$

8.14 $h/d = f(\sigma t^2/\rho d^3, \gamma t^2/\rho d, \mu t/\rho d^2)$

8.16 $\Delta p d^4/\rho Q^2 = f(\mu d/\rho Q, D/d)$

8.18 $\dot{m}/(\sqrt{\rho \Delta p} D^2) = f(\mu D/\dot{m})$

8.20 $F_D/(\rho V^2 S) = f(\omega^2 S/V^2)$

8.22 $V_b/\sqrt{gD} = f[\mu/(\rho_f g^{1/2} D^{3/2}),$
       $(\rho_f - \rho_b)/\rho_f]$

8.24 $F_D/\rho V^2 B^2 = f(\mu/\rho VB, u_{rms}/V, L_x/B)$

8.30 $U_m = 21.4$ m/s, $F_{D,m}/F_{D,p} = 0.500$

8.32 $V_5 = 1.5$ m/s

8.34 $V_m/V_p = 5$

8.36 $F_L = 21.1$ kN

8.38 $p = 27.6$ kbar

8.40   (c)

8.42   Re $= 25{,}200$, $F_D = 20.4 \times 10^{-3}$ N,
       $P = 16.3 \times 10^{-3}$ W

8.44 $V_m = 4.50 \text{ m/s}$

8.46 $V_a = 11.6 \text{ m/s}, \Delta p_\omega = 7.33 \text{ kPa}$

8.48 $\rho_m = 0.024 \text{ slugs/ft}^3, F_p = 500 \text{ lbf}$

8.50 $V_p = 0.215 \text{ m/s}, n_p = 117 \text{ N} \cdot \text{m}$

8.52 $p_m = 808 \text{ kPa}$

8.54 $d = 3.93 \text{ mm}$

8.56 $\nu_m / \nu_p = (L_m/L_p)^3 / ^2$

8.58 $V_p = 12.5 \text{ m/s}, Q_p = 312 \text{ m}^3/\text{s}$

8.60 $V_m/V_p = 1/6, Q_m/Q_p = 1/7776,$
       $Q_m = 0.386 \text{ m}^3/\text{s}$

8.62 $V_p = 39.3 \text{ ft/s}, Q_p = 11,000 \text{ ft}^3/\text{s}$

8.64 $t_p = 5 \text{ min}, Q_p = 312 \text{ m}^3/\text{s}$

8.66 $F_p = 3.83 \text{ MN}$

8.68 $L_m/L_p = 0.0318$

8.70 $V_p = 25 \text{ ft/s}, F_p = 31,200 \text{ lbf}$

8.72 $p_{\text{windward wall}} = 1.08 \text{ kPa}$
       $p_{\text{side wall}} = -2.93 \text{ kPa}$
       $p_{\text{leeward wall}} = -868 \text{ pa}$
       $F_p = 5.65 \text{ MN}$

Chapter 9

9.4 $V = 2.13 \text{ m/s}$

9.6 $\mu = 3.85 \times 10^{-2} \text{ N} \cdot \text{s/m}^2$

9.8   (a) T, (b) F, (c) F, (d) F, (e) T

9.10   (a) F, (b) F, (c) T, (d), F, (e) T

9.12 $T = 1.99 \text{ N} \cdot \text{m}$

9.14 $T = 3.45 \times 10^{-3} \text{ N} \cdot \text{m}$

9.16 $P = 0.00780 \text{ W}$

9.18 $u_{\text{max}} = 0.150 \text{ ft/s}$

9.20 $q = 4.65 \times 10^{-5} \text{ m}^2/\text{s}$

9.22 $dp/ds = -464 \text{ psf/ft}$

9.24 $q = 1.44 \text{ m}^2/\text{hr}$

9.26 $t = 1.024 \, \mu_o U/L$

9.28   4.8%

9.34 $\delta/x = 0.0071$

9.36   (a) is correct

9.38 $F_x = 5.15 \text{ N}$

9.40 $u = U_0 = 1 \text{ m/s}$

9.42 $\delta = 15.8 \text{ mm}, F_s = 0.0943 \text{ N}$

9.46 $F_s = 1.29 \times 10^{-4} \text{ N}$

9.48 $F_{s,\text{wing}} = 230 \text{ N}, P = 12.8 \text{ kW}$
       $x_{cr} = 14.4 \text{ cm}$
       $F_{\text{tripped B.L.}} / F_{\text{normal}} = 1.162$

9.50 $F_s/B = 3.53 \text{ N/m}, du/dy = 9.33 \times 10^4 \text{ s}^{-1}$

9.52 $\delta* = \int_0^\delta [1 - (\rho u)/(\rho_\infty U_\infty)]dy$

9.54 $\delta*/\delta = 0.125$

9.56 $P = 10.4 \text{ kW}$

9.58 $U_0 = 0.805 \text{ m/s}$

9.60 $U_0 = 103 \text{ m/s}$

9.62 $F_s = 26.4 \text{ lbf}$

9.64 $P = 103 \text{ hp}$

9.66 $F_{s100} = 1360 \text{ N}, F_{s200} = 5000 \text{ N}$
       $P_{100} = 37.8 \text{ kW}, P_{200} = 278 \text{ kW}$

9.68 $\delta/W_{\text{min vel}} = 0.0406, \delta/W_{\text{max vel}} = 0.036$

9.70 $F_s = 375 \text{ lbf}$

9.72 $F_{\text{wave}} = 3.72 \times 10^4 \text{ lbf}$

9.74 $\tau_{0,\text{min}} = 106 \text{ N/m}^2$

9.76 $F_D = 287 \text{ kN}, \delta = 0.678 \text{ m}$

Chapter 10

10.2   Flow is turbulent, $L_e = 7.5 \text{ m}$

10.6 $p_{\text{tank}} = 1.75 \text{ kPa gage}$

10.8 $V = 2.19 \text{ m/s}, Q = 0.110 \text{ L/s}$

10.12 $\dot{m} = 0.0141 \text{ kg/s}, f = 0.064,$
       $h_f \neq L = 0.00108 \text{ m per meter of pipe length},$
       $\Delta p/L = 10.6 \text{ Pa per meter of pipe length}$

10.14 $V_2 = 0.215 \text{ m/s}$

10.16 $h_f = 66.4 \text{ ft per 100 ft run of pipe}$

10.18 $V = 0.81 \text{ ft/s}, Q = 2.76 \times 10^{-4} \text{ cfs}$

10.20 $P = 1340 \text{ W}$

10.22   Correct choice is (d)

10.24 $V_2 = 0.0409 \text{ m/s}$

10.26 $\nu = 8.91 \times 10^{-5} \text{ m}^2/\text{s}$

10.28   Flow is downward (from right to left),
       $f = 0.076, \mu = 0.00154 \text{ lbf} \cdot \text{s/ft}^2, \text{laminar}$

10.30 $\Delta p = 684 \text{ Pa}$

10.32 $f = 0.0185$

10.34 $f = 0.0258$

10.36 $\text{Re} = 6.37 \times 10^5, f = 0.023, \tau_o = 51.7 \text{ Pa}$

10.38 $h_f = 182 \text{ ft}$

10.40 $\nu = 2.0 \times 10^{-8} \text{ m}^2/\text{s}$

10.42   (a) $\Delta p = 1.58$ psi, (b) $h_f = 3.64$ ft,
         (c) $P = 0.0675$ hp

10.44 $\Delta p / L = 208$ Pa/m

10.46 $\Delta p / L = 2.48$ psf/foot

10.48 $\Delta p = 48.9$ psf, $P = 349$ hp

10.50   (a) case 1, (b) case 3, (c) case 3

10.52 $D = 0.022$ m

10.54 $V = 3.15$ m/s

10.56 $Q = 6.59 \times 10^{-3}$ m^3/s

10.58 $D = 22$ cm, $P = 45.6$ kW for each
         kilometer of pipe length

10.60 $P = 18.3$ kW

10.62 $t = 23.7$ min

10.64 $t = 46.5$ min

10.66 $P = 30.1$ kW

10.68   Select a pipe with $D = 8$ in.

10.70 $P = 726$ W (clean tube),
         $P = 3.03$ kW (scaled tubes)

10.72 $P = 7.49$ kW

10.74 $P = 10.1 \times 10^{-4}$ hp

10.76   Specify a 12-cm pipe

10.78 $P = 17.4$ MW

10.80   Cavitation could occur in the venturi throat section or just
         downstream of the abrupt contraction.

10.82 $z_1 = 114$ m

10.84 $P_{\text{loss}} = 40.4$ kW

10.86 $V_{\text{trap}} / V_{\text{rect}} = 0.84$

10.88 $Q = 0.25$ m^3/s

10.90 $Q = 4700$ gpm

10.92 $V_A / V_B = 1.26$

10.94 $Q_1 = 2$ cfs

10.96 $(Q_{\text{large}} / Q_{\text{small}}) = 3.86$

10.98 $Q(12$ inch pipe$) = 6.46$ cfs

         $Q(14$ inch pipe$) = 7.75$ cfs

         $Q(16$ inch pipe$) = 10.8$ cfs

         $h_{L_{AB}} = 107$ ft

Chapter 11

11.2   Correct choice is (d)

11.4 $C_D = 2.0$

11.8 $F_D = 2250$ lbf

11.10 $F_D = 198$ lbf

11.12 $F_D = 6.24 \times 10^6$ lbf

11.14 $V = 19.7$ m/s

11.16 $F_D = 18.6$ kN

11.18 $M_o = 3.12$ MN $\cdot$ m

11.20 $T = 142$ N

11.22 $M = 21.2$ kN $\cdot$ m

11.24 $(5.9$ m$)/$s$) \leq V \leq (17.7$ m/s$)$

11.26 $P = 55.5$ kW

11.28   Energy $= 77.2$ kJ $= 18.4$ food calories

11.30   Additional power $= 21.9$ hp

11.32   14.7%

11.34 $P = 47.2$ kW

11.36 $V_c = 12.6$ m/s

11.38   756 N

11.42   The bubble will accelerate as it moves upward. Form drag.

11.44 $V_0 = 1.32$ m/s

11.46 $V_0 = 1.33$ m/s upward

11.48 $V_o = 1.55$ mm/s

11.50 $V_o = 9.13$ m/s

11.52 $V_o = 5.70$ m/s

11.54   Time to reach 99% of the terminal velocity is 0.54 s. The
         corresponding distance of travel is 14.2 cm.

11.56   (a) $F_L = 4.84$ N,
         (b) $A = 5.23 \times 10^3$ mm^2

11.60 $F_L = 2.82$ N

11.62 $b = 20.9$ ft

11.64   The correct answer is (d)

11.66 $V = 29.6$ m/s

11.68 $V_s = 99.8$ m/s, $V_L = 108$ m/s

11.70 $V_0 = 10.5$ m/s, $F_L/$length $= 16,000$ N/m

11.72 $C_L = \sqrt{\pi \Lambda C_{D_0}}$,

         $C_L / C_D = (1/2)\sqrt{\pi \Lambda / C_{D_0}}$

11.74 $F_D = 4000$ N

Chapter 12

12.2   761 mph

12.4 $M = 27.1$

12.6 $c = 1070$ m/s

12.8 $c_{\text{He}} - c_{\text{N}_2} = 656$ m/s

12.10 $c = 1480$ m/s

12.12 $T_t = 218$°C

12.14 $V = 200$ m/s

12.18 $T_t = 51$°C, $p_t = 284.6$ kPa

12.20 $T = 407$ K, $p = 177$ kPa, $V = 346$ m/s

12.22 $T = 291$ K, $p = 487$ kPa, $M = 0.192$,
$\dot{m} = 0.032$ kg/s

12.24 $C_p = 2/(kM^2)[(1 + (k-1)M^2/2)^{(k/k-1)} - 1]$, $C_p(2) = 2.43$,
$C_p(4) = 13.47$, $C_{p,\text{inc}} = 1.0$

12.26 T, T, F

12.28 $M_2 = 0.657$, $p_2 = 208$ kPa,
$T_2 = 316$ K, $\Delta s = 35.6$ J/kg K

12.30 $M = 1.59$

12.32 $V_1 = 1200$ m/s

12.34 $M_2^2 \approx 2 - M_1^2$

12.38 $\dot{m} = 0.0733$ kg/s, (with Bernoulli)
$\dot{m} = 0.0794$ kg/s

12.40 $\dot{m} = 0.100$ kg/s (130 kPa),
$\dot{m} = 0.322$ kg/s (350 kPa)

12.48 $A_e/A_* = 4.00$, $A_T = 29.5$ cm^2

12.50 Underexpanded

12.52 $M = \sqrt{2}$, $A/A_* = 1.123$

12.54 $A_e/A_* = 3.60$, T (ideal) $= 2791$ N,
$T = 2790$ N

12.56 $A/A_* = 2.97$, $x = 5.40$ cm

12.58 $M_3 = 0.336$, $p_3 = 461$ kPa, $p_t = 499$ kPa

Chapter 13

13.4 $V_o = 0.511$ m/s

13.6 $V = 3.96$ m/s

13.8 Percent error $= 0.1\%$

13.10 $Q = 4.26 \times 10^{-3}$ m^3/s

13.12 $V_{\text{mean}} = 4.33$ m/s, $V_{\text{max}}/V_{\text{mean}} = 2$,
$Q = 0.196$ m^3/s, laminar

13.14 (a) $r_m/D = 0.224$, (b) $r_c/D = 0.341$,
(c) $\dot{m} = 9.96$ kg/s

13.16 $Q = 549$ m^3/s

13.18 $V = 0.468$ m/s

13.20 $C_v = 0.975$, $C_c = 0.640$, $C_d = 0.624$

13.24 $h_{\text{mercury}} = 1.54$ ft

13.28 $V_{\text{pipe}} = 1.21$ m/s

13.30 Percent increase in discharge $= 96\%$

13.34 $d = 6.26$ cm

13.36 $d = 0.601$ m

13.38 $Q = KA_0\sqrt{2\Delta p/\rho}$, $Q = 0.290$ m^3/s

13.40 $h = 0.44$ m

13.42 $Q = 1.36$ cfs

13.44 $Q = 11.3$ m^3/s

13.46 $Q = 6.08$ cfs

13.48 $Q = 0.00124$ m^3/s

13.50 $h_L = 64V_0^2/2g$

13.54 (a) $V = (L/\Delta t)[-1 + \sqrt{1 + (c\Delta t/L)^2}$,
(b) $V = c^2\Delta t/(2L)$, (c) $V = 22.5$ m/s

13.58 $Q = 0.0248$ m^3/s

13.60 Correct choice is (b)

13.62 Correct choice is (c)

13.64 $H = 0.53$ ft, $Q = 2.54$ ft^3/s

13.66 $Q = 62.7$ ft^3/s

13.68 Water level is falling

13.72 $Q = 3.96$ ft^3/s

13.74 $h = 1.24$ m

13.78 $\dot{m} = 0.0021$ kg/s

13.80 $\dot{m} = 0.0338$ lbm/s

13.82 $Q = 3.49$ cfs, $U_Q = 0.192$ cfs

Chapter 14

14.4 $F_T = 926$ N, $P = 35.7$ kW

14.6 $N = 1160$ rpm

14.8 $D = 1.71$ m, $V_0 = 89.4$ m/s

14.10 $N = 1170$ rpm

14.12 $a = 0.783$ m/s^2

14.16 $Q = 0.22$ m^3/s, $P = 6.5$ kW

14.18 $Q = 54.6$ cfs, $\Delta H = 21.8$ ft, $P = 169$ hp

14.22 $D = 0.882$ m, $P = 14.2$ kW

14.24 $P = 726$ kW

14.28 $Q = 0.228$ m^3/s

14.30 $H_{1600} = 261$ ft

14.32 $Q = 0.218$ m^3/s

14.38 (a) $Q = 0.218$ m^3/s, (b) $Q = 76.4$ gpm, (c) $Q = 77.4$ gpm

14.40 Mixed flow pump

14.42 Radial flow pump

14.44 Radial flow pump

14.46 $P = 94.4$ kW

14.48 $P = 229$ kW

14.50 $P = 10.6$ MW, $D = 2.85$ m

14.52 $F = (1/2)\rho A V_j^2$

14.56 $\alpha_1 = 6.36°$, $T = 44{,}700$ N-m,
$P = 281$ kW

14.58 $P_{\text{out}} = 271$ hp

14.60 $P_{\text{max}} = 4.69$ kW

14.62 $Q = 289$ gpm

Chapter 15

15.2 $R_h = b/2$

15.4 (c)

15.6 $Q = 448$ m³/s

15.8 Using Darcy-Weisbach, $Q = 243$ cfs;
using Manning, $Q = 214$ cfs

15.10 $Q = 10.6$ ft³/s

15.12 Using Darcy-Weisbach, $V = 5.74$ ft/s and $Q = 758$ cfs;
using Manning, $V = 5.18$ fps and
$Q = 684$ cfs.

15.14 $Q = 546$ cfs

15.16 $d = 4.92$ ft

15.18 Half-hexagon with all three sides having length of 8.57 ft

15.20 Undesirable

15.22 Supercritical

15.24 $\text{Fr}_{0.3} = 7.77$ (supercritical),
$\text{Fr}_{1.0} = 1.27$ (supercritical), and
$\text{Fr}_{2.0} = 0.452$ (subcritical),
$y_c = 1.18$ m

15.26 $y_c = 2.55$ m

15.28 Subcritical

15.30 Alternate depth is $y = 5.38$ m;
sequent depth is $y_2 = 2.33$ m.

15.32 $Q = 187$ cfs

15.34 $Q = 50.5$ cfs

15.36 Elev. $= 101.4$ m

15.38 For upstep, $\Delta y = -0.51$ m, new water elev. is 2.49 m.
For downstep, $\Delta y = 0.40$ m, new water elev. is 3.4.
Before upstream depth change: $z_{\text{step, max}} = 0.43$ m.

15.40 $\Delta z = 0.89$ m

15.42 Ship squat $= 0.30$ m

15.44 $Q = 35.5$ m²/s

15.46 $y = 0.23$ m

15.48 $h_L = 2.30$ ft; $P = 4.70$ hp, and
$F_{\text{ramp,H}} = 51.2$ lbf opposite to direction of flow

15.50 Hydraulic jump can occur; $y_2 = 2.82$ m

15.52 $y_2 = 2.09$ m

15.54 $q = 29.07$ ft²/s

15.56 A hydraulic jump will form at ≈ 29 m downstream of sluice gate.

15.58 $\Delta \text{Elev} = 1.86$ m (increase)

15.60 S3; $\tau_0 = 143$ N/m²

15.62 (d)

15.64 M2

15.68 ≈ 1.51 m

15.70 $Q = 19.2$ m³/s

15.72 Profile progresses from an elevation of 52.2 m to 53.5 m

I N D E X

Table A.6 NOMENCLATURE

Symbol	Dimensions	Description	Symbol	Dimensions	Description
A	L^2	Area	h_L	L	Head loss
A_j	L^2	Jet area	h_p	L	Head supplied by pump
A_0	L^2	Orifice area	h_t	L	Head given up to turbine
A_*	L^2	Nozzle area at M $= 1$	$\bar{I}$	L^4	Area moment of inertia, centroidal
a	L/T^2	Acceleration	$\mathbf{i}$	...	Unit vector in x direction
b	...	Intensive property	$\mathbf{j}$	...	Unit vector in y direction
B	L	Linear measure	$\mathbf{k}$	...	Unit vector in z direction
B	...	Extensive property	K	...	Minor loss coefficient
b	L	Linear measure	k	...	Ratio of specific heats
C_c	...	Coefficient of contraction	k_s	L	Equivalent sand roughness
C_D	...	Coefficient of drag	L	L	Linear measure
C_d	...	Coefficient of discharge	l	L	Linear measure
C_f	...	Average shear stress coefficient	ℓ	L	Linear measure
C_F	...	Force coefficient	M	...	Mach number
C_H	...	Head coefficient	M	ML^2/T^2	Moment
C_L	...	Coefficient of lift	m	M	Mass
C_P	...	Power coefficient	$\dot{m}$	M/T	Mass flow rate
C_p	...	Pressure coefficient	N	T^{-1}	Rotational speed
C_Q	...	Discharge coefficient	N_s	$L^{3/4}/T^{3/2}$	Specific speed
C_T	...	Thrust coefficient	N_{ss}	$L^{3/4}/T^{3/2}$	Suction specific speed
C_v	...	Coefficient of velocity	n	...	Manning's roughness coefficient
c	L/T	Speed of sound	n	T^{-1}	Rotational speed
c_f	...	Local shear stress coefficient	n_s	...	Specific speed
c_p	$L^2/T^2\theta$	Specific heat at constant pressure	n_{ss}	...	Suction specific speed
c_v	$L^2/T^2\theta$	Specific heat at constant volume	p	M/LT^2	Pressure
CP	...	Center of pressure	Δp	M/LT^2	Change in pressure
cs	...	Control surface	P	ML^2/T^3	Power
cv	...	Control volume	p_*	M/LT^2	Pressure at M $= 1$
D	L	Diameter	p_t	M/LT^2	Total pressure
D	L	Hydraulic depth	p_v	M/LT^2	Vapor pressure
D_h	L	Hydraulic diameter	p_z	ML^2/T^2	Piezometric pressure
d	L	Diameter	Q	L^3/T	Discharge, volumetric flow rate
d	L	Depth	Q	ML^2/T^2	Heat transferred
E	ML^2/T^2	Energy	q	L^2/T	Discharge per unit width
E	L	Specific energy	q	M/LT^2	Kinetic pressure
E_v	M/LT^2	Elasticity, bulk	R_h	L	Hydraulic radius
e	L^2/T^2	Energy per unit mass	R	ML/T^2	Reaction or resultant force
Fr	...	Froude number	R	$L^2/\theta T^2$	Gas constant
F	ML/T^2	Force	Re	...	Reynolds number
F_D	ML/T^2	Drag force	r	L	Linear measure in radial direction
F_L	ML/T^2	Lift force	S	L^2	Planform area
F_S	ML/T^2	Surface resistance	St	...	Strouhal number
f	...	Friction factor	S_0	...	Channel slope
G	...	Giga, multiple $= 10^9$	s	$L^2/T^2\theta$	Specific entropy
g	L/T^2	Acceleration due to gravity	S	...	Specific gravity
H	L	Head	s	L	Linear measure
h	L	Piezometric head	T	ML^2/T^2	Torque
h	$L^2/T^2\theta$	Specific enthalpy	T	θ	Temperature
h_f	L	Head loss in pipe			

(Continued)

TABLE F.3   USEFUL CONSTANTS

Name of Constant	Value
Acceleration of gravity	$g = 9.81 \text{ m/s}^2 = 32.2 \text{ ft/s}^2$
Universal gas constant	$R_u = 8.314 \text{ kJ/kmol} \cdot \text{K} = 1545 \text{ ft} \cdot \text{lbf/lbmol} \cdot {}^\circ\text{R}$
Standard atmospheric pressure	$p_{\text{atm}} = 1.0 \text{ atm} = 101.3 \text{ kPa} = 14.70 \text{ psi} = 2116 \text{ psf} = 33.90 \text{ ft of water}$
	$p_{\text{atm}} = 10.33 \text{ m of water} = 760 \text{ mm of Hg} = 29.92 \text{ in of Hg} = 760 \text{ torr} = 1.013 \text{ bar}$

TABLE F.4   PROPERTIES OF AIR [$T = 20^\circ$C (68 $^\circ$F), $p = 1$ atm]

Property	SI Units	Traditional Units
Specific gas constant	$R_{\text{air}} = 287.0 \text{ J/kg} \cdot \text{K}$	$R_{\text{air}} = 1716 \text{ ft} \cdot \text{lbf/slug} \cdot {}^\circ\text{R}$
Density	$\rho = 1.20 \text{ kg/m}^3$	$\rho = 0.0752 \text{ lbm/ft}^3 = 0.00234 \text{ slug/ft}^3$
Specific weight	$\gamma = 11.8 \text{ N/m}^3$	$\gamma = 0.0752 \text{ lbf/ft}^3$
Viscosity	$\mu = 1.81 \times 10^{-5} \text{ N} \cdot \text{s/m}^2$	$\mu = 3.81 \times 10^{-7} \text{ lbf} \cdot \text{s/ft}^2$
Kinematic viscosity	$\nu = 1.51 \times 10^{-5} \text{ m}^2/\text{s}$	$\nu = 1.63 \times 10^{-4} \text{ ft}^2/\text{s}$
Specific heat ratio	$k = c_p/c_v = 1.40$	$k = c_p/c_v = 1.40$
Specific heat	$c_p = 1004 \text{ J/kg} \cdot \text{K}$	$c_p = 0.241 \text{ Btu/lbm} \cdot {}^\circ\text{R}$
Speed of sound	$c = 343 \text{ m/s}$	$c = 1130 \text{ ft/s}$

TABLE F.5   PROPERTIES OF WATER [$T = 15^\circ$C (59 $^\circ$F), $p = 1$ atm]

Property	SI Units	Traditional Units
Density	$\rho = 999 \text{ kg/m}^3$	$\rho = 62.4 \text{ lbm/ft}^3 = 1.94 \text{ slug/ft}^3$
Specific weight	$\gamma = 9800 \text{ N/m}^3$	$\gamma = 62.4 \text{ lbf/ft}^3$
Viscosity	$\mu = 1.14 \times 10^{-3} \text{ N} \cdot \text{s/m}^2$	$\mu = 2.38 \times 10^{-5} \text{ lbf} \cdot \text{s/ft}^2$
Kinematic viscosity	$\nu = 1.14 \times 10^{-6} \text{ m}^2/\text{s}$	$\nu = 1.23 \times 10^{-5} \text{ ft}^2/\text{s}$
Surface tension (water-air)	$\sigma = 0.073 \text{ N/m}$	$\sigma = 0.0050 \text{ lbf/ft}$
Bulk modulus of elasticity	$E_v = 2.14 \times 10^9 \text{ Pa}$	$E_v = 3.10 \times 10^5 \text{ psi}$

TABLE F.6   PROPERTIES OF WATER [$T = 4^\circ$C (39 $^\circ$F), $p = 1$ atm]

Property	SI Units	Traditional Units
Density	1000 kg/m^3	$62.4 \text{ lbm/ft}^3 = 1.94 \text{ slug/ft}^3$
Specific weight	9810 N/m^3	62.4 lbf/ft^3

TABLE F.2 COMMONLY USED EQUATIONS

Specific weight

$\gamma = \rho g$ (Eq. 2.2, p. 16)

Specific gravity

$S = \dfrac{\rho}{\rho_{H_2O \text{ at } 4°C}} = \dfrac{\gamma}{\gamma_{H_2O \text{ at } 4°C}}$ (Eq. 2.3, p. 16)

Ideal gas law

$p = \rho R T$ (Eq. 2.5, p. 17)

Definition of viscosity

$\tau = \mu \dfrac{dV}{dy}$ (Eq. 2.6, p. 19)

Kinematic viscosity

$\nu = \mu / \rho$ (Eq. 2.8, p. 20)

Pressure equation

$p_{abs} = p_{atm} + p_{gage}$ (Eq. 3.3a, p. 35)

$p_{abs} = p_{atm} - p_{vacuum}$ (Eq. 3.3b, p. 35)

Hydrostatic equation

$\dfrac{p_1}{\gamma} + z_1 = \dfrac{p_2}{\gamma} + z_2 = \text{constant}$ (Eq. 3.7a, p. 38)

$p_z = p_1 + \gamma z_1 = p_2 + \gamma z_2 = \text{constant}$ (Eq. 3.7b, p. 38)

$\Delta p = -\gamma \Delta z$ (Eq. 3.7c, p. 38)

Manometer equations

$p_2 = p_1 + \sum_{down} \gamma_i h_i - \sum_{up} \gamma_i h_i$ (Eq. 3.18, p. 45)

$h_1 - h_2 = \Delta h (\gamma_B / \gamma_A - 1)$ (Eq. 3.19, p. 46)

Hydrostatic force equations (flat panels)

$F = \bar{p} A$ (Eq. 3.23, p. 49)

$y_{cp} - \bar{y} = \dfrac{\bar{I}}{\bar{y} A}$ (Eq. 3.28, p. 51)

Buoyant force (Archimedes equation)

$F_B = \gamma \Psi_D$ (Eq. 3.36, p. 56)

The Bernoulli equation

$\dfrac{p_1}{\gamma} + \dfrac{V_1^2}{2g} + z_1 = \dfrac{p_2}{\gamma} + \dfrac{V_2^2}{2g} + z_2$ (Eq. 418b, p. 92)

$p_1 + \dfrac{\rho V_1^2}{2} + \rho g z_1 = p_2 + \dfrac{\rho V_2^2}{2} + \rho g z_2$ (Eq. 418a, p. 92)

Coefficient of pressure

$C_p = \dfrac{p_z - p_{zo}}{\rho V_o^2 / 2} = \dfrac{h - h_o}{V_o^2 / (2g)}$ Eq. 4.50, p. 109

Volume flow rate equation

$Q = \bar{V} A = \dfrac{\dot{m}}{\rho} = \int_A V dA = \int_A \mathbf{V} \cdot d\mathbf{A}$ (Eq. 5.8, p. 131)

Mass flow rate equation

$\dot{m} = \rho A \bar{V} = \rho Q = \int_A \rho V dA = \int_A \rho \mathbf{V} \cdot d\mathbf{A}$ (Eq. 5.9, p. 131)

Continuity equation

$\dfrac{d}{dt} \int_{cv} \rho \, d\Psi + \int_{cs} \rho \mathbf{V} \cdot d\mathbf{A} = 0$ (Eq. 5.24, p. 138)

$\dfrac{d}{dt} M_{cv} + \sum_{cs} \dot{m}_o - \sum_{cs} \dot{m}_i = 0$ (Eq. 5.25, p. 138)

$\rho_1 A_1 V_2 = \rho_2 A_2 V_2$ (Eq. 5.26, p. 142)

Momentum equation

$\sum \mathbf{F} = \dfrac{d}{dt} \int_{cv} \mathbf{v} \rho \, d\Psi + \int_{cs} \mathbf{v} \rho \mathbf{V} \cdot d\mathbf{A}$ (Eq. 6.5, p. 164)

$\sum \mathbf{F} = \dfrac{d}{dt} \int_{cv} \rho \mathbf{v} \, d\Psi + \sum_{cs} \dot{m}_o \mathbf{v}_o - \sum_{cs} \dot{m}_i \mathbf{v}_i$ (Eq. 6.6, p. 164)

Energy equation

$\dfrac{p_1}{\gamma} + \alpha_1 \dfrac{\bar{V}_1^2}{2g} + z_1 + h_p = \dfrac{p_2}{\gamma} + \alpha_2 \dfrac{\bar{V}_2^2}{2g} + z_2 + h_t + h_L$

(Eq. 7.29; p. 225)

The power equation

$P = FV = T\omega$ (Eq. 7.3, p. 218)

$P = \dot{m} g h = \gamma Q h$ (Eq. 7.31, p. 227)

Efficiency of a machine

$\eta = \dfrac{P_{output}}{P_{input}}$ (Eq. 7.32; p. 227)

Reynolds number (pipe)

$Re = \dfrac{VD}{\nu} = \dfrac{\rho VD}{\mu} = \dfrac{4Q}{\pi D \nu} = \dfrac{4\dot{m}}{\pi D \mu}$ (Eq. 10.2, p. 317)

Combined head loss equation

$h_L = \sum_{pipes} f \dfrac{L}{D} \dfrac{V^2}{2g} + \sum_{components} K \dfrac{V^2}{2g}$ (Eq. 10.45, p. 339)

Friction factor f (Resistance coefficient)

$f = \dfrac{64}{Re} \qquad Re \leq 2000$ (Eq. 10.34, p. 326)

$f = \dfrac{0.25}{\left[\log_{10} \left(\dfrac{k_s}{3.7D} + \dfrac{5.74}{Re^{0.9}} \right) \right]^2}$ $(Re \geq 3000)$(Eq. 10.39, p. 331)

Drag force equation

$F_D = C_D A \left(\dfrac{\rho V_0^2}{2} \right)$ (Eq. 11.5, p. 365)

Lift force equation

$F_L = C_L A \left(\dfrac{\rho V_o^2}{2} \right)$ (Eq. 11.17, p. 381)